Nanocarriers in Neurodegenerative Disorders

Due to the lack of secure, efficient, and patient-friendly therapies for neurodegenerative disorders, there is a rising demand for innovative approaches. Despite the limited number of nanocarriers approved for human use, they have demonstrated significant potential in preclinical and, in some instances, clinical trials. In alignment with this objective, the chapters of the book are structured to offer a comprehensive overview of recent advancements in medication and dosage form development, specifically emphasizing the nanoparticulate system for targeting the brain. This book aims to furnish readers with a thorough understanding of the clinical application of nanocarrier systems for treating neurodegenerative disorders, encompassing the latest developments, challenges, safety concerns, toxicity issues, regulatory considerations, prospects, and limitations. Individuals in academia, the scientific community, business, and education seeking a more effective approach to target the brain will find valuable insights in this resource.

Key Features

- Provides a comparative perspective of various nanocarrier systems, therefore facilitating the researcher's selection of appropriate nanoparticulate carriers
- Highlights the related restrictions of brain delivery and current available medicines
- Includes information on the advantages and disadvantages of various biomaterials utilized in the development of nanocarriers for brain targeting
- Emphasizes distinct facets of surface functionalization according to the brain area of interest
- Presents the current advances, preclinical and clinical development, and the future potential of multiple brain-targeting technologies

Dr. Wael Mohamed, a physician neuroscientist, earned his PhD from PSU in the United States and currently holds the position of Professor Madya at IIUM Medical School in Malaysia. With an extensive background, Dr. Mohamed has been invited to deliver over 150 lectures both domestically and internationally. His contributions to neuroscience/psychiatry are notable, with the publication of more than 120 peer-reviewed papers, boasting an h-index of 23. Furthermore, Dr. Mohamed serves as an editor for *Frontiers in Neurology* and *PLOS One* journals, where he has contributed to various journal special issues on brain disorders. In the realm of books, he is actively involved with Springer, CRC Press, Elsevier, and Cambridge in the field of neuroscience. His research endeavors have attracted numerous grants from esteemed national and international organizations, including IBRO, ISN, MJF, STDF, FRGS, and INDO-ASEAN, amounting to a total research funding of half-a-million US dollars. Notably, he is the founder and chair of the AfrAbia-PD-Genomic Consortium funded by MJF and GP2.

Nanocarriers in Neurodegenerative Disorders

Therapeutic Hopes and Hypes

Edited by Wael Mohamed

CRC Press
Taylor & Francis Group
Boca Raton London New York

CRC Press is an imprint of the
Taylor & Francis Group, an **informa** business

Designed cover images: Shutterstock

First edition published 2025
by CRC Press
2385 NW Executive Center Drive, Suite 320, Boca Raton, FL 33431

and by CRC Press
4 Park Square, Milton Park, Abingdon, Oxon, OX14 4RN

CRC Press is an imprint of Taylor & Francis Group, LLC

ISBN: 9781032467955 (hbk)
ISBN: 9781032468051 (pbk)
ISBN: 9781003383376 (ebk)

DOI: 10.1201/9781003383376

Typeset in Times
by Apex CoVantage, LLC

Dedication

I express heartfelt gratitude and deep appreciation as I dedicate this work to my wife, Dr. Rehab Ismaeil. Your unwavering support and unconditional love have been constant pillars throughout my academic journey and beyond. I am truly grateful for the love and inspiration you bring to my life, and I acknowledge that none of this would be possible without you.

Wael Mohamed
Pahang, Malaysia

Contents

Preface

To study the phenomenon of disease without books is to sail an uncharted sea, while to study books without patients is not to go to sea at all.

—William Osler (1849–1919)

Neurodegenerative disorders, including Alzheimer's and Parkinson's, significantly contribute to the global burden of chronic and debilitating illnesses. To elaborate, they stand as the primary cause of disability-adjusted life years (DALYs) and rank second in global mortality. The severity of this issue has escalated due to a lack of effective treatments and public awareness. The intricate anatomy of the brain poses challenges for transporting therapeutic agents, with the blood–brain barrier (BBB) safeguarding the brain's health by restricting the entry of most pathogens through the bloodstream. The brain, however, relies on various active transport systems, such as receptor-mediated transcytosis, carrier-mediated transport, ion channel transporter, and efflux transport, to ensure the supply of nutrients and metabolites. The conventional methods of drug delivery often result in low bioavailability of bioactives in the brain. Consequently, modern medicine has explored the use of advanced nanoparticulate systems to overcome key barriers, like the BBB, and efficiently target medications to the brain.

Transported by nanocarriers, drugs become more soluble and permeable to biological membranes, significantly enhancing their bioavailability. Nanovectors offer numerous acknowledged advantages, including small size, biodegradability, improved release control, prolonged effects, enhanced stability, minimal adverse effects, active targeting, increased bioavailability, and potent therapeutic efficacy. Despite these benefits, there are notable challenges in the clinical adoption of nanotherapies, such as biocompatibility, cellular toxicity, stability, drug loading, release behavior, high cost, and scalability. Hence, the aim of this book is to shed light on various aspects, including the merits and demerits of utilizing nanoparticles as vectors for the targeted delivery of diverse therapeutic and diagnostic substances to the brain in the context of neurodegenerative disorders. The increasing demand for innovative approaches stems from the lack of safe, effective, and patient-friendly therapies for neurodegenerative disorders. Despite the limited licensing of nanocarriers for human use, their considerable promise in preclinical and, at times, clinical trials are evident. With this objective in mind, the chapters of the book are structured to offer a comprehensive overview of recent advancements in medication and dosage form development, specifically focusing on the nanoparticulate system for targeting the brain.

This book endeavors to provide readers with an all-encompassing understanding of the clinical translation of nanocarrier systems for treating neurodegenerative disorders. It covers the latest developments, obstacles, safety concerns, toxicity issues, regulatory factors, future potential, and limitations. Individuals in academia, the scientific community, the business sector, and the educational field seeking a more effective strategy for brain targeting stand to benefit from this resource. The present book offers a comparative analysis of different nanocarrier systems, providing valuable assistance to researchers in choosing appropriate nanoparticulate carriers. It sheds light on the existing challenges related to brain delivery and available medications, offering comprehensive insights into the diverse biomaterials employed in the development of nanocarriers for brain targeting, along with their advantages and disadvantages. Additionally, it emphasizes specific aspects of surface functionalization tailored to the brain area of interest and explores the latest advancements, preclinical and clinical developments, and the future potential of various technologies for targeting the brain. This wealth of information serves as a valuable resource for students, researchers, and emerging scientists in guiding the selection, design, and implementation of their work for the advancement of society.

Drawing from a background rooted in neuropsychiatric health research, Dr. Mohamed, the editor, has chosen to collaborate with colleagues possessing expertise in nanotechnology to create a comprehensive book. In essence, this new publication introduces updated and innovative concepts at the intersection of neurological disorders and nanotechnology. It is anticipated to be highly engaging for researchers and clinical scientists involved in neuropsychiatry, neurology, and biotechnology. The editorial team extends gratitude to all the authors for their significant efforts in crafting excellent chapters for this edition. Additionally, sincere appreciation is extended to each author for their patience throughout the compilation and final editing phases of this book.

Wael Mohamed, MBBCH, MMSc, MD, PhD
Editor
IIUM, Kuantan, Pahang, Malaysia

Acknowledgments

First, I would like to send a great appreciation for all the authors who contributed to this timely project. The high level of devotion and dedication between the authors and editors made writing this book an enjoyable journey. In addition, I also extend our gratefulness for the authors who are in the fields of neurology and neuropsychiatric research for providing years of their experience and work in different areas of psychiatric disorders to deliver such an elegant piece of work. The herein discussed topics and applications are of a great value in the areas of biotechnology, neurological disorders, and neurodegeneration. Finally, I would like to thank the encouragement of many of my friends and colleagues for their unconditional love, encouragement, and inspiration throughout the endeavor of the project. Thank you.

Wael Mohamed, MBBCH, MMSc, MD, PhD
Editor
IIUM, Kuantan, Pahang, Malaysia

Contributors

Heba Abd-El-Azim
Damanhour University
Damanhour, Egypt

Nada Abdel-Monaem
Zagazig University
Zagazig, Egypt

Mohamed Abdelkader
American University in Cairo
Cairo, Egypt

Noha AboQuella
Charité University of Medicine
Berlin, Germany

Nermeen Z. Abuelezz
Instituto de Neurociencias de Alicante
Universidad Miguel Hernández-CSIC
San Juan de Alicante, Spain

Osama M. Ahmed
Beni-Suef University
Beni-Suef, Egypt

Grace Temitope Akingbade
Federal University of Technology
Akure, Ondo State, Nigeria

Victor Ebube Anadu
Federal University of Technology
Akure, Ondo State, Nigeria

Mai M. Anwar
National Organization for Drug Control and Research (NODCAR)/Egyptian Drug Authority (EDA)
Cairo, Egypt

Asmaa Ashraf
Helwan University
Giza, Egypt

Hadeer N. Atalah
American University in Cairo
New Cairo, Egypt

Gamal Abdel Nasser Atia
Suez Canal University
Ismailia, Egypt

Emad B. Basalious
Cairo University
Al-Kasr Al-Aini, Cairo, Egypt

Behloul Soundous Malek
Benbadis University Hospital
Constantine, Algeria

Achouak Boutalbi
University of Batna 2
Algeria

Sudarshan Naidu Chilamakuri
National Institute of Pharmaceutical Education and Research Kolkata
Kolkata, West Bengal, India

Mohamed Dahab
Helwan University
Giza, Egypt

Mona G. El-Sisi
Ain Shams University, Abassia
Cairo, Egypt

Abdullah M. M. Elbadry
Badr University in Cairo (BUC)
Badr, Cairo, Egypt

Michael Oluwatosin Gabriel
University of Warsaw
Warsaw, Poland

María-Salud García-Ayllón
Universidad Miguel Hernández-CSIC
San Juan de Alicante, Spain

Nadia M. Hamdy
Ain Shams University
Abassia, Cairo, Egypt

Hamsa Hassan
Mansoura University
Mansoura, Egypt

Victor Hmingthansanga
National Institute of Pharmaceutical Education and Research Kolkata
Kolkata, West Bengal, India

Olayemi Kafilat Ijomone
The Neuro-Lab, School of Basic Medical Sciences (SBMS)
Federal University of Technology
Akure, Ondo State, Nigeria

Omamuyovwi Meashack Ijomone
The Neuro-Lab, School of Basic Medical Sciences (SBMS)
Federal University of Technology
Akure, Ondo State, Nigeria

Happiness Ochi Inyang
University of Medical Sciences, Ondo
Ondo State, Nigeria

Mona Mohammed Abdel Hakeem Khalifa
Cairo University
Cairo, Egypt

Ahmad S. Kodous
Saveetha Institute of Medical and Technical Sciences
Saveetha Dental College & Hospitals
Chennai, India

Khaled M. Kotb
Mansoura University
Mansoura, Egypt

Nahed S. Lamloum
Beni-Suef University
Beni-Suef, Egypt

Rajukumar Malayandi
National Institute of Pharmaceutical Education and Research Hajipur
Vaishali, Bihar, India

Sivakumar Manickam
Universiti Teknologi Brunei
Bandar Seri Begawan, Brunei

Yasser Mecheri
Benbadis University Hospital
Constantine, Algeria

Adejoke Elizabeth Memudu
Neuroscience Unit Faculty of Basic Medical Sciences
Edo State University Uzairue
Edo State, Nigeria

Shivakumar M. Methri
National Institute of Pharmaceutical Education and Research Kolkata
Kolkata, West Bengal, India

Eman E. Mohamed
Beni-Suef University
Beni-Suef, Egypt

Hend H. Mohamed
Badr University in Cairo
Badr City, Cairo, Egypt

Arab Iram Saba Gaus Mohammed
National Institute of Pharmaceutical Education and Research Kolkata
Kolkata, West Bengal, India

Fayza Eid Nasr
Alexandria University
Alexandria, Egypt

Subramanian Natesan
National Institute of Pharmaceutical Education and Research
Kolkata, West Bengal, India

Haseen Fatima Niazi
University of Central Punjab
Lahore, Pakistan

Islam Omar
South Valley University
Qena, Egypt

Esraa M. Qansuwa
American University in Cairo
New Cairo, Egypt

Priya Ramanathan
Cancer Institute (WIA)
Chennai, Tamilnadu, India

Arunagiri Kuha Deva Magendhra Rao
Cancer Institute (WIA)
Chennai, Tamilnadu, India

V. Ravichandiran
National Institute of Pharmaceutical Education and Research
Kolkata, West Bengal, India

Javier Sáez-Valero
Universidad Miguel Hernández–CSIC
San Juan de Alicante, Spain

Amgad E. Salem
Mansoura University
Mansoura, Egypt

Al-Aliaa Sallam
Ain Shams University
Abassia, Cairo, Egypt

Moshera Samy
University of Deusto, Avda
Bilbao, Spain

Fatma H. Shaker
Ain Shams University
Abassia, Cairo, Egypt

Mohamed N. Sharafeldin
Mansoura University
Mansoura, Egypt

Rahma Soliman
Helwan University
Giza, Egypt

Nouran A. Taha
MARS–Global
London, United Kingdom

Ravichandiran Velayutham
National Institute of Pharmaceutical Education and Research Kolkata
Kolkata, West Bengal, India

Ahmed Yahya
Egypt–Japan University of Science and Technology
Alexandria, Egypt

Mohamed Y. Zaky
Beni-Suef University
Beni-Suef, Egypt

Mariam Zewail
Damanhour University
Damanhour, Egypt

Abbreviations

AAs	amino acids
Aβ	beta-amyloid
AD	Alzheimer's disease
ADP	adenosine diphosphate
ADSC-Exo	adipose-derived stem cell exosomes
AJs	adherens junctions
AKT	protein kinase B
ALS	amyotrophic lateral sclerosis
AMPA	α-amino–hydroxyl-5-methyl-4-isoxazole-propionate
APP	amyloid precursor protein
ASC-Exo	astrocytic exosomes
ATG	autophagy-related gene
ATP	adenosine triphosphate
BACE1	β-site amyloid precursor protein cleaving enzyme-1
BBB	blood–brain barrier
Bcl	B-cell lymphoma
Bcl-2	B-cell lymphoma 2
BME	brain microenvironment
BCNU	O6-benzylguanineand carmustine carmustine
BG	O6-benzylguanine
BRB	blood–retina barrier
BtT	brain-to-tumor
CBCLR	carbohydrate-binding C-type lectin receptors
CBF	cerebral blood flow
CD	cluster of differentiation
CDK	cyclin-dependent kinase
CH	catechin hydrate
CHO	carbohydrates
CNS	central nervous system
CPPs	cell-penetrating peptides
CS	chitosan
CSNPs	chitosan nanoparticles
CSF	blood–cerebrospinal fluid
CTCF	CCCTC-binding factor
Cur	curcumin
CVO	circumventricular organs
DAMPs	mitochondria-derived damage-associated molecular patterns
DB	database
DBP	vitamin D–binding protein
DDS	drug delivery system
DNA	deoxyribonucleic acid
DNMT	DNA methyltransferase
Drp1	dynamin-related protein 1
dsDNA	double-stranded DNA
DOE	design-of-experiment
DOX	doxorubicin
EbD	exosomes-by-design
ECs	endothelial cells
ECM	extracellular matrix
EE%	encapsulation efficacy
EGCG	epigallocatechin-3-gallate
EM	exosome-mimetic
EPR	permeability and retention effect
ETO	etoposide
EVs	extracellular vesicles
Exo	exosomes
FGF2	fibroblast growth factor 2
GABA	gamma-aminobutyric acid
GAL	galantamine
GBM	glioblastoma
GJs	gap junctions
GLUT1	glucose transporter protein 1
GM1	monosialotetrahexosylganglioside
GM3	trisialotetrahexosylganglioside
GSK-3	glycogen synthase kinase-3
HAT	histone acetyltransferase
HD	Huntington's disease
HEs	hybrid exosomes
HOX	homeobox
HSP	heat shock protein
HupA	huperzine A
Htt	huntingtin
IDH	isocitrate dehydrogenase
IONPs	iron oxide nanoparticles
IoNTs	Internet of Nanothings
IN	intranasal administration
LAMs	leukocyte adhesion molecules
LC3	light chain 3
Lf	lactoferrin
Lf-PLGA-PEG NPs	lactoferrin-poly (lactic-co-glycolic acid)-poly(ethylene glycol) nanoparticles
LOP	loperamide
lncRNA	long non-coding RNA
LNP	lipid nanoparticle
MCT1	monocarboxylate transporter 1
MDEs	macrophage-derived exosomes
MEG3	maternally expressed gene 3
MEM	memantine hydrochloride
Mfn	mitofusins
MGMT	O6-methylguanine-DNA methyl transferase
mHtt	mutant huntingtin protein
miRNAs	micro-RNAs
MPEG	methoxy poly (ethylene glycol)
MPEG-PCL	methoxy poly (ethylene glycol)-poly (ε-caprolactone) co-polymer
mRNA	messenger RNA
MSCs	mesenchymal stem cells
mtDNA	mitochondrial DNA
mtDNMT	mitochondrial DNA methyltransferase

mTORC1	mammalian target of rapamycin complex 1
MTT	multi-target therapy
MVBs	multivesicular bodies
NbD	nano-by-design
ncRNA	non-coding RNA
ND	neurodegenerative
NDD	neurodegenerative diseases
NEAT1	nuclear enriched abundant transcript 1
NFT	neurofibrillary tangles
NGNM	next-generation nanomedicine
NK	natural killer
NLRP3	NLR family pyrin domain containing 3
NM	nanomedicine
NMDA	N-methyl-D-aspartate
NPs	nanoparticles
NRF	nuclear respiratory factors
NRSF	neuron-restrictive silencing factor
NT	nanotechnology
NtB	nose-to-brain
NVs	nanovesicles
OXPHOS	oxidative phosphorylation
PBCA	poly (butyl cyanoacrylate)
PBCA-CS	poly (butyl cyanoacrylate)-co-chitosan co-polymer
PCs	pericytes
PCL	poly(ε-caprolactone)
PD	Parkinson's disease
PDCD4	programmed cell death 4
PGC-1α	peroxisome proliferator-activated receptors (PPAR) gamma coactivator-1α
PDGF-B	platelet-derived growth factor B
PDI	polydistribution index
PEG	poly (ethylene glycol)
PEG-PLA	poly (ethylene glycol)-co-poly (lactic acid) co-polymer
PEG–PLGA	poly (ethylene glycol)-poly (lactic-co-glycolic acid) co-polymer
PI3P	phospholipid phosphatidylinositol-3-phosphate
PINK1	phosphatase and tensin homolog
PLA	poly (lactic acid)
PLGA	poly (lactic-co-glycolic acid)
PLGA/CS NPs	poly(lactide-co-glycolides)/chitosan nanoparticles
PNs	polymeric nanoparticles
PM	personalized medicine
polyQ	polyglutamine
PrPC	cellular prion protein
PS	particle size
PS1	presenilin-1
PINK1	(PTEN)-induced putative kinase 1
RAGE	receptor for advanced glycation end products
RMT	receptor-mediated transcytosis
RNPs	rotigotine-loaded chitosan nanoparticles
ROS	reactive oxygen species
rRNA	ribosomal RNA
SEM	scanning electron microscope
SGLT1	sodium glucose cotransporter 1
siRNA	silencing RNA
SIRT1	sirtuin 1
SIT-CS-NPs	sitagliptin loaded chitosan nanoparticles
SNCA	synuclein alpha
SNHG1	small nucleolar RNA host gene 1
SOD-1	Cu/Zn superoxide dismutase
TAR	trans-activating response element
TBE	temozolomide butyl ester
TEM	transmission electron microscope
TDP43	TAR DNA-binding protein 43
TFAM	mitochondrial transcription factor A
TFEB	transcription factor EB
TfR	transferrin receptor
TJs	tight junctions (TJs)
TLR	toll-like receptors
TMZ	temozolomide
TNF-α	tumor necrosis factor
TUG1	taurine upregulated gene 1
Ub	ubiquitin
ULK	unc-51-like autophagy-activating kinases
VEGF	vascular endothelial growth factor
WHO	World Health Organization
Wnt5a	Wnt family member 5a
ZP	zeta potential

1 Fundamentals of Blood–Brain Barrier Anatomy and Global Functions

Yasser Mecheri and Behloul Soundous Malek

1.1 INTRODUCTION

In this introductory chapter, we will explore the fundamental basis of the functioning of the blood–brain barrier (BBB). We will also explore its cellular components and structural characteristics and the different pathways allowing the transport of molecules into the brain. Such knowledge is crucial for the understanding of all the limitations facing the development of drugs intended to pass through the BBB and the consequent necessity of developing nanocarriers to bypass these limitations.

The vascular system of the brain is characterized by a dense capillary network distinguishable from other tissue vasculature by the presence of the BBB (Profaci et al., 2020). The BBB, unique by its composition, structure, and function, has a key role in the protection and the functioning of the central nervous system (CNS), which, because of its functional importance, requires continuous tight control of the cerebral blood flow (CBF) and the metabolism and exchange of substances with the blood (Attwell et al., 2010; Iadecola, 2013; Zhao et al., 2015). BBB serves as an exclusive tight bridge allowing limited passage of these substances between the blood and the neuronal compartments (Dyrna et al., 2013).

The blood–brain barrier was initially described in 1885 by Paul Ehrlich in his book *Das Sauerstoffbedürfnis des Organismus. Eine Farbenanalytische Studie*; afterward, his student Lewandowski followed on his steps and reached the conclusion that "brain capillaries must hold back certain molecules" from penetrating the brain (Dyrna et al., 2013; Lewandowsky, 1909). But it was the insights offered by the electron microscopy that allowed Reese and Karnovsky to finally describe the exact morphology of the BBB and its continuous barrier composed of overlapping endothelial cells (ECs) held tightly to each other by the tight junctions (TJs) system (Reese & Karnovsky, 1967). BBB along with the blood–cerebrospinal fluid (CSF) barrier formed by the ECs in the choroid plexus and the vascular arachnoid epithelium constitute the three barriers separating the brain from the rest of the body (Abbott et al., 2006).

The BBB is a highly selective semi-permeable membrane with a unique tight structure made of many layers interposed between the blood and the brain. The endothelial cells composing cerebral capillary walls are—unlike other organs—closely linked by TJs; in addition, the interposition of the basal lamina and the presence of many pericytes (PCs) and astrocytes all participate in the high selectivity of the BBB (Dong, 2018) and the sealing of the intercellular gap, preventing any inadvertent paracellular leakage (Dyrna et al., 2013).

BBB has special characteristics allowing the strict control and regulation of the movement of metabolites, hormones, drugs, and blood cells between the cerebral interstitium and the blood (Daneman & Prat, 2015) to create an optimal environment for neuronal activity (Zhao et al., 2015), while keeping toxic plasmatic substances and pathogens outside the brain (Montagne et al., 2017). BBB participates also in CNS immune privilege, which is an attenuated immune response to alloantigens and pathogens; this can provide inadvertent protection for certain pathogens when they succeed in penetrating into the brain after having passed the physical and functional BBB (Forrester et al., 2018). BBB dysfunction and disintegration leads to the leakage of toxic substances and cells into the brain with altered clearance of molecules that can contribute to neuronal dysfunction and accelerated neurodegeneration processes in neurodegenerative disorders (Zlokovic, 2011).

1.2 WHY STUDY THE BBB?

Since BBB strictly monitors blood–CNS substances exchange and is crucial to the protection and maintenance of brain functions, its dysfunction and damage are seen in many acute and chronic CNS disorders and aggressions and will negatively influence CNS function (Zhao et al., 2015). Therefore, BBB alterations have been a major focus in CNS research aiming to explain the underlying pathological mechanisms and the possible interventions allowing the maintenance of the BBB and to reverse the fluctuations of vessel permeability in such disorders. This can potentially allow the identification of efficient methods for diagnostics and therapy. Another important area of interest is to understand the pathological mechanisms with the capacity to bypass the BBB and to try replicating them in order to allow the transport of drugs and other molecules incapable of penetrating the BBB because of the restrictions imposed by its normal functioning. The ultimate goal is to develop technologies that are safe, reversible, and able to act on BBB permeability for a defined period without causing irreversible damage.

DOI: 10.1201/9781003383376-1

1.3 BBB COMPOSITION

Physical components of the BBB include several cell types, special cell junctions, and the extracellular matrix. In this section, we will describe the most important BBB elements.

1.3.1 Cellular Components

At least five cell types contribute to the neurovascular unit (NVU) composition and maintenance: endothelial cells, pericytes, astrocytes, neurons, and microglia. Since the BBB is mainly a vascular structure, its cellular composition is parallel to that of other blood vessels, with ECs playing the most important role and other mural cells (muscular smooth cells and pericytes mainly) sitting on top of the abluminal surface and participating in the structural and functional integrity of the BBB (Daneman & Prat, 2015).

1.3.1.1 Endothelial Cells

Epithelial cells are the most important component of the BBB; they are the staple BBB constituent and its selectivity pillar, as they represent its main physical and functional barrier (Armulik et al., 2010). Their interaction with mural cells maintain BBB structure and function. ECs also interact with several other cell types, such as immune cells, neural cells, and glial cells, to regulate, maintain, and protect the BBB (Daneman & Prat, 2015). ECs are first derived from the mesoderm then differentiate into modified simple squamous epithelial cells lacking fenestrations but having a particularly important number of molecular transporters. They form afterward a tight, thin, and continuous barrier at the inner part of blood vessels with a direct contact with blood (Aird, 2007). ECs allow a highly selective control and regulation of molecules, cells, and plasma ions movement through the BBB (Obermeier et al., 2013) thanks to the association of a solid physical barrier with abundant TJs limiting the paracellular transport (Brightman & Reese, 1969; Reese & Karnovsky, 1967), their limited transcellular traffic linked to reduced number of pinocytotic vesicles (Coomber & Stewart, 1985) and a functional barrier maintained by specific transporters and reduced passage of immune cells (Armulik et al., 2010). Efflux pumps and receptor-mediated transcytosis (RMT) constitute the most important membranous selective transporters with a specificity to certain molecules and nutriments, and following a defined direction into or out of the brain, an increased consumption of energy is a natural result of such specific but power-hungry transport mechanisms and therefore follows the need for a high number of mitochondria within the ECs to provide sufficient amounts of ATP. Leukocyte adhesion molecules (LAMs) are reduced in number in ECs, limiting immune cells' diffusion to the CNS (Kolářová et al., 2014).

1.3.1.2 Mural Cells

The capillary wall is formed by two types of mural cells: vascular smooth muscle cells covering partially the ECs, and CNS pericytes, which, like peripheral PCs, control capillary diameter and, therefore, vascular permeability using the contractile proteins in their long cellular extensions in contact with the subluminal part of the endothelial wall (Hall et al., 2014).

1.3.1.3 Pericytes

Pericytes are contractile perivascular cells covering, depending on vessel type, between 22% and 99% of the endothelial cell surface by their extensive cytoplasmic fibroblast-like feet. They are identified by the presence of CD248 and PDGFRβ (platelet-derived growth factor B receptor) as surface markers (Varga & Lafyatis, 2015).

PCs participate in the constitution of the basal membrane and communicate with ECs via paracrine signaling pathways (Caplan et al., 2017) to regulate vascular permeability in BBB and all tissues (Armulik et al., 2011). They are considered a second line of defense, as they have macrophagic functions. They also interact with ECs and play many important roles, including BBB-specific enzymes production, endothelial permeability modulation, microvessel wall stabilization, and angiogenic processes promotion (Daneman & Prat, 2015). This allows them to take part in the BBB function regulation and in the maintenance of BBB by the conservation of ECs integrity. It is admitted that pericytes downregulate the expression of molecules that enhance vascular permeability and facilitate the entrance of immune cells into the CNS (Armulik et al., 2010). However, the regulation of the ECs BBB-specific gene expression is not entirely proven (Daneman et al., 2010). They also play a role in astrocyte functionalities control as they induce the polarization of astrocytes end feet (Armulik et al., 2010).

The CNS PCs are recruited in the embryonic period (Daneman et al., 2010) following the secretion of platelet-derived growth factor B (PDGF-B) by ECs (Andrae et al., 2008; Bjarnegård et al., 2004; Enge et al., 2002; Goel et al., 2011; Hellström et al., 1999). By interacting with ECs, PCs occupy an important role in the modulation of BBB development in early stages of life as they inhibit "leaky" properties by suppressing LAM expression and macromolecules transcytosis (Alonso-Alonso et al., 2022; Armulik et al., 2010; Daneman & Prat, 2015).

The disruption of pericytes leads to BBB dysfunction and CNS neuroinflammation (Daneman et al., 2010). They are also implicated in various brain injuries, like ischemic stroke and Alzheimer's disease (Shah & Abbruscato, 2017). Research performed on mice showed that pericyte disturbance increased the permeability of the BBB to many substances, including water and a number of low-molecular- and high-molecular-mass substances, by promoting endothelial transcytosis (Armulik et al., 2010). Also, recently, it was found that BBB pericytes can be a target of HIV-1 infection and help HIV-1 replication (Bertrand et al., 2019). Furthermore, it has been shown that the total absence of pericytes is lethal for embryonic cells (Obermeier et al., 2013).

1.3.1.4 Astrocytes

Astrocytes represent the most frequent and phenotypically diverse non-neuronal CNS cells (about 50% of all cells) (Emsley & Macklis, 2006). They are a type of glial cells that wrap around synapses, nodes of Ranvier, and blood vessels. One astrocyte has the capacity to extend over a huge number of synapses (up to 2 million) at the same time (Oberheim et al., 2009). They are responsible for neurotransmitters uptake (Rothstein et al., 1996), the support of neuronal metabolism (Banker, 1980), the regulation of ions concentration, the repairs in the CNS following aggressions and damage, the regulation of vascular properties depending on neuronal activity, and the maintenance of BBB (Bush et al., 1999). However, astrocytes cannot be the inductors of BBB formation (Ben-Zvi et al., 2014; Daneman et al., 2010; Sauvageot & Stiles, 2002; Sohet et al., 2015) since they appear starting from the 18th day of embryonic life, three days after the establishment of BBB properties. Ageing-induced transcriptional changes in astrocytes and subsequent activation promoted by microglia induce a phenotype similar to neuroinflammatory reactive astrocytes (Clarke et al., 2018). Also, dysfunctional astrocytes play an important role in inducing CNS age-related neurodegenerative disorders, like dementia (Hallmann et al., 2017) and ALS (Nagai et al., 2007).

1.3.2 BBB Junctional Components

BBB protein complex junctions ensure the most important property of the BBB, which is the high selective permeability by a strict regulation of paracellular molecule passage. These junctional proteins comprise tight junctions (TJs), adherens junctions (AJs), and gap junctions (GJs) and are localized between the cerebral endothelial cells. They have a specific structural disposition and are made following a unique protein expression pattern (Stamatovic et al., 2016). There are many similarities of junctions between epithelial cells and ECs; however, epithelial cells have the particularity to also form desmosomes in addition to TJs and AJs, unlike ECs (Bazzoni et al., 1999). The distribution of AJs and TJs is also different between the two cell types; TJs and AJs are better organized within the epithelial cells, with TJs being more apical to the rim of the intercellular cleft and AJs being located below them. In ECs, this organization is less clear, and TJs and AJs are intermingled (Ichimura et al., 2008). TJs and AJs, despite having different protein structure, have functional similarities and are involved in adhesion between cells, in intracellular anchorage into actin cytoskeleton, and in signaling pathways, including those dedicated to gene transcription regulation (Hartsock & Nelson, 2008).

1.3.2.1 The Tight Junction Complex

Tight junctions of the BBB ensure direct adherence between endothelial cells and are more similar to those found in epithelial cells than to those found on ECs of peripheral blood vessels (Kniesel & Wolburg, 2000). TJs are localized along the lateral membrane of ECs, and as their name indicates, they are highly linked together. The transmembrane adherens protein complex has a parallel series of multiple barriers made of an anastomotic protein strand interacting with their homologues on the other side.

Occludin and claudin are two transmembrane proteins composing TJs and allowing them to perform their two major functions: the first, a fence function, limiting membrane proteins located at the apical part of the membrane from mixing with ones at the laterobasal parts of the membrane; the second, a transport control function, restricting paracellular transport and ion exchange between cells (Hartsock & Nelson, 2008). TJs have also many linked cytoplasmic proteins—ZO-1, ZO-2, ZO-3, cingulin, and 7H6—that allow their anchoring to the actin cytoskeleton and to AJs, which gives them a structural support (Bazzoni & Dejana, 2004; Konari et al., 1995). TJs participate in intracellular signaling (Bazzoni & Dejana, 2004) and use many well-known signaling pathways, including G-proteins, serine, threonine, tyrosine kinases, cAMP levels, proteases, TNF alpha, and calcium levels in the intra- and extracellular space (Kniesel & Wolburg, 2000).

1.3.2.2 Adherens Junctions

Adherens junctions have a similar organization and morphology to TJs despite having different components and being less tight than TJs. The most important proteins involved in AJs' structural composition are intracellular p120-catenin, beta-catenin, and alpha-catenin and transmembrane E-cadherin, which allow AJs to have linking and signaling roles (Bazzoni & Dejana, 2004). E-cadherin, a classical cadherin from the Ca2+-dependent adhesion proteins family, allows AJs to initiate intercellular contacts by binding to another E-cadherin protein from the other side (Gooding et al., 2004; Gumbiner, 2005). However, E-cadherin is not required for the maintenance of mature junctions, while the catenins are more essential for the maturation of the AJs, by creating a direct connection with the actin cytoskeleton. They are also important for the maintenance of functional junctions by the control of the adhesive state of the E-cadherins (Capaldo & Macara, 2007; Gumbiner, 2005; Hartsock & Nelson, 2008). The catenins also play a major role in the AJs' signaling role, by interacting with signaling molecules that regulate the actin cytoskeleton (Gumbiner, 2005).

1.3.2.3 Gap Junctions

Gap junctions are present in all solid tissue cells and not only in excitable cells, like initially described. They are intercellular channels that allow cell-to-cell direct passage of ions and small particles and also the transduction of signals between adjacent cells, allowing direct communication between them (Goodenough & Paul, 2009; Stamatovic et al., 2016). GJs are composed by the docking of two opposing connexons, excluding most other membrane proteins, and only leaving a small extracellular "gap" of 2 nm, which they are named after (Goodenough & Paul, 2009). *Connexons* are hexamers of medium-size families of integral proteins, connexins (Goodenough et al., 1996), and *connexins* are a family of 21 members of transmembrane proteins that form the major part of gap junctions in chordates (Söhl & Willecke, 2004). Connexins may co-oligomerize into the same connexons (i.e. homomeric) or form heteromeric connexons when they mix with another type of connexins (Falk, 2000), particularly three types of them (Cx43, Cx40, and Cx37) are expressed in ECs. Gap junctions Cx43 allow the direct communication between pericytes and ECs (Sweeney et al., 2018; Winkler et al., 2011). Gap junction Cx30 and Cx43 are present in astrocytes and regulate their end feet perivascular polarity (Ezan et al., 2012). Cx43 is essential to maintain immune quiescence in the CNS.

1.3.2.4 Junction Proteins in BBB Dysfunction

Alterations in junction protein structure caused either by the changes in protein expression or by post-translational modifications contribute to BBB dysfunction as they alter BBB permeability; disrupt junction function and, subsequently, transcellular exchange; and change endothelial metabolic processes. These changes will significantly affect BBB properties, particularly barrier permeability, the remodeling of brain ECs surface, and the regulation of brain ECs phenotype (Stamatovic et al., 2016).

1.3.3 Extracellular Matrix and Basement Membrane

The importance of the extracellular matrix (ECM) is often forgotten and underestimated. It is involved in the regulation of BBB properties and is considered a structural support for the vessels. It also constitutes a physical obstacle for cell migration into the brain and a gateway for growth factors. The ECM is composed of several structural proteins: collagens, fibronectin, laminins, integrins, heparin sulfate proteoglycans, and chondroitin sulfate proteoglycans (Correale & Villa, 2009), which create transmembrane heterodimers and interact with ECs cytoskeleton, allowing their anchorage into the ECM (Tietz & Engelhardt, 2015). Integrins are implicated in regulating ECs differentiation, adhesion, migration, and survival, by inducing the expression of ECM ligands and growth factors (Baeten & Akassoglou, 2011). The dysfunction of integrins, laminin 1, and laminin 2 leads to severe BBB abnormalities, including the destruction of the basement membrane, the loss of astrocyte end feet polarity, the absence of TJs or their reduced density, increased permeability, and inefficient filtering of molecules (Menezes et al., 2014; Yamamoto et al., 2015; Yao et al., 2014).

1.4 DEVELOPMENT

Primitive BBB is formed during the early embryonic period from the neuroectoderm vessels. In humans and mammals, it is not yet clear whether BBB is fully functional at birth or not (Daneman et al., 2010; Ek et al., 2012; Saunders et al., 2012), whereas mature cells, including astrocytes and myelinated neurons, are detected in the cerebral cortex after birth (Daneman et al., 2010; Obermeier et al., 2013; Yang et al., 2013). More certain is that the BBB will go through many developmental phases to arrive at its final fully functioning form (Engelhardt, 2003). Three stages are individualized:

- **First Stage:** Studies performed on mice showed that the development of BBB starts around the 10th embryonic day (Ben-Zvi et al., 2014; Daneman et al., 2010). Brain endothelial cells are derived from the angioblasts of permeable vascular plexus; they afterward migrate from the perineural region and colonize the neuroectoderm (Risau et al., 1998). The angiogenic process is stimulated by the vascular endothelial growth factor (VEGF) (Potente et al., 2011) and is regulated by Wnt, which contribute in both angiogenesis and barrier genesis. The resulting intraneural vessels lack BBB (Cho et al., 2017).
- **Second Stage:** The newly formed vessels will become more mature and develop the BBB characteristics and phenotype by the establishment of TJs and the transport systems for hydrophobic compounds. The reason only ECs located in the CNS—but not those from the periphery—are capable of forming TJs is still debatable (Sohet et al., 2015). Lipolysis-stimulated lipoprotein receptor (LSR) was described by Ikenouchi et al. (2005) as protein-forming tricellular TJs necessary to the recruitment of tricellulin to the tricellular TJs of ECs (Masuda et al., 2011; Sohet et al., 2015). LSR knockout mice are uncapable of surviving beyond embryonic day 15.5 (E15.5), but the cause of death is unknown (Mesli et al., 2004).
- **Third Stage:** In this last stage, pericytes will appear and cover the ECs. The basement membrane will also develop and acquire the protective role being an electrostatic selective filter for charged macromolecules. The known differentiating features of ECs—important number of TJs, reduced transcytosis, increased transporter function and expression, and decreased expression of LAMs—will all afterward be maintained for life by contact with astroglia (Obermeier et al., 2013).

Recent research is concentrating on CNS-specific pathways (Wnt/beta-catenin) and molecules (GPR124) necessary to better understand brain angiogenesis mechanisms and the BBB development (Engelhardt & Liebner, 2014). It is important to mention that it has been demonstrated in animals that some peripheral signals could modulate BBB development, properties, and function; for example,

CBF generates a mechanical stimulation on ECs (Cucullo et al., 2011). Also, the alteration of gut flora in the mother increases BBB permeability by the reduction of expression of some molecules (Braniste et al., 2014).

1.5 ORGAN SYSTEMS INVOLVED

The NVUs—structurally and physiologically dynamic structures—are found in most CNS capillaries and postcapillary venules. They are the basis of the brain's restricted and highly regulated exchange system, as neurovascular coupling is the central component of the BBB and the blood–retina barrier (BRB) (Kugler et al., 2021).

The circumventricular organs (CVOs) represent an exception to this rule in the CNS by lacking BBB and having fenestrated capillaries, allowing free passage and exchange of substances between the blood and the brain and with the CSF. This open space allows these CVOs to perform their very specialized and important functions, being specialized neuroepithelial structures with regulated sensory and secretory neuroendocrine functions (Kiecker, 2018).

There are mainly seven major CVOs, grouped around the midline of the brain near the third and fourth ventricles. They are divided according to their function into four secretory CVOs (the medial eminence, the neuro pituitary gland, the pineal gland, and the subcommissural body) and three sensory CVOs (the area postrema, the subfornical organ, and the organum vasculosum of the lamina terminalis) (Kaur & Ling, 2017). There is no common ontological and developmental origin of the CVOs, but mainly a structural and functional homology (Kiecker, 2018). Newly published MRI studies have shown that dura mater lymphatic vessels that are in direct contact with peripheral immune system cells are also implicated in the CNS immune response and also in toxin transport (Absinta et al., 2017).

1.6 FUNCTIONS OF THE BBB

The primary function of the BBB is to provide a strict control of molecules transport and exchange between the blood and the brain. Every cellular type has an important role to maintain a stable homeostasis compatible with normal brain function and also in the protection from aggressions. While capillary ECs and pericytes are the most important for molecular transport and regulation of substances' access to the brain, astrocytes are more involved in maintaining BBB integrity. We should note that not all molecules submit to this passage regulation, as gases and lipophilic molecules smaller than 400 Da can pass the BBB freely (Zeisel et al., 2015).

We can divide the BBB into three distinct barriers based on their functions:

- The "physical" barrier. Its main role is to prevent paracellular diffusion between ECs. It is made of AJs and TJs that are composed of structural proteins, including claudins, occludins, and other transmembrane proteins held together by zona occludins proteins, ensuring the anchorage into the cytoskeleton (Abbott et al., 2006).
- The "transport" barrier includes transport proteins, endocytosis receptors, and ATP-dependent proteins on the ECs, permitting nutrients, ions, and molecules influx and toxins expulsion.
- The "metabolic" barrier is composed of enzymes specialized in the metabolism of substances entering into the ECs, such as P450 and monoamine oxidase (Shlosberg et al., 2010).

1.7 MECHANISMS OF TRANSPORT ACROSS THE BLOOD–BRAIN BARRIER

ECs of the cerebral capillaries have some structural particularities, allowing its special functions. Compared to peripheral ECs, CNS ECs have abundant TJs, which ensure strict molecule transport control. The cytoplasmic vesicles are less abundant than what is usually observed due to limited storage function and limited transcytosis (Oldendorf & Davson, 1967). Also, due to the high energy demand required by active molecules transport, the number of mitochondria is significantly higher (Oldendorf et al., 1977). Every EC's membrane side has a different disposition of membrane transporters and structures, depending on if it is facing the blood or the brain, to adapt to the type of molecules transported (Abbott & Romero, 1996).

1.7.1 Diffusion

Diffusion is a passive mechanism requiring no energy and following the rules of osmosis. Small water-soluble molecules enter the brain through the paracellular space between the TJs. Small lipid-soluble substances like alcohol pass the BBB transcellularly through the cellular membrane lipids (Chen & Liu, 2012).

1.7.2 Efflux Pumps

Efflux pumps ensure the excretion of endogenous and exogenous molecules out of the brain.

Active drug efflux transporters encoded by the ATP-binding cassette (ABC) gene superfamily play a particularly important role in the elimination of drugs, nucleosides, and medications. The superfamily contains 48 transporters divided into 7 sub-families (Dean et al., 2001). The most important members are ABCB1 (or P-glycoprotein), ABCA2, breast cancer resistance protein (BCRP), and multidrug resistance–associated proteins (MRP) 1, 2, 3, 4, and 5 (Begley, 2004; Dauchy et al., 2008). These transporters have the side effect of restraining the distribution of useful drugs into the brain, making the brain particularly vulnerable to certain pathogens and inaccessible to most of medical interventions (Löscher & Potschka, 2005).

1.7.3 Ion transport

Ion transport is mediated by ATP-dependent pumps that allow the transport of ions against the concentration gradient to maintain membrane resting potential and enable the excitability of the neuronal membrane and the production of action potentials. The most important pumps are the abluminal sodium pump (Na-KATPase), the luminal sodium-potassium-chloride cotransporter (Na-K-Cl) (Foroutan et al., 2005), the luminal sodium–hydrogen (Na-H) exchanger that transports protons outside of the brain (Yuen et al., 2014), the sodium–calcium ($Na\text{-}Ca^2$) exchanger cotransporter, and the voltage-gated potassic channel KV1 that allows calcium entry into the brain (Kisler et al., 2017; Millar et al., 2008).

1.7.4 Transport Proteins

Because of BBB high selectivity, some substances, such as glucose and amino acids (AAs), need to be transported by binding to a protein transporter that changes conformation and allows the transport following concentration gradient from high to low. Since the concentration of these molecules is always lower in the brain than in the blood, the favored transportation direction is into the brain (Sweeney et al., 2019). Charged molecules need to be transported actively (using ATP) when the concentration gradient is inverted (Chen & Liu, 2012).

1.7.4.1 Carbohydrate Transporters

The glucose transporter protein 1 (GLUT1) has a single binding site that fixes glucose and other hexoses at the extracellular and intracellular parts of both the abluminal and luminal membranes of ECs (Deng et al., 2014). These transporters are more abundant on the abluminal membrane of ECs and are absent from neuronal membrane (Farrell & Pardridge, 1991). Sodium glucose cotransporter 1 (SGLT1), however, is expressed in both ECs and neurons (Elfeber et al., 2004).

1.7.4.2 Amino Acid Transporters

AA transporters present in the BBB are divided, depending on their specificity, to a certain group of AAs. All neutral essential AAs are transported by large neutral endothelial AA transporters 1 and 2 (LAT1/2) (O'Kane & Hawkins, 2003). The cationic AA transporters 1 and 3 (CAT1/3) assure cationic AA transport (Mann et al., 2003; Stoll et al., 1993), while nonessential AAs are eliminated from the brain at the abluminal membrane by sodium-dependent AA transporters ASCT1/2 and GLYT1 (Hawkins et al., 2006).

1.7.4.3 Monocarboxylate Transporters

Monocarboxylate transporter-1 (MCT1) allows the entry into the brain of lactic acid resulting from muscle activity and of ketonic acids released from fat metabolism (Simpson et al., 2007). Both ketones and lactate are complementary energy sources for the brain (Puchowicz et al., 2007).

1.7.4.4 Fatty Acid Transporters

Fatty acids are necessary for brain development and its normal functioning. Fatty acid transport protein 1 and 4 (FATP1/4) (Mitchell et al., 2011), and the MFSD2A, are the most important transporters present on the laminal part of the BBB. They allow the entry of essential fatty acids, particularly unsaturated ones, like omega-3 (Nguyen et al., 2014).

1.7.4.5 Nucleotide Transporters

Concentrative nucleoside transporter-2 (CNT2) and the equilibrative nucleoside transporter-1 and 2 (ENT1/2) allow nucleotides entry to the brain and their usage in the productions of nucleic acids (Cass et al., 1998).

1.7.5 Receptor-Mediated Transcytosis (RMT)

Transcytosis is an alternative transport mechanism for macromolecules. Receptor-mediated transcytosis (RMT) uses specific receptors allowing the selective binding and crossing of certain molecules, like insulin, transferrin, and lipoproteins. These RMT receptors are contained in specific regions in cerebral ECs membranes called clathrin-coated pits. After binding the substances, these pits invaginate into the cytoplasm and form free coated vesicles called endosomes. Finally, once the endosomal acidification is complete, the molecules detach from the receptor and pass to the other side of the membrane (Pulgar, 2018). The most important RMTs in the BBB are mainly expressed at the luminal membrane (Deane et al., 2003) and allow the influx of these substances into the brain: transferrin receptor (TfR) (Bray, 2015; Jefferies et al., 1984), insulin receptor (IR) (Ito et al., 2017; Pardridge, 2005), leptin receptor (LEP-R), and the receptor for advanced glycation end products (RAGE) (Banks, 2001; Golden et al., 1997). LRP1 and LRP2 are lipoproteins transporters expressed on the abluminal side of the BBB and allow the clearance of amyloid-β outside the brain (Deane et al., 2004; Méresse et al., 1989; Shibata et al., 2000).

1.7.6 Adsorptive-Mediated Transcytosis (AMT)

The second transcytosis mechanism is called adsorptive-mediated transcytosis (AMT), or pinocytosis, and uses the energy produced by the big number of mitochondria in ECs. AMT results from the interaction between a positively charged protein and the negatively charged luminal surface of the ECs membrane and leads to exocytosis to the opposite side of the membrane. The movement can happen in both directions across the plasma membrane (Hervé et al., 2008).

1.7.7 Cell-Mediated Transcytosis (CMT)

Unlike previously described transport modalities that allow only substances with certain characteristics or specificities to move across the BBB, cell-mediated transcytosis (CMT) allows any type of substances and even pathogens to cross the BBB unchecked, as it uses immune cells like macrophages and monocytes as a vessel or, more properly, a Trojan horse. A classic example is the diapedesis HIV-infected cells into the brain (Lorin et al., 2020).

1.8 PATHOPHYSIOLOGY

There is a tight relation between CNS disorders and BBB dysfunction seen in both acute and chronic neurological diseases which are, in a significant part, attributed to cerebrovascular dysfunction and vascular pathology (Sweeney et al., 2018). For example, strokes, neuroinflammatory disorders, and several CNS injuries are caused by the damage of the TJ seal. Also, increased BBB permeability is seen in hypoxic and ischemic injuries, inflammatory mechanisms such as septic encephalopathy, HIV-induced dementia, multiple sclerosis, and Alzheimer's disease (Ballabh et al., 2004). In what follows, we will provide some insights and examples.

1.8.1 Ischemic Strokes

Ischemia and hypoxia are deleterious to the functioning of the brain and cause serious complications, like cerebral edema (intracellular and extracellular edema), hemorrhage transformation, seizure, delirium (Kumar et al., 2010; Langhorne et al., 2000). These severe complications are linked to BBB dysfunction caused by hypoxia-induced disruption of TJs structure and the increase in BBB permeability. The edema formed in the intracellular space is induced by the energy failure and the mitochondrial crisis. The extracellular edema is caused by decreased barrier tightness (Gerriets et al., 2009). Many pathophysiological mechanisms explaining these changes have been modelled in animals and were also observed in stroke patients, such as increase in the release of vascular endothelial growth factor (VEGF), of nitrogen monoxide (Ballabh et al., 2004), of pro-inflammatory cytokines, of IL-1h, and of TNFa (Feuerstein et al., 1994; Tarkowski et al., 1997). Hypoxia also induces an early loss of NVU pericytes, an increase in oxidative stress, an increase in DNA damage, and in protein kinase C activation (Shah & Abbruscato, 2017). VEGF has a positive effect on transcytosis and GJs formation between ECs on the one hand, increasing BBB permeability (Hippenstiel et al., 1998), and cytokines secretion in the other, and the resulting upregulation of endothelial and neutrophil adhesion molecules results in the recruitment of leukocytes to the brain across the BBB, which will induce TJs structural disruption and subsequently lead to TJs disorganization and BBB breakdown (Bolton et al., 1998). It has been noted that TJ seals stay relatively preserved during the early phase, and the increase in BBB permeability is limited to small molecules. Major abnormalities appear few hours later, leading to the leakage of bigger molecules (Zhang et al., 2020).

1.8.2 Intraventricular Hemorrhage

Intraventricular hemorrhage (IVG) is a frequent complication of the prenatal and postnatal periods, leading afterward to many sequelae (epilepsy, neurodevelopmental delay, cerebral palsy, and sometimes death) (Ballabh et al., 2004; Guo et al., 2020). Many factors are involved in the development of IVH, like perinatal hypoxia (Antoniuk & da Silva, 2000), chorioamnionitis (DiSalvo, 1998), seizures, respiratory distress syndrome, post-hemorrhagic hydrocephalus, and periventricular hemorrhagic infarction (Volpe, 1989). IVH usually starts in the germinal matrix formed by neuronal and glial precursors and is the consequence of disturbances of CBF, the immaturity and fragility of germinal matrix abundant microvasculature which have an incomplete BBB coverage due to the paucity of TJs and PCs (Ballabh et al., 2004), the immaturity of basal lamina, and the deficiency of glial fibrillary acidic protein (GFAP) in astrocytes end feet (El-Khoury et al., 2006).

1.8.3 Septic Encephalopathy

Septic encephalopathies are associated with severe inflammatory reaction altered neurotransmitter secretion and ECs and astrocytes dysfunction. Such alterations will lead to cerebral edema, decreased CBF, and subsequent BBB alteration (Papadopoulos et al., 1999).

1.8.4 Epilepsy

BBB dysfunction in epilepsy and consequent increased BBB permeability are a regional phenomenon occurring in the same region as the epileptogenic focus in the brain (Nitsch & Klatzo, 1983). These changes positively influence neuronal excitability (Marchi et al., 2007; Rapoport, 2000) via a quick increase in extracellular K+ levels (Kofuji & Newman, 2004) and in the levels of glutamate (Coulter & Eid, 2012). The increases in BBB permeability and in seizure frequency are closely related without a correlation to the amount of neuronal loss (Han et al., 2017; Rigau et al., 2007). It has been demonstrated that in humans and in rat models of temporal lobe epilepsy, an increased perfusion of the epileptic focus was induced by the local release of VEGF (Rigau et al., 2007). This release had a positive neuroprotective effect (McCloskey et al., 2005) through a fast adaptation of the CBF to the metabolic needs of affected brain regions via an immediate NO-induced vasodilatation (Ahmad et al., 2006). The increased release of VGEF also reduced the number of spontaneous epileptiform discharges, suggesting also an antiepileptic effect (McCloskey et al., 2005). However, recurrent seizures caused by a chronic epileptic focus will entertain persistent angiogenic processes and induce chronic BBB alterations and increased permeability. These changes are considered deleterious to the brain tissue (Hayashi et al., 2006) and may lead to the aggravation of the epileptic state and to the development of drug-resistant epilepsy (van Vliet et al., 2007, 2014). The impaired BBB will lose its TJs, allowing chronic immunoglobulins G (IgG) leakage to the interstitial space along with other serum proteins (like albumin), which will all be captured and accumulated in neurons. This has been suggested as the main mechanism explaining the chronic hypoperfusion-induced neuronal dysfunction–described chronic epilepsy (Rigau et al., 2007; Seiffert et al., 2004).

1.8.5 Alzheimer's Disease

AD is the most frequent neurodegenerative disorder and the most frequent type of dementia ("2022 Alzheimer's Disease Facts and Figures," 2022).

Following the two-hit vascular hypothesis, it is believed that oxidative stress and mitotic signaling disturbances can initiate independently Alzheimer's disease, but their coexistence is an obligatory condition to generate the disease pathology (Zhu et al., 2007). The primum event which is considered the first hit is vascular damage, leading to BBB dysfunction and the dysregulation of brain blood flow; the second hit is B-amyloid accumulation (Zhu et al., 2004). These B-amyloid deposits are the hallmark of AD and are detrimental in the development of neuronal injury, leading ineluctably to neuronal death and apoptosis and to dementia clinical symptoms. Vascular damage is multifactorial, with the associated effect of genetic factors (carriers of APOE-ε4 gene) (Genin et al., 2011; Yamazaki et al., 2021); environmental factors, like toxins and pollution; and most importantly, vascular risk factors, like poorly controlled hypertension, diabetes, and dyslipidemia, which all lead to an associated small vessels disease and a mixed dementia (Sweeney et al., 2018). Moreover, the early loss of pericytes in the NVU has a role in Alzheimer's disease development (Shah & Abbruscato, 2017).

1.8.6 Parkinson's Disease

Parkinson's disease (PD) is linked to the degeneration of dopaminergic neurons in the substantia nigra pars compacta. The histological hallmark is the formation of Lewy bodies by the accumulation of α-synuclein inclusions. PD risk factors are classically divided into genetic factors, like SYNC and LRRK2 mutations, and environmental risk factors, like pesticides, inflammation, and the effect of alimentation. BBB dysfunction was not incriminated previously in PD pathogenesis as it was thought intact, and the peripheral aromatic amino acid decarboxylase (AAAD) inhibitors, like carbidopa, were unable to penetrate BBB and were successfully used to inhibit the peripheral decarboxylase enzyme and increase levodopa plasma concentration without an action on central decarboxylase (Desai et al., 2007). However, more recent works mainly conducted on animal models were in favor of a vascular disruption in PD related to BBB disruption and hypoperfusion, causing neuronal dysfunction and, consequently, cell death. In human PD patients, a small positron emission tomography (PET) study showed BBB transporter system dysfunction; additionally, increased permeability of the BBB in the post-commissural putamen was found upon histological examination (Al-Bachari et al., 2020).

1.8.7 Huntington's Disease

Huntington's disease (HD) is an inherited dominant autosomal severe neurodegenerative disease that manifests with an association of motor symptoms (mainly chorea), cognitive impairment, and psychiatric complaints. The disease is caused by a CAG-repeat expansion in the HTT (huntingtin) gene, resulting in increased expression and accumulation of mutant HTT protein (mHTT) affecting striatal neurons and astrocytes. Astrocytes' resulting dysfunction will cause BBB damage, which will entertain mHTT resulting dysfunction and damage (Hsiao et al., 2015). Postmortem MRI and morphological studies performed on mice models and HD patients found a reduction in blood vessel diameter, a reduced number of TJs, and an increased number of transcytotic vesicles sometimes containing mHTT aggregates, signs of BBB leakage and dysfunction (Drouin-Ouellet et al., 2015).

1.8.8 Traumatic Brain Injury

TBI severity and prognosis are closely related to long-lasting BBB disruptions, appearing hours after the trauma and potentially persisting years after. Extravasation of the serum protein fibrinogen and immunoglobulin G (IgG) and, consequently, edema formation are two hallmarks of BBB disruption in TBI (Pop & Badaut, 2011). Two types of cerebral edema are present in TBI: vasogenic edema by fluid accumulation in the perivascular space, causing cerebral blood flow (CBF) disturbance and increased intracranial pressure (ICP). The second type of edema is cytotoxic edema, linked to increased intracellular entrance of water and its accumulation by the activation of cells ion channels. Uncontrolled, the edema will lead to neural dysfunction and irreversible brain damage (Cash & Theus, 2020).

1.9 CLINICAL SIGNIFICANCE IN DRUG DELIVERY

BBB permeability in normal and pathological states is a very important factor to consider in the development of new medications destined to treat CNS disorders and also in the choice of available therapeutic options, as it plays the role of the policeman controlling all exchanges between the brain tissue and the blood. A small number of medications fulfil the necessary molecular proprieties of size, hydrophilicity, and degree of dissociation to be able to pass the BBB. Furthermore, the majority of macromolecules and most of the micromolecules do not have an effective BBB transportation system. Moreover, BBB disruption following CNS aggressions leads to perivascular accumulation of blood-derived toxic products and the reduction of CBF locally (Cash & Theus, 2020). Such changes lead to decreased therapeutic efficacy by the disruption of endothelial cells' capacity of molecule transport and exchange and by blocking normal flows of substances within the interstitial space (Sweeney et al., 2018).

REFERENCES

2022 Alzheimer's Disease Facts and Figures. (2022). *Alzheimer's & Dementia*, *18*(4), 700–789. https://doi.org/10.1002/alz.12638

Abbott, N. J., & Romero, I. A. (1996). Transporting therapeutics across the blood-brain barrier. *Molecular Medicine Today*, *2*(3), 106–113. https://doi.org/10.1016/1357-4310(96)88720-x

Abbott, N. J., Rönnbäck, L., & Hansson, E. (2006). Astrocyte-endothelial interactions at the blood-brain barrier. *Nature Reviews. Neuroscience*, *7*(1), 41–53. https://doi.org/10.1038/nrn1824

Absinta, M., Ha, S.-K., Nair, G., Sati, P., Luciano, N. J., Palisoc, M., Louveau, A., Zaghloul, K. A., Pittaluga, S., Kipnis, J., & Reich, D. S. (2017). Human and nonhuman primate meninges harbor lymphatic vessels that can be visualized noninvasively by MRI. *ELife*, *6*, e29738. https://doi.org/10.7554/eLife.29738

Ahmad, S., Hewett, P. W., Wang, P., Al-Ani, B., Cudmore, M., Fujisawa, T., Haigh, J. J., le Noble, F., Wang, L., Mukhopadhyay, D., & Ahmed, A. (2006). Direct evidence for endothelial vascular endothelial growth factor receptor-1 function in nitric oxide-mediated angiogenesis. *Circulation Research*, *99*(7), 715–722. https://doi.org/10.1161/01.RES.0000243989.46006.b9

Aird, W. C. (2007). Phenotypic heterogeneity of the endothelium. *Circulation Research*, *100*(2), 174–190. https://doi.org/10.1161/01.RES.0000255690.03436.ae

Al-Bachari, S., Naish, J. H., Parker, G. J. M., Emsley, H. C. A., & Parkes, L. M. (2020). Blood-brain barrier leakage is increased in Parkinson's disease. *Frontiers in Physiology*, *11*, 593026. https://doi.org/10.3389/fphys.2020.593026

Alonso-Alonso, M. L., Sampedro-Viana, A., Fernández-Rodicio, S., Bazarra-Barreiros, M., Ouro, A., Sobrino, T., Campos, F., Castillo, J., Hervella, P., & Iglesias-Rey, R. (2022). Need for a paradigm shift in the treatment of ischemic stroke: The blood-brain barrier. *International Journal of Molecular Sciences*, *23*(16), 9486. https://doi.org/10.3390/ijms23169486

Andrae, J., Gallini, R., & Betsholtz, C. (2008). Role of platelet-derived growth factors in physiology and medicine. *Genes & Development*, *22*(10), 1276–1312. https://doi.org/10.1101/gad.1653708

Antoniuk, S., & da Silva, R. V. (2000). [Periventricular and intraventricular hemorrhage in the premature infants]. *Revista De Neurologia*, *31*(3), 238–243.

Armulik, A., Genové, G., & Betsholtz, C. (2011). Pericytes: Developmental, physiological, and pathological perspectives, problems, and promises. *Developmental Cell*, *21*(2), 193–215. https://doi.org/10.1016/j.devcel.2011.07.001

Armulik, A., Genové, G., Mäe, M., Nisancioglu, M. H., Wallgard, E., Niaudet, C., He, L., Norlin, J., Lindblom, P., Strittmatter, K., Johansson, B. R., & Betsholtz, C. (2010). Pericytes regulate the blood–brain barrier. *Nature*, *468*(7323), Article 7323. https://doi.org/10.1038/nature09522

Attwell, D., Buchan, A. M., Charpak, S., Lauritzen, M., Macvicar, B. A., & Newman, E. A. (2010). Glial and neuronal control of brain blood flow. *Nature*, *468*(7321), 232–243. https://doi.org/10.1038/nature09613

Baeten, K. M., & Akassoglou, K. (2011). Extracellular matrix and matrix receptors in blood–brain barrier formation and stroke. *Developmental Neurobiology*, *71*(11), 1018–1039. https://doi.org/10.1002/dneu.20954

Ballabh, P., Braun, A., & Nedergaard, M. (2004). The blood-brain barrier: An overview: Structure, regulation, and clinical implications. *Neurobiology of Disease*, *16*(1), 1–13. https://doi.org/10.1016/j.nbd.2003.12.016

Banker, G. A. (1980). Trophic interactions between astroglial cells and hippocampal neurons in culture. *Science (New York, N.Y.)*, *209*(4458), 809–810. https://doi.org/10.1126/science.7403847

Banks, W. A. (2001). Leptin transport across the blood-brain barrier: Implications for the cause and treatment of obesity. *Current Pharmaceutical Design*, *7*(2), 125–133. https://doi.org/10.2174/1381612013398310

Bazzoni, G., & Dejana, E. (2004). Endothelial cell-to-cell junctions: Molecular organization and role in vascular homeostasis. *Physiological Reviews*, *84*(3), 869–901. https://doi.org/10.1152/physrev.00035.2003

Bazzoni, G., Dejana, E., & Lampugnani, M. G. (1999). Endothelial adhesion molecules in the development of the vascular tree: The garden of forking paths. *Current Opinion in Cell Biology*, *11*(5), 573–581. https://doi.org/10.1016/s0955-0674(99)00023-x

Begley, D. J. (2004). ABC transporters and the blood-brain barrier. *Current Pharmaceutical Design*, *10*(12), 1295–1312. https://doi.org/10.2174/1381612043384844

Ben-Zvi, A., Lacoste, B., Kur, E., Andreone, B. J., Mayshar, Y., Yan, H., & Gu, C. (2014). Mfsd2a is critical for the formation and function of the blood–brain barrier. *Nature*, *509*(7501), Article 7501. https://doi.org/10.1038/nature13324

Bertrand, L., Cho, H. J., & Toborek, M. (2019). Blood–brain barrier pericytes as a target for HIV-1 infection. *Brain*, *142*(3), 502–511. https://doi.org/10.1093/brain/awy339

Bjarnegård, M., Enge, M., Norlin, J., Gustafsdottir, S., Fredriksson, S., Abramsson, A., Takemoto, M., Gustafsson, E., Fässler, R., & Betsholtz, C. (2004). Endothelium-specific ablation of PDGFB leads to pericyte loss and glomerular, cardiac and placental abnormalities. *Development*, *131*(8), 1847–1857. https://doi.org/10.1242/dev.01080

Bolton, S. J., Anthony, D. C., & Perry, V. H. (1998). Loss of the tight junction proteins occludin and zonula occludens-1 from cerebral vascular endothelium during neutrophil-induced blood-brain barrier breakdown in vivo. *Neuroscience*, *86*(4), 1245–1257. https://doi.org/10.1016/s0306-4522(98)00058-x

Braniste, V., Al-Asmakh, M., Kowal, C., Anuar, F., Abbaspour, A., Tóth, M., Korecka, A., Bakocevic, N., Ng, L. G., Kundu, P., Gulyás, B., Halldin, C., Hultenby, K., Nilsson, H., Hebert, H., Volpe, B. T., Diamond, B., & Pettersson, S. (2014). The gut microbiota influences blood-brain barrier permeability in mice. *Science Translational Medicine*, *6*(263), 263ra158–263ra158. https://doi.org/10.1126/scitranslmed.3009759

Bray, N. (2015). Transferrin' bispecific antibodies across the blood–brain barrier. *Nature Reviews Drug Discovery*, *14*(1), Article 1. https://doi.org/10.1038/nrd4522

Brightman, M. W., & Reese, T. S. (1969). Junctions between intimately apposed cell membranes in the vertebrate brain. *Journal of Cell Biology*, *40*(3), 648–677. https://doi.org/10.1083/jcb.40.3.648

Bush, T. G., Puvanachandra, N., Horner, C. H., Polito, A., Ostenfeld, T., Svendsen, C. N., Mucke, L., Johnson, M. H., & Sofroniew, M. V. (1999). Leukocyte infiltration, neuronal degeneration, and neurite outgrowth after ablation of scar-forming, reactive astrocytes in adult transgenic mice. *Neuron, 23*(2), 297–308. https://doi.org/10.1016/s0896-6273(00)80781-3

Capaldo, C. T., & Macara, I. G. (2007). Depletion of E-cadherin disrupts establishment but not maintenance of cell junctions in Madin-Darby canine kidney epithelial cells. *Molecular Biology of the Cell, 18*(1), 189–200. https://doi.org/10.1091/mbc.e06-05-0471

Caplan, L. R., Biller, J., Leary, M. C., Lo, E. H., Thomas, A. J., Yenari, M. A., & Zhang, J. H. (2017). *Primer on cerebrovascular diseases* (2nd ed.). Academic Press.

Cash, A., & Theus, M. H. (2020). Mechanisms of blood-brain barrier dysfunction in traumatic brain injury. *International Journal of Molecular Sciences, 21*(9), 3344. https://doi.org/10.3390/ijms21093344

Cass, C. E., Young, J. D., & Baldwin, S. A. (1998). Recent advances in the molecular biology of nucleoside transporters of mammalian cells. *Biochemistry and Cell Biology, 76*(5), 761–770. https://doi.org/10.1139/o98-095

Chen, Y., & Liu, L. (2012). Modern methods for delivery of drugs across the blood-brain barrier. *Advanced Drug Delivery Reviews, 64*(7), 640–665. https://doi.org/10.1016/j.addr.2011.11.010

Cho, C., Smallwood, P. M., & Nathans, J. (2017). Reck and Gpr124 are essential receptor cofactors for Wnt7a/Wnt7b-specific signaling in mammalian CNS angiogenesis and blood-brain barrier regulation. *Neuron, 95*(5), 1056–1073.e5. https://doi.org/10.1016/j.neuron.2017.07.031

Clarke, L. E., Liddelow, S. A., Chakraborty, C., Münch, A. E., Heiman, M., & Barres, B. A. (2018). Normal aging induces A1-like astrocyte reactivity. *Proceedings of the National Academy of Sciences of the United States of America, 115*(8), E1896–E1905. https://doi.org/10.1073/pnas.1800165115

Coomber, B. L., & Stewart, P. A. (1985). Morphometric analysis of CNS microvascular endothelium. *Microvascular Research, 30*(1), 99–115. https://doi.org/10.1016/0026-2862(85)90042-1

Correale, J., & Villa, A. (2009). Cellular elements of the blood-brain barrier. *Neurochemical Research, 34*(12), 2067–2077. https://doi.org/10.1007/s11064-009-0081-y

Coulter, D. A., & Eid, T. (2012). Astrocytic regulation of glutamate homeostasis in epilepsy. *Glia, 60*(8), 1215–1226. https://doi.org/10.1002/glia.22341

Cucullo, L., Hossain, M., Puvenna, V., Marchi, N., & Janigro, D. (2011). The role of shear stress in blood-brain barrier endothelial physiology. *BMC Neuroscience, 12*(1), 40. https://doi.org/10.1186/1471-2202-12-40

Daneman, R., & Prat, A. (2015). The blood-brain barrier. *Cold Spring Harbor Perspectives in Biology, 7*(1), a020412. https://doi.org/10.1101/cshperspect.a020412

Daneman, R., Zhou, L., Kebede, A. A., & Barres, B. A. (2010). Pericytes are required for blood–brain barrier integrity during embryogenesis. *Nature, 468*(7323), Article 7323. https://doi.org/10.1038/nature09513

Dauchy, S., Dutheil, F., Weaver, R. J., Chassoux, F., Daumas-Duport, C., Couraud, P.-O., Scherrmann, J.-M., De Waziers, I., & Declèves, X. (2008). ABC transporters, cytochromes P450 and their main transcription factors: Expression at the human blood–brain barrier. *Journal of Neurochemistry, 107*(6), 1518–1528. https://doi.org/10.1111/j.1471-4159.2008.05720.x

Dean, M., Hamon, Y., & Chimini, G. (2001). The human ATP-binding cassette (ABC) transporter superfamily. *Journal of Lipid Research, 42*(7), 1007–1017. https://doi.org/10.1016/S0022-2275(20)31588-1

Deane, R., Du Yan, S., Submamaryan, R. K., LaRue, B., Jovanovic, S., Hogg, E., Welch, D., Manness, L., Lin, C., Yu, J., Zhu, H., Ghiso, J., Frangione, B., Stern, A., Schmidt, A. M., Armstrong, D. L., Arnold, B., Liliensiek, B., Nawroth, P., . . Zlokovic, B. (2003). RAGE mediates amyloid-β peptide transport across the blood-brain barrier and accumulation in brain. *Nature Medicine, 9*(7), Article 7. https://doi.org/10.1038/nm890

Deane, R., Wu, Z., Sagare, A., Davis, J., Yan, S. D., Hamm, K., Xu, F., Parisi, M., LaRue, B., Hu, H. W., Spijkers, P., Guo, H., Song, X., Lenting, P. J., Nostrand, W. E. V., & Zlokovic, B. V. (2004). LRP/amyloid β-peptide interaction mediates differential brain efflux of Aβ isoforms. *Neuron, 43*(3), 333–344. https://doi.org/10.1016/j.neuron.2004.07.017

Deng, D., Xu, C., Sun, P., Wu, J., Yan, C., Hu, M., & Yan, N. (2014). Crystal structure of the human glucose transporter GLUT1. *Nature, 510*(7503), Article 7503. https://doi.org/10.1038/nature13306

Desai, B. S., Monahan, A. J., Carvey, P. M., & Hendey, B. (2007). Blood-brain barrier pathology in Alzheimer's and Parkinson's disease: Implications for drug therapy. *Cell Transplantation, 16*(3), 285–299. https://doi.org/10.3727/000000007783464731

DiSalvo, D. (1998). The correlation between placental pathology and intraventricular hemorrhage in the preterm infant. The Developmental Epidemiology Network Investigators. *Pediatric Research, 43*(1), 15–19. https://doi.org/10.1203/00006450-199801000-00003

Dong, X. (2018). Current strategies for brain drug delivery. *Theranostics, 8*(6), 1481–1493. https://doi.org/10.7150/thno.21254

Drouin-Ouellet, J., Sawiak, S. J., Cisbani, G., Lagacé, M., Kuan, W.-L., Saint-Pierre, M., Dury, R. J., Alata, W., St-Amour, I., Mason, S. L., Calon, F., Lacroix, S., Gowland, P. A., Francis, S. T., Barker, R. A., & Cicchetti, F. (2015). Cerebrovascular and blood-brain barrier impairments in Huntington's disease: Potential implications for its pathophysiology. *Annals of Neurology, 78*(2), 160–177. https://doi.org/10.1002/ana.24406

Dyrna, F., Hanske, S., Krueger, M., & Bechmann, I. (2013). The blood-brain barrier. *Journal of Neuroimmune Pharmacology: The Official Journal of the Society on NeuroImmune Pharmacology, 8*(4), 763–773. https://doi.org/10.1007/s11481-013-9473-5

Ek, C. J., Dziegielewska, K. M., Habgood, M. D., & Saunders, N. R. (2012). Barriers in the developing brain and neurotoxicology. *Neurotoxicology, 33*(3), 586–604. https://doi.org/10.1016/j.neuro.2011.12.009

Elfeber, K., Köhler, A., Lutzenburg, M., Osswald, C., Galla, H.-J., Witte, O. W., & Koepsell, H. (2004). Localization of the Na+-d-glucose cotransporter SGLT1 in the blood-brain barrier. *Histochemistry and Cell Biology, 121*(3), 201–207. https://doi.org/10.1007/s00418-004-0633-9

El-Khoury, N., Braun, A., Hu, F., Pandey, M., Nedergaard, M., Lagamma, E. F., & Ballabh, P. (2006). Astrocyte end-feet in germinal matrix, cerebral cortex, and white matter in developing infants. *Pediatric Research, 59*(5), 673–679. https://doi.org/10.1203/01.pdr.0000214975.85311.9c

Emsley, J. G., & Macklis, J. D. (2006). Astroglial heterogeneity closely reflects the neuronal-defined anatomy of the adult murine CNS. *Neuron Glia Biology, 2*(3), 175–186. https://doi.org/10.1017/S1740925X06000202

Enge, M., Bjarnegård, M., Gerhardt, H., Gustafsson, E., Kalén, M., Asker, N., Hammes, H.-P., Shani, M., Fässler, R., & Betsholtz, C. (2002). Endothelium-specific platelet-derived growth factor-B ablation mimics diabetic retinopathy. *EMBO Journal*, *21*(16), 4307–4316.

Engelhardt, B. (2003). Development of the blood-brain barrier. *Cell and Tissue Research*, *314*(1), 119–129. https://doi.org/10.1007/s00441-003-0751-z

Engelhardt, B., & Liebner, S. (2014). Novel insights into the development and maintenance of the blood-brain barrier. *Cell and Tissue Research*, *355*(3), 687–699. https://doi.org/10.1007/s00441-014-1811-2

Ezan, P., André, P., Cisternino, S., Saubaméa, B., Boulay, A.-C., Doutremer, S., Thomas, M.-A., Quenech'du, N., Giaume, C., & Cohen-Salmon, M. (2012). Deletion of astroglial connexins weakens the blood–brain barrier. *Journal of Cerebral Blood Flow & Metabolism*, *32*(8), 1457–1467. https://doi.org/10.1038/jcbfm.2012.45

Falk, M. M. (2000). Connexins/connexons: Cell-free expression. In R. Bruzzone & C. Giaume (Eds.), *Connexin methods and protocols* (Vol. 154, pp. 91–116). Humana Press. https://doi.org/10.1385/1-59259-043-8:91

Farrell, C. L., & Pardridge, W. M. (1991). Blood-brain barrier glucose transporter is asymmetrically distributed on brain capillary endothelial lumenal and ablumenal membranes: An electron microscopic immunogold study. *Proceedings of the National Academy of Sciences*, *88*(13), 5779–5783. https://doi.org/10.1073/pnas.88.13.5779

Feuerstein, G. Z., Liu, T., & Barone, F. C. (1994). Cytokines, inflammation, and brain injury: Role of tumor necrosis factor-alpha. *Cerebrovascular and Brain Metabolism Reviews*, *6*(4), 341–360.

Foroutan, S., Brillault, J., Forbush, B., & O'Donnell, M. E. (2005). Moderate-to-severe ischemic conditions increase activity and phosphorylation of the cerebral microvascular endothelial cell Na+-K+-Cl– cotransporter. *American Journal of Physiology-Cell Physiology*, *289*(6), C1492–C1501. https://doi.org/10.1152/ajpcell.00257.2005

Forrester, J. V., McMenamin, P. G., & Dando, S. J. (2018). CNS infection and immune privilege. *Nature Reviews. Neuroscience*, *19*(11), 655–671. https://doi.org/10.1038/s41583-018-0070-8

Genin, E., Hannequin, D., Wallon, D., Sleegers, K., Hiltunen, M., Combarros, O., Bullido, M. J., Engelborghs, S., De Deyn, P., Berr, C., Pasquier, F., Dubois, B., Tognoni, G., Fiévet, N., Brouwers, N., Bettens, K., Arosio, B., Coto, E., Del Zompo, M., . . . Campion, D. (2011). APOE and Alzheimer disease: A major gene with semi-dominant inheritance. *Molecular Psychiatry*, *16*(9), 903–907. https://doi.org/10.1038/mp.2011.52

Gerriets, T., Walberer, M., Ritschel, N., Tschernatsch, M., Mueller, C., Bachmann, G., Schoenburg, M., Kaps, M., & Nedelmann, M. (2009). Edema formation in the hyperacute phase of ischemic stroke: Laboratory investigation. *Journal of Neurosurgery*, *111*(5), 1036–1042. https://doi.org/10.3171/2009.3.JNS081040

Goel, S., Duda, D. G., Xu, L., Munn, L. L., Boucher, Y., Fukumura, D., & Jain, R. K. (2011). Normalization of the vasculature for treatment of cancer and other diseases. *Physiological Reviews*, *91*(3), 1071–1121. https://doi.org/10.1152/physrev.00038.2010

Golden, P. L., Maccagnan, T. J., & Pardridge, W. M. (1997). Human blood-brain barrier leptin receptor. Binding and endocytosis in isolated human brain microvessels. *Journal of Clinical Investigation*, *99*(1), 14–18. https://doi.org/10.1172/JCI119125

Goodenough, D. A., & Paul, D. L. (2009). Gap junctions. *Cold Spring Harbor Perspectives in Biology*, *1*(1), a002576. https://doi.org/10.1101/cshperspect.a002576

Goodenough, D. A., Goliger, J. A., & Paul, D. L. (1996). Connexins, connexons, and intercellular communication. *Annual Review of Biochemistry*, *65*, 475–502. https://doi.org/10.1146/annurev.bi.65.070196.002355

Gooding, J. M., Yap, K. L., & Ikura, M. (2004). The cadherin-catenin complex as a focal point of cell adhesion and signalling: New insights from three-dimensional structures. *BioEssays: News and Reviews in Molecular, Cellular and Developmental Biology*, *26*(5), 497–511. https://doi.org/10.1002/bies.20033

Gumbiner, B. M. (2005). Regulation of cadherin-mediated adhesion in morphogenesis. *Nature Reviews. Molecular Cell Biology*, *6*(8), 622–634. https://doi.org/10.1038/nrm1699

Guo, R., Chen, R., Yu, Z., Zhao, X., You, C., Li, H., & Ma, L. (2020). Primary intraventricular hemorrhage in pediatric patients: Causes, characteristics, and outcomes. *World Neurosurgery*, *133*, e121–e128. https://doi.org/10.1016/j.wneu.2019.08.145

Hall, C. N., Reynell, C., Gesslein, B., Hamilton, N. B., Mishra, A., Sutherland, B. A., O'Farrell, F. M., Buchan, A. M., Lauritzen, M., & Attwell, D. (2014). Capillary pericytes regulate cerebral blood flow in health and disease. *Nature*, *508*(7494), 55–60. https://doi.org/10.1038/nature13165

Hallmann, A.-L., Araúzo-Bravo, M. J., Mavrommatis, L., Ehrlich, M., Röpke, A., Brockhaus, J., Missler, M., Sterneckert, J., Schöler, H. R., Kuhlmann, T., Zaehres, H., & Hargus, G. (2017). Astrocyte pathology in a human neural stem cell model of frontotemporal dementia caused by mutant TAU protein. *Scientific Reports*, *7*, 42991. https://doi.org/10.1038/srep42991

Han, H., Mann, A., Ekstein, D., & Eyal, S. (2017). Breaking bad: The structure and function of the blood-brain barrier in epilepsy. *The AAPS Journal*, *19*(4), 973–988. https://doi.org/10.1208/s12248-017-0096-2

Hartsock, A., & Nelson, W. J. (2008). Adherens and tight junctions: Structure, function and connections to the actin cytoskeleton. *Biochimica et Biophysica Acta*, *1778*(3), 660–669. https://doi.org/10.1016/j.bbamem.2007.07.012

Hawkins, R. A., O'Kane, R. L., Simpson, I. A., & Viña, J. R. (2006). Structure of the blood–brain barrier and its role in the transport of amino acids 1, 2, 3. *Journal of Nutrition*, *136*(1), 218S–226S. https://doi.org/10.1093/jn/136.1.218S

Hayashi, T., Deguchi, K., Nagotani, S., Zhang, H., Sehara, Y., Tsuchiya, A., & Abe, K. (2006). Cerebral ischemia and angiogenesis. *Current Neurovascular Research*, *3*(2), 119–129. https://doi.org/10.2174/156720206776875902

Hellström, M., Kalén, M., Lindahl, P., Abramsson, A., & Betsholtz, C. (1999). Role of PDGF-B and PDGFR-β in recruitment of vascular smooth muscle cells and pericytes during embryonic blood vessel formation in the mouse. *Development*, *126*(14), 3047–3055. https://doi.org/10.1242/dev.126.14.3047

Hervé, F., Ghinea, N., & Scherrmann, J.-M. (2008). CNS delivery via adsorptive transcytosis. *The AAPS Journal*, *10*(3), 455–472. https://doi.org/10.1208/s12248-008-9055-2

Hippenstiel, S., Krüll, M., Ikemann, A., Risau, W., Clauss, M., & Suttorp, N. (1998). VEGF induces hyperpermeability by a direct action on endothelial cells. *American Journal of Physiology*, *274*(5), L678–L684. https://doi.org/10.1152/ajplung.1998.274.5.L678

Hsiao, H.-Y., Chen, Y.-C., Huang, C.-H., Chen, C.-C., Hsu, Y.-H., Chen, H.-M., Chiu, F.-L., Kuo, H.-C., Chang, C., & Chern, Y. (2015). Aberrant astrocytes impair vascular reactivity in Huntington disease. *Annals of Neurology*, *78*(2), 178–192. https://doi.org/10.1002/ana.24428

Iadecola, C. (2013). The pathobiology of vascular dementia. *Neuron*, *80*(4), 844–866. https://doi.org/10.1016/j.neuron.2013.10.008

Ichimura, K., Stan, R. V., Kurihara, H., & Sakai, T. (2008). Glomerular endothelial cells form diaphragms during development and pathologic conditions. *Journal of the American Society of Nephrology*, *19*(8), 1463. https://doi.org/10.1681/ASN.2007101138

Ikenouchi, J., Furuse, M., Furuse, K., Sasaki, H., Tsukita, S., & Tsukita, S. (2005). Tricellulin constitutes a novel barrier at tricellular contacts of epithelial cells. *Journal of Cell Biology*, *171*(6), 939–945. https://doi.org/10.1083/jcb.200510043

Ito, S., Yanai, M., Yamaguchi, S., Couraud, P.-O., & Ohtsuki, S. (2017). Regulation of tight-junction integrity by insulin in an in vitro model of human blood–brain barrier. *Journal of Pharmaceutical Sciences*, *106*(9), 2599–2605. https://doi.org/10.1016/j.xphs.2017.04.036

Jefferies, W. A., Brandon, M. R., Hunt, S. V., Williams, A. F., Gatter, K. C., & Mason, D. Y. (1984). Transferrin receptor on endothelium of brain capillaries. *Nature*, *312*(5990), Article 5990. https://doi.org/10.1038/312162a0

Kaur, C., & Ling, E.-A. (2017). The circumventricular organs. *Histology and Histopathology*, *32*(9), 879–892. https://doi.org/10.14670/HH-11-881

Kiecker, C. (2018). The origins of the circumventricular organs. *Journal of Anatomy*, *232*(4), 540–553. https://doi.org/10.1111/joa.12771

Kisler, K., Nelson, A. R., Montagne, A., & Zlokovic, B. V. (2017). Cerebral blood flow regulation and neurovascular dysfunction in Alzheimer disease. *Nature Reviews Neuroscience*, *18*(7), Article 7. https://doi.org/10.1038/nrn.2017.48

Kniesel, U., & Wolburg, H. (2000). Tight junctions of the blood-brain barrier. *Cellular and Molecular Neurobiology*, *20*(1), 57–76. https://doi.org/10.1023/a:1006995910836

Kofuji, P., & Newman, E. A. (2004). Potassium buffering in the central nervous system. *Neuroscience*, *129*(4), 1045–1056. https://doi.org/10.1016/j.neuroscience.2004.06.008

Kolářová, H., Ambrůzová, B., Sviháľková Šindlerová, L., Klinke, A., & Kubala, L. (2014). Modulation of endothelial glycocalyx structure under inflammatory conditions. *Mediators of Inflammation*, *2014*, 694312. https://doi.org/10.1155/2014/694312

Konari, K., Sawada, N., Zhong, Y., Isomura, H., Nakagawa, T., & Mori, M. (1995). Development of the blood-retinal barrier in vitro: Formation of tight junctions as revealed by occludin and ZO-1 correlates with the barrier function of chick retinal pigment epithelial cells. *Experimental Eye Research*, *61*(1), 99–108. https://doi.org/10.1016/s0014-4835(95)80063-8

Kugler, E. C., Greenwood, J., & MacDonald, R. B. (2021). The "neuro-glial-vascular" unit: The role of glia in neurovascular unit formation and dysfunction. *Frontiers in Cell and Developmental Biology*, *9*, 732820. https://doi.org/10.3389/fcell.2021.732820

Kumar, S., Selim, M. H., & Caplan, L. R. (2010). Medical complications after stroke. *Lancet Neurology*, *9*(1), 105–118. https://doi.org/10.1016/S1474-4422(09)70266-2

Langhorne, P., Stott, D. J., Robertson, L., MacDonald, J., Jones, L., McAlpine, C., Dick, F., Taylor, G. S., & Murray, G. (2000). Medical complications after stroke. *Stroke*, *31*(6), 1223–1229. https://doi.org/10.1161/01.STR.31.6.1223

Lewandowsky, M. (1909). Zur lehre der cerebrospinalflussigkeit. *Zeitschrift fur klinische Medizin*, *40*, 480–494.

Lorin, V., Danckaert, A., Porrot, F., Schwartz, O., Afonso, P. V., & Mouquet, H. (2020). Antibody neutralization of HIV-1 crossing the blood-brain barrier. *MBio*, *11*(5), e02424–20. https://doi.org/10.1128/mBio.02424-20

Löscher, W., & Potschka, H. (2005). Blood-brain barrier active efflux transporters: ATP-binding cassette gene family. *NeuroRx: The Journal of the American Society for Experimental NeuroTherapeutics*, *2*(1), 86–98. https://doi.org/10.1602/neurorx.2.1.86

Mann, G. E., Yudilevich, D. L., & Sobrevia, L. (2003). Regulation of amino acid and glucose transporters in endothelial and smooth muscle cells. *Physiological Reviews*, *83*(1), 183–252. https://doi.org/10.1152/physrev.00022.2002

Marchi, N., Angelov, L., Masaryk, T., Fazio, V., Granata, T., Hernandez, N., Hallene, K., Diglaw, T., Franic, L., Najm, I., & Janigro, D. (2007). Seizure-promoting effect of blood-brain barrier disruption. *Epilepsia*, *48*(4), 732–742. https://doi.org/10.1111/j.1528-1167.2007.00988.x

Masuda, S., Oda, Y., Sasaki, H., Ikenouchi, J., Higashi, T., Akashi, M., Nishi, E., & Furuse, M. (2011). LSR defines cell corners for tricellular tight junction formation in epithelial cells. *Journal of Cell Science*, *124*(4), 548–555. https://doi.org/10.1242/jcs.072058

McCloskey, D. P., Croll, S. D., & Scharfman, H. E. (2005). Depression of synaptic transmission by vascular endothelial growth factor in adult rat hippocampus and evidence for increased efficacy after chronic seizures. *Journal of Neuroscience: The Official Journal of the Society for Neuroscience*, *25*(39), 8889–8897. https://doi.org/10.1523/JNEUROSCI.2577-05.2005

Menezes, M. J., McClenahan, F. K., Leiton, C. V., Aranmolate, A., Shan, X., & Colognato, H. (2014). The extracellular matrix protein laminin α2 regulates the maturation and function of the blood–brain barrier. *Journal of Neuroscience*, *34*(46), 15260–15280. https://doi.org/10.1523/JNEUROSCI.3678-13.2014

Méresse, S., Delbart, C., Fruchart, J.-C., & Cecchelli, R. (1989). Low-density lipoprotein receptor on endothelium of brain capillaries. *Journal of Neurochemistry*, *53*(2), 340–345. https://doi.org/10.1111/j.1471-4159.1989.tb07340.x

Mesli, S., Javorschi, S., Bérard, A. M., Landry, M., Priddle, H., Kivlichan, D., Smith, A. J. H., Yen, F. T., Bihain, B. E., & Darmon, M. (2004). Distribution of the lipolysis stimulated receptor in adult and embryonic murine tissues and lethality of LSR-/- embryos at 12.5 to 14.5 days of gestation. *European Journal of Biochemistry*, *271*(15), 3103–3114. https://doi.org/10.1111/j.1432-1033.2004.04223.x

Millar, I. D., Wang, S., Brown, P. D., Barrand, M. A., & Hladky, S. B. (2008). Kv1 and Kir2 potassium channels are expressed in rat brain endothelial cells. *Pflügers Archiv—European Journal of Physiology*, *456*(2), 379–391. https://doi.org/10.1007/s00424-007-0377-1

Mitchell, R. W., On, N. H., Del Bigio, M. R., Miller, D. W., & Hatch, G. M. (2011). Fatty acid transport protein expression in human brain and potential role in fatty acid transport across human brain microvessel endothelial cells: Fatty acid transport protein expression in human brain. *Journal of Neurochemistry*. https://doi.org/10.1111/j.1471-4159.2011.07245.x

Montagne, A., Zhao, Z., & Zlokovic, B. V. (2017). Alzheimer's disease: A matter of blood–brain barrier dysfunction? *Journal of Experimental Medicine*, *214*(11), 3151–3169. https://doi.org/10.1084/jem.20171406

Nagai, M., Re, D. B., Nagata, T., Chalazonitis, A., Jessell, T. M., Wichterle, H., & Przedborski, S. (2007). Astrocytes expressing ALS-linked mutated SOD1 release factors selectively toxic to motor neurons. *Nature Neuroscience*, *10*(5), 615–622. https://doi.org/10.1038/nn1876

Nguyen, L. N., Ma, D., Shui, G., Wong, P., Cazenave-Gassiot, A., Zhang, X., Wenk, M. R., Goh, E. L. K., & Silver, D. L. (2014). Mfsd2a is a transporter for the essential omega-3 fatty acid docosahexaenoic acid. *Nature*, *509*(7501), Article 7501. https://doi.org/10.1038/nature13241

Nitsch, C., & Klatzo, I. (1983). Regional patterns of blood-brain barrier breakdown during epileptiform seizures induced by various convulsive agents. *Journal of the Neurological Sciences*, *59*(3), 305–322. https://doi.org/10.1016/0022-510x(83)90016-3

Oberheim, N. A., Takano, T., Han, X., He, W., Lin, J. H. C., Wang, F., Xu, Q., Wyatt, J. D., Pilcher, W., Ojemann, J. G., Ransom, B. R., Goldman, S. A., & Nedergaard, M. (2009). Uniquely hominid features of adult human astrocytes. *Journal of Neuroscience: The Official Journal of the Society for Neuroscience*, *29*(10), 3276–3287. https://doi.org/10.1523/JNEUROSCI.4707-08.2009

Obermeier, B., Daneman, R., & Ransohoff, R. M. (2013). Development, maintenance and disruption of the blood-brain barrier. *Nature Medicine*, *19*(12), 1584–1596. https://doi.org/10.1038/nm.3407

O'Kane, R. L., & Hawkins, R. A. (2003). Na+-dependent transport of large neutral amino acids occurs at the abluminal membrane of the blood-brain barrier. *American Journal of Physiology-Endocrinology and Metabolism*, *285*(6), E1167–E1173. https://doi.org/10.1152/ajpendo.00193.2003

Oldendorf, W. H., Cornford, M. E., & Brown, W. J. (1977). The large apparent work capability of the blood-brain barrier: A study of the mitochondrial content of capillary endothelial cells in brain and other tissues of the rat. *Annals of Neurology*, *1*(5), 409–417. https://doi.org/10.1002/ana.410010502

Oldendorf, W. H., & Davson, H. (1967). Brain extracellular space and the sink action of cerebrospinal fluid. *Transactions of the American Neurological Association*, *92*, 123–127.

Papadopoulos, M. C., Lamb, F. J., Moss, R. F., Davies, D. C., Tighe, D., & Bennett, E. D. (1999). Faecal peritonitis causes oedema and neuronal injury in pig cerebral cortex. *Clinical Science (London, England: 1979)*, *96*(5), 461–466.

Pardridge, W. M. (2005). Molecular biology of the blood-brain barrier. *Molecular Biotechnology*, *30*(1), 57–69. https://doi.org/10.1385/MB:30:1:057

Pop, V., & Badaut, J. (2011). A neurovascular perspective for long-term changes after brain trauma. *Translational Stroke Research*, *2*(4), 533–545. https://doi.org/10.1007/s12975-011-0126-9

Potente, M., Gerhardt, H., & Carmeliet, P. (2011). Basic and therapeutic aspects of angiogenesis. *Cell*, *146*(6), 873–887. https://doi.org/10.1016/j.cell.2011.08.039

Profaci, C. P., Munji, R. N., Pulido, R. S., & Daneman, R. (2020). The blood-brain barrier in health and disease: Important unanswered questions. *Journal of Experimental Medicine*, *217*(4), e20190062. https://doi.org/10.1084/jem.20190062

Puchowicz, M. A., Xu, K., Sun, X., Ivy, A., Emancipator, D., & LaManna, J. C. (2007). Diet-induced ketosis increases capillary density without altered blood flow in rat brain. *American Journal of Physiology-Endocrinology and Metabolism*, *292*(6), E1607–E1615. https://doi.org/10.1152/ajpendo.00512.2006

Pulgar, V. M. (2018). Transcytosis to cross the blood brain barrier, new advancements and challenges. *Frontiers in Neuroscience*, *12*, 1019. https://doi.org/10.3389/fnins.2018.01019

Rapoport, S. I. (2000). Osmotic opening of the blood-brain barrier: Principles, mechanism, and therapeutic applications. *Cellular and Molecular Neurobiology*, *20*(2), 217–230. https://doi.org/10.1023/a:1007049806660

Reese, T. S., & Karnovsky, M. J. (1967). Fine structural localization of a blood-brain barrier to exogenous peroxidase. *Journal of Cell Biology*, *34*(1), 207–217. https://doi.org/10.1083/jcb.34.1.207

Rigau, V., Morin, M., Rousset, M.-C., de Bock, F., Lebrun, A., Coubes, P., Picot, M.-C., Baldy-Moulinier, M., Bockaert, J., Crespel, A., & Lerner-Natoli, M. (2007). Angiogenesis is associated with blood–brain barrier permeability in temporal lobe epilepsy. *Brain*, *130*(7), 1942–1956. https://doi.org/10.1093/brain/awm118

Risau, W., Esser, S., & Engelhardt, B. (1998). Differentiation of blood-brain barrier endothelial cells. *Pathologie-Biologie*, *46*(3), 171–175.

Rothstein, J. D., Dykes-Hoberg, M., Pardo, C. A., Bristol, L. A., Jin, L., Kuncl, R. W., Kanai, Y., Hediger, M. A., Wang, Y., Schielke, J. P., & Welty, D. F. (1996). Knockout of glutamate transporters reveals a major role for astroglial transport in excitotoxicity and clearance of glutamate. *Neuron*, *16*(3), 675–686. https://doi.org/10.1016/s0896-6273(00)80086-0

Saunders, N. R., Liddelow, S. A., & Dziegielewska, K. M. (2012). Barrier mechanisms in the developing brain. *Frontiers in Pharmacology*, *3*, 46. https://doi.org/10.3389/fphar.2012.00046

Sauvageot, C. M., & Stiles, C. D. (2002). Molecular mechanisms controlling cortical gliogenesis. *Current Opinion in Neurobiology*, *12*(3), 244–249. https://doi.org/10.1016/S0959-4388(02)00322-7

Seiffert, E., Dreier, J. P., Ivens, S., Bechmann, I., Tomkins, O., Heinemann, U., & Friedman, A. (2004). Lasting blood-brain barrier disruption induces epileptic focus in the rat somatosensory cortex. *Journal of Neuroscience: The Official Journal of the Society for Neuroscience*, *24*(36), 7829–7836. https://doi.org/10.1523/JNEUROSCI.1751-04.2004

Shah, K., & Abbruscato, T. (2017). Chapter 6 — The blood–brain barrier: A restricted gateway to the brain. In P. M. Conn (Ed.), *Conn's translational neuroscience* (pp. 141–146). Academic Press. https://doi.org/10.1016/B978-0-12-802381-5.00010-5

Shibata, M., Yamada, S., Kumar, S. R., Calero, M., Bading, J., Frangione, B., Holtzman, D. M., Miller, C. A., Strickland, D. K., Ghiso, J., & Zlokovic, B. V. (2000). Clearance of Alzheimer's amyloid-β_{1-40} peptide from brain by LDL receptor–related protein-1 at the blood-brain barrier. *Journal of Clinical Investigation*, *106*(12), 1489–1499. https://doi.org/10.1172/JCI10498

Shlosberg, D., Benifla, M., Kaufer, D., & Friedman, A. (2010). Blood-brain barrier breakdown as a therapeutic target in traumatic brain injury. *Nature Reviews. Neurology*, *6*(7), 393–403. https://doi.org/10.1038/nrneurol.2010.74

Simpson, I. A., Carruthers, A., & Vannucci, S. J. (2007). Supply and demand in cerebral energy metabolism: The role of nutrient transporters. *Journal of Cerebral Blood Flow & Metabolism*, *27*(11), 1766–1791. https://doi.org/10.1038/sj.jcbfm.9600521

Sohet, F., Lin, C., Munji, R. N., Lee, S. Y., Ruderisch, N., Soung, A., Arnold, T. D., Derugin, N., Vexler, Z. S., & Yen, F. T. (2015). LSR/angulin-1 is a tricellular tight junction protein involved in blood–brain barrier formation. *Journal of Cell Biology*, *208*(6), 703–711.

Söhl, G., & Willecke, K. (2004). Gap junctions and the connexin protein family. *Cardiovascular Research*, *62*(2), 228–232. https://doi.org/10.1016/j.cardiores.2003.11.013

Stamatovic, S. M., Johnson, A. M., Keep, R. F., & Andjelkovic, A. V. (2016). Junctional proteins of the blood-brain barrier: New insights into function and dysfunction. *Tissue Barriers*, *4*(1), e1154641. https://doi.org/10.1080/21688370.2016.1154641

Stoll, J., Wadhwani, K. C., & Smith, Q. R. (1993). Identification of the cationic amino acid transporter (system y+) of the rat blood-brain barrier. *Journal of Neurochemistry*, *60*(5), 1956–1959. https://doi.org/10.1111/j.1471-4159.1993.tb13428.x

Sweeney, M. D., Sagare, A. P., & Zlokovic, B. V. (2018). Blood-brain barrier breakdown in Alzheimer disease and other neurodegenerative disorders. *Nature Reviews. Neurology*, *14*(3), 133–150. https://doi.org/10.1038/nrneurol.2017.188

Sweeney, M. D., Zhao, Z., Montagne, A., Nelson, A. R., & Zlokovic, B. V. (2019). Blood-brain barrier: From physiology to disease and back. *Physiological Reviews*, *99*(1), 21–78. https://doi.org/10.1152/physrev.00050.2017

Tarkowski, E., Rosengren, L., Blomstrand, C., Wikkelsö, C., Jensen, C., Ekholm, S., & Tarkowski, A. (1997). Intrathecal release of pro- and anti-inflammatory cytokines during stroke. *Clinical and Experimental Immunology*, *110*(3), 492–499. https://doi.org/10.1046/j.1365-2249.1997.4621483.x

Tietz, S., & Engelhardt, B. (2015). Brain barriers: Crosstalk between complex tight junctions and adherens junctions. *Journal of Cell Biology*, *209*(4), 493–506. https://doi.org/10.1083/jcb.201412147

van Vliet, E. A., Aronica, E., & Gorter, J. A. (2014). Role of blood–brain barrier in temporal lobe epilepsy and pharmacoresistance. *Neuroscience*, *277*, 455–473. https://doi.org/10.1016/j.neuroscience.2014.07.030

van Vliet, E. A., da Costa Araújo, S., Redeker, S., van Schaik, R., Aronica, E., & Gorter, J. A. (2007). Blood–brain barrier leakage may lead to progression of temporal lobe epilepsy. *Brain*, *130*(2), 521–534. https://doi.org/10.1093/brain/awl318

Varga, J., & Lafyatis, R. (2015). 143—Etiology and pathogenesis of systemic sclerosis. In M. C. Hochberg, A. J. Silman, J. S. Smolen, M. E. Weinblatt, & M. H. Weisman (Eds.), *Rheumatology* (6th ed., pp. 1177–1189). Mosby. https://doi.org/10.1016/B978-0-323-09138-1.00143-1

Volpe, J. J. (1989). Intraventricular hemorrhage in the premature infant—Current concepts. Part I. *Annals of Neurology*, *25*(1), 3–11. https://doi.org/10.1002/ana.410250103

Winkler, E. A., Bell, R. D., & Zlokovic, B. V. (2011). Central nervous system pericytes in health and disease. *Nature Neuroscience*, *14*(11), Article 11. https://doi.org/10.1038/nn.2946

Yamamoto, H., Ehling, M., Kato, K., Kanai, K., van Lessen, M., Frye, M., Zeuschner, D., Nakayama, M., Vestweber, D., & Adams, R. H. (2015). Integrin β1 controls VE-cadherin localization and blood vessel stability. *Nature Communications*, *6*(1), Article 1. https://doi.org/10.1038/ncomms7429

Yamazaki, Y., Liu, C.-C., Yamazaki, A., Shue, F., Martens, Y. A., Chen, Y., Qiao, W., Kurti, A., Oue, H., Ren, Y., Li, Y., Aikawa, T., Cherukuri, Y., Fryer, J. D., Asmann, Y. W., Kim, B. Y. S., Kanekiyo, T., & Bu, G. (2021). Vascular ApoE4 impairs behavior by modulating gliovascular function. *Neuron*, *109*(3), 438–447.e6. https://doi.org/10.1016/j.neuron.2020.11.019

Yang, Y., Higashimori, H., & Morel, L. (2013). Developmental maturation of astrocytes and pathogenesis of neurodevelopmental disorders. *Journal of Neurodevelopmental Disorders*, *5*(1), 22. https://doi.org/10.1186/1866-1955-5-22

Yao, Y., Chen, Z.-L., Norris, E. H., & Strickland, S. (2014). Astrocytic laminin regulates pericyte differentiation and maintains blood brain barrier integrity. *Nature Communications*, *5*(1), Article 1. https://doi.org/10.1038/ncomms4413

Yuen, N., Lam, T. I., Wallace, B. K., Klug, N. R., Anderson, S. E., & O'Donnell, M. E. (2014). Ischemic factor-induced increases in cerebral microvascular endothelial cell Na/H exchange activity and abundance: Evidence for involvement of ERK1/2 MAP kinase. *American Journal of Physiology-Cell Physiology*, *306*(10), C931–C942. https://doi.org/10.1152/ajpcell.00021.2013

Zeisel, A., Muñoz-Manchado, A. B., Codeluppi, S., Lönnerberg, P., La Manno, G., Juréus, A., Marques, S., Munguba, H., He, L., Betsholtz, C., Rolny, C., Castelo-Branco, G., Hjerling-Leffler, J., & Linnarsson, S. (2015). Brain structure. Cell types in the mouse cortex and hippocampus revealed by single-cell RNA-seq. *Science (New York, N.Y.)*, *347*(6226), 1138–1142. https://doi.org/10.1126/science.aaa1934

Zhang, W., Zhu, L., An, C., Wang, R., Yang, L., Yu, W., Li, P., & Gao, Y. (2020). The blood brain barrier in cerebral ischemic injury—Disruption and repair. *Brain Hemorrhages*, *1*(1), 34–53. https://doi.org/10.1016/j.hest.2019.12.004

Zhao, Z., Nelson, A. R., Betsholtz, C., & Zlokovic, B. V. (2015). Establishment and dysfunction of the blood-brain barrier. *Cell*, *163*(5), 1064–1078. https://doi.org/10.1016/j.cell.2015.10.067

Zhu, X., Lee, H., Perry, G., & Smith, M. A. (2007). Alzheimer disease, the two-hit hypothesis: An update. *Biochimica et Biophysica Acta*, *1772*(4), 494–502. https://doi.org/10.1016/j.bbadis.2006.10.014

Zhu, X., Raina, A. K., Perry, G., & Smith, M. A. (2004). Alzheimer's disease: The two-hit hypothesis. *Lancet. Neurology*, *3*(4), 219–226. https://doi.org/10.1016/S1474-4422(04)00707-0

Zlokovic, B. V. (2011). Neurovascular pathways to neurodegeneration in Alzheimer's disease and other disorders. *Nature Reviews Neuroscience*, *12*(12), Article 12. https://doi.org/10.1038/nrn3114

2 Blood–Brain Barrier in Neurodegenerative Disorders

Adejoke Elizabeth Memudu and Michael Oluwatosin Gabriel*

2.1 INTRODUCTION

The blood–brain barrier (BBB), which is a specialized neurovascular unit (NVU) made up of astrocytes, endothelial cells, pericytes, and neurons (Archie et al., 2021), is an important segment of the nervous system that helps maintain homeostasis in the brain milieu (Kadry et al., 2020); hence, it is no news that its disruption triggers or induces lots of neuropathological events in the brain, resulting in neurodegenerative disorders (Verheggen et al., 2020). The main function of the BBB is to regulate the movement of biological substances important for neuronal function, maintenance of homeostasis, and brain metabolic activity between the blood and the brain (Kadry et al., 2020; Archie et al., 2021). The blood–brain barrier (BBB) controls the exchange of compounds between the blood and the brain parenchyma and helps supply nutrients and promote metabolic waste product removal that supports normal functions of the neuron (Verheggen et al., 2020).

According to Archie et al. (2021) and Verheggen et al. (2020), BBB dysfunction predisposes neurodegenerative diseases; however, the correlation between neurological disease and BBB dysfunction is not vivid.

BBB disruption compromises the integrity of the BBB, thereby leading to neurodegenerative diseases (Knox et al., 2022), such as multiple sclerosis, ischemia, Parkinson's and Alzheimer's diseases, epilepsy, brain tumors, and lysosomal storage diseases (Miners et al., 2018; Wu et al., 2021), as a result of the barrier dysfunction or leakages linked to changes in BBB permeability attributed to the opening of the tight junction, breakdown of the BBB, disruption of the basement membrane (Jeon et al., 2020), leading to the uncontrollable passage of neurotoxic plasma proteins, blood cells, and pathogens into the brain. BBB disruption is an initiating factor of the pathological cascade, which triggers white matter pathology (Riphagen et al., 2018; Wong et al., 2019; Hainsworth, 2019). According to Verheggen et al. (2020) and DiBattista et al. (2020), the integrity of the cerebral vasculature is an important factor in the development of neurodegenerative disorders. This review is to consult literature and research works done, to discuss specifically the relationship between BBB and the onset of neurodegenerative disorders. Knox et al. (2022) have reported that there is a connection between BBB and gut microbiota; microbial-derived metabolites help avert BBB disruption. Heart failure, which may result in increased central venous pressure, has been shown to cause BBB disruption (Fulop et al., 2019). A functional BBB is selectively impermeable, while a dysfunctional BBB becomes permeable due to the release of some substances, such as cytokines and chemical mediators secreted by astrocytes, namely, endothelin-1, TNF-α, taurine, NO, glutamate, ATP, aspartate, and macrophages inflammatory protein 2 (MIP2). Vasoactive compounds facilitate permeability of the BBB. Other classes of substance that mediates this permeability are humoral agents, such as bradykinin, serotonin (5HT), histamine, thrombin, substance P, quinolinic acid, platelet-activating factor, and free radicals (Lan et al., 2021), mostly released by the BBB endothelial cells.

BBB disruption has been reported to enable the influx of substances such as lymphocytes, macrophages, and plasma proteins into the brain tissue (Jeon et al., 2020). The influx of peripheral immune cells leads to the induction of microglial activation (Laurent et al., 2017) and neurodegeneration (Mayne et al., 2020). Sweeney et al. (2018) and Werry et al. (2019) made reports on neuroimaging strategies for diagnosing BBB dysfunction in neurodegenerative conditions such as Parkinson's disease, Alzheimer's disease, ischemic stroke, multiple sclerosis, etc. (Wu et al., 2021).

According to Raja et al. (2018), BBB permeability can be detected using a dynamic contrast-enhanced (DCE) MRI that involves the use of a gadolinium-based contrast agent to localize the spread of the contrast from the blood plasma to the brain interstitial fluid, dependent on the extent of BBB disruption.

The alteration in cerebral microvascular function contributes to BBB disruption associated with ageing (Norton et al., 2019). Ageing is linked to a deficiency in nicotinamide adenine dinucleotide (NAD^+) (Csiszar et al., 2019; Yoshino et al., 2018), which is a coenzyme required for endothelial cells proliferation, function, and migration (Bonkowski & Sinclair, 2016; Gomes et al., 2013). NAD^+ deficiency affects the normal function and integrity of endothelial cells, which influences the integrity of the cerebral microvascular function, thereby contributing to BBB damage over time, leading to ND (Verheggen et al., 2020). But restoration of NAD^+ level improves endothelial function and maintenance of cerebral microvasculature in ND (Csiszar et al., 2019; Kiss

* Corresponding author: jokememudu@gmail.com or adejoke.elizabeth@edouniversity.edu.ng

DOI: 10.1201/9781003383376-2

et al., 2019; Tarantini et al., 2019). Brain pericytes express many receptors for endogenous cytokines, pathogenic-like substances such as TNF-α, interleukin (IL)-1β, IL-8, LPS, and α-synuclein, which are implicated in the propagation of inflammatory reactions (Navarro et al., 2016; Dohgu et al., 2019; Perrot et al., 2020). The MMP-9 protein is found in the endothelial cells of BBB (Weekman et al., 2016). Its activation causes tight junction protein disruption and degradation, which leads to endothelial barrier dysfunction (Brilha et al., 2017). Yamazaki et al. (2019) reported that a decline in tight junction proteins correlates to Aβ deposition in AD. VE-cadherin, a gap protein component of endothelial adherens junctions required for vascular integrity maintenance, declines in AD patients and mouse brains (Lee et al., 2018). Endothelial cell dysfunction results from the impairment of endothelial cell migration and proliferation (Ungvari et al., 2018; Tarantini et al., 2019). The disruption of BBB majorly results due to endothelial cell breakdown or shrinkage (Spadoni et al., 2017) and decrease expression of tight junction proteins (Erdő et al., 2017). Aside from disruption of the endothelial cells, degeneration of the pericyte (Banks et al., 2021; Yang et al., 2020), alteration of the basement membrane, and astrocytic end feet detachment from vascular basement membrane (Obermeier et al., 2013), it has also been reported that endothelial cells induce an increase in leukocytes adhesion molecules proteins, causing an increase in leukocytes extravasation in brain tissue in neuropathological conditions (Gerwien et al., 2016). Leukocyte–endothelial cell interaction results in increased permeability of BBB to toxin due to the release of reactive oxygen species, cytokines, and other mediators of barrier disruption (Profaci et al., 2020). Tightening the barrier could reduce the negative effects of the inflammation challenge, or in the presence of any neurodegenerative disease, controlling the integrity of the BBB could slow the progression of the disease and neurodegeneration (Knox et al., 2022).

2.2 EVIDENCE OF BLOOD–BRAIN BARRIER DYSFUNCTION IN NEURODEGENERATIVE DISORDERS

The critical role of the blood–brain barrier (BBB) as a selective interface between the brain and blood implicates it in the pathogenesis and progression of neurodegenerative disorders (Morris et al., 2018). The disruption of the BMs by matrix metalloproteinases is an integral part of BBB dysfunction and posterior leukocyte infiltration, which can be observed in different neurological disorders (Rempe et al., 2016). According to Raja et al. (2018), a disrupted BBB hinders glucose and oxygen supply. The brain tissue is vulnerable to low oxygen and glucose supply. This hypoxic and hypoglycemic condition triggers inflammation (Levit et al., 2020) and neuronal dysfunction in neurodegenerative diseases.

Several methods have been described to investigate BBB damage in neurodegenerative disease, and these include (1) neuroimaging methods using positron emission tomography (PET), magnetic resonance imaging (MRI), dynamic contrast-enhanced magnetic resonance imaging (DCE-MRI), among others; (2) the levels and ratio of peripheral substances like albumin in the cerebrospinal fluid and the serum; (3) pathological presence of blood proteins in the brain; and (4) the occurrence of neuropathological events like cerebral arteriolosclerosis (Arvanitakis et al., 2016; Tao et al., 2022).

The dysregulation of the BBB in neurodegenerative diseases can be detected in animal models and postmortem brain tissues and can be observed clinically by brain imaging of patients (Daneman & Prat, 2015). Several studies have documented the correlation between BBB pathophysiology and Alzheimer's disease, a major neurodegenerative disorder. Some of these studies were based on animal models that provided evidence and mechanisms of BBB disruption in a neurodegenerative state (Giannoni et al., 2016; Haruwaka et al., 2019; Kook et al., 2013). While it has been very difficult to obtain animal models of neurodegenerative diseases that can fully simulate the phenotypes of the conditions in humans, especially BBB breakdown, animal models have continued to provide useful evidence on the existence of BBB dysfunction in neurodegenerative diseases and an understanding of the mechanisms involved (Cirrito et al., 2005; Y. Liu et al., 2020).

In postmortem tissues, BBB breakdown in patients with neurodegenerative diseases was identified by several factors, including degeneration of critical BBB cells, penetration of circulating immune cells, and other cellular and molecular changes that alter the efficiency of the BBB (Sweeney et al., 2018). In patients with early cognitive dysfunction, the DCE-MRI, with a gadolinium-based contrast agent, was used to assess the permeability of the BBB, and it was shown that the BBB was permeable in the region of the hippocampal gray matter of patients (Nation et al., 2019). The study provided evidence that a disruption in the BBB was a marker of early-stage cognitive dysfunction in humans using the DCE-MRI method. Although the tools for brain imaging may be absent in several health facilities, including primary health-care facilities, probing for biomarkers in the blood might offer an alternative approach to predicting BBB alteration in patients with neurodegenerative diseases (Hussain et al., 2021). An increase in BBB permeability in neurodegenerative disease pathology seen in cases such as ischemic stroke and traumatic brain injury is a result of an increased intrinsic cellular pro-inflammatory cytokines (e.g., IL-1, IL-6, TNF-α, interferon-γ), mediators of oxidative stress, vasogenic mediators (e.g., glutamate, serotonin, histamine) that cause abnormal changes in the cerebrovascular unit/neurovascular unit. There are groundbreaking reports on the relationship between BBB integrity and gut microbiota. A study reported that the absence of microbiota results in an increased BBB permeability and alterations in tight junction proteins (Wu et al., 2020) characteristic of impaired BBB. Glucose transport is also affected in AD patients. The expression level of GLUT1 is reduced in brain capillaries in both human AD patients and mouse AD models (Vogelsang et al., 2018). Since glucose is the main energy nutrient for the brain, impaired glucose transport leads to neurodegeneration and brain atrophy (Winkler et al., 2015).

2.3 PATHOLOGICAL MARKERS OF BBB DYSFUNCTION IN NEURODEGENERATIVE DISEASE

Currently, diagnostic and therapeutic options for neurodegenerative diseases have been unsatisfactory partly due to the paucity of biomarkers to detect these pathologies in their early stages (Dokholyan et al., 2022). Since BBB dysfunction has been reported to be an early pathological event in animal models and humans with the neurodegenerative disorder (Nation et al., 2019; Zhong et al., 2008), identifying biomarkers of BBB breakdown may offer a therapeutic pathway for slowing down the progression of neurodegenerative diseases from the onset.

The measurement of the ratio of albumin in the CSF to that of the serum, known as the "CSF/serum albumin quotient" (Q-Alb), has been used as an indirect biomarker to evaluate BBB permeability (Musaeus et al., 2020). This is based on the principle that peripheral albumin, which is not produced in the CNS, is transported to the CSF by passive diffusion, and when the Q-Alb is abnormally increased, it can predict an enhanced BBB permeability (Andersson et al., 1994; Llorens et al., 2015). Although the mechanism that causes pathological CSF levels of albumin and the diagnostic role of Q-Alb in neurodegenerative disorders is not fully understood, an increased Q-Alb has been reported in some neurodegenerative conditions like dementia with Lewy bodies, vascular dementia, and multiple sclerosis (Llorens et al., 2015; Skillbäck et al., 2017; Uher et al., 2016). Since the level of albumin in the CSF can be influenced by several factors, like the amount of CSF or albumin produced or transported per time and albumin degradation in the CNS, the measurement and the use of Q-Alb as a marker to predict BBB breakdown could be flawed; hence, more studies would be needed to clarify potential questions regarding its reliability (LeVine, 2016). Also, the Q-Alb can be used in combination with other markers to predict BBB dysfunction (Tao et al., 2022).

Leukocyte infiltration of the brain which was marked by enhanced matrix metalloproteinase (MMP) activity has been observed in multiple sclerosis using radiolabeled MMP with PET imaging (Gerwien et al., 2016). This study provided a non-invasive approach for monitoring the progression of early multiple sclerosis by focusing on the permeability of the BBB. Other studies have identified some more indicators of BBB dysfunction in neurodegenerative disease, including reduced P-glycoprotein function (Kortekaas et al., 2005), swollen astrocytic end feet and the degeneration of endothelial cells and basement membrane (Garbuzova-Davis et al., 2007), alteration of endothelial cell tight junction proteins and efflux transporters (Pan & Nicolazzo, 2018), etc.

The leakage of the BBB, which is observed in patients with neurodegenerative conditions (van de Haar et al., 2017), makes it easier for some proteins to permeate the BBB into the CNS. For example, an increase in the levels of soluble platelet-derived growth factor receptor β (PDGFRβ) in the CSF, which is used as a biomarker for pericyte injury and BBB dysfunction, was observed in individuals with mild cognitive impairment (Montagne et al., 2015). Blood proteins such as fibrinogen, immunoglobulin G (IgG), thrombin, and hemosiderin that permeate the blood–brain barrier have been reported in postmortem tissues of patients with neurodegenerative disorders (Cullen et al., 2005; Gray & Woulfe, 2015; Loeffler et al., 1995; Pienaar et al., 2015; Ryu & McLarnon, 2009; Sweeney et al., 2018).

2.4 MECHANISMS OF BBB ALTERATIONS IN SELECTED NEURODEGENERATIVE DISEASES

2.4.1 Ischemic Stroke

Ischemic stroke occurs due to a drastic stoppage of arterial supply to the brain tissue, which results in a decline in glucose and oxygen supply to the brain required for energy/ATP production for neural activity (Abdullahi et al., 2018). A decline in ATP level results in abnormal Na/K-ATPase and Ca+2ATPase activity and intracellular ion accumulation. Thereafter, endothelial transporters (Na/H ion-exchanger and Na-K-Cl cotransporter) activity is stimulated, causing an increase in Na+, Cl− and water influx through the BBB into the brain parenchyma. The endothelial cells then begin to accumulate Na+, which contributes to BBB breakdown (Keep et al., 2005), then the cytotoxic edema is followed by the ischemia (Abdullahi et al., 2018). The Na+ cellular uptake depolarizes the cell's membrane, opening voltage-gated ion channels, which enhances Ca2+ uptake, which leads to the prompt release of an excitatory neurotransmitter that is neurotoxic (O'Donnell, 2014). BBB's breakdown in stroke occurs in a two-phase pattern. First, an activation of metalloproteinases (MMP-2), which attack the tight junction proteins of the BBB (Yang et al., 2007). The second phase occurs when pro-inflammatory cytokines such as TNF-alfa and IL-1b activate metalloproteinases MMP-3 and MMP-9 (Abdullahi et al., 2018), resulting in a decline in the expression of occludin and ZO-1 (Rochfort & Cummins, 2015). In this second phase, a compound called cyclooxygenase-2 activates a deleterious opening of the BBB that allows monocytes' and neutrophils' entrance, leading to a local inflammatory response (Abdullahi et al., 2018) involving the activation of astrocytes and pericytes. BBB damage in ischemic stroke is a precursor to the hemorrhagic transformation of the infarcted brain tissue (Kassner & Merali, 2015). In ischemic stroke, BBB damage can occur due to reactive oxygen species (ROS) and oxidative stress. Superoxide anion (O2−) mediates endothelial cell damage after ischemic stroke (Carvalho et al., 2017). Oxidative stress has been implicated in ischemia/reperfusion (I/R) induced neuronal damage by stroke due to events in the neurovascular unit that trigger oxidative stress; these events include depletion of the antioxidant defense system, an increase in vasogenic mediators, and glutamate release (Zhao et al., 2019). In the BBB, the endothelial cells and immune cells release a large amount of ROS during the cerebral ischemia phase, which triggers the activation of nuclear factor-κappa B (NF-κB), inducible endothelial nitric oxide (iNOS), and pro-inflammatory factors, triggering the upregulation of vascular endothelial cell adhesion molecules and causing BBB permeability (Abdul-Muneer et al., 2013). Following these acute CNS injuries described earlier, cerebral edema because of BBB dysfunction develops through several phases, including ionic edema, vasogenic edema, and hemorrhagic conversion (Stokum et al., 2016). Glial cell–derived matrix metalloproteinases (MMPs) contribute to BBB dysfunction by digesting tight junction proteins at this phase. Infiltrating erythrocytes and plasma proteins trigger glial activation and neuronal damage.

2.4.2 Traumatic Brain Injury (TBI)

Traumatic brain injury (TBI) is classified as mild, moderate, and severe TBI. BBB dysfunction is widely observed from mild to severe TBI (O'Keeffe et al., 2020). BBB disruption in mild TBI precedes neuroinflammation responses identified by microglia and astrocyte activation (Wu et al., 2021). BBB disruption occurs within hours after brain injury and can be sustained for years (Hay et al., 2015). Extravasation of plasma proteins, including fibrinogen, IgG, and albumin, has been observed in the brain of patients with TBI in the acute and chronic phases. TBI-induced neuroinflammation responses can be identified by elevated levels of inflammatory mediators within hours post-injury (Corrigan et al., 2016; Lier et al., 2020). Microglia and astrocytes are activated by the extravasation of albumin to release cytokines/chemokines and matrix metalloproteinases (MMPs), which disrupt the BBB (Ralay et al., 2012).

2.4.3 Parkinson's Disease

Parkinson's disease is an ND associated with the loss of dopaminergic neurons and deposition of Lewy bodies in the substantia nigra of the midbrain, leading to motor impairment (Blanchard et al., 2020). The pathogenesis is associated with insoluble α-synuclein (Kali & Kalia, 2015), neuroinflammation, mitochondrial dysfunction, and impaired protein degradation, yet its exact cause is still unknown. α-synuclein is also expressed in endothelial cells and can be transported across the BBB in a bidirectional way (Sui et al., 2014); hence, it can potentiate synaptic failure and activate neuroinflammation through microglia and astrocytes. High serum levels of α-synuclein have been found in PD patients, attributed to an increased efflux to the peripheral blood (Shi et al., 2014); also, postmortem samples have shown cases of α-synuclein aggregates associated with endothelial degeneration and decreased P-gp expression in PD patient (Yang et al., 2017). Experimental studies on PD animal models using neurotoxin models such as 6-hydroxydopamine (6-OHDA) and 1-methyl-4-phenyl-1,2,3,6-tetrahydropyridine (MPTP) administered into the brain parenchyma, resulting in the production of reactive oxygen species, such as hydrogen peroxide, superoxide radicals, and hydroxyl radicals, culminate to dopaminergic neurons loss attributed to oxidative stress-mediated BBB disruption and leakage (Olmedo-Díaz et al., 2017; Kim et al., 2020). In a study involving the co-culture of rat brain endothelial cells and pericyte, the introduction of a monomeric α-synuclein activates the release of inflammatory cytokines from pericytes, leading to an increase in the permeability of brain NVU (Dohgu et al., 2019). This implies that α-synuclein mediates BBB disruption through the activation of pericytes (Elabi et al., 2021). The earlier report assumed that BBB remains normal in PD; however, recent reports in PD patients have identified BBB disruption ([Kim et al., 2020). Neuroimaging technology has shown leakages in the BBB in basal ganglia (Al-Bachari et al., 2017) and cerebral microbleeds (Ham et al., 2014; Janelidze et al., 2015). According to Gray and Woulfe (2015), histology of the basal ganglia of PD patients is characterized by capillary leakages, accumulation of plasma protein associated with the endothelial cell, a decrease of tight junction proteins, and disruption of the basement membrane of the BBB. Another study reported an elevation of albumin and IgG cerebrospinal fluid (CSF)/serum ratios in PD patients (Liguori et al., 2017). Studies using monomeric α-synuclein have implicated pericyte as a target for BBB disruption in animal models of PD (Kuan et al., 2016; Dohgu et al., 2019; Al-Bachari et al., 2020).

2.4.4 Alzheimer's Disease

Alzheimer's disease is the most common cause of dementia, a progressive ND characterized by memory impairment and cognitive decline (Kocahan & Doğan, 2017).

There are two hypotheses for AD pathogenesis associated with neurovascular dysfunction (Nelson et al., 2016) viz: (1) BBB dysfunction leading to the infiltration of neurotoxic substances in brain parenchyma, and (2) neuroinflammation through the process of oxidative stress (Zenaro et al., 2017) resulting in the two main pathological hallmarks of AD, which are amyloid-beta (Aβ) plaque buildup and formation of neurofibrillary tangles (Kocahan & Doğan, 2017; Singh et al., 2021). The accumulation of amyloid beta plaque deposits is linked to an imbalance in the rate of production and clearance of amyloid beta protein (Cai et al., 2018) as a result of leakage in BBB. Aβ impairs the integrity of BBB (Chan et al., 2018; Cuevas et al., 2019). Freeze et al. (2019) mentioned that the buildup of Aβ thickens endothelial walls and narrows the lumen, which engenders hemorrhage and microbleeding into the brain tissue. Ma et al. (2018) made a report on the deposition of Aβ in pericytes of AD patients and APP Swedish mutation mice, indicating that loss of pericyte correlates with an elevated accumulation of Aβ in the human and animal model of AD. A study on animal models has reported a decline and abnormal expression of ABC transporters in the BBB of AD rats, resulting in the efflux of brain-derived amyloid-beta (Aβ) due to BBB permeability (Nelson et al., 2016; Deo et al., 2014). There are scientific hypotheses that suggested that "BBB dysfunction and brain hypoperfusion secondary to blood vessel damage may be the first hit that leads to amyloid-beta (Aβ) accumulation and neuronal injury" (Nelson et al., 2016). Components of BBB have been implicated in AD pathology. For instance, Halliday et al. (2016) mentioned that an accelerated pericyte degeneration results in BBB breakdown in AD-ApoE4 carrier mouse models. Pericytes degeneration activates various pathways of neurodegeneration due to BBB membrane permeability that enables the influx of neurotoxic blood-derived proteins, such as plasminogen, thrombin, and fibrinogen (Halliday et al., 2016; Schultz et al., 2018), which disrupts laminin (a protein matrix), resulting in progressive neuron damage. In recent times, therapeutics that target amyloid plaques proved abortive to avert AD pathogenesis to repair cognitive function and alleviate memory impairment; hence, drugs that target vascular dysfunction and BBB are of great potential (Singh et al., 2021). AD increases glutamate (Huang et al., 2020) that contributes to BBB damage—the permeability of the BBB to plasma protein such as fibrinogen triggers microglial activation (Merlini et al., 2019), followed by the release of inflammatory cytokines, resulting in neuron damage in an AD animal model. Fibrin influx has been seen in various forms of neurodegenerative diseases (Petersen et al., 2018). Elevated concentrations of thrombin activate BBB disruption, neurotoxicity, and memory impairment in AD (Sweeney et al., 2018).

2.4.5 Multiple Sclerosis

Multiple sclerosis is an autoimmune and neuroinflammatory disorder attributed to demyelination and neurodegeneration (Correale et al., 2017). Here, an aberration in the BBB integrity results in the infiltration of T cells and serum protein extravasation into the brain tissue along with fibrinogen buildup (Alvarez et al., 2011). An influx of immune cells activates the release of various kinds of inflammatory mediators that activate astrocytes and microglia, resulting in neuroinflammation and contributing to the progression of MS (Correale et al., 2017).

2.4.6 Huntington's Disease

Huntington's disease (HD) is a neurodegenerative disorder caused by a mutation in *HTT* gene coding for the huntingtin protein, characterized by addition of CAG repeats to the HTT gene, which causes protein folding/aggregation in neurodegeneration (Sweeney et al., 2019). Mutated HTT protein and its aggregates are neurotoxic; oligomers of mutated huntingtin are thought to be more toxic than monomers (Dayalu & Albin, 2015). The mutated huntingtin aggregates are located in brain endothelial cells, perivascular macrophages, and the vascular basal lamina in patients with HD (Drouin-Ouellet et al., 2015). This mutated HTT targets BBB by causing a decline in tight junction proteins and pericytes, resulting in an increase in the membrane permeability of the BBB (Drouin-Ouellet et al., 2015; Di Pardo et al., 2017).

2.4.7 Amyotrophic Lateral Sclerosis (ALS)

Amyotrophic lateral sclerosis (ALS) is a chronic neurodegenerative disease characterized by loss of motor neurons in the spinal cord and brain, leading to progressive paralysis, muscle atrophy, and finally, death (Sweeney et al., 2019; Pape & Grose, 2020; Wu et al., 2021). Some reports claim that ALS pathology is not vivid. However, Wu et al. (2021) suggested a decline in the number of tight junctions, loss of endothelial cells, and pericytes characteristics of BBB breakdown linked to the influx of erythrocytes and serum-derived proteins into the brain tissue. When erythrocytes enter the brain, they degrade and secrete neurotoxins, such as hemoglobin and free iron, causing a buildup of reactive oxygen species (ROS formation), which is toxic to motor neurons, leading to oxidative mediated neuron death through the activation of microglia (Sweeney et al., 2019). A number of studies have demonstrated changes in BBB associated with ALS. Neuropathological changes reported include a decline in tight junction protein, extravasation of erythrocytes leading to the buildup of serum-derived proteins, a reduction in the number of pericytes, dissociation of astrocyte foot processes from the endothelial cell membrane, T cell infiltration in the region of BBB breakdown, and enlargement of the perivascular spaces (Evans et al., 2014; Yamadera et al., 2015; Wu et al., 2021).

2.5 MODELS OF BBB DYSFUNCTION IN NEURODEGENERATIVE DISEASES

Due to the numerous pieces of evidence linking BBB dysfunction to several neurodegenerative diseases and the intricacies and limitations associated with investigating this condition in humans, it has become imperative to explore standard models that can mimic certain patterns of BBB dysfunction in humans. This will promote the understanding of mechanisms connecting BBB breakdown and neurodegenerative diseases for efficient diagnosis and therapy. Both in vivo and in vitro models are relevant for studying the consequences of BBB breakdown; however, in vivo models are very critical for evaluating behavioral and systemic effects, whereas in vitro models are reliable for investigating the drivers of cellular and molecular changes that elucidate in vivo studies (Williams-Medina et al., 2021). The BBB models have been categorized into three major systems that include in vitro, in vivo, and clinical models (Knox et al., 2022):

1. **In vitro models.** In vitro models involve the use of monoculture of brain endothelial cells on Transwell membrane inserts or standard flat tissue culture plates (Ahn et al., 2020). The use of in vitro models to characterize BBB permeability has been around for many years and has continued to advance, strikingly moving from two-dimensional to three-dimensional cultures that more closely recapitulate tissue and organ physiology like the organ-on-a-chip technology (Harris et al., 2022; Pardridge et al., 1990; Wevers et al., 2018; Williams-Medina et al., 2021). In vitro models include monoculture and co-culture; primary, immortalized, and stem cell culture; microfluidic models; dynamic models; etc. to evaluate BBB leakage, trans-endothelial electrical resistance (TEER), changes in cellular and subcellular constituents. Etc. (Knox et al., 2022). In vitro models have been used to connect BBB dysfunction to AD using a 3D AD culture (Shin et al., 2019) and to test the diagnostic and therapeutic prospects of certain peptides, like D-enantiomeric peptides, to inhibit beta-amyloid accumulation in the brain based on their ability to cross the blood–brain barrier (H. Liu et al., 2010; Schartmann et al., 2018; Jagtiani et al., 2022).
   A. **Transwell system.** This is a conventional approach for modelling the BBB to understand its dynamics (Williams-Medina et al., 2021). This system utilized endothelial cells (primary or immortalized) cultured on collagen-coated filters that allow the assessment of the monolayer leakage and TEER and to evaluate the luminal and abluminal sides of the monolayer (Abbott et al., 1992; Erickson et al., 2020). Because this technology was developed with a limited understanding of the dynamics of the BBB, substantial modifications to the materials used have occurred over time but are still based on similar principles (Palmiotti et al., 2014; Williams-Medina et al., 2021). Although this technology provides a simple and affordable alternative for evaluating BBB function and can be adapted for different co-culture modalities in addition to its use for monocultures, its static 2D nature limits its ability to model BBB physiology (Williams-Medina et al., 2021).
   B. **Microfluidic devices.** BBB-on-a-chip provides a more advanced approach for simulating the BBB physiology by incorporating both biochemical and mechanical influences (e.g., stimulation with TNF-α and addition of fluid shear stress, respectively) (Griep et al., 2013). In this study, Griep and colleagues cultured immortalized endothelial cells of the human brain in a microfluidic device and used the result of TEER measurement to show that the addition of fluid shear stress and protein stimulation makes this platform a more realistic approach for investigating BBB function in neurodegenerative diseases. This model was enhanced by co-culturing with astrocytes and pericytes to evaluate the contributions of each cell type as part of the neurovascular unit to neuroinflammation and brain metabolic homeostasis (Herland et al., 2016; Maoz et al., 2018). This is crucial because both neuroinflammation and metabolic dysfunction play critical roles in neurodegenerative diseases (Kshirsagar et al., 2021).
   C. **Organoids.** Although the microfluidic models have undergone different advances and are very beneficial to BBB studies, they are still limited by requiring specialized tools and skillsets (Schreiner et al., 2022). Organoids which are derived from stem cells can bring together multiple cell types into close contact and are relevant to BBB research with high reproducibility; however, the complex interaction between the brain and periphery makes it critical to probe whether an organoid develops vascular networks and functional BBB (Schreiner et al., 2022; Williams-Medina et al., 2021). Human stem cell–derived 3D cerebral organoids were reported to develop distinct brain regions, like the dorsal cortex, prefrontal cortex, hippocampus, ventral cortex, etc., and simulate the cortical pattern of progenitor organization with numerous radial glial cells (Wenzel & Homfray, 2013). A major challenge observed with this organoid is that the absence of a circulatory system produces growth limit and tissue necrosis resulting from insufficient oxygen and nutrients permeating the tissue (Wenzel & Homfray, 2013). To solve this challenge and be able to use this organoid to study the blood–brain barrier, a microvascular unit would need to be incorporated (Caffrey et al., 2021). This has been demonstrated by studies that transplanted organoids into the mouse brain and showed progressive neuronal and glial development with functional neurovasculature (Mansour et al., 2018). Spheroids have also been reported to display certain physiological BBB features, like tight junction proteins, receptor-mediated transcytosis, and the activity of efflux pump, that were better than the Transwell co-culture system and makes them suitable for modelling BBB dysfunction in neurodegenerative diseases (Cho et al., 2017).
2. **In vivo models.** In vivo studies have extensively utilized mammals, and in particular, rodents, to provide a more translational approach to investigating BBB dynamics because they share relatively similar BBB features with humans (Sharma et al., 2018). For instance, rodents have been used to elucidate mechanisms regulating BBB permeability (Vazana et al., 2016; Q. Zhang et al., 2021), demonstrate the complex interactions between BBB dysfunction

and neurodegenerative diseases (Y. Liu et al., 2020; Kook et al., 2012; Zhong et al., 2008), evaluate the penetrability and response of the BBB to some drugs (Rice et al., 2005; Takada et al., 1992; Wu et al., 2020), and reverse BBB leakage in AD (Polis et al., 2019). Smaller vertebrates like the zebra fish have also been used as in vivo models to understand the physiology and molecular architecture of the BBB, because similar tools used to study rodents can also be used to study zebra fish, and the zebra fish share BBB ultrastructural similarities with rodents (Eliceiri et al., 2011). For example, zebra fish cerebral micro-vessels have been reported to possess tight junctions because of the presence of tight junction–associated proteins like claudin-5 and zonular occludens-1 that contribute to the formation of a selective BBB (Jeong et al., 2008; J. Zhang et al., 2012). In addition, both zebra fish and mice express angiogenic factors, like DR6 and TROY, which play an important role in BBB development (Tam et al., 2012). The zebra fish model has been used to test the mechanism of drug delivery across the BBB (Yang et al., 2015) and probe BBB permeability to some bioactive substances, like berberine derivatives, under neurodegenerative conditions (L. Wang et al., 2021). While a huge gap exists between vertebrate and non-vertebrate BBB, certain patterns are conserved, like equivalents of tight junction, efflux transporters, and gap junctions, which are also present in invertebrates like *Drosophila* and make them suitable models for screening drugs for neurodegenerative diseases (Hindle & Bainton, 2014)

3. **Clinical studies.** Clinical studies are useful in understanding BBB dysfunction in ND patients; however, drugs and therapeutics are not screened in clinical studies (Knox et al., 2022). The current clinical therapy against neurodegenerative conditions has not reached a satisfactory level due to the complexities associated with their pathogenesis; efforts to enhance our understanding of the pathological events that induce their occurrence or promote their progression are still in place. One of the areas gaining attention is the connection between vascular diseases and neurodegenerative diseases. Some studies have reported that no significant changes in the permeability of the BBB were seen in patients with AD compared to the control subjects (Caserta et al., 1998; Starr et al., 2009); however, some of these studies utilized small sample sizes that require further investigation. More robust studies have established a link between neurovascular and BBB alterations in normal ageing patients and subjects with cognitive impairment or neurodegenerative disorders, living or autopsied (Arvanitakis et al., 2016; Nation et al., 2019; Toledo et al., 2013). These studies were designed to understand the interplay between cerebrovascular diseases and neurodegenerative disease, since epidemiological studies have shown that a mixed pathology exists in patients (Jellinger & Attems, 2010). The BBB leakage was shown to be an early pathological event in ageing and the onset of AD (Montagne et al., 2015; van de Haar et al., 2017; Iturria-Medina et al., 2016). Although BBB damage in early AD was shown to be independent of Aβ or Tau levels (Nation et al., 2019), BBB damage can influence the Aβ burden in the brain (Toledo et al., 2012; D. Wang et al., 2021). The preceding studies detected the BBB alterations by neuroimaging using MRI, DCE-MRI, documented neuropathological assessment of vascular disease of deceased patients, soluble PDGFRβ in CSF, etc.

Another area of interest is the development of drugs for the treatment of neurodegenerative disorders which has been limited by the delay to develop sufficient drugs that have the capacity to cross the BBB. This delay is due to the presence of multiple alternatives that bypass the BBB to deliver drugs to the brain and the absence of routine measurement of BBB drug transport that is free from artefacts (Pardridge, 2020). Hence, the need for more robust preclinical and in vitro studies to test several BBB permeable drug targets is highly warranted for the treatment of neurodegenerative diseases.

2.6 MEDIATORS OF BBB DISRUPTION IN HEALTH AND DISEASES

Oxidative stress and inflammatory dysfunction of the BBB are evidenced by ROS activity and an increase in inflammatory mediators (Zenaro et al., 2017).

2.6.1 Inflammation

Inflammation has been reported to induce a series of changes in BBB function and physiology (Knox et al., 2022). Inflammation is characterized by activation of glia cells, which, in turn, attract leukocytes across the BBB, resulting in alteration of paracellular and transcellular transport and tight junction protein expressed (Schneider et al., 2019). BBB dysfunction can amplify neuroinflammation and act as a key process in the development of neuroinflammation (Takata et al., 2021). Importantly, systemic administration of lipopolysaccharide (LPS), a Gram-negative bacterial endotoxin, caused increased BBB permeability to sodium fluorescein in parallel with an increased number of activated microglia (Nishioku et al., 2009). Ju et al. (2018) reported that increased BBB permeability induced by mannitol led to microglial activation. BBB dysfunction positively amplifies the development of neuroinflammation, rather than being a pathological result of glial activation and neuroinflammation (Takata et al., 2021).

2.6.2 Oxidative Stress

The brain is susceptible to oxidative stress due to the high reactive oxygen consumed by the brain (Segarra et al., 2021). The endothelial cells that carry blood are more sensitive to oxidative stress when compared with the pericytes (Segarra et al., 2021). Oxidative stress occurs as a result of imbalance in the ratio of oxidant and antioxidant, which results in excess secretion of oxidizing agents (Pun et al., 2009) such as ROS, including superoxide, hydrogen peroxide, peroxynitrite, nitric oxide, and hydroxyl radicals (Knox et al., 2022).

For instance, nitric oxide is the main oxidative species that have been reported to contribute to BBB disruption (Knox et al., 2022). To study the process of BBB disruption due to oxidative stress, cell cultures are modelled using glucose deprivation, hypoxia, or hydrogen peroxide treatment (Song et al., 2020), while animal models can include exogenous treatment with free radicals, stroke, or transgenic mouse models which use normal metabolic processes to produce a pro-oxidative state.

2.6.3 Aging

BBB disruption as a result of aging results in cognitive decline attributed to activation of neuroinflammation (Wen et al., 2020). The dysfunctional BBB results in abnormality in the integrity and function of the BBB (Knox et al., 2022). The alteration and breakdown of the functional structure of BBB result in various neurological disorders, such as cognitive deficit, dementia (Erdő et al., 2017).

2.7 CONCLUDING REMARKS

The blood–brain barrier is an integral part of the brain implicated in most neurodegenerative disorders. We have reviewed the BBB changes in selected neurodegenerative disorders in this section, as well as pathological markers for identifying BBB disruption in ND. The leakage of the brain vessels can be observed in neurodegenerative disorders associated with a decline in tight junction proteins, reduction in endothelial cells, loss of pericyte, and displacement of astrocytes end feet. Hence, the BBB is considered one of the potential treatment targets for NDDs. From another point of view, various experimental models for studying BBB in neurodegenerative diseases were also discussed.

REFERENCES

Abbott, N. J., Hughes, C. C. W., Revest, P. A., & Greenwood, J. (1992). Development and characterisation of a rat brain capillary endothelial culture: Towards an in vitro blood-brain barrier. *Journal of Cell Science*, *103*(1), 23–37. https://doi.org/10.1242/jcs.103.1.23

Abdullahi, W., Tripathi, D., & Ronaldson, P. T. (2018). Blood-brain barrier dysfunction in ischemic stroke: Targeting tight junctions and transporters for vascular protection. *American Journal of Physiology. Cell Physiology*, *315*(3), C343–C356.

Abdul-Muneer, P. M., Schuetz, H., Wang, F., Skotak, M., Jones, J., Gorantla, S., Zimmerman, M. C., Chandra, N., & Haorah, J. (2013). Induction of oxidative and nitrosative damage leads to cerebrovascular inflammation in an animal model of mild traumatic brain injury induced by primary blast. *Free Radical Biology and Medicine*, *60*, 282–291. https://doi.org/10.1016/j.freeradbiomed.2013.02.029

Ahn, S. I., Sei, Y. J., Park, H. J., Kim, J., Ryu, Y., Choi, J. J., et al. (2020). Microengineered human blood–brain barrier platform for understanding nanoparticle transport mechanisms. *Nature Communications*, *11*, 175.

Al-Bachari, S., Naish, J. H., Parker, G. J. M., Emsley, H. C. A., & Parkes, L. M. (2020). Blood-brain barrier leakage is increased in Parkinson's disease. *Frontiers in Physiology*, *11*, 593026.

Al-Bachari, S., Vidyasagar, R., Emsley, H., & Parkes, L. (2017). PO071 MRI assessment of neurovascular changes in idiopathic Parkinson's disease. *Journal of Neurology, Neurosurgery, and Psychiatry*, *88*, A3–A30.

Alvarez, J. I., Cayrol, R., & Prat, A. (2011). Disruption of central nervous system barriers in multiple sclerosis. *Biochimica et Biophysica Acta*, *1812*, 252–264.

Andersson, M., Alvarez-Cermeñio, J., Bernardi, G., Cogato, I., Fredman, P., Frederiksen, J., et al. (1994). Cerebrospinal fluid in the diagnosis of multiple sclerosis: A consensus report. *Journal of Neurology, Neurosurgery and Psychiatry*, *57*(8), 897–902. https://doi.org/10.1136/jnnp.57.8.897

Archie, S. R., Al Shoyaib, A., & Cucullo, L. (2021). Blood-brain barrier dysfunction in CNS disorders and putative therapeutic targets: An overview. *Pharmaceutics*, *13*(11), 1779.

Arvanitakis, Z., Capuano, A. W., Leurgans, S. E., Bennett, D. A., & Schneider, J. A. (2016). Relation of cerebral vessel disease to Alzheimer's disease dementia and cognitive function in elderly people: A cross-sectional study. *Lancet Neurology*, *15*(9), 934–943. https://doi.org/10.1016/S1474-4422(16)30029-1

Banks, W. A., Reed, M., Logsdon, A. F., Rhea, E. M., & Erickson, M. A. (2021). Healthy aging and the blood–brain barrier. *Nature Aging*, *1*(54), 243–254.

Blanchard, J. W., Bula, M., Davila-Velderrain, J., Akay, L. A., Zhu, L., Frank, A., et al. (2020). Reconstruction of the human blood-brain barrier in vitro reveals a pathogenic mechanism of APOE4 in pericytes. *Nature Medicine*, *26*, 952–963.

Bonkowski, M. S., & Sinclair, D. A. (2016). Slowing ageing by design: The rise of NAD+ and sirtuin-activating compounds. *Nature Reviews Molecular Cell Biology*, *17*, 679–690.

Brilha, S., Ong, C. W. M., Weksler, B., Romero, N., Couraud, P. O., & Friedland, J. S. (2017). Matrix metalloproteinase-9 activity and a downregulated hedgehog pathway impair blood-brain barrier function in an in vitro model of CNS tuberculosis. *Scientific Reports*, *7*, 16031.

Caffrey, T. M., Button, E., & Robert, J. (2021). Toward three-dimensional in vitro models to study neurovascular unit functions in health and disease. *Neural Regeneration Research*, *16*(11), 2132–2140. https://doi.org/10.4103/1673-5374.310671

Cai, Z., Qiao, P. F., Wan, C. Q., Cai, M., Zhou, N. K., & Li, Q. (2018). Role of blood-brain barrier in Alzheimer's disease. *Journal of Alzheimer's Disease*, *63*(4), 1223–1234. https://doi.org/10.3233/JAD-180098

Carvalho, A. N., Firuzi, O., Gama, M. J., Horssen, J. V., & Saso, L. (2017). Oxidative stress and antioxidants in neurological diseases: Is there still hope? *Current Drug Targets*, *18*(6), 705–718.

Caserta, M. T., Caccioppo, D., Lapin, G. D., Ragin, A., & Groothuis, D. R. (1998). Blood-brain barrier integrity in

Alzheimer's disease patients and elderly control subjects. *Journal of Neuropsychiatry and Clinical Neurosciences*, *10*(1), 78–84. https://doi.org/10.1176/jnp.10.1.78

Chan, Y., Chen, W., Wan, W., Chen, Y., Li, Y., & Zhang, C. (2018). Aβ1–42 oligomer induces alteration of tight junction scaffold proteins via RAGE-mediated autophagy in bEnd.3 cells. *Experimental Cell Research*, *369*(2), 266–274. https://doi.org/10.1016/j.yexcr.2018.05.025

Cho, C. F., Wolfe, J. M., Fadzen, C. M., Calligaris, D., Hornburg, K., Chiocca, E. A., et al. (2017). Blood-brain-barrier spheroids as an in vitro screening platform for brain-penetrating agents. *Nature Communications*, *8*. https://doi.org/10.1038/ncomms15623

Cirrito, J. R., Deane, R., Fagan, A. M., Spinner, M. L., Parsadanian, M., Finn, M. B., et al. (2005). P-glycoprotein deficiency at the blood-brain barrier increases amyloid-β deposition in an Alzheimer disease mouse model. *Journal of Clinical Investigation*, *115*(11), 3285–3290. https://doi.org/10.1172/JCI25247

Correale, J., Gaitán, M. I., Ysrraelit, M. C., & Fiol, M. P. (2017). Progressive multiple sclerosis: From pathogenic mechanisms to treatment. *Brain*, *140*, 527–546.

Corrigan, F., Mander, K. A., Leonard, A. V., & Vink, R. (2016). Neurogenic inflammation after traumatic brain injury and its potentiation of classical inflammation. *Journal of Neuroinflammation*, *13*, 264.

Csiszar, A., Tarantini, S., Yabluchanskiy, A., Balasubramanian, P., Kiss, T., Farkas, E., Baur, J. A., & Ungvari, Z. (2019). Role of endothelial NAD+ deficiency in age-related vascular dysfunction. *American Journal of Physiology – Heart and Circulatory Physiology*, *316*(6), H1253–H1266. https://doi.org/10.1152/ajpheart.00039.2019

Cuevas, E., Rosas-Hernandez, H., Burks, S. M., Ramirez-Lee, M. A., Guzman, A., Imam, S. Z., et al. (2019). Amyloid beta 25–35 induces blood-brain barrier disruption in vitro. *Metabolic Brain Disease*, *34*, 1365–1374.

Cullen, K. M., Kócsi, Z., & Stone, J. (2005). Pericapillary haem-rich deposits: Evidence for microhaemorrhages in aging human cerebral cortex. *Journal of Cerebral Blood Flow and Metabolism*, *25*(12), 1656–1667. https://doi.org/10.1038/sj.jcbfm.9600155

Daneman, R., & Prat, A. (2015). The blood–brain barrier. *Cold Spring Harbor Perspectives in Biology*, *7*(1). https://doi.org/10.1101/cshperspect.a020412

Dayalu, P., & Albin, R. L. (2015). Huntington disease: Pathogenesis and treatment. *Neurologic Clinics*, *33*, 101–114.

Deo, A. K., Borson, S., Link, J. M., Domino, K., Eary, J. F., Ke, B., Richards, T. L., Mankoff, D. A., Minoshima, S., O'Sullivan, F., Eyal, S., Hsiao, P., Maravilla, K., & Unadkat, J. D. (2014). Activity of P-glycoprotein, a β-amyloid transporter at the blood-brain barrier, is compromised in patients with mild Alzheimer disease. *Journal of Nuclear Medicine*, *55*(7), 1106–1111. https://doi.org/10.2967/jnumed.113.130161

DiBattista, A. M., Sierra, F., & Masliah, E. (2020). *NIA workshop on senescence in brain aging and Alzheimer's disease and its related dementias*. Springer.

Di Pardo, A., Amico, E., Scalabrì, F., Pepe, G., Castaldo, S., Elifani, F., et al. (2017). Impairment of blood-brain barrier is an early event in R6/2 mouse model of Huntington disease. *Scientific Reports*, *7*, 41316.

Dohgu, S., Takata, F., Matsumoto, J., Kimura, I., Yamauchi, A., & Kataoka, Y. (2019). Monomeric α-synuclein induces blood-brain barrier dysfunction through activated brain pericytes releasing inflammatory mediators in vitro. *Microvascular Research*, *124*, 61–66.

Dokholyan, N. V., Mohs, R. C., & Bateman, R. J. (2022). Challenges and progress in research, diagnostics, and therapeutics in Alzheimer's disease and related dementias. *Alzheimer's & Dementia: Translational Research & Clinical Interventions*, *8*(1). https://doi.org/10.1002/trc2.12330

Drouin-Ouellet, J., Sawiak, S. J., Cisbani, G., Lagacé, M., Kuan, W. L., Saint-Pierre, M., et al. (2015). Cerebrovascular and blood-brain barrier impairments in Huntington's disease: Potential implications for its pathophysiology. *Annals of Neurology*, *78*, 160–177.

Elabi, O., Gaceb, A., Carlsson, R., Padel, T., Soylu-Kucharz, R., Cortijo, I., et al. (2021). Human α-synuclein overexpression in a mouse model of Parkinson's disease leads to vascular pathology, blood brain barrier leakage and pericyte activation. *Scientific Reports*, *11*, 1120.

Eliceiri, B. P., Gonzalez, A. M., & Baird, A. (2011). Zebrafish model of the blood-brain barrier: Morphological and permeability studies. *Methods in Molecular Biology*, *686*, 371–378. https://doi.org/10.1007/978-1-60761-938-3_18

Erdő, F., Denes, L., & de Lange, E. (2017). Age-associated physiological and pathological changes at the blood-brain barrier: A review. *Journal of Cerebral Blood Flow & Metabolism*, *37*, 4–24.

Erickson, M. A., Wilson, M. L., & Banks, W. A. (2020). In vitro modeling of blood-brain barrier and interface functions in neuroimmune communication. *Fluids and Barriers of the CNS*, *17*(1). https://doi.org/10.1186/s12987-020-00187-3

Evans, M. C., Serres, S., Khrapitchev, A. A., Stolp, H. B., Anthony, D. C., Talbot, K., et al. (2014). *T2* -weighted MRI detects presymptomatic pathology in the SOD1 mouse model of ALS. *Journal of Cerebral Blood Flow & Metabolism*, *34*, 785–793.

Freeze, W. M., Bacskai, B. J., Frosch, M. P., Jacobs, H. I. L., Backes, W. H., Greenberg, S. M., et al. (2019). Blood-brain barrier leakage and microvascular lesions in cerebral amyloid angiopathy. *Stroke*, *50*, 328–335.

Fulop, G. A., Ahire, C., Csipo, T., Tarantini, S., Kiss, T., Balasubramanian, P., Yabluchanskiy, A., Farkas, E., Toth, A., Nyúl-Tóth, Á., Toth, P., Csiszar, A., & Ungvari, Z. (2019). Cerebral venous congestion promotes blood-brain barrier disruption and neuroinflammation, impairing cognitive function in mice. *GeroScience*, *41*(5), 575–589. https://doi.org/10.1007/s11357-019-00110-1

Garbuzova-Davis, S., Haller, E., Saporta, S., Kolomey, I., Nicosia, S. V., & Sanberg, P. R. (2007). Ultrastructure of blood-brain barrier and blood-spinal cord barrier in SOD1 mice modeling ALS. *Brain Research*, *1157*(1), 126–137. https://doi.org/10.1016/j.brainres.2007.04.044

Gerwien, H., Hermann, S., Zhang, X., Korpos, E., Song, J., Kopka, K., et al. (2016). Imaging matrix metalloproteinase activity in multiple sclerosis as a specific marker of leukocyte penetration of the blood-brain barrier. *Science Translational Medicine*, *8*, 364. https://doi.org/10.1126/scitranslmed.aaf8020

Giannoni, P., Arango-Lievano, M., Neves, I. D., Rousset, M. C., Baranger, K., Rivera, S., et al. (2016). Cerebrovascular pathology during the progression of experimental Alzheimer's disease. *Neurobiology of Disease*, *88*, 107–117. https://doi.org/10.1016/j.nbd.2016.01.001

Gomes, A. P., Gomes, A. P., Price, N. L., Price, N. L., Ling, A. J. Y., Ling, A. J. Y., et al. (2013). Declining NAD(+) induces a pseudohypoxic state disrupting nuclear-mitochondrial communication during aging. *Cell*, *155*(7), 1624–1638. http://eutils.ncbi.nlm.nih.gov/entrez/eutils/elink.fcgi?dbfrom=pu

bmed&id=24360282&retmode=ref&cmd=prlinks%5Cnpapers3://publication/doi/10.1016/j.cell.2013.11.037
Gray, M. T., & Woulfe, J. M. (2015). Striatal blood-brain barrier permeability in Parkinson's disease. *Journal of Cerebral Blood Flow & Metabolism, 35*, 747–750.
Griep, L. M., Wolbers, F., De Wagenaar, B., Ter Braak, P. M., Weksler, B. B., Romero, I. A., et al. (2013). BBB on CHIP: Microfluidic platform to mechanically and biochemically modulate blood-brain barrier function. *Biomedical Microdevices, 15*(1), 145–150. https://doi.org/10.1007/s10544-012-9699-7
Hainsworth, A. (2019). White matter lesions in cerebral small vessel disease: Underperfusion or leaky vessels? *Neurology, 92*.
Halliday, M. R., Rege, S. V., Ma, Q., Zhao, Z., Miller, C. A., Winkler, E. A., et al. (2016). Accelerated pericyte degeneration and blood-brain barrier breakdown in apolipoprotein E4 carriers with Alzheimer's disease. *Journal of Cerebral Blood Flow & Metabolism, 36*, 216–227.
Ham, J. H., Yi, H., Sunwoo, M. K., Hong, J. Y., Sohn, Y. H., & Lee, P. H. (2014). Cerebral microbleeds in patients with Parkinson's disease. *Journal of Neurology, 261*, 1628–1635.
Harris, W. J., Asselin, M. C., Hinz, R., Parkes, L. M., Allan, S., Schiessl, I., et al. (2022). In vivo methods for imaging blood–brain barrier function and dysfunction. *European Journal of Nuclear Medicine and Molecular Imaging*. https://doi.org/10.1007/s00259-022-05997-1
Haruwaka, K., Ikegami, A., Tachibana, Y., Ohno, N., Konishi, H., Hashimoto, A., et al. (2019). Dual microglia effects on blood brain barrier permeability induced by systemic inflammation. *Nature Communications, 10*(1). https://doi.org/10.1038/s41467-019-13812-z
Hay, J. R., Johnson, V. E., Young, A. M., Smith, D. H., & Stewart, W. (2015). Blood-brain barrier disruption is an early event that may persist for many years after traumatic brain injury in humans. *Journal of Neuropathology & Experimental Neurology, 74*, 1147–1157.
Herland, A., Van Der Meer, A. D., FitzGerald, E. A., Park, T. E., Sleeboom, J. J. F., & Ingber, D. E. (2016). Distinct contributions of astrocytes and pericytes to neuroinflammation identified in a 3D human blood-brain barrier on a chip. *PLoS ONE, 11*(3). https://doi.org/10.1371/journal.pone.0150360
Hindle, S. J., & Bainton, R. J. (2014). Barrier mechanisms in the Drosophila blood-brain barrier. *Frontiers in Neuroscience, 8*. https://doi.org/10.3389/fnins.2014.00414
Huang, Z., Wong, L. W., Su, Y., Huang, X., Wang, N., Chen, H., et al. (2020). Blood-brain barrier integrity in the pathogenesis of Alzheimer's disease. *Frontiers in Neuroendocrinology, 59*, 100857.
Hussain, B., Fang, C., & Chang, J. (2021). Blood–brain barrier breakdown: An emerging biomarker of cognitive impairment in normal aging and dementia. *Frontiers in Neuroscience, 15*. https://doi.org/10.3389/fnins.2021.688090
Iturria-Medina, Y., Sotero, R. C., Toussaint, P. J., Mateos-Pérez, J. M., & Evans, A. C. (2016). Early role of vascular dysregulation on late-onset Alzheimer's disease based on multifactorial data-driven analysis. *Nature Communications, 7*. Retrieved from www.embase.com/search/results?subaction=viewrecord&from=export&id=L610904515%0Ahttp://dx.doi.org/10.1038/ncomms11934
Jagtiani, E., Yeolekar, M., Naik, S., & Patravale, V. (2022). In vitro blood brain barrier models: An overview. *Journal of Controlled Release, 343*, 13–30.
Janelidze, S., Lindqvist, D., Francardo, V., Hall, S., Zetterberg, H., Blennow, K., et al. (2015). Increased CSF biomarkers of angiogenesis in Parkinson disease. *Neurology, 85*, 1834–1842.
Jellinger, K. A., & Attems, J. (2010). Prevalence of dementia disorders in the oldest-old: An autopsy study. *Acta Neuropathologica, 119*(4), 421–433. https://doi.org/10.1007/s00401-010-0654-5
Jeon, J., Lourenco, J., Kaiser, E. E., Waters, E. S., Scheulin, K. M., Fang, X., et al. (2020). Dynamic changes in the gut microbiome at the acute stage of ischemic stroke in a pig model. *Frontiers in Neuroscience, 14*, 587986.
Jeong, J. Y., Kwon, H. B., Ahn, J. C., Kang, D., Kwon, S. H., Park, J. A., et al. (2008). Functional and developmental analysis of the blood-brain barrier in zebrafish. *Brain Research Bulletin, 75*(5), 619–628. https://doi.org/10.1016/j.brainresbull.2007.10.043
Ju, F., Ran, Y., Zhu, L., Cheng, X., Gao, H., Xi, X., et al. (2018). Increased BBB permeability enhances activation of microglia and exacerbates loss of dendritic spines after transient global cerebral ischemia. *Frontiers in Cellular Neuroscience, 12*, 236.
Kadry, H., Noorani, B., & Cucullo, L. (2020). A blood–brain barrier overview on structure, function, impairment, and biomarkers of integrity. *Fluids Barriers CNS, 17*, 69.
Kalia, L. V., & Kalia, S. K. (2015). α-synuclein and Lewy pathology in Parkinson's disease. *Current Opinion in Neurology, 28*, 375–381.
Kassner, A., & Merali, Z. (2015). Assessment of blood-brain barrier disruption in stroke. *Stroke, 46*(11), 3310–3315. https://doi.org/10.1161/STROKEAHA.115.008861
Keep, R. F., Andjelkovic, A. V., Stamatovic, S. M., Shakui, P., & Ennis, S. R. (2005). Ischemia-induced endothelial cell dysfunction. *Acta Neurochirurgica Supplement, 95*, 399–402.
Kim, H., Shin, J. Y., Lee, Y. S., Yun, S. P., Maeng, H. J., & Lee, Y. (2020). Brain endothelial P-glycoprotein level is reduced in Parkinson's disease via a vitamin D receptor-dependent pathway. *International Journal of Molecular Sciences, 21*, 8538.
Kiss, T., Balasubramanian, P., Valcarcel-Ares, M. N., Tarantini, S., Yabluchanskiy, A., Csipo, T., Lipecz, A., Reglodi, D., Zhang, X. A., Bari, F., Farkas, E., Csiszar, A., & Ungvari, Z. (2019). Nicotinamide mononucleotide (NMN) treatment attenuates oxidative stress and rescues angiogenic capacity in aged cerebromicrovascular endothelial cells: A potential mechanism for the prevention of vascular cognitive impairment. *GeroScience, 41*(5), 619–630. https://doi.org/10.1007/s11357-019-00074-2
Knox, E. G., Aburto, M. R., Clarke, G., Cryan, J. F., & O'Driscoll, C. M. (2022). The blood-brain barrier in aging and neurodegeneration. *Molecular Psychiatry, 27*(6), 2659–2673. https://doi.org/10.1038/s41380-022-01511-z
Kocahan, S., & Doğan, Z. (2017). Mechanisms of Alzheimer's disease pathogenesis and prevention: The brain, neural pathology, N-methyl-D-Aspartate receptors, tau protein and other risk factors. *Clinical Psychopharmacology and Neuroscience, 15*(1), 1–8. https://doi.org/10.9758/cpn.2017.15.1.1
Kook, S.-Y., Hong, H. S., Moon, M., Ha, C. M., Chang, S., & Mook-Jung, I. (2012). Aβ1-$_{42}$-RAGE interaction disrupts tight junctions of the blood-brain barrier via Ca2+-calcineurin signaling. *Journal of Neuroscience, 32*(26), 8845–8854. Retrieved from http://www.embase.com/search/results?subaction=viewrecord&from=export&id=L365108513%0Awww.jneurosci.org/content/32/26/8845.full.pdf+html%0Ahttp://dx.doi.org/10.1523/JNEUROSCI.6102-11.2012

Kook, S.-Y., Seok Hong, H., Moon, M., & Mook-Jung, I. (2013). Disruption of blood-brain barrier in Alzheimer disease pathogenesis. *Tissue Barriers*, *1*(2), e23993. https://doi.org/10.4161/tisb.23993

Kortekaas, R., Leenders, K. L., Van Oostrom, J. C. H., Vaalburg, W., Bart, J., Willemsen, A. T. M., et al. (2005). Blood-brain barrier dysfunction in Parkinsonian midbrain in vivo. *Annals of Neurology*, *57*(2), 176–179. https://doi.org/10.1002/ana.20369

Kshirsagar, V., Thingore, C., & Juvekar, A. (2021). Insulin resistance: A connecting link between Alzheimer's disease and metabolic disorder. *Metabolic Brain Disease*, *36*(1), 67–83. https://doi.org/10.1007/s11011-020-00622-2

Kuan, W. L., Bennett, N., He, X., Skepper, J. N., Martynyuk, N., Wijeyekoon, R., et al. (2016). α-Synuclein pre-formed fibrils impair tight junction protein expression without affecting cerebral endothelial cell function. *Experimental Neurology*, *285*, 72–81.

Lan, G., Wang, P., Chan, R. B., Liu, Z., Yu, Z., Liu, X., et al. (2021). Astrocytic VEGFA: An essential mediator in blood-brain-barrier disruption in Parkinson's disease. *Glia*, *70*, 337–353.

Laurent, C., Dorothée, G., Hunot, S., Martin, E., Monnet, Y., Duchamp, M., et al. (2017). Hippocampal T cell infiltration promotes neuroinflammation and cognitive decline in a mouse model of tauopathy. *Brain*, *140*, 184–200.

Lee, D., Cho, S. J., Lim, H. J., Seok, J., Jo, C., Jo, S. A., et al. (2018). Alteration of vascular endothelial cadherin in Alzheimer's disease patient and mouse model. *BioRxiv*, 430140.

LeVine, S. M. (2016). Albumin and multiple sclerosis. *BMC Neurology*, *16*(1). https://doi.org/10.1186/s12883-016-0564-9

Levit, A., Hachinski, V., & Whitehead, S. N. (2020). Neurovascular unit dysregulation, white matter disease, and executive dysfunction: The shared triad of vascular cognitive impairment and Alzheimer disease. *GeroScience*, 1–21.

Lier, J., Ondruschka, B., Bechmann, I., & Dreßler, J. (2020). Fast microglial activation after severe traumatic brain injuries. *International Journal of Legal Medicine*, *134*, 2187–2193.

Liguori, C., Olivola, E., Pierantozzi, M., Cerroni, R., Galati, S., Saviozzi, V., et al. (2017). Cerebrospinal-fluid Alzheimer's disease biomarkers and blood-brain barrier integrity in a natural population of cognitive intact Parkinson's disease patients. *CNS & Neurological Disorders – Drug Targets*, 16, 339–345.

Liu, H., Funke, S. A., & Willbold, D. (2010). Transport of Alzheimer disease amyloid-β-binding d-amino acid peptides across an in vitro blood-brain barrier model. *Rejuvenation Research*, *13*(2–3), 210–213. https://doi.org/10.1089/rej.2009.0926

Liu, Y., Huber, C. C., & Wang, H. (2020). Disrupted blood-brain barrier in 5×FAD mouse model of Alzheimer's disease can be mimicked and repaired in vitro with neural stem cell-derived exosomes. *Biochemical and Biophysical Research Communications*, *525*(1), 192–196. https://doi.org/10.1016/j.bbrc.2020.02.074

Llorens, F., Schmitz, M., Gloeckner, S. F., Kaerst, L., Hermann, P., Schmidt, C., et al. (2015). Increased albumin CSF/serum ratio in dementia with Lewy bodies. *Journal of the Neurological Sciences*, *358*(1–2), 398–403. https://doi.org/10.1016/j.jns.2015.10.011

Loeffler, D. A., Connor, J. R., Juneau, P. L., Snyder, B. S., Kanaley, L., DeMaggio, A. J., et al. (1995). Transferrin and iron in normal, Alzheimer's disease, and Parkinson's disease brain regions. *Journal of Neurochemistry*, *65*(2), 710–716. https://doi.org/10.1046/j.1471-4159.1995.65020710.x

Ma, Q., Zhao, Z., Sagare, A. P., Wu, Y., Wang, M., Owens, N. C., et al. (2018). Blood-brain barrier-associated pericytes internalize and clear aggregated amyloid-B42 by LRP1-dependent apolipoprotein E isoform-specific mechanism. *Molecular Neurodegeneration*, *13*, 57.

Mansour, A. A., Gonçalves, J. T., Bloyd, C. W., Li, H., Fernandes, S., Quang, D., et al. (2018). An in vivo model of functional and vascularized human brain organoids. *Nature Biotechnology*, *36*(5), 432–441. https://doi.org/10.1038/nbt.4127

Maoz, B. M., Herland, A., Fitzgerald, E. A., Grevesse, T., Vidoudez, C., Pacheco, A. R., et al. (2018). A linked organ-on-chip model of the human neurovascular unit reveals the metabolic coupling of endothelial and neuronal cells. *Nature Biotechnology*, *36*(9), 865–877. https://doi.org/10.1038/nbt.4226

Mayne, K., White, J. A., McMurran, C. E., Rivera, F. J., & de la Fuente, A. G. (2020). Aging and neurodegenerative disease: Is the adaptive immune system a friend or foe? *Frontiers in Aging Neuroscience*, *12*, 572090.

Merlini, M., Rafalski, V. A., Rios Coronado, P. E., Gill, T. M., Ellisman, M., Muthukumar, G., et al. (2019). Fibrinogen induces microglia-mediated spine elimination and cognitive impairment in an Alzheimer's disease model. *Neuron*, *101*, 1099–1108.

Miners, J. S., Schulz, I., & Love, S. (2018). Differing associations between Aβ accumulation, hypoperfusion, blood-brain barrier dysfunction and loss of PDGFRB pericyte marker in the precuneus and parietal white matter in Alzheimer's disease. *Journal of Cerebral Blood Flow & Metabolism*, *38*, 103–115.

Montagne, A., Barnes, S. R., Sweeney, M. D., Halliday, M. R., Sagare, A. P., Zhao, Z., et al. (2015). Blood-brain barrier breakdown in the aging human hippocampus. *Neuron*, *85*(2), 296–302. https://doi.org/10.1016/j.neuron.2014.12.032

Morris, G., Fernandes, B. S., Puri, B. K., Walker, A. J., Carvalho, A. F., & Berk, M. (2018). Leaky brain in neurological and psychiatric disorders: Drivers and consequences. *Australian and New Zealand Journal of Psychiatry*, *52*(10), 924–948. https://doi.org/10.1177/0004867418796955

Musaeus, C. S., Gleerup, H. S., Høgh, P., Waldemar, G., Hasselbalch, S. G., & Simonsen, A. H. (2020). Cerebrospinal fluid/plasma albumin ratio as a biomarker for blood-brain barrier impairment across neurodegenerative dementias. *Journal of Alzheimer's Disease*, *75*(2), 429–436. https://doi.org/10.3233/JAD-200168

Nation, D. A., Sweeney, M. D., Montagne, A., Sagare, A. P., D'Orazio, L. M., Pachicano, M., et al. (2019). Blood–brain barrier breakdown is an early biomarker of human cognitive dysfunction. *Nature Medicine*, *25*(2), 270–276. https://doi.org/10.1038/s41591-018-0297-y

Navarro, R., Compte, M., Alvarez-Vallina, L., & Sanz, L. (2016). Immune regulation by pericytes: Modulating innate and adaptive immunity. *Frontiers in Immunology*, *7*, 480.

Nelson, A. R., Sweeney, M. D., Sagare, A. P., & Zlokovic, B. V. (2016). Neurovascular dysfunction and neurodegeneration in dementia and Alzheimer's disease. *Biochimica et Biophysica Acta*, *1862*(5), 887–900.

Nishioku, T., Dohgu, S., Takata, F., Eto, T., Ishikawa, N., Kodama, K. B., et al. (2009). Detachment of brain pericytes from the basal lamina is involved in disruption of the blood-brain barrier caused by lipopolysaccharide-induced sepsis in mice. *Cellular and Molecular Neurobiology*, *29*, 309–316.

Norton, E., Bridges, L., Kenyon, L., Esiri, M., Bennett, D., & Hainsworth, A. (2019). Cell senescence and cerebral small vessel disease in the brains of people aged 80 years and older. *Journal of Neuropathology & Experimental Neurology*, *78*.

Obermeier, B., Daneman, R., & Ransohoff, R. M. (2013). Development, maintenance and disruption of the blood-brain barrier. *Nature Medicine*, *19*, 1584–1596.

O'Donnell, M. E. (2014). Blood-brain barrier Na transporters in ischemic stroke. *Advances in Pharmacology*, *71*, 113–146.

O'Keeffe, E., Kelly, E., Liu, Y., Giordano, C., Wallace, E., Hynes, M., et al. (2020). Dynamic blood-brain barrier regulation in mild traumatic brain injury. *Journal of Neurotrauma*, *37*, 347–356.

Olmedo-Díaz, S., Estévez-Silva, H., Orädd, G., af Bjerkén, S., Marcellino, D., & Virel, A. (2017). An altered blood–brain barrier contributes to brain iron accumulation and neuroinflammation in the 6-OHDA rat model of Parkinson's disease. *Neuroscience*, *362*, 141–151.

Palmiotti, C. A., Prasad, S., Naik, P., Abul, K. M. D., Sajja, R. K., Achyuta, A. H., et al. (2014). In vitro cerebrovascular modeling in the 21st century: Current and prospective technologies. *Pharmaceutical Research*, *31*(12), 3229–3250. https://doi.org/10.1007/s11095-014-1464-6

Pan, Y., & Nicolazzo, J. A. (2018). Impact of aging, Alzheimer's disease and Parkinson's disease on the blood-brain barrier transport of therapeutics. *Advanced Drug Delivery Reviews*, *135*, 62–74. https://doi.org/10.1016/j.addr.2018.04.009

Pape, J. A., & Grose, J. H. (2020). The effects of diet and sex in amyotrophic lateral sclerosis. *Revue Neurologique*, *176*, 301–315.

Pardridge, W. M. (2020). Treatment of Alzheimer's disease and blood–brain barrier drug delivery. *Pharmaceuticals*, *13*(11), 1–25. https://doi.org/10.3390/ph13110394

Pardridge, W. M., Triguero, D., Yang, J., & Cancilla, P. A. (1990). Comparison of in vitro and in vivo models of drug transcytosis through the blood-brain barrier. *Journal of Pharmacology and Experimental Therapeutics*, *253*(2), 884–891.

Perrot, C. Y., Herrera, J. L., Fournier-Goss, A. E., & Komatsu, M. (2020). Prostaglandin E2 breaks down pericyte-endothelial cell interaction via EP1 and EP4-dependent downregulation of pericyte N-cadherin, connexin-43, and R-Ras. *Scientific Reports*, *10*, 11186.

Petersen, M. A., Ryu, J. K., & Akassoglou, K. (2018). Fibrinogen in neurological diseases: Mechanisms, imaging and therapeutics. *Nature Reviews Neuroscience*, *19*, 283–301.

Pienaar, I. S., Lee, C. H., Elson, J. L., McGuinness, L., Gentleman, S. M., Kalaria, R. N., et al. (2015). Deep-brain stimulation associates with improved microvascular integrity in the subthalamic nucleus in Parkinson's disease. *Neurobiology of Disease*, *74*(1), 392–405. https://doi.org/10.1016/j.nbd.2014.12.006

Polis, B., Gurevich, V., Assa, M., & Samson, A. O. (2019). Norvaline restores the BBB integrity in a mouse model of Alzheimer's disease. *International Journal of Molecular Sciences*, *20*(18). https://doi.org/10.3390/ijms20184616

Profaci, C. P., Munji, R. N., Pulido, R. S., & Daneman, R. (2020). The blood-brain barrier in health and disease: Important unanswered questions. *Journal of Experimental Medicine*, *217*, e20190062.

Pun, P. B. L., Lu, J., & Moochhala, S. (2009). Involvement of ROS in BBB dysfunction. *Free Radical Research*, *43*, 348–364.

Raja, R., Rosenberg, G. A., & Caprihan, A. (2018). MRI measurements of blood-brain barrier function in dementia: A review of recent studies. *Neuropharmacology*, *134*, 259–271.

Ralay, R., Hodge, H., Choi, J. N., & Wainwright, M. S. (2012). Albumin induces upregulation of matrix metalloproteinase-9 in astrocytes via MAPK and reactive oxygen species-dependent pathways. *Journal of Neuroinflammation*, *9*, 68.

Rempe, R. G., Hartz, A. M. S., & Bauer, B. (2016). Matrix metalloproteinases in the brain and blood-brain barrier: Versatile breakers and makers. *Journal of Cerebral Blood Flow and Metabolism*, *36*(9), 1481–1507. https://doi.org/10.1177/0271678X16655551

Rice, A., Liu, Y., Michaelis, M. L., Himes, R. H., Georg, G. I., & Audus, K. L. (2005). Chemical modification of paclitaxel (taxol) reduces P-glycoprotein interactions and increases permeation across the blood-brain barrier in vitro and in situ. *Journal of Medicinal Chemistry*, *48*(3), 832–838. https://doi.org/10.1021/jm040114b

Riphagen, J. M., Gronenschild, E. H., Salat, D. H., Freeze, W. M., Ivanov, D., Clerx, L., Verhey, F. R. J., Aalten, P., & Jacobs, H. I. L. (2018). Shades of white: Diffusion properties of T1- and FLAIR-defined white matter signal abnormalities differ in stages from cognitively normal to dementia. *Neurobiology of Aging*, *68*, 48–58. https://doi.org/10.1016/j.neurobiolaging.2018.03.029

Rochfort, K. D., & Cummins, P. M. (2015). Cytokine-mediated dysregulation of zonula occludens-1 properties in human brain microvascular endothelium. *Microvascular Research*, *100*, 48–53.

Ryu, J. K., & McLarnon, J. G. (2009). A leaky blood-brain barrier, fibrinogen infiltration and microglial reactivity in inflamed Alzheimer's disease brain. *Journal of Cellular and Molecular Medicine*, *13*(9A), 2911–2925. https://doi.org/10.1111/j.1582-4934.2008.00434.x

Schartmann, E., Schemmert, S., Niemietz, N., Honold, D., Ziehm, T., Tusche, M., et al. (2018). In vitro potency and preclinical pharmacokinetic comparison of all-d-enantiomeric peptides developed for the treatment of Alzheimer's disease. *Journal of Alzheimer's Disease*, *64*(3), 859–873. https://doi.org/10.3233/JAD-180165

Schneider, C. A., Figueroa Velez, D. X., Azevedo, R., Hoover, E. M., Tran, C. J., Lo, C., et al. (2019). Imaging the dynamic recruitment of monocytes to the blood–brain barrier and specific brain regions during Toxoplasma gondii infection. *Proceedings of the National Academy of Sciences of the United States of America*, *116*, 24796–24807.

Schreiner, T. G., Creangă-Murariu, I., Tamba, B. I., Lucanu, N., & Popescu, B. O. (2022). In vitro modeling of the blood–brain barrier for the study of physiological conditions and Alzheimer's disease. *Biomolecules*, *12*(8). https://doi.org/10.3390/biom12081136

Schultz, N., Brännström, K., Byman, E., Moussaud, S., Nielsen, H. M., Olofsson, A., et al. (2018). Amyloid-beta 1–40 is associated with alterations in NG2+ pericyte population ex vivo and in vitro. *Aging Cell*, *17*, e12728.

Segarra, M., Aburto, M. R., & Acker-Palmer, A. (2021). Blood-brain barrier dynamics to maintain brain homeostasis. *Trends in Neurosciences*, *44*, 3931–39405.

Sharma, B., Luhach, K., & Kulkarni, G. T. (2018). In vitro and in vivo models of BBB to evaluate brain targeting drug delivery. *Brain Targeted Drug Delivery Systems: A Focus on Nanotechnology and Nanoparticulates*, 53–101. https://doi.org/10.1016/B978-0-12-814001-7.00004-4

Shi, M., Liu, C., Cook, T. J., Bullock, K. M., Zhao, Y., Ginghina, C., et al. (2014). Plasma exosomal α-synuclein is likely CNS-derived and increased in Parkinson's disease. *Acta Neuropathologica*, *128*, 639–650.

Shin, Y., Choi, S. H., Kim, E., Bylykbashi, E., Kim, J. A., Chung, S., et al. (2019). Blood–brain barrier dysfunction in a 3D in vitro model of Alzheimer's disease. *Advanced Science*, *6*(20). https://doi.org/10.1002/advs.201900962

Singh, C. S. B., Choi, K. B., Munro, L., Wang, H. Y., Pfeifer, C. G., & Jefferies, W. A. (2021). Reversing pathology in a preclinical model of Alzheimer's disease by hacking cerebrovascular neoangiogenesis with advanced cancer therapeutics. *eBioMedicine*, *71*, 103503.

Skillbäck, T., Delsing, L., Synnergren, J., Mattsson, N., Janelidze, S., Nägga, K., et al. (2017). CSF/serum albumin ratio in dementias: A cross-sectional study on 1861 patients. *Neurobiology of Aging*, *59*, 1–9. https://doi.org/10.1016/j.neurobiolaging.2017.06.028

Song, K., Li, Y., Zhang, H., An, N., Wei, Y., Wang, L., et al. (2020). Oxidative stress-mediated blood-brain barrier (BBB) disruption in neurological diseases. *Oxidative Medicine and Cellular Longevity*, 4356386.

Spadoni, I., Fornasa, G., & Rescigno, M. (2017). Organ-specific protection mediated by cooperation between vascular and epithelial barriers. *Nature Reviews Immunology*, *17*, 761–773.

Starr, J. M., Farrall, A. J., Armitage, P., McGurn, B., & Wardlaw, J. (2009). Blood-brain barrier permeability in Alzheimer's disease: A case-control MRI study. *Psychiatry Research—Neuroimaging*, *171*(3), 232–241. https://doi.org/10.1016/j.pscychresns.2008.04.003

Stokum, J. A., Gerzanich, V., & Simard, J. M. (2016). Molecular pathophysiology of cerebral edema. *Journal of Cerebral Blood Flow & Metabolism*, *36*, 513–538.

Sui, Y. T., Bullock, K. M., Erickson, M. A., Zhang, J., & Banks, W. A. (2014). Alpha synuclein is transported into and out of the brain by the blood-brain barrier. *Peptides*, *62*, 197–202.

Sweeney, M. D., Sagare, A. P., & Zlokovic, B. V. (2018). Blood-brain barrier breakdown in Alzheimer disease and other neurodegenerative disorders. *Nature Reviews Neurology*, *14*(3), 133–150. https://doi.org/10.1038/nrneurol.2017.188

Sweeney, M. D., Zhao, Z., Montagne, A., Nelson, A. R., & Zlokovic, B. V. (2019). Blood-brain barrier: From physiology to disease and back. *Physiological Reviews*, *99*, 21–78.

Takada, Y., Greig, N. H., Vistica, D. T., Rapoport, S. I., & Smith, Q. R. (1992). Affinity of antineoplastic amino acid drugs for the large neutral amino acid transporter of the blood-brain barrier. *Cancer Chemotherapy and Pharmacology*, *30*(2), 164. https://doi.org/10.1007/BF00686413

Takata, F., Nakagawa, S., Matsumoto, J., & Dohgu, S. (2021). Blood-brain barrier dysfunction amplifies the development of neuroinflammation: Understanding of cellular events in brain microvascular endothelial cells for prevention and treatment of BBB dysfunction. *Frontiers in Cellular Neuroscience*, *15*, 661838.

Tam, S. J., Richmond, D. L., Kaminker, J. S., Modrusan, Z., Martin-McNulty, B., Cao, T. C., et al. (2012). Death receptors DR6 and TROY regulate brain vascular development. *Developmental Cell*, *22*(2), 403–417. https://doi.org/10.1016/j.devcel.2011.11.018

Tao, Q. Q., Lin, R. R., Chen, Y. H., & Wu, Z. Y. (2022). Discerning the role of blood brain barrier dysfunction in Alzheimer's disease. *Aging and Disease*, *13*(5), 1391–1404. https://doi.org/10.14336/AD.2022.0130-1

Tarantini, S., Yabluchanskiy, A., Csipo, T., Fulop, G., Kiss, T., Balasubramanian, P., DelFavero, J., Ahire, C., Ungvari, A., Nyúl-Tóth, Á., Farkas, E., Benyo, Z., Toth, A., Csiszar, A., & Ungvari, Z. (2019). Treatment with the poly(ADP-ribose) polymerase inhibitor PJ-34 improves cerebromicrovascular endothelial function, neurovascular coupling responses and cognitive performance in aged mice, supporting the NAD + depletion hypothesis of neurovascular aging. *FASEB Journal*, *34*(S1), 1–1. https://doi.org/10.1096/fasebj.2020.34.s1.04293

Toledo, J. B., Arnold, S. E., Raible, K., Brettschneider, J., Xie, S. X., Grossman, M., et al. (2013). Contribution of cerebrovascular disease in autopsy confirmed neurodegenerative disease cases in the National Alzheimer's coordinating centre. *Brain*, *136*(9), 2697–2706. https://doi.org/10.1093/brain/awt188

Toledo, J. B., Toledo, E., Weiner, M. W., Jack, C. R., Jagust, W., Lee, V. M. Y., et al. (2012). Cardiovascular risk factors, cortisol, and amyloid-β deposition in Alzheimer's disease neuroimaging initiative. *Alzheimer's and Dementia*, *8*(6), 483–489. https://doi.org/10.1016/j.jalz.2011.08.008

Uher, T., Horakova, D., Tyblova, M., Zeman, D., Krasulova, E., Mrazova, K., et al. (2016). Increased albumin quotient (QAlb) in patients after first clinical event suggestive of multiple sclerosis is associated with development of brain atrophy and greater disability 48 months later. *Multiple Sclerosis*, *22*(6), 770–781. https://doi.org/10.1177/1352458515601903

Ungvari, Z., Tarantini, S., Donato, A. J., Galvan, V., & Csiszar, A. (2018). Mechanisms of vascular aging. *Circulation Research*, *123*, 849–867.

van de Haar, H. J., Burgmans, S., Jansen, J. F. A., van Osch, M. J. P., van Buchem, M. A., Muller, M., et al. (2017). Blood-brain barrier leakage in patients with early Alzheimer disease. *Radiology*, *282*(2), 615–615. https://doi.org/10.1148/radiol.2017164043

Vazana, U., Veksler, R., Pell, G. S., Prager, O., Fassler, M., Chassidim, Y., et al. (2016). Glutamate-mediated blood–brain barrier opening: Implications for neuroprotection and drug delivery. *Journal of Neuroscience*, *36*(29), 7727–7739. https://doi.org/10.1523/JNEUROSCI.0587-16.2016

Verheggen, I. C. M., de Jong, J. J. A., van Boxtel, M. P. J., Gronenschild, E. H. B. M., Palm, W. M., Postma, A. A., Jansen, J. F. A., Verhey, F. R. J., & Backes, W. H. (2020). Increase in blood–brain barrier leakage in healthy, older adults. *GeroScience*, *42*(4), 1183–1193. https://doi.org/10.1007/s11357-020-00211-2

Vogelsang, P., Giil, L. M., Lund, A., Vedeler, C. A., Parkar, A. P., Nordrehaug, J. E., et al. (2018). Reduced glucose transporter-1 in brain derived circulating endothelial cells in mild Alzheimer's disease patients. *Brain Research*, *1678*, 304–309.

Wang, D., Chen, F., Han, Z., Yin, Z., Ge, X., & Lei, P. (2021). Relationship between amyloid-β deposition and blood–brain barrier dysfunction in Alzheimer's disease. *Frontiers in Cellular Neuroscience*, *15*. https://doi.org/10.3389/fncel.2021.695479

Wang, L., Sheng, W., Tan, Z., Ren, Q., Wang, R., Stoika, R., et al. (2021). Treatment of Parkinson's disease in Zebrafish model with a berberine derivative capable of crossing blood brain barrier, targeting mitochondria, and convenient for bioimaging experiments. *Comparative Biochemistry and Physiology Part—C: Toxicology and Pharmacology*, *249*. https://doi.org/10.1016/j.cbpc.2021.109151

Weekman, E. M., & Wilcock, D. M. (2016). Matrix metalloproteinase in blood-brain barrier breakdown in Dementia. *Journal of Alzheimer's Disease*, *49*, 893–903.

Wen, J., Ding, Y., Wang, L., & Xiao, Y. (2020). Gut microbiome improves postoperative cognitive function by decreasing permeability of the blood-brain barrier in aged mice. *Brain Research Bulletin*, *164*, 249–256.

Wenzel, D., & Homfray, T. (2013). Cerebral organoids model human brain development and microcephaly ① 「マウス

で見る」 ということを ② 神経発生 ・ 神経培養について の ・・・ といいな, *29*(August), 25740.

Werry, E. L., Bright, F. M., Piguet, O., Ittner, L. M., Halliday, G. M., Hodges, J. R., et al. (2019). Recent developments in TSPO PET imaging as a biomarker of neuroinflammation in neurodegenerative disorders. *International Journal of Molecular Sciences*, *20*, 3161. https://doi.org/10.3390/ijms20133161

Wevers, N. R., Kasi, D. G., Gray, T., Wilschut, K. J., Smith, B., Vught, R., et al. (2018). A perfused human blood-brain barrier on-a-chip for high-throughput assessment of barrier function and antibody transport. *Fluids and Barriers of the CNS*, *15*(1). https://doi.org/10.1186/s12987-018-0108-3

Williams-Medina, A., Deblock, M., & Janigro, D. (2021). In vitro models of the blood–brain barrier: Tools in translational medicine. *Frontiers in Medical Technology*, *2*. https://doi.org/10.3389/fmedt.2020.623950

Winkler, E. A., Nishida, Y., Sagare, A. P., Rege, S. V., Bell, R. D., Perlmutter, D., et al. (2015). GLUT1 reductions exacerbate Alzheimer's disease vasculo-neuronal dysfunction and degeneration. *Nature Neuroscience*, *18*, 521–530.

Wong, S. M., Jansen, J. F., Zhang, C. E., Hoff, E. I., Staals, J., van Oostenbrugge, R. J., et al. (2019). Blood-brain barrier impairment and hypoperfusion are linked in cerebral small vessel disease. *Neurology*, *92*, e1669–e1677.

Wu, Q., Yuan, X., Li, B., Han, R., Zhang, H., & Xiu, R. (2020). Salvianolic acid alleviated blood-brain barrier permeability in spontaneously hypertensive rats by inhibiting apoptosis in pericytes via P53 and the Ras/Raf/MEK/ERK pathway. *Drug Design, Development and Therapy*, *14*, 1523–1534. https://doi.org/10.2147/DDDT.S245959

Wu, Y.-C., Sonninen, T.-M., Peltonen, S., Koistinaho, J., & Lehtonen, Š. (2021). Blood-brain barrier and neurodegenerative diseases-modeling with iPSC-derived brain cells. *International Journal of Molecular Sciences*, *22*, 7710.

Yamadera, M., Fujimura, H., Inoue, K., Toyooka, K., Mori, C., Hirano, H., et al. (2015). Microvascular disturbance with decreased pericyte coverage is prominent in the ventral horn of patients with amyotrophic lateral sclerosis. *Amyotrophic Lateral Sclerosis (ALS) and Frontotemporal Degeneration*, *16*, 393–401.

Yamazaki, Y., Shinohara, M., Shinohara, M., Yamazaki, A., Murray, M. E., Liesinger, A. M., et al. (2019). Selective loss of cortical endothelial tight junction proteins during Alzheimer's disease progression. *Brain*, *142*, 1077–1092.

Yang, A. C., Stevens, M. Y., Chen, M. B., Lee, D. P., Stähli, D., Gate, D., et al. (2020). Physiological blood–brain transport is impaired with age by a shift in transcytosis. *Nature*, *583*, 425–430.

Yang, P., Min, X.-L., Mohammadi, M., Turner, C., Faull, R., Waldvogel, H., et al. (2017). Endothelial degeneration of Parkinson's disease is related to alpha-synuclein aggregation. *Journal of Alzheimer's Disease and Parkinsonism*, *7*.

Yang, T., Martin, P., Fogarty, B., Brown, A., Schurman, K., Phipps, R., et al. (2015). Exosome delivered anticancer drugs across the blood-brain barrier for brain cancer therapy in Danio Rerio. *Pharmaceutical Research*, *32*(6), 2003–2014. https://doi.org/10.1007/s11095-014-1593-y

Yoshino, J., Baur, J. A., & Imai, S.-I. (2018). NAD+ intermediates: The biology and therapeutic potential of NMN and NR. *Cell Metabolism*, *27*, 513–528.

Zenaro, E., Piacentino, G., & Constantin, G. (2017). The blood-brain barrier in Alzheimer's disease. *Neurobiology of Disease*, *107*, 41–56.

Zhang, J., Liss, M., Wolburg, H., Blasig, I. E., & Abdelilah-Seyfried, S. (2012). Involvement of claudins in zebrafish brain ventricle morphogenesis. *Annals of the New York Academy of Sciences*, *1257*(1), 193–198. https://doi.org/10.1111/j.1749-6632.2012.06507.x

Zhang, Q., Zheng, M., Betancourt, C. E., Liu, L., Sitikov, A., Sladojevic, N., et al. (2021). Increase in blood-brain barrier (BBB) permeability is regulated by MMP3 via the erk signaling pathway. *Oxidative Medicine and Cellular Longevity*, *2021*. https://doi.org/10.1155/2021/6655122

Zhao, X. Y., Lu, M. H., Yuan, D. J., Xu, D. E., Yao, P. P., Ji, W. L., et al. (2019). Mitochondrial dysfunction in neural injury. *Frontiers in Neuroscience*, *4*, 13–30. https://doi.org/10.3389/fnins.2019.00030

Zhong, Z., Deane, R., Ali, Z., Parisi, M., Shapovalov, Y., O'Banion, M. K., et al. (2008). ALS-causing SOD1 mutants generate vascular changes prior to motor neuron degeneration. *Nature Neuroscience*, *11*(4), 420–422. https://doi.org/10.1038/nn2073

3 Neurodegenerative Disorders
Available Therapies and Their Limitations

Mohamed Y. Zaky, Eman E. Mohamed, and Osama M. Ahmed*

3.1 INTRODUCTION

The phrase "neurodegeneration" consists of two words: *neuro* refers to nerve cells, and *degeneration* is another word meaning "degradation." The decrease of neural function is a sign of neurodegenerative diseases (Finkbeiner, 2011; Rahman et al., 2020; Rahman et al., 2022). Neurodegeneration is a feature of both acute and chronic neurological disorders, such as Alzheimer's disease (AD), Parkinson's disease (PD), neurotropic viral infections, paraneoplastic disorders, stroke, multiple sclerosis (MS), and traumatic brain injury (TBI) (Amor et al., 2010). Degenerative nerve disorders can affect many different bodily processes, including mobility, memory, eyesight, intellect, breathing, speech, heart function, and a variety of others. The progressive deterioration of a neuron's structure or functionality, known as neurodegeneration, is the underlying cause of neurodegenerative disorders. Neuronal circuitry can be impacted by neurodegeneration at the molecular, cellular, and systemic levels (Berman and Bayati, 2018). The majority of NDDs are characterized by aberrant protein buildup, as amyloid-β (Aβ) in AD, α-synuclein in PD, and TAR DNA-binding protein 43 in ALS (Peng et al., 2020). Inflammasomes can be activated when these proteins accumulate abnormally in neuroinflammation and neurodegeneration (Voet et al., 2019). In recent years, neurodegeneration and neuroprogression have received more attention as possible pathogenic components associated to mood disorders (Kapczinski et al., 2017; Tang et al., 2017).

3.2 NEURODEGENERATIVE DISORDERS (NDDS)

Neurodegeneration is defined by gradual, chronic alterations in the populations of motor, sensory, or cognitive neurons that initially affect how the neurons operate. A pathological reorganization of the central nervous system (CNS) is referred to by way of neuroprogression, and this includes minor degenerative processes and diminished plasticity (Kapczinski et al., 2017; Berk et al., 2011; Tang et al., 2017).

These neurodegenerative and neuroprogressive pathways include multiple and complex interactions between the CNS, endocrine, and immunological systems, including inflammatory processes, oxidative stress, neurotoxic protein buildup, and neuroprotection deficiency. These interactions would be associated to recurring and chronic mood disorder trajectories because they would result in failure or incorrect reorganization in neuroplasticity (Kapczinski et al., 2017; Berk et al., 2011; Tang et al., 2017). However, neuroinflammation can have effects on the immune system, the body's biochemistry, and the body's physiological processes. According to reports, neuroinflammation has a part in the etiology of a variety of different brain disorders, including AD, PD, and Huntington's disease (HD) (Kumar et al., 2021), MS, and ALS (Guzman-Martinez et al., 2019; Kumar et al., 2021). NDDs are distinguished by apoptosis/necrosis and neuronal cell malfunction, which results in a malignant influence on the nervous system. The brain is especially prone to oxidative stress since it is the most active portion of the body. Because the brain has a greater oxygen requirement, it consumes 20% more oxygen than the rest of the body (McQuillen and Ferriero, 2004). Oxidative stress has been hypothesized as a regulatory factor in ageing and a variety of neurological diseases (Singh et al., 2019).

3.2.1 Role of Oxidative Stress in NDDs

Although ageing is the main cause of the majority of NDDs, oxidative stress is the primary cause of efficient decline as we age because it causes lipid oxidation, proteins, and DNA in the presence of reactive oxygen or nitrogen species (ROS or RNS) (Liguori et al., 2018). Oxidative imbalance has also been identified as a distinguishing hallmark of NDDs (Buizza et al., 2012; Cheignon et al., 2018; Chinta and Andersen, 2008; Covarrubias-Pinto et al., 2015; Musgrove et al., 2019; Xie et al., 2013; Daixing et al., 2013). NDDs are characterized visually by progressive cell loss in susceptible neuronal cells, which is frequently coupled with cytoskeletal protein aggregates generating neuronal inclusions and/or glial cells. The formation of aberrant aggregated proteins and the disturbance of metal ion balance are strongly linked to oxidative stress (Chen et al., 2012). An increase in the

* Corresponding author: mohamedzaki448@science.bsu.edu.eg

DOI: 10.1201/9781003383376-3

degree of nitrosative stress may have an impact on mitochondrial respiration and disturb the dynamics of the mitochondria, which can lead to synaptic injury and neuronal death (Cho et al., 2009), given that growing mitochondrial dysfunction is a hallmark of AD (Abtahi et al., 2020; Wang et al., 2020).

3.2.2 Role of Oxidoreductase in NDDs

An *oxidoreductase* is an enzyme that converts electrons from an oxidant to a reductant during oxidation–reduction reactions. An oxidase is composed of an oxygenase, a dehydrogenase, a peroxidase, a hydroxylase, and reductase. GSH-Px (glutathione peroxidase) is a biomarker for oxidative stress, and its decreased activity correlates with increased inflammation and together have an association with cognitive impairment in the elderly (Melo et al., 2011). Increased the level of Nrf2 and decreased level of the superoxide dismutase 1 (SOD1) are linked to mental decline (Mota et al., 2015). The expression of inflammatory and metabolism-related genes is regulated by Nrf2, which responds to oxidative stress and toxic insults (Butterfield and Keller, 2012).

3.3 DIAGNOSIS OF NDDs

Early accurate differential diagnosis of NDDs is critical for two reasons. First, combined with current treatment, early diagnosis can delay hospitalization, according to studies (Feldman et al., 2009), and in the future, once disease-modifying drugs become available, it will become increasingly important to diagnose diseases early (Siemers et al., 2016). Second, creating novel medicines necessitates the early and perfect designation of appropriate goal groups. It has been proposed that overly varied research populations may explain the failure of several earlier medicinal studies (Falahati et al., 2014). In addition to computed tomography and magnetic resonance imaging (MRI), modern neuroimaging techniques like positron emission tomography (PET) and single-photon emission computed tomography (SPECT) have been supplanted structural neuroimaging techniques due to their limited specificity (Camprodon and Stern, 2013). AD is associated with structural MRI evidence of hippocampus and entorhinal cortex atrophy (Appel et al., 2009). Apolipoprotein E polymorphisms, presenilin gene mutations, and amyloid precursor proteins are known genetic markers for AD, whereas mutations of PrP or α-synuclein protein, which encodes the familiar type of prion disease, are known genetic markers for PD (García and Bustos, 2018). New diagnostic studies are aimed at developing biomarkers that can be easily detected in blood or saliva in order to discriminate between different types of neurological disorders (Malamud, 2011). Researchers are increasingly interested in amino acid profiling, particularly its potential application in the early identification of a variety of disorders. A number of research studies have revealed differences in the amino acid profile of patients with different illnesses and a healthy group, like people with AD (Van Assche et al., 2015; Gong et al., 2015; Corso et al., 2017), PD (Blesa et al., 2015; Figura et al., 2018), and elderly patients (Adachi et al., 2018). In the early stages of neurodegenerative changes, metabolic variables may be important mediators, at least in part (Bredesen, 2015).

Glutamine contributes to the primary metabolic system. A major function of glutamate in the brain is as a precursor to amino acids, such as glutamate and aspartate, as well as to the inhibitory amino acid and γ-amino butyric acid. According to several researches, the metabolism of neurotransmitters and their metabolites in body fluids varies considerably between healthy people and people with NDDs like AD (Zhao et al., 2017; Gong et al., 2015). Extensive neurotransmitter and metabolite study and certification would have a significant clinical impact in the progress of NDDs diagnosis.

3.4 RISK FACTORS FOR MOST NDDs

3.4.1 Ageing and Risk of NDDs

The majority of NDDs, including as AD and PD, are caused predominantly by ageing. AD affects one out of every ten adults over the age of 65, and its incidence increases with age. There are few or no effective medicines available for age-related NDDs, which tend to progress irreversibly and are associated with severe economic and personal costs (Hou et al., 2019). NDDs are prevalent, and disease-free brains are uncommon, particularly in the elderly; brain ageing may create a continuum with NDDs. The development of NDDs is determined by human genetic and environmental variables (Wyss-Coray, 2016). Ageing is a well-known risk factor for NDDs, a collection of incurable, disabling, and always-deadly diseases that include a variety of disorders, such as AD and other dementias, PD, HD, motor neuron disease, and prion disease (Walker et al., 2019). It has been discovered through molecular research that older people's brain tissue has abnormal accumulations of aggregated proteins, like hyperphosphorylated tau (p-tau), Aβ, and α-synuclein. However, it is still unknown whether the presence of these deposits correlates with the severity of cognitive decline (Elobeid et al., 2016). According to several research, the risk of NDDs is connected with early developmental problems, suggesting that brain structural alterations may occur far earlier than cognitive impairment (Dean et al., 2014).

3.4.2 Insomnia and Risk of NDDs

Insomnia is defined as the inability to fall asleep or maintain sleep, waking up earlier than expected, or having sleep that does not contribute to the restoration of the body (Shamim et al., 2019). Insomnia, which often appears as a subjective complaint of difficulty initiating or maintaining sleep or nonrestorative sleep, is one of the most prevalent sleep disorders seen in the geriatric clinic population. It can cause serious daytime symptoms, like difficulty concentrating and mood problems (Patel et al., 2018). Insomnia is always accompanied by daytime drowsiness and weariness (Roth, 2007). Although insomnia is a frequent symptom of many NDDs, it was only recently discovered that it is a risk factor for AD and PD (Shamim et al., 2019). A person is termed a short sleeper if they can only sleep for less than seven hours each night, regardless of their symptoms in the morning (Schutte-Rodin et al., 2008). If a person has a regular sleep period, efficiency, and quality but does not receive a restorative sleep, the body does not recover (Wilkinson and Shapiro, 2012). NDDs and insomnia are often related. AD causes fragmented sleep, and poor sleep increases the buildup of Aβ and hyperphosphorylated tau protein in brain tissue of human, as well as the risk of developing AD (Minakawa et al., 2019).

3.5 TYPES OF NDDs

3.5.1 Alzheimer's Disease (AD)

AD is the primary cause of dementia and one of the most pressing medical challenges of our day. Dementia is estimated to affect 40 million people globally, with the number anticipated to rise every 20 years till 2050 (Prince et al., 2013). AD is the most prevalent kind of dementia and NDDs, characterized by memory problems, particularly losing recent knowledge, recall capacity impairment, and loss of time tracking, problem-solving, language, and recognition issues. AD is also a worldwide disease; however, despite all scientific efforts, treating AD remains difficult (Sadeghmousavi et al., 2020). AD is an age-related NDD. Although a buildup of brain β-amyloid peptide and hyperphosphorylated tau protein has been linked to the development of AD, the disease's etiology is unknown (Shoshan-Barmatz et al., 2018). AD was once thought to be a clinicopathological entity, which meant that if a patient met the clinical criteria for amnestic dementia and all other diseases were checked out, AD pathology may be assumed to be the cause (Petersen, 2017). The global incidence of AD is linked with that of all-cause dementia (Nelson et al., 2011; Boyle et al., 2018). Although AD is the most common cause of dementia (Schneider et al., 2009), dementia can be induced by a variety of neurodegenerative or cerebrovascular disorders, especially in the elderly (Kapasi et al., 2017; Karanth et al., 2020).

3.5.1.1 Biomarkers of Neuroinflammation in AD

Inflammation normally disappears on its own and is necessary for the healing process. Pathologic (i.e., chronic) inflammation, on the other hand, can develop if inflammation is protracted, resulting in adverse effects on brain function owing to excessive or continuous production of cytotoxic substances. This chronic over-activation of pro-inflammatory reactions has been linked to a variety of NDDs, including AD (Chaney et al., 2019). Neuroinflammation is a natural immune response of the nervous system that includes microglia, astrocytes, cytokines, and chemokines, and it plays a key role in the early stages of AD pathogenesis (Heneka et al., 2015; Businaro et al., 2018). At the molecular level, the presence of amyloid-β aggregates, including oligomers and fibrils, promotes inflammation (Glass et al., 2010; Cameron and Landreth, 2010; Forloni and Balducci, 2018). Amyloid-β induced an increase in the production of cytokines and chemokines by microglia, neurons, and astrocytes, which triggers amyloid formation (Solfrizzi et al., 2006).

IL-1, a well-studied pro-inflammatory cytokine, has been linked to ptau and amyloid-β deposition (Sheng et al., 2001). IL-1β is produced as a result of mitochondrial malfunction, which leads to cell death (Ramesh et al., 2018). Tumor necrosis factor (TNF-α), IL-10, and IL-12 are also important mediators of the inflammatory process (Baird et al., 2015; Leung et al., 2013) and the acute reactant protein C (Cheng et al., 2018).

There is also a significant neurobiological relationship in the AD between the lack of anti-inflammatory cytokines like TGF-β1 and the early pro-inflammatory process (Caraci et al., 2018). TGF-β1 is a neurotrophic factor whose deficiency is linked to AD, and the pathway is selectively impaired in early AD (Ueberham et al., 2006; Tesseur et al., 2006).

3.5.2 Parkinson's Disease (PD)

PD is the most commonly occurring NDD, and its occurrence is expected to more than double in the next 30 years. PD commonly manifests in older individuals with both motor and non-motor dysfunctions. PD is a progressive NDD caused by dopaminergic neuronal cell loss in the substantia nigra pars compacta area of the midbrain. PD is traditionally characterized as a motor system disorder with four cardinal symptoms: bradykinesia (slowness of movement), rigidity (stiffness of the limbs and trunk), postural instability (impaired balance and coordination), and tremor (trembling in hands, arms, legs, and face) Santens et al., 2003; Jankovic, 2005; Berganzo et al., 2016; Ferrazzoli et al., 2020). Many of these symptoms are caused by dopamine-producing neuron loss in the substantia nigra pars compacta, and bradykinesia and stiffness, in particular, initially react well to symptomatic dopamine replacement therapy. Non-motor symptoms, such as constipation and hyposmia, are also common in PD and are unrelated to dopamine depletion (Kalia & Lang, 2015). PD has historically been diagnosed mainly on motor symptoms. Despite the identification of cardinal indications of PD in clinical examinations, most of the rating scales used to determine disease severity have not been extensively examined and validated (Jankovic, 2008). Although non-motor symptoms (e.g., cognitive alterations, such as difficulties with attention and planning, sleep disturbances, sensory abnormalities, such as olfactory dysfunction) are present in many individuals prior to the development of PD, they are not always present (Jankovic, 2008; Tremblay et al., 2017),

They lack specificity, are difficult to determine, and/or produce variation from patient to patient (Zesiewicz et al., 2006). As a result, non-motor symptoms do not yet allow for the independent diagnosis of PD (Braak et al., 2003), while some have been employed as supporting diagnostic criteria (Postuma et al., 2015).

3.5.2.1 Inflammatory Biomarkers in PD

Inflammatory indicators such as cytokines such as IL and TNF-α, chemokines such as chemokine ligand (CCL) and CX3 chemokine ligand (CX3CL), and CRP have been identified as key immune activation signaling molecules with effects in the CNS and periphery (Tabas and Glass, 2013). Furthermore, peripheral inflammation may contribute to the etiology and progression of PD. Less-intrusive indicators found in peripheral blood and CSF can help researchers better understand the etiology of PD and identify possible biomarkers for the condition; nevertheless, their efficacy varies widely among studies due to variances in study locations and equipment. Previous studies have shown that PD patients have higher levels of inflammatory markers in their peripheral blood and CSF than healthy people (Zimmermann and Brockmann, 2022; Chen et al., 2018).

3.5.3 Multiple Sclerosis (MS)

MS is a CNS immune-mediated chronic inflammatory disease that produces demyelination and widespread neurodegeneration in both brain and spinal cord gray and white matter (Thomson et al., 2018; Lassmann et al., 2007). MS is a primary cause of impairment in young adults, affecting an estimated 2.8 million individuals globally (Walton et al., 2020). The etiology of MS is unknown; however, risk factors for the disease have been identified, including geographic location, genetic composition, and biological sex, among others. North America and Europe have the highest MS prevalence, with greater latitudes being associated with higher MS incidence (Baecher-Allan et al., 2018; Simpson et al., 2019; Walton et al., 2020). In genome-wide association research, more than 230 risk variants have been found, the majority of which are linked to immune system function (International Multiple Sclerosis Genetics Consortium, 2019). Furthermore, females are two to three times more likely than men to be diagnosed with MS, implying a hormonal component to disease initiation (Walton et al., 2020). MS is a multifaceted, diverse disease with widely varying clinical symptoms across all MS subtypes due to the possibility of inflammatory or demyelinating lesions affecting any location of the CNS. MS is also a chronic disorder with no recognized cure or final therapy, while innovative medicines for disease management have developed in recent years. Since the first acceptance of interferon beta-lb in 1993, significant progress has been achieved in the pathogenesis, diagnosis, and therapy of MS (Pardo and Jones, 2017; Chataway, 2018). Although clinical results are difficult to predict, cumulative axonal damage is often regarded as the most important predictor of long-term prognosis. Brain volume loss, or atrophy, has lately emerged as a popular magnetic resonance imaging (MRI) diagnostic of neurodegeneration in MS, pathologically coinciding with demyelination, reduced axonal count, and neuronal cell death (Miller et al., 2002; Minagar et al., 2013). The average of brain volume reduction in healthy persons is between 0.2 and 0.5% each year, depending on the participants' age (Zivadinov et al., 2016b), whereas in those with MS, this process appears to be three to five times faster (Barkhof, 2016). Brain atrophy is prevalent and clinically significant in all subtypes, from patients with a clinically isolated condition to those with long-term progressive disease (Sastre-Garriga et al., 2017; Zivadinov et al., 2016a), becoming more evident in the latter phases but appearing to move at a comparable rate over the full clinical range (De Stefano et al., 2010). The progression of brain atrophy, particularly in the gray matter, has been identified as an independent predictor of current and future physical disability, mood disorders, and cognitive impairment, exceeding traditional MRI measures of white matter inflammation (Sastre-Garriga et al., 2017; Rocca et al., 2017).

To describe how progressive MS is initiated, several ideas have been put forth. One hypothesis states that while inflammatory processes similar to those seen in relapsing-remitting multiple sclerosis (RRMS) are thought to be the cause of brain damage, as the disease progresses, a microenvironment within the CNS is created that favors the homing and retention of inflammatory cells, rendering disease-modifying therapies ineffective (Fischer et al., 2013). Another hypothesis is that MS begins as an inflammatory condition before switching, over time, to a neurodegenerative process that is unrelated to inflammatory reactions (Meuth et al., 2008). Finally, inflammation may be a secondary reaction to NDDs in MS, aggravating progressive stages (Barnett and Prineas, 2004; Kassmann et al., 2007).

3.5.3.1 Biomarkers of MS

By detecting a blood vessel in the center of a T2 hyperintense lesion, the central vein sign (CVS) has been proposed to distinguish MS lesions from MRI imitators, such as microvascular disorders and migraine (Sati et al., 2016; Tallantyre et al., 2011). Several studies have examined the frequency, sensitivity, and specificity of CVS in discriminating MS from non-MS lesions using various MRI sequences that identify field inhomogeneities, including as T2-weighted, susceptibility-weighted, and phase imaging (Clarke et al., 2019; Dworkin et al., 2018; Maggi et al., 2019; Mistry et al., 2015; Sinnecker et al., 2019).

Glial fibrillary acidic protein (GFAP) levels in the blood are higher in people with progressive MS and are linked to illness duration and clinical disability (Högel et al., 2020). In progressive MS, GFAP levels correlate with Nf-L, lesion load, and brain volume in both white and gray matter, indicating that GFAP may be a marker of disease progression (Axelsson et al., 2011; Ayrignac et al., 2020; Högel et al., 2020). Although previous research has suggested a weak to moderate link between serum GFAP and higher disability, it should be noted that the strength of this correlation has not been determined (Martínez et al., 2015; Norgren et al., 2004).

The latest modification of the McDonald criteria includes oligoclonal immunoglobulin bands (OCBs) in CSF to help in the diagnosis of MS by identifying B cell activity (Obermeier et al., 2011; Thompson et al., 2018). The presence of unique OCBs in the CSF but not in the serum (both taken from the same patient on the same day) supports intrathecal synthesis of immunoglobulins directed against ubiquitous intracellular antigens in cellular debris (Winger and Zamvil, 2016).

3.5.4 Amyotrophic Lateral Sclerosis (ALS)

ALS, also known as Lou Gehrig's disease, causes the loss of muscle control in the brain and spinal cord as a result of nerve cell damage. Weak muscles, behavioral changes, speech problems, and cognitive problems are some of the most common clinical symptoms of ALS (Cho and Shukla, 2020).

ALS was previously only understood to be a kind of motor neuron degeneration that eventually results in paralysis, the potential for respiratory failure, and consequences from immobility. It is currently recognized as a multisystem of NDDs, with the disease being characterized at the clinical, neuropathological, and genetic levels (Masrori and Van Damme, 2020). ALS is an NDD characterized by progressive muscular paralysis caused by loss of motor neurons in the primary motor cortex, brain stem, and spinal cord. The body location of onset, the percentage of upper motor neuron (UMN) and lower motor neuron (LMN) involvement, the rate of progression and cognitive impairment all play a part in the highly diverse phenotypic expression of ALS (Machtoub and Kasugai, 2016).

A variety of factors are involved in the pathophysiology of the ALS, including oxidative stress, mitochondrial dysfunction, and mutations in mitochondrial-related genes (Petrov et al., 2016; Richardson et al., 2013). SOD1 gene is mutated in 15–20% of familial ALS patients, which plays an important role in oxidative stress resistance (Pansarasa et al., 2018). There have been more than 150 mutations in the SOD1 gene that are related to ALS, and they cause protein misfolding, aggregation, ROS production, and redox system disequilibrium (Pansarasa et al., 2018; Huai and Zhang, 2019). ALS is associated with several interconnected risk factors, like neuroinflammation, excitotoxicity, mitochondrial dysfunction, and stress on the endoplasmic reticulum (Obrador et al., 2020; Blasco et al., 2017; Bellezza et al., 2018). ALS patients have elevated levels of oxidative stress biomarkers like MDA and 8-hydroxyguanosine (Wang et al., 2019b). According to Kazama et al. (2020), cystine/glutamate antiporter upregulation in sporadic ALS patients results in enhanced oxidative stress and extracellular glutamate buildup. Additionally, motor neuron death is influenced by dysregulation of the retinoic acid signaling system, a vitamin A product (Kolarcik and Bowser, 2012).

3.5.4.1 Laboratory Investigations and Differential Diagnosis of ALS

Although there is no accurate diagnostic test for ALS, symptoms of UMN and LMN can be used to detect ALS by recognizing disease progression in different regions of the body. Some LMN symptoms include weakness and atrophy, whereas UMN symptoms include spasticity and hyperreflexia (Goutman, 2017). The most crucial technical test to confirm the diagnosis of ALS up to this point has been electromyography (EMG). While imaging and laboratory tests, such as MRI, may not always provide diagnostically useful information, they are important to rule out alternative conditions that could be more curable (Machtoub and Kasugai, 2016). The Revised ALS Functional Rating Scale (ALSFRS-R) is a method of evaluation that evaluates the degree of functional impairment in ALS patients. The examination includes 12 tasks that measure bulbar, motor, and respiratory function and are scored from 0 to 4, with 0 indicating not functional at all and 4 indicating totally functional (Ohashi et al., 2001). A person with ALS may experience mild symptoms initially, such as muscle twitching, cramping, stiffness, weakness, jerking movements that are involuntarily triggered, tremors, and inability to control bowels or bladder, or difficulty opening their eyes (Machtoub and Kasugai, 2016).

3.5.5 Huntington's Disease (HD)

HD is a hereditary NDD that causes devastating cognitive, behavioral, and motor symptoms. HD was a difficult problem to solve, as the gene encoded one of the biggest non-membrane proteins in the human proteome, measuring 3,144 amino acids or 350 kDa. Over the next 25 years, the normal function of huntingtin also eluded strict definition. Cell biology and biochemical research indicated that the protein lacked enzymatic activity and no conserved functional patterns, with the exception that proteins with polyglutamine tracts were frequently transcriptionally active, and huntingtin included HEAT repeats seen in various protein scaffolds (Takano and Gusella, 2002). Huntingtin was discovered to be at the center of numerous complexes in interactome research (Brandi et al., 2018), and it has been localized to the nucleus (Davies et al., 1997), endoplasmic reticulum (Atwal et al., 2007), and early, recycling, and late endosomes (Kim et al., 1999), the neuronal postsynaptic density (Marcora et al., 2003), mitotic spindle (Atwal et al., 2011; Godin et al., 2010), and the primary cilium (Maiuri et al., 2013; Keryer et al., 2011).

The mutation that causes HD is an extension of the huntingtin (HTT) gene's cytosine-adenine-guanine (CAG) repeat. A clinical diagnosis of symptomatic HD involves the evidence of motor abnormalities, such as chorea and motor impersistence (i.e., the inability to maintain basic voluntary motions), as well as a family history of the disease (Paulsen et al., 2008). Protein aggregation in the brain cells of HD patients includes mutant HTT, polyQ-expanded ataxins, and α-synuclein, just like other neurodegenerative diseases (Bernis et al., 2015; Jansen et al., 2017).

The pathophysiology of HD, like that of the majority of NDDs, is primarily related to the CNS and the brain as an organ where significant degradation takes place. As HD advances, the disease's pathophysiology causes a huge loss of medium spiny neurons in the striatum and neurons in the cortex (Bates et al., 2015). Clinical HD symptoms include motor impairments, psychological impairment, and cognitive deterioration (Ghosh and Tabrizi, 2018). In addition to NDDs, overall (brain and peripheral) expression of mutant HTT may contribute to additional symptoms, such as weight loss, glucose metabolism, cardiomyopathy, muscle and endocrine dysfunction (Van der Burg et al., 2009; Bozzi and Sciandra, 2020). The peripheral nerve system acts as a direct signal communicator between the CNS and peripheral organs, but different types of indirect signals can also be transmitted via the immune system. Neuroinflammation causes several similar alterations in a larger set of NDDs (Ransohoff, 2016). HD patients have persistent inflammation that may be identified in the peripheral for years before the disorder start (Valadão et al., 2020).

3.5.5.1 Inflammatory Biomarkers of HD

Evidence suggests that microglia activation during the asymptomatic period correlates with disease severity in HD patients (Tai et al., 2007). Increased inflammatory mediators released by microglia, such as IL-6, IL-8, IL-10, matrix metallopeptidase 9 (MMP-9), and chemokine C-C motif ligand 2 (CCL2) mRNA, have been found in the brain tissue of postmortem HD patients (Saba et al., 2022). Increased levels of IL-4, IL-6, IL-8, and TNF-α have been seen in HD patients' plasma and CSF. When compared to controls, plasma levels of the chemokines eotaxin-3, macrophage inflammatory protein-1 (MIP-1), eotaxin, monocyte chemotactic protein-1 (MCP-1), and MCP-4 are considerably higher in HD (Wild et al., 2011).

3.6 THERAPEUTIC STRATEGIES FOR COMBATTING NDDs

Gene therapy is rapidly gaining popularity as a potent treatment option for a wide range of NDDs, including AD, PD, and HD. Gene therapy for NDDs has made significant development over the last few decades. Increased knowledge of the pathogenetic processes of these disorders has enabled several improvements in important technologies to converge, including the identification of novel treatment targets and routes (Dunbar et al., 2018). The main causes of NDDs with both single-gene and complex etiologies have been specifically targeted by several genetic therapies as a result of this greater understanding. Gene therapy's long-term, even permanent, therapeutic effects are especially appealing for compartmentalized organs like the eye, cochlea, or CNS, which are hard to treat due to most agents not being able to penetrate physiological barriers, like the blood–cerebrospinal fluid barrier (BCSFB), blood–retina barrier (BRB), and blood–brain barrier (BBB) (Hudry and Vandenberghe, 2019; Wang et al., 2019a). In addition, gene therapy, which has the ability to handle both gain-of-function mutations and loss-of-function mutations, may be able to address some genetic targets that are resistant to treatment with conventional drugs.

However, the pathophysiology of the illness and the temporal progression of the clinical phenotype determine the appropriate gene therapy parameters. Additionally, the geographical and temporal specificity of gene expression must be taken into account. While geographical specificity describes if and how the gene expression is restricted to a certain brain area or cell type, temporal specificity in gene therapy relates to whether or not the gene expression is constitutive or controlled. By delivering gene therapy vectors through focused neurosurgery, the blood–brain barrier can be crossed, and the therapy may be targeted to anatomical regions of interest, preventing exposure to other parts of the brain where transgene expression is either unnecessary or undesirable (Richardson et al., 2009).

Phytochemicals have been demonstrated to be effective in the treatment of NDDs, such as AD and PD. Natural plants and phytochemicals may slow its progression, shorten its duration, and promote recovery by targeting many pathogenic causes. They regulate mitochondrial stress, apoptosis, free radical scavenging, and neurotrophic factors, among other things. Amyloidoses, tauopathies, alpha-synucleinopathies, and TDP-43 proteinopathies are among the most common NDDs (Dugger and Dickson, 2017). Many of these conditions are hereditary, but they can also be brought on by poisons, chemicals, or viruses. NDDs have a 50% likelihood of recurrence when inherited in an autosomal dominant form (Roberts et al., 2020). Vitamin C (ascorbic acid) is a crucial antioxidant molecule in the brain because it helps maintain the integrity and functionality of many different brain functions (Shohag et al., 2022).

3.6.1 Therapeutic Approaches for AD

3.6.1.1 Antibody-Targeting Amyloid-Beta (Aβ) Plaques

AD is a complex NDD, and in addition to anti-amyloid treatments, several medications are under development that target the disease's other characteristics, including tau tangles, synaptic dysfunction, neuroinflammation, and vascular pathology (Scheltens et al., 2021; Cummings et al., 2021). The fact that two additional amyloid-targeting antibodies are currently on the same accelerated approval track as aducanumab may have benefits, lowering costs and accelerating the need for clinical trials to demonstrate efficacy, but it is crucial that the current focus on amyloid immunotherapy does not detract from efforts in other drug discovery areas, because the AD drug development pipeline goes well beyond Aβ (Cummings et al., 2021).

Acuranumab (Aduhelm), the first medicine licenzed for the treatment of Alzheimer's disease, received FDA approval in June 2021 (Dunn et al., 2021). Aducanumab is an IgG1 monoclonal antibody that binds to and aids in the removal of extracellular Aβ plaques in the brain. Every four weeks, it is given as an intravenous infusion, lasting about an hour (Walsh et al., 2021). Aducanumab is provisionally licensed, although clinical results indicate a decrease in the burden of plaques without any correlation to patients' cognitive performance. It will need further clinical research to make a firm determination about whether the medication improves cognitive abilities. However, optimism among AD sufferers and advocacy groups has also increased as a result of aducanumab's approval. They think it will lead the way for comparable treatments in the near future in addition to being the first treatment to focus on changing the histology of the disease (Lamptey et al., 2022).

3.6.1.2 Cholinesterase Inhibitors

AD is marked by programmed cell death of cholinergic neuron and a resulting reduction of acetylcholine-mediated neurotransmission, leading in neurodegeneration. Acetylcholinesterase and butyrylcholinesterase have potential as medicinal targets in the treatment of AD because inhibition of these enzymes can be employed to restore synaptic acetylcholine concentrations. The three most often used cholinesterase inhibitors in clinical use are donepezil, rivastigmine, and galantamine (Okello and Mather, 2020). In AD, there is a loss of cholinergic neurons and a decrease in the amount of acetylcholine in the cortical areas of the brain. Several investigations have found that increasing the availability of acetylcholine in dementia patients helps slow cognitive deterioration. Cholinesterase inhibitors slow the breakdown of acetylcholine, and the patient benefits from greater cholinergic activity (Hampel et al., 2019).

Although numerous drugs have been developed for AD, cholinesterase inhibitors remain the only option available to patients. The recently approved drug aducanumab is contentious in its efficacy and exorbitant in pricing. However, cholinesterase inhibitors show limited efficacy. They achieve a modest improvement in patients' cognitive ability, and they are labeled as symptomatic treatment options rather than altering the pathology (Liu et al., 2019).

3.6.1.3 Glutamate Regulators

Glutamate is a significant excitatory neurotransmitter in the mammalian brain that plays a role in memory, learning, cognition, and motor activity (Danbolt, 2001; Johnson and Kotermanski, 2006). Under typical physiological conditions, glutamate is involved in the development of dendrites and synapse. Proton magnetic resonance spectroscopy can be used to detect glutamatergic activity (Hall et al., 2012) that reflects neural integrity. The use of proton magnetic resonance spectroscopy has also emerged as a potential method in the examination of neurological and mental diseases, including AD (Wang et al., 2015; Su et al., 2016).

3.6.2 Therapeutic Approaches for PD

Although PD is the second most prevalent NDD, there is a dearth of effective medication that specifically modifies the pathophysiology of the disease. Instead, some alternatives treat motor-related and non-motor-related problems independently in order to provide symptomatic alleviation to patients. The primary method to controlling PD is to refill the brain's decreasing dopamine levels in the substantia nigra area. Several ways have been implemented in order to refill dopamine levels (Emamzadeh and Surguchov, 2018).

3.6.2.1 Levodopa and Carbidopa Combination

Current treatments for PD are palliative, with levodopa/carbidopa medication remaining the preferred option despite its inability to slow the course of neurodegeneration. Because of the complexities of such a multifaceted condition and the considerable limitations of therapy, various genetic techniques have proven useful in identifying genes and pathways involved in this disease (Behl et al., 2021). Levodopa is an immediate precursor of dopamine that aids in the restoration of motor capabilities lost due to dopamine deficiency. Carbidopa is used with levodopa to prevent levodopa from being broken down in the peripheral nervous system before it enters the brain (Emamzadeh and Surguchov, 2018).

The use of levodopa to ameliorate PD motor symptoms is the gold standard. Casimir Funk, a Polish scientist, discovered levodopa in 1911. The United States Food and Drug Administration authorized levodopa as a dopamine replacement therapy to treat PD motor symptoms in 1970 (Abbott, 2010). Levodopa activates dopamine receptors in the central nervous system and peripheral circulation by converting dopamine via dopamine decarboxylase. Carbidopa works as a decarboxylase inhibitor to increase levodopa availability in the brain (Guebila and Thiele, 2016). Levodopa has been used to treat PD for more than 50 years. Unfortunately, one study found that dyskinesia was caused by levodopa medication. The mechanism underlying long-term levodopa therapy–induced dyskinesia is unknown (Sharma et al., 2015; Wu et al., 2019).

3.6.2.2 N-Acetylcysteine (NAC)

NAC is an antioxidant that increases glutathione levels (Coles et al., 2018). NAC has long been thought to offer potential as a disease-modifying therapy for Parkinson's disease. NAC has been found in preclinical research in Parkinson's disease models to be protective against dopaminergic cell death caused by rotenone (Chen et al., 2007). There have also been various clinical trials using NAC for neurological diseases, such as PD (Slattery et al., 2015; Monti et al., 2016; Reyes et al., 2016; Monti et al., 2019). NAC has been proven to reach the cerebrospinal fluid after oral ingestion (Katz et al., 2015; Reyes et al., 2016). NAC has abilities to increase glutathione S-transferase activity, replenishing glutathione, scavenging free radicals, and stabilizing protein structures by cross-linking cysteine disulfide molecules, in addition to its antioxidant, anti-inflammatory, and mucolytic effects (Schwalfenberg, 2021).

3.6.2.3 β-Agonists

While aggregated α-synuclein is not required for the development of neurodegeneration or clinical Parkinsonism (Engelender and Isacson, 2017), it is definitely of great potential significance, as proven by rare people with PD who have α-synuclein gene duplications (Chartier-Harlin et al., 2004). Nonetheless, postmortem examination of the majority of PD patients' brains reveals multiple forms of misfolded proteins, not just α-synuclein, and an unambiguous proof that α-synuclein aggregation is harmful rather than compensatory has yet to be shown (Paumier et al., 2015). One example is beta 2-adrenoreceptor, which is a putative regulator of the α-synuclein gene. The usage of beta 2-adrenoreceptor agonists has been linked to a lower incidence of PD and decreased α-synuclein transcription (Mittal et al., 2017), while preliminary research appeared to indicate a symptomatic advantage.

3.6.3 Therapeutic Approaches for MS

MS treatment is progressing from non-specific medicines with undetermined pharmacodynamics to molecular-specific therapies (Inojosa et al., 2022).

3.6.3.1 Disease-Modifying Therapies (DMTs)

DMTs, acute relapse therapy, comorbidity management, symptom control, psychological support, rehabilitative techniques, and lifestyle adjustments are all included in MS treatment (McGinley et al., 2021). Currently, the European Medicines Agency approved 17 DMTs to treat the disorder. Various effectiveness patterns for medicines administered subcutaneously, intramuscularly, intravenously, and orally have been documented. Their modes of action, bad effects, and safety profiles, as well as clinical experience with each of them, differ greatly across medications, and their precise use in the course of the illness is still not well standardized and understood (Inojosa et al., 2022). DMTs for MS minimize the incidence of relapses and short-term impairment. In 1993, the FDA authorized the first DMT, interferon beta-1b. Prior to the licensure of interferon beta-1b, various broad-spectrum immunosuppressive agents were utilized, including azathioprine, methotrexate, mycophenolate mofetil, intravenous immunoglobulin, and corticosteroids. The effectiveness of these earlier medicines was based on studies that were not conclusive due to small sample numbers. Because DMTs are now accessible, these treatments are being recommended less frequently (McGinley et al., 2021)

3.6.3.2 T Lymphocyte Targeting

MS is a chronic disease that has a varied course. RRMS begins with clinical relapses, followed by near or complete recovery, although recovery may be partial over time, and impairment frequently increases. Around 20% of RRMS patients suffer significant neurologic impairment later in the illness and move to secondary progressive MS (SPMS). A limited percentage of people (15%) have primary progressive MS (PPMS) from the time of start (Cree et al., 2016).

Therapy aimed at modulating the composition and differentiation of T cells in progressive MS generally aims to minimize inflammation and minimize any involvement of T cells in the disease process. T cells can be discovered in the spinal cord parenchyma of PPMS patients; however, they are present at lower levels and are more diffuse in the normal-appearing white matter (NAWM) and normal-appearing gray matter (NAGM) (Androdias et al., 2010). The level of inflammation markers in CSF in patients with progressive MS is comparable to that of those with RRMS, despite T cells playing a minor role in inflammation in progressive MS lesions (Komori et al., 2015). Patients with progressive MS have low levels of T cells in the parenchyma, whereas their meninges contain higher numbers of T cells. Axonal damage in the NAWM seems to be associated with the severity of meningeal T cell inflammation (Androdias et al., 2010). It is possible that long-range soluble factors are responsible for T cells' detrimental role in progressive MS. The T cells accumulate slowly in the Virchow–Robin space and the meninges in progressive MS (Lassmann, 2019).

3.6.3.3 Antioxidants and MS

Another treatment strategy for progressive MS is to inhibit the production of ROS and RNS from myeloid cells (Chechneva et al., 2011). It is believed that progressive MS is characterized by chronic inflammation behind a closed blood–brain barrier with activation of microglia and prolonged participation of T cells and B cells. Neurodegeneration is caused by the release of ROS and RNS, which ultimately impair mitochondria and axons (Faissner et al., 2019). Low-dose dextromethorphan therapy improved MS by inhibiting the ROS-generating enzyme NADPH oxidase 2 (NOX2) (Chechneva et al., 2011). In addition, NOX2 is present at high levels in microglia and macrophages, and inhibiting it with dextromethorphan increases the proliferation of oligodendroglial progenitor cells (Lisak et al., 2014) and reduces microglial activation, resulting in lower generation of nitric oxide and superoxide free radicals (Liu et al., 2003).

3.6.4 Therapeutic Approaches for ALS

ALS is a motor neuron disease that causes symptoms of frontotemporal dementia, behavioral abnormalities, and cognitive deterioration as the disease progresses. Within three to five years of the onset of symptoms, ALS patients die of respiratory failure (Bucchia et al., 2015). There are two drugs authorized for ALS patients: riluzole and edaravone. Riluzole, a glutamate-receptor antagonist, was licensed in 1995 as a 100 mg/day oral tablet. Riluzole treatment extends the lives of ALS patients by three to four months as compared to a placebo group in clinical studies.

Edaravone, a free radical scavenger, was recently licensed as an intravenous infusion with a dosage of 60 mg/day in 2017, and it helps halt disease development (Jaiswal, 2019). In addition to these medications, patients are treated symptomatically to improve their quality of life.

3.6.5 Therapeutic Approaches for HD

The HD drugs development strategy has changed dramatically in recent years. Therapies that target certain components of HD biology are being developed and tested for the first time in order to putatively halt the course of the illness, in addition to efforts centered on symptomatic therapy. In this review, we give an update on the developments in the symptomatic management of HD and outline the expanding efforts in disease-modifying medicines, with a particular focus on interventions designed to reduce the level of mHTT in the brain (Dash and Mestre, 2020).

3.6.5.1 Therapies Related to mutHTT Modification

Various proteases are known to proteolytically cleave the mutHTT protein (Gafni et al., 2012; Miller et al., 2010), releasing N-terminal polyQ-containing fragments that can bind with other abnormal proteins, resulting in the development of mutHTT aggregates (Bates et al., 2015). Notably, whereas mutations in the HTT gene cause HD pathogenesis, this gene is required for proper neuronal development as a knockout (Nasir et al., 1995). In addition to symptomatic therapy for HD improvement, novel therapeutic techniques that target the mutHTT protein are being developed (Kumar et al., 2020). Therapeutic efforts to reduce mutHTT content at the transcription and translational stages, as well as therapeutic ways to increase mutHTT proteosomal destruction, are critical. Furthermore, post-translational alteration of mutHTT as a pharmaceutical strategy is being researched. Treatments connected with mutHTT modification and degradation, treatments associated with signaling pathways, therapies associated with lower mutHTT content, stem cell therapies, and pharmaceutical therapies are some of the therapeutic options for HD (Huang et al., 2016; Alkanli et al., 2022).

3.7 NANOTECHNOLOGY AND NDDs

CNS is regarded as the most important organ in the human body. The essential contacts between the CNS and the periphery are the CNS barriers. The BBB is the most significant biological barrier to medication delivery to the brain. It functions as a semi-permeable diffusion barrier to keep unpleasant chemicals and other hazardous substances out of the bloodstream and controls the supply of nutrients to the brain to ensure appropriate brain function (Akhtar et al., 2021).

Drug delivery is the process of conveying pharmaceutical substance into the body for therapeutic purposes using the appropriate methods, formulations, technologies, and procedures. It includes methods of providing medicinal substances to people and animals in order to achieve therapeutic efficacy. There have been major advances in drug delivery systems in the past few years focusing on smart drug delivery systems which provide optimal timing of drug administration, dose, and place for optimum safety and efficacy (Thakur et al., 2015). For drug diffusion from the blood to the brain, it is crucial for a biologically active molecule to cross lipid membranes. It is difficult for many medications to cross the BBB because they lack the necessary physicochemical properties, including high lipid solubility, small molecular sizes, and positive charges. This is why numerous ways, both invasive and non-invasive, have been developed to circumvent the BBB. Direct delivery of medications or antibodies to the brain interstitium is the most direct technique to bypass the BBB. Through the delivery of medications directly to an intracranial target, intramuscular medication administration could theoretically achieve high concentrations of CNS drugs with low systemic exposure and toxicity. Furthermore, this method allows for the maintenance of intracranial drug concentrations, which is critical in the treatment of several chemotherapeutic drugs (Baratchi et al., 2009).

The structure and size range of nanoparticles (NPs) are diverse, ranging from nanocapsules to nanospheres with continuous matrix structures and ranging from 1 to 1,000 nm in size (Blasi et al., 2007; Silva, 2007). They are cutting-edge drug delivery systems that cover the BBB restriction, maintain drug stability, and have the potential to delay drug release, hence reducing medication toxicity. Affordable, nontoxic, non-immunogenic, biodegradable, and non-thrombogenic carrier nanoparticles are ideal (Silva, 2007; Rieux et al., 2006; Blasi et al., 2007). Because NPs provide us the freedom to manipulate drugs or alter the BBB's permeability, they have recently attracted a lot of attention as a mechanism for drug delivery. Their expansive surface area, which may be modified to enhance targeting, biocompatibility, solubility, and controlled drug release, is another advantage (Silva, 2007; Rieux et al., 2006; Blasi et al., 2007). NP-based drug delivery allows drugs to be released more slowly, giving the medication time to work therapeutically and respond to stimuli such as pH, light, heat, and enzymes (Yoo et al., 2019). Organic NPs are a suitable model for delivering calcitonin, insulin, dalargin, and cyclosporine (Yadav et al., 2011).

One promising method for delivering various drugs is the nanomicelle, particularly polymeric nanomicelle (Sharma and Singh, 2020; Baba et al., 2015). According to a new investigation, using chitosan nanomicelles in the proper amounts can successfully transfect brain cells (Kim et al., 2013). According to Layek and Singh (2013), the advantages of chitosan nanomicelles are their biodegradability, lack of toxicity at the use concentration, and flexibility for surface modification. Due to these benefits, conjugated chitosan nanoparticles were found to be considerably reduced in vitro by α-synuclein aggregation and to have strong neuroprotective effects in PD models, making them a good carrier for delivering medicines, proteins to the brain (Xue et al., 2020). To improve transport across the BBB, chitosan can be conjugated with other polymers. In comparison to bare PLGA nanoparticles, those coated with chitosan demonstrated a higher BBB absorption (Jaruszewski et al., 2012).

Nanomedicine offers promise for NDDs and has the capacity to get beyond limitations of existing and conventional therapy modalities (Zhu et al., 2021). We emphasized publications using nanoparticles to combat several NDDs, which might lead to the development of nanomedicine. Understandably, the process of developing a curative medicine takes time, but early research in the area may provide a foundation for the eventual eradication of NDDs. More in vitro and in vivo evidence must be generated, nevertheless, in order to demonstrate the effectiveness against NDDs. To evaluate the effectiveness of nanoparticles, rigorous in vitro and in vivo studies and the formation of their correlation are needed. This would aid the scientific community in discovering more or more efficient nanoparticles for use in diagnostic or therapeutic procedures (Lamptey et al., 2022).

3.8 LIMITATIONS OF THE CURRENT THERAPEUTIC STRATEGIES OF NDDs

Administering medications in the form of oral capsules or tablets is typical for current treatments of PD and AD. Medications have inadequate access to the brain, considerable first-pass metabolism reduces their half-life, and they may have unwanted side effects when they reach peripheral tissues that are not their intended targets, making systemic delivery of medications to the CNS a significant problem. For the treatment of NDDs, several endogenous proteins and antibody-based biopharmaceuticals have been developed and tested; nevertheless, their therapeutic efficacy is still insufficient (Sevigny et al., 2016).

3.9 CONCLUSION

NDD is a progressive condition brought on by environmental, physiological, genetic, and epigenetic causes. It has been discovered that the oxidoreductase-induced production of free radicals and the degeneration of the antioxidant system allow the CNS to accumulate misfolded proteins, which then drive mitochondrial dysfunctions and neuroinflammations, which ultimately result in NDDs. Most often, measuring the specific receptor binding of neurodegenerative illnesses, cellular metabolic alterations, or anatomical structure changes is used to diagnose them. By using radiolabeled tracers, neuroimaging techniques like PET, SPECT, MRI, etc. have been widely employed to diagnose the metabolic state and receptor activity of damaged neuronal cells under pathological situations. Metabolomics applications are yet another, more recent method for the diagnosis and prognosis of NDDs. Brain mapping techniques, on the other hand, can be used to characterize regional anatomical alterations in the brain caused by disease. Because of these technological developments, it is now considerably simpler to diagnose neurodegeneration, and for the majority of the major NDDs, an early diagnosis is also somewhat feasible. Although a complete treatment for NDDs or other neurodegenerative diseases has not yet been found, medications that can stop NDDs from developing in the first place are being researched. There are few medical medicines on the market for the treatment of NDDs, other than prevention. However, other approaches, including immunomodulatory therapy, neurotrophic factor therapy, epigenetic modulation, targeting oxidoreductases, and insulin therapy, are currently being studied and tested in clinical settings. The current advancements in biochemical, pathological, and pharmaceutical research may guarantee a better future for neuronal health globally, but to prevent the formation of such NDDs, a healthier lifestyle must be adopted from an early age.

REFERENCES

Abbott, A. (2010). Levodopa: The story so far. Nature, 466(7310), S6–S7.

Abtahi, S. L., Masoudi, R., & Haddadi, M. (2020). The distinctive role of Tau and amyloid beta in mitochondrial dysfunction through alteration in Mfn2 and Drp1 mRNA levels: A comparative study in drosophila melanogaster. Gene, 754, 144854.

Adachi, Y., Ono, N., Imaizumi, A., Muramatsu, T., Andou, T., Shimodaira, Y., Nagao, K., Kageyama, Y., Mori, M., Noguchi, Y., Hashizume, N., & Nukada, H. (2018). Plasma amino acid profile in severely frail elderly patients in Japan. International Journal of Gerontology, 12(4), 290–293.

Akhtar, A., Andleeb, A., Waris, T. S., Bazzar, M., Moradi, A. R., Awan, N. R., & Yar, M. (2021). Neurodegenerative diseases and effective drug delivery: A review of challenges and novel therapeutics. Journal of Controlled Release, 330, 1152–1167.

Alkanli, S. S., Alkanli, N., Ay, A., & Albeniz, I. (2022). CRISPR/Cas9 mediated therapeutic approach in Huntington's disease. Molecular Neurobiology, 1–13.

Amor, S., Puentes, F., Baker, D., & Van Der Valk, P. (2010). Inflammation in neurodegenerative diseases. Immunology, 129(2), 154–169.

Androdias, G., Reynolds, R., Chanal, M., Ritleng, C., Confavreux, C., & Nataf, S. (2010). Meningeal T cells associate with diffuse axonal loss in multiple sclerosis spinal cords. Annals of Neurology, 68(4), 465–476.

Appel, J., Potter, E., Shen, Q., Pantol, G., Greig, M. T., Loewenstein, D., & Duara, R. (2009). A comparative analysis of structural brain MRI in the diagnosis of Alzheimer's disease. Behavioural Neurology, 21(1–2), 13–19.

Atwal, R. S., Desmond, C. R., Caron, N., Maiuri, T., Xia, J., Sipione, S., & Truant, R. (2011). Kinase inhibitors modulate huntingtin cell localization and toxicity. Nature Chemical Biology, 7(7), 453–460.

Atwal, R. S., Xia, J., Pinchev, D., Taylor, J., Epand, R. M., & Truant, R. (2007). Huntingtin has a membrane association signal that can modulate huntingtin aggregation, nuclear entry and toxicity. Human Molecular Genetics, 16(21), 2600–2615.

Axelsson, M., Malmeström, C., Nilsson, S., Haghighi, S., Rosengren, L., & Lycke, J. (2011). Glial fibrillary acidic protein: A potential biomarker for progression in multiple sclerosis. Journal of Neurology, 258(5), 882–888.

Ayrignac, X., Le Bars, E., Duflos, C., Hirtz, C., Maleska Maceski, A., Carra-Dallière, C., Charif, M., Pinna, F., Prin, P., Menjot de Champfleur, N., Deverdun, J., Kober, T., Marechal, B., Fartaria, M. J., Corredor Jerez, R., Labauge, P., & Lehmann, S. (2020). Serum GFAP in multiple sclerosis: Correlation with disease type and MRI markers of disease severity. Scientific Reports, 10(1), 10923.

Baba, M., Itaka, K., Kondo, K., Yamasoba, T., & Kataoka, K. (2015). Treatment of neurological disorders by introducing mRNA in vivo using polyplex nanomicelles. Journal of Controlled Release, 201, 41–48.

Baecher-Allan, C., Kaskow, B. J., & Weiner, H. L. (2018). Multiple sclerosis: Mechanisms and immunotherapy. Neuron, 97(4), 742–768.

Baird, A. L., Westwood, S., & Lovestone, S. (2015). Blood-based proteomic biomarkers of Alzheimer's disease pathology. Frontiers in Neurology, 6, 236.

Baratchi, S., Kanwar, R. K., Khoshmanesh, K., Vasu, P., Ashok, C., Hittu, M., & Kanwar, J. R. (2009). Promises of nanotechnology for drug delivery to brain in neurodegenerative diseases. Current Nanoscience, 5(1), 15–25.

Barkhof, F. (2016). Brain atrophy measurements should be used to guide therapy monitoring in MS–NO. Multiple Sclerosis Journal, 22(12), 1524–1526.

Barnett, M. H., & Prineas, J. W. (2004). Relapsing and remitting multiple sclerosis: Pathology of the newly forming lesion. Annals of Neurology, 55(4), 458–468.

Bates, G. P., Dorsey, R., Gusella, J. F., Hayden, M. R., Kay, C., Leavitt, B. R., Nance, M., Ross, C. A., Scahill, R. I., Wetzel, R., & Tabrizi, S. J. (2015). Huntington disease. Nature Reviews Disease Primers, 1(1), 1–21.

Behl, T., Kaur, G., Fratila, O., Buhas, C., Judea-Pusta, C. T., Negrut, N., Bustea, C., & Bungau, S. (2021). Cross-talks among GBA mutations, glucocerebrosidase, and α-synuclein in GBA-associated Parkinson's disease and their targeted therapeutic approaches: A comprehensive review. Translational Neurodegeneration, 10, 1–13.

Bellezza, I., Grottelli, S., Costanzi, E., Scarpelli, P., Pigna, E., Morozzi, G., Mezzasoma, L., Peirce, M. J., Moresi, V., Adamo, S., & Minelli, A. (2018). Peroxynitrite activates the NLRP3 inflammasome cascade in SOD1 (G93A) mouse model of amyotrophic lateral sclerosis. Molecular Neurobiology, 55, 2350–2361.

Berganzo, K., Tijero, B., González-Eizaguirre, A., Somme, J., Lezcano, E., Gabilondo, I., Fernandez, M., Zarranz, J. J., & Gómez-Esteban, J. C. (2016). Síntomas no motores y motores en la enfermedad de Parkinson y su relación con la calidad de vida y los distintos subgrupos clínicos. Neurología, 31(9), 585–591.

Berk, M., Kapczinski, F., Andreazza, A. C., Dean, O. M., Giorlando, F., Maes, M., Yücel, M., Gama, C. S., Dodd, S., Dean, B., Magalhães, P. V. S., Amminger, P., McGorry, P., & Malhi, G. S. (2011). Pathways underlying neuroprogression in bipolar disorder: Focus on inflammation, oxidative stress and neurotrophic factors. Neuroscience & Biobehavioral Reviews, 35(3), 804–817.

Berman, T., & Bayati, A. (2018). What are neurodegenerative diseases and how do they affect the brain? Frontiers for Young Minds, 6, 1–6.

Bernis, M. E., Babila, J. T., Breid, S., Wüsten, K. A., Wüllner, U., & Tamgüney, G. (2015). Prion-like propagation of human brain-derived alpha-synuclein in transgenic mice expressing human wild-type alpha-synuclein. Acta Neuropathologica Communications, 3, 75.

Blasco, H., Garcon, G., Patin, F., Veyrat-Durebex, C., Boyer, J., Devos, D., Vourc'h, P., Andres, C. R., & Corcia, P. (2017). Panel of oxidative stress and inflammatory biomarkers in ALS: A pilot study. Canadian Journal of Neurological Sciences, 44(1), 90–95.

Blasi, P., Giovagnoli, S., Schoubben, A., Ricci, M., & Rossi, C. (2007). Solid lipid nanoparticles for targeted brain drug delivery. Advanced Drug Delivery Reviews, 59(6), 454–477.

Blesa, J., Trigo-Damas, I., Quiroga-Varela, A., & Jackson-Lewis, V. R. (2015). Oxidative stress and Parkinson's disease. Frontiers in Neuroanatomy, 9, 1–9.

Boyle, P. A., Yu, L., Wilson, R. S., Leurgans, S. E., Schneider, J. A., & Bennett, D. A. (2018). Person-specific contribution of neuropathologies to cognitive loss in old age. Annals of Neurology, 83(1), 74–83.

Bozzi, M., & Sciandra, F. (2020). Molecular mechanisms underlying muscle wasting in Huntington's disease. International Journal of Molecular Sciences, 21(21), 8314.

Braak, H., Del Tredici, K., Rüb, U., De Vos, R. A., Steur, E. N. J., & Braak, E. (2003). Staging of brain pathology related to sporadic Parkinson's disease. Neurobiology of Aging, 24(2), 197–211.

Brandi, V., Di Lella, V., Marino, M., Ascenzi, P., & Polticelli, F. (2018). A comprehensive in silico analysis of huntingtin and its interactome. Journal of Biomolecular Structure and Dynamics, 36(12), 3155–3171.

Bredesen, D. E. (2015). Metabolic profiling distinguishes three subtypes of Alzheimer's. Aging, 7(8), 595–600.

Bucchia, M., Ramirez, A., Parente, V., Simone, C., Nizzardo, M., Magri, F., Dametti, S., & Corti, S. (2015). Therapeutic development in amyotrophic lateral sclerosis. Clinical Therapeutics, 37(3), 668–680.

Buizza, L., Cenini, G., Lanni, C., Ferrari-Toninelli, G., Prandelli, C., Govoni, S., Buoso, E., Racchi, M., Barcikowska, M., Styczynska, M., Szybinska, A., Butterfeld, D. A., Memo, M., & Uberti, D. (2012). Conformational altered p53 as an early marker of oxidative stress in Alzheimer's disease. PLoS ONE, 7(1), e29789.

Businaro, R., Corsi, M., Asprino, R., Di Lorenzo, C., Laskin, D., Corbo, R. M., Ricci, S., & Pinto, A. (2018). Modulation of inflammation as a way of delaying Alzheimer's disease progression: The diet's role. Current Alzheimer Research, 15(4), 363–380.

Butterfield, D. A., & Keller, J. N. (2012). Antioxidants and antioxidant treatment in disease. Biochimica et Biophysica Acta-Molecular Basis of Disease, 1822(5), 615.

Cameron, B., & Landreth, G. E. (2010). Inflammation, microglia, and Alzheimer's disease. Neurobiology of Disease, 37(3), 503–509.

Camprodon, J. A., & Stern, T. A. (2013). Selecting neuroimaging techniques: A review for the clinician. The Primary Care Companion for CNS Disorders, 15(4), 26837.

Caraci, F., Spampinato, S. F., Morgese, M. G., Tascedda, F., Salluzzo, M. G., Giambirtone, M. C., Caruso, G., Munafò, A., Torrisi, S. A., Leggio, G. M., & Trabace, L. (2018). Neurobiological links between depression and AD: The role of TGF-β1 signaling as a new pharmacological target. Pharmacological Research, 130, 374–384.

Chaney, A., Williams, S. R., & Boutin, H. (2019). In vivo molecular imaging of neuroinflammation in Alzheimer's disease. Journal of Neurochemistry, 149(4), 438–451.

Chartier-Harlin, M. C., Kachergus, J., Roumier, C., Mouroux, V., Douay, X., Lincoln, S., Levecque, C., Larvor, L., Andrieux, J., Hulihan, M., & Waucquier, N. (2004). α-synuclein locus duplication as a cause of familial Parkinson's disease. Lancet, 364(9440), 1167–1169.

Chataway, J. (2018). Evolving diagnostic criteria for multiple sclerosis. Lancet Neurology, 17(2), 118.

Chechneva, O. V., Mayrhofer, F., Daugherty, D. J., Pleasure, D. E., Hong, J. S., & Deng, W. (2011). Low dose dextromethorphan attenuates moderate experimental autoimmune encephalomyelitis by inhibiting NOX_2 and reducing peripheral immune cells infiltration in the spinal cord. Neurobiology of Disease, 44(1), 63–72.

Cheignon, C., Tomas, M., Bonnefont-Rousselot, D., Faller, P., Hureau, C., & Collin, F. (2018). Oxidative stress and the amyloid beta peptide in Alzheimer's disease. Redox Biology, 14, 450–464.

Chen, C. M., Yin, M. C., Hsu, C. C., & Liu, T. C. (2007). Antioxidative and anti-inflammatory effects of four cysteine-containing agents in striatum of MPTP-treated mice. Nutrition, 23(7–8), 589–597.

Chen, X., Guo, C., & Kong, J. (2012). Oxidative stress in neurodegenerative diseases. Neural Regeneration Research, 7(5), 376.

Chen, X., Hu, Y., Cao, Z., Liu, Q., & Cheng, Y. (2018). Cerebrospinal fluid inflammatory cytokine aberrations in Alzheimer's disease, Parkinson's disease and amyotrophic lateral sclerosis: A systematic review and meta-analysis. Frontiers in Immunology, 9, 2122.

Cheng, Z., Yin, J., Yuan, H., Jin, C., Zhang, F., Wang, Z., Liu, X., Wu, Y., Wang, T., & Xiao, S. (2018). Blood-derived plasma protein biomarkers for Alzheimer's disease in Han Chinese. Frontiers in Aging Neuroscience, 10, 414.

Chinta, S. J., & Andersen, J. K. (2008). Redox imbalance in Parkinson's disease. Biochimica et Biophysica Acta (BBA)-General Subjects, 1780(11), 1362–1367.

Cho, D. H., Nakamura, T., Fang, J., Cieplak, P., Godzik, A., Gu, Z., & Lipton, S. A. (2009). S-nitrosylation of Drp1 mediates β-amyloid-related mitochondrial fission and neuronal injury. Science, 324(5923), 102–105.

Cho, H., & Shukla, S. (2020). Role of edaravone as a treatment option for patients with amyotrophic lateral sclerosis. Pharmaceuticals, 14(1), 29.

Clarke, M. A., Samaraweera, A. P. R., Falah, Y., Pitiot, A., Allen, C. M., Dineen, R. A., Tench, C. R., Morgan, P. S., & Evangelou, N. (2019). Single test to ARrive at multiple sclerosis (STAR-MS) diagnosis: A prospective pilot study assessing the accuracy of the central vein sign in predicting multiple sclerosis in cases of diagnostic uncertainty. Multiple Sclerosis, 26(4), 433–441.

Coles, L. D., Tuite, P. J., Öz, G., Mishra, U. R., Kartha, R. V., Sullivan, K. M., Cloyd, J. C., & Terpstra, M., (2018). Repeated-dose oral N-acetylcysteine in Parkinson's disease: Pharmacokinetics and effect on brain glutathione and oxidative stress. Journal of Clinical Pharmacology, 58(2), 158–167.

Corso, G., Cristofano, A., Sapere, N., la Marca, G., Angiolillo, A., Vitale, M., Fratangelo, R., Lombardi, T., Porcile, C., Intrieri, M., & Di Costanzo, A. (2017). Serum amino acid profiles in normal subjects and in patients with or at risk of Alzheimer dementia. Dementia and Geriatric Cognitive Disorders Extra, 7(1), 143–159.

Covarrubias-Pinto, A., Moll, P., Solís-Maldonado, M., Acuña, A. I., Riveros, A., Miró, M. P., Papic, E., Beltrán, F. A., Cepeda, C., Concha, I. I., & Brauchi, S. (2015). Beyond the redox imbalance: Oxidative stress contributes to an impaired GLUT3 modulation in Huntington's disease. Free Radical Biology and Medicine, 89, 1085–1096.

Cree, B. A., Gourraud, P. A., Oksenberg, J. R., Bevan, C., Crabtree-Hartman, E., Gelfand, J. M., Goodin, D. S., Graves, J., Green, A. J., & Mowry, E. (2016). Long-term evolution of multiple sclerosis disability in the treatment era. Annals of Neurology, 80(4), 499–510.

Cummings, J., Lee, G., Zhong, K., Fonseca, J., & Taghva, K. (2021). Alzheimer's disease drug development pipeline: 2021. Alzheimer's & dementia: Translational Research & Clinical Interventions, 7(1), e12179.

Daixing, Z., Chengye, Z., Qiang, Z., & Shusheng, L. (2013). Upregulation of sestrin-2 expression via P53 protects against 1-methyl-4-phenylpyridinium (MPP+) neurotoxicity. Journal of Molecular Neuroscience, 51, 967–975.

Danbolt, N. C. (2001). Glutamate uptake. Progress in Neurobiology, 65(1), 1–105.

Dash, D., & Mestre, T. A. (2020). Therapeutic update on Huntington's disease: Symptomatic treatments and emerging disease-modifying therapies. Neurotherapeutics, 17(4), 1645–1659.

Davies, S. W., Turmaine, M., Cozens, B. A., DiFiglia, M., Sharp, A. H., Ross, C. A., Scherzinger, E., Wanker, E. E., Mangiarini, L., & Bates, G. P. (1997). Formation of neuronal intranuclear inclusions underlies the neurological dysfunction in mice transgenic for the HD mutation. Cell, 90(3), 537–548.

Dean, D. C., Jerskey, B. A., Chen, K., Protas, H., Thiyyagura, P., Roontiva, A., O'Muircheartaigh, J., Dirks, H., Waskiewicz, N., Lehman, K., & Siniard, A. L. (2014). Brain differences in infants at differential genetic risk for late-onset Alzheimer disease: A cross-sectional imaging study. JAMA Neurology, 71(1), 11–22.

De Stefano, N., Giorgio, A., Battaglini, M., Rovaris, M., Sormani, M. P., Barkhof, F., Korteweg, T., Enzinger, C., Fazekas, F., Calabrese, M., & Dinacci, D. (2010). Assessing brain atrophy rates in a large population of untreated multiple sclerosis subtypes. Neurology, 74(23), 1868–1876.

Dugger, B. N., & Dickson, D. W. (2017). Pathology of neurodegenerative diseases. Cold Spring Harbor Perspectives in Biology, 9(7), a028035.

Dunbar, C. E., High, K. A., Joung, J. K., Kohn, D. B., Ozawa, K., & Sadelain, M. (2018). Gene therapy comes of age. Science, 359(6372), eaan4672.

Dunn, B., Stein, P., & Cavazzoni, P. (2021). Approval of aducanumab for Alzheimer disease—the FDA's perspective. JAMA Internal Medicine, 181(10), 1276–1278.

Dworkin, J. D., Sati, P., Solomon, A., Pham, D. L., Watts, R., Martin, M. L., Ontaneda, D., Schindler, M. K., Reich, D. S., & Shinohara, R. T. (2018). Automated integration of multimodal MRI for the probabilistic detection of the central vein sign in white matter lesions. AJNR American Journal of Neuroradiology, 39(10), 1806–1813.

Elobeid, A., Libard, S., Leino, M., Popova, S. N., & Alafuzoff, I. (2016). Altered proteins in the aging brain. Journal of Neuropathology & Experimental Neurology, 75(4), 316–325.

Emamzadeh, F. N., & Surguchov, A. (2018). Parkinson's disease: Biomarkers, treatment, and risk factors. Frontiers in Neuroscience, 12, 612.

Engelender, S., & Isacson, O. (2017). The threshold theory for Parkinson's disease. Trends in Neurosciences, 40(1), 4–14.

Faissner, S., Plemel, J. R., Gold, R., & Yong, V. W. (2019). Progressive multiple sclerosis: From pathophysiology to therapeutic strategies. Nature Reviews Drug Discovery, 18(12), 905–922.

Falahati, F., Westman, E., & Simmons, A. (2014). Multivariate data analysis and machine learning in Alzheimer's disease with a

focus on structural magnetic resonance imaging. Journal of Alzheimer's Disease, 41(3), 685–708.
Feldman, H. H., Pirttila, T., Dartigues, J. F., Everitt, B., Van Baelen, B., Schwalen, S., & Kavanagh, S. (2009). Treatment with galantamine and time to nursing home placement in Alzheimer's disease patients with and without cerebrovascular disease. International Journal of Geriatric Psychiatry: A Journal of the Psychiatry of Late Life and Allied Sciences, 24(5), 479–488.
Ferrazzoli, D., Ortelli, P., Cucca, A., Bakdounes, L., Canesi, M., & Volpe, D. (2020). Motor-cognitive approach and aerobic training: A synergism for rehabilitative intervention in Parkinson's disease. Neurodegenerative Disease Management, 10(1), 41–55.
Figura, M., Kuśmierska, K., Bucior, E., Szlufik, S., Koziorowski, D., Jamrozik, Z., & Janik, P. (2018). Serum amino acid profile in patients with Parkinson's disease. PLoS ONE, 13(1), e0191670.
Finkbeiner, S. (2011). Huntington's disease. Cold Spring Harbor Perspectives in Biology, 3(6), a007476.
Fischer, M. T., Wimmer, I., Höftberger, R., Gerlach, S., Haider, L., Zrzavy, T., Hametner, S., Mahad, D., Binder, C. J., Krumbholz, M., & Bauer, J. (2013). Disease-specific molecular events in cortical multiple sclerosis lesions. Brain, 136(6), 1799–1815.
Forloni, G., & Balducci, C. (2018). Alzheimer's disease, oligomers, and inflammation. Journal of Alzheimer's Disease, 62(3), 1261–1276.
Gafni, J., Papanikolaou, T., DeGiacomo, F., Holcomb, J., Chen, S., Menalled, L., Kudwa, A., Fitzpatrick, J., Miller, S., Ramboz, S., & Tuunanen, P. I. (2012). Caspase-6 activity in a BACHD mouse modulates steady-state levels of mutant huntingtin protein but is not necessary for production of a 586 amino acid proteolytic fragment. Journal of Neuroscience, 32(22), 7454–7465.
García, J. C., & Bustos, R. H. (2018). The genetic diagnosis of neurodegenerative diseases and therapeutic perspectives. Brain Sciences, 8(12), 222.
Ghosh, R., & Tabrizi, S. J. (2018). Huntington disease. Handbook of Clinical Neurology, 147, 255–278.
Glass, C. K., Saijo, K., Winner, B., Marchetto, M. C., & Gage, F. H. (2010). Mechanisms underlying inflammation in neurodegeneration. Cell, 140(6), 918–934.
Godin, J. D., Colombo, K., Molina-Calavita, M., Keryer, G., Zala, D., Charrin, B. C., Dietrich, P., Volvert, M. L., Guillemot, F., Dragatsis, I., & Bellaiche, Y. (2010). Huntingtin is required for mitotic spindle orientation and mammalian neurogenesis. Neuron, 67(3), 392–406.
Gong, Y., Liu, Y., Zhou, L., Di, X., Li, W., Li, Q., & Bi, K. (2015). A UHPLC-TOF/MS method based metabonomic study of total ginsenosides effects on Alzheimer disease mouse model. Journal of Pharmaceutical and Biomedical Analysis, 115, 174–182.
Goutman, S. A. (2017). Diagnosis and clinical management of amyotrophic lateral sclerosis and other motor neuron disorders. CONTINUUM: Lifelong Learning in Neurology, 23(5), 1332–1359.
Guebila, M. B., & Thiele, I. (2016). Model-based dietary optimization for late-stage, levodopa-treated, Parkinson's disease patients. NPJ Systems Biology and Applications, 2(1), 1–8.
Guzman-Martinez, L., Maccioni, R. B., Andrade, V., Navarrete, L. P., Pastor, M. G., & Ramos-Escobar, N. (2019). Neuroinflammation as a common feature of neurodegenerative disorders. Frontiers in Pharmacology, 10, 1008.
Hall, H., Cuellar-Baena, S., & Dahlberg, C, In't Zandt, R., Denisov, V., & Kirik, D. (2012). Magnetic resonance spectroscopic methods for the assessment of metabolic functions in the diseased brain. Current Topics in Behavioral Neurosciences, 11, 169–198.
Hampel, H., Mesulam, M. M., Cuello, A. C., Khachaturian, A. S., Vergallo, A., Farlow, M. R., Snyder, P., Giacobini, E., Khachaturian, Z., & Cholinergic System Working Group, and for the Alzheimer Precision Medicine Initiative (APMI). (2019). Revisiting the cholinergic hypothesis in Alzheimer's disease: Emerging evidence from translational and clinical research. Journal of Prevention of Alzheimer's Disease, 6, 2–15.
Heneka, M. T., Carson, M. J., El Khoury, J., Landreth, G. E., Brosseron, F., Feinstein, D. L., Jacobs, A. H., Wyss-Coray, T., Vitorica, J., Ransohoff, R. M., & Herrup, K. (2015). Neuroinflammation in Alzheimer's disease. Lancet Neurology, 14(4), 388–405.
Högel, H., Rissanen, E., Barro, C., Matilainen, M., Nylund, M., Kuhle, J., & Airas, L. (2020). Serum glial fibrillary acidic protein correlates with multiple sclerosis disease severity. Multiple Sclerosis, 26(2), 210–219.
Hou, Y., Dan, X., Babbar, M., Wei, Y., Hasselbalch, S. G., Croteau, D. L., & Bohr, V. A. (2019). Ageing as a risk factor for neurodegenerative disease. Nature Reviews Neurology, 15(10), 565–581.
Huai, J., & Zhang, Z. (2019). Structural properties and interaction partners of familial ALS-associated SOD1 mutants. Frontiers in Neurology, 10, 527.
Huang, W. J., Chen, W. W., & Zhang, X. (2016). Huntington's disease: Molecular basis of pathology and status of current therapeutic approaches. Experimental and Therapeutic Medicine, 12(4), 1951–1956.
Hudry, E., & Vandenberghe, L. H. (2019). Therapeutic AAV gene transfer to the nervous system: A clinical reality. Neuron, 101(5), 839–862.
Inojosa, H., Proschmann, U., Akgün, K., & Ziemssen, T. (2022). The need for a strategic therapeutic approach: Multiple sclerosis in check. Therapeutic Advances in Chronic Disease, 13, 20406223211063032.
International Multiple Sclerosis Genetics Consortium. (2019). Multiple sclerosis genomic map implicates peripheral immune cells and microglia in susceptibility. Science, 365(6460), eaav7188.
Jaiswal, M. K. (2019). Riluzole and Edaravone: A tale of two amyotrophic lateral sclerosis drugs. Medicinal Research Reviews, 39(2), 733–748.
Jankovic, J. (2005). Motor fluctuations and dyskinesias in Parkinson's disease: Clinical manifestations. Movement Disorders: Official Journal of the Movement Disorder Society, 20(S11), S11–S16.
Jankovic, J. (2008). Parkinson's disease: Clinical features and diagnosis. Journal of Neurology, Neurosurgery, and Psychiatry, 79, 368–376.
Jansen, A. H., Batenburg, K. L., Pecho-Vrieseling, E., & Reits, E. A. (2017). Visualization of prion-like transfer in Huntington's disease models. Biochimica et Biophysica Acta (BBA)-Molecular Basis of Disease, 1863(3), 793–800.
Jaruszewski, K. M., Ramakrishnan, S., Poduslo, J. F., & Kandimalla, K. K. (2012). Chitosan enhances the stability and targeting of immuno-nanovehicles to cerebro-vascular deposits of Alzheimer's disease amyloid protein. Nanomedicine: Nanotechnology, Biology and Medicine, 8(2), 250–260.

Johnson, J. W., & Kotermanski, S. E. (2006). Mechanism of action of memantine. Current Opinion in Pharmacology, 6(1), 61–67.

Kalia, L. V., & Lang, A. E. (2015). Parkinson's disease. Lancet, 386(9996), 896–912.

Kapasi, A., DeCarli, C., & Schneider, J. A. (2017). Impact of multiple pathologies on the threshold for clinically overt dementia. Acta Neuropathologica, 134, 171–186.

Kapczinski, N. S., Mwangi, B., Cassidy, R. M., Librenza-Garcia, D., Bermudez, M. B., Kauer-Sant'anna, M., Kapczinski, F., & Passos, I. C. (2017). Neuroprogression and illness trajectories in bipolar disorder. Expert Review of Neurotherapeutics, 17(3), 277–285.

Karanth, S., Nelson, P. T., Katsumata, Y., Kryscio, R. J., Schmitt, F. A., Fardo, D. W., Cykowski, M. D., Jicha, G. A., Van Eldik, L. J., & Abner, E. L. (2020). Prevalence and clinical phenotype of quadruple misfolded proteins in older adults. JAMA Neurology, 77(10), 1299–1307.

Kassmann, C. M., Lappe-Siefke, C., Baes, M., Brügger, B., Mildner, A., Werner, H. B., Natt, O., Michaelis, T., Prinz, M., Frahm, J., & Nave, K. A. (2007). Axonal loss and neuroinflammation caused by peroxisome-deficient oligodendrocytes. Nature Genetics, 39(8), 969–976.

Katz, M., Won, S. J., Park, Y., Orr, A., Jones, D. P., Swanson, R. A., & Glass, G. A. (2015). Cerebrospinal fluid concentrations of N-acetylcysteine after oral administration in Parkinson's disease. Parkinsonism & Related Disorders, 21(5), 500–503.

Kazama, M., Kato, Y., Kakita, A., Noguchi, N., Urano, Y., Masui, K., Niida-Kawaguchi, M., Yamamoto, T., Watabe, K., Kitagawa, K., & Shibata, N. (2020). Astrocytes release glutamate via cystine/glutamate antiporter upregulated in response to increased oxidative stress related to sporadic amyotrophic lateral sclerosis. Neuropathology, 40(6), 587–598.

Keryer, G., Pineda, J. R., Liot, G., Kim, J., Dietrich, P., Benstaali, C., Smith, K., Cordelières, F. P., Spassky, N., Ferrante, R. J., Dragatsis, I., & Saudou, F. (2011). Ciliogenesis is regulated by a huntingtin-HAP1-PCM1 pathway and is altered in Huntington disease. Journal of Clinical Investigation, 121(11), 4372–4382.

Kim, J. Y., Choi, W. I., Kim, Y. H., & Tae, G. (2013). Brain-targeted delivery of protein using chitosan-and RVG peptide-conjugated, pluronic-based nano-carrier. Biomaterials, 34(4), 1170–1178.

Kim, M., Velier, J., Chase, K., Laforet, G., Kalchman, M. A., Hayden, M. R., Won, L., Heller, A., Aronin, N., & DiFiglia, M. (1999). Forskolin and dopamine D1 receptor activation increase huntingtin's association with endosomes in immortalized neuronal cells of striatal origin. Neuroscience, 89(4), 1159–1167.

Kolarcik, C. L., & Bowser, R. (2012). Retinoid signaling alterations in amyotrophic lateral sclerosis. American Journal of Neurodegenerative Disease, 1(2), 130.

Komori, M., Blake, A., Greenwood, M., Lin, Y. C., Kosa, P., Ghazali, D., Winokur, P., Natrajan, M., Wuest, S. C., Romm, E., & Panackal, A. A. (2015). Cerebrospinal fluid markers reveal intrathecal inflammation in progressive multiple sclerosis. Annals of Neurology, 78(1), 3–20.

Kumar, A., Kumar, V., Singh, K., Kumar, S., Kim, Y. S., Lee, Y. M., & Kim, J. J. (2020). Therapeutic advances for Huntington's disease. Brain Sciences, 10(1), 43.

Kumar, V., Singh, C., & Singh, A. (2021). Zebrafish an experimental model of Huntington's disease: Molecular aspects, therapeutic targets and current challenges. Molecular Biology Reports, 1–14.

Lamptey, R. N., Chaulagain, B., Trivedi, R., Gothwal, A., Layek, B., & Singh, J. (2022). A review of the common neurodegenerative disorders: Current therapeutic approaches and the potential role of nanotherapeutics. International Journal of Molecular Sciences, 23(3), 1851.

Lassmann, H. (2019). Pathogenic mechanisms associated with different clinical courses of multiple sclerosis. Frontiers in Immunology, 9, 3116.

Lassmann, H., Brück, W., & Lucchinetti, C. F. (2007). The immunopathology of multiple sclerosis: An overview. Brain Pathology, 17(2), 210–218.

Layek, B., & Singh, J. (2013). Amino acid grafted chitosan for high performance gene delivery: Comparison of amino acid hydrophobicity on vector and polyplex characteristics. Biomacromolecules, 14(2), 485–494.

Leung, R., Proitsi, P., Simmons, A., Lunnon, K., Güntert, A., Kronenberg, D., & Lovestone, S. (2013). Inflammatory proteins in plasma are associated with severity of Alzheimer's disease. PLoS One, 8(6), e64971.

Liguori, I., Russo, G., Curcio, F., Bulli, G., Aran, L., Della-Morte, D., Gargiulo, G., Testa, G., Cacciatore, F., Bonaduce, D., & Abete, P. (2018). Oxidative stress, aging, and diseases. Clinical Interventions in Aging, 757–772.

Lisak, R. P., Nedelkoska, L., & Benjamins, J. A. (2014). Effects of dextromethorphan on glial cell function: Proliferation, maturation, and protection from cytotoxic molecules. GLIA, 62, 751–762.

Liu, P. P., Xie, Y., Meng, X. Y., & Kang, J. S. (2019). History and progress of hypotheses and clinical trials for Alzheimer's disease. Signal Transduction and Targeted Therapy, 4(1), 29.

Liu, Y., Qin, L., Li, G., Zhang, W., An, L., Liu, B., & Hong, J. S. (2003). Dextromethorphan protects dopaminergic neurons against inflammation-mediated degeneration through inhibition of microglial activation. Journal of Pharmacology and Experimental Therapeutics, 305(1), 212–218.

Machtoub, L., & Kasugai, Y. (2016). Amyotrophic lateral sclerosis: Advances and perspectives of neuronanomedicine

Maggi, P., Absinta, M., Sati, P., Perrotta, G., Massacesi, L., Dachy, B., Pot, C., Meuli, R., Reich, D. S., Filippi, M., Pasquier, R. D., & Théaudin, M. (2019). The "central vein sign" in patients with diagnostic "red flags" for multiple sclerosis: A prospective multicenter 3T study. Multiple Sclerosis, 26(4), 421–432.

Maiuri, T., Woloshansky, T., Xia, J., & Truant, R. (2013). The huntingtin N17 domain is a multifunctional CRM1 and Ran-dependent nuclear and cilial export signal. Human Molecular Genetics, 22(7), 1383–1394.

Malamud, D. (2011). Saliva as a diagnostic fluid. Dental Clinics, 55(1), 159–178.

Marcora, E., Gowan, K., & Lee, J. E. (2003). Stimulation of NeuroD activity by huntingtin and huntingtin-associated proteins HAP1 and MLK2. Proceedings of the National Academy of Sciences, 100(16), 9578–9583.

Martínez, M. A. M., Olsson, B., Bau, L., Matas, E., Calvo, A. C., Andreasson, U., Blennow, K., Romero-Pinel, L., Martínez-Yélamos, S., & Zetterberg, H. (2015). Glial and neuronal markers in cerebrospinal fluid predict progression in multiple sclerosis. Multiple Sclerosis, 21(5), 550–561.

Masrori, P., & Van Damme, P. (2020). Amyotrophic lateral sclerosis: A clinical review. European Journal of Neurology, 27(10), 1918–1929.

McGinley, M. P., Goldschmidt, C. H., & Rae-Grant, A. D. (2021). Diagnosis and treatment of multiple sclerosis: A review. JAMA, 325(8), 765–779.

McQuillen, P. S., & Ferriero, D. M. (2004). Selective vulnerability in the developing central nervous system. Pediatric Neurology, 30(4), 227–235.

Melo, A., Monteiro, L., Lima, R. M., de Oliveira, D. M., de Cerqueira, M. D., & El-Bachá, R. S. (2011). Oxidative stress in neurodegenerative diseases: Mechanisms and therapeutic perspectives. Oxidative Medicine and Cellular Longevity, 2011, 467180.

Meuth, S. G., Simon, O. J., Grimm, A., Melzer, N., Herrmann, A. M., Spitzer, P., Landgraf, P., Wiendl, H. (2008). CNS inflammation and neuronal degeneration is aggravated by impaired CD200–CD200R-mediated macrophage silencing. Journal of Neuroimmunology, 194, 62–69.

Miller, D. H., Barkhof, F., Frank, J. A., Parker, G. J., & Thompson, A. J. (2002). Measurement of atrophy in multiple sclerosis: Pathological basis, methodological aspects and clinical relevance. Brain, 125(8), 1676–1695.

Miller, J. P., Holcomb, J., Al-Ramahi, I., De Haro, M., Gafni, J., Zhang, N., Kim, E., Sanhueza, M., Torcassi, C., Kwak, S., & Botas, J. (2010). Matrix metalloproteinases are modifiers of huntingtin proteolysis and toxicity in Huntington's disease. Neuron, 67(2), 199–212.

Minagar, A., Barnett, M. H., Benedict, R. H., Pelletier, D., Pirko, I., Sahraian, M. A., Frohman, E., & Zivadinov, R. (2013). The thalamus and multiple sclerosis: Modern views on pathologic, imaging, and clinical aspects. Neurology, 80(2), 210–219.

Minakawa, E. N., Wada, K., & Nagai, Y. (2019). Sleep disturbance as a potential modifiable risk factor for Alzheimer's disease. International Journal of Molecular Sciences, 20(4), 803.

Mistry, N., Abdel-Fahim, R., Samaraweera, A., Mougin, O., Tallantyre, E., Tench, C., Jaspan, T., Morris, P., Morgan, P. S., & Evangelou, N. (2015). Imaging central veins in brain lesions with 3-T T2*-weighted magnetic resonance imaging differentiates multiple sclerosis from microangiopathic brain lesions. Multiple Sclerosis, 22(10), 1289–1296.

Mittal, S., Bjørnevik, K., Im, D. S., Flierl, A., Dong, X., Locascio, J. J., Abo, K. M., Long, E., Jin, M., Xu, B., & Xiang, Y. K. (2017). β2-adrenoreceptor is a regulator of the α-synuclein gene driving risk of Parkinson's disease. Science, 357(6354), 891–898.

Monti, D. A., Zabrecky, G., Kremens, D., Liang, T. W., Wintering, N. A., Cai, J., Wei, X., Bazzan, A. J., Zhong, L., Bowen, B., & Intenzo, C. M. (2016). N-acetyl cysteine may support dopamine neurons in Parkinson's disease: Preliminary clinical and cell line data. PLoS ONE, 11(6), e0157602.

Monti, D. A., Zabrecky, G., Kremens, D., Liang, T. W., Wintering, N. A., Bazzan, A. J., Zhong, L., Bowens, B. K., Chervoneva, I., Intenzo, C., & Newberg, A. B. (2019). N-acetyl cysteine is associated with dopaminergic improvement in Parkinson's disease. Clinical Pharmacology & Therapeutics, 106(4), 884–890.

Mota, S. I., Costa, R. O., Ferreira, I. L., Santana, I., Caldeira, G. L., Padovano, C., Fonseca, A. C., Baldeiras, I., Cunha, C., Letra, L., & Oliveira, C., & Rego, A. C. (2015). Oxidative stress involving changes in Nrf2 and ER stress in early stages of Alzheimer's disease. Biochimica et Biophysica Acta (BBA)-Molecular Basis of Disease, 1852(7), 1428–1441.

Musgrove, R. E., Helwig, M., Bae, E. J., Aboutalebi, H., Lee, S. J., Ulusoy, A., & Di Monte, D. A. (2019). Oxidative stress in vagal neurons promotes parkinsonian pathology and intercellular α-synuclein transfer. Journal of Clinical Investigation, 129(9), 3738–3753.

Nasir, J., Floresco, S. B., O'Kusky, J. R., Diewert, V. M., Richman, J. M., Zeisler, J., Borowski, A., Marth, J. D., Phillips, A. G., & Hayden, M. R. (1995). Targeted disruption of the Huntington's disease gene results in embryonic lethality and behavioral and morphological changes in heterozygotes. Cell, 81, 811–823.

Nelson, P. T., Head, E., Schmitt, F. A., Davis, P. R., Neltner, J. H., Jicha, G. A., Abner, E. L., Smith, C. D., Van Eldik, L. J., Kryscio, R. J., & Scheff, S. W. (2011). Alzheimer's disease is not "brain aging": Neuropathological, genetic, and epidemiological human studies. Acta Neuropathologica, 121, 571–587.

Norgren, N., Sundström, P., Svenningsson, A., Rosengren, L., Stigbrand, T., & Gunnarsson, M. (2004). Neurofilament and glial fibrillary acidic protein in multiple sclerosis. Neurology, 63(9), 1586–1590.

Obermeier, B., Lovato, L., Mentele, R., Brück, W., Forne, I., Imhof, A., Lottspeich, F., Turk, K. W., Willis, S. N., Wekerle, H., Hohlfeld, R., Hafler, D. A., O'Connor, K. C., & Dornmair, K. (2011). Related B cell clones that populate the CSF and CNS of patients with multiple sclerosis produce CSF immunoglobulin. Journal of Neuroimmunology, 233(1), 245–248.

Obrador, E., Salvador, R., López-Blanch, R., Jihad-Jebbar, A., Vallés, S. L., & Estrela, J. M. (2020). Oxidative stress, neuroinflammation and mitochondria in the pathophysiology of amyotrophic lateral sclerosis. Antioxidants, 9(9), 901.

Ohashi, Y., Tashiro, K., Itoyama, Y., Nakano, I., Sobue, G., Nakamura, S., Sumino, S., & Yanagisawa, N. (2001). Study of functional rating scale for amyotrophic lateral sclerosis: Revised ALSFRS (ALSFRS-R) Japanese version. No to Shinkei= Brain and Nerve, 53(4), 346–355.

Okello, E. J., & Mather, J. (2020). Comparative kinetics of acetyl- and butyryl-cholinesterase inhibition by green tea catechinsl relevance to the symptomatic treatment of Alzheimer's disease. Nutrients, 12(4), 1090.

Pansarasa, O., Bordoni, M., Diamanti, L., Sproviero, D., Gagliardi, S., & Cereda, C. (2018). SOD1 in amyotrophic lateral sclerosis:"Ambivalent" behavior connected to the disease. International Journal of Molecular Sciences, 19(5), 1345.

Pardo, G., & Jones, D. E. (2017). The sequence of disease-modifying therapies in relapsing multiple sclerosis: Safety and immunologic considerations. Journal of Neurology, 264(12), 2351–2374.

Patel, D., Steinberg, J., & Patel, P. (2018). Insomnia in the elderly: A review. Journal of Clinical Sleep Medicine, 14(6), 1017–1024.

Paulsen, J. S., Langbehn, D. R., Stout, J. C., Aylward, E., Ross, C. A., Nance, M., Guttman, M., Johnson, S., MacDonald, M., Beglinger, L. J., & Duff, K. (2008). Detection of Huntington's disease decades before diagnosis: The Predict-HD study. Journal of Neurology, Neurosurgery & Psychiatry, 79(8), 874–880.

Paumier, K. L., Luk, K. C., Manfredsson, F. P., Kanaan, N. M., Lipton, J. W., Collier, T. J., Steece-Collier, K., Kemp, C. J., Celano, S., Schulz, E., & Sandoval, I. M. (2015). Intrastriatal injection of pre-formed mouse α-synuclein fibrils into rats triggers α-synuclein pathology and bilateral nigrostriatal degeneration. Neurobiology of Disease, 82, 185–199.

Peng, C., Trojanowski, J. Q., & Lee, V. M. Y. (2020). Protein transmission in neurodegenerative disease. Nature Reviews Neurology, 16(4), 199–212.

Petersen, R. C. (2018). How early can we diagnose Alzheimer disease (and is it sufficient)? The 2017 Wartenberg lecture. Neurology, 91(9), 395–402.

Petrov, D., Daura, X., & Zagrovic, B. (2016). Effect of oxidative damage on the stability and dimerization of superoxide dismutase 1. Biophysical Journal, 110(7), 1499–1509.

Postuma, R. B., Berg, D., Stern, M., Poewe, W., Olanow, C. W., Oertel, W., Obeso, J., Marek, K., Litvan, I., Lang, A. E., & Halliday, G. (2015). MDS clinical diagnostic criteria for Parkinson's disease. Movement Disorders, 30(12), 1591–1601.

Prince, M., Bryce, R., Albanese, E., Wimo, A., Ribeiro, W., & Ferri, C. P. (2013). The global prevalence of dementia: A systematic review and metaanalysis. Alzheimer's & Dementia, 9(1), 63–75.

Rahman, M. M., Ferdous, K. S., & Ahmed, M. (2020). Emerging promise of nanoparticle-based treatment for Parkinson's disease. Biointerface Research in Applied Chemistry, 10, 7135–7151.

Rahman, M. M., Islam, M. R., Islam, M. T., Harun-Or-Rashid, M., Islam, M., Abdullah, S., Uddin, M. B., Das, S., Rahaman, M. S., Ahmed, M., & Alhumaydhi, F. A. (2022). Stem cell transplantation therapy and neurological disorders: Current status and future perspectives. Biology, 11(1), 147.

Ramesh, S., Govindarajulu, M., Jones, E., Knowlton, S., Weeks, L., Suppiramaniam, V., & Dhanasekaran, M. (2018). Current and Novel Biomarkers for Alzheimer's Disease. Reno, NV: MedDocs Publishers LLC, pp. 1–11.

Ransohoff, R. M. (2016). How neuroinflammation contributes to neurodegeneration. Science, 353(6301), 777–783.

Reyes, R. C., Cittolin-Santos, G. F., Kim, J. E., Won, S. J., Brennan-Minnella, A. M., Katz, M., Glass, G. A., & Swanson, R. A. (2016). Neuronal glutathione content and antioxidant capacity can be normalized in situ by N-acetyl cysteine concentrations attained in human cerebrospinal fluid. Neurotherapeutics, 13, 217–225.

Richardson, K., Allen, S. P., Mortiboys, H., Grierson, A. J., Wharton, S. B., Ince, P. G., Shaw, P. J., & Heath, P. R. (2013). The effect of SOD1 mutation on cellular bioenergetic profile and viability in response to oxidative stress and influence of mutation-type. PLoS ONE, 8(6), e68256.

Richardson, R. M., Varenika, V., Forsayeth, J. R., & Bankiewicz, K. S. (2009). Future applications: Gene therapy. Neurosurgery Clinics, 20(2), 219–224.

Rieux, A., Fievez, V., Garinot, M., Schneider, Y. J., & Préat, V. (2006). Nanoparticles as potential oral delivery systems of proteins and vaccines: A mechanistic approach. Journal of Controlled Release, 116(1), 1–27.

Roberts, J. S., Patterson, A. K., & Uhlmann, W. R. (2020). Genetic testing for neurodegenerative diseases: Ethical and health communication challenges. Neurobiology of Disease, 141, 104871.

Rocca, M. A., Battaglini, M., Benedict, R. H., De Stefano, N., Geurts, J. J., Henry, R. G., Horsfield, M. A., Jenkinson, M., Pagani, E., & Filippi, M. (2017). Brain MRI atrophy quantification in MS: From methods to clinical application. Neurology, 88(4), 403–413.

Roth, T. (2007). Insomnia: Definition, prevalence, etiology, and consequences. Journal of Clinical Sleep Medicine, 3(5 Suppl.), S7–S10.

Saba, J., Couselo, F. L., Bruno, J., Carniglia, L., Durand, D., Lasaga, M., & Caruso, C. (2022). Neuroinflammation in Huntington's disease: A starring role for astrocyte and microglia. Current Neuropharmacology, 20(6), 1116–1143.

Sadeghmousavi, S., Eskian, M., Rahmani, F., & Rezaei, N. (2020). The effect of insomnia on development of Alzheimer's disease. Journal of Neuroinflammation, 17(1), 1–20.

Santens, P., Boon, P., Van Roost, D., & Caemaert, J. (2003). The pathophysiology of motors symptoms in Parkinson's disease. Acta Neurologica Belgica, 103(3), 129–134.

Sastre-Garriga, J., Pareto, D., & Rovira, À. (2017). Brain atrophy in multiple sclerosis: Clinical relevance and technical aspects. Neuroimaging Clinics, 27(2), 289–300.

Sati, P., Oh, J., Constable, R. T., Evangelou, N., Guttmann, C. R. G., Henry, R. G., Klawiter, E. C., Mainero, C., Massacesi, L., McFarland, H., Nelson, F., Ontaneda, D., Rauscher, A., Rooney, W. D., Samaraweera, A. P. R., Shinohara, R. T., Sobel, R. A., Solomon, A. J., Treaba, C. A., & NAIMS Cooperative. (2016). The central vein sign and its clinical evaluation for the diagnosis of multiple sclerosis: A consensus statement from the North American Imaging in Multiple Sclerosis Cooperative. Nature Reviews Neurology, 12(12), 714–722.

Scheltens, P., De Strooper, B., Kivipelto, M., Holstege, H., Chételat, G., Teunissen, C. E., Cummings, J., & van der Flier, W. M. (2021). Alzheimer's disease. Lancet, 397(10284), 1577–1590.

Schneider, J. A., Arvanitakis, Z., Leurgans, S. E., & Bennett, D. A. (2009). The neuropathology of probable Alzheimer disease and mild cognitive impairment. Annals of Neurology: Official Journal of the American Neurological Association and the Child Neurology Society, 66(2), 200–208.

Schutte-Rodin, S., Broch, L., Buysse, D., Dorsey, C., & Sateia, M. (2008). Clinical guideline for the evaluation and management of chronic insomnia in adults. Journal of Clinical Sleep Medicine, 4(5), 487–504.

Schwalfenberg, G. K. (2021). N-acetylcysteine: A review of clinical usefulness (an old drug with new tricks). Journal of Nutrition and Metabolism, 2021.

Sevigny, J., Chiao, P., Bussière, T., Weinreb, P. H., Williams, L., Maier, M., Dunstan, R., Salloway, S., Chen, T., Ling, Y., & O'Gorman, J. (2016). The antibody aducanumab reduces Aβ plaques in Alzheimer's disease. Nature, 537(7618), 50–56.

Shamim, S. A., Warriach, Z. I., Tariq, M. A., Rana, K. F., & Malik, B. H. (2019). Insomnia: Risk factor for neurodegenerative diseases. Cureus, 11(10).

Sharma, D., & Singh, J. (2020). Long-term glycemic control and prevention of diabetes complications in vivo using oleic acid-grafted-chitosan-zinc-insulin complexes incorporated in thermosensitive copolymer. Journal of Controlled Release, 323, 161–178.

Sharma, S., Singh, S., Sharma, V., Singh, V. P., & Deshmukh, R. (2015). Neurobiology of l-DOPA induced dyskinesia and the novel therapeutic strategies. Biomedicine & Pharmacotherapy, 70, 283–293.

Sheng, J. G., Jones, R. A., Zhou, X. Q., McGinness, J. M., Van Eldik, L. J., Mrak, R. E., & Griffin, W. S. T. (2001). Interleukin-1 promotion of MAPK-p38 overexpression in experimental animals and in Alzheimer's disease: Potential significance for tau protein phosphorylation. Neurochemistry International, 39(5–6), 341–348.

Shohag, S., Akhter, S., Islam, S., Sarker, T., Sifat, M. K., Rahman, M., & Sharma, R. (2022). Perspectives on the molecular mediators of oxidative stress and antioxidant strategies in the context of neuroprotection and neurolongevity: An extensive review. Oxidative Medicine and Cellular Longevity, 2022.

Shoshan-Barmatz, V., Nahon-Crystal, E., Shteinfer-Kuzmine, A., & Gupta, R. (2018). VDAC1, mitochondrial dysfunction, and Alzheimer's disease. Pharmacological Research, 131, 87–101.

Siemers, E. R., Sundell, K. L., Carlson, C., Case, M., Sethuraman, G., Liu-Seifert, H., & Demattos, R. (2016). Phase 3

solanezumab trials: Secondary outcomes in mild Alzheimer's disease patients. Alzheimer's & Dementia, 12(2), 110–120.
Silva, G. A. (2007). Nanotechnology approaches for drug and small molecule delivery across the blood brain barrier. Surgical Neurology, 67(2), 113–116.
Simpson, S., Jr., Wang, W., Otahal, P., Blizzard, L., van der Mei, I. A. F., & Taylor, B. V. (2019). Latitude continues to be significantly associated with the prevalence of multiple sclerosis: An updated meta-analysis. Journal of Neurology, Neurosurgery, and Psychiatry, 90(11), 1193–1200.
Singh, A., Kukreti, R., Saso, L., & Kukreti, S. (2019). Oxidative stress: A key modulator in neurodegenerative diseases. Molecules, 24(8), 1583.
Sinnecker, T., Clarke, M. A., Meier, D., Enzinger, C., Calabrese, M., De Stefano, N., Pitiot, A., Giorgio, A., Schoonheim, M. M., Paul, F., Pawlak, M. A., Schmidt, R., Kappos, L., Montalban, X., Rovira, À., Evangelou, N., Wuerfel, J., & MAGNIMS Study Group. (2019). 22 of 25 MEY et al., 76(12), 1446–1456.
Slattery, J., Kumar, N., Delhey, L., Berk, M., Dean, O., Spielholz, C., & Frye, R. (2015). Clinical trials of N-acetylcysteine in psychiatry and neurology: A systematic review. Neuroscience & Biobehavioral Reviews, 55, 294–321.
Solfrizzi, V., D'Introno, A., Colacicco, A. M., Capurso, C., Todarello, O., Pellicani, V., Capurso, S. A., Pietrarossa, G., Santamato, V., Capurso, A., & Panza, F. (2006). Circulating biomarkers of cognitive decline and dementia. Clinica Chimica Acta, 364(1–2), 91–112.
Su, L., Blamire, A. M., Watson, R., He, J., Hayes, L., & O'Brien, J. T. (2016). Whole-brain patterns of 1H-magnetic resonance spectroscopy imaging in Alzheimer's disease and dementia with Lewy bodies. Translational Psychiatry, 6(8), e877.
Tabas, I., & Glass, C. K. (2013). Anti-inflammatory therapy in chronic disease: Challenges and opportunities. Science, 339, 166–172.
Tai, Y. F., Pavese, N., Gerhard, A., Tabrizi, S. J., Barker, R. A., Brooks, D. J., & Piccini, P. (2007). Imaging microglial activation in Huntington's disease. Brain Research Bulletin, 72(2–3), 148–151.
Takano, H., & Gusella, J. F. (2002). The predominantly HEAT-like motif structure of huntingtin and its association and coincident nuclear entry with dorsal, an NF-kB/Rel/dorsal family transcription factor. BMC Neuroscience, 3, 1–13.
Tallantyre, E. C., Dixon, J. E., Donaldson, I., Owens, T., Morgan, P. S., Morris, P. G., & Evangelou, N. (2011). Ultra-high-field imaging distinguishes MS lesions from asymptomatic white matter lesions. Neurology, 76(6), 534–539.
Tang, S. W., Helmeste, D. M., & Leonard, B. E. (2017). Neurodegeneration, neuroregeneration, and neuroprotection in psychiatric disorders. Neuroprogression in Psychiatric Disorders, 31, 107–123.
Tesseur, I., Zou, K., Esposito, L., Bard, F., Berber, E., Van Can, J., Lin, A. H., Crews, L., Tremblay, P., Mathews, P., & Mucke, L. (2006). Deficiency in neuronal TGF-β signaling promotes neurodegeneration and Alzheimer's pathology. Journal of Clinical Investigation, 116(11), 3060–3069.
Thakur, A., Roy, A., Chatterjee, S., Chakraborty, P., Bhattacharya, K., & Mahata, P. P. (2015). Recent Trends in Targeted Drug Delivery. SMGroup.
Thompson, A. J., Banwell, B. L., Barkhof, F., Carroll, W. M., Coetzee, T., Comi, G., Correale, J., Fazekas, F., Filippi, M., Freedman, M. S., Fujihara, K., Galetta, S. L., Hartung, H. P., Kappos, L., Lublin, F. D., Marrie, R. A., Miller, A. E., Miller, D. H., Montalban, X., & Cohen, J. A. (2018). Diagnosis of multiple sclerosis: 2017 revisions of the McDonald criteria. Lancet Neurology, 17(2), 162–173.
Thomson, A. J., Baranzini, S. E., Geurts, J., Hemmer, B., & Ciccarelli, O. (2018). Multiple sclerosis. Lancet, 391, 1622–1636.
Tremblay, C., Martel, P. D., & Frasnelli, J. (2017). Trigeminal system in Parkinson's disease: A potential avenue to detect Parkinson-specific olfactory dysfunction. Parkinsonism & Related Disorders, 44, 85–90.
Ueberham, U., Ueberham, E., Gruschka, H., & Arendt, T. (2006). Altered subcellular location of phosphorylated Smads in Alzheimer's disease. European Journal of Neuroscience, 24(8), 2327–2334.
Valadão, P. A. C., Santos, K. B. S., e Vieira, T. H. F., e Cordeiro, T. M., Teixeira, A. L., Guatimosim, C., & de Miranda, A. S. (2020). Inflammation in Huntington's disease: A few new twists on an old tale. Journal of Neuroimmunology, 348, 577380.
Van Assche, R., Temmerman, L., Dias, D. A., Boughton, B., Boonen, K., Braeckman, B. P., Schoofs, L., & Roessner, U. (2015). Metabolic profiling of a transgenic Caenorhabditis elegans Alzheimer model. Metabolomics, 11, 477–486.
van der Burg, J. M., Björkqvist, M., & Brundin, P. (2009). Beyond the brain: Widespread pathology in Huntington's disease. Lancet Neurology, 8(8), 765–774.
Voet, S., Srinivasan, S., Lamkanfi, M., & van Loo, G. (2019). Inflammasomes in neuroinflammatory and neurodegenerative diseases. EMBO Molecular Medicine, 11(6), e10248.
Walker, L., McAleese, K. E., Erskine, D., & Attems, J. (2019). Neurodegenerative diseases and ageing. Biochemistry and Cell Biology of Ageing: Part II Clinical Science, 75–106.
Walsh, S., Merrick, R., Milne, R., & Brayne, C. (2021). Aducanumab for Alzheimer's Disease? BMJ, 374.
Walton, C., King, R., Rechtman, L., Kaye, W., Leray, E., Marrie, R. A., Robertson, N., La Rocca, N., Uitdehaag, B., van der Mei, I., Wallin, M., Helme, A., Angood Napier, C., Rijke, N., & Baneke, P. (2020). Rising prevalence of multiple sclerosis worldwide: Insights from the Atlas of MS, third edition. Multiple Sclerosis, 26(14), 1816–1821.
Wang, D., Tai, P. W., & Gao, G. (2019a). Adeno-associated virus vector as a platform for gene therapy delivery. Nature Reviews Drug Discovery, 18(5), 358–378.
Wang, W., Zhao, F., Ma, X., Perry, G., & Zhu, X. (2020). Mitochondria dysfunction in the pathogenesis of Alzheimer's disease: Recent advances. Molecular Neurodegeneration, 15, 1–22.
Wang, X. L., Li, C. J., Xing, Y., Yang, Y. H., & Jia, J. P. (2015). Hypervalinemia and hyperleucine-isoleucinemia caused by mutations in the branched-chain-amino-acid aminotransferase gene. Journal of Inherited Metabolic Disease, 38, 855–861.
Wang, Z., Bai, Z., Qin, X., & Cheng, Y. (2019b). Aberrations in oxidative stress markers in amyotrophic lateral sclerosis: A systematic review and meta-analysis. Oxidative Medicine and Cellular Longevity, 2019.
Wild, E., Magnusson, A., Lahiri, N., Krus, U., Orth, M., Tabrizi, S. J., & Björkqvist, M. (2011). Abnormal peripheral chemokine profile in Huntington's disease. PLoS Currents, 3, RRN1231.
Wilkinson, K., & Shapiro, C. (2012). Nonrestorative sleep: Symptom or unique diagnostic entity? Sleep Medicine, 13(6), 561–569.
Winger, R. C., & Zamvil, S. S. (2016). Antibodies in multiple sclerosis oligoclonal bands target debris. Proceedings of the National Academy of Sciences of the United States of America, 113(28), 7696–7698.

Wu, J., Lim, E. C., Nadkarni, N. V., Tan, E. K., & Kumar, P. M. (2019). The impact of levodopa therapy-induced complications on quality of life in Parkinson's disease patients in Singapore. Scientific Reports, 9(1), 1–6.

Wyss-Coray, T. (2016). Ageing, neurodegeneration and brain rejuvenation. Nature, 539(7628), 180–186.

Xie, H., Hou, S., Jiang, J., Sekutowicz, M., Kelly, J., & Bacskai, B. J. (2013). Rapid cell death is preceded by amyloid plaque-mediated oxidative stress. Proceedings of the National Academy of Sciences, 110(19), 7904–7909.

Xue, Y., Wang, N., Zeng, Z., Huang, J., Xiang, Z., & Guan, Y. Q. (2020). Neuroprotective effect of chitosan nanoparticle gene delivery system grafted with acteoside (ACT) in Parkinson's disease models. Journal of Materials Science & Technology, 43, 197–207.

Yadav, S. C., Kumari, A., & Yadav, R. (2011). Development of peptide and protein nanotherapeutics by nanoencapsulation and nanobioconjugation. Peptides, 32(1), 173–187.

Yoo, J., Park, C., Yi, G., Lee, D., & Koo, H. (2019). Active targeting strategies using biological ligands for nanoparticle drug delivery systems. Cancers, 11(5), 640.

Zesiewicz, T. A., Sullivan, K. L., & Hauser, R. A. (2006). Nonmotor symptoms of Parkinson's disease. Expert Review of Neurotherapeutics, 6, 1811–1822.

Zhao, X. E., He, Y., Li, M., Chen, G., Wei, N., Wang, X., Sun, J., Zhu, S., & You, J. (2017). Analysis of amino acid and monoamine neurotransmitters and their metabolites in rat urine of Alzheimer's disease using in situ ultrasound-assisted derivatization dispersive liquid-liquid microextraction with UHPLC–MS/MS. Journal of Pharmaceutical and Biomedical Analysis, 135, 186–198.

Zhu, F. D., Hu, Y. J., Yu, L., Zhou, X. G., Wu, J. M., Tang, Y., & Wu, A. G. (2021). Nanoparticles: A hope for the treatment of inflammation in CNS. Frontiers in Pharmacology, 12, 683935.

Zimmermann, M., & Brockmann, K. (2022). Blood and cerebrospinal fluid biomarkers of inflammation in Parkinson's disease. Journal of Parkinson's Disease, 1–18.

Zivadinov, R., Dwyer, M. G., & Bergsland, N. (2016a). Brain atrophy measurements should be used to guide therapy monitoring in MS–YES. Multiple Sclerosis Journal, 22(12), 1522–1524.

Zivadinov, R., Jakimovski, D., Gandhi, S., Ahmed, R., Dwyer, M. G., Horakova, D., Weinstock-Guttman, B., Benedict, R. R., Vaneckova, M., Barnett, M., & Bergsland, N. (2016b). Clinical relevance of brain atrophy assessment in multiple sclerosis. Implications for its use in a clinical routine. Expert Review of Neurotherapeutics, 16(7), 777–793.

4 Challenges in Brain Targeting and Mechanisms of Drug Transfer across the BBB

Ahmad S. Kodous, Priya Ramanathan, Arunagiri Kuha Deva Magendhra Rao, and Gamal Abdel Nasser Atia*

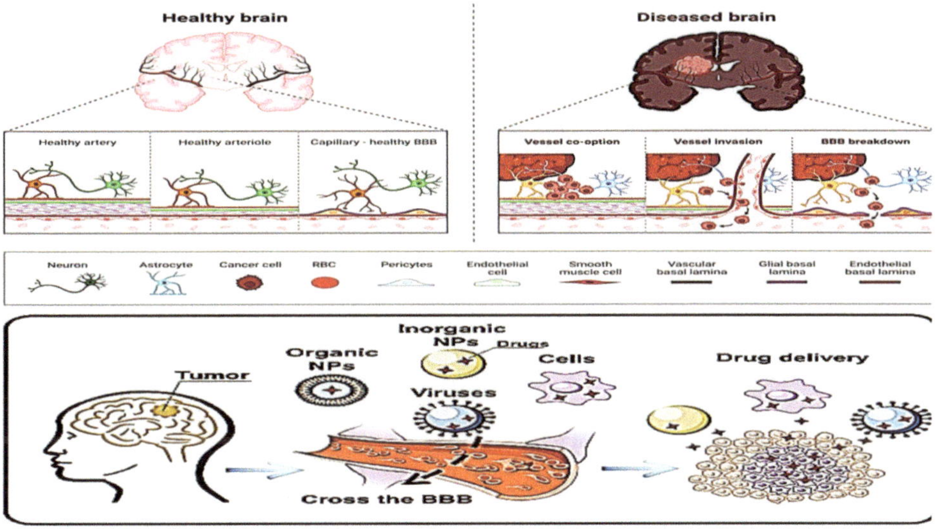

FIGURE 4.1A Challenges in brain targeting and mechanism of drug transfer across BBB.

4.1 INTRODUCTION

The brain is a component of the central nervous system (CNS), which is regarded as the most significant component of the human body as a whole and is sometimes called the body's command headquarters. The CNS's neurons are essential parts, and these neural networks are in charge of controlling neuronal signaling by using a variety of both biological and electrical impulses to control the ionic conditions at terminals and fibers [1]. An advanced biological system has a major "blood–brain barrier" (BBB) that was found, and given that term in the early 20th century, between the CNS and the outside portion of the body. The transit of numerous chemicals from the blood to the brain is rigorously regulated by the blood–brain barrier (BBB) [2, 3]. Due to the extremely poor BBB infiltration capacity of most medications, it is difficult to achieve therapeutic concentrations in the brain parenchyma for successful therapy [4, 5].

Different types of highly upregulated ligands and carriers on the BBB perform crucial functions in feeding the brain with vital nutrients and energy [6–8]. This physiological process gives receptor-mediated transcytosis (RMT) and carrier-mediated transportation (CMT) an intriguing prospective in the administration of drugs that target the brain [9, 10]. Due to their excellent qualities in improving medication durability and transferring and lowering side effects, among other things, nano-biomaterials have been used largely and widely as medication carriers [11, 12]. The use of nanotechnology in conjunction with RMT and CMT may increase BBB entry for systemic administration of drugs with a focus on the brain [13, 14]. Various ligands for these channels and carriers, as well as their messenger molecules, have been widely employed during the past 20 years to increase the effectiveness of delivery to the brain [15–17].

Nearly all brain disorders, including chronic Alzheimer's disease (AD) and Parkinson's disease (PD), acute ischemic stroke and traumatic brain injury (TBI), epileptic convulsions, and brain tumors, cause pathological BBB disruption with structural and functional abnormalities in afflicted

* Corresponding author: ahmadkmp11@gmail.com

DOI: 10.1201/9781003383376-4

brain areas [18, 19]. These vascular modifications frequently involve higher levels of luminal adhesion compounds, raised leukocyte adhesion and migration, decreased tight junction (TJ) expression, boosted endothelial transcytosis, affected receptor and transporter expression, endothelial degeneration, pericyte degeneration, perivascular retention of harmful substances, irritation, and an allergic reaction [20]. It is amazing how frequently BBB developments occur throughout the start and course of brain disorders [21]. It is still unclear, nevertheless, whether BBB failure is one of the fundamental factors that trigger brain disorders or only one of their aftereffects [22]. For instance, as the BBB is crucial in preserving brain homeostasis, it is becoming more and more apparent that its improper functioning may be a pathogen in the pathogenesis of neurological illnesses such as AD and PD [23].

On the other hand, a breach in the blood–brain barrier (BBB) enables the admission of potentially harmful blood-derived cells, particulates, and microbes into the brain and is related to pro-inflammatory and immunological reactions that may trigger a number of neurodegenerative processes [24]. Inflammatory cytokines and chemokines, on the other hand, have a role where junction proteins are broken down, downregulated, and distributed differently in space, which results in an impairment of TJ function. The accumulation and cerebral distribution of the majority of drugs are impacted by BBB evolutions or changes. A breakdown of the BBB brought on by disease may offer a way to treat neurons that are being impacted. Therefore, medication distribution throughout the damaged BBB is distinct from and more complicated than administration via the healthy BBB for brain disorders. This study aims to quickly describe the BBB developments in different brain illnesses and the associated approaches for developing nanoscopic drug delivery devices that target the brain.

4.2 ANATOMY, PHYSIOLOGY, AND MOLECULAR CONSTITUENTS OF THE BLOOD–BRAIN BARRIER

Capillary endothelial cells (ECs) make up the adult blood–brain barrier (BBB), which is made up of closely linked tight junctions (TJs) and adherens junctions (AJs) that block paracellular transit [25] and possess a diminished level of pinocytosis [26, 27], albeit only a small amount of transcellular transfer happens [28]. Additionally, pericytes, microglia, and closely related perivascular astrocytic end feet all have an impact on the BBB, as shown in Figure 4.1 [29].

The growth, immigration, and vascular sprouting of the brain endothelial cells are controlled by these several cellular lineages, which have crucial roles in the development and protection of the BBB. Additionally, the pericytes and endothelial cells are supported structurally by the basement membrane. The astrocyte's end feet's plasma membranes and the basal lamina are connected [30]. The transendothelial electrical resistance (TEER) of the BBB, which is roughly 1,500 $\Omega\cdot cm^2$, offers significant ion movement resistance [27], a 100-fold increase over peripheral micro-vessels [31, 32]. Tight junctions (TJs) and adherens junctions (AJs) link the capillary endothelial cells (ECs) of the brain to one another. The basal membrane that envelops the endothelial cells also covers the pericytes that join them. A crucial component of the barrier function is also provided by the astrocyte end feet that surround the brain micro-vessels. Microglia cells and linking neurons are other subtypes of supporting cells.

4.2.1 Endothelial Cells

BBB ECs differ from peripheral vasculature in that they contain more mitochondria, have less pinocytotic activity,

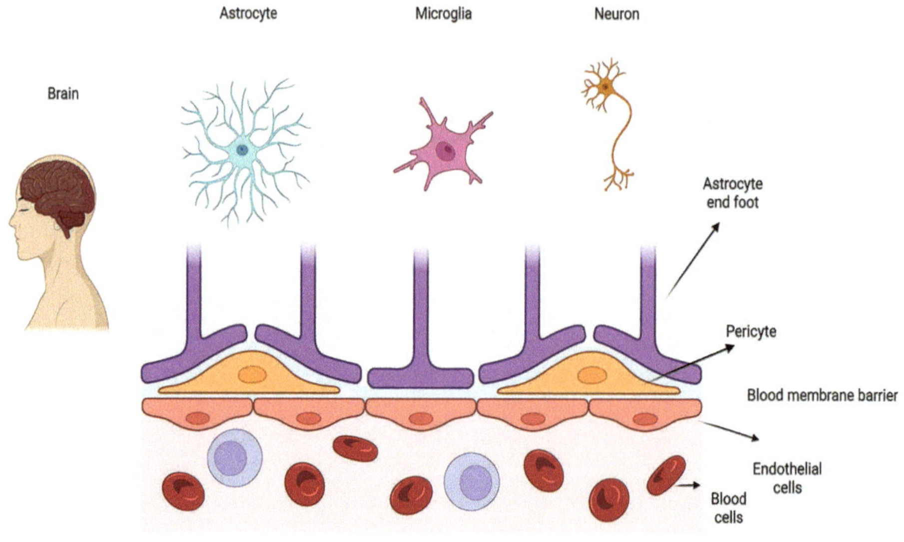

FIGURE 4.1 Blood–brain barrier building blocks: endothelial cells, basement membrane, pericytes, and astrocytes.

and do not have fenestrations [33–36]. Two intercellular molecular binding structures, the AJs and the TJs, provide the justification for the reduced paracellular permeability of the capillary EC layer. TJs are dynamic complexes containing several protein components, such as membrane-associated guanylate kinase (MAGUK)-like proteins (ZO-1, -2, and -3), occludin, claudins (i.e., claudin-1, -3, and -5), junctional adhesion molecules (JAMs), and occludin [30]. Vascular endothelium (VE), cadherin, actin, nectin, and catenin are some of the proteins that make up AJs [37]. Diverse transporter protein types are expressed on the apical and basolateral surfaces of ECs. The function of a barrier is also crucially provided by the astrocyte end feet that surround the brain micro-vessels. Microglia cells and interconnecting neurons are other subtypes of supporting cells.

4.2.2 Astrocytes

In order to preserve the homeostatic equilibrium within the brain microenvironment, astrocytes, which are glial cells, regulate the levels of neurotransmitters and ions. They also modulate synaptic communication, govern immunological responses [38], and communicate with endothelial cells via their end feet extensions, which ring the basolateral side of brain capillaries [39].

4.2.3 Pericytes

Pericytes control the following processes: (1) BBB integrity, which includes tight or adherens junctions and transcytosis across the BBB; (2) angiogenesis; (3) phagocytosis; (4) cerebral blood circulation; (5) neuroinflammation; and (6) multipotent stem cell behavior. Phagocytosis is the process of clearing harmful substances from the central nervous system [40].

4.2.4 Microglia

The major immunological cells of the brain, called microglia, are widely dispersed throughout the central nervous system (CNS) and become active in reaction to a variety of CNS pathosis, inflammatory processes, and trauma [35, 41–43].

4.3 BBB DYSFUNCTION

In a disordered state, BBB disruption and malfunction cause blood components to leak into the CNS, infiltrate different cells, and abnormally transfer and clear particles and ions that reduce and dysregulate cerebral blood flow (CBF), further leading to a variety of deficiencies in the brain [18].

4.3.1 Acute Neurological Diseases Due to BBB Dysfunction

4.3.1.1 Bacterial Meningitis

Bacterial meningitis is an irritation of the subarachnoid cavity and meninges triggered by a variety of microbial infections, including those by *Streptococcus pneumoniae* and *Neisseria meningitides* [44]. The spinal cord and cerebral cortex are also affected by the inflammation [45]. Bacterial invasion into circulation, survivability in the blood, entry into the subarachnoid region, and colonization are the four steps that all help germs move to the brain [46]. Pathogenic and host variables combine to cause the ensuing inflammation and CNS damage. Among the major symptoms of bacterial meningitis are migraine, pyrexia, cephalalgia, photophobia, and lack of awareness [47]. Bacteria enter the meninges mainly through the choroid plexus (CPs) and CNS barrier vessels. *H. influenzae*, which conforms to ECs, connects with the basolateral side of CP's ependymal monolayer, transcytoses to the cytoplasm, and is discharged in the ventricles [48].

The two primary pathogens that cause newborn meningitis are *E. coli* K1 and group B streptococcus (GBS), but they each have a unique way of invading the subarachnoid region. Cytotoxic necrotizing factor-1 (CNF1), invasion of brain endothelial cells (IBE), and other *E. coli* K1 factors facilitate bacterial invasion of ECs that activate RHO group GTPases [49]. Different host cell signaling proteins are triggered when bacterial substances communicate with host cell receptors [50]. Bacteria can adhere to host cells thanks to factors including outer-membrane protein A (OmpA), FimH, and the lipoprotein NLPI, which facilitate adhesion through N-acetyl glucosamine of the glycoprotein (GP96) [50]. The human nasopharynx is where *N. meningitidis* and *S. pneumoniae* naturally inhabit. From there, they go through the respiratory system, enter the circulation, and result in septicemia [51].

The ability of *N. meningitidis* to survive in the blood and resist destruction by host effector cells is due to its infectious nature. Pili type IV are necessary for adherence to ECs. *N. meningitidis* attaches to the CNS capillaries and effectively breaches the blood–brain barrier by causing cell–cell junctions to come apart [52]. Via the lungs, intravascular spaces found inside the tissues, or spaces of the inner ear, *Streptococcus pneumoniae* can enter the circulation. *S. pneumoniae* can enter the meninges through the olfactory neuron once it has become bloodborne. Pneumococcal surface proteins (PsP), pneumolysin, and capsule are pathogenic variables that support *S. pneumoniae* survival and growth in the blood and brain parenchyma. These components are also present in several other meningitis-causing bacteria [53]. *S. pneumoniae* that has been enveloped is more resistant to phagocytosis. PsP allows bacteria to adhere to ECs using pneumococcal pilus, in addition to PECAM-1, a molecule that adheres platelets to the endothelium, and plgR, a protein that facilitates translocation through the BBB [54]. Many meningeal diseases directly interfere with AJs or TJs, which compromise BBB integrity. A pore-forming toxin is directly secreted by GBS and *S. pneumoniae*, changing the integrity of the BBB. Nitric oxide generation from ECs is increased by *E. coli* K1, *S. pneumoniae*, and GBS, compromising BBB integrity [55]. Furthermore, pathogen-derived toxicities can have a deleterious influence on BBB integrity by increasing the generation of inflammation-related chemical messengers or mediators in reaction to contamination by the host [56].

4.3.1.2 Epilepsy

The erratic and uncontrolled activity of either the entire CNS or just a portion of it is what defines *epilepsy*. When CNS excitability exceeds a specific critical threshold, an epileptic patient has attacks [53]. Episodes and epilepsy vary primarily in that the former occur just once, while the second kind is characterized by two or more induced episodes [57]. There are three types of epilepsy: grand mal, petit mal, and focal mal. The whole brain experiences neuronal discharges as a symptom of grand mal epilepsy. The victim passes into unconsciousness, which lasts for three to four minutes. The entire brain may experience an increase in voltage and repetition of electrical impulses [58]. Muscle spasms around the head for 3–30 seconds are symptoms of petit mal epilepsy. The spike-and-dome rhythm can be used to show the cerebral waveform in petit mal epilepsy [59]. The cause of grand mal epilepsy, according to in vivo investigations, is thought to be excitatory and inhibitory thalamic and corticothalamic neurons oscillations [60]. A localized organic lesion or functional anomaly, like brain scar tissue that drags on neighboring neuronal tissue, a lesion that compresses the brain area, and congenitally faulty regional circulation, can cause focal mal epilepsy [58].

Intense psychological stimulation, severe injuries, and over breathing-induced alkalosis can all lead to epilepsy. In addition, patients have been shown to have increased microvascular density and altered GABAergic processes [61], TJ loss, IgG leak in the hippocampal resections [62], an instantaneous amnesic period [63], abrupt breathlessness, and apprehension fits of irrational wrath, jerks, and shock movements [64]. The incidences of seizures and BBB failure are categorically unrelated, and neither is neuronal loss. Pro-inflammatory molecules like TNF, IL-1, and high-mobility group box 1 (HMGB1) are produced and released by epileptogenic injuries and seizures, which lower the threshold for seizures. Seizures also result due to rapid shifts of glutamate and aminobutyric acid (GABA) sensor phosphorylation. Additionally, this results in channelopathies that alter intrinsic brain excitability [65]. BBB disruption can result from seizures as well. When the BBB is artificially opened, rat neuronal activity synchronizes, causing albumin and immunoglobulin G (IgG) neuropil eruptions. Neuronal hyperexcitability is caused by albumin altering astrocytes' ability to buffer K+ [66]. TGF produced by different cell types has an impact on BBB integrity as well. When tissue plasminogen activator (tPA) is suppressed by plasminogen activator inhibitor-1 (PAI-1), astrocytes produce TGF, which causes the BBB to close [67].

4.3.1.3 Traumatic Brain Injury (TBI)

Traumatic brain injury (TBI), often referred to as a result of brain damage or brain damage as a result of sudden trauma, damages the brain and interferes with its normal function, further leading to significant impairment and fatality. It often happens as a result of an intense blow or jolt to the cranium. When the head is hit, the CSF shields the brain against external influences and acts as a cushion. However, when the shock goes beyond the threshold, as in TBI, pro-inflammatory protein concentrations of IL-6/8/10 rise in the CSF [68]. Consciousness loss, headaches, post-injury amnesia, seizures, blurred vision, migraines, dizziness, light sensitivity, anxiety, aggressiveness changes, memory blurs, insufficient control over impulses, poor decision-making, and post-traumatic stress disorder are common symptoms. The BBB limits the entrance of medicinal substances into the brain. However, when a brain injury occurs, broken micro-vessels cause the barrier to depolarize, allowing blood components to pass through the brain parenchyma [69].

The walls of these micro-vessels are further disrupted by primary damage brought on by dysfunction, such as neuro-inflammation and apoptosis, which starts a process of coagulation and results in secondary injury [70]. As a result of ischemia, which is brought on by intravascular coagulation that restricts blood flow to tissues, fibrinogen, thrombin, and albumin, among other blood-borne substances, can enter the brain [71]. Ischemia affects the blood's constituent parts and may set off inflammatory cascades that prevent ECs, ECM, pericytes, and astroglia cells from operating properly. Following a TBI, the BBB can become dysfunctional in two stages. In the first stage, micro-vessels may sustain shear damage shortly after the TBI. After three days of TBI, the activation of inflammatory cells is the next phase, which leads to the permeability changes in BBB. According to in vivo research, BBB disruption has a role in changing the neurovascular network's transcriptional processes. It also leads to the emergence of degenerative disorders, such as epilepsy, psychological issues, dementia, and Alzheimer's disease [70].

4.3.2 Chronic Neurological Diseases

4.3.2.1 Alzheimer's Disease (AD)

The increasing degeneration and loss of brain cells are side effects of AD, a brain condition. It causes a steady loss in behavioral, cognitive, and other social abilities, interfering with the individual's capacity for independence. Dementia, anxiety, memory loss, restlessness, exhaustion, and vertigo are a few symptoms of AD [70]. Numerous other genes, such as apolipoprotein (APOE4), presenilin-1 (PSEN1), presenilin-2 (PSEN2), microtubule-associated protein tau (MAPT), and amyloid-beta precursor protein (APP) are also connected with a larger or reduced likelihood of early and delayed onset of AD [72]. Neuronal loss, plaques, hyperphosphorylated tau neurofibrillary tangles (NFTs), and cerebrovascular dysfunction are all key factors in the pathogenesis of AD and cognitive impairment [70]. NFT is associated with chromosome 17q21 and can potentially be stimulated by the deposition of phosphorylated tau protein in the neuron. Tau is a protein that is connected to micro-tubules and facilitates axonal transport, which is required for neuronal signaling and trafficking [73]. A single tau molecule normally contains two to three phosphates; however, in tauopathy patients, the phosphoryl concentration is increased by many times. When tau protein is hyperphosphorylated, it separates from the microtubule, causing unbound microtubules to proliferate and phosphorylated tau protein to gradually build up until NFTs are formed [74]. On the contrary, PSEN1 and PSEN2 are the enzyme secretase's catalytic constituents [75]. It has been demonstrated that human PSEN1 mutations encourage BBB disintegration and cerebrovascular impairment [76]. Less than 5% of all instances of AD have PSEN2 mutations [77]. Additionally, APOE4 is one of the main genetic predisposing causes for sporadic and late-onset AD. APOE4 alleles double the risk of AD compared to APOE3 alleles and lead to toxicity in the neurons and the cerebrovascular system [78]. The APOE4 carriers also cause BBB breakdown and pericyte degeneration [79], early neurovascular dysfunction [80], and low glucose absorption through the BBB. Proteins like thrombin, fibrinogen, IgG, hemosiderin, and albumin have been shown to leak from vascular capillaries in the hippocampus, prefrontal cortex, and entorhinal cortex of AD patients. There are over 40 APP mutations, and proteins that co-localize with A [80] related to the development of AD have been found. This results in BBB disintegration, cerebral amyloid angiopathy (CAA), and cerebrovascular disease [81]. CAA is brought on by BBB disruption at the arterial and arteriolar levels and smooth muscle cell (SMC) vascular degeneration [82]. Additionally, APOE4 transporters speed up BBB thinning by activating the proinflammatory CypA-MMP-9 pathway. As a result, endothelial TJs and basement membrane proteins degrade, worsening the BBB's vulnerability [83].

4.3.2.2 Amyotrophic Lateral Disease (ALS)

Neurons in charge of voluntary motions become dysfunctional as a result of amyotrophic lateral sclerosis (ALS). It is also referred to as Lou Gehrig's disease, or motor neuron disease. Patients with ALS have trouble walking, speaking, breathing, and losing the use of their hands and arms. The typical time to survive after being diagnosed with ALS is around three years, and respiratory failure is frequently the cause of death [84]. Particular neurons impacted by ALS include the lower motor neuron (LMN) and the upper motor neuron (UMN). While the LMN originates from the brain stem to the skeletal muscles to produce motion, the UMN contains motor information extending from the cerebral cortex, or brain stem [85]. Either sporadic or familial ALS can occur. Around 90% of instances of ALS are sporadic, which is random, whereas familial ALS is inherited. Angiogenin (ANG), transactive response DNA-binding protein (TARDBP), optineurin (OPTN), and RNA-binding protein (FUS) are among the 15 genes that have dominant inheritance in 10% of instances [86]. Dyspnea (difficulties breathing), orthopnea (difficulties breathing while lying flat), a faint cough, being very sleepy, cognitive impairment, etc. are some of the disease's respiratory symptoms. Erythrocytes and plasma components cannot enter the central nervous system (CNS) due to the BBB and BSCB. Assemblages of proteins like fibrin and thrombin in the spinal cord or motor cortex of ALS patients have been linked to the collapse of these barriers [87]; decreased quantities of erythrocytes, hemosiderin, hemoglobin, and TJ proteins all contribute to the production of ROS, which is harmful to motor neurons. In ALS, there are higher amounts of adhesion agents, like vascular cell adhesion molecules (VCAM) and intercellular adhesion molecules (ICAM-1/2). Both types of this illness have been shown to have a lower expression of BSCB TJs proteins and an enhanced ratio of CSF to serum albumin [88]. Pericytes play a crucial role in preserving the blood–CNS boundaries, which is widely known.

4.3.2.3 Huntington's Disease (HD)

A degenerative ailment called *Huntington's disease* (HD) is described as muscular flicks that develop into severe bodily distortions. Along with physical dysfunctions, significant cognitive issues also manifest. The repeating 36 CAG sequence in exon 1 of the HTT gene results in mutant Huntington protein (mHtt) aggregation, leading to neurodegeneration, and is involved in the genetics of HD abnormalities [88]. It may be said that Huntington's is a polyglutamine illness, since CAG encodes for glutamine in the aberrant neural cell protein. A typical person has 10–35 CAG repeats, and a larger number of CAG repeats causes symptoms to appear in the person [89], also known as the putamen and caudate nucleus [89]. Chorea is characterized by involuntary jerky motions that affect the hip, shoulders, and face; dystonia, known to be with abnormal muscle tone that causes muscular spasm and incorrect posture); and cognitive loss. Behavioral issues are all indications of HD [90]. Greater BBB permeability has been seen in HD patients, and this has been linked to decreased expression claudin-5 and occludin involved in the development of BBB TJs IL-8 and metalloproteinase 1 tissue inhibitor [91, 92]. Seeing a 2.5-fold increase in the extravascular fibrin and BBB disruption in HD patients has been demonstrated [93] in contrast to controls. Recent research on HD patients using dynamic contrast-enhanced (DCE) MRI shows increased BBB permeability in the caudate nucleus in addition to a rise in gray matter cerebral blood levels.

4.3.2.4 Parkinson's Disease (PD)

The buildup of oligomeric-synuclein (-syn) and the subsequent degeneration of dopaminergic neurons in the substantia nigra pars compacta (SNpc) are hallmarks of Parkinson's disease (PD) [94]. Since the pars compacta follows a nigrostriatal network that aids in triggering and initiating movement in the cerebral cortex, the decline of pars compacta neurons eventually ends with a poor state of motion. Patients with PD struggle to do daily activities like walking, jogging, etc. because of muscle rigidity in their limbs. Dementia, dizziness, loss of postural reflexes, and diminished facial reflexes and cognitive issues are other symptoms. In PD patients, basal ganglia vascular dysfunction results in BBB degradation and malfunction. The MRI showed microbleeds and decreased xenobiotic active discharge, in addition to numerous possible poisons being observed by verapamil PET [95]. Additionally, greater 11C-verapamil uptake was seen in frontal white matter areas compared to controls [72]. Neurotoxic fibrinogen, thrombin, plasminogen, and RBC extravasation build up as a result of BBB degradation. ROS are produced by the release of Hb and Fe2+, which damage dopaminergic neurons. Pro-inflammatory cytokines secreted during neurodegeneration include IL-1, IFN-, and TNF, as well as different MMPs. Several recent investigations have emphasized the potential syn infiltration into the BBB and accumulation in the brain. LRP1-mediated transcytosis then facilitates additional clearance of the brain through the BBB. Leucine-rich repeat kinase 2 (LRRK2) missense mutations are linked to delayed onset of Parkinson's disease [96]. Furthermore, in vivo investigations emphasized mutations of MDR1 gene linked to both familial and sporadic PD. The strong expression signature of MDR1 at the BBB endothelium causes deregulation of ABCB1 (P-glycoprotein) in PD patients [93].

4.3.2.5 Multiple Sclerosis (MS)

A neurological condition called multiple sclerosis (MS) causes the BBB to become damaged, eventually allowing B cells, CD4+ T cells, and peripheral macrophages to enter the central nervous system (CNS). This sets off an inflammatory cascade that causes demyelination and axonal damage [97]. The brain, spinal cord, and optic nerves of the CNS are all impacted by MS's demyelination and inflammation of the nerves. The major histocompatibility complex (MHC) contains several genes that increase a person's susceptibility to MS. Leukocytes, especially T cells, have been thought to move to the BBB [97]. The symptoms involve tingling, numbness, and visual issues; changes in the optic nerve; sensory, bladder, and bowel function issues; as well as cognitive deficits. The hallmarks of MS include BBB collapse at an early stage, fibrinogen buildup, TJs downregulation, and deterioration of endothelium. BBB dysregulation and migration of activated leukocytes are early cerebrovascular complexities, which result in the production of inflammatory mediators and chemokines [98]. Vascular endothelial cadherin (VE-adherin), occludin, and phenotypes in AJs and TJs of cerebral endothelial cells (CECs) are altered by Th1 cytokines via numerous inflammatory genes [83]. No effective medication against MS has been created so far, since the majority of medicines are unable to get through the BBB. Nevertheless, researchers have found that dNP2 peptide, when combined with cytotoxic T cells, can pass through the BBB and effectively combat MS [99].

4.3.3 Autoimmune Neurological Diseases

Numerous illnesses, particularly autoimmune conditions linked to the CNS, have been found to have commonalities and to breach the BBB.

4.3.3.1 Susac's Syndrome

It is an uncommon condition that affects the CNS, causes hearing loss, and causes visual problems by occluding the branch retinal arteries (BRAO). Upon MRI examination, lesions are frequently seen. There are links between a number of neuropsychiatric illnesses, including personality disorders. Seizures, mood swings, disorientation, headaches, and other symptoms are seen. Additionally, there might be a total loss of eyesight, clouded vision, photopsia, etc. [100–102]. Typically, women are reported to have this condition more frequently. If this condition is detected early on, patients may recover [102, 103].

4.3.3.2 Chronic Inflammatory Demyelinating Polyneuropathy (CIDP)

Peripheral nerve demyelination occurs as a result of chronic inflammatory demyelinating polyneuropathy (CIDP), which primarily impacts the PNS. Autoantibodies generated towards the antigens expressed on the peripheral nerves are responsible for it [104, 105]. Most of these instances result in elevated CSF protein levels, which are more frequently seen in males. First affected are the lower limbs, and weakness makes standing up difficult [106]. According to the clinical symptoms, it has been further divided into various subtypes [105].

The autoimmune diseases include Fisher syndrome, Bickerstaff brain stem encephalitis (BBE), and Guillain–Barré syndrome. All have similar symptoms that influence both the CNS and PNS. Furthermore, given that all three of these illnesses contain the same autoantibodies, they may potentially share a common etiology [107]. Patients with FS, BBE, and GBS exhibit anti-ganglioside antibodies. Ataxia, hypo- or areflexia, ophthalmoplegia, and reflex weakness are frequently seen in FS patients. Reduced awareness, ataxia, and ophthalmoplegia are hallmarks of BBE. Consciousness disturbance is a characteristic of BBE that sets it apart from FS. The main symptoms of GBS are flexia and symmetrical limb weakness [108].

4.3.3.3 Cerebral Ischemia

Brain ischemia is brought on by the stimulation of neural inflammation, where blood vessel injury affects the design and functionality of the BBB [109]. The stroke is a type of cerebral ischemia in which the brain's restricted blood flow might cause neurological illnesses. Immune cells are activated by inflammation release cytokines and effector cells that restore tissue architecture and function. Due to BBB leakage brought on by excessive microglial cell activation, interleukins, interferons, and TNF-α are released, which may further lead to neuronal death [110–112].

4.3.3.4 Neuromyelitis Optica (NMO)

By creating lesions, neuromyelitis optica (NMO) harms both the spinal cord and the optic nerves. It is a condition of the autoimmune CNS. The individuals specifically exhibit serum IgG against aquaporin 4. When antibodies reach the central nervous system (CNS), they attach to astrocyte aquaporin 4, which activates the traditional complement pathway [113, 114]. It has been proposed that a number of elements that facilitate BBB access play a role in the development of such antibodies. Sjögren's syndrome and neuropsychiatric systemic lupus erythematosus (NPSLE) autoantibodies and self-reactive lymphocytes are developed as a result of NPSLE and Sjögren's disease. The symptoms of NPSLE, an inflammatory autoimmune illness, are extremely similar to those of other neuropsychiatric diseases. Seizures, aseptic meningitis, cognitive dysfunction, disorientation, and psychosis are only a few of the neuropsychiatric symptoms that are connected [115]. It frequently results in BBB malfunction, which contributes to the development of the illness. Additionally, immune system effectors such as cytokines and other antibodies that induce tissue injury and neuroinflammation also have contributing roles [116]. Sjögren's syndrome is an inflammatory condition that results in inflammatory reactions of the salivary and lacrimal glands, which, in turn, cause the mucosa to dry up [117].

4.3.3.5 CNS Vasculitis

The blood vessels in the CNS become inflamed and necrotic as a result of CNS vasculitis. C-reactive protein (CRP) and erythrocyte sedimentation rate (ESR) elevations are also associated with this condition, which is also connected to a number of other conditions, including arthritis, brain damage, and Sjögren's syndrome. Corticosteroids are used during therapy [118]. For the advancement of therapies for CNS illnesses and the transport of systemically delivered medications to the brain, the BBB presents a tremendous hurdle [71]. The BBB is believed to exclude adeno-associated viruses (AAV), which are under investigation for use as gene delivery vehicles, generic proteins and peptides, antibodies, and nearly all large-molecule therapies [119, 120]. It has been calculated that only 0.1% of antibody therapies given intravenously reach the brain, necessitating substantially greater circulation levels or prolonged administration to reach therapeutic amounts at their target, both of which are linked to a greater likelihood of systemic adverse effects [121]. Levodopa, a lipid-insoluble precursor of dopamine, serves as an excellent example of the need for specialized techniques to promote brain absorption. Although levodopa is converted to dopamine in the CNS to treat Parkinson's disease (PD), a greater systemic dose is necessary to reach therapeutic CNS concentrations. But this levodopa conversion results in nausea and vomiting in the circulatory system. Carbidopa, a decarboxylase inhibitor, is used along with levodopa to avoid this peripheral conversion and reduce levodopa adverse reactions since it is polar and cannot penetrate the blood–brain barrier [122]. The action of efflux carriers, such as Pgp for digoxin and cyclosporine A, can further diminish the therapeutic concentrations of lipophilic medicines in the brain [122]. Additionally, peptidases and cholinesterases are two enzymes that reduce the concentration of xenobiotics in the brain through metabolism [123]. Several brain disorders are characterized by a protracted preliminary period, a lack of indicators, an elevated level of unpredictability, and comorbidity, in addition to these difficulties in medication administration [124]. As a result, numerous neurological illnesses, including AD, only have symptomatic therapies available. This calls for greater expenditure on drug discovery and methods to get beyond the BBB's barrier to drug delivery. The latter methods may typically be divided into invasive and non-invasive categories, each of which has certain benefits and drawbacks.

4.4 BRAIN DRUG DELIVERY STRATEGIES FOR OVERCOMING THE BBB AND FOR CNS DISEASES

To get medications and diagnostic tools into the brain, great lengths have been taken. Several innovative tactics have been utilized in conjunction with current developments in the BBB investigation. The works released during the last five years are included in this review, as revealed in Table 4.1.

The BBB's transport mechanisms allow endogenous amino acids to reach the brain. This understanding is used to connect medications with amino acids with potential to cross the BBB in one appealing method of drug delivery to the brain. Peura et al. synthesized three amino acids, the pro-agents of dopamine, to improve absorption across through amino acid transporters. Utilizing a brain perfusion approach in rats, they tested three prodrugs. Compared to other prodrugs, phenylalanine demonstrated stronger receptor binding and absorption [140]. A methotrexate (MTX)–lysine combination was found to improve absorption across the BBB through the endogenous lysine transport pathway. The effective biodistribution and pharmacokinetics of the MTX–lysine conjugation showed that this prodrug was specifically transported to the brain [141]. This type of prodrug method, however, could only be appropriate for tiny compounds due to the size of amino acids. Amino acids may be too tiny to alter the absorption route of macromolecules like proteins and siRNAs, or macromolecules may be too large to transit via transportation networks, as shown in Table 4.1.

4.4.1 Structural Challenges

4.4.1.1 Viral Vectors

Nucleic acids can naturally infect cells using viral vectors. Over two decades have been spent researching the use of viral vectors for the administration of genes to individuals with neurological diseases. Viral vectors have high transfection efficiency of ~80% [142]. Gene transfer has been accomplished in the brain using lentivirus, herpes simplex virus, adenovirus, and adeno-associated virus (AAV) vectors. The difficulty in synthesis, the increased expenditure, and most significantly, the biocompatibility of viral vectors due to patient deaths in clinical trials are some of the drawbacks of employing viral vectors for drug

TABLE 4.1
Brain drug delivery strategies for overcoming the BBB and for CNS diseases

Strategies	Advantages	Limitations	Reference
Intrathecal administration	• High drug concentration in CSF	• Highly invasive • Poor patient compliance	[125]
Convention-enhanced non-invasive delivery	• Enhanced brain drug uptake and distribution by bulk flow • Potential open BBB	• Invasive • Difficulty in monitoring drug in vivo	[126]
Interstitial wafers and implant	• Sustained and controlled release	• Invasive • Drug dosage limited by implant size	[127]
Lipidization	• Non-invasive	• Fast clearance • Inhomogeneous drug	[128]
Substrates for RMT	• Non-invasive • Highly specific	• Off-targets side effects • Interference by endogenous ligands	[129]
Substrates for CMT	• Non-invasive • Highly specific	• Stereospecific • Limited to smaller cargoes	[130]
Viral-mediated (Viral vectors) BBB delivery	• Specific cellular tropism	• Highly invasive, brain direct injection • Risk of oncogenesis	[131]
Exosome-mediated BBB delivery	• Non-immunologic • Potential to cross BBB	• Loading procedures • Exosomes donor cells	[132]
Low-intensity ultrasound	• Positive bio-effect • Reproducible	• Need sophisticated equipment • Need precise mechanisms	[133]
Brain permeability enhancer (osmotic BBB opening)	• Open BBB transiently • Clinically applicable	• Neurotoxic • Nonselective	[134]
BBB active transporters delivery	• Potential to cross BBB by intravenous injection	• Mainly for small molecules	[135]
Nanoparticles	• Specific brain-targeting • delivery	• Crossing BBB	[136]
Permeable BBB delivery (disease conditions)	• Potential to cross BBB	• Limited knowledge about BBB dynamic changes	[137]
Intranasal delivery	• Non-invasive • Practical and effective	• Suitable only for low doses	[138]
Brain imaging/diagnostics nanoparticles	• Hyper-permeable BBB in case of disease conditions • Highly efficient brain imaging	• Limited knowledge about BBB dynamic changes and cross BBB	[139]

delivery [143, 144]. The secure use of viral vectors has to be established before they can be used in therapeutic settings. Though there is still some worry about immunogenicity, AVV vectors have so far been safe and capable of delivering genes to the brain [145]. In light of this, the AVV vector is widely used in ongoing clinical studies of genetic therapies for brain illnesses. In order to either mechanically or physiologically circumvent the BBB, a number of delivery methods, including stereotaxic and CSF injection, have been investigated [146]. Despite the fact that these ways of delivering medication are exceedingly specialized, patients face severe dangers from very invasive neurosurgery. For instance, 12 cortical sites were required for the delivery of AAV2CUhCLN2, which was used to treat late infantile neuronal ceroid lipofuscinosis, through six burr holes. In the clinical experiment, serious negative consequences were documented, although it was not obvious if those negative effects might be directly linked to the AAV2CUhCLN2 vector [147].

Several viral vectors are now being developed for the management of different brain illnesses; however, most methods still rely on direct brain injection [144]. Therefore, the primary goal in the creation of innovative viral vectors is safety and systemic distribution. Numerous adeno-associated virus (AAV) serotypes have shown the ability to target CNS cells while evading the BBB [23, 26, 27]. Foust et al. were the first to show that intravenous treatment of AAV9 at a high dosage of 2×10^4 vg/kg have resulted in widespread transduction of neurons and astrocytes in mice [28]. About nine recombinant AAVs (rAAVs) were evaluated for CNS transduction to determine the best AAVs to cross the BBB. Furthermore, in establishing the capacity of rAAVs to penetrate the CNS in newborn mice, rAAVrh.10 performed similarly to rAAV9 [26]. Mice with Huntington's illness were intravenously injected with AAV9-gfaABC1D-glutamate transporter 1 (GLT1) by Vagner et al., and 30% and 49% of the genes were successfully transduced in the striatum and cortex, respectively, at a dosage of 1.86×10^{12} vg/kg [23]. Despite these successes, additional research on novel vectors is required to further lower the dose and enable the translation of AAV-based medicines to people. Still, when employing viral vectors for CNS gene therapy, systemic administration is one area with a significant unmet demand.

The most effective way of brain medications is intranasal mode. In this method, medicines reach the brain via the olfactory pathway instead of the BBB. Due to the first-pass effect, many medications for HIV have poor bioavailability and low BBB permeability. According to reports, the most significant HIV reservoir location is the CNS. Efavirenz was encapsulated with lipid nanoparticles using high pressure to increase bioavailability and brain absorption. When administered intranasally instead of orally, efavirenz nanoparticles increase the levels of the drug in the brain by 150 folds [186]. Yet due to the regulated dose volume, intranasal administration has limitations in terms of the overall quantity of medication that might be transported into the brain. The intranasal route would be a good method for delivering medications to the brain if they were extremely powerful.

4.4.1.2 Non-Viral Nanoparticles

4.4.1.2.1 Drug Delivery

Nanoparticles have been suggested as an exciting method to possibly improve medicine transport across the BBB since the development of nanotechnologies. You may find comprehensive evaluations elsewhere [148–150]. Generally, nanoparticles have benefits in terms of multifunctionality, capacity to transport therapeutic encapsulation, management of drug liberation, and alteration of the drug's pharmacokinetics. Additionally, the increased permeability and retention (EPR) result suggests that nanoparticles, due to their nanosize (200 nm), might enter tumor tissue that is "leaky" to help with medication administration [151]. Although employing nanoparticles for brain medication delivery can result in higher drug concentrations in the brain, this does not essentially mean that the nanoparticles' tiny size allows them to pass across a healthy BBB. When compared to standard formulations, nanoparticles may provide drugs with additional chances to infiltrate the BBB by lengthening their blood circulation duration. Alternatively, nanoparticles may boost drug levels at the surface of BBB cells. For instance, the healthy BBB prevents the passage of polyethylene glycol–polylactic acid (PEG-PLA) protein complex nanoparticles [152]. Yet the complex nanoparticles increased the effectiveness of cerebral artery blockage in mouse model for stroke by delivering brain-derived neurotrophic factor (BDNF) to the brain [135]. The PEG-PLA-BDNF complex reached the brain through the stroke-induced BBB damage [152]. Additionally, the complex most likely lengthened BDNF's plasma half-life in order to account for the higher therapeutic effectiveness shown in mice. There are also several discussions on whether or not nanoparticles or their encapsulated payloads may traverse the BBB. Additionally, the complex most likely lengthened BDNF's plasma half-life in order to account for the higher therapeutic effectiveness shown in mice. There are also several discussions on whether or not nanoparticles or their encapsulated payloads may traverse the BBB [148, 150, 153]. In order to monitor nanoparticles at the cellular level, Medina et al. recently created "barcoded" poly (lactic-co-glycolic acid) (PLGA) nanoparticles with quantum dots (QDs) [137]. The QD-PLGA nanoparticles was not found to pass the BBB in the healthy brain, but they did reach the CNS.

Size, surface characteristics, and shape are just a few of the variables that affect how well nanoparticles can penetrate the porous BBB [149]. Intravenously administered surfactants polysorbate 80 and poloxamer 188 have demonstrated that improving the function of nanoparticles causes them to carry medications over the BBB. Further, polysorbate 80 or poloxamer 188 was shown to absorb Apo A-I and/or Apo E from the blood and aid in crossing the BBB through receptor-mediated endocytosis [154]. Each form of nanoparticle, including those made of lipids, polymers, dendrimers, and inorganic materials, most likely has a distinct destiny. Additionally, it is challenging to isolate each attribute and focus just on studying size in an experimental context. Nevertheless, given the mounting data, it is not enough to merely assert that nanoparticles' tiny size allows them to pass through the BBB. With the advancement of novel multispectral techniques, including the use of QD-labeled PLGA nanoparticles, it will be possible to understand the unique function of nanoparticles in the brain.

4.4.1.2.2 Nanoparticles for Brain Imaging and Diagnosis

In-depth research has been done on nanoparticles for tumor imaging and diagnosis. The difficulty of the BBB may be the reason there has not been much study on brain imaging for CNS illnesses. The development of diagnostic tools, particularly MRI and computed tomography (CT), has improved the management and prognostication of neurodegenerative illnesses. DCE-MRI can quantitatively measure BBB breakdown in ischemic stroke patients [155]. The dynamic course of the damage and BBB breakdown after intracerebral hemorrhage, a primary factor in morbidity and mortality, has also been monitored using multipurpose MRI [156]. Numerical and optical evaluation of enhanced BBB permeability, alongside diagnostics and monitoring therapy impacts, may aid in the selection of suitable therapeutic measures outside of the predetermined time window. Patients with an acute stroke receive tPA as their initial therapy. However, tPA may cause bleeding. Hemorrhage transformation was shown to be connected with higher BBB permeability, according to studies employing CT angiography. Before receiving tPA therapy, CT may be used as a technique to evaluate the risk of hemorrhage [157]. Furthermore, Gd-micelles, a contrast reagent for MRI tumor imaging that was created [157], were employed to test the permeability of the BBB in a rat model [158]. The ischemic hemisphere showed a substantial contrast region in the MRI scans, showing the BBB's permeability to macromolecules. The Gd-micelles were able to stay in the ischemic hemisphere due to their high molecular mass. In order to estimate the danger of hemorrhage, the Gd-micelle MRI contrast agent allowed the BBB opening to be seen. These studies imply that CT and MRI brain imaging might be used to evaluate individuals for brain disorders. Patients now undergoing CT procedures are exposed to a sizable dose of radiation. For the purpose of improving CT or MRI for tumor diagnosis, nanoparticles have been created. It is conceivable to use tumor diagnostic nanoparticles, such as Gd-micelles, for brain illnesses because of the BBB's leakiness in specific disease situations. Enhanced imageable nanoparticles could reduce the quantity of contrast chemicals, which would make CT or MRI procedures safer.

4.4.1.3 Exosomes

Tiny extracellular vesicles called exosomes are produced by cells. Exosomes have a lengthy and steady circulation due to their non-immunogenic nature, which is a key advantage over other artificial nanoparticles. Exosome components extracted from brain ECs serve as regulators for chemical exchange over the BBB and uphold cellular interplay in the brain [159]. Exosomes have been used to transfer numerous pharmaceutical agents over the BBB. You may find a thorough evaluation here [159]. Getting siRNA into the brain is a major challenge. Exosomes were recovered from brain EC culture medium and transfected with VEGF siRNA. Exosomes helped siRNA penetrate the BBB and reduce VEGF in xenotransplanted, brain tumor–bearing zebra fish [140]. Exosomes as therapeutic transporters, nevertheless, face important challenges and limitations before they can be used to their fullest extent in clinical settings. These challenges and limitations include the selection of cells secreting the exosomes loading procedure optimization, assessment of the siRNA encapsulation capacity, pharmaceutics, and biocompatibility and pharmacological investigations. We must continue to investigate and eagerly anticipate the answers to these concerns.

Exosomes are small vesicles that can pass through the blood–brain barrier (BBB) and be used to transfer small pharmaceutical molecules to the brain tumor and glioblastoma (Figure 4.2) [160]. Exosomes have been used to transfer numerous pharmaceutical agents over the BBB, including siRNAs (Figure 4.2) [161]. Although siRNA has immense potential as a medicine, getting it into the brain remains a major challenge [160]. Here are some insights from the search results on the use of exosomes for delivering therapeutic agents to the brain. The main advantages of using exosomes for drug delivery are its ability to cross BBB, its use in transporting even small pharmaceutical molecules [160], and their ability to be engineered to carry surface molecules for targeting brain cells [161], as well as being considered a successful delivery system that transfers drugs and genes to target cells [163]. Several natural brain-targeted exosomes, which could target the brain by homing, have been widely used for delivering drugs to the brain [164]. Exosome drug delivery techniques faced challenges, like some type of exosomes getting degraded, a few of the exosomes not crossing the BBB, and immune cells possibly limiting the access of exosomes to the injured target sites [162], moderate to markedly [162]. As revealed, exosomes have shown promise as delivery tools for overcoming the BBB and glioblastoma therapy. They can transfer various therapeutic agents to the brain, including siRNAs, which have immense potential as a medicine. However, there are still challenges to overcome, such as the limited ability of some exosomes to cross the BBB and the potential for immune cells to limit their access to the injured target sites.

Exosomes cross the BBB. Passive diffusion and active transport selectively move molecules over the BBB. Transport relies on molecular weight, lipid solubility, plasma protein binding, and cerebral blood flow. Passive transport is confined to tiny lipid-soluble molecules like gaseous oxygen and carbon dioxide. However, carrier-mediated transporters, efflux pumps, receptor-mediated transporters, and adsorption-mediated transcytosis are needed to cross the BBB [165, 166].

Exosomes traverse the BBB from the blood to the brain and back [167]. However, how exosomes move from peripheral circulation to the brain is yet unknown [167]. Exosomes cross the BBB through transcytosis, like immune cells and infectious pathogens. Exosomes reach the intracellular compartment through transcytosis, unlike brain cells' paracellular routes [167]. However, studies suggest that exosomes may enter the brain either by crossing the endothelial cell barrier or by being trapped within it [167, 168].

Liposomes, SLN, dendrimers, nanogels, nanoemulsions, and NLC have been utilized to transfer target molecules into the CNS [169]. Liposomes load hydrophilic and lipophilic medications. Liposome clearance through the reticuloendothelial system is the principal disadvantage [170]. SLNs promote loading protein stability and biocompatibility. These particles emit compounds over extended durations [171]. SLNs' noticeable crystalline shape lowers drug loading efficiency and induces early burst release [172]. SLNs eject dissolved medicines because drug molecules are positioned between glycerides or fatty acid molecule [173]. Hydrogel are nanosized hydrogels consisting cross-linked polymer [174]. Nanogels are biocompatible, flexible, and water-rich [175]. Due to solvent removal, surfactant toxicity, weight scaling up, and mean size control challenges, nanogel manufacture is expensive [176]. Star-shaped dendrimers are nanoscale. Dendrimers carry medications because of their polyvalence, small size, and stability [177]. Dendrimers directly interact with proteins, organelles, and membranes, inducing cell lysis [178]. Synthetic nanocarriers carry target molecules with varied disadvantages. Exosomes transfer biological substances, like genetic components, as natural drug delivery vehicles [179]. Exosomes are appropriate across biological barriers because of their stability or prolonged half-life, compatibility, immune tolerance, and tissue-specific ligands [180]. Exosomes may transfer compounds through the BBB to develop novel brain disease remedies [181]. Nucleic acids treat numerous illnesses.

Delivering these molecules to designated cells or tissues is the toughest hurdle. Nucleic acids are supplied to target cells using viral vectors, liposomes, or polymeric nanoparticles [182]. Exosomes are possible transporters of nucleic acids (miRNA and siRNA), proteins, and tiny chemicals like curcumin and doxorubicin [183]. Transfection, incubation, and electroporation are the most popular exosome encapsulation techniques [184]. Sonication and extrusion of catalase into the exosomal lumen boosted loading and cellular uptake [185]. Exosomes may load foreign DNA under 1,000 kb [186]. Hybrid exosomes transfected larger plasmids. To promote exosome–liposome fusion and plasmid encapsulation, exosomes were treated with liposomes and target plasmids at specified periods. Electroporation generates low-rate plasmid transfection into exosomes [187]. Transient electrical pulses insert nucleic acid plasmids into exosomes in the CNP biochip [188]. Source cells create big exosomes with massive mRNAs [188]. CRISPR-Cas9 technology can also effectively insert mRNAs into exosomes [189]. Cells release exosomes in normal and abnormal conditions. Thus, metabolic activity and environmental stimuli alter cell capacity. Exosome testing from diverse sources has been done for decades. Therapeutic exosome isolation involves immature dendritic cells. Dendritic cell exosomes lack surface CD86, CD40, MHC-I, and MHC-II ligands, lowering immunogenicity. Peripheral blood includes many cell type–specific exosomes. The exosomes of mesenchymal stem cell are also of interest due to their immunosuppressive properties [190]. Cancer cell–derived exosomes may potentially induce T cell immunological responses. Cancer-derived exosomes may potentially interact with tissue ligands owing to their high tetraspanin content [191], as shown in Figure 4.2. These exosomes are contested for typical use. Cancer cell–derived exosomes may be harmful to parental cells and inaccessible to non-cancer cells [192]. Fruits and plants produce therapeutic exosomes [193]. Grapefruit exosomes target and suppress inflammatory tumors [194]. Inflammatory conditions may disrupt BBB integrity and exacerbate brain–blood chemical and substance leakage [195]. However, transfer of CSF material into the circulation can lower the half-life of CSF exosome [196]. The specificity of exosomes for matching receptors on the EC surface revealed that the transcellular route may transport exosomes into brain tissue. Consider whether transcellular and paracellular pathways transfer exosomes from the circulation to the CSF. Specific modification of the signaling pathways and the affinity for receptors significantly improve BBB–safe exosome dispersion. Pathological injuries do not indicate the severity of BBB leaks. Additionally, endosomal flow towards transcytosis or lysosomal breakdown in inflammatory cells near the BBB must be assessed. Inflammatory cells near the BBB may affect exosome trafficking [197]. Activated macrophages phagocytose and destroy exosomal cargo [198]. Moderately to highly active immune cells may inhibit exosomes from reaching injured target sites [198].

4.4.1.4 Non-Invasive Focused Ultrasound (FUS) Drug Delivery

In the past few years, using ultrasound to help pharmaceuticals traverse the BBB has grown in popularity. By boosting BBTB permeability in gliomas, the non-invasive technology known as microbubble-enhanced diagnostic ultrasonography (MEUS) has successfully assisted medicines in crossing the BBTB. Prominent TJ proteins include occludin, JAMs, and claudins in the BBB [199]. Microbubbles

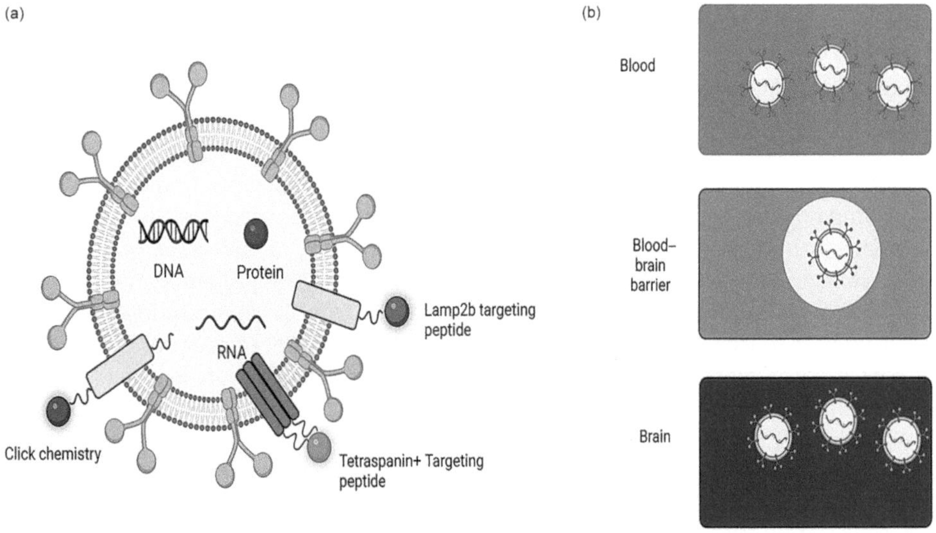

FIGURE 4.2 Techniques for targeted delivery of therapeutic exosomes to the brain: (a) Exosomes may be delivered specifically to the brain by labeling their surface with a variety of targeting moieties. Therapeutic exosomes may be genetically modified to express targeting peptides fused with exosomal membrane-associated components, such as Lamp2b and tetraspanins, or they can be chemically modified, such as via click chemistry, to express different targeting moieties. (b) By labeling the targeting peptides on the surface of exosomes, RMT may be employed to transport exosomes to the brain [161].

and ultrasonic irradiation have the potential to temporarily breach the BBB without causing harm to healthy brain tissue by reducing the expression of these TJ proteins [88]. Additionally, Ningaraj et al. demonstrated that MEUS elevated KCa channel expression in gliomas, which enhanced pinocytosis and ultimately raised BBTB permeability [199]. In addition to the BBTB, the BBB continues to act as a barrier to the transport of drugs to brain tumors. Focused ultrasound (FUS) and microbubbles have the potential to disrupt the BBB in surrounding tissue and increase the permeability of the BBTB in brain tumors. Further, DCE-MRI, FUS, and microbubbles were used in conjunction with doxorubicin administration. It was also shown that the combination improved BBB and BBTB crossing and prolonged drug retention duration in the tissue over 24 h [200]. It's also intriguing that MEUS could short-circuit P-gp expression. MEUS might be used to suppress P-gp for up to 48 h before restoring it after 72 h [201]. To learn more about the physiological changes in the brain caused by the BBB opening generated by FUS, it was tested on non-human primates at varying sonic pressures. The pharmacokinetic research proved that FUS effectively aided medication distribution by temporarily and locally opening the BBB. The acoustic pressure and inhomogeneity of the brain determined the degree of BBB opening affecting the medication levels in the brain [202]. You may find the fundamental ideas and a thorough explanation of the opportunities of ultrasound-facilitated medication delivery here [203]. The transcranial magnetic stimulation (TMS) boosts glutamate release and also enhances neuronal activity, resulting in easier transfer of drugs over the BBB. In 10 out of 15 patients with malignant brain tumors, a pilot clinical trial found that TMS increased BBB permeability, suggesting the possibility of employing TMS in clinical trials to promote medication administration into the brain [204].

4.4.1.5 Liposomal Drug Delivery

Liposomal transport is a drug delivery method that utilizes liposomes as a vehicle to deliver drugs to the target site. Liposomes are tiny spherical-in-shape synthetic vesicles formulated with cholesterol and phospholipids with hydrophobic and hydrophilic properties as well [205]. The properties of liposomes may differ depending on various factors, such as size, charge, composition of bilayer, method of preparation, saturation, and surface modifications [205]. In ischemic stroke, liposomal accumulation in the mouse brain correlated with biphasic blood–brain barrier (BBB) breakdown [205]. Similarly, PEGylated liposomes conjugated with ZL006 and HAIPRH targeted the transferrin receptors (T7) [206]. The results showed extensive brain and endothelium distribution of T7-PEGylated liposomes [206]. These studies provide evidence for the neuroprotective effect of liposomes during two therapeutic windows [205]. Liposomes are a promising tool for treating strokes, as they can shuttle diverse cargo to target sites. In conclusion, liposomes are a promising drug delivery system due to their unique structure and properties. Liposomes are highly versatile nanoparticles and have been considered for many biomedical applications [205–207].

Most liposomes, including those carrying temozolomide (TMZ), are able to cross over the BBB by exploiting the porous vascular network present in cancer microenvironments. Endothelial cells in healthy arteries are tightly interconnected, blocking the chemical communication between the brain and the blood. Under cancer conditions, the EPR effect allows external chemicals, particles, and other things to enter the brain via holes in the endothelium wall. This effect is magnified when lymph fluid leaks from the area around the tumor. Classical liposomes, which have not had their surface chemistry altered, benefit from passive targeting [208, 209]. The BTB is produced during tumor angiogenesis and serves to partition cancer tissue from the newly formed arteries. The BTB is more hospitable to liposomes than the BBB because of the EPR effect and the presence of interendothelial gaps [209]. Tumor cells with epidermal growth factor receptors take in liposomes. The second strategy for breaching the BBB is active targeting. Using the chemical features of the intact membrane to selectively absorb liposomes, this transport is more efficient than passive targeting. A liposomal activation of a route via membrane facilities is possible [208]. Figure 4.3 shows potential active transport mechanisms in engineered TMZ-liposomes.

Glucose transporters [210], adsorptive transcytosis through electrostatic interactions with cationic liposomes [211], and transferrin (Tf) and Angiopep-2 (An2)-decorated TMZ-liposomes for receptor-mediated transcytosis [212] all mediate transcytosis. Cross-BBB transport is possible, albeit pinpointing its limits is tricky. Due to the multifaceted nature of biology, there is a possibility that some of these strategies may cross over. Based on several studies, it seems that multifunctional (targeted) liposomes may be more selectively absorbed by cancer cells if they accumulate at tumor sites via precise interaction between targets and ligands [213, 214]. Liposomes protect TMZ from degrading, lessen its side effects on unintended organs, and boost its therapeutic efficacy at lower doses [213].

Since liposomes and other nanocarriers have the same disadvantages, overestimation of EPR effect, underestimation of tumor and tissue penetration, overestimation of active targeting, propensity to design complex materials without feasible applicability, failure to take tumor biology into account in drug formulation design, formulations designed for accumulation at the tumor site rather than metastases, and underestimation of nanocarriers' potential in personalized therapy were all found. Some nanomedicines, most notably liposome-based ones, made it to market despite the limitations, gaps, and misinterpretations previously mentioned [215–217]. The benefits of nanocarrier-based formulations were overshadowed by those of medication encapsulation in a carrier (such as a liposome). Due to their composition of all-natural phospholipids, liposomes pose no health risks. Some of the research found that uncoated liposomes were not harmful to cell viability,

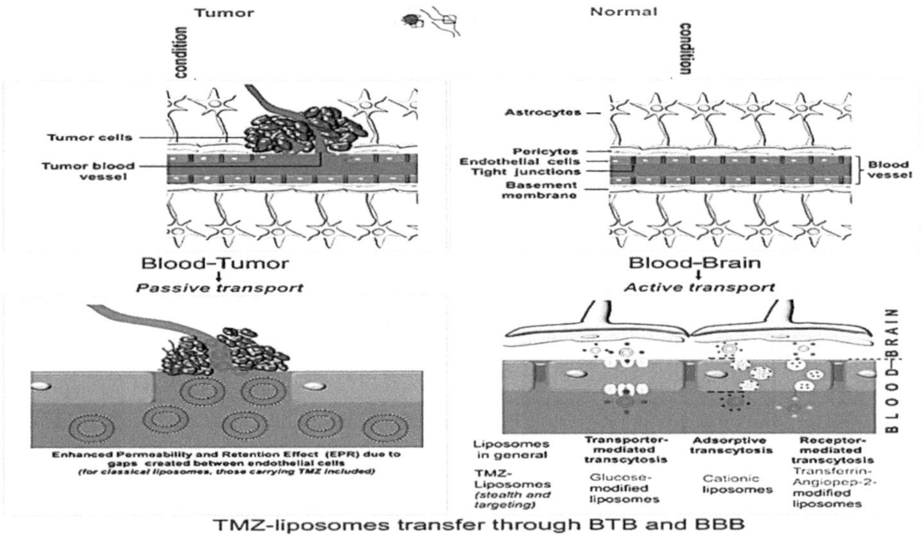

FIGURE 4.3 Strategies of crossing the blood–brain barrier (BBB) and the blood–tissue barrier (BTB) using TMZ-liposomes. (Adopted from Amarandi et al. [219].)

lending credence to their use as safe nanocarriers [218]. However, the efficiency of liposomes is not improved by utilizing natural or natural-identic phospholipids. Because the extra component in the liposomal formulation might be toxic, doing so would be a gamble [219]. As a result, the safety of liposomes should be evaluated with each new formulation to account for any changes in their chemical makeup.

4.4.2 Chemical Challenges

4.4.2.1 Enzyme Modulation

The development of new drugs for disorders of the central nervous system (CNS) presents unique challenges when compared with other disease areas [220]. The BBB is a major challenge for the treatment of CNS-related disorders, as it restricts the entry of many drugs into the brain [220]. However, the BBB contains receptors that determine the expression of drug-metabolizing enzymes [221]. Therefore, drugs that target CNS disorders must be designed to overcome these challenges.

Enzyme modulation is another approach to developing drugs for CNS disorders. Enzymes play a critical role in the metabolism of drugs, and modulating their activity can affect drug efficacy and toxicity. CYP enzyme modulation: CYP enzymes are responsible for the metabolism of many drugs, and modulating their activity can affect drug efficacy and toxicity. The BBB contains nuclear receptors that influence the expression of drug-metabolizing enzymes, like CYP enzymes [221]. Dopamine beta-hydroxylase (DBH) is an enzyme that converts dopamine to norepinephrine, and inhibiting DBH can increase dopamine levels in the brain and is being developed for the treatment of CNS disorders, such as depression and attention-deficit hyperactivity disorder (ADHD) [220].

Drug-metabolizing enzymes, such as CYP enzymes, play a crucial role in the metabolism of drugs, and their expression is regulated by nuclear receptors in the blood–brain barrier (BBB). Upregulation of these receptors has been linked to higher CYP enzyme expression and poorer medication penetration across the BBB, particularly in CNS illnesses, such as treatment-resistant epilepsy. However, regulating these receptors can improve drug accessibility and lower the BBB's chemical barrier [222–224]. Different CYP enzyme expression in the BBB is regulated by several nuclear receptors, including the pregnane X receptor (PXR), the constitutive androstane receptor (CAR) [225], and the glucocorticoid receptor (GR). Studies have shown that downregulation of GR in brain microvascular endothelial cells leads to increased phenytoin transport and decreased CYP3A4 protein levels [223].

Other central nervous system (CNS) illnesses, such as Alzheimer's disease, have been linked to GR dysregulation in addition to epilepsy [226]. Clinical trials have investigated the safety and efficacy of ketoconazole (NCT00860275, NCT00931073) as supplementing treatments for Alzheimer's disease [227]. Further evidence for the therapeutic potential of GR antagonism comes from studies using a mouse model of Alzheimer's disease and the drug mifepristone to diminish amyloid levels in the central nervous system [228]. Although targeting these receptors seems to hold promise as a treatment for certain CNS disorders, the widespread expression of GR and other nuclear receptors in the human body means that there is a chance of unintended side effects.

4.4.2.2 Receptor Agonists and Antagonists

Receptor agonists and antagonists are drugs that target specific receptors in the brain and CNS. They work by either activating or blocking the receptor, respectively. Here are some examples of receptor agonists and antagonists. These drugs target the muscarinic receptor class of acetylcholine agonists and antagonists and are used therapeutically for various CNS disorders [221]. Adenosine receptor signaling has long been a target for drug development, with adenosine itself or its derivatives being used clinically since the 1940s [229]. Cannabinoid CB2 receptor agonists and antagonists: these drugs target the CB2 receptor and are being developed for the treatment of neurodegenerative disorders [230]. Dopamine D3 receptor antagonists: these drugs target the D3 receptor and are being developed for the treatment of neurological diseases, such as schizophrenia and drug abuse [231]. The widespread expression of GR and other nuclear receptors in the human body means that there is a chance of unintended side effects when targeting these receptors. Overall, understanding the regulation of drug-metabolizing enzymes and their expression in the BBB is crucial for developing effective treatments for CNS disorders [222–225].

Developing drugs for CNS disorders is a challenging task due to the BBB and the complex nature of the CNS. Receptor agonists and antagonists and enzyme modulation are two approaches to developing drugs for CNS disorders.

4.2.3 Receptor- (or Transporter-) Mediated Transcytosis

Over 30 years have been spent researching actively targeted delivery, which involves enhancing nanoparticle absorption across the BBB by utilizing ligands of transporters or receptors. In this strategy, the ligand serves as a vehicle to administer medication enclosed in nanoparticles rather than a medication [232]. This method's preferred mechanism is receptor/transporter-mediated transcytosis of nanoparticles between the basolateral and apical surfaces of brain ECs. For instance, low-density lipoproteins are released at the basolateral surface of the brain via receptor-mediated transcytosis through the ECs, skipping the lysosomal compartment [233]. Nevertheless, it is debatable if therapeutic agents are transported to the brain by nanoparticles. After being injected intravenously into mice, brain-targeted pluronic-based nanoparticles made with RVG29 showed improved brain delivery [234]. In order to create nanoparticles, RVG29 was additionally complexed with dendrimers and subsequently conjugated with DNA. After being injected into the bloodstream of mice, RVG29-DNA nanoparticles tended to gather in the brain, where they crossed brain capillary ECs with the help of clathrin and caveolae [235]. However, other studies have shown that nanoparticles may enhance drug absorption, as they are not brain-specific [148, 236, 237]. RVG29-coated PLGA filled with camptothecin helped get the medicine into the brain by making it easier for the nanoparticles to stick to the GABA surface receptors. They came to the conclusion that the medicine was transmitted non-specifically since they did not see the nanoparticles internalize in the target tissue [153]. In response to this, Chen et al. used Förster resonance energy transfer imaging to investigate the release of hydrophobic fluorescent probes encapsulated in fluorescently labeled polymeric micelles. Tumor cells internalize micelles before hydrophobic probes were discharged in the extracellular space, indicating a membrane-aided mechanism for cellular absorption [236]. Together, these results show that, despite having the same targeted ligand (RVG29), different nanoparticles have different methods for improving drug absorption in the brain. These contentious findings highlight the difficulties, variety of nanoparticles, and crucial factors to take into account while designing and testing actively targeted nanoparticles that can cross the BBB.

Chemical characteristics of endothelial cells, which make up a portion of the BBB, must be overcome in order to transport medicinal substances across it. Receptor-mediated transcytosis (RMT) is a medication delivery strategy that has the ability to sidestep chemical characteristics, including enzymatic degradation. When a medication binds to a ligand-recognition receptor on the luminal side of BBB endothelial cells, the RMT process is initiated. Subsequently, the endothelial cell endocytoses the drug-bound receptor, and the exosome carries the drug-receptor conjugate and fuses with the plasma membrane to release the drug [238–240]. The BBB is not compromised in any way during RMT, and the chemical characteristics of endothelial cells are circumvented, making RMT a viable therapy option for a wide range of central nervous system (CNS) problems.

Transferrin receptor and insulin receptor are only two examples of BBB endothelial cell surface receptors targeted in CNS disorders. To transport a bioactive peptide, scientists have recently identified a nanobody against the mouse anti-transferrin receptor that can do so through RMT. The animals were injected with these nanobodies on the periphery (intraperitoneally, intravenously, or subcutaneously) [241]. This is only one of several studies demonstrating the feasibility of RMT as a means of delivering drugs over the BBB's chemical barrier. RMT medication delivery through the transferrin receptor [242] and the insulin receptor [243] is now being tested in human studies to prevent neurodegeneration in a rare hereditary illness. The use of RMT to transport medications over the BBB and into the sick brain is expected to increase in the near future.

Adopting actively targeted techniques, it is also necessary to take the state and course of the illness into account while designing the nanoparticles. However, as the illness progresses, the receptor expression and transport pathways change. Examples of commonly targeted receptors include insulin and transferrin receptors, which have been exploited to create functionalized nanoparticles. Research has shown that neuroinflammatory conditions and illness development affect the activity of the receptors [244].

4.2.3.1 *Transferrin Receptor*

The iron regulatory protein (IRP) system is in charge of controlling the transferrin receptor's expression. IRP depletion contributes to neurodegeneration, which results in a disease called neuronal iron shortage. IRP2-Null mice exhibit decreased transferrin receptor expression [245]. Therefore, transferrin receptor targeting is not an efficient mechanism to transport medications to the brain. Additionally, the impact of illness circumstances on the expression of receptors is highly particular and requires individual evaluation. Ho et al. showed that, while seeing decreased insulin receptor signaling, the overall amount of insulin receptor in the brains of nondiabetic sporadic AD patients did not significantly alter [245]. Transferrin receptor concentrations were evaluated in brain specimens from AD patients and AD animal models, and no declines were seen [246]. The control of the expression of receptors or carriers in the BBB during diseases is poorly understood. A thorough investigation on this aspect must be done before selecting the receptor for drug delivery.

The transferrin receptor (TfR) is a well-studied RMT cargo transporter through brain endothelial cells and into brain parenchyma [247]. RMT is a physiological mechanism that carries macromolecule nutrients, including as iron-bound transferrin, insulin, and leptin, across the blood–brain barrier (BBB) through vesicular trafficking of the ligand–receptor complexes [248] via endogenous receptors expressed on the luminal side of the BBB. One of the most widely used receptors for RMT, TfR has been crucial in the efficient transport of therapeutic proteins and antibodies [249]. Transcytosis across the BBB is possible through the selectivity of the TfR-mediated RMT mechanism [249]. Binding to the extracellular domain of the receptor is followed by endocytosis and transcytosis to the interstitial space 2 on the luminal (brain) side of the capillary endothelium. Inside the brain, the bispecific antibody may find and attach to the target protein, releasing the therapeutic payload [248]. Using transcytosis endothelial cells with intrinsic receptor-mediated phagocytosis [250], the TfR has been employed to improve antibody distribution through the BBB. Recent evidence indicates that TfR efficiently penetrates the BBB and is distributed throughout the brain parenchyma [251]. The TfR has been shown to be a reliable receptor for RMT-mediated BBB crossing in the treatment of glioblastoma and other brain disorders. Therapeutics such liposomes [252–254] and chitosan nanospheres [255] have been used for TfR-mediated brain delivery. TfR-mediated brain transport of medicines has also showed encouraging outcomes in various clinical studies. Since the brain takes in more iron than Tf [256], the process of Tf transcytosis remains mysterious. Researchers then determined that only a limited amount of Tf is able to cross the lipid membrane of ES cells lining cerebral blood vessels and deposit itself in the brain [257]. Despite worries about the distribution mechanism, Tf ligand brain targeting has happened. Shin et al. [258] demonstrate that Tf-iodinated IgG3 antibody may be delivered to the brain. Human IgG3 was used as an antibody model to assess the most efficient brain delivery Tf linkage points. Tf binding to the IgG3 hinge region increased brain absorption by 0.3%. Recently, Tf-targeted PEGylated albumin nanoparticles (PEG-NP) have been used to administer azidothymidine (AZT). After four hours, the brains of rats were shown to have recovered 21.1% of the drug when Tf-targeted PEG-NP was administered, but only 9.3% when non-targeted PEG-NP was [259].

Because endogenous Tf persists in the plasma at 25 M [260], Tf is not an effective brain delivery route, because the TfR is practically saturated. Efficacy of anti-TfR antibodies has been shown. OX26 is the mouse monoclonal antibody (MAb) most studied in relation to rat TfR. This antibody binds to a distinct TfR epitope, hence minimizing competition between the vector and the natural ligand [261]. Therapeutic cargo may be delivered to the brain and pharmacological effects produced via the OX26 BBB-targeting vector. Symptoms of stroke have been treated using the neuroprotective substance BDNF [262]. BDNF was conjugated to OX26 through biotin (B) and streptavidin (SA) using PEG-hydrazide. Intravenous BDNF or OX26-BDNF was given to rats with permanent blockage of the middle cerebral artery (MCAO). Stroke treatment potential was shown in rats when the OX26-BDNF conjugate enhanced motor performance by 243% compared to BDNF alone [262].

OX26 can non-invasively transport several drugs in vivo. Blood pressure in the brain is controlled by a protein called VIP. While unconjugated VIP did not affect hemispheric blood flow in awake adult rats, the OX26–VIP conjugate (SA/B) did so by 60%. FGF-2 is neuroprotective when injected intracerebroventricularly after brain ischemia [263]. OX26-conjugated bFGF (SA/B) administered intravenously decreased infarct volume by 80% in rats with chronic MCAO. The same amount of unconjugated biotinylated bFGF was used in each experiment [264]. In a rat model of HD, Kordower et al. [265] found that OX26 conjugated to NGF mitigated the loss of cholinergic striatal acetyltransferase-immunoreactive neurons. OX26 can transport not just protein but also plasmid DNA containing therapeutic genes. Plasmid DNA encoding for galactosidase was delivered to the brain of rats in a non-invasive manner via PEGylated liposomes that were targeted towards the OX26 receptor. TfR is highly expressed in the brain parenchyma, liver, and spleen. Methods comparable to those used to distribute daunomycin [266, 267] and antisense oligonucleotides (ODN) and peptide nucleic acids (PNA) [268, 269] have also been developed.

In mice, the BBB has been targeted with the use of another monoclonal antibody, 8D3 [261]. Exogenous plasmid DNA encoding galactosidase and luciferase may be delivered to mice brains through 8D3-targeted PEGylated liposomes [270]. The brain and other TfR-rich organs, like lung, liver, and spleen, expressed galactosidase activity using the constitutive SV40 promoter. Gene expression was low in peripheral organs but high in astrocytes when the brain-specific glial fibrillary acidic protein (GFAP) promoter controlled enzyme expression [270]. In order to transport galactosidase to the central nervous system and other tissues, rather than the gene itself, we used a SA/B linkage to couple it to 8D3. SA/B linkage was used to attach 8D3 to a radiolabeled 16-mer antisense PNA for high-definition imaging. The exon 1 [271] of the huntingtin gene was designed to be the hybridization target for the antisense PNA [271]. Radioactivity testing on brain tissue from transgenic mice revealed that they retained three times as much 8D3-PNA as littermates. The 8D3 method was used to visualize beta-amyloid plaques in transgenic mice with Alzheimer's disease. Amyloid deposition in transgenic mice was visualized and measured by intravenous infusion of 8D3-conjugated, radiolabeled A1–40 [272]. The monoclonal antibody R17–127, which targets mouse TfR, may undergo RMT via the mouse BBB. In comparison to 8D3, which has a higher brain uptake and percent given dose/g brain [261], R17–127 is more brain-selective. A BBB TfR epitope that is inaccessible to the liver may be recognized by R17–127 [261]. Thus, targeting vectors serve distinct purposes even though they all target the same receptor.

Using the OX26 and 8D3 antibodies, the TfR has been widely investigated as a potential RMT mechanism for non-invasively transporting medications to the brain. Firstly, only 0.44% of the injected dose of OX26 reaches the rat brain after 24 h [273], due to the widespread TfR expression on peripheral organs limiting its potential on brain. After 1 h, 1.5% of 8D3 and 0.8% (1.6%) of R17–127 reached the brains of mice, respectively [261]. There is no doubt that drugs have their intended effects, but use of a brain-specific promoter to stimulate tissue-specific gene expression might lessen the medication's unwanted effects in places outside the brain [270]. In contrast to other antibodies, these do not identify the human TfR [274]. Immunogenicity in humans may result from the usage of these non-human antibodies.

4.2.3.2 Lipoprotein Receptors

Nowadays, some novel techniques using nanoparticles have been indicated to particularly distribute medications to the brain. Also, propionylated amylose helix was used to create nanoclusters that contained the hydrophobic medication propofol, which was used to produce calming properties in rabbits. The brain ECs have more phosphophatidylethanolamine (POPE) than other ECs. The complex was bound to the POPE bilayer's surface due to H-bonding with propionylated amylose, which also caused the complex to break apart and liberate the drug. This created a localized high-concentration gradient that made it easier for the hydrophobic drug to cross the BBB. The drug's encapsulation and release were modelled using molecular dynamics modelling, which was used to choose the helix [275]. As a result, this novel strategy focused on brain targeting, not on BBB crossing, by using a distinct, biochemically specific circumstance to initiate drug release in the BBB. Since they are natural, lipoproteins have long been investigated as potential drug delivery systems. As a substitute, high-density lipoprotein (HDL) synthetic mimics made from recombinant apolipoproteins and natural or synthetic lipids have been produced. In order to inhibit the production of A oligomers and hasten A cellular breakdown, Song et al. created Apo E-reconstituted HDLs. The findings demonstrated that Apo E-reconstituted HDLs focused on A deposits and aided BBB penetration [51]. The self-assembled HDL-imitating nanoparticles were made from natural lipids, an HDL-like structure that successfully encapsulated nerve growth factor (NGF). The NGF HDL-imitating nanoparticles maintained NGF biological activity for promoting neurite development and also sustained NGF [276, 277]. Porphyrin-lipid nanoparticles mimicked HDL in order to target glioblastoma. Apo E3 was then coated on the nanoparticle surface [278].

The low-density lipoprotein receptor (LDLR) family endocytoses LDLs, such ApoB and ApoE. The LDL particle comprises an apolipoprotein encapsulated in fat molecules, including phospholipids, cholesterol, and triglycerides, which carry fatty acids to cells. LDLR is a brain-expressed receptor that efficiently transports medicines. ApoB- and ApoE-derived peptides may transport lysosomal enzymes across the BBB [279]. Liposomes, HDL, and polymersomes functionalized with ApoE-derived peptides promote BBB bridging by LDLR- and LRP1-mediated transcytosis in BCECs [280, 281]. ApoB-derived peptides target brain parenchyma delivery of siRNA and nanoparticles [36, 37, 282]. Angiopep-2 is a promising targeting moiety for brain delivery and glioma-targeting nanoparticles [283, 284]. LRP1 is abundantly expressed in human and mouse BCECs [285] and gliomas [286].

The Apo E3 porphyrin-lipid nanoparticles engage with overexpressed low-density lipoprotein receptor on glioblastoma cells. The unique Apo E3 porphyrin-lipid nanoparticles may be employed in theranostics since porphyrin has dual functions for fluorescence imaging and photodynamic treatment. Although investigating the imaging in an orthotopic tumor mouse model, scientists only showed the photodynamic therapy's sensitivity in glioblastoma cells. Angiopep-2 is intended to attach to LRP-1, a protein that mediates transcytosis through the low-density lipoprotein receptor. A combination of angiopep-2 and paclitaxel, known as ANG1005, was used in phase II tests to show that angiopep-2 had the ability to enter the brain [287]. A cell-penetrating peptide was coupled with the angiopep-2 paclitaxel conjugate to increase penetration in gliomas. The transfer of paclitaxel into tumor tissues was greatly enhanced by the novel conjugate, which combined one for brain penetration and the other for tumor penetration, and raised the survival rate [288].

4.2.3.3 Insulin Receptor (INSR) and Glucose Transporter 1 (GLUT1)

Like the TfR, the insulin receptor resides majorly on the luminal membrane of brain capillary endothelial cells [289]. It may undergo RMT across the BBB endothelium. The insulin receptor is an integral membrane glycoprotein (~300 kDa) [290]. When insulin binds, a conformation shift allows tyrosine kinase activity and receptor internalization [290]. Due to two issues, insulin has not been used as a vector. Insulin has a serum half-life of ten minutes and may produce hypoglycemia if disrupted [291]. Like the TfR, insulin receptor antibodies can target the BBB. RMT delivery vector study has used the 83–14 mouse MAb against the human insulin receptor (HIR) [292]. The antibody 83–14's transport through the monkey BBB is tenfold higher than that of an anti-human TfR MAb [293]; 83–14, like transferrin antibodies, is a mouse antibody that may cause immunogenic reactions in humans. Engineered 83–14 MAbs have been designed to improve its humanness. First, a chimeric MAb with 85% human and 15% mouse variable fragment binding sites was generated [294]. The chimeric antibody reacted to the HIR of human brain capillaries like the parent mouse 83–14 antibody [294, 295]. Fully humanized antibodies are considerably less immunogenic than chimeric antibodies [295]. Only the antigen-binding CDR loops were grafted onto B43 IgG heavy chain and human REI kappa light chain [295]. After intravenous injection, the humanized antibody reached all areas of the Rhesus monkey brain [295]. Therefore, 83–14 MAb travels into the animal model brain and might be used clinically in humans after humanization.

Many investigations employing the differently humanized 83–14 MAb to transport drug cargo were comparable to those using TfR antibodies. Table I lists a few high-impact studies. When conjugated to 83–14 and intravenously injected into Rhesus monkeys, radiolabeled Aβ1–40 increased brain concentration similar in the mouse with 8D3 [296]. The emulsion autoradiography revealed that the radiolabeled conjugate was attached to autopsied Alzheimer brain tissue slices neuritic plaques [297]. The 83-14-decorated liposomes enclosed a β-galactosidase-expressing plasmid. After intravenous injection, the foreign gene crossed the BBB and was expressed at high levels in neuron plasma membranes [298]. The expression plasmid regulated by the ocular-specific opsin promoter confined β-galactosidase gene expression to the monkey eye, indicating tissue-specific gene targeting [299]. Recently, the intravenous injection of human BDNF single polypeptide and 83–14 MAb chimera into Rhesus monkeys showed an increase in BDNF levels more than tenfold [300].

The mannose 6-phosphate (M6P) receptor, also known as the insulin-like growth factor II (IGF-II) receptor, has been examined in neonatal mouse BBB transport [301]. M6P receptors attach to the M6P component of lysosomal enzymes, internalizing and sorting them to the lysosome. A 312 kDa enzyme β-glucuronidase (GUS) was previously found to be active in the brain in the first two weeks of development of N14. This enzyme deficiency causes mucopolysaccharidosis type VII [301]. In newborn mice, radiolabeled and phosphorylated-GUS (131I-P-GUS) reached the brain, whereas nonphosphorylated GUS did not. Then, 131I-P-GUS transfer decreased at seven weeks. Through the M6P/IGF-II receptor, P-GUS undergoes RMT early in postnatal development [301]. Thus, lysosomal storage disorders may be treated in neonates by RMT using enzyme therapeutics, and while in adults through M6P receptors [301].

4.2.3.3.1 Age-Related Changes to the Blood–Brain Barrier

The impact of ageing on medication transport to the brain has received little attention in literature and investigation. Several results from the literature are included in this part of the chapter. These BBB elements may be impacted by ageing. For instance, research revealed that ageing astrocytes have elevated levels of genes associated with scarring and inflammation [302]. Astrocytic activities crucial for recovering from a stroke were affected by ageing in male and female rats [303, 304]. Additionally, as people age, astrocytes produce fewer trophic elements that suppress brain degeneration [304, 305]. In these investigations, AAV5-GFP-hIGF-1 injections were used to deliver the insulin-like growth factor 1 (IGF-1) gene to astrocytes in middle-aged female rats by injecting it into the cortex and striatum. The findings demonstrated that IGF-1 aided in repair caused by stroke, like BBB permeability and neuroinflammation [306]. They nevertheless failed to examine this IGF-1 gene therapy in rats of various ages. As we get older, astrocytes also diminish the brain's ability to absorb nutrients like glucose and the BBB's expression of the appropriate receptors like GLUT-1 [307]. As a result, when given to seniors, also with age, pericytes decline, paralleling the rise in BBB permeability, in addition to astrocytes [308]. Age changes the permeability of molecules because ageing affects the BBB structure. In one experiment, hypoxic-ischemic brain injury was compared between newborn, neonatal, and adult healthy rats in terms of their permeability to NGF. The findings indicated that NGF penetration was much higher in the newborn rats under hypoxic conditions. In addition, NGF exhibited increasingly greater permeability in neonatal rats for the ageing impact [309]. Although BBB impairment develops early in the pathophysiology, frequent pressures from illnesses will further change BBB function in elderly people. Also, lipopolysaccharide caused BBB breakdown in aged animals, simulating the typical sepsis stress. Additionally, BBB rupture was connected to the overexpression of gp91phox, the suppression of protein kinase activity, claudin-5, and occludin degradation [310]. Because of problems and limited studies, it is currently unknown how and to what extent ageing affects the BBB. Because numerous CNS problems are more common in older people, it is important to thoroughly comprehend how genes behave and BBB permeability in individuals of all ages. When developing new medication delivery methods for CNS illnesses, investigators studying drug delivery must take ageing factors into account.

4.4.3 Brain Permeability Challenges

Numerous compounds have shown the capacity to temporarily crack through the blood–brain barrier (BBB) and let large amounts of systemically delivered chemotherapeutics enter the brain [311]. One of the explanations for these compounds' ability to breach the BBB is dependent on the temporary breakdown of the BBB caused by a reduction in the production of TJ proteins, such as occludin, tricellulin, and claudin-1. They initially used intra-arterial mannitol together with chemotherapeutic drugs to treat brain tumors [287]. At present, it has been demonstrated that cereport (a bradykinin analogue) increases BBB permeability, which enhances the anticancer effectiveness of cancer-fighting medicines administered together in animal models. Clinical investigations, nevertheless, did not demonstrate the advantages of combined use in individuals with gliomas [312]. The adenosine receptor agonist regadenoson also improved BBB permeability in animal tests, but it had no effect on the infiltration of simultaneously delivered contrastors in people [311]. For these compounds, there seems to be a variation between findings in rats and humans. The authors suggested more research on dosages, timing, and combination therapies. In recent investigations, borneol has been shown to improve oral medication absorption as well as drug penetration into the brain in experimental animals [313–315]. Four distinct oral forms of borneol and puerarin were compared by Yi et al. Among these, compared to previous preparations, puerarin and borneol produced noticeably greater AUCs in both the plasma and the brain [315]. As demonstrated in the earlier cereport and regadenoson trials, it is extremely probable that co-administration of a medication and a distribution promoter is not enough to reap the advantages of the enhancer in people. Due to the temporary nature of the enhancers' interactions with the BBB, it may be crucial to transport both the enhancer and the medication on the same carrier in order to permit the drug to go through the BBB while the booster accesses the BBB. Medication delivery across the accessible blood–brain barrier in the presence of illnesses: the BBB has long been acknowledged as a significant barrier to the transport of drugs to the brain. Although it is known that several illness situations change the BBB leakiness, further specific information, such as the magnitude and duration of the BBB opening, is not well recognized. With more sophisticated research, different processes have been found. For instance, ECs' release of glutamate encourages BBB permeability [204]. Recent studies using cutting-edge techniques for brain imaging have provided increasingly thorough data on BBB leakiness. The current section revisits an "old" idea in light of new research on BBB permeability in pathological settings. In disorders including inflammatory, painful, and deteriorating circumstances, BBB opening has been seen. BBB opening is frequently the defining clinical manifestation [316]. In an optimally functioning brain, interendothelial connections play a crucial role in preserving tissue–fluid balance. When interendothelial connections are damaged due to certain diseases, proteinaceous fluid penetrates the interstitium and causes edema [317, 318]. On the contrary, as illnesses advance, brain damage progressively modifies BBB permeability [319]. MRI is now becoming a popular non-invasive method for examining BBB impairment in patients [155, 320–322]. One of the main reasons for BBB disruption is inflammation. Cardiopulmonary bypass, which causes an overall inflammation response, was used on patients. Under cardiac bypass, dynamic contrast-enhanced magnetic resonance imaging (DCE-MRI) and diffusion-weighted imaging techniques were used to assess BBB damage and its integrity time to be restored [320]. Following the operation, the BBB's integrity was compromised. Despite the fact that BBB permeability returned after a few days, the brief BBB rupture was associated with postoperative neurocognitive impairment. Using DCE-MRI, Wong et al. showed that patients with cerebrovascular illness may have their mild BBB leakage evaluated with good to exceptional repeatability [322]. Understanding the dramatic shift in the permeability and BBB processes in ischemic stroke is another key area of emphasis. It is debatable when BBB disruption happens following a stroke. According to several studies, stroke affects the BBB many hours after it starts [323]. Nevertheless, a new investigation using improved MRI revealed that BBB leakage in individuals immediately following acute ischemic stroke (AIS) increased with time [324]. The BBB's permeability peaked 6–48 h following the initiation of AIS. Following AIS, BBB deterioration may result in pathologic events, such as edema and hemorrhagic transformation [325]. A primate cerebral artery occlusion paradigm was used in conjunction with imaging technology to analyze the change in permeability, depending on the CSF/serum albumin ratio. The findings indicated that the BBB breakdown happened quickly following ischemia and that its severity was strongly connected with the progression of the illness [326]. PEGylated liposomes transferred to the ischemic brain area during the onset of reperfusion by concentrating on the broken BBB [327]. The aforementioned team additionally blended tissue plasminogen activator (tPA) with the neuroprotectant drug Fasudil in liposomes to manage ischemic stroke. Prior to tPA, liposomal Fasudil therapy showed notable neuroprotective benefits [328]. Due to tumor blood vessel development, which is a result of extremely porous capillaries and a lack of lymphatic drainage, the EPR effect vanishes in brain tumors [199, 329, 330]. It is crucial to understand the link between nanoparticle attributes such as size and the dimensions of the BBTB pores in order to utilize the impact of EPR on brain tumors. Dendrimers are artificial nanoparticles, and through nanoparticle creation, it is possible to design and regulate their size, porosity, and surface charge. Polyamidoamine (PAMAM) dendrimers (generations 1–8) were prepared to explore the impact of its dimensions on particulate aggregation in tumor-bearing rat glioma cells [331]. This study results showed that it required Gd-chelated dendrimers at a size threshold of 12 nm to pass through the BBTB, and particles larger than this size were not able to do so. In the animal approach, nanoparticles with a diameter of 4 to 10 nm also produced elevated levels in the blood for an extended period [332]. As a result, sizes smaller than 12 nm may be able to pass the BBTB in dendrimers. For additional kinds of nanoparticles, such as lipid nanoparticles, it is difficult to synthesize nanoparticles of different sizes using the same ingredients; as a result, it is challenging to completely rule out the impact of other features (such as surface properties and shape) on these nanoparticles. As a result, it is currently unclear how particle size affects a nanoparticle's ability to traverse the BBB or BBTB. Additionally, research has revealed that several linked therapies for brain disorders can repair the BBB leak. A humanized monoclonal antibody called bevacizumab prevents VEGF from functioning biologically. In the advanced stages of malignant gliomas, bevacizumab is routinely prescribed. This therapy improves the brain tumor vasculature while repairing the BBB's leakiness [333]. The absorption of any other medications during the treatment regimen might be a challenge when bevacizumab is employed in combination treatment. Understanding BBB dynamics may shed light on potential therapies for degenerative brain conditions in the future. If patients could be treated during the crucial period of its opening, the BBB might not be such a big obstacle. Consequently, by making use of certain clinical situations, existing unresolved problems in brain medication delivery might be resolved.

4.5 CONCLUSION AND FUTURE DIRECTIONS

The previous five years' worth of the administration of medicines to the brain have been covered in this chapter. To develop effective pharmaceutical delivery systems for brain disorders, a full knowledge of BBB disruption is necessary. Recent developments in studies have not only shown that the BBB is permeable in cases of brain damage but have also identified the processes by which the BBB is regulated. Innovative non-surgical methods like MEUS and TMS have been investigated, along with commonly employed technology, like viral vectors and nanoparticles, to rapidly expose the BBB and enhance brain medicine absorption. Innovative delivery methods may facilitate brain illness diagnostics. Using knowledge of the leaky BBB, brain illnesses might be treated using previously created nanoparticles that target tumors via the EPR effect. Highly diverse ranges exist in gliomas, and permeability is typical on the periphery of these ranges. Therefore, it may be necessary to take into account a combination of techniques that penetrate both permeable and normal BBBs. Additionally, new studies on the mechanisms of disturbances to the BBB will be released, which will promote the development of suitable administration methods that make use of BBB leakage. The impact of ageing on BBB dysfunctions is another relevant subject that merits research. Seldom can you find in the literature brain medication delivery mechanisms that have taken the influences of ageing into account and have been tested in animals of different ages. In conclusion, the intricacy of the BBB requires thorough research into administration methods, but it could additionally present special chances to develop effective delivery methods for the treatment of different brain illnesses.

REFERENCES

1. Abbott, N.J., et al., *Structure and function of the blood–brain barrier.* Neurobiology of Disease, 2010. **37**(1): pp. 13–25.
2. Gürsoy-Özdemir, Y. and Y.C. Tas, Anatomy and physiology of the blood–brain barrier, in *Nanotechnology methods for neurological diseases and brain tumors.* Elsevier, 2017, pp. 3–13.
3. Abbott, N.J., *Blood–brain barrier structure and function and the challenges for CNS drug delivery.* Journal of Inherited Metabolic Disease, 2013. **36**: pp. 437–449.
4. Pardridge, W.M., *Drug transport across the blood–brain barrier.* Journal of Cerebral Blood Flow & Metabolism, 2012. **32**(11): pp. 1959–1972.
5. Ndemazie, N.B., et al., *Multi-disciplinary approach for drug and gene delivery systems to the brain.* American Association of Pharmaceutical Scientists PharmSciTech, 2022. **23**: pp. 1–21.
6. Kadry, H., B. Noorani, and L. Cucullo, *A blood–brain barrier overview on structure, function, impairment, and biomarkers of integrity.* Fluids and Barriers of the CNS, 2020. **17**(1): pp. 1–24.
7. Li, J., et al., *Development of novel therapeutics targeting the blood–brain barrier: From barrier to carrier.* Advanced Science, 2021. **8**(16): p. 2101090.
8. Keaney, J. and M. Campbell, *The dynamic blood–brain barrier.* The FEBS Journal, 2015. **282**(21): pp. 4067–4079.
9. Lajoie, J.M. and E.V. Shusta, *Targeting receptor-mediated transport for delivery of biologics across the blood-brain barrier.* Annual Review of Pharmacology and Toxicology, 2015. **55**: pp. 613–631.
10. Ruan, S., et al., *Rethinking CRITID procedure of brain targeting drug delivery: Circulation, blood brain barrier recognition, intracellular transport, diseased cell targeting, internalization, and drug release.* Advanced Science, 2021. **8**(9): p. 2004025.
11. Teleanu, D.M., et al., *Blood-brain delivery methods using nanotechnology.* Pharmaceutics, 2018. **10**(4): p. 269.
12. Zhou, Y., et al., *Crossing the blood-brain barrier with nanoparticles.* Journal of Controlled Release, 2018. **270**: pp. 290–303.
13. Botti, G., A. Dalpiaz, and B. Pavan, *Targeting systems to the brain obtained by merging prodrugs, nanoparticles, and nasal administration.* Pharmaceutics, 2021. **13**(8): p. 1144.
14. Saeedi, M., et al., *Applications of nanotechnology in drug delivery to the central nervous system.* Biomedicine & Pharmacotherapy, 2019. **111**: pp. 666–675.
15. Pinheiro, R.G.R., et al., *Nanoparticles for targeted brain drug delivery: What do we know?* International Journal of Molecular Sciences, 2021. **22**(21): p. 11654.
16. Sartaj, A., et al., *An insight to brain targeting utilizing polymeric nanoparticles: Effective treatment modalities for neurological disorders and brain tumor.* Frontiers in Bioengineering and Biotechnology, 2022. **10**: p. 19.
17. Lu, C.-T., et al., *Current approaches to enhance CNS delivery of drugs across the brain barriers.* International Journal of Nanomedicine, 2014. **9**: p. 2241.
18. Sweeney, M.D., et al., *Blood-brain barrier: From physiology to disease and back.* Physiological Reviews, 2019. **99**(1): pp. 21–78.
19. Archie, S.R., A. Al Shoyaib, and L. Cucullo, *Blood-brain barrier dysfunction in CNS disorders and putative therapeutic targets: An overview.* Pharmaceutics, 2021. **13**(11).

20. Sweeney, M.D., A.P. Sagare, and B.V. Zlokovic, *Blood-brain barrier breakdown in Alzheimer disease and other neurodegenerative disorders.* Nature Reviews Neurology, 2018. **14**(3): pp. 133–150.
21. Knox, E.G., et al., *The blood-brain barrier in aging and neurodegeneration.* Molecular Psychiatry, 2022. **27**(6): pp. 2659–2673.
22. Erickson, M.A. and W.A. Banks, *Blood-brain barrier dysfunction as a cause and consequence of Alzheimer's disease.* Journal of Cerebral Blood Flow Metabolism, 2013. **33**(10): pp. 1500–1513.
23. Carvey, P.M., B. Hendey, and A.J. Monahan, *The blood-brain barrier in neurodegenerative disease: A rhetorical perspective.* Journal of Neurochemistry, 2009. **111**(2): pp. 291–314.
24. Han, L. and C. Jiang, *Evolution of blood-brain barrier in brain diseases and related systemic nanoscale brain-targeting drug delivery strategies.* Acta Pharmaceutica Sinica B, 2021. **11**(8): pp. 2306–2325.
25. Abbott, N.J., et al., *Structure and function of the blood-brain barrier.* Neurobiology of Disease, 2010. **37**(1): pp. 13–25.
26. Nahirney, P.C., P. Reeson, and C.E. Brown, *Ultrastructural analysis of blood-brain barrier breakdown in the peri-infarct zone in young adult and aged mice.* Journal of Cerebral Blood Flow Metabolism, 2016. **36**(2): pp. 413–425.
27. Redzic, Z., *Molecular biology of the blood-brain and the blood-cerebrospinal fluid barriers: Similarities and differences.* Fluids and Barriers of the CNS, 2011. **8**(1): p. 3.
28. Kadry, H., B. Noorani, and L. Cucullo, *A blood–brain barrier overview on structure, function, impairment, and biomarkers of integrity.* Fluids and Barriers of the CNS, 2020. **17**(1): p. 69.
29. Kaur, C. and E.A. Ling, *Blood brain barrier in hypoxic-ischemic conditions.* Current Neurovascular Research, 2008. **5**(1): pp. 71–81.
30. Hawkins, B.T. and T.P. Davis, *The blood-brain barrier/neurovascular unit in health and disease.* Pharmacological Reviews, 2005. **57**(2): pp. 173–185.
31. Crone, C. and O. Christensen, *Electrical resistance of a capillary endothelium.* Journal of General Physiology, 1981. **77**(4): pp. 349–371.
32. Gorlé, N., et al., *The effect of aging on brain barriers and the consequences for Alzheimer's disease development.* Mammalian Genome, 2016. **27**(7–8): pp. 407–420.
33. Oldendorf, W.H., M.E. Cornford, and W.J. Brown, *The large apparent work capability of the blood-brain barrier: A study of the mitochondrial content of capillary endothelial cells in brain and other tissues of the rat.* Annals of Neurology, 1977. **1**(5): pp. 409–417.
34. Fenstermacher, J., et al., *Structural and functional variations in capillary systems within the brain.* Annals of the New York Academy of Sciences, 1988. **529**: pp. 21–30.
35. Sanchez-Covarrubias, L., et al., *Transporters at CNS barrier sites: Obstacles or opportunities for drug delivery?* Current Pharmaceutical Design, 2014. **20**(10): pp. 1422–1449.
36. Takakura, Y., K.L. Audus, and R.T. Borchardt, *Blood-brain barrier: Transport studies in isolated brain capillaries and in cultured brain endothelial cells.* Advances in Pharmacology, 1991. **22**: pp. 137–165.
37. Vorbrodt, A.W. and D.H. Dobrogowska, *Molecular anatomy of intercellular junctions in brain endothelial and epithelial barriers: Electron microscopist's view.* Brain Research Reviews, 2003. **42**(3): pp. 221–242.
38. Rodríguez-Arellano, J.J., et al., *Astrocytes in physiological aging and Alzheimer's disease.* Neuroscience, 2016. **323**: pp. 170–182.
39. Abbott, N.J., L. Rönnbäck, and E. Hansson, *Astrocyte–endothelial interactions at the blood–brain barrier.* Nature Reviews Neuroscience, 2006. **7**(1): pp. 41–53.
40. De Lange, E., et al., Maturation, barrier function, aging and breakdown of the blood-brain barrier, in *Aging: Exploring a Complex Phenomenon.* Ahmad, S.I., ed. CRC Press, 2018, p. 271.
41. Kettenmann, H., et al., *Physiology of microglia.* Physiological Reviews, 2011. **91**(2): pp. 461–553.
42. Kofler, J. and C.A. Wiley, *Microglia: Key innate immune cells of the brain.* Toxicologic Pathology, 2011. **39**(1): pp. 103–114.
43. Harry, G.J., *Microglia during development and aging.* Pharmacology & Therapeutics, 2013. **139**(3): pp. 313–326.
44. Fowler, M.I., et al., *Different meningitis-causing bacteria induce distinct inflammatory responses on interaction with cells of the human meninges.* Cellular Microbiology, 2004. **6**(6): pp. 555–567.
45. Davis, L.E., *Acute bacterial meningitis.* Continuum: Lifelong Learning in Neurology, 2018. **24**(5): pp. 1264–1283.
46. Coureuil, M., et al., *A journey into the brain: Insight into how bacterial pathogens cross blood-brain barriers.* Nature Reviews Microbiology, 2017. **15**(3): pp. 149–159.
47. van de Beek, D., et al., *Community-acquired bacterial meningitis.* Nature Reviews Disease Primers, 2016. **2**(1): p. 16074.
48. Häuser, S., et al., *Capsule and fimbriae modulate the invasion of Haemophilus influenzae in a human blood-cerebrospinal fluid barrier model.* International Journal of Medical Microbiology, 2018. **308**(7): pp. 829–839.
49. Al-Obaidi, M.M.J. and M.N.M. Desa, *Mechanisms of blood brain barrier disruption by different types of bacteria, and bacterial-host interactions facilitate the bacterial pathogen invading the brain.* Cellular and Molecular Neurobiology, 2018. **38**(7): pp. 1349–1368.
50. Prasadarao, N.V., et al., *A novel interaction of outer membrane protein A with C4b binding protein mediates serum resistance of Escherichia coli K1.* Journal of Immunology, 2002. **169**(11): pp. 6352–6360.
51. Al-Obaidi, M.M.J. and M.N.M. Desa, *Mechanisms of blood brain barrier disruption by different types of bacteria, and bacterial–host interactions facilitate the bacterial pathogen invading the brain.* Cellular and Molecular Neurobiology, 2018. **38**: pp. 1349–1368.
52. Coureuil, M., et al., *A journey into the brain: Insight into how bacterial pathogens cross blood–brain barriers.* Nature Reviews Microbiology, 2017. **15**(3): pp. 149–159.
53. Iovino, F., et al., *pIgR and PECAM-1 bind to pneumococcal adhesins RrgA and PspC mediating bacterial brain invasion.* Journal of Experimental Medicine, 2017. **214**(6): pp. 1619–1630.
54. Yau, B., et al., *Blood–brain barrier pathology and CNS outcomes in Streptococcus pneumoniae meningitis.* International Journal of Molecular Sciences, 2018. **19**(11).
55. Mittal, R. and N.V. Prasadarao, *gp96 expression in neutrophils is critical for the onset of Escherichia coli K1 (RS218) meningitis.* Nature Communications, 2011. **2**(1): p. 552.
56. Doran, K.S., G.Y. Liu, and V. Nizet, *Group B streptococcal beta-hemolysin/cytolysin activates neutrophil signaling pathways in brain endothelium and contributes to development of meningitis.* Journal of Clinical Investigation, 2003. **112**(5): pp. 736–744.

57. Pack, A.M., *Epilepsy overview and revised classification of seizures and epilepsies*. Continuum: Lifelong Learning in Neurology, 2019. **25**(2): pp. 306–321.
58. Stafstrom, C.E. and L. Carmant, *Seizures and epilepsy: An overview for neuroscientists*. Cold Spring Harbor Perspectives in Medicine, 2015. **5**(6): p. a022426.
59. Millichap, J.G. and F. Aymat, *Treatment and prognosis of petit mal epilepsy*. Pediatric Clinics of North America, 1967. **14**(4): pp. 905–920.
60. Miyamoto, H., et al., *Impaired cortico-striatal excitatory transmission triggers epilepsy*. Nature Communications, 2019. **10**(1): p. 1917.
61. Khazipov, R., *GABAergic synchronization in epilepsy*. Cold Spring Harbor Perspectives in Medicine, 2016. **6**(2): p. a022764.
62. Rigau, V., et al., *Angiogenesis is associated with blood–brain barrier permeability in temporal lobe epilepsy*. Brain, 2007. **130**(7): pp. 1942–1956.
63. Tramoni-Negre, E., et al., *Long-term memory deficits in temporal lobe epilepsy*. Revue Neurologique, 2017. **173**(7–8): pp. 490–497.
64. Thijs, R.D., et al., *Epilepsy in adults*. Lancet, 2019. **393**(10172): pp. 689–701.
65. Kim, S.Y., et al., *Blood–brain barrier dysfunction–induced inflammatory signaling in brain pathology and epileptogenesis*. Epilepsia, 2012. **53**: pp. 37–44.
66. David, Y., et al., *Astrocytic dysfunction in epileptogenesis: Consequence of altered potassium and glutamate homeostasis?* Journal of Neuroscience, 2009. **29**(34): pp. 10588–10599.
67. Kim, J.A., et al., *Astrocyte regulation of human brain capillary endothelial fibrinolysis*. Thrombosis Research, 2003. **112**(3): pp. 159–165.
68. Zetterberg, H. and K. Blennow, *Fluid markers of traumatic brain injury*. Molecular and Cellular Neuroscience, 2015. **66**: pp. 99–102.
69. Habgood, M.D., et al., *Changes in blood–brain barrier permeability to large and small molecules following traumatic brain injury in mice*. European Journal of Neuroscience, 2007. **25**(1): pp. 231–238.
70. Shlosberg, D., et al., *Blood–brain barrier breakdown as a therapeutic target in traumatic brain injury*. Nature Reviews Neurology, 2010. **6**(7): pp. 393–403.
71. Sweeney, M.D., et al., *Blood-brain barrier: From physiology to disease and back*. Physiological Reviews, 2018. **99**(1): pp. 21–78.
72. Kortekaas, R., et al., *Blood–brain barrier dysfunction in parkinsonian midbrain in vivo*. Annals of Neurology, 2005. **57**(2): pp. 176–179.
73. Gao, Y., et al., *Tau in Alzheimer's disease: Mechanisms and therapeutic strategies*. Current Alzheimer Research, 2018. **15**(3): pp. 283–300.
74. Boimel, M., et al., *Efficacy and safety of immunization with phosphorylated tau against neurofibrillary tangles in mice*. Experimental Neurology, 2010. **224**(2): pp. 472–485.
75. Szaruga, M., et al., *Qualitative changes in human γ-secretase underlie familial Alzheimer's disease*. Journal of Experimental Medicine, 2015. **212**(12): pp. 2003–2013.
76. Montagne, A., Z. Zhao, and B.V. Zlokovic, *Alzheimer's disease: A matter of blood–brain barrier dysfunction?* Journal of Experimental Medicine, 2017. **214**(11): pp. 3151–3169.
77. Marín-Muñoz, J., et al., *Novel mutation (Gly212Val) in the PS2 gene associated with early-onset familial Alzheimer's disease*. Journal of Alzheimer's Disease, 2016. **53**(1): pp. 73–78.
78. Zlokovic, B.V., *Cerebrovascular effects of apolipoprotein E: Implications for Alzheimer disease*. JAMA Neurology, 2013. **70**(4): pp. 440–444.
79. Halliday, M.R., et al., *Relationship between cyclophilin a levels and matrix metalloproteinase 9 activity in cerebrospinal fluid of cognitively normal apolipoprotein e4 carriers and blood-brain barrier breakdown*. JAMA Neurology, 2013. **70**(9): pp. 1198–1200.
80. Suri, S., et al., *Reduced cerebrovascular reactivity in young adults carrying the APOE ε4 allele*. Alzheimer's & Dementia, 2015. **11**(6): pp. 648–657.
81. Zarranz, J.J., et al., *Iowa APP mutation-related hereditary cerebral amyloid angiopathy (CAA): A new family from Spain*. Journal of the Neurological Sciences, 2016. **363**: pp. 55–56.
82. Saito, S. and M. Ihara, *Interaction between cerebrovascular disease and Alzheimer pathology*. Current Opinion in Psychiatry, 2016. **29**(2): pp. 168–173.
83. Barrier, B.-B., *From physiology to disease and back*. Physiological Reviews, 2019. **99**.
84. Niedermeyer, S., M. Murn, and P.J. Choi, *Respiratory failure in amyotrophic lateral sclerosis*. Chest, 2019. **155**(2): pp. 401–408.
85. Grad, L.I., et al., *Clinical spectrum of amyotrophic lateral sclerosis (ALS)*. Cold Spring Harbor Perspectives in Medicine, 2017. **7**(8): p. a024117.
86. Al-Chalabi, A., L.H. Van Den Berg, and J. Veldink, *Gene discovery in amyotrophic lateral sclerosis: Implications for clinical management*. Nature Reviews Neurology, 2017. **13**(2): pp. 96–104.
87. Winkler, E.A., et al., *Blood–spinal cord barrier disruption contributes to early motor-neuron degeneration in ALS-model mice*. Proceedings of the National Academy of Sciences, 2014. **111**(11): pp. E1035–E1042.
88. Henkel, J.S., et al., *Decreased mRNA expression of tight junction proteins in lumbar spinal cords of patients with ALS*. Neurology, 2009. **72**(18): pp. 1614–1616.
89. Drouin-Ouellet, J., et al., *Cerebrovascular and blood–brain barrier impairments in Huntington's disease: Potential implications for its pathophysiology*. Annals of Neurology, 2015. **78**(2): pp. 160–177.
90. Walker, F.O., *Huntington's disease*. Lancet, 2007. **369**(9557): pp. 218–228.
91. Lim, R.G., et al., *Huntington's disease iPSC-derived brain microvascular endothelial cells reveal WNT-mediated angiogenic and blood-brain barrier deficits*. Cell Reports, 2017. **19**(7): pp. 1365–1377.
92. Silvestroni, A., et al., *Distinct neuroinflammatory profile in post-mortem human Huntington's disease*. Neuroreport, 2009. **20**(12): pp. 1098–1103.
93. Sengillo, J.D., et al., *Deficiency in mural vascular cells coincides with blood–brain barrier disruption in Alzheimer's disease*. Brain Pathology, 2013. **23**(3): pp. 303–310.
94. Li, J.-Q., L. Tan, and J.-T. Yu, *The role of the LRRK2 gene in Parkinsonism*. Molecular Neurodegeneration, 2014. **9**(1): pp. 1–17.
95. Forsythe, P., *Mast cells in neuroimmune interactions*. Trends in Neurosciences, 2019. **42**(1): pp. 43–55.
96. Hongge, L., et al., *The role of LRRK2 in the regulation of monocyte adhesion to endothelial cells*. Journal of Molecular Neuroscience, 2015. **55**: pp. 233–239.
97. Ortiz, G.G., et al., *Role of the blood–brain barrier in multiple sclerosis*. Archives of Medical Research, 2014. **45**(8): pp. 687–697.

98. Minagar, A. and J.S. Alexander, *Blood-brain barrier disruption in multiple sclerosis*. Multiple Sclerosis Journal, 2003. **9**(6): pp. 540–549.
99. Lim, S., et al., *dNP2 is a blood–brain barrier-permeable peptide enabling ctCTLA-4 protein delivery to ameliorate experimental autoimmune encephalomyelitis*. Nature Communications, 2015. **6**(1): p. 8244.
100. Hahn, J.S., W.C. Lannin, and M.M. Sarwal, *Microangiopathy of brain, retina, and inner ear (Susac's syndrome) in an adolescent female presenting as acute disseminated encephalomyelitis*. Pediatrics, 2004. **114**(1): pp. 276–281.
101. Bitra, R.K. and E. Eggenberger, *Review of Susac syndrome*. Current Opinion in Ophthalmology, 2011. **22**(6): pp. 472–476.
102. Vishnevskia-Dai, V., et al., *Susac syndrome: Clinical characteristics, clinical classification, and long-term prognosis*. Medicine, 2016. **95**(43).
103. Kayser, M.S. and J. Dalmau, *The emerging link between autoimmune disorders and neuropsychiatric disease*. Journal of Neuropsychiatry and Clinical Neurosciences, 2011. **23**(1): pp. 90–97.
104. Hughes, R.A.C., et al., *Pathogenesis of chronic inflammatory demyelinating polyradiculoneuropathy*. Journal of the Peripheral Nervous System, 2006. **11**(1): pp. 30–46.
105. Kuwabara, S. and S. Misawa, *Chronic inflammatory demyelinating polyneuropathy*. Myelin: Basic and Clinical Advances, 2019: pp. 333–343.
106. Whitesell, J., *Inflammatory neuropathies*. © Thieme Medical Publishers.
107. Kamm, C. and U.K. Zettl, *Autoimmune disorders affecting both the central and peripheral nervous system*. Autoimmunity Reviews, 2012. **11**(3): pp. 196–202.
108. Shahrizaila, N. and N. Yuki, *Bickerstaff brainstem encephalitis and Fisher syndrome: Anti-GQ1b antibody syndrome*. Journal of Neurology, Neurosurgery & Psychiatry, 2013. **84**(5): pp. 576–583.
109. Rosenberg, G.A., *Neurological diseases in relation to the blood-brain barrier*. Journal of Cerebral Blood Flow Metabolism, 2012. **32**(7): pp. 1139–1151.
110. Lee, R.H.C., et al., *Cerebral ischemia and neuroregeneration*. Neural Regeneration Research, 2018. **13**(3): pp. 373–385.
111. Xanthos, D.N. and J. Sandkühler, *Neurogenic neuroinflammation: Inflammatory CNS reactions in response to neuronal activity*. Nature Reviews Neuroscience, 2014. **15**(1): pp. 43–53.
112. Chodobski, A., B.J. Zink, and J. Szmydynger-Chodobska, *Blood-brain barrier pathophysiology in traumatic brain injury*. Translational Stroke Research, 2011. **2**(4): pp. 492–516.
113. Shimizu, F., et al., *Sera from neuromyelitis optica patients disrupt the blood–brain barrier*. Journal of Neurology, Neurosurgery & Psychiatry, 2012. **83**(3): pp. 288–297.
114. Papadopoulos, M.C., J.L. Bennett, and A.S. Verkman, *Treatment of neuromyelitis optica: State-of-the-art and emerging therapies*. Nature Reviews Neurology, 2014. **10**(9): pp. 493–506.
115. Jeltsch-David, H. and S. Muller, *Neuropsychiatric systemic lupus erythematosus: Pathogenesis and biomarkers*. Nature Reviews Neurology, 2014. **10**(10): pp. 579–596.
116. Bertsias, G.K. and D.T. Boumpas, *Pathogenesis, diagnosis and management of neuropsychiatric SLE manifestations*. Nature Reviews Rheumatology, 2010. **6**(6): pp. 358–367.
117. Brito-Zerón, P., et al., *Sjögren syndrome*. Nature Reviews Disease Primers, 2016. **2**(1): pp. 1–20.
118. Byram, K., Hajj-Ali, R.A. and Calabrese, L., *CNS vasculitis: an approach to differential diagnosis and management*. Current Rheumatology Reports, 2018. **20**: pp. 1–7.
119. Pardridge, W.M., *The blood-brain barrier: Bottleneck in brain drug development*. NeuroRx, 2005. **2**: pp. 3–14.
120. Neuwelt, E., et al., *Strategies to advance translational research into brain barriers*. Lancet Neurology, 2008. **7**(1): pp. 84–96.
121. Golde, T.E., *Open questions for Alzheimer's disease immunotherapy*. Alzheimer's Research & Therapy, 2014. **6**(1): pp. 1–7.
122. Haddad, F., et al., *Dopamine and levodopa prodrugs for the treatment of Parkinson's disease*. Molecules, 2017. **23**(1): p. 40.
123. Zlokovic, B.V., *The blood-brain barrier in health and chronic neurodegenerative disorders*. Neuron, 2008. **57**(2): pp. 178–201.
124. Li, C. and J. Götz, *Tau-based therapies in neurodegeneration: Opportunities and challenges*. Nature Reviews Drug Discovery, 2017. **16**(12): pp. 863–883.
125. Soderquist, R.G. and M.J. Mahoney, *Central nervous system delivery of large molecules: Challenges and new frontiers for intrathecally administered therapeutics*. Expert Opinion on Drug Delivery, 2010. **7**(3): pp. 285–293.
126. Wu, S.-K., et al., *Focused ultrasound and microbubbles-mediated drug delivery to brain tumor*. Pharmaceutics, 2020. **13**(1): p. 15.
127. Akiyama, Y., et al., *Advantages and disadvantages of combined chemotherapy with carmustine wafer and bevacizumab in patients with newly diagnosed glioblastoma: A single-institutional experience*. World Neurosurgery, 2018. **113**: pp. e508–e514.
128. Stockwell, J., et al., *Novel central nervous system drug delivery systems*. Chemical Biology & Drug Design, 2014. **83**(5): pp. 507–520.
129. Aleynik, A., et al., *Stem cell delivery of therapies for brain disorders*. Clinical and Translational Medicine, 2014. **3**: pp. 1–10.
130. Pandit, R., L. Chen, and J. Götz, *The blood-brain barrier: Physiology and strategies for drug delivery*. Advanced Drug Delivery Reviews, 2020. **165**: pp. 1–14.
131. Chen, W., et al., *BBB-crossing adeno-associated virus vector: An excellent gene delivery tool for CNS disease treatment*. Journal of Controlled Release, 2021. **333**: pp. 129–138.
132. Liu, Y., et al., *Targeted exosome-mediated delivery of opioid receptor Mu siRNA for the treatment of morphine relapse*. Scientific Reports, 2015. **5**(1): p. 17543.
133. Zhang, S., et al., *Ultrasound-assisted brain delivery of nanomedicines for brain tumor therapy: Advance and prospect*. Journal of Nanobiotechnology, 2022. **20**(1): pp. 1–27.
134. He, Q., et al., *Towards improvements for penetrating the blood–brain barrier—recent progress from a material and pharmaceutical perspective*. Cells, 2018. **7**(4): p. 24.
135. Burgess, A., et al., *Focused ultrasound-mediated drug delivery through the blood–brain barrier*. Expert Review of Neurotherapeutics, 2015. **15**(5): pp. 477–491.
136. Neganova, M.E., et al., *Benefits and limitations of nanomedicine treatment of brain cancers and age-dependent neurodegenerative disorders*. In Seminars in Cancer Biology, 2022. **86**: pp. 805–833.
137. Akhtar, A., et al., *Neurodegenerative diseases and effective drug delivery: A review of challenges and novel*

therapeutics. Journal of Controlled Release, 2021. **330**: pp. 1152–1167.
138. Ali, J., et al., *Potential of nanoparticulate drug delivery systems by intranasal administration*. Current Pharmaceutical Design, 2010. **16**(14): pp. 1644–1653.
139. Lisik, K. and A. Krokosz, *Application of carbon nanoparticles in oncology and regenerative medicine*. International Journal of Molecular Sciences, 2021. **22**(15): p. 8341.
140. Yang, T., et al., *Delivery of small interfering RNA to inhibit vascular endothelial growth factor in zebrafish using natural brain endothelia cell-secreted exosome nanovesicles for the treatment of brain cancer*. The AAPS Journal, 2017. **19**: pp. 475–486.
141. Singh, V.K. and B.B. Subudhi, *Development and characterization of lysine-methotrexate conjugate for enhanced brain delivery*. Drug Delivery, 2016. **23**(7): pp. 2327–2337.
142. Ma, K., et al., *Progress of cationic gene delivery reagents for non-viral vector*. Applied Microbiology and Biotechnology, 2021. **105**: pp. 525–538.
143. Mashel, T.V., et al., *Overcoming the delivery problem for therapeutic genome editing: Current status and perspective of non-viral methods*. Biomaterials, 2020. **258**: p. 120282.
144. Choudhury, S.R., et al., *Viral vectors for therapy of neurologic diseases*. Neuropharmacology, 2017. **120**: pp. 63–80.
145. Sayed, N., et al., *Gene therapy: Comprehensive overview and therapeutic applications*. Life Sciences, 2022: p. 120375.
146. Dong, X., *Current strategies for brain drug delivery*. Theranostics, 2018. **8**(6): p. 1481.
147. Sondhi, D., et al., *Slowing late infantile Batten disease by direct brain parenchymal administration of a rh.10 adeno-associated virus expressing CLN2*. Science Translational Medicine, 2020. **12**(572).
148. Masserini, M., *Nanoparticles for brain drug delivery*. International Scholarly Research Notices, 2013. **2013**.
149. Saraiva, C., et al., *Nanoparticle-mediated brain drug delivery: Overcoming blood–brain barrier to treat neurodegenerative diseases*. Journal of Controlled Release, 2016. **235**: pp. 34–47.
150. Wohlfart, S., S. Gelperina, and J. Kreuter, *Transport of drugs across the blood–brain barrier by nanoparticles*. Journal of Controlled Release, 2012. **161**(2): pp. 264–273.
151. Huang, L. and Y. Liu, *In vivo delivery of RNAi with lipid-based nanoparticles*. Annual Review of Biomedical Engineering, 2011. **13**: pp. 507–530.
152. Jiang, Y., et al., *SOD1 nanozyme salvages ischemic brain by locally protecting cerebral vasculature*. Journal of Controlled Release, 2015. **213**: pp. 36–44.
153. Cook, R.L., et al., *A critical evaluation of drug delivery from ligand modified nanoparticles: Confounding small molecule distribution and efficacy in the central nervous system*. Journal of Controlled Release, 2015. **220**: pp. 89–97.
154. Kreuter, J., *Mechanism of polymeric nanoparticle-based drug transport across the blood-brain barrier (BBB)*. Journal of Microencapsulation, 2013. **30**(1): pp. 49–54.
155. Villringer, K., et al., *DCE-MRI blood–brain barrier assessment in acute ischemic stroke*. Neurology, 2017. **88**(5): pp. 433–440.
156. Yang, J., et al., *Multimodality MRI assessment of grey and white matter injury and blood-brain barrier disruption after intracerebral haemorrhage in mice*. Scientific Reports, 2017. **7**(1): pp. 1–12.
157. Rosenberg, G.A., *Neurological diseases in relation to the blood–brain barrier*. Journal of Cerebral Blood Flow & Metabolism, 2012. **32**(7): pp. 1139–1151.
158. Shiraishi, K., et al., *A polymeric micelle magnetic resonance imaging (MRI) contrast agent reveals blood–brain barrier (BBB) permeability for macromolecules in cerebral ischemia-reperfusion injury*. Journal of Controlled Release, 2017. **253**: pp. 165–171.
159. Haqqani, A.S., et al., *Method for isolation and molecular characterization of extracellular microvesicles released from brain endothelial cells*. Fluids and Barriers of the CNS, 2013. **10**: pp. 1–13.
160. Khatami, S.H., et al., *Exosomes: Promising delivery tools for overcoming blood-brain barrier and glioblastoma therapy*. Molecular Neurobiology, 2023: p. 1–20.
161. Choi, H., et al., *Strategies for targeted delivery of exosomes to the brain: Advantages and challenges*. Pharmaceutics, 2022. **14**(3): p. 672.
162. Heidarzadeh, M., et al., *Exosomal delivery of therapeutic modulators through the blood–brain barrier: Promise and pitfalls*. Cell & Bioscience, 2021. **11**: pp. 1–28.
163. Abdelsalam, M., et al., *Insights into exosome transport through the blood–brain barrier and the potential therapeutical applications in brain diseases*. Pharmaceuticals, 2023. **16**(4): p. 571.
164. Chen, J., et al., *Brain-targeted exosomes-based drug delivery system to overcome the treatment bottleneck of brainstem glioma*. Advanced Functional Materials, 2023: p. 2302378.
165. Barar, J., et al., *Blood-brain barrier transport machineries and targeted therapy of brain diseases*. BioImpacts: BI, 2016. **6**(4): p. 225.
166. Bellettato, C.M. and M. Scarpa, *Possible strategies to cross the blood–brain barrier*. Italian Journal of Pediatrics, 2018. **44**(2): pp. 127–133.
167. Banks, W.A., et al., *Transport of extracellular vesicles across the blood-brain barrier: Brain pharmacokinetics and effects of inflammation*. International Journal of Molecular Sciences, 2020. **21**(12): p. 4407.
168. Console, L., M. Scalise, and C. Indiveri, *Exosomes in inflammation and role as biomarkers*. Clinica Chimica Acta, 2019. **488**: pp. 165–171.
169. Spandana, K.M.A., et al., *A comprehensive review of nano drug delivery system in the treatment of CNS disorders*. Journal of Drug Delivery Science and Technology, 2020. **57**: p. 101628.
170. Khalid, M. and H.S. El-Sawy, *Polymeric nanoparticles: Promising platform for drug delivery*. International Journal of Pharmaceutics, 2017. **528**(1–2): pp. 675–691.
171. Barman, R.K., et al., *In vivo pharmacokinetic and hemocompatible evaluation of lyophilization induced nifedipine solid-lipid nanoparticle*. Pharmacology & Pharmacy, 2014. **5**(5): pp. 455–461.
172. Makwana, V., et al., *Solid lipid nanoparticles (SLN) of Efavirenz as lymph targeting drug delivery system: Elucidation of mechanism of uptake using chylomicron flow blocking approach*. International Journal of Pharmaceutics, 2015. **495**(1): pp. 439–446.
173. Ghasemiyeh, P. and S. Mohammadi-Samani, *Solid lipid nanoparticles and nanostructured lipid carriers as novel drug delivery systems: Applications, advantages and disadvantages*. Research in Pharmaceutical Sciences, 2018. **13**(4): p. 288.
174. Maya, S., et al., *Smart stimuli sensitive nanogels in cancer drug delivery and imaging: A review*. Current Pharmaceutical Design, 2013. **19**(41): pp. 7203–7218.

175. Hamidi, M., A. Azadi, and P. Rafiei, *Hydrogel nanoparticles in drug delivery*. Advanced Drug Delivery Reviews, 2008. **60**(15): pp. 1638–1649.
176. Kendre, P.N. and T.S. Satav, *Current trends and concepts in the design and development of nanogel carrier systems*. Polymer Bulletin, 2019. **76**: pp. 1595–1617.
177. Rai, A.K., et al., *Dendrimers: A potential carrier for targeted drug delivery system*. Pharmaceutical and Biological Evaluations, 2016. **3**(3): pp. 275–287.
178. Mecke, A., et al., *Lipid bilayer disruption by polycationic polymers: The roles of size and chemical functional group*. Langmuir, 2005. **21**(23): pp. 10348–10354.
179. Patil, S.M., S.S. Sawant, and N.K. Kunda, *Exosomes as drug delivery systems: A brief overview and progress update*. European Journal of Pharmaceutics and Biopharmaceutics, 2020. **154**: pp. 259–269.
180. Perets, N., et al., *Golden exosomes selectively target brain pathologies in neurodegenerative and neurodevelopmental disorders*. Nano Letters, 2019. **19**(6): pp. 3422–3431.
181. Fraietta, J.A., et al., *Disruption of TET2 promotes the therapeutic efficacy of CD19-targeted T cells*. Nature, 2018. **558**(7709): pp. 307–312.
182. Jang, S.C., et al., *Bioinspired exosome-mimetic nanovesicles for targeted delivery of chemotherapeutics to malignant tumors*. ACS Nano, 2013. **7**(9): pp. 7698–7710.
183. Shi, R., et al., *Maternal exosomes in diabetes contribute to the cardiac development deficiency*. Biochemical and Biophysical Research Communications, 2017. **483**(1): pp. 602–608.
184. Haney, M.J., et al., *Exosomes as drug delivery vehicles for Parkinson's disease therapy*. Journal of Controlled Release, 2015. **207**: pp. 18–30.
185. Lamichhane, T.N., R.S. Raiker, and S.M. Jay, *Exogenous DNA loading into extracellular vesicles via electroporation is size-dependent and enables limited gene delivery*. Molecular Pharmaceutics, 2015. **12**(10): pp. 3650–3657.
186. Lin, Y., et al., *Exosome–liposome hybrid nanoparticles deliver CRISPR/Cas9 system in MSCs*. Advanced Science, 2018. **5**(4): p. 1700611.
187. Gallego-Pérez, D., et al., *Topical tissue nano-transfection mediates non-viral stroma reprogramming and rescue*. Nature Nanotechnology, 2017. **12**(10): pp. 974–979.
188. Yang, Z., et al., *Large-scale generation of functional mRNA-encapsulating exosomes via cellular nanoporation*. Nature Biomedical Engineering, 2020. **4**(1): pp. 69–83.
189. Usman, W.M., et al., *Efficient RNA drug delivery using red blood cell extracellular vesicles*. Nature Communications, 2018. **9**(1): p. 2359.
190. Rajabi, H., et al., *Current status of used protocols for mesenchymal stem cell differentiation: A focus on insulin producing, osteoblast-like and neural cells*. Current Stem Cell Research & Therapy, 2019. **14**(7): pp. 570–578.
191. Rana, S., et al., *Toward tailored exosomes: The exosomal tetraspanin web contributes to target cell selection*. The International Journal of Biochemistry & Cell Biology, 2012. **44**(9): pp. 1574–1584.
192. Qiao, L., et al., *Tumor cell-derived exosomes home to their cells of origin and can be used as Trojan horses to deliver cancer drugs*. Theranostics, 2020. **10**(8): p. 3474.
193. Ju, S., et al., *Grape exosome-like nanoparticles induce intestinal stem cells and protect mice from DSS-induced colitis*. Molecular Therapy, 2013. **21**(7): pp. 1345–1357.
194. Wang, Q., et al., *Grapefruit-derived nanovectors use an activated leukocyte trafficking pathway to deliver therapeutic agents to inflammatory tumor sites hijacked leukocyte pathway for targeted delivery*. Cancer Research, 2015. **75**(12): pp. 2520–2529.
195. Cheng, Y., et al., *TNFα disrupts blood brain barrier integrity to maintain prolonged depressive-like behavior in mice*. Brain, Behavior, and Immunity, 2018. **69**: pp. 556–567.
196. Suzuki, H., T. Terasaki, and Y. Sugiyama, *Role of efflux transport across the blood-brain barrier and blood-cerebrospinal fluid barrier on the disposition of xenobiotics in the central nervous system*. Advanced Drug Delivery Reviews, 1997. **25**(2–3): pp. 257–285.
197. Cheng, L., W. Zhao, and A.F. Hill, *Exosomes and their role in the intercellular trafficking of normal and disease associated prion proteins*. Molecular Aspects of Medicine, 2018. **60**: pp. 62–68.
198. Feng, D., et al., *Cellular internalization of exosomes occurs through phagocytosis*. Traffic, 2010. **11**(5): pp. 675–687.
199. Zhang, J., et al., *Increasing of blood-brain tumor barrier permeability through transcellular and paracellular pathways by microbubble-enhanced diagnostic ultrasound in a C6 glioma model*. Frontiers in Neuroscience, 2017. **11**: p. 86.
200. Park, J., et al., *Evaluation of permeability, doxorubicin delivery, and drug retention in a rat brain tumor model after ultrasound-induced blood-tumor barrier disruption*. Journal of Controlled Release, 2017. **250**: pp. 77–85.
201. Aryal, M., et al., *Effects on P-glycoprotein expression after blood-brain barrier disruption using focused ultrasound and microbubbles*. PLoS One, 2017. **12**(1): p. e0166061.
202. Samiotaki, G., et al., *Pharmacokinetic analysis and drug delivery efficiency of the focused ultrasound-induced blood-brain barrier opening in non-human primates*. Magnetic Resonance Imaging, 2017. **37**: pp. 273–281.
203. Dasgupta, A., et al., *Ultrasound-mediated drug delivery to the brain: Principles, progress and prospects*. Drug Discovery Today: Technologies, 2016. **20**: pp. 41–48.
204. Xhima, K., D. Weber-Adrian, and J. Silburt, *Glutamate induces blood–brain barrier permeability through activation of N-methyl-D-aspartate receptors*. Journal of Neuroscience, 2016. **36**(49): pp. 12296–12298.
205. Al-Ahmady, Z.S., et al., *Selective liposomal transport through blood brain barrier disruption in ischemic stroke reveals two distinct therapeutic opportunities*. ACS Nano, 2019. **13**(11): pp. 12470–12486.
206. Yang, Y., Y.S. Phang, and Y. Zhao, Nanotechnology for stroke treatment, in *Nanotechnology for hematology, blood transfusion, and artificial blood*. Elsevier, 2022, pp. 339–369.
207. Gajbhiye, K.R., et al., *PEGylated nanocarriers: A promising tool for targeted delivery to the brain*. Colloids and Surfaces B: Biointerfaces, 2020. **187**: p. 110770.
208. Mojarad-Jabali, S., et al., *An update on actively targeted liposomes in advanced drug delivery to glioma*. International Journal of Pharmaceutics, 2021. **602**: p. 120645.
209. Ding, Z., Y. Jiang, and X. Liu, *Nanotechnology-based targeted drug delivery system for brain tumors*. Academic Press, 2018.
210. Zhang, Y., H. Qu, and X. Xue, *Blood–brain barrier penetrating liposomes with synergistic chemotherapy for glioblastoma treatment*. Biomaterials Science, 2022. **10**(2): pp. 423–434.
211. Hu, J., et al., *Pharmacokinetics and antitumor efficacy of DSPE-PEG2000 polymeric liposomes loaded with quercetin and temozolomide: Analysis of their effectiveness in enhancing the chemosensitization of drug-resistant glioma cells*. International Journal of Molecular Medicine, 2016. **37**(3): pp. 690–702.

212. Kim, S.-S., et al., *Encapsulation of temozolomide in a tumor-targeting nanocomplex enhances anti-cancer efficacy and reduces toxicity in a mouse model of glioblastoma.* Cancer Letters, 2015. **369**(1): pp. 250–258.
213. Ismail, M., et al., *Targeted liposomes for combined delivery of artesunate and temozolomide to resistant glioblastoma.* Biomaterials, 2022. **287**: p. 121608.
214. Lam, F.C., et al., *Enhanced efficacy of combined temozolomide and bromodomain inhibitor therapy for gliomas using targeted nanoparticles.* Nature Communications, 2018. **9**(1): p. 1991.
215. Bregoli, L., et al., *Nanomedicine applied to translational oncology: A future perspective on cancer treatment.* Nanomedicine: Nanotechnology, Biology and Medicine, 2016. **12**(1): pp. 81–103.
216. Gabizon, A.A., R.T.M. de Rosales, and N.M. La-Beck, *Translational considerations in nanomedicine: The oncology perspective.* Advanced Drug Delivery Reviews, 2020. **158**: pp. 140–157.
217. Torchilin, V.P., *Recent advances with liposomes as pharmaceutical carriers.* Nature Reviews Drug Discovery, 2005. **4**(2): pp. 145–160.
218. Jain, K.K. and K.K. Jain, *Nanopharmaceuticals.* The Handbook of Nanomedicine, 2017: pp. 201–271.
219. Amarandi, R.-M., et al., *Liposomal-based formulations: A path from basic research to temozolomide delivery inside glioblastoma tissue.* Pharmaceutics, 2022. **14**(2): p. 308.
220. Danon, J.J., T.A. Reekie, and M. Kassiou, *Challenges and opportunities in central nervous system drug discovery.* Trends in Chemistry, 2019. **1**(6): pp. 612–624.
221. Broadley, K.J. and D.R. Kelly, *Muscarinic receptor agonists and antagonists.* Molecules, 2001. **6**(3): pp. 142–193.
222. Ghosh, C., et al., *Overexpression of pregnane X and glucocorticoid receptors and the regulation of cytochrome P450 in human epileptic brain endothelial cells.* Epilepsia, 2017. **58**(4): pp. 576–585.
223. Ghosh, C., et al., *Modulation of glucocorticoid receptor in human epileptic endothelial cells impacts drug biotransformation in an in vitro blood–brain barrier model.* Epilepsia, 2018. **59**(11): pp. 2049–2060.
224. Hossain, M., et al., *Heat shock proteins accelerate the maturation of brain endothelial cell glucocorticoid receptor in focal human drug-resistant epilepsy.* Molecular Neurobiology, 2020. **57**: pp. 4511–4529.
225. Ferguson, S.S., et al., *Human CYP2C8 is transcriptionally regulated by the nuclear receptors constitutive androstane receptor, pregnane X receptor, glucocorticoid receptor, and hepatic nuclear factor 4α.* Molecular Pharmacology, 2005. **68**(3): pp. 747–757.
226. Canet, G., et al., *Central role of glucocorticoid receptors in Alzheimer's disease and depression.* Frontiers in Neuroscience, 2018. **12**: p. 739.
227. Loose, D.S., E.P. Stover, and D. Feldman, *Ketoconazole binds to glucocorticoid receptors and exhibits glucocorticoid antagonist activity in cultured cells.* Journal of Clinical Investigation, 1983. **72**(1): pp. 404–408.
228. Baglietto-Vargas, D., et al., *Mifepristone alters amyloid precursor protein processing to preclude amyloid beta and also reduces tau pathology.* Biological Psychiatry, 2013. **74**(5): pp. 357–366.
229. Chen, J.-F., H.K. Eltzschig, and B.B. Fredholm, *Adenosine receptors as drug targets—what are the challenges?* Nature Reviews Drug Discovery, 2013. **12**(4): pp. 265–286.
230. Navarro, G., et al., *Targeting cannabinoid CB2 receptors in the central nervous system: Medicinal chemistry approaches with focus on neurodegenerative disorders.* Frontiers in Neuroscience, 2016. **10**: p. 406.
231. Maramai, S., et al., *Dopamine D3 receptor antagonists as potential therapeutics for the treatment of neurological diseases.* Frontiers in Neuroscience, 2016. **10**: p. 451.
232. Georgieva, J.V., D. Hoekstra, and I.S. Zuhorn, *Smuggling drugs into the brain: An overview of ligands targeting transcytosis for drug delivery across the blood–brain barrier.* Pharmaceutics, 2014. **6**(4): pp. 557–583.
233. Candela, P., et al., *Physiological pathway for low-density lipoproteins across the blood-brain barrier: Transcytosis through brain capillary endothelial cells in vitro.* Endothelium, 2008. **15**(5–6): pp. 254–264.
234. Kim, J.-Y., et al., *Brain-targeted delivery of protein using chitosan and RVG peptide-conjugated, pluronic-based nano-carrier.* Biomaterials, 2013. **34**(4): pp. 1170–1178.
235. Liu, Y., et al., *Brain-targeting gene delivery and cellular internalization mechanisms for modified rabies virus glycoprotein RVG29 nanoparticles.* Biomaterials, 2009. **30**(25): pp. 4195–4202.
236. Chen, H., et al., *Release of hydrophobic molecules from polymer micelles into cell membranes revealed by Förster resonance energy transfer imaging.* Proceedings of the National Academy of Sciences, 2008. **105**(18): pp. 6596–6601.
237. Xu, P., et al., *Intracellular drug delivery by poly (lactic-co-glycolic acid) nanoparticles, revisited.* Molecular Pharmaceutics, 2009. **6**(1): pp. 190–201.
238. Tashima, T., *Smart strategies for therapeutic agent delivery into brain across the blood–brain barrier using receptor-mediated transcytosis.* Chemical and Pharmaceutical Bulletin, 2020. **68**(4): pp. 316–325.
239. Wevers, N.R., et al., *A perfused human blood–brain barrier on-a-chip for high-throughput assessment of barrier function and antibody transport.* Fluids and Barriers of the CNS, 2018. **15**(1): pp. 1–12.
240. Freskgård, P.-O. and E. Urich, *Antibody therapies in CNS diseases.* Neuropharmacology, 2017. **120**: pp. 38–55.
241. Wouters, Y., et al., *Identification and in vivo characterization of a brain-penetrating nanobody.* Fluids and Barriers of the CNS, 2020. **17**: pp. 1–10.
242. Okuyama, T., et al., *A phase 2/3 trial of pabinafusp alfa, IDS fused with anti-human transferrin receptor antibody, targeting neurodegeneration in MPS-II.* Molecular Therapy, 2021. **29**(2): pp. 671–679.
243. Giugliani, R., et al., *Neurocognitive and somatic stabilization in pediatric patients with severe Mucopolysaccharidosis Type I after 52 weeks of intravenous brain-penetrating insulin receptor antibody-iduronidase fusion protein (valanafusp alpha): An open label phase 1–2 trial.* Orphanet Journal of Rare Diseases, 2018. **13**: pp. 1–11.
244. Schenk, G.J. and H.E. de Vries, *Altered blood–brain barrier transport in neuro-inflammatory disorders.* Drug Discovery Today: Technologies, 2016. **20**: pp. 5–11.
245. Jeong, S.Y., et al., *Iron insufficiency compromises motor neurons and their mitochondrial function in Irp2-null mice.* PLoS One, 2011. **6**(10): p. e25404.
246. Bien-Ly, N., et al., *Lack of widespread BBB disruption in Alzheimer's disease models: Focus on therapeutic antibodies.* Neuron, 2015. **88**(2): pp. 289–297.
247. Pulgar, V.M., *Transcytosis to cross the blood brain barrier, new advancements and challenges.* Frontiers in Neuroscience, 2019. **12**: p. 1019.
248. Zhao, P., N. Zhang, and Z. An, *Engineering antibody and protein therapeutics to cross the blood–brain barrier.* Antibody Therapeutics, 2022. **5**(4): pp. 311–331.

249. Sonoda, H., et al., *Treatment of neuronopathic mucopolysaccharidoses with blood–brain barrier-crossing enzymes: Clinical application of receptor-mediated transcytosis.* Pharmaceutics, 2022. **14**(6): p. 1240.
250. Edavettal, S., et al., *Enhanced delivery of antibodies across the blood-brain barrier via TEMs with inherent receptor-mediated phagocytosis.* Medicine, 2022. **3**(12): pp. 860–882.
251. Arguello, A., et al., *Molecular architecture determines brain delivery of a transferrin receptor–targeted lysosomal enzyme.* Journal of Experimental Medicine, 2022. **219**(3): p. e20211057.
252. Sharma, G., et al., *Cell penetrating peptide tethered bi-ligand liposomes for delivery to brain in vivo: Biodistribution and transfection.* Journal of Controlled Release, 2013. **167**(1): pp. 1–10.
253. Zhang, Y., et al., *Lysosomal enzyme replacement of the brain with intravenous non-viral gene transfer.* Pharmaceutical Research, 2008. **25**: pp. 400–406.
254. Zhang, Y. and W.M. Pardridge, *Near complete rescue of experimental Parkinson's disease with intravenous, non-viral GDNF gene therapy.* Pharmaceutical Research, 2009. **26**: pp. 1059–1063.
255. Karatas, H., et al., *A nanomedicine transports a peptide caspase-3 inhibitor across the blood–brain barrier and provides neuroprotection.* Journal of Neuroscience, 2009. **29**(44): pp. 13761–13769.
256. Taylor, E.M. and E.H. Morgan, *Developmental changes in transferrin and iron uptake by the brain in the rat.* Developmental Brain Research, 1990. **55**(1): pp. 35–42.
257. Crowe, A. and E.H. Morgan, *Iron and transferrin uptake by brain and cerebrospinal fluid in the rat.* Brain Research, 1992. **592**(1–2): pp. 8–16.
258. Shin, S.-U., et al., *Transferrin-antibody fusion proteins are effective in brain targeting.* Proceedings of the National Academy of Sciences, 1995. **92**(7): pp. 2820–2824.
259. Mishra, V., et al., *Targeted brain delivery of AZT via transferrin anchored pegylated albumin nanoparticles.* Journal of Drug Targeting, 2006. **14**(1): pp. 45–53.
260. Qian, Z.M., et al., *Targeted drug delivery via the transferrin receptor-mediated endocytosis pathway.* Pharmacological Reviews, 2002. **54**(4): pp. 561–587.
261. Lee, H.J., et al., *Targeting rat anti-mouse transferrin receptor monoclonal antibodies through blood-brain barrier in mouse.* Journal of Pharmacology and Experimental Therapeutics, 2000. **292**(3): pp. 1048–1052.
262. Zhang, Y. and W.M. Pardridge, *Blood–brain barrier targeting of BDNF improves motor function in rats with middle cerebral artery occlusion.* Brain Research, 2006. **1111**(1): p. 227–229.
263. Lyons, M.K., R.E. Anderson, and F.B. Meyer, *Basic fibroblast growth factor promotes in vivo cerebral angiogenesis in chronic forebrain ischemia.* Brain Research, 1991. **558**(2): pp. 315–320.
264. Song, B.-W., et al., *Enhanced neuroprotective effects of basic fibroblast growth factor in regional brain ischemia after conjugation to a blood-brain barrier delivery vector.* Journal of Pharmacology and Experimental Therapeutics, 2002. **301**(2): pp. 605–610.
265. Kordower, J.H., et al., *Intravenous administration of a transferrin receptor antibody-nerve growth factor conjugate prevents the degeneration of cholinergic striatal neurons in a model of Huntington disease.* Proceedings of the National Academy of Sciences, 1994. **91**(19): pp. 9077–9080.
266. Huwyler, J., D. Wu, and W.M. Pardridge, *Brain drug delivery of small molecules using immunoliposomes.* Proceedings of the National Academy of Sciences, 1996. **93**(24): pp. 14164–14169.
267. Huwyler, J., J. Yang, and W.M. Pardridge, *Receptor mediated delivery of daunomycin using immunoliposomes: Pharmacokinetics and tissue distribution in the rat.* Journal of Pharmacology and Experimental Therapeutics, 1997. **282**(3): pp. 1541–1546.
268. Shi, N., R.J. Boado, and W.M. Pardridge, *Antisense imaging of gene expression in the brain in vivo.* Proceedings of the National Academy of Sciences, 2000. **97**(26): pp. 14709–14714.
269. Suzuki, T., et al., *Imaging endogenous gene expression in brain cancer in vivo with 111In-peptide nucleic acid antisense radiopharmaceuticals and brain drug-targeting technology.* Journal of Nuclear Medicine, 2004. **45**(10): pp. 1766–1775.
270. Shi, N., et al., *Brain-specific expression of an exogenous gene after iv administration.* Proceedings of the National Academy of Sciences, 2001. **98**(22): pp. 12754–12759.
271. Lee, H.J., et al., *Imaging gene expression in the brain in vivo in a transgenic mouse model of Huntington's disease with an antisense radiopharmaceutical and drug-targeting technology.* Journal of Nuclear Medicine, 2002. **43**(7): pp. 948–956.
272. Lee, H.J., et al., *Imaging brain amyloid of Alzheimer disease in vivo in transgenic mice with an Aβ peptide radiopharmaceutical.* Journal of Cerebral Blood Flow & Metabolism, 2002. **22**(2): pp. 223–231.
273. Friden, P.M., et al., *Anti-transferrin receptor antibody and antibody-drug conjugates cross the blood-brain barrier.* Proceedings of the National Academy of Sciences, 1991. **88**(11): pp. 4771–4775.
274. Friden, P.M., et al., *Characterization, receptor mapping and blood-brain barrier transcytosis of antibodies to the human transferrin receptor.* Journal of Pharmacology and Experimental Therapeutics, 1996. **278**(3): pp. 1491–1498.
275. Gao, W., et al., *Rapid and efficient crossing blood-brain barrier: Hydrophobic drug delivery system based on propionylated amylose helix nanoclusters.* Biomaterials, 2017. **113**: pp. 133–144.
276. Prathipati, P., J. Zhu, and X. Dong, *Development of novel HDL-mimicking α-tocopherol-coated nanoparticles to encapsulate nerve growth factor and evaluation of biodistribution.* European Journal of Pharmaceutics and Biopharmaceutics, 2016. **108**: pp. 126–135.
277. Zhu, J. and X. Dong, *Preparation and characterization of novel HDL-mimicking nanoparticles for nerve growth factor encapsulation.* Journal of Visualized Experiments: Jove, 2017. **123**: p. e55584.
278. Rajora, M.A., et al., *Tailored theranostic apolipoprotein E3 porphyrin-lipid nanoparticles target glioblastoma.* Chemical Science, 2017. **8**(8): pp. 5371–5384.
279. Sorrentino, N.C., et al., *A highly secreted sulphamidase engineered to cross the blood-brain barrier corrects brain lesions of mice with mucopolysaccharidoses type IIIA.* EMBO Molecular Medicine, 2013. **5**(5): pp. 675–690.
280. Wang, Q., et al., *ApoE mimetic peptide targeted nanoparticles carrying a BRD4 inhibitor for treating Medulloblastoma in mice.* Journal of Controlled Release, 2020. **323**: pp. 463–474.
281. Ouyang, J., et al., *Doxorubicin delivered via ApoE-directed reduction-sensitive polymersomes potently inhibit*

orthotopic human glioblastoma xenografts in nude mice. International Journal of Nanomedicine, 2021. **16**: p. 4105.
282. Spencer, B., et al., *Systemic peptide mediated delivery of an siRNA targeting α-syn in the CNS ameliorates the neurodegenerative process in a transgenic model of Lewy body disease.* Neurobiology of Disease, 2019. **127**: pp. 163–177.
283. He, C., et al., *LRP1-mediated pH-sensitive polymersomes facilitate combination therapy of glioblastoma in vitro and in vivo.* Journal of Nanobiotechnology, 2021. **19**: pp. 1–17.
284. Xu, H., et al., *Angiopep-2-modified calcium Arsenite-loaded liposomes for targeted and pH-responsive delivery for anti-glioma therapy.* Biochemical and Biophysical Research Communications, 2021. **551**: pp. 14–20.
285. Uchida, Y., et al., *Quantitative targeted absolute proteomics of human blood–brain barrier transporters and receptors.* Journal of Neurochemistry, 2011. **117**(2): pp. 333–345.
286. Baum, L., et al., *Low density lipoprotein receptor related protein gene amplification and 766T polymorphism in astrocytomas.* Neuroscience Letters, 1998. **256**(1): pp. 5–8.
287. Thomas, F.C., et al., *Uptake of ANG1005: A novel paclitaxel derivative, through the blood-brain barrier into brain and experimental brain metastases of breast cancer.* Pharmaceutical Research, 2009. **26**: pp. 2486–2494.
288. Li, Y., et al., *Delivery of a peptide-drug conjugate targeting the blood brain barrier improved the efficacy of paclitaxel against glioma.* Oncotarget, 2016. **7**(48): p. 79401.
289. Simonneau, C., et al., *Investigating receptor-mediated antibody transcytosis using blood–brain barrier organoid arrays.* Fluids and Barriers of the CNS, 2021. **18**(1): pp. 1–17.
290. Ellis, L., J. Sissom, and A. Levitan, *Truncation of the ectodomain of the human insulin receptor results in secretion of a soluble insulin binding protein from transfected CHO cells.* Journal of Molecular Recognition, 1988. **1**(1): pp. 25–31.
291. Bickel, U., T. Yoshikawa, and W.M. Pardridge, *Delivery of peptides and proteins through the blood–brain barrier.* Advanced Drug Delivery Reviews, 2001. **46**(1–3): pp. 247–279.
292. Pardridge, W.M., et al., *Human insulin receptor monoclonal antibody undergoes high affinity binding to human brain capillaries in vitro and rapid transcytosis through the blood–brain barrier in vivo in the primate.* Pharmaceutical Research, 1995. **12**: pp. 807–816.
293. Sheff, J., et al., *Epitope mapping of a blood–brain barrier crossing antibody targeting the cysteine-rich region of IGF1R using hydrogen-exchange mass spectrometry enabled by electrochemical reduction.* Journal of Biochemistry, 2023. **173**(2): pp. 95–105.
294. Boado, R.J., et al., *AGT-181: Expression in CHO cells and pharmacokinetics, safety, and plasma iduronidase enzyme activity in Rhesus monkeys.* Journal of Biotechnology, 2009. **144**(2): pp. 135–141.
295. Alata, W., et al., *Targeting insulin-like growth factor-1 receptor (IGF1R) for brain delivery of biologics.* FASEB Journal, 2022. **36**(3): p. e22208.
296. Bolcaen, J., et al., *A perspective on the radiopharmaceutical requirements for imaging and therapy of glioblastoma.* Theranostics, 2021. **11**(16): p. 7911.
297. Jokar, S., et al., *Synthesis, biological evaluation and preclinical study of a novel 99mTc-peptide: A targeting probe of amyloid-β plaques as a possible diagnostic agent for Alzheimer's disease.* Bioorganic Chemistry, 2020. **99**: pp. 103857.
298. Zhang, Y., F. Schlachetzki, and W.M. Pardridge, *Global non-viral gene transfer to the primate brain following intravenous administration.* Molecular Therapy, 2003. **7**(1): pp. 11–18.
299. Zhang, Y., et al., *Organ-specific gene expression in the Rhesus monkey eye following intravenous non-viral gene transfer.* Molecular Vision, 2003. **9**: pp. 465–472.
300. Boado, R.J., et al., *Genetic engineering, expression, and activity of a fusion protein of a human neurotrophin and a molecular Trojan horse for delivery across the human blood–brain barrier.* Biotechnology and Bioengineering, 2007. **97**(6): pp. 1376–1386.
301. Wang, Y., et al., *Insulin-Like Growth Factor-II/Cation-Independent Mannose 6-Phosphate Receptor in Neurodegenerative Diseases.* Molecular Neurobiology, 2017. **54**(4): pp. 2636–2658.
302. Buga, A.M., et al., *The genomic response of the ipsilateral and contralateral cortex to stroke in aged rats.* Journal of Cellular and Molecular Medicine, 2008. **12**(6b): pp. 2731–2753.
303. Latour, A., et al., *Omega-3 fatty acids deficiency aggravates glutamatergic synapse and astroglial aging in the rat hippocampal CA 1.* Aging Cell, 2013. **12**(1): pp. 76–84.
304. Lewis, D.K., et al., *Age-related severity of focal ischemia in female rats is associated with impaired astrocyte function.* Neurobiology of Aging, 2012. **33**(6): p. 1123–e1.
305. Chisholm, N.C. and F. Sohrabji, *Astrocytic response to cerebral ischemia is influenced by sex differences and impaired by aging.* Neurobiology of Disease, 2016. **85**: pp. 245–253.
306. Wang, J., et al., *Gene therapy of adeno-associated virus (AAV) vectors in preclinical models of ischemic stroke.* CNS Neuroscience & Therapeutics, 2023. **29**(12): pp. 3725–3740.
307. Okoreeh, A.K., S. Bake, and F. Sohrabji, *Astrocyte-specific insulin-like growth factor-1 gene transfer in aging female rats improves stroke outcomes.* Glia, 2017. **65**(7): pp. 1043–1058.
308. Bell, R.D., et al., *Pericytes control key neurovascular functions and neuronal phenotype in the adult brain and during brain aging.* Neuron, 2010. **68**(3): pp. 409–427.
309. Zhou, W., et al., *Permeability and distribution of nerve growth factor in the brain of neonatal rats by periphery venous injection in hypoxic-ischemic state.* SpringerPlus, 2016. **5**: pp. 1–9.
310. Wang, X., et al., *Melatonin alleviates lipopolysaccharide-compromised integrity of blood–brain barrier through activating AMP-activated protein kinase in old mice.* Aging Cell, 2017. **16**(2): pp. 414–421.
311. Jackson, S., et al., *The effect of regadenoson on the integrity of the human blood–brain barrier: A pilot study.* Journal of Neuro-Oncology, 2017. **132**: pp. 513–519.
312. Prados, M.D., et al., *A randomized, double-blind, placebo-controlled, phase 2 study of RMP-7 in combination with carboplatin administered intravenously for the treatment of recurrent malignant glioma.* Neuro-Oncology, 2003. **5**(2): pp. 96–103.
313. Cai, Z., et al., *Preparation and evaluation of sustained-release solid dispersions co-loading gastrodin with borneol as an oral brain-targeting enhancer.* Acta Pharmaceutica Sinica B, 2014. **4**(1): pp. 86–93.
314. Zhang, Q., et al., *Improved blood–brain barrier distribution: Effect of borneol on the brain pharmacokinetics of kaempferol in rats by in vivo microdialysis sampling.* Journal of Ethnopharmacology, 2015. **162**: pp. 270–277.

315. Yi, T., et al., *Enhancing both oral bioavailability and brain penetration of puerarin using borneol in combination with preparation technologies*. Drug Delivery, 2017. **24**(1): pp. 422–429.
316. Reinhold, A.K. and H.L. Rittner, *Barrier function in the peripheral and central nervous system—A review*. Pflügers Archiv-European Journal of Physiology, 2017. **469**(1): pp. 123–134.
317. Komarova, Y.A., et al., *Protein interactions at endothelial junctions and signaling mechanisms regulating endothelial permeability*. Circulation Research, 2017. **120**(1): pp. 179–206.
318. Sahin, D., et al., *Changes in electroencephalographic characteristics and blood-brain barrier permeability in WAG/Rij rats with cortical dysplasia*. Epilepsy & Behavior, 2017. **67**: pp. 70–76.
319. Danielski, L.G., et al., *Brain barrier breakdown as a cause and consequence of neuroinflammation in sepsis*. Molecular Neurobiology, 2018. **55**: pp. 1045–1053.
320. Abrahamov, D., et al., *Blood–brain barrier disruption after cardiopulmonary bypass: Diagnosis and correlation to cognition*. The Annals of Thoracic Surgery, 2017. **104**(1): pp. 161–169.
321. Renú, A., et al., *Vessel wall enhancement and blood–cerebrospinal fluid barrier disruption after mechanical thrombectomy in acute ischemic stroke*. Stroke, 2017. **48**(3): pp. 651–657.
322. Wong, S.M., et al., *Measuring subtle leakage of the blood–brain barrier in cerebrovascular disease with DCE-MRI: Test–retest reproducibility and its influencing factors*. Journal of Magnetic Resonance Imaging, 2017. **46**(1): pp. 159–166.
323. Harris, N.M., et al., *Nano-particle delivery of brain derived neurotrophic factor after focal cerebral ischemia reduces tissue injury and enhances behavioral recovery*. Pharmacology Biochemistry and Behavior, 2016. **150**: pp. 48–56.
324. Merali, Z., et al., *Evolution of blood-brain-barrier permeability after acute ischemic stroke*. PLoS One, 2017. **12**(2): p. e0171558.
325. Sandoval, K.E. and K.A. Witt, *Blood-brain barrier tight junction permeability and ischemic stroke*. Neurobiology of Disease, 2008. **32**(2): pp. 200–219.
326. Zhang, Y., et al., *Temporal analysis of blood–brain barrier disruption and cerebrospinal fluid matrix metalloproteinases in rhesus monkeys subjected to transient ischemic stroke*. Journal of Cerebral Blood Flow & Metabolism, 2017. **37**(8): pp. 2963–2974.
327. Ishii, T., et al., *Amelioration of cerebral ischemia–reperfusion injury based on liposomal drug delivery system with asialo-erythropoietin*. Journal of Controlled Release, 2012. **160**(1): pp. 81–87.
328. Fukuta, T., et al., *Combination therapy with liposomal neuroprotectants and tissue plasminogen activator for treatment of ischemic stroke*. FASEB Journal, 2017. **31**(5): pp. 1879–1890.
329. Karki, K., J.R. Ewing, and M.M. Ali, *Targeting glioma with a dual mode optical and paramagnetic nanoprobe across the blood-brain tumor barrier*. Journal of Nanomedicine & Nanotechnology, 2016. **7**(4).
330. Warren, M.S., et al., *Comparative gene expression profiles of ABC transporters in brain microvessel endothelial cells and brain in five species including human*. Pharmacological Research, 2009. **59**(6): pp. 404–413.
331. Sarin, H., et al., *Effective transvascular delivery of nanoparticles across the blood-brain tumor barrier into malignant glioma cells*. Journal of Translational Medicine, 2008. **6**(1): pp. 1–15.
332. Sarin, H., *Recent progress towards development of effective systemic chemotherapy for the treatment of malignant brain tumors*. Journal of Translational Medicine, 2009. **7**(1): pp. 1–14.
333. Stegmayr, C., et al., *Influence of Bevacizumab on blood–brain barrier permeability and O-(2–18F-Fluoroethyl)-l-Tyrosine uptake in rat gliomas*. Journal of Nuclear Medicine, 2017. **58**(5): pp. 700–705.

5 Introduction to Nanomedicine, Nanotechnology, and Nanopharmaceuticals

Abdullah M. M. Elbadry and Hend H. Mohamed*

5.1 INTRODUCTION

For many years, nanomedicine has promised to transform how millions of people receive treatment due to the fact that they have an impact on cells and molecules within organs and tissues. Currently, research is focused on how nanotechnology can be applied in medicine. It is used for developing medical devices, encapsulating and delivering medications, diagnostics, genes, and cell and tissue engineering (Linkov et al., 2008). The term "nanoparticles (NPs)" refers to the smallest particles, typically between 1 and 100 nm (nanometers) in size. Particles are created by manipulating metals or chemicals, either naturally or artificially. In various industries, including cosmetics, agriculture, and pharmaceuticals, synthesized NPs have been used. Different types of NPs, with different physical and chemical characteristics, have so far been created, including solid lipid (SL), polylactide-co-glycoside (PLGA), liposomal, polymeric, fullerene, nanoemulsions, metal and metal oxides, and others. The synthesized NPs have been used in various industries, including cosmetics, agriculture, and pharmaceuticals. The discovery of NPs appears to be significant and has the potential to improve the efficacy and efficiency of future medications. The current increase in NPs presents significant potential for treating neurodegenerative disorders (Ngowi et al., 2021). Typically, conventional drug delivery systems are unable to penetrate the blood–brain barrier (BBB); this leads to massive restrictions on the ability to diagnose and treat different neurodegenerative diseases, for example, Parkinson's disease (PD), Alzheimer's disease (AD), and various neurological disorders as well. The multifunctional and variable structures of NPs make them suitable for the transport of drugs to the brain. However, the most important factors to consider before applying them are shape, size, charge, hydrophobicity, and surface chemistry. An ideal NP should have low toxicity, good biocompatibility, and the capacity to bind and deliver medications. It should not degrade easily in vivo in order to regulate the release of medicines over extended periods and can pass across the BBB. These characteristics result in those NPs with high potency in BBB penetration (Kumar et al., 2020). These diseases are characterized by a gradual decline (and even death) of neuronal structure and function and impose a substantial burden on the individual and society. Every year, neurodegenerative diseases affect millions of individuals worldwide. The prevalence of Alzheimer's disease rose from 26.6 million in 2006 to 36 million in 2014 (Rubinsztein, 2006; Sharma et al., 2015). Neurodegenerative diseases, characterized by neuronal dysfunction and progressive degeneration, are caused by protein degradation (Rubinsztein, 2006), various environmental factors (Brown et al., 2005), mitochondrial defects, familial history (Mosconi et al., 2007), abnormal protein accumulation in neurons, etc. (Barnham et al., 2004); however, ageing is regarded as one of the most significant issues in neurodegenerative diseases (Hung et al., 2010). A wide range of cutting-edge technologies have been developed over the past few decades to transport a vast array of compounds and bioactive molecules to mitochondria. These technologies have made it possible to significantly improve the pharmacokinetic profile, intracellular penetration, target site distribution, and pharmacological effects of drugs (Couvreur & Vauthier, 2006). Due to the need to break the BBB for drugs to access the brain, the development of effective and easy-to-administer chronic treatments for neurodegenerative disorders remains a significant challenge. This barrier regulates the passage of foreign blood substances, potentially dangerous molecules, and pathogens, preventing them from freely penetrating the extracellular fluid of the gray matter of the brain. In addition, the BBB protects brain cells by regulating their environment (cerebrovascular homeostasis, pH) against changes in hormones and neurotransmitters, for example. Thus, it provides strong resistance to ion movement, allowing for the maintenance of the brain's chemical equilibrium and the modification of its environment to guarantee optimal signal transmission between neurons. Moreover, the presence of tight junctions (TJs) between each endothelial cell substantially

* Corresponding author: abdullah.elbadry1@gmail.com

DOI: 10.1201/9781003383376-5

restricts the passage of numerous substances into the brain (98% of substances cannot cross it) (A. Cunha et al., 2021). Numerous attempts have been made to develop nano-drug delivery systems with these characteristics. Even though nano-drug delivery systems have produced some encouraging results, this is still an emerging field of study that requires additional research and analysis.

5.2 CATEGORIES AND APPLICATIONS OF NANOCARRIERS IN NEURODEGENERATIVE DISEASE

5.2.1 Metal Nanoparticles

Metal or metallic nanoparticles (MNPs) have gained much interest in medication development due to their optical properties, specifically surface plasmon resonance (SPR) and their ability to control the optical field, making them potential candidates in the field of biomedical applications. MNPs' small size makes it easier for them to pass across biological or physiological membranes, which are often impenetrable to other macromolecules (Chandrakala et al., 2022). MNPs have been widely used in animal and mammalian cells to deliver a variety of substances intracellularly, including DNA, medicines, peptides, and others (Jat et al., 2020). Pure metal-based NPs make up the first category of nanomaterials (e.g., zinc, magnesium, iron, titanium, platinum, copper, gold, and silver, and NPs). The other categories of nanomaterials include metal oxide nanoparticles, which include silver oxide, zinc oxide, titanium dioxide, and others (Yaqoob et al., 2020). Table 5.1 shows the different applications of metal nanoparticles in neurodegenerative diseases.

5.2.1.1 Gold Nanoparticles (AuNPs)

Gold nanoparticles (Au NPs) are one of the most stable, easily synthesized nanoparticles, and they exhibit several excellent characteristics, such as the ability to assemble different types and the quantum size effect (Daniel & Astruc, 2004; Amina & Guo, 2020). They have a negative charge, which makes it simple for biomolecules like medicines, genes, and targeting ligands to functionalize them (Kong et al., 2017). AuNPs have been shown to have some toxic properties, despite some research indicating that they are nontoxic (Aghaie et al., 2019). Additionally, they can penetrate the BBB and have antioxidant and anti-inflammatory properties that induce therapeutic applications for the treatment of neurodegenerative disorders (Silveira et al., 2023).

5.2.1.2 Silver Nanoparticles (AgNPs)

Silver nanoparticles (AgNPs) are also potential nanosystems for the treatment of NDs due to their physicochemical properties, such as their high surface area, simple synthesis method, and flexibility, as well as their capacity to pass the BBB and trigger an immunological response in the brain; nevertheless, AgNPs may cause neurological problems by inducing an inflammatory response in neuronal cells. According to Lin et al., exposure to AgNPs may cause the progression of neurodegenerative diseases that are related to Alzheimer's disease (Ribeiro et al., 2022; Lin et al., 2016). These harmful restrictions can be overcome by employing green synthesis methods and targeting specific cells with the use of coatings and ligands that can decrease the toxicity of AgNPs (Ribeiro et al., 2022).

5.2.2 Lipid-Based Nanoparticles

Lipid-based nanoparticles (LBNPs) have gained much interest in medication development. These NPs have a long half-life and a regulated release of the medication, which prolongs the duration of pharmacological activity. They can carry hydrophobic and hydrophilic molecules and have very little or no toxicity. Major types of these nanoparticles include liposomes, solid lipid nanoparticles (SLN), and nanostructured lipid carriers (NLC). These nanosystems may contain chemical modifications (such as gangliosides or polyethylene glycol [PEG]) to escape immune system

TABLE 5.1
Metal nanoparticle application in neurodegenerative disease

Nanocarrier system	Particle size		Drug	Disease	Ref.
PEGylated gold nanoparticles	AuNP14a AuNP14b	250 nm 150 nm	siRNA targets APOE4	Alzheimer's disease	(Okła et al., 2023)
Chitosan (CTS)@gold nanoparticles (GNP)	CTS@GNP	5 nm 10 nm	Plasmid DNA (pDNA)—nerve growth factor (NGF)	Parkinson's disease	(Hu et al., 2018)
PEGylated polysorbate 80–coated gold nanoparticle	50–90 nm		Donepezil	Alzheimer's disease	(Sivaji & Kannan, 2019)
Silver nanoparticles (green synthesis)	*Lampranthus coccineus* aqueous nanoextract *Malephora lutea* aqueous nanoextract	136 nm 206.7 nm	Extract of *Lampranthus coccineus* and *Malephora lutea*	Alzheimer's disease	(Youssif et al., 2019)

detection or to increase a drug's solubility (García-Pinel et al., 2019). Table 5.2 shows the different applications of lipid-based nanoparticles in neurodegenerative diseases.

5.2.2.1 Liposomes

Liposomes consist of enclosed vesicles of lipid bilayers of cholesterol and phospholipids that self-assemble in concentric layers that mimic natural cell membranes. Due to their excellent biocompatibility, safety, and entrapment efficiency, they have long been used as drug carriers and non-viral vectors for gene therapy. They are often divided into four types: multilamellar vesicles (MLV), multivesicular vesicles (MVV), small unilamellar vesicles (SUV), and large unilamellar vesicles (LUV), depending on the size and structure of bilayers (He et al., 2019; Vemana et al., 2021). They play an important role as therapeutic carriers for different diseases, including neurodegenerative disorders. Several liposomes with surface modifications of BBB-targeting ligands have been developed to penetrate the BBB through different mechanisms, like transcytosis (Seo & Park, 2021).

5.2.2.2 Solid Lipid Nanoparticles

Solid lipid nanoparticles (SLNs) are small colloidal carriers that range from 40 to 1,000 nm in size. SLNs are made of a hydrophobic lipid core, such as fatty acids and triglycerides that remain solid at room temperature and body temperature, where medication coated with a monolayer of long-chain acids is dissolved or dispersed. The ability of SLNs to enclose both hydrophilic and lipophilic drug molecules, regulate drug release, and direct drug delivery to particular cells and tissues has shown potential application (Yaghmur & Mu, 2021). SLNs with particles 120–200 nm in size have better blood circulation through bypassing spleen or liver filtration (Ganesan et al., 2018). By avoiding the use of organic solvents and harmful monomers, they surmount the limitations of polymeric nanoparticles. Lipid in the system also shows improved transport efficiency with prolonged interaction with BBB. The lipids employed in the manufacturing of SLN are biodegradable and biocompatible (made from physiologically acceptable lipids), making them a safer delivery vehicle for drugs to the brain (Kaur et al., 2008).

5.2.2.3 Nanostructured Lipid Carriers

To overcome many drawbacks of SLNs, such as low drug loading, for the next generation, it was discovered that adding a liquid lipid to the core solid lipid matrix of the nanoparticle leads to an increase in the number of defects in the core solid matrix, therefore enabling to incorporate more drug while maintaining the physical stability of the nanocarriers. This new SLN with an unstructured matrix was designed to be nanostructured lipid carriers (NLCs). NLCs are composed of a surfactant or a combination of surfactants present in the aqueous phase and the unstructured solid lipid matrix, which is composed of a combination of solid and liquid lipids. Usually, solid and liquid lipids are combined in a 70:30 to 99.9:0.1 ratio, with a surfactant content of 1.5–5 % (w/v) (Beloqui et al., 2016). NLC's lipophilic properties and small particle/droplet sizes make it easier for drugs to cross the BBB, and the drug's encapsulation in the lipid matrix shields the formulation from enzymatic breakdown, allowing active drugs to reach the brain at therapeutic concentrations (S. Cunha et al., 2021).

5.2.3 Polymeric Nanoparticles

Polymeric nanoparticles are flexible carriers commonly used for the delivery of drugs and other therapeutic applications (Beach et al., 2024). They are classified into two categories biodegradable and non-biodegradable polymers (Banik et al., 2016). Biodegradable polymers include polycaprolactone (PCL), polylactic acid (PLA), chitosan, gelatin, and poly(lactic-co-glycolic acid) (PLGA). However, non-biodegradable polymers include polymethyl methacrylate) (PMMA), polyacrylamide, polyacrylates, and polystyrene (Banik et al., 2016; Kumari et al., 2010).

5.2.3.1 PLA Nanoparticles

Polylactic acid (PLA) is a biodegradable polymer, that is frequently utilized in nanomedicine (Beach et al., 2024). PLA NPs are a promising tool in the development of advanced therapies for conditions including neurodegenerative diseases because of their biocompatibility, biodegradability, and ability to cross the BBB. Consequently, they provide an efficient platform for delivering therapeutic agents in the treatment of neurodegenerative and different diseases (Liu et al., 2008; Masoudi Asil et al., 2020; Sahini, 2023). They have also been used to reduce drug elimination while enhancing the absorption of medications through the nasal mucosa and improving the paracellular absorption of most hydrophilic drugs in the nasal cavity (Pan et al., 2018; Rabiee et al., 2021).

5.2.3.2 PLGA Nanoparticles

Poly(lactic-co-glycolic acid) (PLGA) is a biodegradable and biocompatible copolymer made from lactic acid and glycolic acid (Y. Lu et al., 2023). PLGA NPs are commonly used in drug delivery, especially for their ability to penetrate the BBB and transport therapeutic substances to treat neurodegenerative diseases (Amo et al., 2021; Pinto et al., 2022). In addition to being non-toxic and biocompatible, these NPs have the advantage of enhanced cellular internalization, protection against enzymatic degradation, improved targeting efficiency, and improved drug solubility (A. Cunha et al., 2021).

5.2.3.3 Chitosan Nanoparticles

Chitosan is a biodegradable and biocompatible polysaccharide polymer derived from chitin. The structure comprises of β-(1→4)-linked D-glucosamine and N-acetyl-D-glucosamine units (Pellis et al., 2022). Based on its molecular weight range, chitosan can be classified into

TABLE 5.2
Lipid-based nanoparticle applications in neurodegenerative disease

Nanocarrier system	Particle size		Drug	Disease	Ref.
Cationic liposomes (liposomes containing cationic lipids DLinDMA or DOTAP)	DLinDMA liposomes DOTAP liposomes	141.20 ± 10.45 nm 110.36 ± 7.83 nm	EPM2A (laforin)	Lafora disease	(Vemana et al., 2021)
Liposomes	Nanoceria Empty liposomes Nanoceria-loaded liposomes	12 ± 1 nm 165 ± 42 nm 230 ± 10 nm	Cerium oxide nanoparticles (nanoceria)	Neurodegenerative disorders	(Grillone et al., 2017)
Solid lipid nanoparticles	82.5 ± 4.07 nm		Rivastigmine	Alzheimer's disease	(Shah et al., 2015)
Solid lipid nanoparticles	SLNs Nar-SLNs	176 nm 210.1 nm	Naringenin	Neurodegenerative disorders	(Nouri et al., 2022)
Solid lipid nanoparticles	SLNs SLNs-CUR	176.0 ± 0.2 nm 153.3 ± 0.2 nm	Curcumin	Alzheimer's disease	(Campisi et al., 2022)
Solid lipid nanoparticles	SLNs CIT-SLNs	141 ± 11 nm 201 ± 24 nm	Citicoline	Neurodegenerative disorders	(Margari et al., 2022)
Nanostructured lipid carriers	239.7±4.04nm		Silibinin	Neurodegenerative disorders	(Makhdoomi et al., 2022)
Nanostructured lipid carriers	Rapa anionic-NLC Rapa transferrin decorated-NLC	130 ± 10 nm 150 ± 9 nm	Rapamycin	Alzheimer's disease	(Khonsari et al., 2022)
Nanostructured lipid carriers	—		Alpha-mangostin (αM)	Parkinson's disease (PD)	(Sakamula et al., 2022)

three groups: low molecular weight chitosan (LMWC, less than 150 kDa), medium molecular weight chitosan (MMWC, 150–700 kDa), and high molecular weight chitosan (HMWC, >700 kDa) (Román-Doval et al., 2023). Extensive research has been conducted on its potential to deliver nano-drugs to treat different diseases (Herdiana et al., 2023; Zaiki et al., 2023). Chitosan NPs are often used in drug delivery due to their unique properties, including biocompatibility, biodegradability, mucoadhesion, low toxicity, and the ability to improve drug solubility and stability (Kuen & Masarudin, 2022; Yadav et al., 2023). These NPs are suitable carriers for targeted and controlled delivery of drugs as they can be utilized to carry a wide range of therapeutic agents, such as small molecules, proteins, peptides, and nucleic acids (Garg et al., 2019; Rostami, 2020). Moreover, they can improve drug permeability through the BBB by influencing the tight junction. Chitosan NPs, with a positive charge, can adhere to the negatively charged cell membrane, enhancing their presence on the nasal mucosa, and facilitating the transfer of medication from the nasal cavity to the brain. These features make chitosan NPs particularly well-suited for delivering drugs to the brain (Yu et al., 2019).

5.2.3.4 Gelatin Nanoparticles

Gelatin is a flexible natural biopolymer derived from collagen protein breakdown through hydrolysis. It is widely used in the pharmaceutical industry due to its biocompatibility, biodegradability, cost-effectiveness, and numerous active groups for attaching targeted compounds. These benefits result in its use to produce NPs for gene and drug delivery (Elzoghby, 2013). Gelatin NPs exhibit considerable promise in treating neurodegenerative conditions because of their low cytotoxicity, high drug binding capacity, and capability to penetrate the BBB without causing harm to the barrier. These properties enable therapeutic applications for the treatment of neurodegenerative disorders (C. T. Lu et al., 2015; Mahmoudi Saber, 2019; Zhao et al., 2016).

5.2.4 Nanotheranostics

Nanotheranostics combine diagnostic and therapeutic agents in a single formulation linked and integrated by nanoparticles (Ladju et al., 2022). Theranostics could be used in many treatment areas, including personalized treatment, by fusing molecular imaging and molecular therapy, including early disease detection, disease staging, therapy selection, treatment planning, identifying side effects at an early stage of the treatment, and scheduling follow-up therapies (Kim et al., 2013). They have several advantages, as theranostic systems provide drug choice, optimization, and selective diagnosis together with real-time biopharmaceutical factor and drug release pattern analysis for desired therapeutic impact. They reduce the duration of patient care and keep track of how medicinal drugs are delivered throughout the body. Better assessment methods like biomarker sensitivity and tracking activity help with accurate medication identification, dose calculation, and risk reduction for drug overdose or side effects in clinical studies (Shende & Trivedi, 2021).

5.2.5 Stimuli-Responsive Nanocarriers

Stimuli-responsive nanocarriers are nanocarriers that respond to external and internal stimuli. They are primarily used to deliver, release, and activate drugs to targeted tissues that enhance the biological activity, effectiveness, and specificity of drug delivery (Mi, 2020).

5.2.5.1 Intrinsic Stimuli-Responsive Nanocarriers

Intrinsic stimuli-responsive nanocarriers are nanocarriers that respond to internal stimuli, including pH, glutathione (GSH) content, overexpression of particular enzymes, and hyperthermia. The biological factors that differ between normal and diseased tissues trigger novel approaches for designing nanovehicles for drug delivery with improved target site properties and cellular absorption efficiency. Multiple stimuli-responsive triggers could be used synergistically for better intracellular absorption of nanocarriers in order to produce more efficient drug delivery. This would allow for more effective drug delivery (Liu et al., 2017).

5.2.5.2 Extrinsic Stimuli-Responsive Nanocarriers

Extrinsic stimuli-responsive nanocarriers have received much interest in controlled drug delivery by light, magnetic fields, ultrasound, and electric fields. These systems enable control of the dosage, location, and timing of the release of encapsulated substances (Li et al., 2020).

5.2.5.2.1 Light/Photostimuli-Responsive Nanocarriers

Light-responsive nanocarriers are non-invasive drug delivery systems that can be controlled remotely to target specific sites at specific time in response to applying a certain light wavelength like ultraviolet (UV), visible or near-infrared (NIR), etc. (Majumder & Minko, 2021). Like numerous ideas, photo-responsive systems were inspired by nature, which has developed numerous complex biological systems capable of utilizing light as an external source of power and information. As an example, the cis-trans isomerization of retinal molecule produced by light, attached to the protein opsin's membrane, causes a number of processes, such as changes in the protein's structure and membrane permeability, resulting in a neural signal and, eventually, the perception of light (Ercole et al., 2010).

5.2.5.2.2 Magnetic Stimuli-Responsive Nanocarriers

The benefit of employing a magnetic field depends on the variety of forms that the magnetic response can take, such as magnetic guidance under a permanent magnetic field, temperature rise under an alternating magnetic field, or both when employed alternately. Therefore, a variety of drug delivery pathways is possible with magnetically responsive systems. Additionally, it is possible to perform magnetic resonance imaging and so combine diagnostics and treatment in a single theranostic device (Yang et al., 2011; Mura et al., 2013).

5.2.5.2.3 Ultrasound Stimuli-Responsive Nanocarriers

Ultrasound (US) consists of pressure waves called acoustic waves with frequencies greater than the limit of human hearing (>20 kHz); unlike light or electromagnetic waves, US requires a medium to move through (Entzian & Aigner, 2021). US provides a successful way to provide spatiotemporal control over medication release at the desired site, preventing serious side effects on healthy tissues. The use of ultrasounds is also attractive because of its non-invasiveness, lack of ionizing radiation, and simple control of tissue penetration depth through adjusting frequency, duty cycles, and exposure time. Through the thermal and/or mechanical effects caused by cavitation phenomena or radiation forces, US waves can lead to drug release from various nanocarriers (Mura et al., 2013).

5.2.5.2.4 Electric Stimuli-Responsive Nanocarriers

Electric stimuli-responsive nanocarriers have been employed as external stimuli in drug administration because they have several advantages, such as portability, avoiding big equipment, the ability to run cycles for a long time, and remote control via a sensor or microchip (Gopalan et al., 2020). Various actuation techniques can produce pulsed or sustained medication release using weak electric fields (usually around 1 V). For instance, conductive polymer-based nanoparticles with controlled drug-release profiles included electrochemical reduction-oxidation and the movement of charged molecules triggered by an electric field (Ge et al., 2012; Mura et al., 2013).

5.3 ADVANTAGES OF NANOCARRIERS

Researchers, scientists, and medical professionals are paying close attention to nanomedicines as a means of treating neurological problems to improve the pharmacokinetic features of poorly soluble medicines by enhancing solubility in NPs. Additionally, for decreased toxicity and adverse effects, these nanocarriers provide focused medication administration by inherent passive transport. As a result, formulations based on nanotechnology can enhance the drug's safety, effectiveness, pharmacokinetics, and bioavailability. By dissolving into the hydrophobic or hydrophilic compartment and reaching the targeted region, nanoformulation-based therapy increases the solubility of poorly soluble bioactive and reduces metabolic rate. In addition, nanocarriers can target specific tissues or organs to improve drug access to the site of action after internalization with a physiological membrane for passive absorption. However, tissue targeting via nanocarriers requires the use of advanced active targeting techniques, ligand-investigated carriers, surface functionalization, and prodrug production. In certain neurodegenerative diseases, Due to their unique properties (ease of traversing tight cell junctions, across the BBB, enhanced bioavailability, drug protection, and high drug encapsulation efficiency), nanoformulation-based drug delivery systems are generally a compelling option for the treatment of AD and PD (Ashique et al., 2023).

5.4 CHALLENGES TO CLINICAL TRANSLATION

NPs enable genetically independent wireless neural stimulation. Despite this, their potential use in the treatment of neurodegenerative diseases in humans continues to face obstacles. NPs are capable of stimulating cells in response to external stimuli; plasmonic nanoparticles, for instance, can be used for photothermal stimulation, iron oxide nanoparticles for magnetic stimulation, and silicon nanostructures or organic nanoparticles for photostimulation. In addition, NPs can be bioconjugated to take advantage of photochemistry for neurotransmitter or drug delivery (i.e., they can release a molecule in response to light or ultrasound). Nanoparticles can establish a coupling interface with neurons and receive energy from external sources; therefore, they can be used for neurostimulation (Couvreur & Vauthier, 2006). Building an interface with neurons involves the delivery and dispersion of nanoparticles as well as the biochemical modification of their surface properties to permit interaction with the neuron. Although there is widespread accord regarding the neurostimulation potential of nanomedicine, its clinical application remains limited. The greatest barrier to clinical implementation is regulating the fate and toxicity of nanoparticles in vivo. Once injected, nanoparticles can leave the area requiring therapeutic action and accumulate in other organs, where they are toxic. In addition, determining the fate of nanoparticles in vivo remains difficult (Carvalho-de-Souza et al., 2015).

5.5 CONCLUSION

Nanotechnology will be employed in all phases of drug development, ranging from formulations for optimal delivery to diagnostic applications in clinical trials, but there are no nanoparticles available optimally for all aspects of drug discovery. The choice may depend on the specifications. The toxicity of certain pharmaceuticals and traditional formulations, as well as the difficulty of treating certain diseases, has prompted the creation of new nanomaterials. Due to their biocompatibility, biodegradability, and versatility, NPs are garnering a great deal of attention, opening the door to a vast array of materials that could possess the necessary properties for a particular application. Nanomaterials were discovered to be promising agents in medical applications due to their exceptional biocompatibility, excellent water solubility, and low toxicity. At this point, the enhancement of pinpointing the precise site of action to achieve therapeutic effect reduces the incidence of adverse drug reactions dramatically. This would indicate a novel nanomedicine strategy for the use of treatments for a variety of pathologies. The efforts applied in the development of nanomaterials may include the specific requirements for a particular delivery system, the enhanced understanding of disease mechanisms among scientists, and the collaborative research conducted across all scientific fields, which will improve the current state of nanomaterials' use in the medical field, resulting in more effective and safer treatments.

REFERENCES

Aghaie, T., Jazayeri, M. H., Manian, M., Khani, L., Erfani, M., Rezayi, M., Ferns, G. A., & Avan, A. (2019). Gold nanoparticle and polyethylene glycol in neural regeneration in the treatment of neurodegenerative diseases. *Journal of Cellular Biochemistry*, *120*(3), 2749–2755.

Amina, S. J., & Guo, B. (2020). A review on the synthesis and functionalization of gold nanoparticles as a drug delivery vehicle. *International Journal of Nanomedicine*, *15*, 9823–9857. https://doi.org/10.2147/IJN.S279094

Amo, L. D., Cano, A., Ettcheto, M., Souto, E. B., Espina, M., Camins, A., García, M. L., & Sánchez-López, E. (2021). Surface functionalization of PLGA nanoparticles to increase transport across the BBB for Alzheimer's disease. *Applied Sciences*, *11*(9), 4305. https://doi.org/10.3390/APP11094305

Ashique, S., Afzal, O., Yasmin, S., Hussain, A., Altamimi, M. A., Webster, T. J., & Altamimi, A. S. A. (2023). Strategic nanocarriers to control neurodegenerative disorders: Concept, challenges, and future perspective. *International Journal of Pharmaceutics*, *633*. https://doi.org/10.1016/J.IJPHARM.2023.122614

Banik, B. L., Fattahi, P., & Brown, J. L. (2016). Polymeric nanoparticles: The future of nanomedicine. *Wiley Interdisciplinary Reviews: Nanomedicine and Nanobiotechnology*, *8*(2), 271–299. https://doi.org/10.1002/WNAN.1364

Barnham, K. J., Masters, C. L., & Bush, A. I. (2004). Neurodegenerative diseases and oxidative stress. *Nature.com*. Retrieved May 12, 2023, from www.nature.com/articles/nrd1330

Beach, M. A., Nayanathara, U., Gao, Y., Zhang, C., Xiong, Y., Wang, Y., & Such, G. K. (2024). Polymeric nanoparticles for drug delivery. *Chemical Reviews*, *124*(9), 5505–5616. https://doi.org/10.1021/ACS.CHEMREV.3C00705

Beloqui, A., Solinís, M. Á., Rodríguez-Gascón, A., Almeida, A. J., & Préat, V. (2016). Nanostructured lipid carriers: Promising drug delivery systems for future clinics. *Nanomedicine: Nanotechnology, Biology and Medicine*, *12*(1), 143–161. https://doi.org/https://doi.org/10.1016/j.nano.2015.09.004

Brown, R. C., Lockwood, A. H., & Sonawane, B. R. (2005). Neurodegenerative diseases: An overview of environmental risk factors. *Environmental Health Perspectives*, *113*(9), 1250–1256. https://doi.org/10.1289/EHP.7567

Campisi, A., Sposito, G., Pellitteri, R., Santonocito, D., Bisicchia, J., Raciti, G., Russo, C., Nardiello, P., Pignatello, R., Casamenti, F., & Puglia, C. (2022). Effect of unloaded and curcumin-loaded solid lipid nanoparticles on tissue transglutaminase isoforms expression levels in an experimental model of Alzheimer's disease. *Antioxidants*, *11*(10). https://doi.org/10.3390/antiox11101863

Carvalho-de-Souza, J., Treger, J., Dang, B., Kent, S. B. H., Pepperberg, D. R., & Bezanilla, F. (2015). Photosensitivity of neurons enabled by cell-targeted gold nanoparticles. *Neuron*. Retrieved May 13, 2023, from www.sciencedirect.com/science/article/pii/S0896627315001488

Chandrakala, V., Aruna, V., & Angajala, G. (2022). Review on metal nanoparticles as nanocarriers: Current challenges and perspectives in drug delivery systems. *Emergent Materials*, *5*(6), 1593–1615. https://doi.org/10.1007/s42247-021-00335-x

Couvreur, P., & Vauthier, C. (2006). Nanotechnology: Intelligent design to treat complex disease. *Pharmaceutical Research*, *23*(7), 1417–1450. https://doi.org/10.1007/S11095-006-0284-8

Cunha, A., Gaubert, A., Latxague, L., & Dehay, B. (2021). PLGA-based nanoparticles for neuroprotective drug delivery in neurodegenerative diseases. *Pharmaceutics*, *13*(7), 1042. https://doi.org/10.3390/PHARMACEUTICS13071042

Cunha, S., Forbes, B., Sousa Lobo, J. M., & Silva, A. C. (2021). Improving drug delivery for Alzheimer's disease through nose-to-brain delivery using nanoemulsions, nanostructured lipid carriers (NLC) and in situ hydrogels. *International Journal of Nanomedicine*, *16*, 4373–4390. https://doi.org/10.2147/IJN.S305851

Daniel, M.-C., & Astruc, D. (2004). Gold nanoparticles: Assembly, supramolecular chemistry, quantum-size-related properties, and applications toward biology, catalysis, and nanotechnology. *Chemical Reviews*, *104*(1), 293–346.

Elzoghby, A. O. (2013). Gelatin-based nanoparticles as drug and gene delivery systems: Reviewing three decades of research. *Journal of Controlled Release*, *172*(3), 1075–1091. https://doi.org/10.1016/J.JCONREL.2013.09.019Entzian, K., & Aigner, A. (2021). Drug delivery by ultrasound-responsive nanocarriers for cancer treatment. *Pharmaceutics*, *13*(8), 1135.

Ercole, F., Davis, T. P., & Evans, R. A. (2010). Photo-responsive systems and biomaterials: Photochromic polymers, light-triggered self-assembly, surface modification, fluorescence modulation and beyond. *Polymer Chemistry*, *1*(1), 37–54.

Ganesan, P., Ramalingam, P., Karthivashan, G., Ko, Y. T., & Choi, D.-K. (2018). Recent developments in solid lipid nanoparticle and surface-modified solid lipid nanoparticle delivery systems for oral delivery of phyto-bioactive compounds in various chronic diseases. *International Journal of Nanomedicine*, *13*, 1569–1583. https://doi.org/10.2147/IJN.S155593

García-Pinel, B., Porras-Alcalá, C., Ortega-Rodríguez, A., Sarabia, F., Prados, J., Melguizo, C., & López-Romero, J. M. (2019). Lipid-based nanoparticles: Application and recent advances in cancer treatment. *Nanomaterials (Basel, Switzerland)*, *9*(4). https://doi.org/10.3390/nano9040638

Garg, U., Chauhan, S., Nagaich, U., & Jain, N. (2019). Current advances in chitosan nanoparticles based drug delivery and targeting. *Advanced Pharmaceutical Bulletin*, *9*(2), 195. https://doi.org/10.15171/APB.2019.023Ge, J., Neofytou, E., Cahill III, T. J., Beygui, R. E., & Zare, R. N. (2012). Drug release from electric-field-responsive nanoparticles. *ACS Nano*, *6*(1), 227–233.

Gopalan, D., Pandey, A., Udupa, N., & Mutalik, S. (2020). Receptor specific, stimuli responsive and subcellular targeted approaches for effective therapy of Alzheimer: Role of surface engineered nanocarriers. *Journal of Controlled Release*, *319*, 183–200. https://doi.org/https://doi.org/10.1016/j.jconrel.2019.12.034

Grillone, A., Li, T., Battaglini, M., Scarpellini, A., Prato, M., Takeoka, S., & Ciofani, G. (2017). Preparation, characterization, and preliminary in vitro testing of nanoceria-loaded liposomes. *Nanomaterials*, *7*(9). https://doi.org/10.3390/nano7090276

He, H., Lu, Y., Qi, J., Zhu, Q., Chen, Z., & Wu, W. (2019). Adapting liposomes for oral drug delivery. *Acta Pharmaceutica Sinica B*, *9*(1), 36–48. https://doi.org/https://doi.org/10.1016/j.apsb.2018.06.005

Herdiana, Y., Wathoni, N., Gozali, D., Shamsuddin, S., & Muchtaridi, M. (2023). Chitosan-based nano-smart drug delivery system in breast cancer therapy. *Pharmaceutics*, *15*(3), 879. https://doi.org/10.3390/PHARMACEUTICS15030879

Hu, K., Chen, X., Chen, W., Zhang, L., Li, J., Ye, J., Zhang, Y., Zhang, L., Li, C.-H., Yin, L., & Guan, Y.-Q. (2018). Neuroprotective effect of gold nanoparticles composites in Parkinson's disease model. *Nanomedicine: Nanotechnology, Biology and Medicine*, *14*(4), 1123–1136. https://doi.org/https://doi.org/10.1016/j.nano.2018.01.020

Hung, C., Chen, Y., Hsieh, W., Chiou, S. H., & Kao, C.-L. (2010). Ageing and neurodegenerative diseases. *Ageing Research Reviews*, *9*, S36–S46. Retrieved May 12, 2023, from www.sciencedirect.com/science/article/pii/S1568163710000632

Jat, S. K., Bhattacharya, J., & Sharma, M. K. (2020). Nanomaterial based gene delivery: A promising method for plant genome engineering. *Journal of Materials Chemistry B*, *8*(19), 4165–4175. https://doi.org/10.1039/D0TB00217H

Kaur, I. P., Bhandari, R., Bhandari, S., & Kakkar, V. (2008). Potential of solid lipid nanoparticles in brain targeting. *Journal of Controlled Release*, *127*(2), 97–109. https://doi.org/https://doi.org/10.1016/j.jconrel.2007.12.018

Khonsari, F., Heydari, M., Sharifzadeh, M., Valizadeh, H., Dinarvand, R., & Atyabi, F. (2022). Transferrin decorated-nanostructured lipid carriers (NLCs) are a promising delivery system for rapamycin in Alzheimer's disease: An in vivo study. *Biomaterials Advances*, *137*, 212827. https://doi.org/10.1016/j.bioadv.2022.212827

Kim, T. H., Lee, S., & Chen, X. (2013). Nanotheranostics for personalized medicine. *Expert Review of Molecular Diagnostics*, *13*(3), 257–269.

Kong, F.-Y., Zhang, J.-W., Li, R.-F., Wang, Z.-X., Wang, W.-J., & Wang, W. (2017). Unique roles of gold nanoparticles in drug delivery, targeting and imaging applications. *Molecules*, *22*(9). https://doi.org/10.3390/molecules22091445

Kuen, C. Y., & Masarudin, M. J. (2022). Chitosan nanoparticle-based system: A new insight into the promising controlled release system for lung cancer treatment. *Molecules*, *27*(2). https://doi.org/10.3390/MOLECULES27020473

Kumar, A., Chaudhary, R. K., Singh, R., Singh, S. P., Wang, S. -Y., Hoe, Z.-Y., Pan, C.-T., Shiue, Y.-L., Wei, D.-Q., Kaushik, A. C., & Dai, X. (2020). Nanotheranostic applications for detection and targeting neurodegenerative diseases. *Frontiers in Neuroscience*, *14*.

Kumari, A., Yadav, S. K., & Yadav, S. C. (2010). Biodegradable polymeric nanoparticles based drug delivery systems. *Colloids and Surfaces B: Biointerfaces*, *75*(1), 1–18. https://doi.org/10.1016/J.COLSURFB.2009.09.001

Ladju, R. B., Ulhaq, Z. S., & Soraya, G. V. (2022). Nanotheranostics: A powerful next-generation solution to tackle hepatocellular carcinoma. *World Journal of Gastroenterology*, *28*(2), 176.

Li, F., Qin, Y., Lee, J., Liao, H., Wang, N., Davis, T. P., Qiao, R., & Ling, D. (2020). Stimuli-responsive nano-assemblies for remotely controlled drug delivery. *Journal of Controlled Release*, *322*, 566–592. https://doi.org/https://doi.org/10.1016/j.jconrel.2020.03.051

Lin, H.-C., Huang, C.-L., Huang, Y.-J., Hsiao, I.-L., Yang, C.-W., & Chuang, C.-Y. (2016). Transcriptomic gene-network analysis of exposure to silver nanoparticle reveals potentially neurodegenerative progression in mouse brain neural cells.

Toxicology in Vitro, *34*, 289–299. https://doi.org/10.1016/j.tiv.2016.04.014

Ling Tan, J. S., Roberts, C. J., & Billa, N. (2018). Mucoadhesive chitosan-coated nanostructured lipid carriers for oral delivery of amphotericin B. *Pharmaceutical Development and Technology*, *24*(4), 504–512. https://doi.org/10.1080/10837450.2018.1515225

Linkov, I., Satterstrom, F. K., & Corey, L. M. (2008). Nanotoxicology and nanomedicine: Making hard decisions. *Nanomedicine: Nanotechnology, Biology and Medicine*, *4*(2), 167–171. https://doi.org/10.1016/J.NANO.2008.01.001

Liu, M., Du, H., Zhang, W., & Zhai, G. (2017). Internal stimuli-responsive nanocarriers for drug delivery: Design strategies and applications. *Materials Science and Engineering: C*, *71*, 1267–1280. https://doi.org/10.1016/j.msec.2016.11.030

Liu, M., Li, H., Luo, G., Liu, Q., & Wang, Y. (2008). Pharmacokinetics and biodistribution of surface modification polymeric nanoparticles. *Archives of Pharmacal Research*, *31*(4), 547–554. https://doi.org/10.1007/S12272-001-1191-8/METRICS

Lu, C. T., Jin, R. R., Jiang, Y. N., Lin, Q., Yu, W. Z., Mao, K. L., Tian, F. R., Zhao, Y. P., & Zhao, Y. Z. (2015). Gelatin nanoparticle-mediated intranasal delivery of substance P protects against 6-hydroxydopamine-induced apoptosis: An in vitro and in vivo study. *Drug Design, Development and Therapy*, *9*, 1955–1962. https://doi.org/10.2147/DDDT.S77237

Lu, Y., Cheng, D., Niu, B., Wang, X., Wu, X., & Wang, A. (2023). Properties of poly (lactic-co-glycolic acid) and progress of poly (lactic-co-glycolic acid)-based biodegradable materials in biomedical research. *Pharmaceuticals*, *16*(3). https://doi.org/10.3390/PH16030454

Mahmoudi Saber, M. (2019). Strategies for surface modification of gelatin-based nanoparticles. *Colloids and Surfaces B: Biointerfaces*, *183*, 110407. https://doi.org/10.1016/J.COLSURFB.2019.110407

Majumder, J., & Minko, T. (2021). Multifunctional and stimuli-responsive nanocarriers for targeted therapeutic delivery. *Expert Opinion on Drug Delivery*, *18*(2), 205–227.

Makhdoomi, S., Mahboobian, M. M., Haddadi, R., Komaki, A., & Mohammadi, M. (2022). Silibinin-loaded nanostructured lipid carriers (NLCs) ameliorated cognitive deficits and oxidative damages in aluminum chloride-induced neurotoxicity in male mice. *Toxicology*, *477*, 153260. https://doi.org/10.1016/j.tox.2022.153260

Margari, A., Monteduro, A. G., Rizzato, S., Capobianco, L., Crestini, A., Rivabene, R., Piscopo, P., D'Onofrio, M., Manzini, V., Trapani, G., Quarta, A., Maruccio, G., Ventra, C., Lieto, L., & Trapani, A. (2022). The encapsulation of citicoline within solid lipid nanoparticles enhances its capability to counteract the 6-Hydroxydopamine-induced cytotoxicity in human neuroblastoma SH-SY5Y cells. *Pharmaceutics*, *14*(9). https://doi.org/10.3390/pharmaceutics14091827

Masoudi Asil, S., Ahlawat, J., Guillama Barroso, G., & Narayan, M. (2020). Nanomaterial based drug delivery systems for the treatment of neurodegenerative diseases. *Biomaterials Science*, *8*(15), 4109. https://doi.org/10.1039/D0BM00809E

Mi, P. (2020). Stimuli-responsive nanocarriers for drug delivery, tumor imaging, therapy and theranostics. *Theranostics*, *10*(10), 4557–4588. https://doi.org/10.7150/thno.38069

Mosconi, L., Brys, M., Switalski, R., Mistur, R., Glodzik, L., Pirraglia, E., Tsui, W., De Santi, S., & De Leon, M. J. (2007). Maternal family history of Alzheimer's disease predisposes to reduced brain glucose metabolism. *Proceedings of the National Academy of Sciences of the United States of America*, *104*(48), 19067–19072. https://doi.org/10.1073/PNAS.0705036104

Mura, S., Nicolas, J., & Couvreur, P. (2013). Stimuli-responsive nanocarriers for drug delivery. *Nature Materials*, *12*(11), 991–1003. https://doi.org/10.1038/nmat3776

Nabar, G., Mahajan, K., Calhoun, M. A., Duong, A. D., Souva, M. S., Xu, J., Czeisler, C., Puduvalli, V. K., Otero, J. J., Wyslouzil, B. E., & Winter, J. O. (2018). Micelle-templated, poly (lactic-co-glycolic acid) nanoparticles for hydrophobic drug delivery. *International Journal of Nanomedicine*. Retrieved May 12, 2023, from www.ncbi.nlm.nih.gov/pmc/articles/PMC5768187/

Ngowi, E. E., Wang, Y.-Z., Qian, L., Helmy, Y. A. S. H., Anyomi, B., Li, T., Zheng, M., Jiang, E.-S., Duan, S.-F., Wei, J.-S., Wu, D.-D., & Ji, X.-Y. (2021). The application of nanotechnology for the diagnosis and treatment of brain diseases and disorders. *Frontiers in Bioengineering and Biotechnology*, *9*.

Nouri, Z., Sajadimajd, S., Hoseinzadeh, L., Bahrami, G., Arkan, E., Moradi, S., Abdi, F., & Farzaei, M. H. (2022). Neuroprotective effect of naringenin-loaded solid lipid nanoparticles against streptozocin-induced neurotoxicity through autophagy blockage. *Journal of Food Biochemistry*, *46*(12), e14408. https://doi.org/10.1111/jfbc.14408

Okła, E., Białecki, P., Kędzierska, M., Pędziwiatr-Werbicka, E., Miłowska, K., Takvor, S., Gómez, R., de la Mata, F. J., Bryszewska, M., & Ionov, M. (2023). Pegylated gold nanoparticles conjugated with siRNA: Complexes formation and cytotoxicity. *International Journal of Molecular Sciences*, *24*, 7. https://doi.org/10.3390/ijms24076638

Pan, L., Zhou, J., Ju, F., & Zhu, H. (2018). Intranasal delivery of α-asarone to the brain with lactoferrin-modified mPEG-PLA nanoparticles prepared by premix membrane emulsification. *Drug Delivery and Translational Research*, *8*(1), 83–96. https://doi.org/10.1007/S13346-017-0438-8/FIGURES/10

Pellis, A., Guebitz, G. M., & Nyanhongo, G. S. (2022). Chitosan: Sources, processing and modification techniques. *Gels*, *8*(7), 393. https://doi.org/10.3390/GELS8070393

Pinto, M., Silva, V., Barreiro, S., Silva, R., Remião, F., Borges, F., & Fernandes, C. (2022). Brain drug delivery and neurodegenerative diseases: Polymeric PLGA-based nanoparticles as a forefront platform. *Ageing Research Reviews*, *79*, 101658. https://doi.org/10.1016/J.ARR.2022.101658

Rabiee, N., Ahmadi, S., Afshari, R., Khalaji, S., Rabiee, M., Bagherzadeh, M., Fatahi, Y., Dinarvand, R., Tahriri, M., Tayebi, L., Hamblin, M. R., & Webster, T. J. (2021). Polymeric nanoparticles for nasal drug delivery to the brain: Relevance to Alzheimer's disease. *Advanced Therapeutics*, *4*(3), 2000076. https://doi.org/10.1002/ADTP.202000076

Ribeiro, T. C., Sábio, R. M., Carvalho, G. C., Fonseca-Santos, B., & Chorilli, M. (2022). Exploiting mesoporous silica, silver and gold nanoparticles for neurodegenerative diseases treatment. *International Journal of Pharmaceutics*, *624*, 121978. https://doi.org/10.1016/j.ijpharm.2022.121978

Román-Doval, R., Torres-Arellanes, S. P., Tenorio-Barajas, A. Y., Gómez-Sánchez, A., & Valencia-Lazcano, A. A. (2023). Chitosan: Properties and its application in agriculture in context of molecular weight. *Polymers*, *15*(13), 2867. https://doi.org/10.3390/POLYM15132867

Rostami, E. (2020). Progresses in targeted drug delivery systems using chitosan nanoparticles in cancer therapy: A mini-review. *Journal of Drug Delivery Science and Technology*, *58*, 101813. https://doi.org/10.1016/J.JDDST.2020.101813

Rubinsztein, D. C. (2006). The roles of intracellular protein-degradation pathways in neurodegeneration. *Nature.com*.

Retrieved May 12, 2023, from www.nature.com/articles/nature05291

Sahini, M. G. (2023). Polylactic acid (PLA)-based materials: A review on the synthesis and drug delivery applications. *Emergent Materials*, *6*(5), 1461–1479. https://doi.org/10.1007/S42247-023-00551-7

Sakamula, R., Yata, T., & Thong-Asa, W. (2022). Nanostructure lipid carriers enhance alpha-mangostin neuroprotective efficacy in mice with rotenone-induced neurodegeneration. *Metabolic Brain Disease*, *37*(5), 1465–1476. https://doi.org/10.1007/s11011-022-00967-w

Seo, M.-W., & Park, T.-E. (2021). Recent advances with liposomes as drug carriers for treatment of neurodegenerative diseases. *Biomedical Engineering Letters*, *11*(3), 211–216. https://doi.org/10.1007/s13534-021-00198-5

Shah, B., Khunt, D., Bhatt, H., Misra, M., & Padh, H. (2015). Application of quality by design approach for intranasal delivery of rivastigmine loaded solid lipid nanoparticles: Effect on formulation and characterization parameters. *European Journal of Pharmaceutical Sciences*, *78*, 54–66. https://doi.org/10.1016/j.ejps.2015.07.002

Sharma, M., Deogaonkar, M., & Rezaia, A. (2015). Assessment of potential targets for deep brain stimulation in patients with Alzheimer's disease. *Journal of Clinical Medicine Research*. Retrieved May 12, 2023, from www.ncbi.nlm.nih.gov/pmc/articles/PMC4432890/

Shende, P., & Trivedi, R. (2021). Nanotheranostics in epilepsy: A perspective for multimodal diagnosis and strategic management. *Nano Select*, *2*(7), 1277–1290.

Silveira, P. C. L., Rodrigues, M. S., Gelain, D. P., & de Oliveira, J. (2023). Gold nanoparticles application to the treatment of brain dysfunctions related to metabolic diseases: Evidence from experimental studies. *Metabolic Brain Disease*, *38*(1), 123–135. https://doi.org/10.1007/s11011-022-00929-2

Sivaji, K., & Kannan, R. R. (2019). Polysorbate 80 coated gold nanoparticle as a drug carrier for brain targeting in Zebrafish odel. *Journal of Cluster Science*, *30*(4), 897–906. https://doi.org/10.1007/s10876-019-01548-1

Vemana, H. P., Saraswat, A., Bhutkar, S., Patel, K., & Dukhande, V. V. (2021). A novel gene therapy for neurodegenerative Lafora disease via EPM2A-loaded DLinDMA lipoplexes. *Nanomedicine*, *16*(13), 1081–1095. https://doi.org/10.2217/nnm-2020-0477

Yadav, M., Kaushik, B., Rao, G. K., Srivastava, C. M., & Vaya, D. (2023). Advances and challenges in the use of chitosan and its derivatives in biomedical fields: A review. *Carbohydrate Polymer Technologies and Applications*, *5*, 100323. https://doi.org/10.1016/J.CARPTA.2023.100323

Yaghmur, A., & Mu, H. (2021). Recent advances in drug delivery applications of cubosomes, hexosomes, and solid lipid nanoparticles. *Acta Pharmaceutica Sinica B*, *11*(4), 871–885. https://doi.org/10.1016/j.apsb.2021.02.013

Yang, H.-W., Hua, M.-Y., Liu, H.-L., Huang, C.-Y., Tsai, R.-Y., Lu, Y.-J., Chen, J.-Y., Tang, H.-J., Hsien, H.-Y., Chang, Y.-S., Yen, T.-C., Chen, P.-Y., & Wei, K.-C. (2011). Self-protecting core-shell magnetic nanoparticles for targeted, traceable, long half-life delivery of BCNU to gliomas. *Biomaterials*, *32*(27), 6523–6532. https://doi.org/10.1016/j.biomaterials.2011.05.047

Yaqoob, A. A., Ahmad, H., Parveen, T., Ahmad, A., Oves, M., Ismail, I. M. I., Qari, H. A., Umar, K., & Mohamad Ibrahim, M. N. (2020). Recent advances in metal decorated nanomaterials and their various biological applications: A review. *Frontiers in Chemistry*, *8*.

Youssif, K. A., Haggag, E. G., Elshamy, A. M., Rabeh, M. A., Gabr, N. M., Seleem, A., Salem, M. A., Hussein, A. S., Krischke, M., Mueller, M. J., & Abdelmohsen, U. R. (2019). Anti-Alzheimer potential, metabolomic profiling and molecular docking of green synthesized silver nanoparticles of *Lampranthus coccineus* and *Malephora lutea* aqueous extracts. *PloS One*, *14*(11), e0223781.

Yu, S., Xu, X., Feng, J., Liu, M., & Hu, K. (2019). Chitosan and chitosan coating nanoparticles for the treatment of brain disease. *International Journal of Pharmaceutics*, *560*, 282–293. https://doi.org/10.1016/J.IJPHARM.2019.02.012

Zaiki, Y., Iskandar, A., & Wong, T. W. (2023). Functionalized chitosan for cancer nano drug delivery. *Biotechnology Advances*, *67*, 108200. https://doi.org/10.1016/J.BIOTECHADV.2023.108200

Zhao, Y. Z., Jin, R. R., Yang, W., Xiang, Q., Yu, W. Z., Lin, Q., Tian, F. R., Mao, K. L., Lv, C. Z., Wáng, Y. X. J., & Lu, C. T. (2016). Using gelatin nanoparticle mediated intranasal delivery of neuropeptide substance P to enhance neuro-recovery in hemiparkinsonian rats. *PloS One*, *11*(2), e0148848. https://doi.org/10.1371/JOURNAL.PONE.0148848

6 Regulations and Ethics of Nanomedicine and Public Acceptance

Nada Abdel-Monaem

6.1 INTRODUCTION

Nanomedicine is one of the applications of nanotechnology in medicine. It has emerged as a promising field with the potential to revolutionize health care. It offers unique opportunities for targeted drug delivery, improved imaging techniques, and enhanced diagnostics. The use of nanomaterials and nanoscale devices enables precise interactions at the molecular level, leading to more effective and personalized treatments with reduced side effects (Satalkar et al., 2016). However, the development and application of nanomedicine also raise significant ethical, legal, and regulatory considerations that must be addressed to ensure its responsible implementation and public acceptance (Foulkes et al., 2020).

Due to the distinct properties and behaviors of nanoscale particles and materials, the traditional approaches to regulation and ethical considerations in medicine may not be sufficient. Nanomedicine introduces novel challenges in terms of safety evaluation, risk assessment, and manufacturing standards (Satalkar et al., 2016). Regulatory agencies such as the US Food and Drug Administration (FDA) and the European Medicines Agency (EMA) have recognized the need for specific guidelines tailored to nanomedicine products. These guidelines address aspects such as the characterization and assessment of nanomaterials, preclinical and clinical evaluation, and post-market surveillance (Foulkes et al., 2020).

The development of regulatory frameworks for nanomedicine involves striking a balance between fostering innovation and ensuring patient safety. It requires interdisciplinary collaboration among scientists, policymakers, ethicists, and industry stakeholders to address the unique challenges posed by nanomedicine. The regulatory landscape is still evolving, and ongoing research and collaboration are essential to establish robust frameworks that can keep pace with the rapid advancements in nanotechnology (Foulkes et al., 2020).

Ethical deliberations are crucial in the development and application of nanomedicine. The unique properties of nanomaterials raise concerns regarding unintended harm to human health and the environment. Ethical principles such as beneficence, nonmaleficence, autonomy, and justice should guide the research, development, and deployment of nanomedicine interventions (AlKahtani, 2018). Addressing issues related to privacy, informed consent, access to nanomedicine technologies, equitable distribution, and responsible research conduct is essential (Wasti et al., 2023).

Ethical frameworks provide a foundation for navigating the complex ethical challenges associated with nanomedicine. They facilitate the integration of ethical considerations into research and development processes, clinical practice, and policymaking. Transparent and inclusive decision-making processes involving stakeholders and affected communities can help ensure that ethical concerns are adequately addressed (AlKahtani, 2018; Wasti et al., 2023).

Public acceptance and perception of nanomedicine play a critical role in its successful integration into healthcare systems. The level of public awareness and understanding of nanomedicine directly impacts public support and trust. Educating the public about the benefits, risks, and ethical implications of nanomedicine is paramount. Open dialogue, effective risk communication, and engagement with the public can help address misconceptions, alleviate concerns, and build public confidence (AlKahtani, 2018; Wasti et al., 2023). Public acceptance is influenced by various factors, including familiarity with nanotechnology, perceived risks and benefits, trust in regulatory systems, and cultural attitudes. Public engagement initiatives such as public consultations, community involvement, and public deliberation can foster a sense of ownership and enable citizens to contribute to decision-making processes (Wasti et al., 2023).

6.2 REGULATORY FRAMEWORKS FOR NANOMEDICINE

Regulatory frameworks for nanomedicine are essential to address the unique challenges posed by these emerging technologies and ensure their safe and effective development. At the moment, the field of nanomedicine lacks standards and guidelines, which impedes advancement and creates safety concerns (Sechi et al., 2014). When nanomedicines come into contact with biomolecules and genetic materials, they may become mutagenic and genotoxic (Engelmann et al., 2021). When immune cells respond inflammatorily, reactive oxygen and nitrogen species are produced, which causes oxidative and nitrosative stress, which is often the mechanism by which nanomedicines cause toxicity (Allon et al., 2017).

DOI: 10.1201/9781003383376-6

The accumulation of free radicals generated by nanomedicines can cause extensive damage to the body, including oxidative DNA damage, protein denaturation, lipid peroxidation, mitochondrial membrane damage, and transcription of genes responsible for carcinogenesis and fibrosis (Sechi et al., 2014). When nanomedicines are administered intravenously, they can accumulate in organs such as the liver and translocate to other systems, including the central nervous, cardiovascular, and renal systems. The lack of understanding of the precise interactions between nanomedicines and biological systems makes it challenging to assess their physicochemical and toxicological properties (Engelmann et al., 2021). Furthermore, the unique properties observed at the nanoscale, such as nanoparticle type, surface properties, administration route, and morphology, contribute to the complexity of regulating nanomedicines (Sechi et al., 2014).

The current absence of standardized regulatory guidance in nanomedicine prevents progress in addressing the safety concerns associated with these technologies (Sechi et al., 2014). It is crucial to develop regulatory frameworks that consider the specific challenges and risks posed by nanomedicines, taking into account their unique properties and potential impacts on human health and the environment (Allon et al., 2017). To establish effective regulatory frameworks for nanomedicine, international collaboration and stakeholder engagement are crucial (Allon et al., 2017; Atalla et al., 2016). The involvement of academics, clinicians, regulatory authorities, and other relevant stakeholders can contribute to the development of guidelines that ensure the safe and ethical use of nanomedicines (Allon et al., 2017; Silva Costa et al., 2011). In addition to addressing safety concerns, regulatory frameworks for nanomedicine should also consider ethical, privacy, and intellectual property issues (Atalla et al., 2016). The unique characteristics and potential societal impacts of nanomedicine require careful consideration of ethical implications, protection of privacy rights, and fair management of intellectual property (Atalla et al., 2016; Wiesing & Clausen, 2014).

6.2.1 Regulatory Agencies and Guidelines

The oversight of the application of nanotechnology in medical products is largely the responsibility of regulatory bodies. Regarding the special characteristics and behaviors of products utilizing nanotechnology, the Food and Drug Administration (FDA) in the United States has released a number of guidance documents. These papers address issues like how nanotechnology is used in FDA-regulated goods, whether nanomaterials are safe in food and cosmetic products, and how to apply a risk-based framework to nanotechnology (Engelmann et al., 2021). In order to address concerns about the regulatory status, safety, efficacy, or other characteristics of products containing nanomaterials, the FDA encourages industry consultation in advance of the agency's regulatory authority over nanotechnology products (Soares et al., 2018)

In the European Union, the European Medicines Agency (EMA) has established an expert group on nanomedicines to provide scientific information and develop or review guidelines. Additionally, the movement towards formalized nanomedicine regulations in the EU is being led by the REFINE Nanomed project, which aims to define the criteria for regulatory needs for nanomedicines and nanomaterials for clinical use (Alphandéry, 2019). The regulatory frameworks for nanomedicine vary across different countries and regions, and there is a lack of global consensus on definitions and clarity around critical safety attributes. However, initiatives and guidelines are in place to provide indications of what to expect in the future (Engelmann et al., 2021). Existing laws and individual authorities serve as the foundation for the regulation and oversight of nanomaterials and applications of nanotechnology. Federal agencies strive to assess the applications' safety, efficacy, impact on public health, and environmental implications by utilizing the best available scientific and technical data (Soares et al., 2018) (Figure 6.1).

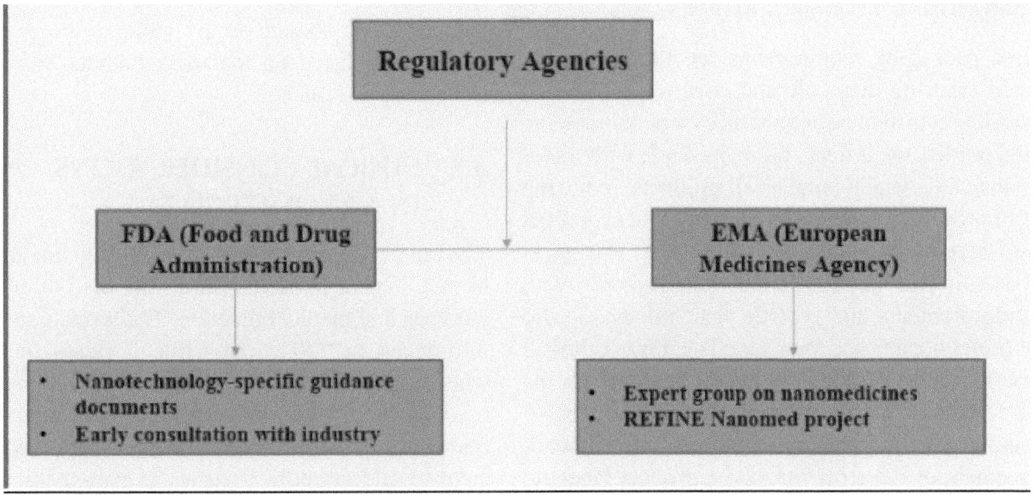

FIGURE 6.1 Diagram for regulatory agencies.

6.2.2 Safety Assessment and Risk Management

Assessing the safety and risk management of nanomedicines is a critical aspect of their regulation. The regulatory safety evaluation of nanomedical products involves addressing key issues to refine the safety assessment and risk management. Nanotechnologies have rapidly developed, enabling significant improvements in medicinal products and medical devices (Soares et al., 2018). However, the introduction of new nanomedicinal entities can be challenging with regard to the preclinical safety evaluation before clinical trials and market authorization. The lack of global consensus on definitions and clarity around critical safety attributes presents regulatory challenges and gaps in currently available testing methods for the safety evaluation of nanomedicines (Alphandéry, 2019).

The US Food and Drug Administration (FDA) recommends taking into account the distinct qualities and behaviors of nanotechnology products when assessing their safety, efficacy, impact on public health, or regulatory status. A risk-based framework, particular guidelines for nonclinical and clinical trials, manufacturing quality and control standards, and unique environmental considerations for drug and biologic products containing nanomaterials have all been delineated by the FDA's Nanotechnology Task Force (Soares et al., 2018).

In the European Union, the movement toward formalized nanomedicine regulations is being led by the REFINE Nanomed project. This project addresses issues specifically for nanomedicines and contributes to establishing science-based regulatory requirements for safety (Alphandéry, 2019). The safety assessment for medical applications in nanomedicine relies heavily on nano-specific toxicological concepts and findings and on a multidisciplinary collaborative approach involving material scientists, physicians, and toxicologists. Detailed physicochemical characterization of nanomaterials is crucial for safety assessment and risk management in nanomedicine (Schillmeier et al., 2016).

6.2.3 Labeling and Packaging Requirements

Labeling and packaging requirements for nanomedicines have a role in ensuring their safe and effective use. While the FDA has not published clear guidance on nanomedicine approval and marketing, it treats products made with nanotechnology the same way it handles all products, requiring companies to prove safety and efficacy before their product can come to market (Schillmeier et al., 2016). The packaging and labeling of nanomedicines must comply with regulatory requirements and provide clear information to healthcare professionals and patients. The ISO technical specification ISO/TS 13830:2013 provides guidance on the format and content of voluntary labels for consumer products containing manufactured nano-objects, recommending the placement of the term "nano" on product labels to enable consumers to make informed purchasing decisions (Soares et al., 2018). While there are no mandatory labeling requirements for nanomaterials in consumer products and adhering to best practices, utilizing standardized terminology, providing safety information, and staying informed about scientific research are key aspects of effective nanomaterial labeling (Alphandéry, 2019).

6.2.4 Clinical Trials and Post-Market Surveillance

Clinical trials and post-market surveillance play a crucial role in evaluating the safety and efficacy of nanomedicines. The European Medicines Agency (EMA) has established an expert group on nanomedicines to provide scientific information and develop or review guidelines for the development and approval of nanomedicines (Soares et al., 2018). Similarly, the FDA's Nanotechnology Task Force has outlined specific requirements for nonclinical and clinical trials, manufacturing quality and controls, and special environmental considerations for drug and biologic products containing nanomaterials. However, challenges and inconsistencies in the regulatory frameworks for nanomedicines, particularly in terms of clinical trials and post-market surveillance, still exist (Alphandéry, 2019).

The FDA has approved the commercialization of 100 nanomedicine applications and products, demonstrating the significant role of nanotechnology in biomedical science. Despite this, there are concerns about the ethical issues, costs, commercial development, and possible market failure associated with nanopharmaceuticals (Soares et al., 2018). The long-term exposure effects on humans, animals, and the environment due to nanomedicines are still not fully understood. Moreover, the lack of toxicity information and data currently available makes it difficult to create adequate regulations for maintaining patient safety and regulating the use of nanomedicines in a clinical setting (Soares et al., 2018). A crucial step in guaranteeing the continued safety and effectiveness of nanotherapeutics is post-marketing surveillance, which tries to monitor the medication after marketing approval (Alphandéry, 2019). However, because of their limited understanding of nanomedicines, regulatory agencies have difficulty determining whether a particular nanomedicine should be available as an over-the-counter product or placed on the market under strict supervision (Alphandéry, 2019).

6.3 ETHICAL CONSIDERATIONS IN NANOMEDICINE

The ethical considerations surrounding nanomedicine have been a subject of discussion in the field of emerging technologies and medical practice. The term "nano-ethics" was introduced in 2004, and while it raises questions about whether nanomedicine presents new ethical dilemmas or requires a new type of ethics, it is generally accepted that existing ethical principles and approaches are sufficient to address ethical issues in nanomedicine (Mazahir et al., 2023). However, there are valid reasons to support the need for specialized expertise in nano-ethics. Given the complexity and incomplete knowledge in nanoscience, a comprehensive understanding and analysis of the ethical implications of nanomedicine require continuous study, firsthand experience, and collaboration with stakeholders in the field (Mazahir et al., 2023).

Nanomedicine has the potential to interfere with various biological processes in the human body, leading to both advantages and challenges. The heterogeneity of nanomedicine complicates the formulation of general statements, and an approach that considers each case individually is favored over generalizations (Mazahir et al., 2023). Nanomedicine encompasses a wide range of applications from simple patches containing nanoparticles to complex computer–brain interfaces, each posing different levels of complexity (Mazahir et al., 2023). When analyzing the ethical implications of nanomedicine, it is essential to draw from the principles and approaches commonly used in bioethics and medical ethics. No single principle or theory alone is sufficient to address the complexity of ethical issues in nanomedicine. Instead, a circular approach that combines inductive and deductive reasoning is preferred, as it allows for making general assumptions based on individual cases and vice versa (Mazahir et al., 2023).

6.3.1 Informed Consent and Privacy

Informed consent is a critical ethical consideration in nanomedicine research and development. It ensures that individuals participating in clinical trials and other applications fully understand the potential risks and benefits associated with the use of nanotechnology. The principle of respect for autonomy is particularly relevant in nanomedicine due to the unique properties and effects of nanoparticles on human health (Vandemeulebroucke et al., 2022). Researchers and clinicians provided clear explanations of the risks, benefits, and uncertainties of nanomedicine treatments to patients and their families. This approach helps ensure that patients provide truly informed consent, which is essential for upholding trust and respecting individual autonomy (de Kanter et al., 2023).

6.3.2 Access, Equity, and Justice

Ensuring access to nanomedicine for all individuals is a significant ethical concern. Regardless of their socioeconomic status, nanomedicine has the potential to improve health outcomes and enhance the quality of life for millions of people. Promoting social justice and equity requires strategies that enable affordable and accessible nanomedicine products, investments in health infrastructure, and the promotion of health education and awareness (Vandemeulebroucke et al., 2022). Additionally, the societal implications of nanomedicine, such as the use of nanotechnology for physical enhancement rather than therapy, need to be carefully considered to address concerns related to social justice and equity (Malik et al., 2023).

6.3.3 Enhancement Technologies and Ethical Boundaries

The development and application of nanomedicine raise ethical questions concerning the use of enhancement technologies and the boundaries of ethical permissibility. As nanotechnology advances, it becomes possible to enhance human abilities and characteristics, ranging from cognitive function to strength and mobility (Bawa & Johnson, 2019). While these enhancements can offer significant benefits, they also introduce concerns related to fairness, privacy, and unintended consequences. Establishing clear and consistent guidelines for the development and use of

TABLE 6.1
Ethical considerations in nanomedicine

Ethical considerations	Description	Key points	Ref.
Informed consent and privacy	Obtaining informed consent from participants in nanomedicine research and clinical trials is crucial to ensure understanding of potential risks and benefits.	Respect for autonomy and clear explanations of risks, benefits, and uncertainties are essential.	(de Kanter et al., 2023; Vandemeulebroucke et al., 2022)
Access, equity, and justice	Ensuring affordable and accessible nanomedicine products for all individuals is critical.	Strategies include investments in health infrastructure, education, and awareness.	(Vandemeulebroucke et al., 2022; Bawa & Johnson, 2019)
Enhancement technologies and ethical boundaries	Establishing clear guidelines for nanomedicine enhancement technologies is necessary to address concerns about fairness, privacy, and unintended consequences.	Ethical permissibility and responsible development practices are paramount.	(Bawa & Johnson, 2019; Wasti et al., 2023)
Responsible conduct of research	Adherence to rigorous scientific and ethical standards is essential to ensure the safety and efficacy of nanomedicine products.	Protecting privacy, confidentiality, and research integrity is a key consideration.	(Hock et al., 2011; de Kanter et al., 2023)

nanomedicine enhancement technologies is essential to ensure responsible and ethical practices that promote the well-being of all individuals (Wasti et al., 2023).

6.3.4 Responsible Conduct of Research

Ensuring the safety and efficacy of nanomedicine products requires responsible research conduct that complies with strict scientific and ethical guidelines. This includes getting participants' informed permission, maintaining participant privacy and confidentiality, and truthfully and openly disclosing study results. Beyond what is covered by the common rule, research on nanomedicine raises ethical questions about dangers to participants' families, research personnel, workers in nanotechnology, the environment, and society (de Kanter et al., 2023). Therefore, in order to safeguard research personnel as well as the environment, investigators and research institutions should abide by occupational health and environmental laws. Furthermore, because the field of nanomedicine research is new and evolving, it may be more challenging to successfully obtain real informed consent from participants in trials involving nanomedical devices and nanochemical drugs. By promoting responsible conduct of research, nanomedicine can advance in a manner that is scientifically and ethically sound (Hock et al., 2011).

6.4 PUBLIC ACCEPTANCE OF NANOMEDICINE

The public acceptance of nanomedicine has been a subject of interest in recent research. A study on the knowledge, perceptions, and attitudes of medical residents toward nanomedicine found that a positive attitude toward nanomedicine was correlated with a higher willingness to use it to diagnose and treat patients (Nassani et al., 2019). Additionally, more than 50 nanomedicine formulations have been approved for clinical use, indicating the growing maturity of nanomedicine for global health care (Germain et al., 2020). However, there are concerns about limited understanding of nanomedicine among medical residents, with 66% of them believing they do not have adequate access to information about nanotechnology (Nassani et al., 2019). This highlights the need for improved education and awareness about nanomedicine among medical professionals and the public. The public acceptance of nanomedicine has been a topic of discussion for over a decade, with questions arising about whether patient consumers will be reluctant to embrace nanomedicine (Gupta et al., 2015). Limited understanding of the subject can lead to misinformed biases and associations, emphasizing the importance of accurate information and education about nanomedicine (Rosaria et al., 2018).

6.4.1 Factors Influencing Public Acceptance

Public acceptance of nanomedicine is crucial for its continued advancement and widespread adoption. Several factors influence public acceptance, including awareness, education, perception of risks and benefits, and trust in regulatory systems and healthcare institutions (Kamarulzaman, et al., 2020; Hock, et al., 2011). Awareness and education play a significant role in shaping public acceptance of nanomedicine. Currently, there is a substantial knowledge gap between the general public's understanding and knowledge of nanomedicine (Chhikara, 2017). Insufficient awareness can lead to misconceptions and mistrust, highlighting the need for enhanced education and public relations strategies (Hock et al., 2011).

The perception of risks and benefits associated with nanomedicine is another influential factor in public acceptance. The public's perception of risk versus benefit is crucial in determining the level of social acceptance of nanotechnology (Hock et al., 2011). It is essential to ensure that the public understands the potential advantages of nanomedicine, such as improved diagnostics, treatment, and patient outcomes (Chhikara, 2017). By effectively communicating the benefits and addressing potential risks, a positive attitude towards nanomedicine can be fostered. Trust in regulatory systems and healthcare institutions also plays a vital role in public acceptance of nanomedicine (Kamarulzaman et al., 2020). The public must have confidence that regulatory systems and healthcare institutions are reliable and effective in overseeing the development and use of nanomedicine (Kamarulzaman et al., 2020). Transparent and accountable regulatory processes can help build trust and promote the adoption of nanomedicine.

Trust and communication are closely intertwined with public acceptance. Effective communication of nanomedicine's risks and benefits is crucial in building trust and ensuring public acceptance (Kamarulzaman et al., 2020). It is essential to address misconceptions and provide accurate information about nanomedicine to the public (Kamarulzaman et al., 2020). Engaging with the public through various channels, such as public forums, educational campaigns, and online platforms, can facilitate dialogue and build trust (Kamarulzaman et al., 2020).

6.4.2 Stakeholder Engagement and Communication

Public acceptance of nanomedicine is crucial for its continued advancement, and stakeholder engagement and communication play a significant role in shaping this acceptance. Transparency and open dialogue are fundamental principles that should underpin the development and application of nanomedicine. Providing the public with access to accurate and reliable information about nanomedicine is essential in dispelling misconceptions and fostering trust in the field (Malviya et al., 2021). Openly sharing information regarding the benefits, risks, and regulatory processes associated with nanomedicine can help create an informed public opinion and facilitate the acceptance of this emerging technology.

Engaging with various stakeholders, including patients, healthcare providers, and the public, is crucial for

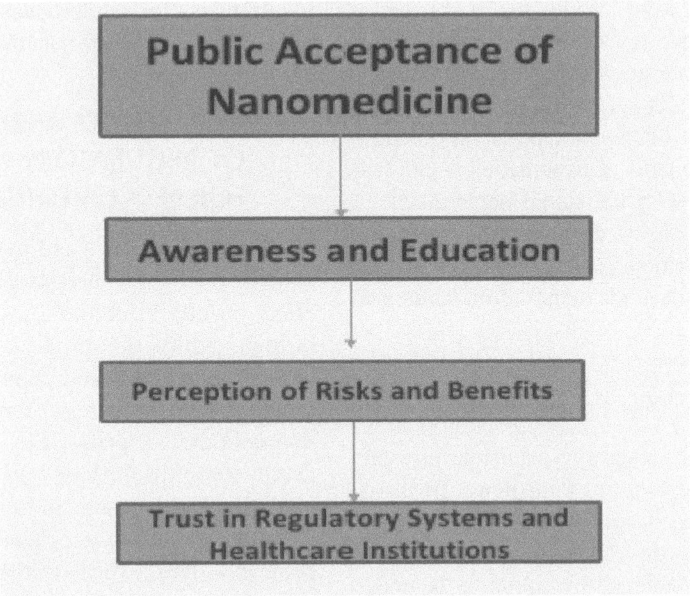

FIGURE 6.2 Factors influencing public acceptance of nanomedicine.

understanding their concerns, expectations, and attitudes towards nanomedicine. This stakeholder engagement allows policymakers and regulators to gather valuable insights that can inform the development of appropriate regulations, guidelines, and communication strategies (Ragucci et al., 2023). By actively involving these stakeholders, regulators can ensure that the regulatory frameworks for nanomedicine align with their needs and aspirations. The influence of media framing can also be taken into account when considering public acceptance of nanomedicine. The way media portrays nanotechnology and its associated risks and benefits can significantly impact public perception (Rathore & Mahesh, 2021). Therefore, accurate and responsible reporting of nanomedicine is essential in shaping a positive public opinion and fostering a better understanding of the technology.

The advice and opinions of healthcare professionals, backed by clinical research, have a role in public acceptance of new medicines, including nanomedicine. Patients often rely on their doctor's advice, clinical evidence, information about side effects, and regulatory approvals, such as those from the FDA (Nassani et al., 2019). Therefore, it is essential to ensure that healthcare professionals are well-informed about nanomedicine and its potential benefits. This can enhance public confidence in the technology and contribute to its acceptance.

Addressing concerns and misconceptions is another crucial aspect of promoting public acceptance of nanomedicine. Clear communication and providing accurate information can help dispel common misconceptions and alleviate concerns regarding the safety, efficacy, and ethical implications of nanomedicine (Malviya et al., 2021). By addressing these concerns head-on, regulators can foster a more positive public perception and facilitate a better understanding of the potential benefits of nanomedicine.

6.4.3 Addressing Public Concerns and Misconceptions

Addressing public concerns and misconceptions is critical for ensuring the acceptance and successful integration of nanomedicine into health care. A number of key concerns have been identified, including safety and long-term effects, ethical implications and moral objections, and managing nanotechnology hype while setting realistic expectations. Recent scientific papers offer valuable insights into addressing these concerns.

Public perception of nanotechnology can be influenced by biases and associations based on previous experiences with biotechnologies (Gupta et al., 2015). However, there is evidence to suggest that emphasizing the medical applications of nanotechnology, particularly in terms of prolonging life, helps rekindle interest and trust in the field (Gupta et al., 2015). Therefore, it is crucial to emphasize the safety and long-term effects of nanomedicine through rigorous safety assessments, ongoing research, and transparent communication about the potential benefits and risks (Gupta et al., 2015).

Ethical implications and moral objections are frequently raised by the public and can significantly impact the acceptance of nanotechnology (Rosaria et al., 2018). Addressing these concerns requires open dialogue and transparent communication among stakeholders. It is important to provide clear ethical justifications for the use of nanomedicine, such as improved patient outcomes and enhanced quality of life (Rosaria et al., 2018). By highlighting the ethical foundations of nanomedicine and engaging in ethical discussions, stakeholders can help alleviate public concerns and promote acceptance.

Managing the hype surrounding nanotechnology and establishing realistic expectations are crucial for public acceptance (Sechi et al., 2014). While the public generally holds a positive attitude toward nanotechnology, it is essential to provide accurate information about its current state and limitations (Sechi et al., 2014). Transparent communication regarding the timeline for potential advancements and the practical benefits of nanomedicine can help align public expectations with the actual potential of the technology (Rosaria et al., 2018). By emphasizing tangible benefits and ethical justifications, stakeholders can foster a more informed and realistic understanding of nanomedicine (Sechi et al., 2014).

6.4.4 Case Studies and Best Practices

Case studies and best practices play a vital role in establishing regulatory frameworks for nanomedicine. By examining successful examples, it can identify key strategies and approaches to navigate the challenges associated with this emerging field. One noteworthy initiative is undertaken by the US Food and Drug Administration (FDA), Environmental Protection Agency (EPA), and the Health and Consumer Protection Directorate of the European Commission (Silva et al., 2011). These organizations have developed regulatory frameworks specific to nanomedicine to address potential risks and ensure the safety and efficacy of nanomedical products. Additionally, the REFINE project serves as a local initiative aiming to define regulatory criteria for nanomedicines and nanomaterials intended for clinical use (Silva Costa et al., 2011). Such projects contribute to the establishment of standardized guidelines and procedures that can be adopted globally, facilitating the responsible development and regulation of nanomedicine.

Ethical considerations are of paramount importance in nanomedicine. However, comprehensive reviews of ethical and legal concerns related to nanomedical technology are limited (Atalla et al., 2016). To address this gap, it is crucial to identify and address key ethical and legal issues associated with nanomedicine. This can be achieved through a consultative process involving academics, clinicians, and other stakeholders (Silva et al., 2011). By formulating guidelines and principles that govern the ethical conduct of nanomedicine research and development, it can ensure that this technology is used in a responsible and socially acceptable manner.

Public acceptance campaigns and strategies are essential for the continued advancement of nanotechnology in medicine. Public acceptance plays a critical role in determining the adoption and success of any new technology. In the case of nanomedicine, it is crucial to address common misconceptions and concerns related to social and ethnic issues associated with emerging nanotechnologies (Ventola, 2012). The limited understanding of nanotechnology can lead to misinformed biases and associations, potentially resulting in public opposition (Ventola, 2012). To mitigate these challenges, the FDA considers the need for increased patient and consumer education and engagement. Furthermore, an evaluation of current labeling requirements for nanoproducts is conducted to ensure they align with public sentiment (Wiesing & Clausen, 2014).

6.5 CHALLENGES AND CONSIDERATIONS OF REGULATORY FRAMEWORKS FOR NANOMEDICINE

Numerous obstacles stand in the way of the regulation of nanomaterials and nanomedicines, which may have an effect on nanomedicine's future and impede the creation of high-quality, safe products. These challenges include a lack of consistency and definitions, responsibility for formulating guidelines, regulatory framework and harmonization, and the adequacy of existing regulatory frameworks (Sechi et al., 2014). One of the primary challenges in nanomedicine is the design and engineering of nanostructures with appropriate physicochemical properties and biological behaviors. This involves optimizing reproducible manufacturing processes and selecting reliable analysis methods for adequate characterization (Ventola, 2012). For instance, dynamic light scattering (DLS) is commonly used for nanoparticle size analysis in liquid solutions, but it has limitations in accurately estimating the particle size distribution (Joubert et al., 2019). Nanoparticle tracking analysis (NTA) is emerging as an alternative technique that overcomes the limitations of DLS by measuring particle size based on the changing position of scattering diffusing particles through image analysis (Joubert et al., 2019).

Nanomedicine products must have favorable pharmacology and toxicity profiles to ensure their safety and efficacy in clinical trials. Comprehensive preclinical studies are necessary to assess these factors and establish the potential benefits and risks associated with nanomedicine (Allon et al., 2017). The challenges lie in understanding the behavior of nanomedicine in biological systems and ensuring its compatibility with the human body. By addressing these challenges, nanomedicine can demonstrate its clinical utility and pave the way for further development (Joubert et al., 2019).

Thehosphopment of nanomedicine requires a multidisciplinary approach involving experts from various fields, such as biology, chemistry, nanotechnology, imaging, and medicine (Ventola, 2012). Collaboration between academia, large pharmaceutical companies, smaller companies, and hospitals is crucial for the successful translation of nanomedicine from bench to bedside. The integration of diverse resources and expertise can accelerate the development of translatable therapeutics and overcome the complexities associated with nanomedicine (Ventola, 2012; Silva Costa et al., 2011).

Nanomedicine presents unique regulatory challenges due to its complexity and multicomponent nature. Health authorities need to develop and refine definitions, quality standards, guidelines, and regulatory frameworks specific to nanomedicine (Atalla et al., 2016). The regulatory landscape should address the additional variables and

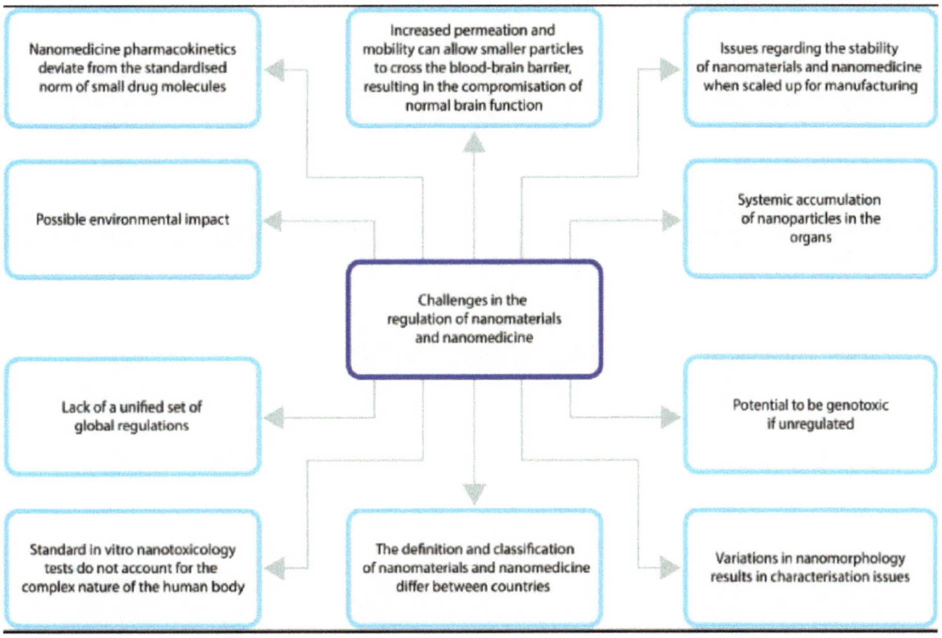

FIGURE 6.3 Diagram illustrating the main difficulties in regulating nanomedicine. (From Foulkes et al., 2020.)

considerations associated with controlling processes and monitoring the behavior of nanomedicine in biological systems (Wiesing & Clausen, 2014). Establishing appropriate regulatory frameworks will ensure the safety, efficacy, and ethical use of nanomedicine in clinical practice (Sechi et al., 2014).

6.5 RECOMMENDATIONS FOR POLICYMAKERS, RESEARCHERS, AND HEALTHCARE PROFESSIONALS

Regulatory frameworks for nanomedicine are crucial to ensure the safe and responsible development and implementation of this innovative field. Policymakers, researchers, and healthcare professionals have key roles in addressing the ethical, social, and regulatory challenges associated with nanomedicine. Risk assessment and management in clinical trials involving nanomedicine should be a priority for policymakers (Kamarulzaman et al., 2020). Clear guidelines need to be established to evaluate the toxicological, pharmacological, and immunological properties of different nanomaterials before their clinical use (Kamarulzaman et al., 2020). This approach ensures the safety and efficacy of nanomedicine interventions.

Public engagement and education are essential to gain public support and address concerns related to nanomedicine (Neme et al., 2021). Researchers and healthcare professionals should proactively communicate the benefits and risks of nanomedicine to the public (Neme et al., 2021). Transparent communication and public forums for discussion can help foster public trust and understanding of nanomedicine (Neme et al., 2021). The regulatory framework for nanomedicine should be adapted to address the unique challenges posed by this field (Foulkes et al., 2020). Policymakers need to evaluate whether current regulations are adaptable to newly emerging nanomedicine products or whether new frameworks are necessary (Foulkes et al., 2020). This assessment ensures that regulatory measures keep pace with technological advancements and adequately address safety and efficacy concerns. Informed consent is a critical aspect of nanomedical research (AlKahtani, 2018). Researchers should ensure that participants provide proper informed consent, which includes clear explanations of the research objectives, potential risks, and privacy concerns related to nanomedicine (AlKahtani, 2018). Respecting participants' autonomy and privacy is essential for ethical research practices.

Establishing oversight bodies can help address public concerns and provide expertise in nanomedicine research (Wasti et al., 2023). For instance, a Secretary's Advisory Committee can serve as a platform for addressing ethical and regulatory issues related to nanomedicine (Wasti et al., 2023). Additionally, a coordinated approach to federal and institutional oversight of nanomedicine human subjects research should be developed to ensure comprehensive governance (Wasti et al., 2023). Adopting ethical guidelines for the development and use of nanomedicine is crucial (Hock et al., 2011). These guidelines should ensure that nanomachines are specialized, are non-self-replicating, and do not pose potential harm to society (Hock et al., 2011). Policymakers and researchers should work collaboratively to establish and adhere to these guidelines, promoting responsible and ethical practices in nanomedicine.

6.6 CONCLUSION

Nanomedicine is a rapidly developing field that has the potential to revolutionize health care. However, the ethical implications of nanotechnology in medicine have not been widely discussed. The ethical considerations of nonmaleficence, beneficence, and respect for autonomy are in conflict, and it is important to determine which considerations should have the most weight. Nanomedicine regulation is a complicated topic, and developing appropriate regulations is challenging in the absence of extensive knowledge about the field. Authorities in charge of regulations must determine whether a particular nanomedicine should be sold under close supervision or as over-the-counter goods. An international problem is the absence of official regulation of the production of nanomaterials and nanomedicines for use in health-related applications. In evaluating nanomedicine, ethics plays two roles; first, it supports policy decisions, and second, it aids in patient care by providing clinical ethics advice.

Note: This chapter outline provides a general structure for discussing regulations, ethics, and public acceptance of nanomedicine. Each section can be further expanded with relevant examples, case studies, and current research to provide a comprehensive understanding of the topic.

REFERENCES

AlKahtani, R. N. (2018). The implications and applications of nanotechnology in dentistry: A review. Saudi Dental Journal, 30(2), 107–116.

Allon, I., Ben-Yehudah, A., Dekel, R., Solbakk, J. H., Weltring, K. M., & Siegal, G. (2017). Ethical issues in nanomedicine: Tempest in a teapot? Medicine, Health Care and Philosophy, 20, 3–11.

Alphandéry, E. (2019). A discussion on existing nanomedicine regulation: Progress and pitfalls. Applied Materials Today, 17, 193–205.

Atalla, K., Chaudhary, A., Eshaghian, M., Gupta, A., Mehta, R., Nayak, A., Prajogi, A., Ravicz, K., Shiroma, B., & Trivedi, P. (2016). Ethical, privacy, and intellectual property issues in nanomedicine. Wireless Computing in Medicine: From Nano to Cloud with Ethical and Legal Implications, 567–600. https://doi.org/10.1002/9781118993620.ch20.

Bawa, R., & Johnson, S. (2019). Overview of ethical issues in nanomedicine. In The Road from Nanomedicine to Precision Medicine (pp. 1013–1025). Jenny Stanford Publishing.

Chhikara, B. S. (2017). Current trends in nanomedicine and nanobiotechnology research. Journal of Materials Nano Science, 4(1), 19–24.

De Kanter, A. F. J., Jongsma, K. R., Verhaar, M. C., & Bredenoord, A. L. (2023). The ethical implications of tissue engineering for regenerative purposes: A systematic review. Tissue Engineering Part B: Reviews, 29(2), 167–187.

Engelmann, W., von Hohendorff, R., & Leal, D. W. S. (2021). Regulations for using nanotechnology in food and medical products. In Biopolymer-Based Nano Films (pp. 387–411). Elsevier.

Foulkes, R., Man, E., Thind, J., Yeung, S., Joy, A., & Hoskins, C. (2020). The regulation of nanomaterials and nanomedicines for clinical application: Current and future perspectives. Biomaterials Science, 8. https://doi.org/10.1039/D0BM00558D.

Germain, M., Caputo, F., Metcalfe, S., Tosi, G., Spring, K., Åslund, A. K. O., Pottier, A., Schiffelers, R., Ceccaldi, A., & Schmid, R. (2020). Delivering the power of nanomedicine to patients today. Journal of Controlled Release: Official Journal of the Controlled Release Society, 326, 164–171.

Gupta, N., Fischer, A. R. H., & Frewer, L. J. (2015). Ethics, risk and benefits associated with different applications of nanotechnology: A comparison of expert and consumer perceptions of drivers of societal acceptance. Nano Ethics, 9, 93–108.

Hock, S. C., Ying, Y. M., & Wah, C. L. (2011). A review of the current scientific and regulatory status of nanomedicines and the challenges ahead. PDA Journal of Pharmaceutical Science and Technology, 65(2), 177–195.

Joubert, I., Geppert, M., Ess, S., Nestelbacher, R., Gadermaier, G., Duschl, A., Bathke, A., Himly, M. (2019). Public perception and knowledge on nanotechnology: A study based on a citizen science approach. NanoImpact, 17, 100201. https://doi.org/10.1016/j.impact.2019.100201.

Kamarulzaman, N. A., Lee, K. E., Siow, K. S., & Mokhtar, M. (2020). Public benefit and risk perceptions of nanotechnology development: Psychological and sociological aspects. Technology in Society, 62, 101329.

Malik, S., Muhammad, K., & Waheed, Y. (2023). Emerging applications of nanotechnology in healthcare and medicine. Molecules (Basel, Switzerland), 28(18), 6624.

Malviya, R., Fuloria, S., Verma, S., Subramaniyan, V., Sathasivam, K. V., Kumarasamy, V., Hari Kumar, D., Vellasamy, S., Meenakshi, D. U., Yadav, S., Sharma, A., & Fuloria, N. K. (2021). Commercial utilities and future perspective of nanomedicines. PeerJ, 9, e12392. https://doi.org/10.7717/peerj.12392

Massaro, S., & Lorenzoni, G. (2021). Nanomedicine: A sociotechnical system. Technological Forecasting and Social Change, 173, 121066. 10.1016/j.techfore.2021.121066.

Mazahir, F., Bhogale, D., Palai, A. K., & Yadav, A. K. (2023). Nanomedicine: Principles, properties, and regulatory issues. In Smart Polymeric Nano-Constructs in Drug Delivery (pp. 523–565). Academic Press.

Nassani, N., El-Douaihy, Y., Khotsyna, Y., Shwe, T., & El-Sayegh, S. (2019). Knowledge, perceptions, and attitudes of medical residents towards nanomedicine: Defining the gap. Medical Science Educator, 30(1), 179–186.

Neme, K., Nafady, A., Uddin, S., & Tola, Y. B. (2021). Application of nanotechnology in agriculture, postharvest loss reduction and food processing: Food security implications and challenges. Heliyon, 7(12).

Ragucci, F., Sireci, F., Cavallieri, F., Rossi, J., Biagini, G., Tosi, G., Lucchi, C., Molina-Pena, R., Ferreira, N. H., Zarur, M., Ferreiros, A., Bourgeois, W., Berger, F., Abal, M., Rousseau, A., Boury, F., Alvarez-Lorenzo, C., Garcion, E., Pisanello, A., Pavesi, G., . . . Valzania, F. (2023). Insights into healthcare professionals' perceptions and attitudes toward nanotechnological device application: What is the current situation in glioblastoma research? Biomedicines, 11(7), 1854.

Rathore, A., & Mahesh, G. (2021). Public perception of nanotechnology: A contrast between developed and developing countries. Technology in Society, 67, 101751.

Rosaria, V., De Devitiis, B., Carlucci, D., Gianluca, N., & Fabio, S. (2018). On consumers' acceptance of nanotechnologies: An Italian case study. International Journal on Food System Dynamics, 9(4), 321–330.

Satalkar, P., Elger, B. S., & Shaw, D. M. (2016). Defining nano, nanotechnology and nanomedicine: Why should it matter? Science and Engineering Ethics, 22, 1255–1276.

Schillmeier, M., & Cornier, J., Owen, A., Kwade, A., Voorde, M. (2016). Social studies of nanopharmaceutical research. https://doi.org/10.1002/9783527800681.ch21.

Sechi, G., Bedognetti, D., Sgarrella, F., Eperen, L. V., Marincola, F. M., Bianco, A., & Delogu, L. G. (2014). The perception of nanotechnology and nanomedicine: A worldwide social media study. Nanomedicine, 9(10), 1475–1486.

Silva Costa, H., Sethe, S., Pêgo, A. P., & Olsson, I. A. S. (2011). Scientists' perception of ethical issues in nanomedicine: A case study. Nanomedicine, 6(4), 681–691.

Soares, S., Sousa, J., Pais, A., & Vitorino, C. (2018). Nanomedicine: Principles, properties, and regulatory issues. Frontiers in Chemistry, 360.

Vandemeulebroucke, T., Denier, Y., Mertens, E., & Gastmans, C. (2022). Which framework to use? A systematic review of ethical frameworks for the screening or evaluation of health technology innovations. Science and Engineering Ethics, 28(3), 26.

Ventola, C. L. (2012). The nanomedicine revolution: Part 1: Emerging concepts. P & T: A Peer-Reviewed Journal for Formulary Management, 37(9), 512–525.

Wasti, S., Lee, I. H., Kim, S., Lee, J. H., & Kim, H. (2023). Ethical and legal challenges in nanomedical innovations: A scoping review. Frontiers in Genetics, 14, 1163392.

Wiesing, U., & Clausen, J. (2014). The clinical research of nanomedicine: A new ethical challenge? Nanoethics, 8(1), 19–28.

7 Safety of Nanoparticles in Medicine

Mona Mohammed Abdel Hakeem Khalifa

7.1 INTRODUCTION

The application of nanoparticles for medicinal and diagnostic purposes is known as *nanomedicine*. Over the last 20 years, an increasing number of nanomedicines have been approved by regulatory bodies, and many more appear to have great potential. Small molecules, proteins, peptides, and nucleic acids are among the medicinal and imaging agents that are delivered by nanoparticles. The three main classes of therapeutically authorized nanomedicines are protein-, polymer-, and lipid-based particles. For example, more than 20 liposomal medications are enrolled in clinical studies, yet only 14 liposomal medications are marketed in the USA, Europe, or China (Koudelka & Turanek, 2012).

Drug release that is controlled and prolonged can be made possible by nanocarriers that shield the payload from deterioration. Moreover, nanoparticles may enhance medication accumulation and decrease clearance in sick tissue, improving treatment effectiveness and lowering adverse effects. The body contains several barriers that must be overcome for a drug to reach an intended location (Ferrari et al., 2013). These barriers consist of the vascular endothelium, renal and enzymatic degradation, immune clearance, and mechanical degradation membrane pumps, lysosomes, cell membranes, and extracellular matrix. Essentially, these challenges can be overcome by optimizing the characteristics of nanoparticles specifically; a multimodal strategy is made possible by nanoplatforms, where different materials and compartments can be merged to effectively address the difficulties of targeted distribution. For instance, the multistage vector, which releases cargo one step at a time, is well-suited to deal with the different in vivo circumstances and compartments that arise after systemic injection (Godin et al., 2011). In addition, certain nanoparticles have special optical and electrical qualities that can be used in medical applications. For example, sick tissue can be thermally ablated using metal nanoparticles and external energy. For instance, infrared light can be used to heat gold nanoparticles (Shen et al., 2012) and radio waves (Glazer et al., 2010), while iron oxide particles can generate heat when placed in a magnetic field. When combined, these benefits imply that nanoparticles may be a useful tool in the fight against various illnesses. Given the tremendous potential of the subject of nanomedicine, it is crucial to create safety assays that can precisely forecast the possible toxicity of nanotherapeutics. Appropriate assays for assessing nanoparticle toxicity should be used, particularly as these particles have particular characteristics that are not predictable from studying the bulk material. However, in contrast to small molecule medications, it might not be required to set more stringent regulations for the licensing of nanoparticles. Instead, there may be situations where alternative approaches are taken to evaluate safety. The integration of nanoparticles into medical applications holds great promise for revolutionizing health care, offering targeted drug delivery, imaging, and diagnostics. However, ensuring the safety of these nanoparticles is crucial for their successful integration into medical practice. Several safety concerns mitigate potential risks and maximize the benefits of nanoparticle-based medical technologies (Johannsen et al., 2005).

1. *Biocompatibility*. One of the primary safety concerns when integrating nanoparticles into medical applications is their biocompatibility. Nanoparticles must not induce toxic effects or provoke an immune response when introduced into the human body. Biocompatibility encompasses various aspects, including the physicochemical properties of nanoparticles, such as size, shape, surface charge, and surface chemistry. These factors can influence the interaction of nanoparticles with biological systems and determine their biocompatibility. For instance, nanoparticles with certain surface coatings or modifications may trigger adverse reactions in the body, highlighting the need for thorough biocompatibility assessments before clinical use.
2. *Potential Toxicity*. Another critical safety consideration is the potential toxicity of nanoparticles. Due to their small size and unique physicochemical properties, nanoparticles may exhibit unexpected toxicological effects that differ from those of bulk materials. For example, certain nanoparticles could induce oxidative stress, inflammation, or genotoxicity in cells and tissues. Moreover, their ability to penetrate biological barriers and accumulate in specific organs raises concerns about long-term systemic toxicity. Therefore, comprehensive toxicity evaluations are essential to identify any potential adverse effects associated with nanoparticle exposure in medical settings.
3. *Long-Term Biodistribution and Clearance*. Understanding the long-term biodistribution and

DOI: 10.1201/9781003383376-7

clearance of nanoparticles is vital for assessing their safety in medical applications. Once administered, nanoparticles can distribute throughout the body via various routes, potentially accumulating in vital organs, such as the liver, spleen, or kidneys. Prolonged retention of nanoparticles in these organs may raise concerns about chronic toxicity and interfere with normal physiological functions. Additionally, efficient clearance mechanisms are crucial to prevent nanoparticle accumulation and associated adverse effects over time. Therefore, investigating the fate of nanoparticles in vivo and their clearance pathways is imperative for ensuring their long-term safety in medical contexts.

4. *Potential Environmental Impact.* In addition to human health considerations, the potential environmental impact of nanoparticle-based medical technologies should not be overlooked. As these nanoparticles may enter the environment through various routes, such as wastewater discharge or improper disposal, their impact on ecosystems and living organisms warrants careful assessment. Understanding the environmental fate, persistence, and potential bioaccumulation of medical nanoparticles is essential for minimizing ecological risks associated with their use.

In addressing these safety concerns, rigorous testing methodologies and standardized protocols are indispensable for evaluating the safety profile of nanoparticles intended for medical applications. This includes comprehensive physicochemical characterization, in vitro studies using relevant cell lines and tissues, as well as in vivo investigations to assess biodistribution, toxicity, and long-term effects.

Furthermore, regulatory agencies play a pivotal role in establishing guidelines and frameworks for assessing the safety of nanoparticle-based medical products. Robust regulatory oversight ensures that these innovative technologies meet stringent safety standards before entering clinical practice.

In conclusion, while nanoparticle integration holds immense potential for advancing medical applications, addressing safety concerns is paramount to realize their successful integration into healthcare. By thoroughly evaluating biocompatibility, toxicity, biodistribution, clearance mechanisms, and environmental impact, researchers and regulatory bodies can ensure that nanoparticle-based medical technologies are safe and effective for patient care.

7.2 NANOPARTICLES IN MEDICINE

Nanoparticles have gained significant attention in the field of medicine due to their unique properties and potential applications. These tiny particles, typically ranging from 1 to 100 nm in size, offer a wide range of opportunities for medical advancements, including drug delivery, imaging, and diagnostics.

There are various applications of nanoparticles in medicine, such as drug delivery, imaging, and diagnostics:

1. *Drug Delivery.* Nanoparticles are being extensively researched for their potential in targeted drug delivery. By functionalizing nanoparticles with specific ligands or antibodies, they can be designed to selectively target diseased cells or tissues while minimizing the impact on healthy cells. This targeted approach can enhance the efficacy of drugs and reduce their side effects. Additionally, nanoparticles can encapsulate drugs, protecting them from degradation and improving their bioavailability. This has the potential to revolutionize the treatment of various diseases, including cancer, by delivering drugs directly to tumor sites (Nel & Mädler, 2009).
2. *Imaging.* Nanoparticles are also utilized in medical imaging techniques, such as magnetic resonance imaging (MRI), computed tomography (CT), and positron emission tomography (PET). Nanoparticles are ideal for these applications with large surface areas that can be functionalized to introduce multiple modality reporters and targeting vectors to modify localization properties. The particle size, morphology, and surface chemistry can also be modified to influence their behavior in vivo (Cai & Chen, 2008).
3. *Diagnostics.* In the field of diagnostics, nanoparticles play a crucial role in the development of sensitive and specific diagnostic assays. Nanoparticles can be functionalized with biomolecules, such as antibodies or nucleic acids, to create biosensors capable of detecting biomarkers associated with various diseases. These nanoparticle-based biosensors offer high sensitivity and specificity, making them valuable tools for early disease detection and monitoring (Jokerst & Gambhir, 2011).
4. *Therapeutics.* Nanoparticles are also being explored for their potential therapeutic applications beyond drug delivery. For example, gold nanoparticles have been investigated for photothermal therapy, where they absorb light and convert it into heat to selectively destroy cancer cells. Similarly, silver nanoparticles have shown antimicrobial properties and are being studied for applications in combating infections (Dreaden et al., 2012).
5. *Regenerative Medicine.* Nanoparticle synthesis for regenerative medicines mainly focuses on development of entrapment and delivery systems for genetic material, biomolecules (growth and differentiation factors), bone morphogenetic proteins and for reinforcing the bioactivity of 3D scaffolds

for tissue engineering. Microspheres, microcapsules, liposomes, micelles and dendrimers are the nanoparticles mainly used in delivery systems. As per the requirements, solid, hollow or porous nanoparticles are synthesized through molecular self-assembly, nanomanipulation, bio aggregation and photochemical patterning (Allemann et al., 1993; Cade et al., 2004).

The diverse applications of nanoparticles in medicine hold great promise for advancing health care by improving drug delivery, enhancing medical imaging techniques, enabling sensitive diagnostics, exploring new therapeutic modalities, and contributing to regenerative medicine.

7.3 SAFETY ASSESSMENT OF NANOPARTICLES

Assessing the safety of nanoparticles in a medical context is crucial due to their potential applications in drug delivery, imaging, and diagnostics. Nanoparticles are particles with at least one dimension less than 100 nm, and their unique properties make them promising candidates for medical use. However, their small size and large surface area raise concerns about their potential toxicity and adverse effects on human health. Therefore, various methods and techniques have been developed to evaluate the safety of nanoparticles In a medical context:

1. *In Vitro Testing.* One of the primary methods used to assess the safety of nanoparticles is in vitro testing. In vitro studies involve exposing cell cultures to nanoparticles to evaluate their cytotoxicity, genotoxicity, and potential for causing inflammation. These studies provide valuable insights into the cellular responses to nanoparticles and help identify potential risks associated with their use in medical applications (Nel et al., 2006).
2. *In Vivo Studies.* In addition to in vitro testing, in vivo studies play a crucial role in assessing the safety of nanoparticles. In these studies, nanoparticles are administered to animal models to evaluate their biodistribution, pharmacokinetics, and potential systemic toxicity. By examining the interactions between nanoparticles and living organisms, researchers can gain a better understanding of their safety profile and potential long-term effects (Fadeel & Garcia-Bennett, 2010).
3. *Advanced Imaging Techniques.* Furthermore, advanced imaging techniques such as transmission electron microscopy (TEM) and scanning electron microscopy (SEM) are used to visualize the interactions between nanoparticles and biological systems at the nanoscale. These techniques allow researchers to observe the internalization of nanoparticles by cells and tissues, providing valuable information about their behavior within the body (Zhang & Monteiro-Riviere, 2009).
4. *Computational Modeling.* Computational modeling and simulation have emerged as powerful tools for predicting the behavior of nanoparticles in biological environments. By utilizing computational approaches, researchers can simulate the interactions between nanoparticles and biological molecules, as well as predict their potential toxicological effects based on physicochemical properties (Singh & Nalwa, 2007).
5. *Standardized Regulatory Guidelines.* Lastly, standardized regulatory guidelines and frameworks have been established to ensure the safe development and use of nanoparticles in medicine. These guidelines provide a structured approach for evaluating the safety of nanoparticles, including considerations for physicochemical characterization, toxicological testing, and risk assessment (Oberdörster et al., 2005).

In conclusion, assessing the safety of nanoparticles in a medical context involves a multidisciplinary approach that encompasses in vitro testing, in vivo studies, advanced imaging techniques, computational modeling, and regulatory guidelines. By employing these methods and techniques, researchers can gain valuable insights into the safety profile of nanoparticles and facilitate their responsible integration into medical applications.

7.3.1 Importance of Nanoparticles Being Biocompatible

Nanoparticles are those that do not cause harm to living organisms. They are tiny particles with dimensions in the nanometer range. They have gained significant attention in various fields, including medicine, electronics, and environmental science, due to their unique properties. In the context of biomedical applications, the biocompatibility of nanoparticles is of paramount importance. Biocompatible organisms make them suitable for use in medical diagnostics, drug delivery, and therapeutic interventions (Nel et al., 2006).

The importance of nanoparticles being biocompatible lies in their potential to interact with biological systems without eliciting adverse reactions. When nanoparticles are introduced into the body for medical purposes, such as targeted drug delivery or imaging, it is crucial that they do not induce toxicity or provoke immune responses. Biocompatible nanoparticles minimize the risk of tissue damage, inflammation, and other harmful effects, thereby enhancing their safety and efficacy in biomedical applications (Dobson, 2008).

Biocompatible nanoparticles can also be designed to specifically target diseased tissues or cells while sparing healthy ones. This targeted approach reduces off-target effects and improves the precision of medical treatments (Fadeel & Garcia-Bennett, 2010).

Additionally, biocompatible nanoparticles can be engineered to degrade or be eliminated from the body after fulfilling their intended function, minimizing long-term accumulation and potential toxicity (Zhang et al., 2008).

The importance of nanoparticles being biocompatible stems from their potential to advance medical technologies while ensuring safety and minimizing adverse effects on living organisms (Albanese et al., 2012).

7.3.2 The Potential Risks Associated with Nanoparticles

Nanoparticles are tiny particles with dimensions ranging from 1 to 100 nm. Due to their unique properties, nanoparticles have found applications in various fields, such as medicine, electronics, and environmental remediation. However, the increasing use of nanoparticles has raised concerns about their potential risks to human health and the environment. Some of the potential risks associated with nanoparticles include toxicity, inflammation, and immune responses.

1. *Toxicity.* Toxicity is a major concern with nanoparticles, as their small size and large surface area can lead to increased reactivity and potential harm to living organisms. When nanoparticles enter the body through inhalation, ingestion, or skin contact, they can interact with biological systems and cause adverse effects. Studies have shown that certain types of nanoparticles, such as metal oxides and carbon-based nanoparticles, can induce toxic effects in cells and tissues. The mechanisms of nanoparticle toxicity may involve oxidative stress, DNA damage, and disruption of cellular functions (Nel et al., 2006).
2. *Inflammation.* In addition to toxicity, nanoparticles have been associated with inflammation in the body. When nanoparticles are inhaled or come into contact with biological tissues, they can trigger immune responses that lead to inflammation. Chronic inflammation caused by prolonged exposure to nanoparticles has been linked to various diseases, including respiratory disorders and cardiovascular problems. Furthermore, the ability of nanoparticles to penetrate biological barriers such as the blood–brain barrier raises concerns about their potential to induce neuroinflammation and neurological disorders (Fadeel & Garcia-Bennett, 2010).
3. *Immune System Responses.* Moreover, nanoparticles have been shown to interact with the immune system in complex ways. The unique physicochemical properties of nanoparticles can influence immune cell function and modulate immune responses. Some nanoparticles have been reported to elicit immune reactions that may lead to hypersensitivity or autoimmune conditions. On the other hand, certain nanoparticles have been explored for their immunomodulatory effects in medical applications. Understanding the interactions between nanoparticles and the immune system is crucial for assessing their safety and potential therapeutic uses. (Dobrovolskaia & McNeil, 2007).

Overall, while nanoparticles offer promising opportunities for technological advancements, it is essential to carefully evaluate their potential risks to human health and the environment. Research efforts focused on understanding the toxicity, inflammation, and immune responses associated with nanoparticles are critical for developing safe nanomaterials and ensuring responsible use in various applications (Oberdörster et al., 2005).

7.4 REGULATORY GUIDELINES AND AGENCIES RESPONSIBLE FOR OVERSEEING NANOPARTICLE SAFETY IN MEDICINE

The use of nanoparticles in medicine has shown great promise for improving drug delivery, imaging, and diagnostics. However, the unique properties of nanoparticles also raise concerns about their safety and potential health risks. As a result, regulatory guidelines and agencies play a crucial role in overseeing nanoparticle safety in medicine.

1. *Regulatory Guidelines.* In the United States, the Food and Drug Administration (FDA) is the primary regulatory agency responsible for overseeing the safety and efficacy of medical products, including those involving nanoparticles. The FDA provides guidance on the development and evaluation of nanotechnology-based products through its Center for Drug Evaluation and Research (CDER) and Center for Devices and Radiological Health (CDRH). These guidelines aim to ensure that nanoparticle-based medical products meet the necessary safety standards before they can be approved for clinical use.
2. *European Medicines Agency.* EMA in Europe plays a pivotal role in regulating medicinal products containing nanoparticles. The EMA evaluates the quality, safety, and efficacy of these products through its Committee for Medicinal Products for Human Use (CHMP) and Committee for Advanced Therapies (CAT), providing specific guidelines for manufacturers to follow during the development and approval process.
3. *Agencies Responsible for Oversight.* Apart from the FDA and EMA, several other agencies worldwide are involved in overseeing nanoparticle safety in medicine. For instance, Health Canada regulates health products containing nanoparticles through its Health Products and Food Branch (HPFB),

ensuring that these products comply with safety standards before entering the Canadian market.

4. *Pharmaceuticals and Medical Devices Agency*. In Japan, PMDA oversees the approval and post-marketing safety of pharmaceuticals and medical devices, including those utilizing nanotechnology. The PMDA evaluates the safety and effectiveness of nanoparticle-based medical products to safeguard public health.
5. *Challenges and Considerations*. One of the challenges in regulating nanoparticle safety in medicine is the need to establish specific criteria for evaluating the potential risks associated with different types of nanoparticles. Due to their diverse physicochemical properties, nanoparticles may exhibit varying degrees of toxicity, biodistribution, and biocompatibility. As a result, regulatory agencies must consider these factors when assessing the safety profiles of nanoparticle-based medical products.

Furthermore, there is ongoing research into developing standardized testing methods to assess the safety of nanoparticles, including their interactions with biological systems. This includes understanding how nanoparticles behave in vivo, their potential to accumulate in specific tissues or organs, and any long-term effects on human health.

In conclusion, regulatory guidelines and agencies play a critical role in ensuring the safety of nanoparticle-based medical products. The FDA, EMA, Health Canada, PMDA, and other regulatory bodies work to establish comprehensive guidelines and oversight mechanisms to evaluate the safety and efficacy of these innovative technologies. By addressing the unique challenges posed by nanoparticles and considering their potential risks, these agencies contribute to advancing safe and effective nanoparticle applications in medicine.

7.5 IMPORTANCE OF COMPLYING WITH THESE REGULATIONS FOR APPROVAL AND COMMERCIALIZATION

Complying with regulations is of utmost importance for the approval and commercialization of products, especially in industries such as pharmaceuticals, medical devices, food and beverages, and many others. Regulatory compliance ensures that products meet the required standards for safety, efficacy, quality, and ethical considerations. Here are several reasons that complying with regulations is crucial for approval and commercialization:

1. *Safety and Efficacy*. Regulatory compliance ensures that products are safe for use and effective for their intended purpose. For instance, in the pharmaceutical industry, regulatory bodies such as the Food and Drug Administration (FDA) in the United States or the European Medicines Agency (EMA) in Europe set stringent guidelines to ensure that drugs undergo rigorous testing to demonstrate their safety and efficacy before they can be approved for commercialization.
2. *Quality Assurance*. Compliance with regulations ensures that products meet specific quality standards. This is essential to maintain consumer confidence and protect public health. For example, in the food industry, adherence to regulations such as Good Manufacturing Practices (GMP) is crucial to ensure that food products are manufactured, processed, and stored in a manner that prevents contamination and ensures their safety.
3. *Ethical Considerations*. Many regulations also encompass ethical considerations, particularly in industries involving human subjects or animal testing. For instance, in the field of medical research, compliance with regulations such as the Declaration of Helsinki is essential to ensure that research involving human subjects is conducted ethically and with respect for the participants' rights and well-being.
4. *Market Access*. Compliance with regulations is often a prerequisite for market access. Without regulatory approval, products may not be allowed to be commercialized or sold in certain markets. Therefore, adherence to regulatory requirements is critical for companies seeking to enter new markets and expand their commercial reach.
5. *Legal Obligations*. Non-compliance with regulations can lead to legal repercussions, including fines, product recalls, or even legal action. By ensuring compliance with regulations from the outset, companies can avoid costly legal issues that may arise from non-compliance.
6. *Reputation Management*. Regulatory compliance also plays a significant role in shaping a company's reputation. Adhering to regulations demonstrates a commitment to quality, safety, and ethical business practices, which can enhance a company's standing in the eyes of consumers, investors, and other stakeholders.

In conclusion, complying with regulations is paramount for the approval and commercialization of products across various industries. It not only ensures safety, efficacy, and quality but also upholds ethical standards while facilitating market access and safeguarding against legal and reputational risks.

7.6 NANOPARTICLES IN MEDICINE AND THEIR SAFETY PROFILES

Nanoparticles have gained significant attention in the field of medicine due to their unique properties and potential applications in drug delivery, imaging, diagnostics, and therapeutics. These tiny particles, typically ranging from

1 to 100 nm in size, offer distinct advantages, such as high surface-area-to-volume ratio, tunable surface chemistry, and the ability to encapsulate or carry therapeutic agents to specific targets within the body. However, the use of nanoparticles in medicine raises concerns about their safety profiles, including potential toxicity and long-term effects on human health. we will explore specific examples of nanoparticles used in medicine and their safety profiles, as well as notable successes and challenges in ensuring their safety

7.6.1 Examples of Nanoparticles Used in Medicine and Their Safety Profiles

7.6.1.1 Liposomal Nanoparticles

Liposomal nanoparticles are lipid-based vesicles that have been extensively studied for drug delivery applications. These nanoparticles can encapsulate a wide range of therapeutic agents, including chemotherapeutic drugs, nucleic acids, and imaging agents. One notable example is Doxil®, a liposomal formulation of doxorubicin approved for the treatment of various cancers. Studies have shown that while liposomal nanoparticles can improve the pharmacokinetics and biodistribution of drugs, they may also induce unique adverse effects related to their composition and interactions with biological systems. Research has focused on understanding the potential toxicity of liposomal nanoparticles, particularly in terms of immune responses, organ accumulation, and long-term effects on healthy tissues.

7.6.1.2 Gold Nanoparticles

Gold nanoparticles have been investigated for their applications in cancer therapy, imaging, and diagnostics. These nanoparticles exhibit favorable optical properties that make them suitable for photothermal therapy and imaging contrast enhancement. Additionally, gold nanoparticles can be functionalized with targeting ligands to specifically deliver therapeutic payloads to tumor sites. Despite their promising applications, concerns have been raised regarding the potential toxicity of gold nanoparticles, especially when considering their accumulation in vital organs, such as the liver and spleen. Studies have aimed to elucidate the mechanisms of gold nanoparticle clearance and metabolism in vivo to assess their long-term safety profiles.

7.6.1.3 Quantum Dots

Quantum dots are semiconductor nanoparticles with unique optical properties, making them valuable for bioimaging and diagnostic purposes. These nanoparticles can emit bright fluorescence at specific wavelengths, enabling sensitive detection of biomolecules and cellular processes. However, the use of quantum dots in medicine has raised concerns about their potential cytotoxicity due to factors such as heavy metal content (e.g., cadmium) and surface modifications. Efforts have been made to develop biocompatible coatings for quantum dots and understand their interactions with biological systems to mitigate potential safety risks.

7.6.2 Notable Successes and Challenges in Ensuring Nanoparticle Safety

7.6.2.1 Successes

- *Improved Understanding.* Significant progress has been made in elucidating the interactions between nanoparticles and biological systems at the molecular and cellular levels. This improved understanding has led to the development of safer nanoparticle formulations with reduced toxicity.
- *Regulatory Guidelines.* Regulatory agencies have established guidelines for evaluating the safety of nanoparticle-based products, leading to more rigorous preclinical assessments and clinical trials to assess their potential risks.
- *Targeted Delivery Strategies.* Advances in nanoparticle engineering have enabled the design of targeted delivery systems that minimize off-target effects and enhance therapeutic efficacy while mitigating safety concerns.

7.6.2.2 Challenges

- *Long-Term Effects.* Assessing the long-term effects of nanoparticle exposure remains a challenge due to the complex interactions between nanoparticles and biological systems over extended periods.
- *Standardized Testing Methods.* The lack of standardized testing methods for evaluating nanoparticle safety hinders comparability across studies and complicates regulatory decision-making.
- *Multifactorial Toxicity.* Nanoparticle toxicity is influenced by multiple factors, such as size, shape, surface chemistry, and dose, making it challenging to predict their safety profiles accurately.

In conclusion, while nanoparticles hold immense promise for revolutionizing medicine through targeted therapies and advanced diagnostics, ensuring their safety remains a critical consideration. Ongoing research efforts continue to address these challenges by advancing our understanding of nanoparticle interactions with biological systems and developing robust safety assessment strategies.

7.7 IMPROVING THE SAFETY OF NANOPARTICLES IN MEDICINE

The ongoing research and innovations aimed at improving the safety of nanoparticles in medicine are crucial for advancing the field of nanomedicine. One area of focus is the development of novel nanoparticle formulations that exhibit enhanced biocompatibility and reduced toxicity.

Researchers are exploring various surface modifications and coatings to mitigate potential adverse effects associated with nanoparticle exposure. Additionally, efforts are underway to optimize the design of nanoparticles to ensure efficient targeting of specific tissues or cells while minimizing off-target interactions.

Another promising direction in nanoparticle safety research involves the advancement of sophisticated imaging and analytical techniques to track the biodistribution and fate of nanoparticles in living organisms. By gaining a deeper understanding of how nanoparticles behave in biological systems, scientists can better assess their potential risks and develop strategies to enhance their safety profiles.

Furthermore, ongoing investigations seek to elucidate the long-term effects of nanoparticle exposure on human health. Longitudinal studies and epidemiological research are essential for comprehensively evaluating the safety implications of prolonged nanoparticle exposure, especially in the context of therapeutic applications.

In conclusion, it is evident that the field of nanomedicine holds tremendous promise for revolutionizing health care through targeted drug delivery, imaging, and diagnostics. However, ensuring the safety of nanoparticles remains a paramount concern. The key takeaways from this chapter underscore the need for continued vigilance in assessing and mitigating potential risks associated with nanoparticle-based interventions. As research progresses, it is imperative to maintain a proactive approach towards addressing safety concerns and implementing robust regulatory frameworks to safeguard public health.

REFERENCES

Albanese, A., Tang, P. S., & Chan, W. C. W. (2012). The effect of nanoparticle size, shape, and surface chemistry on biological systems. Annual Review of Biomedical Engineering, 14(1), 1–16.

Allemann, E., Gurny, R., & Doelker, E. (1993). Drug loaded nanopaticles~methods and drug targeting issues. European Journal of Pharmaceutics and Biopharmaceutics, 39, 173–191.

Cade, D., Ramus, E., Rinaudo, M., et al. (2004). Tailoring of biocompatible polymers for elaboration of sugar-functionalized nanoparticles. Biomacromolecules, 5, 922–927.

Cai, W. B., & Chen, X. Y. (2008). Multimodality molecular imaging of tumor angiogenesis. Journal of Nuclear Medicine, 49 2(1), 113S–128S.

Dobrovolskaia, M. A., & McNeil, S. E. (2007). Immunological properties of engineered nanomaterials. Nature Nanotechnology, 2(8), 469–478.

Dobson, J. (2008). Nanoscale biocompatibility: A new frontier for diagnostics and therapeutics. Expert Review of Medical Devices, 5(4), 469–471.

Dreaden, E. C., Alkilany, A. M., Huang, X., Murphy, C. J., & El-Sayed, M. A. (2012). The golden age: Gold nanoparticles for biomedicine. Chemical Society Reviews, 41(7), 2740–2779.

Fadeel, B., & Garcia-Bennett, A. E. (2010). Better safe than sorry: Understanding the toxicological properties of inorganic nanoparticles manufactured at large scale. Nanomedicine: Nanotechnology, Biology and Medicine, 5(6), 873–879.

Ferrari, M. (2010). Frontiers in cancer nanomedicine: Directing mass transport through biological barriers. Trends in Biotechnology, 28(4), 181–188.

Glazer, E. S., Zhu, C., Massey, K. L., et al. (2010). Noninvasive radiofrequency field destruction of pancreatic adenocarcinoma xenografts treated with targeted gold nanoparticles. Clinical Cancer Research, 16(23), 5712–5721.

Godin, B., Tasciotti, E., Liu, X., Serda, R. E., & Ferrari, M. (2011). Multistage nanovectors: From concept to novel imaging contrast agents and therapeutics. Accounts of Chemical Research, 44(10), 979–989.

Johannsen, M., Thiesen, B., Jordan, A., et al. (2005). Magnetic fluid hyperthermia (MFH)reduces prostate cancer growth in the orthotopic Dunning R3327 rat model. Prostate, 64(3), 283–292.

Jokerst, J. V., & Gambhir, S. S. (2011). Molecular imaging with theragnostic nanoparticles. *Accounts of Chemical Research*, 44, 1050–1060.

Koudelka, S., & Turanek, J. (2012). Liposomal paclitaxel formulations. *Journal of Controlled Release*, 163(3), 322–334.

Nel, A. E., Mädler, L., Velegol, D., Xia, T., & Hoek, E. M. (2009). Understanding biophysicochemical interactions at the nano-bio interface. Nature Materials, 8(7), 543–557.

Nel, A., Xia, T., Mädler, L., & Li, N. (2006). Toxic potential of materials at the nanolevel. Science, 311(5761), 622–627.

Oberdörster, G., Oberdörster, E., & Oberdörster, J. (2006). Nanotoxicology: An emerging discipline evolving from studies of ultrafine particles. Environmental Health Perspectives, 13(7), 823–839.

Shen, H., You, J., Zhang, G., et al. (2012). Cooperative, nanoparticle-enabled thermal therapy of breast cancer. Advanced Healthcare Materials, 1(1), 84–89.

Singh, S., & Nalwa, H. S. (2007). Nanotechnology and health safety—toxicity and risk assessments of nanostructured materials on human health. Journal of Nanoscience and Nanotechnology, 7(9), 3048–3070.

Zhang, L., Gu, F., Chan, J., Wang, A., Langer, R., & Farokhzad, O. (2008). Nanoparticles in medicine: Therapeutic applications and developments. Clinical Pharmacology & Therapeutics, 83(5), 761–769.

Zhang, L. W., & Monteiro-Riviere, N. A. (2009). Assessment of quantum dot penetration into intact human skin using an in vitro model. Journal of Investigative Dermatology, 21(3), 166–180.

8 Enhancing Brain Drug Delivery with Lipid Nanocarriers for Bioactive Compounds

*Shivakumar M. Methri, Sivakumar Manickam, V. Ravichandiran, and Subramanian Natesan**

8.1 INTRODUCTION

The brain is an incredibly intricate and specialized organ responsible for controlling vital bodily functions, such as memory, emotions, sensory perception (touch, vision, etc.), and hunger (Brain Anatomy and How the Brain Works, 2021). However, its unique structure and physiology pose significant challenges for drug delivery. The presence of the blood–brain barrier (BBB) acts as a protective shield for the brain, preventing the entry of most drugs and bioactive substances. This selective barrier limits the effectiveness of therapies for neurological disorders, including Alzheimer's disease, Parkinson's disease, and brain tumors (Pardridge, 2012). Biologics, encompassing recombinant proteins, therapeutic antibodies, receptors, neurotrophins, enzyme therapies, peptide therapeutics, and therapeutic nucleic acids, are classified as large molecule therapeutics that face a significant hurdle in crossing the BBB. Consequently, delivering these biologics for treating various brain disorders presents a formidable challenge. To enhance their therapeutic effectiveness, there is a pressing need for an efficient delivery mechanism capable of facilitating the passage of these large-molecule medications through the BBB (Pardridge, 2014). Central nervous system (CNS) disorders have a substantial impact on public health, affecting approximately one in six individuals and resulting in nearly seven million deaths annually (Rehabilitation, 2021). Addressing these neurological disorders requires innovative approaches to drug delivery. Lipid carriers, including liposomes, solid lipid nanoparticles (SLNs), and nanostructured lipid carriers (NLCs), have emerged as promising vehicles for brain-targeted drug delivery. They offer the ability to encapsulate drugs, protect bioactive compounds from degradation, enhance solubility, and improve pharmacokinetics (Micheli et al., 2012). Among these lipid-based formulations, SLNs and NLCs have garnered significant attention in the biomedical field, particularly as targeted delivery systems for conditions like brain cancer and neurodegenerative diseases (Tapeinos et al., 2017). Recent research efforts have been dedicated to overcoming the challenges in brain-targeted drug delivery. This has led to substantial progress in the development and application of lipid nanocarriers, paving the way for preclinical and clinical studies involving various drugs, such as curcumin, to create more precise and effective delivery systems with enhanced therapeutic outcomes (Scioli Montoto et al., 2020).

This chapter offers a comprehensive overview of recent advancements in utilizing lipid nanocarriers for delivering bioactive compounds to the brain. This overview encompasses various aspects, including the types of nanocarriers employed, formulation strategies, characterization techniques, mechanisms facilitating brain delivery, and their applications in treating neurological disorders. We have also explored the existing challenges and potential opportunities within this domain while exploring the prospects of developing safe and efficient lipid nanocarriers tailored for brain drug delivery.

8.1.1 Background and Significance of Brain Delivery of Bioactives

BBB is a highly intricate system that responds to local changes and the specific demands arising from various cell types and mechanisms. It plays a pivotal role in both normal physiological processes and pathological conditions. Disruptions in the structural integrity of the BBB can lead to neuroinflammation and neurodegeneration. While the BBB offers essential protective functions by preventing the entry of circulating chemicals into the brain and maintaining brain homeostasis, it also poses a formidable challenge for drug delivery into the CNS (Balda & Matter, 2009; Abbott et al., 2006). The brain endothelium and its associated cells are vital contributors to the day-to-day functions of the brain. As our understanding of BBB modulation expands, emerging opportunities exist to target the brain endothelium to support its health and facilitate recovery following injuries or infections. This emphasizes the critical role of the BBB in influencing approaches to brain drug therapy (Abbott et al., 2006). For instance, diseases such as Alzheimer's, Parkinson's, and multiple sclerosis are

* Corresponding author: snatesan@niperkolkata.ac.in

DOI: 10.1201/9781003383376-8

characterized by the accumulation of toxic proteins, neuroinflammation, and neurodegeneration in specific regions of the brain (Khalid et al., 2016). Various therapies have been developed to treat these conditions, but their effectiveness is often limited by BBB, which prevents the drugs from reaching their target sites. Therefore, there is a need for a new, more specific, and precise delivery system that can overcome the limitations and deliver the bioactives to the brain (Terstappen et al., 2021). Moreover, the brain is a frequently affected site for various forms of cancer, encompassing both primary brain tumors and metastatic tumors originating from other regions of the body. Conventional cancer treatments, such as chemotherapy and radiation therapy, encounter significant limitations when dealing with brain tumors. The formidable barrier posed by the BBB restricts the access of therapeutic drugs to cancer cells within the brain (Aldape et al., 2019). Consequently, there is a pressing need for the development of innovative and effective approaches to deliver bioactive substances for the treatment of brain disorders, including cancer.

8.1.2 Challenges in the Brain Drug Delivery System

BBB. As previously indicated, BBB acts as a formidable protective barrier for the brain (Abbott et al., 2010). However, its selective nature presents a significant obstacle to drug delivery. Many drugs struggle to traverse the BBB, resulting in low drug concentrations at the intended target site.

Brain Heterogeneity. The brain is a highly heterogeneous organ comprising various cell types and regions with distinct physiological and biochemical properties (Alnaes et al., 2019). This heterogeneity complicates uniform drug distribution and can influence the efficacy of drugs in different brain regions.

Drug Stability and Degradation. The brain is a sensitive organ, and many drugs can undergo rapid degradation by brain enzymes or lose their potency due to fluctuations in brain pH (Löscher & Potschka, 2005). Hence, ensuring drug stability and preventing degradation are vital when designing drug delivery systems for the brain.

Target Specificity. Achieving specificity in drug targeting, that is, ensuring that drugs reach particular cells or regions within the brain, is crucial for efficient and effective drug delivery (Rosenblum et al., 2018). However, attaining this specificity is challenging due to the brain's complexity and heterogeneity.

Toxicity. Drug toxicity is another critical consideration in brain delivery, as drugs can potentially cause severe side effects in the brain and other organs. It is essential to mitigate such toxic effects during drug delivery to the brain.

8.1.3 Lipid Nanocarriers as a Promising Approach

The effectiveness of recent neurotherapeutic agents in treating brain disorders is hampered by several challenges, including poor solubility, lack of specificity, toxicity, and short half-life, which can result in various side effects (Haider et al., 2020). To address these issues, pharmaceutical industries have applied nanoparticles, particularly lipid nanocarriers, to achieve optimized drug delivery profiles characterized by improved solubility, drug loading, specificity, and stability (Elmowafy & Al-Sanea, 2021). The ideal selection criteria for lipid-based carriers include their smaller size, biocompatibility, and lipophilicity, facilitating their passage through the BBB into the brain (Denora et al., 2009).

Various lipid nanocarriers, viz., liposomes, solid lipid nanoparticles, and nanostructured lipid carriers, have been developed for brain drug delivery systems (Carbone et al., 2013), demonstrating the significant potential for delivering diverse drugs effectively to the brain.

8.2 TYPES OF LIPID NANOCARRIERS FOR BRAIN DELIVERY

8.2.1 Liposomes

Liposomes are minuscule spherical artificial vesicles that can be crafted from cholesterol and natural phospholipids, including phosphatidylcholine and phosphatidylethanolamine (Akbarzadeh et al., 2013). The global market demand for liposomes has already reached an impressive valuation of US$6.7 billion, and projections indicate that it will experience a compound annual growth rate (CAGR) of 6%, potentially reaching a valuation of approximately US$13 billion by 2033 (Rising Demand for Targeted Drug Delivery Systems Boosting Liposomes Market, 2023). This remarkable growth can be attributed to the outstanding biocompatibility, biodegradability, and low immunogenicity of liposomes.

Liposomes are widely recognized as the primary nanocarriers for various hydrophobic and hydrophilic compounds. Moreover, liposomes have demonstrated their ability to enhance drug solubility, facilitate controlled drug release, and offer surface modification options for targeted, extended, and sustained drug delivery (Akbarzadeh et al., 2013).

8.2.1.1 Composition

Liposomes are constructed from non-toxic phospholipids (Figure 8.1), often with the addition of cholesterol to enhance stability.

Phospholipids. Liposomes can be synthesized from a combination of both natural and synthetic phospholipids. The lipid composition used in liposomes directly impacts various properties, such as particle size, rigidity, fluidity, stability, and electrical charge. Natural lipids are used in liposome formation and are typically derived from soybean and egg yolk sources. These natural lipids can be classified into different types, including phosphatidylcholine (PC), phosphatidic acid (PA), and phosphatidylglycerol. It is worth noting that natural lipids tend to be less stable in liposomes than synthetic phospholipids, primarily due to unsaturated characteristics in their hydrocarbon chains (Nsairat et al., 2022). Synthetic phospholipids are typically generated through straightforward chemical modifications of the polar and non-polar regions of natural phospholipids. The primary synthetic phospholipids are saturated and derived from fatty acids, such as stearic and palmitic (Li et al., 2015). It is also possible to synthesize them using a combination of mixed fatty acids, including unsaturated fatty acids, either in a single hydrocarbon chain or multiple chains (Monteiro et al., 2014). These modifications allow for greater control over the properties of liposomes, such as their stability and fluidity.

Cholesterol. Indeed, synthetic phospholipids, when combined with phospholipids, contribute to the stability of liposomes. They offer the added benefit of regulating membrane fluidity, reducing the permeation of encapsulated active compounds, eliminating the pre-transition temperature peak, and enhancing the overall stability of liposomes in plasma (Isalomboto Nkanga et al., 2019). This makes them valuable in liposome formulations, allowing for better control over liposome properties and drug delivery characteristics.

Surfactants. Surfactants play a crucial role in the synthesis of liposomes as they help enhance encapsulation and release properties. Some commonly used surfactants include Span 60, Span 80, Tween 20, and sodium cholate (Nsairat et al., 2022). These surfactants aid in developing and stabilizing liposomes, making them suitable for various drug delivery applications.

8.2.1.2 Classification

Liposomes can be classified based on several factors, including their size, number of vesicles, composition, application, and preparation method (Figure 8.2) (Isalomboto Nkanga et al., 2019). These classifications help categorize liposomes and determine their suitability for specific drug delivery purposes.

1. Classification of liposomes based on size and the number of vesicles:
   a. Small unilamellar vesicles (SUV): 20–100 nm
   b. Large unilamellar vesicles (LUV): >100 nm
   c. Giant unilamellar vesicles (GULV): >1,000 nm
   d. Oligolamellar large vesicles (OLV): 100–1,000 nm
   e. Multilamellar large vesicles (MLV): >500 nm
   f. Multivesicular vesicles (MVV): >1,000 nm

2. Classification based on composition and intended application:
   a. Conventional liposomes
   b. PEGylated liposomes
   c. Multifunctional liposomes
   d. Ligand-targeted liposomes

Conventional Liposomes are considered first-generation liposomes and can be formulated with or without cholesterol. These liposomes consist of a bilayer of phospholipids surrounding an aqueous chamber and serve as the foundation for subsequent liposome developments. However, conventional liposomes have some limitations, including a short blood circulation time due to their susceptibility to clearance by the mononuclear phagocyte system. They tend to accumulate quickly in organs such as the liver and spleen, hindering their transport to target areas and

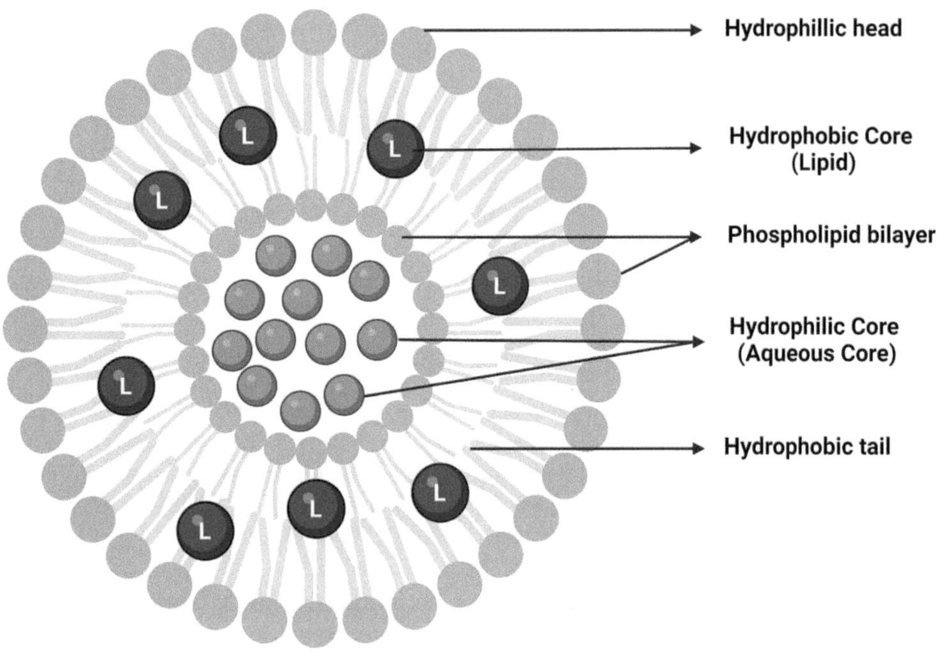

FIGURE 8.1 Structure of liposome.

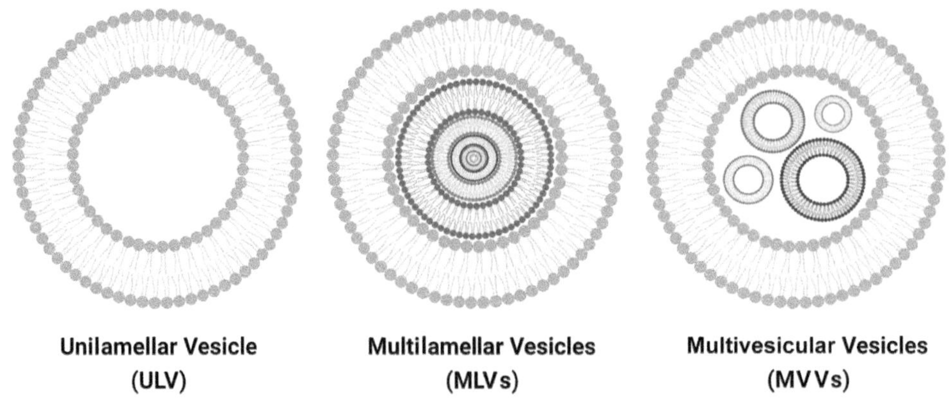

FIGURE 8.2 Types of liposomes based on size and number of vesicles: (a) ULV, (b) MLV, (c) MVV.

limiting their circulation in other body parts. Furthermore, conventional liposomes exhibit relatively limited *in vitro* and *in vivo* stability. To address these issues, stealth-stable liposomes were developed to enhance their stability *in vivo* and prolong their circulation time. This includes the use of charged lipids, such as oleic acid and N-[1(2,3-dioleoyloxy) propyl]-N,N,N-trimethylammonium chloride (DOTAP), leading to the generation of cationic and anionic liposomes. These modifications aimed to improve the performance of liposomes for drug delivery. Cationic liposomes comprise basic lipids with a positively charged head group, which can be tertiary or quaternary ammonium groups or polyamines, combined with a hydrophobic domain of alkyl chains or cholesterol. These liposomes have shown promise as carriers targeting angiogenic endothelial cells in tumors. One of their notable advantages is their ability to pass through the BBB using receptor-mediated drug delivery or absorptive-mediated transcytosis. However, it is important to note that the higher positive charge on the surface of cationic liposomes can have drawbacks. It can lead to interactions with anionic species in the bloodstream, affecting blood circulation and lowering the encapsulated drug's bioavailability. Anionic liposomes, in contrast to neutral and cationic liposomes, tend to be less stable and are associated with higher clearance rates. They are primarily utilized for transdermal drug delivery to enhance drug permeation through the stratum corneum, the outermost layer of the skin (Chauhan et al., 2020; González-Rodríguez & Rabasco, 2011).

PEGylated Liposomes. They are a liposomal formulation incorporating polyethylene glycol (PEG) to enhance stability (Figure 8.3), prolong residence time, improve drug encapsulation, and increase circulation in *in vivo* studies (Eleraky et al., 2020; Harris & Chess, 2003). PEGylation has the added benefit of inhibiting the adsorption of protein opsonins onto liposome surfaces in the circulation. This can reduce clearance by mononuclear phagocytic cells in the liver and spleen. An example of a PEGylated liposomal formulation available in the market is Droxil®, loaded with doxorubicin (DXR) and approved for treating Kaposi's sarcoma (Harris & Chess, 2003).

Multifunctional Liposomes. Ligand-targeted liposomes, also known as theragnostic liposomes, are designed with surface modifications involving antibodies and other targeting molecules. This modification enables the creation of targeted drug carriers, which can specifically deliver therapeutic agents to particular cells or tissues, improving the precision and efficacy of drug delivery.

Ligand-Targeted Liposomes. These are second-generation liposomes featuring surface modifications with various targeting ligands. These modifications can include synthetic polymers, polysaccharides, glycoproteins, and other surface-active ligands. This approach enhances the versatility and specificity of liposomal drug delivery systems, allowing for multiple functions and potential applications (Chen et al., 2021).

8.2.1.3 Formulation of Liposomes

The preparation of liposomes aims to achieve several key objectives, including forming monodisperse particles with a narrow size distribution, a high degree of lamellarity, efficient drug loading, and long-term stability (Xiang & Cao, 2017). The preparation process typically involves six main steps: (1) dissolving lipids in an organic solvent; (2) elimination of the organic phase; (3) hydration of the lipid surface through stirring or agitation; (4) size reduction or adjustment of lamellarity; (5) post-formulation processing, including purification and sterilization; and (6) characterization of liposomes to ensure their quality and properties (Otake et al., 2001).

8.2.1.3.1 Conventional Methods

Thin-Film Hydration (TFH) Method. Thin-film hydration method is one of the oldest, simplest, and most widely used techniques for preparing liposomes (Figure 8.4). Following is the description of the THF method:

Dissolution. Phospholipids are dissolved in an organic solvent, creating a homogeneous mixture.

Evaporation. The organic solvent is evaporated using a rotary evaporator, forming a thin lipid film inside the flask.

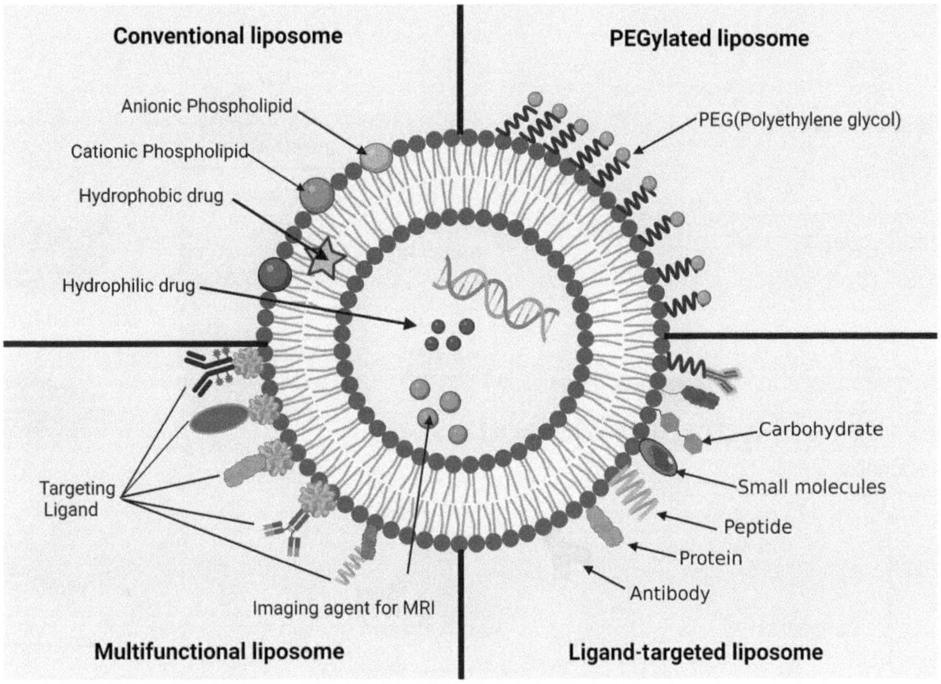

FIGURE 8.3 Classification of liposomes based on composition and application.

Solvent Removal. Dry nitrogen or argon gas eliminates any residual traces of the organic solvent in a fume hood.

Hydration. The thin lipid film is hydrated by introducing buffer solutions at temperatures ranging from 60 to 70°C. This hydration process typically lasts for 1 to 2 h.

Vesicle Formation. During hydration, the lipid layer swells and forms multilamellar vesicles (MLVs) with sizes exceeding 500 nm.

Agitation. Agitation or stirring may be employed to help detach the swollen lipid vesicles from the internal surface of the flask.

This method effectively produces liposomes suitable for various applications (Torres-Flores et al., 2020).

Detergent Removal/Depletion Method. This method processes the lipid layer using a detergent solution, typically including agents like Triton X-100, sodium deoxycholate, sodium cholate, and alkyl glycosides. The following are the steps involved in this method:

Solubilization. The lipid layer is hydrolyzed or solubilized by adding a detergent solution. Common detergents include Triton X-100, sodium deoxycholate, sodium cholate, and alkyl glycosides. When mixed, the phospholipids interact with the detergent, resulting in self-association and the formation of micelles (Figure 8.5).

Micelle Formation. The direct interaction between phospholipids and the detergent solution leads to the creation of micelles. These micelles comprise phospholipids and detergent molecules and serve as intermediate structures.

Progressive Washing. The solution is gradually washed with a buffer to remove the detergent. As the detergent is removed, the mixed micelles become enriched with lipids and undergo a transition, ultimately forming unilamellar vesicles (ULVs). This method generates unilamellar vesicles suitable for various applications by controlling the detergent removal (Schubert, 2003).

Solvent Injection Method. The method involves introducing lipids, typically dissolved in an organic solvent, into an aqueous phase. Specifically, ethanol and ethers are the two primary organic solvents employed for creating liposomal nanoformulations. The solvent injection technique encompasses two main steps: first, the lipids are dissolved in the organic solvent, and then the resulting solution is injected into the aqueous phase.

Ethanol Injection Method. This method involves several steps. First, the aqueous phase, typically consisting of a buffer or tris-HCl, is heated to around 60°C. Then, a solution of lipids dissolved in ethanol is rapidly injected into the heated aqueous phase. The quick injection results in ethanol dilution, which carries the dissolved lipid into the aqueous layer. This dilution reduces the ethanol concentration below a critical level, promoting the self-assembly of the dissolved lipid in the aqueous layer. This self-assembly leads to the precipitation of lipids and the subsequent formation of bilayer planar fragments. Finally, the organic solvents, in this case,

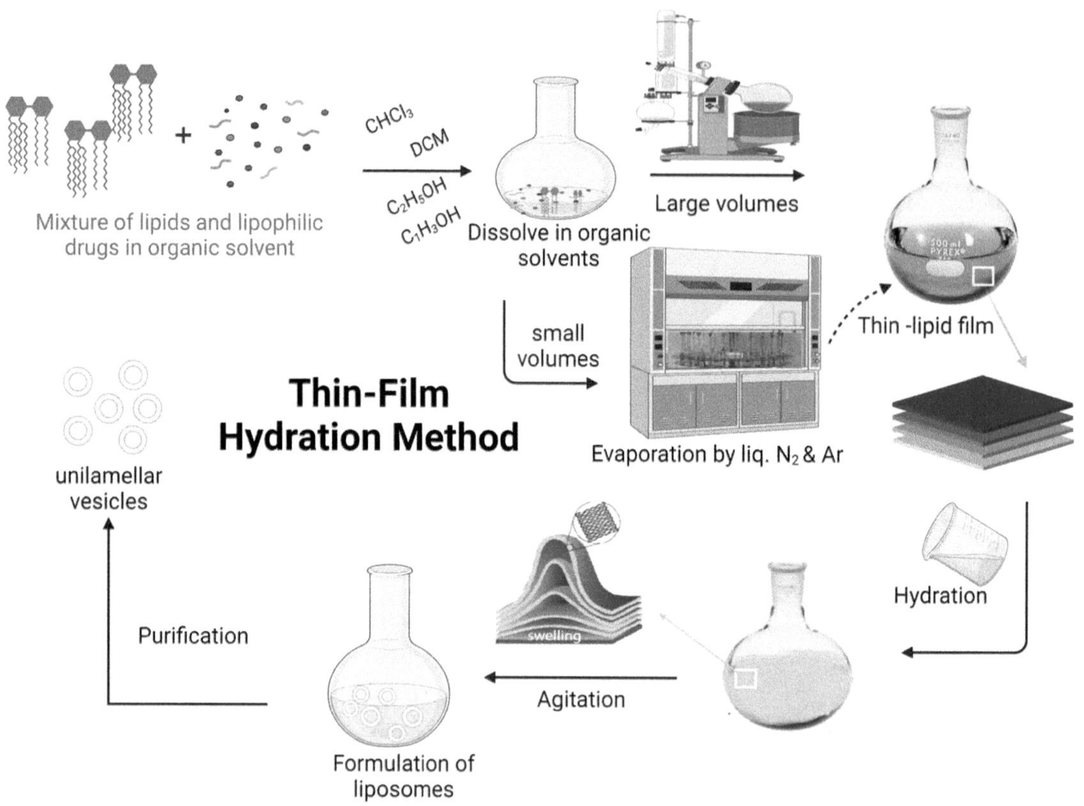

FIGURE 8.4 Preparation of liposomes by the thin-film hydration method.

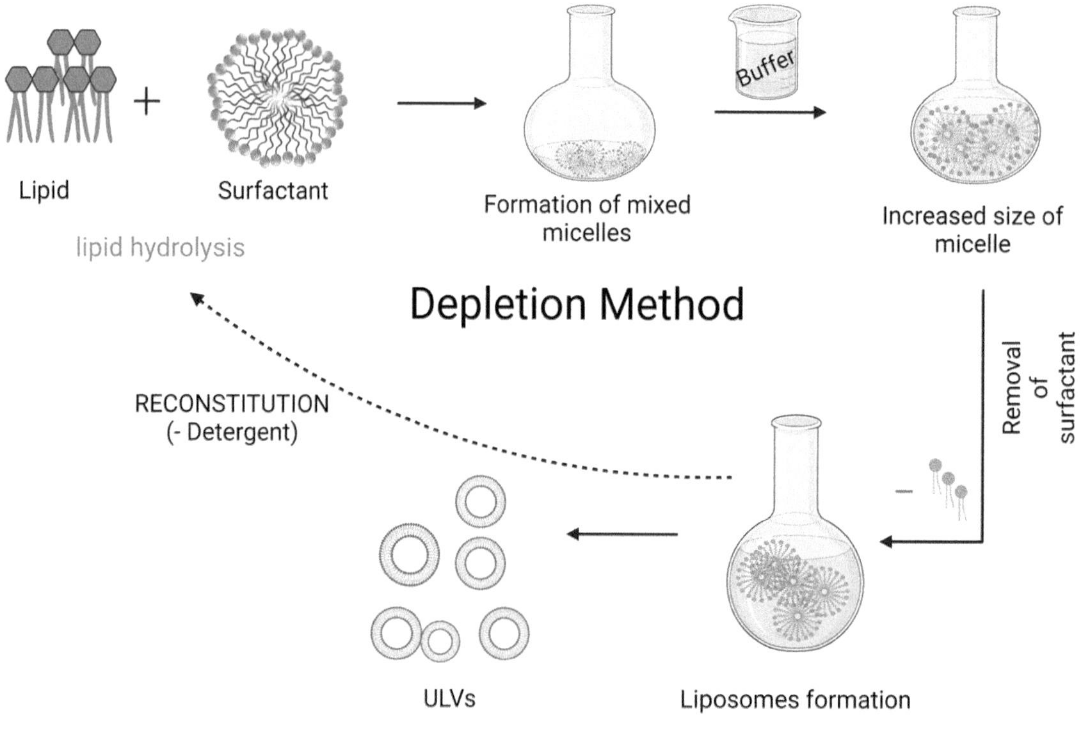

FIGURE 8.5 Preparation of liposomes by the depletion method.

ethanol, are evaporated through continuous stirring. This evaporation causes the fusion of lipid fragments, ultimately forming unilamellar vesicles (ULVs) (Gouda et al., 2021). This method offers several advantages, including its simplicity, high level of reproducibility, use of a non-harmful solvent like ethanol, and ease of scalability for larger production quantities. However, it does have some drawbacks, such as the challenge of removing residual ethanol (which forms an azeotrope with water) and the tendency for the final formulation to be somewhat heterogeneous (Figure 8.6).

Ether Injection Method. This method is similar to the previous one, preheating the aqueous solution to 65–75°C and then rapidly injecting a lipid solution dissolved in ether. The rapid injection leads to the self-assembly of lipids in the aqueous layer, forming a precipitate. One advantage of this method over the ethanol injection method is the efficient removal of the ether solvent (Figure 8.6). Additionally, this method favors the formation of small unilamellar vesicles (SUVs) with diameters ranging from 50 to 200 nm, high entrapment efficiency, and a narrow particle distribution (Samatha, 2018).

Reverse-Phase Evaporation Method. In this method, lipids are initially dissolved in an organic solvent, a mixture of diethyl ether, isopropyl ether, chloroform, and methanol in a 2:1 ratio. This mixture of lipids and organic solvent promotes the formation of inverted micelles. An aqueous phase, typically a buffer, is added to the organic phase. The lipids reorganize at the interface between the water and oil phases, forming a water-in-oil (W/O) microemulsion. This microemulsion is emulsified using mechanical or sonication methods to achieve a homogenous formulation. An excess buffer is added to the formulation to increase the encapsulation efficiency of the liposomes. The organic solvent is then evaporated slowly, disrupting inverted micelles and promoting liposome formation. The concentrated gel collapses at a critical point, and an excess of phospholipid in the solution redistributes itself around the micelles, ultimately forming a lipid bilayer and liposome (Lombardo & Kiselev, 2022).

8.2.2 Solid Lipid Nanoparticles (SLNs)

The concept of solid lipid nanoparticles (SLNs) represents one of the earliest developments in lipid-based colloidal carriers. In this approach, drugs are intricately integrated into biocompatible lipids at the submicron level. These lipid–drug complexes are dispersed within an aqueous surfactant solution, forming an oily or lipidic core enveloped by a single shell layer (Ramsden, 2011). The initial generation of lipid-based carrier systems was designed to facilitate the delivery of active compounds through various administration routes for treating diverse diseases. SLNs were introduced to encapsulate drugs within solid lipids, typically employing either cold or hot homogenization techniques, depending on the thermal stability of the active ingredient. SLNs offer several advantages, including protecting the drug against unfavorable environmental conditions, an enlarged surface area, low toxicity, controlled release, and enhanced targeting capabilities (Salvi & Pawar, 2019).

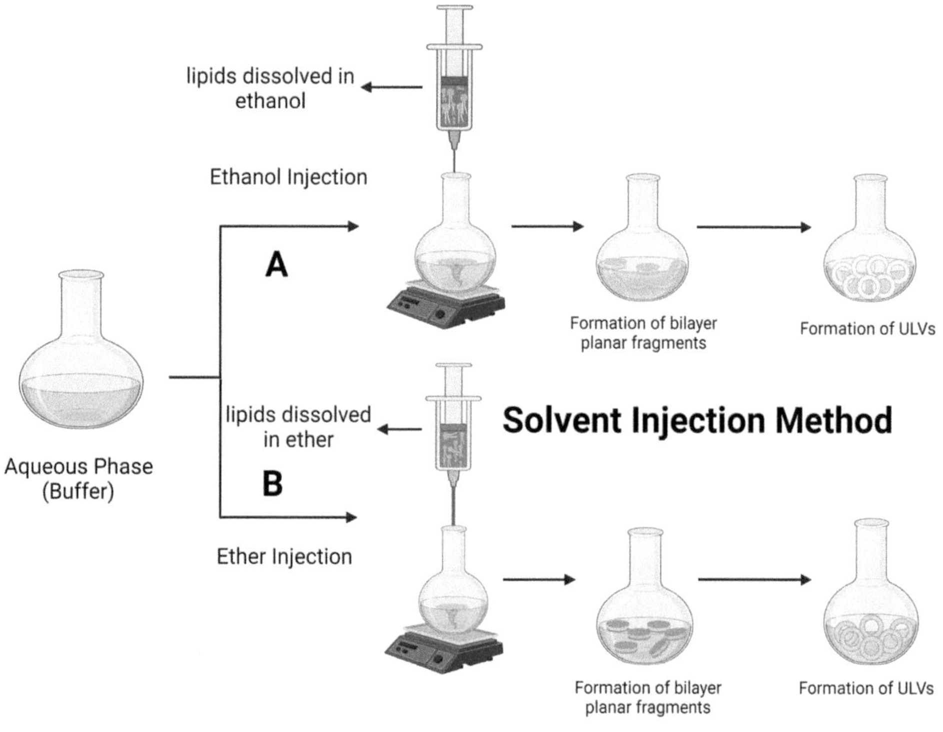

FIGURE 8.6 Preparation of liposomes by the solvent injection method.

8.2.2.1 Preparation Methods

8.2.2.1.1 High-Energy Methods

High-Pressure Homogenization (Hot and Cold). It involves the fusion of the lipid and the drug, along with the surfactants, at a temperature exceeding the lipid's melting point. A hot pre-emulsion is generated after homogenization (typically 2–3 cycles at 500–1,500 bar). SLNs are then formed by cooling the emulsion to room temperature through recrystallisation (Singh, 2019). At higher temperatures, the size of SLNs tends to be smaller due to reduced viscosity, potentially leading to lipid degradation. The molten lipid is initially cooled to room temperature, suspending partially melted lipids and drugs. Subsequently, rapid cooling is achieved by using dry ice and liquid nitrogen to promote uniform drug distribution within the lipid matrix. The resulting dispersion is then subjected to high-pressure homogenization, typically at room or lower temperature, to minimize thermal exposure (Rajpoot, 2019). The gravitational force is sufficient to transform the lipid microparticles into SLNs with sizes ranging from 50 to 1,000 nm.

Ultrasonic/High-Speed Homogenization. SLNs can be produced using high-speed stirring or sonication involving the cavitation phenomenon, including the creation, expansion, and collapse of microbubbles. The size of the SLNs in this method is influenced by factors such as stirring speed, emulsion duration, and cooling temperature (Cheaburu-Yilmaz et al., 2019).

Supercritical Fluid-Based Method. This method is employed to enhance the solubility of a compound by using a fluid that becomes a supercritical fluid at elevated temperatures and pressures, surpassing its critical values (Hirlekar et al., 2011). This method yields small-sized particles with irregular morphology, characterized by high diffusivity, low viscosity, and high compressibility. Supercritical carbon dioxide is frequently chosen as the solvent for this process due to its non-toxic, non-flammable nature and ready availability.

8.2.2.1.2 Low-Energy Methods

Coacervation Method. In this method, the drug is introduced into a solution containing lipids and surfactants, which are dissolved in a non-aqueous solvent. This solution combines an aqueous phase with a coacervation agent, forming SLNs (Battaglia et al., 2017). The concentration of micelles and the extent of polymer utilization for stabilization primarily influence the size of the SLNs.

Microemulsion-Based Method. The lipid is melted and combined with a hot surfactant solution. Gentle shaking is performed until a microemulsion is created. The hot microemulsion is diluted in a significant quantity of cold water (2–10°C) with continuous stirring (Chandana et al., 2019). Temperature gradients induce lipid crystallization and prevent aggregation; hence, this method is called the microemulsion dilution method. The particle size of the SLNs typically falls within the range of 50 to 800 nm. This method is straightforward, reproducible, and easily scalable. It involves the preparation of an oil-in-water (O/W) microemulsion by melting an emulsifying wax (typically at temperatures between 37 and 55°C) and adding water at the same temperature with minimal agitation to create a uniform milky slurry (Battaglia et al., 2017). A clear and stable liquid matrix O/W microemulsion is achieved by adding specific quantities of a suitable, pharmaceutically acceptable polymeric surfactant to water. The undiluted O/W microemulsion is then cooled to either room temperature or 4°C, causing SLNs to precipitate from the O/W microemulsion.

Double Emulsion Method. One of the most popular methods for producing nanoparticles encapsulated with hydrophilic active substances using stabilizers or surfactants. This method, also known as the "multiple emulsion method," consists of three essential steps (Jain & Thareja, 2020). First, the formation of a water-in-oil emulsion or reverse emulsion takes place. Second, the W1/O emulsion is added to an aqueous surfactant solution to create a W1/O/W2 emulsion with continuous agitation, often achieved through sonication or homogenization. Finally, the solvent is evaporated, or the multiple emulsion is filtered to form SLNs.

Phase Inversion Method by Temperature. The temperature-induced phase inversion approach involves repeated phase inversions from an O/W emulsion to a W/O emulsion using heating and cooling cycles until an emulsion is achieved. In this O/W correlation, each inversion reduces the droplet's size, ultimately allowing for the production of nanoparticles through a final dilution step in cold water. This technique can produce SLNs ranging between 30 and 100 nm (Ganesan & Narayanasamy, 2017).

8.2.2.1.3 Organic Solvent Approaches

Solvent Injection Technique. A water-miscible solvent dissolves both the lipid and the active drug. With gentle mechanical agitation, the mixture is dispersed in an aqueous solution containing a surfactant to create a suspension of lipid nanoparticles. The agitation speed can influence the particle size of the SLNs, with higher speeds resulting in smaller particles. Conversely, more lipophilic solvents tend to produce larger particles, which can be a concern (Qushawy & Nasr, 2019). Indeed, this method offers several advantages, including the ability to work at low temperatures, low cutting pressure requirements, ease of handling, and a relatively quick production process that does not necessitate advanced equipment like high-pressure homogenizers. The resulting SLNs typically have a particle size ranging from 100 to 500 nm.

Solvent Emulsion and Diffusion Method. This method achieves a thermodynamic equilibrium between an initially saturated organic phase and an aqueous phase containing a stabilizer. The drug is dissolved with the help of a homogenizer in the saturated solution. Subsequently, an O/W emulsion is created by dispersing an aqueous solution with an emulsifier. To facilitate the diffusion of the solvent into the water, an appropriate amount of water is added to the emulsion in the correct proportion with moderate stirring. This process leads to nanoprecipitation and the formation of SLNs (Patravale & Mirani, 2019). The particle size of the SLNs produced using this method typically falls within the range of 30 to 100 nm.

8.2.3 Nanostructured Lipid Carriers (NLCs)

Nanostructured lipid carriers (NLCs) represent a promising advancement in colloidal nanocarrier systems for drug delivery. They offer improved drug localization in areas such as inflammation, infection, and tumors. NLCs are essentially modified SLNs where the lipid phase comprises solid and liquid lipids. These carriers are known for their physiological compatibility, biodegradability, non-toxicity, and lack of irritation. In recent developments, targeted drug delivery strategies have enhanced the precision of NLCs in delivering drugs to specific target cells. NLCs are particularly useful for encapsulating lipophilic drugs and unstable molecules, safeguarding them against chemical degradation. These carriers are created by combining solid and liquid lipids, resulting in submicron-sized biocompatible lipid particles with an average size ranging from 100 to 500 nm (Müller et al., 2007).

8.2.3.1 Components of NLCs

Various components play essential roles in the formulation of NLCs, which include:

Solid Lipids. Solid lipids form the matrix of NLCs. They contribute to the solid structure of the nanoparticles. Solid lipids commonly used in NLC formulations include stearic acid, cetyl palmitate, cholesterol, beeswax, and others.

Liquid Lipids. Liquid lipids enhance drug loading efficiency, improve stability, and regulate drug release from the NLC formulation. Examples of liquid lipids used in NLCs include castor oil, olive oil, soybean oil, coconut oil, etc.

Surfactants. Surfactants are crucial for stabilizing NLCs, providing nanoparticle flexibility, and reducing particle size. Common surfactants used in NLC formulations include Tween 20, Tween 80, Span 20, glyceryl monooleate, and others combined in specific ratios and formulations; these components result in NLCs suitable for delivering various drugs efficiently and effectively (Figure 8.7).

8.2.3.2 Classification of NLCs

NLCs are classified based on their structural arrangement:

Imperfect Type. NLCs have an imperfectly organized solid lipid matrix. They are formed by blending solid and liquid lipids to achieve a high drug payload. Due to the imperfect lipid matrix organization, drug ejection can occur in imperfect NLCs during the crystallization process.

Amorphous Type. Amorphous NLCs have a structureless solid amorphous matrix. These NLCs are created by combining solid lipids with special lipids, such as isopropyl myristate or medium-chain triglycerides (Li et al., 2017). The unique structure of the lipid matrix in amorphous NLCs prevents drug ejection during the crystallization process, even during storage.

Multiple Types. Multiple NLCs contain solid lipids encapsulating oil nano-compartments. In this type, the drug is loaded into the oil compartments. High concentrations of liquid lipids are mixed into the lipid matrix to prevent drug ejection during crystallization. When the oil reaches its solubility limit, it precipitates into nano-compartments during the cooling process (Elmowafy & Al-Sanea, 2021; Aryasomayajula et al., 2017).

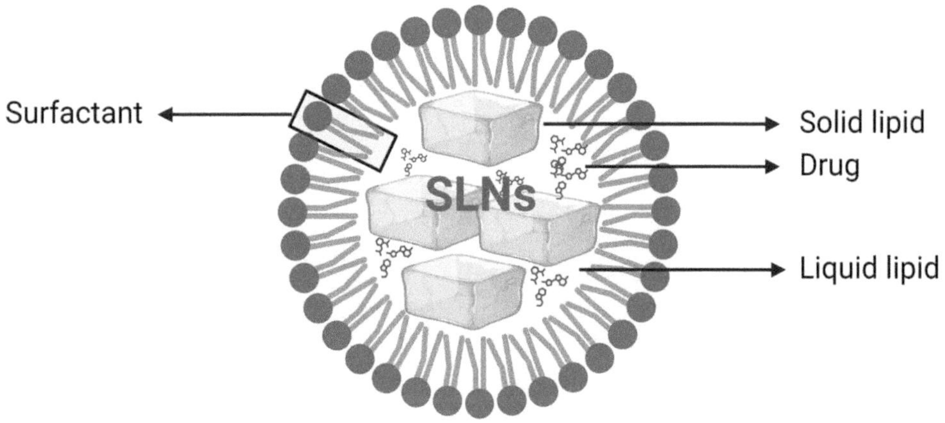

FIGURE 8.7 Structure of NLCs.

8.2.3.3 Preparation Methods of NLCs

The methods of preparing SLNs and NLCs share many similarities. SLNs and NLCs are commonly produced using techniques such as cold homogenization, hot homogenization, hot emulsification, and ultrasonication. However, their primary distinction lies in their core or matrix composition. In SLNs, the drug is either solubilized or directly incorporated into the solid lipids. In contrast, NLCs involve solubilizing or melting the drug in a mixture of liquid and solid lipids, followed by dispersion in an aqueous phase containing a surfactant (Li et al., 2017).

Commonly employed methods for manufacturing NLCs include

1. High-pressure homogenization
   - Hot homogenization
   - Cold homogenization
2. Microemulsion-based
3. Solvent emulsification diffusion
4. Solvent evaporation
5. Solvent injection
6. Double emulsion
7. Ultrasonication
8. Solvent emulsification

8.3 FORMULATION OF LIPID NANOCARRIERS FOR BRAIN DELIVERY

In the preceding section, we deliberated on the formulation of various lipid nanocarriers, such as liposomes, NLCs, and SLNs. The critical factors that assume significance in the choice of excipients, lipids, and drugs are pivotal in facilitating the delivery of drugs to the brain, particularly in the context of transcending the physiological barriers of the brain, notably the BBB.

8.3.1 Selection of Lipids and Excipients

Creating effective lipid nanocarriers for brain drug delivery demands carefully selecting lipids and excipients. The ensuing strategies can be employed to design an efficient lipid nanocarrier:

Lipids. The selection of appropriate lipids is paramount in achieving stability, enhancing drug loading capacity, and enabling the ability to traverse the BBB (de Boer & Gaillard, 2007). Lipids are chosen based on their ability to dissolve the drug to the maximum extent possible (Eleraky et al., 2020). Some examples of lipids that have demonstrated success in brain drug delivery systems encompass phospholipids, triglycerides, and cholesterol.

Excipients. They play a key role in stabilizing the formulation, augmenting drug solubility, and facilitating the uptake of the lipid carrier into the brain (Rautio et al., 2008). Various excipients employed in lipid-based brain drug delivery systems encompass surfactants (Tween 80 and Poloxamer 188), co-surfactants (propylene glycol and glycerol), and co-solvents (ethanol and propylene glycol) (Yang, 2010).

Surface Modifications. They can be harnessed to enhance the lipid carrier's capacity to traverse the BBB. The surface modifications incorporate targeting elements like antibodies, peptides, or ligands that exhibit specificity for receptors situated on the BBB (Micheli et al., 2012).

Particle Size and Charge. It is crucial to investigate particle size and charge while developing lipid carriers for brain drug delivery. The optimal particle size falls within the range of 100 –200 nm, ensuring that the lipid carrier can efficiently traverse the BBB (Dhaval et al., 2021; Khan et al., 2020). Furthermore, the lipid carrier's charge significantly impacts its ability to cross the BBB, with a preference for neutral or slightly negatively charged nanoparticles.

Drug Properties. The physicochemical characteristics of the drug can also significantly influence the choice of lipids and excipients (Demetriou et al., 2019). For instance, hydrophobic drugs may necessitate lipids with higher melting points or increased saturation levels to ensure stability. In contrast, hydrophilic drugs may require the addition of co-solvents or surfactants to enhance solubility.

In summary, the selection of lipids and excipients for the development of lipid carriers for brain drug delivery should be guided by several key considerations, including the drug's physicochemical properties, the method of preparation, the lipid carrier's ability to cross the BBB, and the stability and safety of the formulation.

8.3.2 Optimization of Formulations

Optimizing various parameters during the development process is essential to ensure the effectiveness and safety of lipid carriers for brain drug delivery. The following are the key parameters that can be optimized:

Lipid Composition. Optimizing the selection of lipids based on the drug-to-lipid ratio can improve stability, drug-loading capacity, and brain uptake of the lipid carrier. Experimenting with different combinations of phospholipids, triglycerides, and cholesterol can help identify the best lipid composition for brain-specific drug delivery (Nguyen et al., 2021).

Excipient Selection. The choice of excipients can be optimized to enhance the lipid carrier's stability, solubility, and uptake. Testing various surfactants, co-surfactants, and co-solvents can help determine the best combination for brain-specific drug delivery (Dhaval et al., 2021).

Particle Size and Charge. The particle size and charge of the lipid carrier impact its ability to cross the BBB and distribute within the brain. Utilizing dynamic light scattering or zeta potential analysis, different sizes and charges can be tested to identify the optimal parameters (Sharma et al., 2019).

Surface Modifications. Adding targeting moieties to the surface of lipid nanocarriers can enhance their ability to cross the BBB and provide targeted drug release within the brain. Experimenting with different targeting moieties or modifications can help identify the most effective strategy for brain-specific drug delivery (Gaudin et al., 2014).

Drug Loading and Release. Optimizing the drug's loading capacity and release kinetics is crucial for sustained release and maximum therapeutic effect. Exploring various methods of drug loading, such as passive or active loading, can aid in identifying the best strategy tailored to the specific drug and lipid carrier (Ramasamy et al., 2015).

In summary, achieving the ideal formulation for lipid carriers in brain drug delivery systems requires a systematic and iterative approach. This involves testing different lipid compositions, excipients, particle sizes and charges, surface modifications, drug loading, and release methods to determine the most effective strategy for brain delivery.

8.4 CHARACTERIZATION OF LIPID NANOCARRIERS FOR BRAIN DRUG DELIVERY

Lipid nanocarriers (LNCs) represent a highly promising drug delivery approach for brain targeting, given their capability to traverse the BBB and effectively transport drugs to the brain. The characterization of LNCs for brain drug delivery typically encompasses assessing their physicochemical attributes, *in vitro* release profiles, and *in vivo* pharmacokinetic investigations.

8.4.1 Physicochemical Properties

The physicochemical properties of LNCs encompass a range of physical and chemical attributes, including dimensions, morphology, surface charge, stability, and drug-carrying capacity. These properties directly impact the capability of LNCs to traverse the BBB, engage with brain cells, and discharge the drug payload at the desired site. Consequently, it is paramount to assess these properties through methodologies like dynamic light scattering, zeta potential analysis, transmission electron microscopy, and drug loading assays to guarantee the integrity and safety of the LNCs.

8.4.1.1 Size

Size is a critical factor influencing the ability of lipid carriers to traverse the BBB. Smaller particles exhibit faster drug release than larger particles due to an increased surface-area-to-volume ratio and shorter diffusion path, as observed by Khosa et al. (2018). Additionally, Loo et al. demonstrated in their study that smaller-sized NLCs tend to possess greater physical stability when compared to larger-sized particles. The optimal particle size should not exceed 200 nm, allowing for an extended circulation time in the bloodstream, ensuring that the drug remains in contact with the BBB for the maximum possible duration.

8.4.1.2 Shape

LNCs typically exhibit a spherical shape, which holds significant importance in facilitating drug delivery to the intended target site, as indicated by Teng et al. (2018). The surface morphology and shape of these carriers can be examined and characterized using techniques like transmission electron microscopy (TEM).

8.4.1.3 Surface Charge and Stability

Surface charge is indeed a critical factor in assessing the stability of LNCs. Particles with a neutral or negative charge tend to reduce the absorption of serum proteins, which, in turn, enhances their circulation in the bloodstream. This characteristic is significant for maintaining the stability of LNCs and promoting their effectiveness in drug delivery.

8.4.2 *In Vitro* Studies

In vitro release studies involve examining the release of the drug payload from LNCs under simulated physiological conditions, including variations in pH and temperature. Techniques like dialysis, centrifugation, and chromatography, as described by Chauhan et al. (2020), are commonly used for this purpose. These *in vitro* release studies offer valuable insights into the kinetics of drug release and the stability of LNCs.

To generate release profiles, the percentage of cumulative drug release over time is calculated, typically using Equation (8.1):

$$\textit{Percentage Release}(\%) = \frac{\textit{Amount of drug released in buffer}}{\textit{Total amount of drug in the formulation}} \times 100 \tag{8.1}$$

8.4.3 *In Vivo* Pharmacokinetic Studies

This study focuses on assessing the *in vivo* performance of LNCs within the body and their capacity to reach the brain. Various methods, including biodistribution studies, imaging techniques, and pharmacokinetic analysis, as suggested by Teng et al. (2018), are employed for this purpose. *In vivo* studies yield crucial information concerning the pharmacokinetics, pharmacodynamics, and safety of LNCs, which are vital considerations when advancing LNCs toward clinical applications.

8.5 MECHANISMS OF BRAIN DELIVERY BY LIPID NANOCARRIERS

More than 98% of newly discovered chemical compounds face the formidable challenge of breaching the BBB, making drug delivery to the brain one of the most challenging routes of administration. Nanoparticle drug delivery systems have proven effective in circumventing this barrier, successfully transporting drugs such as loperamide, dopamine, and doxorubicin into the brain (Gastaldi et al., 2014). As previously indicated, the BBB serves as a protective shield around the brain and restricts the free passage of drugs, including bioactive compounds. Therefore, developing efficient methods for delivering drugs to the brain is imperative for treating numerous CNS disorders. LNCs have emerged as a promising drug delivery system capable of efficiently delivering bioactive substances through the BBB while reducing side effects. The process of brain drug delivery using LNCs involves several key steps.

BBB Penetration. LNCs are designed to be sufficiently small to traverse the BBB. They can traverse this barrier through various mechanisms, including passive diffusion, receptor-mediated transcytosis, and adsorptive-mediated transcytosis (Pardridge, 2022).

Interaction with Brain Cells. Once past the BBB, LNCs interact with neurons and glial cells in the brain (Neves et al., 2017). The drug payload is released into the brain when the lipid bilayer of the nanocarriers fuses with the cell membrane.

Intracellular Drug Delivery. Drug-loaded LNCs can target specific sites within brain cells, leading to therapeutic effects, such as reduced inflammation or increased neurotransmitter levels (Witika et al., 2022).

Clearance. Following the delivery of bioactive compounds, LNCs are eliminated from the body and the brain through various mechanisms, including metabolism, excretion, and phagocytosis (Poudel & Park, 2022).

In essence, LNCs are adept at traversing the BBB, engaging with brain cells, releasing their drug cargo, and being cleared from the body (Figure 8.8). This approach holds significant potential for markedly enhancing the therapeutic efficacy of treatments for CNS diseases.

8.6 APPLICATIONS OF LIPID NANOCARRIERS FOR BRAIN DELIVERY OF BIOACTIVES

8.6.1 Neurodegenerative Diseases and Neuroprotection

Alzheimer's disease, Parkinson's disease, and dementia are chronic and progressive neurodegenerative conditions primarily marked by neuronal loss and other associated ailments. Neurodegeneration leads to structural and functional impairments in the BBB. In order to concentrate on precise drug therapies for the CNS, researchers have identified therapeutic advancements aimed at gaining a deeper understanding of the pathways associated with the onset of neurodegenerative disorders. Current evidence indicates that the disruption of lipid metabolism may have a pivotal role in developing these conditions. Lipid transporters, such as liposomes, exosomes, and lipid-based nanoparticles, have demonstrated significant promise in treating neurodegenerative disorders. Liposomes can rapidly traverse the BBB and target drug molecules, specifically in the brain. Numerous studies have demonstrated that liposomes

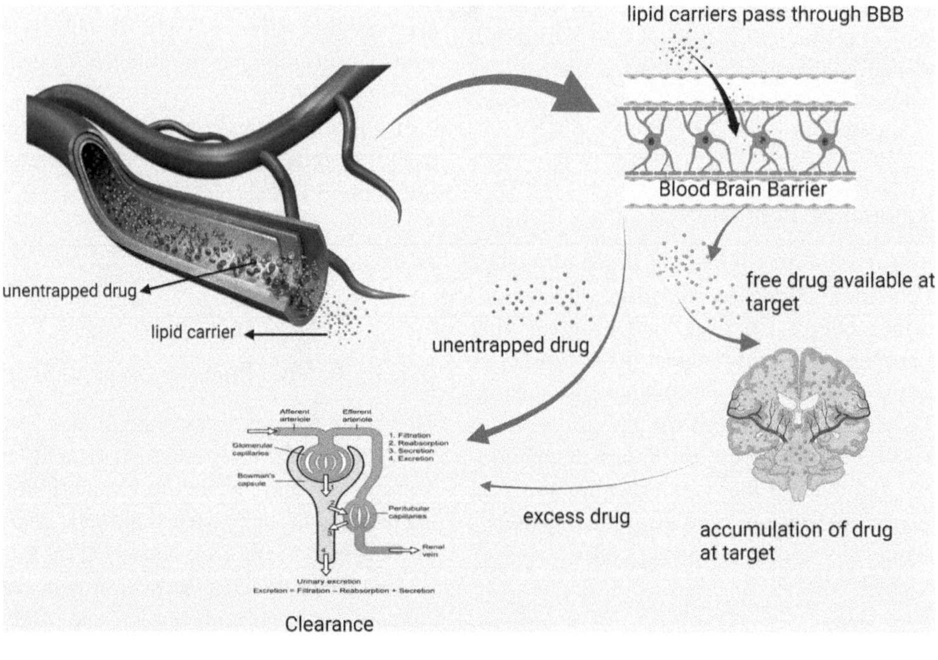

FIGURE 8.8 Mechanism of lipid carriers in brain drug delivery system.

containing neuroprotective agents like curcumin, resveratrol, and quercetin can mitigate oxidative stress, reduce inflammation, and enhance cognitive function in animal models of Alzheimer's and Parkinson's disease (Rahman et al., 2020; Khan et al., 2019). Exosomes, small extracellular vesicles released by various cell types, including neurons, contain bioactive components, such as proteins, lipids, and nucleic acids. They can also permeate the BBB and play a crucial role in intercellular communication. Recent studies have indicated that exosomes derived from mesenchymal stem cells (MSCs) have the potential to enhance cognitive function and reduce neuroinflammation in models of Alzheimer's and Parkinson's disease by delivering neuroprotective miRNAs (Chen et al., 2020; Abrishamdar et al., 2023). Lipid nanoparticles (LNPs) can encapsulate nucleic acids like siRNA and mRNA, and they hold substantial promise in gene therapy for neurodegenerative disorders. For instance, in preclinical investigations, an LNP formulation containing siRNA targeting the huntingtin gene, responsible for Huntington's disease (HD), successfully lowered huntingtin protein levels and enhanced motor function in HD animal models (Germain et al., 2023).

In summary, lipid carriers, including liposomes, exosomes, and LNPs, exhibit significant potential in treating neurodegenerative disorders. These carriers can efficiently breach the BBB and directly deliver neuroprotective agents or gene therapies to the brain, reducing oxidative stress, neuroinflammation, and neuronal degeneration (Figure 8.9). Nonetheless, further investigations are essential to refine their efficacy and safety for clinical application.

8.6.2 Brain Cancer Therapy

Lipid carriers have gained increasing prominence in cancer therapy because of their capacity to enhance drug bioavailability and therapeutic effectiveness. Lipid-based nanoparticle systems involve nanosized lipid carriers that encapsulate hydrophobic drugs and transport them to cancer cells. Several studies on these LNPs have demonstrated enhanced effectiveness of chemotherapeutic drugs like paclitaxel and doxorubicin, accompanied by reduced toxicity (Zhang et al., 2007, 2022). Liposomes, spherical vesicles containing a lipid bilayer capable of encapsulating hydrophobic and hydrophilic drugs, are widely used in cancer therapy due to their biocompatibility and ability to target tumor cells. The FDA has approved liposomal doxorubicin (Doxil) for treating ovarian and breast cancer (Gu et al., 2020). SLNs, composed of solid lipids, have exhibited substantial potential in cancer therapy as they can encapsulate hydrophobic drugs and transport them to tumor cells. SLNs have been employed to enhance the bioavailability and therapeutic efficacy of drugs such as curcumin and docetaxel (Wang et al., 2018). As a result, lipid carriers have demonstrated remarkable potential in treating tumors by enhancing drug bioavailability, increasing efficacy, mitigating toxicity, and improving drug targeting to tumor cells.

8.6.3 Applications in Pain and Management

Pain and neuroinflammation are intricate conditions that can pose challenges for effective treatment. Recent research has revealed that lipid carriers, including liposomes and nanoparticles, hold significant promise in delivering anti-inflammatory and analgesic agents to the brain. Liposomes are vesicular structures composed of a phospholipid bilayer, enabling them to encapsulate hydrophilic and hydrophobic drugs. They are biocompatible, biodegradable, and readily customizable. Numerous studies have indicated anti-inflammatory liposomes containing substances like curcumin,

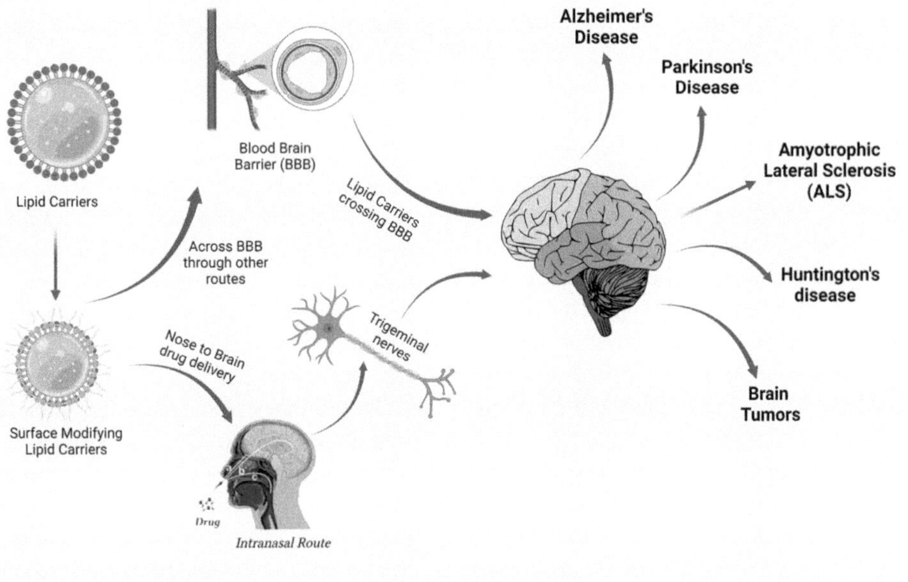

FIGURE 8.9 Application of lipid carriers in various neurological disorders.

resveratrol, and ibuprofen can diminish neuroinflammation and enhance pain management in animal models (Chinraj & Raman, 2022; Costa et al., 2016). Furthermore, liposomes containing opioids such as morphine have shown promising outcomes for targeted drug delivery and reducing opioid dependence (Chen & Liu, 2012). Nanoparticles represent another promising lipid carrier for the treatment of neuroinflammation and pain. Their small size facilitates efficient drug delivery while minimizing toxicity. Numerous studies have demonstrated the successful delivery of anti-inflammatory agents and opioids to the brain using nanoparticles. For example, a study employing poly(lactic-co-glycolic acid) (PLGA) nanoparticles loaded with curcumin reported a significant reduction in pain-related behavior in rats (Bei et al., 2020). Another study by Kashyap and Shukla (2019) using PLGA nanoparticles loaded with the opioid antagonist naloxone showed that these LNPs effectively reduced opioid tolerance and dependence.

In conclusion, lipid carriers like liposomes and nanoparticles hold substantial promise for addressing neuroinflammation and pain management. These carriers can readily traverse the BBB and directly deliver therapeutic agents to the brain, diminishing inflammation and alleviating pain. Nevertheless, additional research is necessary to enhance their effectiveness, safety, and clinical applicability for brain-targeted drug delivery.

8.7 FUTURE PROSPECTS AND CHALLENGES

8.7.1 Translational Potential of Lipid Carriers

Despite the promising outcomes observed in preclinical studies, translating lipid carriers for brain drug delivery to clinical applications encounters several hurdles. This section examines the potential prospects and challenges associated with the clinical translation of lipid carriers. One of the most formidable challenges lies in achieving effective and safe delivery of therapeutic agents to the human brain. While lipid carriers have demonstrated encouraging results in animal models, their translation to human clinical trials remains intricate due to the intricate and diverse nature of the human BBB (Pardridge, 2014). Overcoming the barrier posed by the human BBB and ensuring the successful delivery of therapeutic agents to the brain remain a major challenge. Toxicity and safety concerns with lipid carriers represent another issue. Although lipid carriers are inherently biocompatible and biodegradable, specific formulations can induce toxicity and trigger immune responses (Pahnke et al., 2014). To mitigate the risk of adverse effects in patients, it is imperative to conduct thorough assessments of the safety of lipid carriers. Additionally, the manufacturing and scalability of lipid carriers present challenges. Successful clinical translation necessitates consistent and standardized properties in lipid carriers. Therefore, optimizing manufacturing processes is crucial to ensure batch-to-batch consistency and uphold stringent quality control standards.

Notwithstanding these obstacles, lipid carriers hold considerable promise for future applications in brain drug delivery. Nanotechnology and material science advances have enabled the development of novel lipid-based carriers with improved attributes, including enhanced BBB permeability, increased drug loading capacity, and targeted drug delivery (Griffin et al., 2016). Furthermore, combining lipid carriers with gene therapy introduces new avenues for treating neurodegenerative diseases.

In conclusion, the clinical translation of lipid carriers for brain drug delivery remains formidable. However, with ongoing research and development efforts, lipid carriers offer substantial potential for delivering therapeutic agents to the brain effectively and safely. Further studies are required to optimize their efficacy, safety, and clinical applicability.

8.7.2 Safety and Regulatory Aspects

The use of lipid carriers for drug delivery to the brain has demonstrated substantial promise in treating various neurological disorders. However, the challenges related to safety and regulatory considerations in lipid carriers for brain delivery systems must be effectively addressed before they can be translated into clinical applications. A key challenge is the absence of clear regulatory guidelines for developing and approving lipid-based drug delivery systems. Regulatory authorities need to establish well-defined guidelines for their development and safety assessment to ensure the safe and effective utilization of lipid carriers in human subjects. Another concern pertains to the potential toxicity and immunogenicity of lipid carriers. While lipid carriers are generally safe, conducting thorough investigations into their toxicity and immunogenicity is crucial to minimize the risk of adverse effects in patients (Pardridge, 2012). Toxicity assessments should encompass not only the lipid carrier itself but also any potential degradation products or impurities. Manufacturing lipid carriers presents challenges, particularly in ensuring batch-to-batch consistency and quality control. Manufacturing processes must be rigorously controlled and supervised by regulatory agencies, and lipid carrier manufacturers must take steps to ensure their products align with regulatory requirements. Despite these hurdles, the utilization of lipid carriers in brain drug delivery systems holds promising prospects. Nanotechnology and material science advances have led to the development of novel lipid-based carriers with enhanced attributes, including increased drug loading capacity and precise drug delivery (Patel & Patel, 2021).

In summary, addressing the safety and regulatory considerations of lipid carriers for brain drug delivery systems is challenging but indispensable to ensure their secure and effective application in human subjects. Clear regulatory guidelines, comprehensive safety evaluation, and optimized manufacturing processes are essential to surmount these obstacles and facilitate the clinical translation of lipid carriers for brain drug delivery (Satapathy et al., 2021).

8.7.3 Technological Innovations and Emerging Trends

The lipid carrier–based brain drug delivery field is witnessing rapid advancements, driven by emerging trends and technological innovations that offer immense potential to enhance the effectiveness of drug delivery and precision in targeting. Nonetheless, substantial challenges can be overcome in translating these innovations into clinical practice. One notable trend involves the utilization of functionalized lipid carriers designed to target specific receptors or transporters expressed on the BBB or within brain cells (Farshbaf et al., 2022). This approach aims to improve drug delivery to specific brain regions while minimizing off-target effects. For example, liposomes functionalized with antibodies that target the transferrin receptor have shown enhanced brain uptake of drugs dependent on this receptor for crossing the BBB. Another promising trend entails using lipid nanoparticles to deliver drugs with gene therapies, seeking synergistic therapeutic effects (Ezzati Nazhad Dolatabadi et al., 2015). This strategy has the potential to overcome drug resistance and amplify treatment outcomes. Research has indicated that lipid nanoparticles containing both a chemotherapeutic drug and a gene therapy vector can significantly enhance the effectiveness of brain tumor treatment. Technological advancements, such as developing microfluidic platforms for lipid nanoparticle synthesis, promise increased scalability and reproducibility in lipid carrier production. Microfluidic synthesis allows for precise particle size and composition control, producing more uniform and stable lipid nanoparticles (Carvalho et al., 2022). However, several challenges must be addressed in developing lipid carrier–based brain drug delivery systems. One significant challenge lies in the absence of effective methods for assessing *in vivo* drug efficacy and safety. Existing imaging techniques, like magnetic resonance imaging, provide valuable information on nanoparticle distribution in the brain but fall short in real-time drug release and pharmacodynamics monitoring. Developing new imaging techniques to track drug release and therapeutic effects is vital for optimizing brain drug delivery (Üner 2016). Furthermore, concerns regarding potential toxicity and immunogenicity must be thoroughly investigated, particularly with repeated use of lipid carriers. Strategies for reducing toxicity, such as using biodegradable lipid carriers or surface modifications to reduce immune recognition, require further exploration.

In summary, lipid carrier–based drug delivery systems for the brain are a rapidly evolving field characterized by numerous emerging trends and technological breakthroughs. Targeted delivery, combination therapies, and microfluidic synthesis promise to achieve precise brain-targeted drug delivery. Nevertheless, further research and development are essential to address existing challenges and ensure the safety and efficacy of these innovative approaches.

8.8 CONCLUSION

The persistently high mortality and morbidity rates associated with complex neuropathologies, coupled with the intricate mechanisms underlying central nervous system (CNS) disorders, continue to present unresolved challenges. Recent strides in biomedical research have illuminated the blood–brain barrier (BBB) as a potential target for drug delivery to the brain, offering new hope for addressing a range of CNS disorders. Lipid nanocarriers have emerged as a promising strategy for delivering bioactive compounds to the brain. These lipid carrihosphorlizsing liposomes, solid lipid nanoparticles (SLNs), and nanostructured lipid carriers (NLCs), offer several advantages, including their reduced size, facilitating passage through the BBB. The selection of lipids used in their formulation significantly influences drug loading capacity and release characteristics. Moreover, various excipients are incorporated in the formulation, and their choice plays a pivotal role in determining the behavior of the drug delivery system. The attainment of a uniform and homogeneous formulation hinges on carefully selecting the lipid-to-drug ratio and the preparation method. Once formulated, rigorous characterization and analysis are imperative to evaluate the safety and efficacy of the drug in treating CNS disorders. The smaller size, surface charge, and lipophilicity of these carriers confer an advantage in traversing the BBB and targeting specific sites within the brain, thereby enhancing therapeutic outcomes. Despite substantial progress in lipid nanocarrier research, there remains a substantial journey ahead to achieve clinical success in brain drug delivery.

REFERENCES

Abbott, N. J., Patabendige, A. A., Dolman, D. E., Yusof, S. R., & Begley, D. J. (2010, January). Structure and function of the blood–brain barrier. Neurobiology of Disease, 37(1), 13–25. https://doi.org/10.1016/j.nbd.2009.07.030

Abbott, N. J., Rönnbäck, L., & Hansson, E. (2006, January 1). Astrocyte–endothelial interactions at the blood–brain barrier. Nature Reviews Neuroscience, 7(1), 41–53. https://doi.org/10.1038/nrn1824

Abrishamdar, M., Jalali, M. S., & Yazdanfar, N. (2023, March 23). The role of exosomes in pathogenesis and the therapeutic efficacy of mesenchymal stem cell-derived exosomes against Parkinson's disease. Neurological Sciences, 44(7), 2277–2289. https://doi.org/10.1007/s10072-023-06706-y

Akbarzadeh, A., Rezaei-Sadabady, R., Davaran, S., Joo, S. W., Zarghami, N., Hanifehpour, Y., Samiei, M., Kouhi, M., & Nejati-Koshki, K. (2013, February 22). Liposome: Classification, preparation, and applications. Nanoscale Research Letters, 8(1). https://doi.org/10.1186/1556-276x-8-102

Aldape, K., Brindle, K. M., Chesler, L., Chopra, R., Gajjar, A., Gilbert, M. R., Gottardo, N., Gutmann, D. H., Hargrave, D., Holland, E. C., Jones, D. T. W., Joyce, J. A., Kearns, P., Kieran, M. W., Mellinghoff, I. K., Merchant, M., Pfister, S. M., Pollard, S. M., Ramaswamy, V., . . . Gilbertson, R. J. (2019, February 7). Challenges to curing primary brain tumors. Nature Reviews Clinical Oncology, 16(8), 509–520. https://doi.org/10.1038/s41571-019-0177-5

Alnaes, D., Kaufmann, T., van der Meer, D., Córdova-Palomera, A., Rokicki, J., Moberget, T., Bettella, F., Agartz, I., Barch, D. M., Bertolino, A., Brandt, C. L., Cervenka, S., Djurovic, S., Doan, N. T., Eisenacher, S., Fatouros-Bergman, H., Flyckt, L., Di Giorgio, A., Haatveit, B., . . . Westlye, L. T. (2019, July 1). Brain heterogeneity in schizophrenia and its association with polygenic risk. JAMA Psychiatry, 76(7), 7 39. https://doi.org/10.1001/jamapsychiatry.2019.0257

Aryasomayajula, B., Salzano, G., & Torchilin, V. P. (2017). Multifunctional liposomes. Methods in Molecular Biology, 41–61. https://doi.org/10.1007/978-1-4939-6646-2_3

Balda, M. S., & Matter, K. (2009, April). Tight junctions and the regulation of gene expression. Biochimica Et Biophysica Acta (BBA)—Biomembranes, 1788(4), 761–7 67. https://doi.org/10.1016/j.bbamem.2008.11.024

Battaglia, L., Muntoni, E., Chirio, D., Peira, E., Annovazzi, L., Schiffer, D., Mellai, M., Riganti, C., Salaroglio, I. C., Lanotte, M., Panciani, P., Capucchio, M. T., Valazza, A., Biasibetti, E., & Gallarate, M. (2017, March). Solid lipid nanoparticles by coacervation loaded with a methotrexate prodrug: A preliminary study for glioma treatment. Nanomedicine, 12(6), 639–656. https://doi.org/10.2217/nnm-2016-0380

Bei, W., Jing, L., & Chen, N. (2020, January). Cardioprotective role of wogonin loaded nanoparticle against isoproterenol-induced myocardial infarction by moderating oxidative stress and inflammation. Colloids and Surfaces B: Biointerfaces, 185, 1106 35. https://doi.org/10.1016/j.colsurfb.2019.110635

Brain Anatomy and How the Brain Works. (2021, July 14). Brain Anatomy and How the Brain Works Johns Hopkins Medicine. www.hopkinsmedicine.org/health/conditions-and-diseases/anatomy-of-the-brain

Carbone, C., Cupri, S., Leonardi, A., Puglisi, G., & Pignatello, R. (2013). Lipid-based nanocarriers for drug delivery and targeting: A patent survey of methods of production hosphorlizedntion. Pharmaceutical Patent Analyst, 2(5), 665–6 77. https://doi.org/10.4155/ppa.13.43

Carvalho, B. G., Ceccato, B. T., Michelon, M., Han, S. W., & de la Torre, L. G. (2022, January 7). Advanced microfluidic technologies for lipid nano-microsystems from synthesis to biological application. Pharmaceutics, 14(1), 1 41. https://doi.org/10.3390/pharmaceutics14010141

Chandana, K. V., Gupta, N. V., & Kanna, S. (2019, May 14). Nanostructured lipid carriers: The frontiers in drug delivery. Asian Journal of Pharmaceutical and Clinical Research, 8–12. https://doi.org/10.22159/ajpcr.2019.v12i7.33595

Chauhan, I., Yasir, M., Verma, M., & Singh, A. P. (2020). Nanostructured lipid carriers: A groundbreaking approach for transdermal drug delivery. Advanced Pharmaceutical Bulletin, 10(2), 150–165. https://doi.org/10.34172/apb.2020.021

Cheaburu-Yilmaz, C. N., Karasulu, H. Y., & Yilmaz, O. (2019). Nanoscaled dispersed systems used in drug-delivery applications. Polymeric Nanomaterials in Nanotherapeutics, 437–468. https://doi.org/10.1016/b978-0-12-813932-5.00013-3

Chen, H. X., Liang, F. C., Gu, P., Xu, B. L., Xu, H. J., Wang, W. T., Hou, J. Y., Xie, D. X., Chai, X. Q., & An, S. J. (2020, April 27). Exosomes derived from mesenchymal stem cells repair a Parkinson's disease model by inducing autophagy. Cell Death & Disease, 11(4). https://doi.org/10.1038/s41419-020-2473-5

Chen, M., Cui, Y., Hao, W., Fan, Y., Zhang, J., Liu, Q., Jiang, M., Yang, Y., Wang, Y., & Gao, C. (2021, January 1). Ligand-modified homologous targeted cancer cell membrane biomimetic nanostructured lipid carriers for glioma therapy. Drug Delivery, 28(1), 2241–2255. https://doi.org/10.1080/10717544.2021.1992038

Chen, Y., & Liu, L. (2012, May). Modern methods for delivery of drugs across the blood–brain barrier. Advanced Drug Delivery Reviews, 64(7), 640–6 65. https://doi.org/10.1016/j.addr.2011.11.010

Chinraj, V., & Raman, S. (2022, July 5). Neuroprotection by resveratrol: A review on brain delivery strategies for Alzheimer's and Parkinson's disease. Journal of Applied Pharmaceutical Science, 1– 17. https://doi.org/10.7324/japs.2022.120701

Classic Rehabilitation. (2021, July 8). Facts about neurological disorders—Classic Rehabilitation. Classic Rehabilitati on. https://classicrehabilitation.com/facts-about-neurological-disorders

Costa, L. G., Garrick, J. M., Roquè, P. J., & Pellacani, C. (2016). Mechanisms of neuroprotection by quercetin: Counteracting oxidative stress and more. Oxidative Medicine and Cellular Longevity, 1– 10. https://doi.org/10.1155/2016/2986796

de Boer, A., & Gaillard, P. (2007, February 1). Drug targeting to the brain. Annual Review of Pharmacology and Toxicology, 47(1), 323–355. https://doi.org/10.1146/annurev.pharmtox.47.120505.1052372

Demetriou, E., Story, H. E., Bofinger, R., Hailes, H. C., Tabor, A. B., & Golay, X. (2019, August 26). Effect of liposomal encapsulation on the chemical exchange properties of diamagnetic CEST agents. Journal of Physical Chemistry B, 123(35), 7545–7557. https://doi.org/10.1021/acs.jpcb.9b02280

Denora, N., Trapani, A., Laquintana, V., Lopedota, A., & Trapani, G. (2009, February 1). Recent advances in medicinal chemistry and pharmaceutical technology-strategies for drug delivery to the brain. Current Topics in Medicinal Chemistry, 9(2), 182–1 96. https://doi.org/10.2174/156802609787521571

Dhaval, M., Vaghela, P., Patel, K., Sojitra, K., Patel, M., Patel, S., Dudhat, K., Shah, S., Manek, R., & Parmar, R. (2021, October 5). Lipid-based emulsion drug delivery systems — A comprehensive review. Drug Delivery and Translational Research, 12(7), 1616–1639. https://doi.org/10.1007/s13346-021-01071-9

Eleraky, N. E., Omar, M. M., Mahmoud, H. A., & Abou-Taleb, H. A. (2020, May 14). Nanostructured lipid carriers to mediate brain delivery of temazepam: Design and in vivo study. Pharmaceutics, 12(5), 4 51. https://doi.org/10.3390/pharmaceutics12050451

Elmowafy, M., & Al-Sanea, M. M. (2021, September). Nanostructured lipid carriers (NLCs) as drug delivery platform: Advances in formulation and delivery strategies. Saudi Pharmaceutical Journal, 29(9), 999–1012. https://doi.org/10.1016/j.jsps.2021.07.015

Ezzati Nazhad Dolatabadi, J., Valizadeh, H., & Hamishehkar, H. (2015, June 1). Solid lipid nanoparticles as efficient drug and gene delivery systems: Recent breakthroughs. Advanced Pharmaceutical Bulletin, 5(2), 151–1 59. https://doi.org/10.15171/apb.2015.022

Farshbaf, M., Mojarad-Jabali, S., Hemmati, S., Khosroushahi, A. Y., Motasadizadeh, H., Zarebkohan, A., & Valizadeh, H. (2022, May). Enhanced BBB and BBTB penetration and improved anti-glioma behavior of Bortezomib through dual-targeting nanostructured lipid carriers. Journal of Controlled Release, 345, 371–3 84. https://doi.org/10.1016/j.jconrel.2022.03.019

Ganesan, P., & Narayanasamy, D. (2017, December). Lipid nanoparticles: Different preparation techniquhosphorlizedntion, hurdles, and strategies for the production of solid lipid nanoparticles and nanostructured lipid carriers for oral drug delivery. Sustainable Chemistry and Pharmacy, 6, 37– 56. https://doi.org/10.1016/j.scp.2017.07.002

Gastaldi, L., Battaglia, L., Peira, E., Chirio, D., Muntoni, E., Solazzi, I., Gallarate, M., & Dosio, F. (2014, August). Solid lipid nanoparticles as vehicles of drugs to the brain: Current state of the art. European Journal of Pharmaceutics and Biopharmaceutics, 87(3), 433–4 44. https://doi.org/10.1016/j.ejpb.2014.05.004

Gaudin, A., Yemisci, M., Eroglu, H., Lepetre-Mouelhi, S., Turkoglu, O. F., Dönmez-Demir, B., Caban, S., Sargon, M. F., Garcia-Argote, S., Pieters, G., Loreau, O., Rousseau, B., Tagit, O., Hildebrandt, N., Le Dantec, Y., Mougin, J., Valetti, S., Chacun, H., Nicolas, V., . . . Couvreur, P. (2014, November 24). Squalenoyl adenosine nanoparticles provide neuroprotection after stroke and spinal cord injury. Nature Nanotechnology, 9(12), 1054–10 62. https://doi.org/10.1038/nnano.2014.274

Germain, N. D., Chung, W. K., & Sarmiere, P. D. (2023, June). RNA interference (RNAi)-based therapeutics for treatment of rare neurologic diseases. Molecular Aspects of Medicine, 91, 101148. https://doi.org/10.1016/j.mam.2022.101148

González-Rodríguez, M., & Rabasco, A. (2011, May 11). Charged liposomes as carriers enhance the permeation through the skin. Expert Opinion on Drug Delivery, 8(7), 857–8 71. https://doi.org/10.1517/17425247.2011.574610

Gouda, A., Sakr, O. S., Nasr, M., & Sammour, O. (2021, February). Ethanol injection technique for liposomes formulation: An insight into the development, influencing factors, challenges, and applications. Journal of Drug Delivery Science and Technology, 61, 1021 74. https://doi.org/10.1016/j.jddst.2020.102174

Griffin, B. T., Guo, J., Presas, E., Donovan, M. D., Alonso, M. J., & O'Driscoll, C. M. (2016, November). Pharmacokinetic, pharmacodynamic, and biodistribution following oral administration of nanocarriers containing peptide and protein drugs. Advanced Drug Delivery Reviews, 106, 367–3 80. https://doi.org/10.1016/j.addr.2016.06.006

Gu, Z., Da Silva, C., Van der Maaden, K., Ossendorp, F., & Cruz, L. (2020, November 4). Liposome-based drug delivery systems in cancer immunotherapy. Pharmaceutics, 12(11), 10 54. https://doi.org/10.3390/pharmaceutics12111054

Haider, M., Abdin, S. M., Kamal, L., & Orive, G. (2020, March 23). Nanostructured lipid carriers for delivery of chemotherapeutics: A review. Pharmaceutics, 12(3), 2 88. https://doi.org/10.3390/pharmaceutics12030288

Harris, J. M., & Chess, R. B. (2003, March). Effect of pegylation on pharmaceuticals. Nature Reviews Drug Discovery, 2(3), 214–2 21. https://doi.org/10.1038/nrd1033

Hirlekar, R., Garse, H., & Kadam, V. (2011, November 1). Solid lipid nanoparticles and nanostructured lipid carriers: A review. Current Drug Therapy, 6(4), 240–2 50. https://doi.org/10.2174/157488511798109637

Isalomboto Nkanga, C., Murhimalika Bapolisi, A., Ikemefuna Okafor, N., & Werner Maçedo Krause, R. (2019, September 4). General perception of liposomes: Formation, manufacturing and applications. Liposomes—Advances and Perspectiv es. https://doi.org/10.5772/intechopen.84255

Jain, A. K., & Thareja, S. (2020). Solid lipid nanoparticles. Nanomaterials and Environmental Biotechnology, 221–249. https://doi.org/10.1007/978-3-030-34544-0_13

Kashyap, K., & Shukla, R. (2019, December 27). Drug delivery and targeting to the brain through nasal route: Mechanisms, applications, and challenges. Current Drug Delivery, 16(10), 887–901. https://doi.org/10.2174/15672018166661910291 22740

Khalid, A., Persano, S., Shen, H., Zhao, Y., Blanco, E., Ferrari, M., & Wolfram, J. (2016, October 11). Strategies for improving drug delivery: Nanocarriers and microenvironmental priming. Expert Opinion on Drug Delivery, 14(7), 865–877. https://doi.org/10.1080/17425247.2017.1243527

Khan, A. A., Allemailem, K. S., Almatroodi, S. A., Almatroudi, A., & Rahmani, A. H. (2020, March 10). Recent strategies towards the surface modification of liposomes: An innovative approach for different clinical applications. 3 Biotech, 10(4). https://doi.org/10.1007/s13205-020-2144-3

Khan, H., Ullah, H., Aschner, M., Cheang, W. S., & Akkol, E. K. (2019, December 30). Neuroprotective effects of quercetin in Alzheimer's disease. Biomolecules, 10(1), 59. https://doi.org/10.3390/biom10010059

Khosa, A., Reddi, S., & Saha, R. N. (2018, July). Nanostructured lipid carriers for site-specific drug delivery. Biomedicine & Pharmacotherapy, 103, 598–6 13. https://doi.org/10.1016/j.biopha.2018.04.055

Li, J., Wang, X., Zhang, T., Wang, C., Huang, Z., Luo, X., & Deng, Y. (2015, April). A review on phospholipids and their main applications in drug delivery systems. Asian Journal of Pharmaceutical Sciences, 10(2), 81– 98. https://doi.org/10.1016/j.ajps.2014.09.004

Li, Q., Cai, T., Huang, Y., Xia, X., Cole, S. P. C., & Cai, Y. (2017). A review of the structure, preparation, and application of NLCs, PNPs, and PLNs. *Nanomaterials (Basel, Switzerland)*, *7*(6), 1 22. https://doi.org/10.3390/nano7060122

Lombardo, D., & Kiselev, M. A. (2022, February 28). Methods of liposomes preparation: Formation and control factors of versatile nanocarriers for biomedical and nanomedicine application. Pharmaceutics, 14(3), 5 43. https://doi.org/10.3390/pharmaceutics14030543

Löscher, W., & Potschka, H. (2005, July 15). Drug resistance in brain diseases and the role of drug efflux transporters. Nature Reviews Neuroscience, 6(8), 591–6 02. https://doi.org/10.1038/nrn1728

Micheli, M. R., Bova, R., Magini, A., Polidoro, M., & Emiliani, C. (2012, April 1). Lipid-based nanocarriers for CNS-targeted drug delivery. Recent Patents on CNS Drug Discovery, 7(1), 71–86. https://doi.org/10.2174/157488912798842241

Monteiro, N., Martins, A., Reis, R. L., & Neves, N. M. (2014, December 6). Liposomes in tissue engineering and regenerative medicine. Journal of the Royal Society Interface, 11(101), 20140459. https://doi.org/10.1098/rsif.2014.0459

Müller, R. H., Petersen, R. D., Hommoss, A., & Pardeike, J. (2007). Nanostructured lipid carriers (NLC) in cosmetic dermal products. Advanced Drug Delivery Reviews, 59(6), 522–530. https://doi.org/10.1016/j.addr.2007.04.012

Neves, A. R., Queiroz, J. F., Lima, S. A. C., & Reis, S. (2017, April 5). Apo E-functionalization of solid lipid nanoparticles enhances brain drug delivery: Uptake mechanism and transport pathways. Bioconjugate Chemistry, 28(4), 995–1004. https://doi.org/10.1021/acs.bioconjchem.6b00705

Nguyen, T. T., Dung Nguyen, T. T., Vo, T. K., Tran, N. M. A., Nguyen, M. K., Van Vo, T., & Van Vo, G. (2021, November). Nanotechnology-based drug delivery for central nervous system disorders. Biomedicine & Pharmacotherapy, 143, 112117. https://doi.org/10.1016/j.biopha.2021.112117

Nsairat, H., Khater, D., Sayed, U., Odeh, F., Al Bawab, A., & Alshaer, W. (2022, May). Liposomes: Structure, composition, types, and clinical applications. Heliyon, 8(5), e09394. https://doi.org/10.1016/j.heliyon.2022.e09394

Otake, K., Imura, T., Sakai, H., & Abe, M. (2001, June 1). Development of a new preparation method of liposomes using supercritical carbon dioxide. Langmuir, 17(13), 3898–3901. https://doi.org/10.1021/la010122k

Pahnke, J., Fröhlich, C., Paarmann, K., Krohn, M., Bogdanovic, N., Årsland, D., & Winblad, B. (2014, November). Cerebral ABC transporter-common mechanisms may modulate neurodegenerative diseases and depression in elderly subjects. Archives of Medical Research, 45(8), 738–743. https://doi.org/10.1016/j.arcmed.2014.10.010

Pardridge, W. M. (2012, August 29). Drug transport across the blood–brain barrier. Journal of Cerebral Blood Flow & Metabolism, 32(11), 1959–1972. https://doi.org/10.1038/jcbfm.2012.126

Pardridge, W. M. (2014, August 20). Blood–brain barrier drug delivery of IgG fusion proteins with a transferrin receptor monoclonal antibody. Expert Opinion on Drug Delivery, 12(2), 207–222. https://doi.org/10.1517/17425247.2014.952627

Pardridge, W. M. (2022, June 16). A historical review of brain drug delivery. Pharmaceutics, 14(6), 12 83. https://doi.org/10.3390/pharmaceutics14061283

Patravale, V. B., & Mirani, A. G. (2019). Preparation and characterization of solid lipid nanoparticles-based gel for topical delivery. Pharmaceutical Nanotechnology, 293–302. https://doi.org/10.1007/978-1-4939-9516-5_20

Patel, M., & Patel, P. (2021, October 22). Toxicity and safety evaluation of lipid-based nanoparticles for brain delivery. Nanotechnology in Medicine, 233–247. https://doi.org/10.1002/9781119769897.ch10

Poudel, P., & Park, S. (2022, April 11). Recent advances in the treatment of Alzheimer's disease using nanoparticle-based drug delivery systems. Pharmaceutics, 14(4), 835. https://doi.org/10.3390/pharmaceutics14040835

Qushawy, M., & Nasr, A. (2019, November 7). Solid lipid nanoparticles (SLNs) as nano drug delivery carriers: Preparation, characterization, and application. International Journal of Applied Pharmaceutics, 1–9. https://doi.org/10.22159/ijap.2020v12i1.35312

Rahman, M. H., Akter, R., Bhattacharya, T., Abdel-Daim, M. M., Alkahtani, S., Arafah, M. W., Al-Johani, N. S., Alhoshani, N. M., Alkeraishan, N., Alhenaky, A., Abd-Elkader, O. H., El-Seedi, H. R., Kaushik, D., & Mittal, V. (2020, December 30). Resveratrol and neuroprotection: Impact and its therapeutic potential in Alzheimer's disease. Frontiers in Pharmacology, 11. https://doi.org/10.3389/fphar.2020.619024

Rajpoot, K. (2019, December 17). Solid lipid nanoparticles: A promising nanomaterial in drug delivery. Current Pharmaceutical Design, 25(37), 3943–3959. https://doi.org/10.2174/1381612825666190903155321

Ramasamy, T., Ruttala, H. B., Choi, J. Y., Tran, T. H., Kim, J. H., Ku, S. K., Choi, H. G., Yong, C. S., & Kim, J. O. (2015). Engineering of a lipid-polymer nanoarchitectural platform for highly effective combination therapy of doxorubicin and irinotecan. Chemical Communications, 51(26), 5758–5761. https://doi.org/10.1039/c5cc00482a

Ramsden, J. (2011, January 1). Nanotechnology: An Introduction. William Andrew.

Rautio, J., Laine, K., Gynther, M., & Savolainen, J. (2008, February 5). Prodrug approaches for CNS delivery. The AAPS Journal, 10(1), 92–102. https://doi.org/10.1208/s12248-008-9009-8

Rising Demand for Targeted Drug Delivery Systems Boosting Liposomes Market. (2023, February 1). Persistence Market Research. www.persistencemarketresearch.com/market-research/liposomes-market.asp

Rosenblum, D., Joshi, N., Tao, W., Karp, J. M., & Peer, D. (2018, April 12). Progress and challenges towards targeted delivery of cancer therapeutics. Nature Communications, 9(1). https://doi.org/10.1038/s41467-018-03705-y

Salvi, V. R., & Pawar, P. (2019, June). Nanostructured lipid carriers (NLC) system: A novel drug targeting carrier. Journal of Drug Delivery Science and Technology, 51, 255–267. https://doi.org/10.1016/j.jddst.2019.02.017

Samatha, G. (2018, December 5). Fenofibrate niosomes by modified ether injection method of proper diet to help lower "bad". Gastroenterology Pancreatology and Hepatobiliary Disorders, 2(1), 1–3. https://doi.org/10.31579/2641-5194/005

Satapathy, M. K., Yen, T. L., Jan, J. S., Tang, R. D., Wang, J. Y., Taliyan, R., & Yang, C. H. (2021, July 31). Solid Lipid Nanoparticles (SLNs): An advanced drug delivery system targeting brain through BBB. Pharmaceutics, 13(8), 1183. https://doi.org/10.3390/pharmaceutics13081183

Schubert, R. (2003). Liposome preparation by detergent removal. Methods in Enzymology, 46–70. https://doi.org/10.1016/s0076-6879(03)67005-9

Scioli Montoto, S., Muraca, G., & Ruiz, M. E. (2020, October 30). Solid lipid nanoparticles for drug delivery: Pharmacological and biopharmaceutical aspects. Frontiers in Molecular Biosciences, 7. https://doi.org/10.3389/fmolb.2020.587997

Sharma, G., Sharma, A. R., Lee, S. S., Bhattacharya, M., Nam, J. S., & Chakraborty, C. (2019, March). Advances in nanocarriers enabled brain-targeted drug delivery across blood-brain barrier. International Journal of Pharmaceutics, 559, 360–372. https://doi.org/10.1016/j.ijpharm.2019.01.056

Singh, R. (2019, March 15). Preparation of solid lipid nanoparticles through various methods using different precursors. Journal of Drug Delivery and Therapeutics. https://doi.org/10.22270/jddt.v9i2.2461

Tapeinos, C., Battaglini, M., & Ciofani, G. (2017, October). Advances in the design of solid lipid nanoparticles and nanostructured lipid carriers for targeting brain diseases. Journal of Controlled Release, 264, 306–332.https://doi.org/10.1016/j.jconrel.2017.08.033

Teng, Z., Yu, M., Ding, Y., Zhang, H., Shen, Y., Jiang, M., Liu, P., Opoku-Damoah, Y., Webster, T. J., & Zhou, J. (2018, December). Preparation and characterization of nimodipine-loaded nanostructured lipid systems for enhanced solubility and bioavailability. International Journal of Nanomedicine, 14, 119–133. https://doi.org/10.2147/ijn.s186899

Terstappen, G. C., Meyer, A. H., Bell, R. D., & Zhang, W. (2021, March 1). Strategies for delivering therapeutics across the blood–brain barrier. Nature Reviews Drug Discovery, 20(5), 362–383. https://doi.org/10.1038/s41573-021-00139-y

Torres-Flores, G., Gonzalez-Horta, A., Vega-Cantu, Y. I., Rodriguez, C., & Rodriguez-Garcia, A. (2020, October 10). Preparation hosphorlizedntion of liposomal everolimus by thin-film hydration technique. Advances in Polymer Technology, 1–9. https://doi.org/10.1155/2020/5462949

Üner, M. (2016). Characterisation and imaging of solid lipid nanoparticles and nanostructured lipid carriers. Handbook of Nanoparticles, 117–141. https://doi.org/10.1007/978-3-319-15338-4_3

Wang, W., Chen, T., Xu, H., Ren, B., Cheng, X., Qi, R., Liu, H., Wang, Y., Yan, L., Chen, S., Yang, Q., & Chen, C. (2018, June 29). Curcumin-loaded solid lipid nanoparticles enhanced anticancer efficiency in breast cancer. Molecules, 23(7), 1578. https://doi.org/10.3390/molecules23071578

Witika, B. A., Poka, M. S., Demana, P. H., Matafwali, S. K., Melamane, S., Malungelo Khamanga, S. M., & Makoni, P. A. (2022, April 11). Lipid-based nanocarriers for neurological disorders: A review of the state-of-the-art and therapeutic success to date. Pharmaceutics, 14(4), 836. https://doi.org/10.3390/pharmaceutics14040836

Xiang, B., & Cao, D. Y. (2017, December 4). Preparation of drug liposomes by thin-film hydration and homogenization. Liposome-Based Drug Delivery Systems, 1–11. https://doi.org/10.1007/978-3-662-49231-4_2-1

Yang, H. (2010, July 1). Nanoparticle-mediated brain-specific drug delivery, imaging, and diagnosis. Pharmaceutical Research, 27(9), 1759–1771. https://doi.org/10.1007/s11095-010-0141-7

Zhang, L., Gu, F., Chan, J., Wang, A., Langer, R., & Farokhzad, O. (2007, October 24). Nanoparticles in medicine: Therapeutic applications and developments. Clinical Pharmacology & Therapeutics, 83(5), 761–769. https://doi.org/10.1038/sj.clpt.6100400

Zhang, Z., Yao, S., Hu, Y., Zhao, X., & Lee, R. J. (2022). Application of lipid-based nanoparticles in cancer immunotherapy. Frontiers in Immunology, 13, 967505. https://doi.org/10.3389/fimmu.2022.967505

9 Blood–Brain Barrier Disruption and Nanomedicine in Neurodegenerative Disorders

Adejoke Elizabeth Memudu, Michael Oluwatosin Gabriel, Grace Temitope Akingbade, Happiness Ochi Inyang, and Omamuyovwi Meashack Ijomone*

9.1 INTRODUCTION

Neurodegenerative diseases (NDs), such as Alzheimer's disease (AD), Parkinson's disease (PD), Huntington's disease (HD), multiple sclerosis (MS), etc. (Naqvi et al., 2020; Hanif et al., 2020), occur due to the progressive degeneration of the neurons in the brain or the spinal cord (Silveira et al., 2021; Zhu et al., 2021), leading to abnormalities in neural activity that culminate in cell death and cognitive impairments (Hanif et al., 2020; Jagaran and Singh, 2021).

The neuronal dysfunction and death are attributed to neurotoxic mediated oxidative stress (Kumar et al., 2020), neuroinflammation, abnormal protein accumulation, and blood–brain barrier (BBB) disruption (Kumar et al., 2020). In 2020, a public health report states that more than 55 million people across the globe are living with NDs, and this number is expected to double every 20 years, reaching 78 million in 2030 and 139 million in 2050 (Patterson, 2018; Hanif et al., 2020; Foidl and Humpel, 2020). NDs contribute to high morbidity and mortality among adult, hence the need to develop novel rapid brain-acting therapeutic strategies for their treatments (Hanif et al., 2020). One of the novel therapeutic interventions for NDs is to target neuroinflammatory processes (Chiang et al., 2021) associated with activation of microglial, astrocytes, and secretion of pro-inflammatory mediators (Zhao et al., 2020; Zhu et al., 2021). For instance, patients with AD are given drugs such as donepezil (acetylcholinesterase inhibitors) or galantamine (a nicotinic acetylcholine receptor modulators) or memantine (a NMDA receptor antagonists), which have not been proven to be effective in halting disease progression in AD patients (Hanif et al., 2020); hence, most medications remain palliative (Pardridge, 2020; Jagaran and Singh, 2021). The major limitations of presently available therapeutic agents for NDs include delivery and specificity, hence the need to develop new measures to combat this menace (Naqvi et al., 2020), which involve developing drugs that can cross the blood–brain barrier (Ngowi et al., 2021). The blood–brain barrier (BBB) protects the brain parenchyma from harmful substances and poses as the major challenge in delivering drugs into the brain for treatment of NDs (Naqvi et al., 2020). The BBB limits adequate entering of any substance, including drugs, into the brain (Silveira et al., 2021). This impact can be mitigated by nanomedicine strategies using nanoparticles (NPs) that can improve BBB penetration and exert a neuroprotective effect (Zhu et al., 2021) in NDs.

Nanomedicine is a field that implements nanotechnology and medicine via the use of nanoparticles to diagnose and treat diseases, including NDs (Jagaran and Singh, 2021). This new strategy has great ability to suppress issues associated with conventional/traditional treatment approaches (Naqvi et al., 2020); hence, nanoparticles can improve BBB penetration of beneficial therapeutic agents to exert neuroprotective effects (Zhu et al., 2021). Some of these nanoparticles/nanocarriers are liposomes, micelles, nanogels, nanotubes, nanofibers (Tsou et al., 2017; Poupot and Bergozza, 2018), solid lipid nanoparticles (SLNPs), dendrimers, and nanoemulsions for delivery. Nanoformulation of various neurotherapeutic agents with neuroprotective potential, such as curcumin (Naqvi et al., 2020), has been reported in some research. The advantage of nanoparticles is based on its ability to penetrate the BBB (Hassanzadeh et al., 2017) because of its shape, small size (Saraf et al., 2015), circulating half-life, structural stability, and cell-targeting ability to enable targeted treatment of NS pathogenesis (Silva et al., 2017) and the stability of the nanocarriers to carry lipophilic and hydrophilic molecules (Alexander et al., 2016; Kumar et al., 2017).

In this chapter, we highlight how nanomedicine impacts the BBB to make it permeable for therapeutic agents that can target specific brain regions to improve treatment of NDs.

9.2 CHEMICAL DISRUPTION OF THE BBB

A common approach to transiently disrupt the BBB and deliver compounds to the brain involves using an osmotic gradient. In this system, hypertonic solutions such as mannitol

* Corresponding author: jokememudu@gmail.com or adejoke.elizabeth@edouniversity.edu.ng

DOI: 10.1201/9781003383376-9

are infused into cerebral vessels to produce an osmotic gradient that drags water from endothelial cells, dehydrates and shrinks the cells, and eventually results in transient damage to endothelial cells' tight junctions (D'Amico et al., 2020). This method was used in several clinical studies and considered relatively safe for patients with no permanent neurological deficits (Angelov et al., 2009; Boockvar et al., 2011; Burkhardt et al., 2012; Chakraborty et al., 2016; Fortin et al., 2005; Miyagami et al., 1990). Various hyperosmotic agents have been tested; however, mannitol is preferred for clinical tests because its use in patients has been established and it has produced a higher survival rate in experimental animals than other tested agents (Karmur et al., 2020). Another agent for BBB disruption are bradykinin analogues, such as RMP-7, which, similar to endogenous bradykinin, transiently increase BBB permeability through interaction with endothelial cells' B_2 receptors to alter tight junctions (Boddy & Thomas, 1997; Karmur et al., 2020). The increase in permeability of BBB caused by the bradykinin analogue is reported to be more selective for tumor capillaries and common in smaller vessels than in big vessels (Boddy & Thomas, 1997; Chakraborty et al., 2016). Other agents used to chemically modify the BBB are minoxidil sulfate (that selectively activates ATP-mediated potassium channels), NS1619 (an activator of calcium-mediated potassium channels), and cyclic guanosine monophosphate (cGMP), as reviewed by Mo et al. (2021).

9.3 PHYSICAL DISRUPTION OF THE BBB

Several decades ago, it was shown that animals exposed to ultrasound waves develop BBB disruption. Then it was recommended as a strategy for delivering therapeutic agents to the brain (Hynynen, 2008). Although this recommendation has been investigated and shown to produce tissue damage (Kullervo Hynynen, 2008), other studies have reported the possibility of using it without the complication of tissue damage (Vykhodtseva et al., 1995). Ultrasound BBB disruption involves using ultrasound waves of preferably low frequency transmitted intracranially to disrupt the BBB (Rodriguez et al., 2015). This system has advanced and is used clinically as an approach to disrupt the BBB, guided by magnetic resonance imaging (MRI) for efficient targeting of the region of interest (Rodriguez et al., 2015). MRI-guided focused ultrasound (MRIgFUS) can also be used in conjunction with intravenously injected microbubble contrast agents, which, when exposed to low-intensity pulsed ultrasound, undergo expansion and contraction oscillations that break apart tight junctions and induce a transient BBB disruption (Fishman & Fischell, 2021; Hynynen et al., 2001). Several animal experiments have used this strategy to deliver therapeutic agents for neurodegenerative conditions and to deliver viruses, genes, and cells to the brain (Ahmed et al., 2021; Jordão et al., 2010, 2013; Kofoed et al., 2022). In patients with neurodegenerative diseases, MRIgFUS was used to induce BBB disruption and was reported to be safe with no clinical side effects, and it was reversible with 24 h closure of the BBB (Gasca-Salas et al., 2021; Lipsman et al., 2018; Park et al., 2021; Rezai et al., 2020). Despite the beneficial applications of FUS and the evidence of the safety of the procedure (K. Hynynen et al., 2001; Sheikov et al., 2008), some other studies reported evidence of changes in inflammatory patterns and thermal ablation of tissue (Elias et al., 2013; Kovacsa et al., 2017; Singh et al., 2020). Apart from animal and clinical studies, in vitro models of BBB have also utilized low pulsed electric fields that are lower than the threshold for electroporation to induce transient BBB disruption by changing the functions of junctional proteins of cells; however, more studies are needed to further elucidate the mechanisms (Sharabi et al., 2019).

9.4 MECHANISMS OF BLOOD–BRAIN BARRIER DISRUPTION

9.4.1 Oxidative Stress

There is mounting evidence suggesting that oxidative stress plays a pivotal role in causing changes to the BBB. There are diverse pathological mechanisms underlying the disruption of the BBB due to oxidative stress.

Firstly, oxidative stress influences tight junction (TJ) structure and functions. The structural stability of the BBB is believed to be crucially maintained by proteins of the tight junction, as indicated by various lines of evidence. Tight junctions function as molecular barriers within the intercellular space, preventing the passage of ions, blood-borne contaminants, water-soluble molecules, pharmaceuticals, and pathogens via the BBB channel (Bauer & Traweger, 2016). The TJ complex found in the cerebral endothelium comprises integral membrane proteins, such as occludin, claudins, ZO (zonula occludens) proteins, and JAMs (junctional adhesion molecules) (Berndt et al., 2019, Haseloff et al., 2015). One study reported that the formation of ROS can regulate the transcriptional activity of claudin-5, heighten solutes diffusibility, and impact the overall structural stability of the BBB (Irudayanathan et al., 2017). The claudin protein has the potential to modify the selective permeability of tight junctions (Yang et al., 2019).

Furthermore, ROS can modify the BBB's porosity through the modulation of ZO protein localization. The process of phosphorylation and the subsequent distribution of TJ proteins are crucial determinants of blood–brain barrier (BBB) permeability. Consequently, any alteration of these parameters induced by reactive oxygen species (ROS) has the potential to undermine the BBB's integrity.

Secondly, matrix metalloproteinases (MMPs) are a frequent cause of BBB injury induced by oxidative stress. MMPs are proteolytic enzymes with the ability to break down proteins of the extracellular matrix and split cell surface receptors. Following an oxidative insult, MMP-2 and MMP-9 are the MMPs most closely associated with barrier permeability (Turner & Sharp, 2016). MMPs, specifically MMP-9, are activated or upregulated in reaction to oxidative stress (Banks & Rhea, 2021; Gu et al., 2002). MMPs may break down the extracellular matrix and epithelial basement membrane, thereby compromising the BBB's integrity (Reinhard et al., 2015). It has been proposed that inhibiting MMPs can impede the degradation of basement membrane proteins and TJs, thereby averting BBB compromise (Trivedi et al., 2019). Consequently, MMP activity is the primary mediator of BBB permeability. The release of MMP-9 and other factors by pericytes and astrocytes, degradation of the basement membrane, modification of expression levels of proteins of the TJ and their distribution, and disruption of the BBB are attributed to oxidative stress (da Fonseca, 2014).

Thirdly, autophagic responses following oxidative stress can also impact BBB integrity. Autophagy is an essential and highly conserved lysosomal degradation pathway that is crucial in maintaining cellular homeostasis. Its primary function is to facilitate the removal of organelles and proteins during various pathological processes, thereby preventing excessive cellular damage and dysfunction in different organs and cells (Ikeda et al., 2022; Yang et al., 2021; De Meyer, 2015). Recent findings suggest that the impairment of the BBB may be attributed to oxidative stress, which triggers lysosomal dysfunction and initiates autophagy in the pericytes, hippocampus, and astrocytes (Tan et al., 2018). Research has proven that oxidative stress damages the frontal cortex and hippocampus neurons. This damage may be caused by AKT/mTOR signaling pathway activation to control autophagy and suppress neuronal apoptosis (Wang et al., 2019). Under conditions of oxidative stress, ROS may promote autophagy, and autophagy can mitigate oxidative stress–induced damage. Therefore, ROS and autophagy can simultaneously preserve the intracellular environment, the structural and functional integrity of brain cells, and the BBB (Singh et al., 2019; Yin et al., 2017).

9.4.2 Inflammation

Several physiologic and functional alterations in the BBB are induced by inflammation. Inflammation is known to trigger glia activation, leukocytes recruitment across the BBB. Also, inflammation weakens the BBB due to changes in paracellular transport, transcellular transport, and decreased expression of tight junction protein (Knox et al., 2022). These events culminate in increase in BBB permeability. Therefore, microglia activation and leukocyte recruitment are key players in the inflammatory events that result in BBB disruption (Knox et al., 2022; Lee & Funk, 2023). BBB disruption can both cause and result from CNS inflammation, which affects the severity and course of many mental and neurological illnesses. Inflammatory events in the brain may induce damage to endothelial cells and increase in BBB permeability (Kim et al., 2012; Małkiewicz et al., 2019). It is well-documented that neurodegeneration induced by neuroinflammation frequently involves BBB disruption. For instance, patients with neurodegenerative illnesses such as AD, PD, ALS, and MS exhibit BBB disruption (Takata et al., 2021).

The first line of defense against external invaders is innate immunity. Recent studies suggest that BBB disruption is closely linked to increased immune responses in the CNS (Jeon et al., 2021). Neuroinflammation and BBB disruption are also triggered by innate immune cells. Glia and neurons express pathogen recognition receptors (PRRs), which are activated upon binding to their ligands, pathogen-associated molecular patterns (PAMPs). The PRRs participate in brain sterile inflammation as they also identify endogenous substances like damage-associated molecular patterns (DAMPs). As PRRs bind to DAMPs, glial cells become activated. Thus, inflammatory mediators and cytotoxic factors associated with neurodegeneration are released (Chitnis & Weiner, 2017; Liddelow et al., 2017). Inflammatory mediators such as cytokines, chemokines, reactive oxygen species (ROS), and lipid mediators influence BBB integrity (Takata et al., 2021). The increased expression of cytokines that accompany continued microglial activation further contributes to the junctional breakdown in BBB disruption (da Fonseca et al., 2014).

Within the CNS, albumin activates microglia to cause the expression of IL-1β, TNF-α, TGF-β, and MCP-1, which alters BBB permeability as well as TJ distribution (Cash & Theus, 2020). Additionally, as albumin enters the brain via transcytosis, astrocytes are activated to release MMPs, which weaken the basement membrane and increase BBB permeability (Ranaivo et al., 2012). The cytokines released by activation, in turn, induce downstream pathways, including Rho/ROCK, PKC, and MAPK, which influence the phosphorylation of TJ proteins and promote increased paracellular permeability of the BBB (Cash & Theus, 2020). Notably, TNF-α and IL-1β released by activated microglia control BBB permeability by interacting with the endothelial cells of the BBB (Jeon et al., 2021). This interaction causes the modulation of BBB permeability by inducing increased expression of the cell adhesion molecules in leucocytes such as like P-selectin and E-selectin and ICAM-1 (Lee & Funk, 2023), resulting in more severe inflammation. This condition also upregulates MMPs that contribute to the downregulation of extracellular matrix components of BBB (Jeon et al., 2021; Labus et al., 2014). Additionally, upregulated MMPs disrupt TJ adhesion by reorganizing TJ proteins, including claudin-5 and occludin (Lee and Funk, 2023).

Structurally, pericyte loss, swollen and detached astrocytic end feet, decreased angiogenesis and vascular density, and glycocalyx degradation are inflammatory responses that play significant roles in the neurovascular unit's loss of integrity (Hussain et al., 2021; Knox et al., 2022; Lee and Funk, 2023). It is noteworthy that pericytes also respond to injury by mediating inflammation in the BBB (Rustenhoven et al., 2017). The IL-8 and MMP-9 generated by reactive pericytes responding to inflammation could enable leukocytes to extravasate into the brain through the BBB (Takata et al., 2021). Furthermore, the expressions of P-glycoprotein (Pgp) and glucose transport protein (GLUT1) are influenced by inflammatory astrocytic activation (Abbott et al.,

2006), resulting in molecular dyshomeostasis in the brain. For instance, nutrient deficiency and accumulation of metabolic wastes are caused by the downregulation of GLUT1 and efflux transporters Pgp and LRP-1 (Knox et al., 2022; Lee & Funk, 2023). These changes promote neurodegenerative processes as they ultimately culminate in BBB disruption characterized by increase in BBB permeability, imbalance of the brain microenvironment, as well as pathogens and neurotoxins extravasation into the brain (Funk et al., 2021; Lotz et al., 2021).

9.4.3 Pathogenic or Microbial Disruption of the BBB

The parenchyma of the central nervous system is effectively protected from invasion by bacterial, fungal, and viral pathogens by the meninges (the dura, arachnoid, and pia maters). Thus, only some parasites, neurotropic viruses, and molds may penetrate the BBB to cause CNS infection. Some bacteria that enter the meninges' subarachnoid space may also infiltrate the perivascular spaces of the postcapillary venules, thereby disrupting BBB and causing parenchymal infection (Salimi & Klein, 2019).

Examples of microorganisms such as viruses, bacteria, and fungi are capable of penetrating the BBB using different mechanisms that include transcellular traversal of BBB with no apparent damage to tight junctions or paracellular traversal, with the possibility of altering the BBB or the Trojan horse system, where organisms can penetrate the BBB via infected phagocytes (Kim, 2008). This is evident in some neurological disorders, such as bacterial meningitis, where the bacteria in the blood penetrate the BBB, disrupting the internal milieu of the brain (Al-Obaidi & Desa, 2018). Some of these bacteria can stimulate the release of toxins and pro-inflammatory substances that enhance the permeability of the BBB (Kim, 2003). These microorganisms, including *Streptococcus pneumonia*, *Listeria monocytogenes*, *Haemophilus influenza*, *Neisseria meningitides*, etc., have been discussed (Al-Obaidi & Desa, 2018). Although many of these studies discussing the microbial disruption of the BBB were based on outcomes of disease conditions, they provided an understanding of mechanisms by which these microorganisms can penetrate the BBB and are being investigated for therapeutic administration through the BBB. Against this background, the viruses offered a promising direction for the delivery of therapeutics to the brain to treat neurological disorders in both clinical and preclinical studies because of their ability to cross the BBB with high efficiency (Marino & Holt, 2022). To address the safety issues around the use of viruses, particularly adeno-associated viruses (AAVs), and to improve tissue tropism and BBB penetration, modifications are normally made to the viral capsid (Fischell & Fishman, 2021; Marino & Holt, 2022).

Bacterial infections in the CNS are known to typically produce meningitis and are restricted to the CSF compartment. BBB may be disrupted by bacterial invasion into the CNS parenchyma due to host inflammatory reactions and bacterial products. Several species have pili surface proteins that interact with BMECs by binding to pIgR and/or LR. This results in the activation of endothelial cells and overexpression of PAFr, CD31, and/or intercellular adhesion molecules. PAFr activation causes arterial dilatation, platelet aggregation, and enhanced BBB permeability, all of which are terminal stages of bacterial meningitis (Al-Obaidi & Desa 2018; Salimi & Klein 2019).

Blood–brain barrier disruption is caused by mouse adenovirus type-1 (MAV-1) infection without inflammation, indicating viral infection rather than inflammatory reactions as the primary driver of the barrier breakdown. In vitro, MAV-1 suppresses the production of TJ proteins while infecting brain vascular endothelium in vivo. Actually, BBB disruption by neurotropic viruses like JEV, WNV, and HIV-1 *in vivo* is associated with reduced TJ and AJ protein expressions. Viruses achieve this by either reducing TJ mRNA transcription levels or encouraging protein degradation (Dallasta et al., 1999; Gralinski et al., 2009; Kajon et al., 1998; Roe et al., 2012; Wang et al., 2018). According to evidence, neurotropic viruses increase BBB permeability by altering the junctions and adhesions between the endothelial cells. This is mostly accomplished by increasing the production of ROS in the CNS, which then activates several tyrosine kinases, MMPs, and the small GTPase RhoA. These actions culminate in the loss of BBB function (Salimi and Klein, 2019). As immune cells infiltration into the brain is essential for the viral clearance, it can also disrupt the BBB and cause neuronal damage by inducing neuroinflammation. Mice infected with VEEV and TBEV have shown that immune cell infiltration occurs prior to BBB breakdown. This is related to elevated levels of ICAM-1, TNF-, IL-6, IL-1, etc. in brain tissues (Cain et al., 2017; Klein et al., 2005; Phares et al., 2006; Ramos et al., 2012; Roe et al., 2012; Růžek et al., 2011). Also, monocytes and infiltrating lymphocytes secrete inflammatory cytokines in contribution to BBB disruption (Salimi and Klein, 2019).

It is also known that neurotropic parasites, including *Trypanosoma*, *Acanthamoeba*, and *Balamuthia* species, generate and release a range of proteases that attack TJ proteins and the BBB's basal membrane, causing barrier failure. These proteases include cysteine and serine proteases and MMPs (Alsam et al., 2005; Iqbal et al., 2014; Lonsdale-Eccles & Grab, 2002; Sissons et al., 2006). By triggering phospholipase C (PLC), proteases interact with protease-activated receptors on BMEC to increase calcium release from intracellular reserves. Increased calcium levels cause myosin light chain to get activated by calmodulin, which, in turn, amplifies intracellular contraction and causes TJs between brain ECs to be disrupted (Grab et al., 2009; Shen et al., 2006).

9.4.4 Apoptosis

Apoptosis is a thoroughly controlled process of intentional cellular death that is crucial for maintaining tissue homeostasis (Kalkan, 2021). Recent studies have suggested that apoptotic activities in the BBB can lead to BBB damage, which may contribute to the pathogenesis of neurological disorders. Multiple processes have been proposed for how apoptosis can lead to BBB damage, including disruption of TJs, loss of pericytes, activation of the immune response, etc. Several studies demonstrated that apoptotic activity may impair BBB tight junctions and increase endothelial permeability. A recent study found that under hyperglycemic conditions, IL-1β triggered pericyte apoptosis through the activation of NF-κB (Yun, 2021). This study observed that IL-1β levels are elevated in the retina of mice with diabetes. The origin of this cytokine appears to be microglia and endothelial cells (EC). However, it did not elicit apoptosis in endothelial cells. Furthermore, the influence of IL-1β on permeability in the endothelial cell monolayer was not observed. Nonetheless, in the presence of EC and pericytes, IL-1β elicited an increase in endothelial cell permeability by downregulating the expression of tight junction protein in endothelial cells. Consequently, diabetic retinopathy pericyte loss may be prevented by blocking IL-1β/NF-B signaling (Yun, 2021). A recent study examined the impact of EGb761 (extract of Ginkgo biloba 761) on cell apoptosis triggered by Aβ 1–42 and the production of ROS. The results indicated that EGb761 exhibited a protective effect against cellular damage caused by Aβ 1–42. The administration of EGb761 resulted in a notable reduction in cell apoptosis and ROS production in SH-SY5Y cells. Also, the administration of EGb761 resulted in a significant decrease in BBB permeability in the Aβ 1–42-induced SH-SY5Y cells. Additionally, the expression levels of caspase 3 and Bax were observed to decrease, while that of Bcl-2 was markedly increased in these cells. The administration of EGb761 to Aβ 1–42-induced SH-SY5Y cells resulted in an upregulation of p-Akt, Nrf2 (nucleus), and HO-1 expression levels (Wang et al., 2021).

In an investigation aimed at examining the involvement of X-linked inhibitor of apoptosis (XIAP) in the early brain damage that occurs after subarachnoid hemorrhage (SAH), the presence of XIAP was identified through real-time RT-PCR and Western blotting in a SAH model involving autologous blood injection. The investigation of XIAP's function in the context of initial brain damage was conducted. Proteins associated with the process of apoptosis were detected. The findings indicate a notable increase in XIAP expression within the hippocampus and cortex regions, with neuronal cells being the primary site of XIAP expression following SAH. The exacerbation of BBB disruption, neurological deficits, and brain edema was observed upon the inhibition of endogenous XIAP. The administration of recombinant XIAP resulted in the preservation of the BBB, improvement of neurological scores, and reduction of brain edema. Furthermore, the administration of recombinant XIAP resulted in a reduction in the levels of cleaved caspases 9, 8, and 3. However, no significant changes were observed in the expression of cytochrome c, p53, or apoptosis-inducing factors. The findings indicate that XIAP functions as an innate neuroprotective and anti-apoptotic factor following SAH (Gao et al., 2017), thus associating anti-apoptotic activities to maintaining BBB integrity.

9.5 NANOTECHNOLOGY/NANOMEDICINE IN BBB DISRUPTION

BBB has a tight small opening that hinders macromolecules from entering the brain (Zhang et al., 2016); this is averted by drug-loaded nanocarriers (Cano et al., 2020) for brain-targeted drug delivery (Zheng et al., 2018; Ozdemir-Kaynak et al., 2018).

Nanomedicine basically targets the immune cells in a brain tissue (Hanif et al., 2020) and enables permeability of the barrier and targeting efficiency (Ghosh et al., 2018) by acting on selective targeting receptors, such as proteins, peptides, and aptamers, or cell membrane–derived vesicles. Nanomedicine implements nanomaterials such as carbon nanotube, silicon nanowires, gold nanoparticle, magnetic nanoparticle, radiolabeled nanoparticles, nanoprobes, and photoresponsive nanomaterial, amongst many others (Viola et al., 2015; Hanif et al., 2020). Nanodelivery strategies combine immunotherapy and nanomedicine (Haabnen et al., 2017; Hanif et al., 2020) to activate innate and adaptive cellular immune responses for the treatment of NDs (Ghosh et al., 2018). These involve strategies that use lipid-, polymer-, or metal-based nanocarriers or nanoparticles (Neganova et al., 2022), characterized by their small (mostly <200 nm) size, positive charge, suitable surface functionalization, targeting ligands (cell-penetrating peptides and polyethylene glyco) for improved circulation time (Cenã et al., 2018; Jagaran & Singh, 2021), and modulation of the release of drugs, to cross or permeate the BBB. The nanocarrier surface is designed to contain an active ligand which can be modified to target specific brain tissue cell. These cells then absorb the nanocarriers, such as metal-transferrin drug-loaded systems or polymer- (nanocapsules and nanospheres) or lipid- (sulfatide-containing nanoliposomes) based nanovehicles (Neganova et al., 2022). The mechanisms of disruption of BBB using nanotechnologies in NDs are not fully understood. However, studies using nanomedicine-based therapy have identified the involvement of several of the mechanisms of BBB disruption we have described earlier (Neganova et al., 2022). According to reports made by Naqvi et al. (2020) and Joseph et al. (2018), glial activation and neuroinflammatory response in NDs results in an imbalance between pro-inflammatory (IL-1β, TNF-α, and IL-6) and anti-inflammatory cytokines (IL-4, IL-10, etc.). The administration of nanoformulations, such as immunoliposomes, dendrimers, micelles, liposomes, loaded with drugs crosses the BBB due to their small size and specifically targets the brain cells and glial to stabilize and restore normal function by reducing neuroinflammation and glial cell activation. Nanoparticle, fullerenols, act as an antioxidant and free radical scavenger because they act by reducing glutamate, N-methyl-d-aspartate (NMDA), amino-3-hydroxyl-5-methyl-4-isoxazole-propionate (AMPA), and kainite-induced excitotoxicity and apoptosis (Naqvi et al., 2020). Another study reported curcumin-loaded PLGA-PEG (cur-PLGA-NPs) nanoparticles have neuroprotection against impaired BBB (Joseph et al., 2018). Also, silicon nanowires have been formulated to detect plaque formation in ND in forms such as nanosensor and silicon nanoprobes (Peng et al., 2014; Hanif et al., 2020). Gold nanoparticle, which has a larger surface area to bind with other targeting ligands, has been reported to help in proliferation of neurites and neuron stimulation in treating NDs (Nakatsuji et al., 2015; Paviolo & Stoddart, 2017). Gold-NP-decorated wheat germ agglutinin disrupts the BBB by the dissolution of its surfactant molecules, resulting in distorting the endothelial cell junctions and allowing for the entry of the NPs into the brain (Zhang et al., 2016). Super paramagnetic iron oxide nanoparticles (SPIONs) help identify neural lesions or disruption for effective drug delivery to target site (Krupa et al., 2014), as seen in glioblastoma. Furthermore, nanocarriers have been indicated as highly probable potent diagnostic and treatment measures in AD (Derakhshankhah et al., 2020). The use of exosome nanocarriers in AD acts to target and modify Aβ senile plaques. Exosome nanocarriers were used to deliver BACE1-siRNA into the brain, to silence BACE expression, thereby limiting overproduction of Aβ peptide, and aid the clearance of already-formed Aβ plaques (Cano et al., 2020).

Molecules can be nanoengineered to carry out multiple specific functions, such as crossing the BBB, targeting specific cell or signaling pathway, responding to endogenous stimuli, and acting as a vehicle for gene delivery, nerve regeneration support, and cell survival (Naqvi et al., 2020). However, the limitation of nanomedicines is its toxicity profile, which is restricting their use at bedside, even with the promising therapeutic potential against NDs (Neganova et al., 2022).

9.6 CONCLUDING REMARKS

In summary, the disruption of the blood–brain barrier plays a significant role in the pathogenesis of various neurological disorders. This disruption triggers a cascade of events, including increased neuroinflammation, oxidative stress, and neuronal damage, which contribute to the progression and severity of neurological diseases.

Understanding the underlying mechanisms and developing targeted interventions to restore BBB integrity are essential for improving disease outcomes and facilitating the development of effective therapies for CNS disorders.

REFERENCES

Abbott, N. J., Rönnbäck, L., & Hansson, E. (2006). Astrocyte–endothelial interactions at the blood–brain barrier. *Nature Reviews Neuroscience, 7*, 41–53.

Ahmed, N., Gandhi, D., Melhem, E. R., & Frenkel, V. (2021). MRI guided focused ultrasound-mediated delivery of therapeutic cells to the brain: A review of the state-of-the-art methodology and future applications. *Frontiers in Neurology*, 12. https://doi.org/10.3389/fneur.2021.669449

Alexander, A., Ajazuddin, R. J., Patel, S., & Saraf, S. S. (2016). Recent expansion of pharmaceutical nanotechnologies and targeting strategies in the field of phytopharmaceuticals for the delivery of herbal extracts and bioactives. *Journal of Controlled Release, 241*, 110–124.

Al-Obaidi, M. M. J., & Desa, M. N. M. (2018). Mechanisms of blood brain barrier disruption by different types of bacteria, and bacterial–host interactions facilitate the bacterial pathogen invading the brain. *Cellular and Molecular Neurobiology, 38*(7), 1349–1368. https://doi.org/10.1007/s10571-018-0609-2

Alsam, S., Goldsworthy, G., Lightfoot, M., Jarroll, E. L., & Khan, N. A. (2006). Identification and properties of proteases from an Acanthamoeba isolate capable of producing granulomatous encephalitis. *BMC Microbiology, 6*, 42. https://doi.org/10.1186/1471-2180-6-42

Alsam, S., Sissons, J., Jayasekera, S., & Khan, N. A. (2005). Extracellular proteases of acanthamoeba castellanii (encephalitis isolate belonging to T1 genotype) contribute to increased permeability in an in vitro model of the human blood–brain barrier. *Journal of Infection, 51*, 150–156.

Angelov, L., Doolittle, N. D., Kraemer, D. F., Siegal, T., Barnett, G. H., Peereboom, D. M., . . . Neuwelt, E. A. (2009). Blood-brain barrier disruption and intra-arterial methotrexate-based therapy for newly diagnosed primary CNS lymphoma: A multi-institutional experience. *Journal of Clinical Oncology, 27*(21), 3503–3509. https://doi.org/10.1200/JCO.2008.19.3789

Banks, W. A., & Rhea, E. M. (2021). The blood–brain barrier, oxidative stress, and insulin resistance. *Antioxidants, 10*(11), 1695.

Bauer, H., & Traweger, A. (2016). Tight junctions of the blood-brain barrier–a molecular gatekeeper. *CNS & Neurological Disorders-Drug Targets (Formerly Current Drug Targets-CNS & Neurological Disorders), 15*(9), 1016–1029.

Berndt, P., Winkler, L., Cording, J., Breitkreuz-Korff, O., Rex, A., Dithmer, S., . . . & Haseloff, R. F. (2019). Tight junction proteins at the blood–brain barrier: Far more than claudin-5. *Cellular and Molecular Life Sciences, 76*, 1987–2002.

Boddy, A. V., & Thomas, H. D. (1997). RMP-7. Potential as an adjuvant to the drug treatment of brain tumours. *CNS Drugs, 7*(4), 257–263. https://doi.org/10.2165/00023210-199707040-00001

Boockvar, J. A., Tsiouris, A. J., Hofstetter, C. P., Kovanlikaya, I., Fralin, S., Kesavabhotla, K., . . . Riina, H. A. (2011). Safety and maximum tolerated dose of superselective intraarterial cerebral infusion of bevacizumab after osmotic blood-brain barrier disruption for recurrent malignant glioma: Clinical article. *Journal of Neurosurgery, 114*(3), 624–632. https://doi.org/10.3171/2010.9.JNS101223

Burkhardt, J. K., Riina, H., Shin, B. J., Christos, P., Kesavabhotla, K., Hofstetter, C. P., . . . Boockvar, J. A. (2012). Intra-arterial delivery of bevacizumab after blood-brain barrier disruption for the treatment of recurrent glioblastoma: Progression-free survival and overall survival. *World Neurosurgery, 77*(1), 130–134. https://doi.org/10.1016/j.wneu.2011.05.056

Cain, M. D., Salimi, H., Gong, Y., Yang, L., Hamilton, S. L., Heffernan, J. R., Hou, J., Miller, M. J., & Klein, R. S. (2017). Virus entry and replication in the brain precedes blood-brain barrier disruption during intranasal alphavirus infection. *Journal of Neuroimmunology, 308*, 118–130.

Cano, A., Sánchez-López, E., Ettcheto, M., López-Machado, A., Espina, M., Souto, E. B., Galindo, R., Camins, A., García,

M. L., & Turowski, P. (2020). Current advances in the development of novel polymeric nanoparticles for the treatment of neurodegenerative diseases. *Nanomedicine (London)*, *15*(12), 1239–1261.

Cash, A., & Theus, M. H. (2020). Mechanisms of blood–brain barrier dysfunction in traumatic brain injury. *International Journal of Molecular Sciences*, *21*, 3344.

Cenã, V., & Játiva, P. (2018). Nanoparticle crossing of blood–brain barrier: A road to new therapeutic approaches to central nervous system diseases. *Nanomedicine, 13*, 1513–1516. https://doi.org/10.2217/nnm-2018-0139

Chakraborty, S., Filippi, C. G., Wong, T., Ray, A., Fralin, S., Tsiouris, A. J., Praminick, B., Demopoulos, A., McCrea, H. J., Bodhinayake, I., Ortiz, R., Langer, D. J., & Boockvar, J. A. (2016). Superselective intraarterial cerebral infusion of cetuximab after osmotic blood/brain barrier disruption for recurrent malignant glioma: Phase I study. *Journal of Neuro-Oncology*, *128*(3), 405–415. https://doi.org/10.1007/s11060-016-2099-8

Chiang, M. C., Nicol, C. J. B., Lin, C. H., Chen, S. J., Yen, C., & Huang, R. N. (2021). Nanogold induces anti-inflammation against oxidative stress induced in human neural stem cells exposed to amyloid-beta peptide. *Neurochemistry International*, *145*, 104992. https://doi.org/10.1016/j.neuint.2021.104992

Chitnis, T., & Weiner, H. L. (2017). CNS inflammation and neurodegeneration. *Journal of Clinical Investigation*, *127*, 3577–3587.

D'Amico, R. S., Khatri, D., Reichman, N., Patel, N. V., Wong, T., Fralin, S. R., Li, M., Ellis, J. A., Ortiz, R., Langer, D. J., & Boockvar, J. A. (2020). Super selective intra-arterial cerebral infusion of modern chemotherapeutics after blood–brain barrier disruption: Where are we now, and where we are going. *Journal of Neuro-Oncology*, *147*(2), 261–278. https://doi.org/10.1007/s11060-020-03435-6da

Fonseca, A. C., Matias, D., Garcia, C., Amaral, R., Geraldo, L. H., Freitas, C., & Lima, F. R. S. (2014). The impact of microglial activation on blood-brain barrier in brain diseases. *Frontiers in Cellular Neuroscience*, *8*, 362.

Dallasta, L. M., Pisarov, L. A., Esplen, J. E., Werley, J. V., Moses, A. V., Nelson, J. A., & Achim, C. L. (1999). Blood-brain barrier tight junction disruption in human immunodeficiency virus-1 encephalitis. *American Journal of Pathology*, *155*, 1915–1927.

De Meyer, G. R., Grootaert, M. O., Michiels, C. F., Kurdi, A., Schrijvers, D. M., & Martinet, W. (2015). Autophagy in vascular disease. *Circulation research*, *116*(3), 468–479.

Derakhshankhah, H., Sajadimajd, S., Jafari, S., Izadi, Z., Sarvari, S., Sharifi, M., Falahati, M., Moakedi, F., Muganda, W. C. A., Müller, M., Raoufi, M., & Presley, J. F. (2020). Novel therapeutic strategies for Alzheimer's disease: Implications from cell-based therapy and nanotherapy. *Nanomedicine: Nanotechnology, Biology and Medicine*, *24*, 102149.

Elias, W. J., Huss, D., Voss, T., Loomba, J., Khaled, M., Zadicario, E., Frysinger, R. C., Sperling, S. A., Wylie, S., Monteith, S. J., Druzgal, J., Shah, B. B., Harrison, M., & Wintermark, M. (2013). A pilot study of focused ultrasound thalamotomy for essential tremor. *New England Journal of Medicine*, *369*(7), 640–648. https://doi.org/10.1056/nejmoa1300962

Fischell, J. M., & Fishman, P. S. (2021). A multifaceted approach to optimizing AAV delivery to the brain for the treatment of neurodegenerative diseases. *Frontiers in Neuroscience*, 15. https://doi.org/10.3389/fnins.2021.747726

Fishman, P. S., & Fischell, J. M. (2021). Focused ultrasound mediated opening of the blood-brain barrier for neurodegenerative diseases. *Frontiers in Neurology*, 12. https://doi.org/10.3389/fneur.2021.749047

Foidl, B. M., & Humpel, C. (2020). Can mouse models mimic sporadic Alzheimer's disease? *Neural Regeneration Research*, *15*(3), 401–406. https://doi.org/10.4103/1673-5374.2660

Fortin, D., Desjardins, A., Benko, A., Niyonsega, T., & Boudrias, M. (2005). Enhanced chemotherapy delivery by intraarterial infusion and blood-brain barrier disruption in malignant brain tumors: The Sherbrooke experience. *Cancer*, *103*(12), 2606–2615. https://doi.org/10.1002/cncr.21112

Funk, K. E., Arutyunov, A. D., Desai, P., White, J. P., Soung, A. L., Rosen, S. F., Diamond, M. S., & Klein, R. S. (2021). Decreased antiviral immune response within the central nervous system of aged mice is associated with increased lethality of west Nile virus encephalitis. *Aging Cell, 20*, e13412.

Gao, C., Yu, H., Yan, C., Zhao, W., Liu, Y., Zhang, D., Li, J., & Liu, N. (2017). X-linked inhibitor of apoptosis inhibits apoptosis and preserves the blood-brain barrier after experimental subarachnoid hemorrhage. *Scientific Reports*, *7*(1), 1–12. https://doi.org/10.1038/srep44918

Gasca-Salas, C., Fernández-Rodríguez, B., Pineda-Pardo, J. A., Rodríguez-Rojas, R., Obeso, I., Hernández-Fernández, F., Del Álamo, M., Mata, D., Guida, P., Ordás-Bandera, C., Montero-Roblas, J. I., Martínez-Fernández, R., Foffani, G., Rachmilevitch, I., & Obeso, J. A. (2021). Blood-brain barrier opening with focused ultrasound in Parkinson's disease dementia. *Nature Communications*, *12*(1). https://doi.org/10.1038/s41467-021-21022-9

Ghosh, D., Nandi, S., & Bhattacharjee, S. (2018). Combination therapy to checkmate glioblastoma: Clinical challenges and advances. *Clinical and Translational Medicine*, *7*, 33.

Grab, D. J., Garcia-Garcia, J. C., Nikolskaia, O. V., Kim, Y. V., Brown, A., Pardo, C. A., Zhang, Y., Becker, K. G., Wilson, B. A., & de A. Lima, A. P. C. (2009). Protease activated receptor signaling is required for African trypanosome traversal of human brain microvascular endothelial cells. *PLoS Neglected Tropical Diseases*, *3*, e479.

Gralinski, L. E., Ashley, S. L., Dixon, S. D., & Spindler, K. R. (2009). Mouse adenovirus type 1-induced breakdown of the blood-brain barrier. *Journal of Virology*, *83*, 9398–9410.

Gu, Z., Kaul, M., Yan, B., Kridel, S. J., Cui, J., Strongin, A., Smith, J. W., Liddington, R. C., & Lipton, S. A. (2002). S-nitrosylation of matrix metalloproteinases: Signaling pathway to neuronal cell death. *Science*, *297*(5584), 1186–1190. https://doi.org/10.1126/science.1073634

Hanif, S., Muhammad, P., Chesworth, R., Rehman, F. U., Qian, R. J., Zheng, M., & Shi, B. Y. (2020). Nanomedicine-based immunotherapy for central nervous system disorders. *Acta Pharmacologica Sinica*, *41*, 936–953. https://doi.org/10.1038/s41401-020-0429-z

Haseloff, R. F., Dithmer, S., Winkler, L., Wolburg, H., & Blasig, I. E. (2015). Transmembrane proteins of the tight junctions at the blood–brain barrier: Structural and functional aspects. In *Seminars in cell & developmental biology* (Vol. 38, pp. 16–25). Academic Press.

Hassanzadeh, P., Atyabi, F., & Dinarvand, R. (2017). Application of modelling and nanotechnology-based approaches: The emergence of breakthroughs in theranostics of central nervous system disorders. *Life Sciences*, *182*, 93–103.

Hussain, B., Fang, C., & Changm, J. (2021). Blood–brain barrier breakdown: An emerging biomarker of cognitive impairment in normal aging and dementia. *Frontiers in Neuroscience*, *15*, 688090.

Hynynen, K. (2008). Ultrasound for drug and gene delivery to the brain. *Advanced Drug Delivery Reviews*, *60*(10), 1209–1217. https://doi.org/10.1016/j.addr.2008.03.010

Hynynen, K., McDannold, N., Vykhodtseva, N., & Jolesz, F. A. (2001). Noninvasive MR imaging-guided focal opening of the blood-brain barrier in rabbits. *Radiology*, *220*(3), 640–646. https://doi.org/10.1148/radiol.2202001804

Ikeda, S., Zablocki, D., & Sadoshima, J. (2022). The role of autophagy in death of cardiomyocytes. *Journal of Molecular and Cellular Cardiology*, *165*, 1–8.

Iqbal, J., Naeem, K., Siddiqui, R., & Khan, N. A. (2014). In vitro inhibition of protease-activated receptors 1, 2 and 4 demonstrates that these receptors are not involved in an Acanthamoeba Castellanii Keratitis isolate-mediated disruption of the human brain microvascular endothelial cells. *Experimental Parasitology*, *145*, S78–S83.

Irudayanathan, F. J., Wang, N., Wang, X., & Nangia, S. (2017). Architecture of the paracellular channels formed by claudins of the blood–brain barrier tight junctions. *Annals of the New York Academy of Sciences*, *1405*(1), 131–146.

Jagaran, K., & Singh, M. (2021). Nanomedicine for neurodegenerative disorders: Focus on Alzheimer's and Parkinson's diseases. *International Journal of Molecular Sciences*, *22*(16), 9082. https://doi.org/10.3390/ijms22169082

Jeon, M. T., Kim, K. S., Kim, E. S., Lee, S., Kim, J., Hoe, H. S., & Kim, D. G. (2021). Emerging pathogenic role of peripheral blood factors following BBB disruption in neurodegenerative disease. *Ageing Research Reviews*, *68*, 101333.

Jordão, J. F., Ayala-Grosso, C. A., Markham, K., Huang, Y., Chopra, R., McLaurin, J. A., Hynynen, K., & Aubert, I. (2010). Antibodies targeted to the brain with image-guided focused ultrasound reduces amyloid-β plaque load in the TgCRND8 mouse model of Alzheimer's disease. *PLoS ONE*, *5*(5). https://doi.org/10.1371/journal.pone.0010549

Jordão, J. F., Thévenot, E., Markham-Coultes, K., Scarcelli, T., Weng, Y. Q., Xhima, K., O'Reilly, M., Huang, Y., McLaurin, J., Hynynen, K., & Aubert, I. (2013). Amyloid-β plaque reduction, endogenous antibody delivery and glial activation by brain-targeted, transcranial focused ultrasound. *Experimental Neurology*, *248*, 16–29. https://doi.org/10.1016/j.expneurol.2013.05.008

Joseph, A., Wood, T., Chen, C.-C., Corry, K, Snyder, J. M., Juul, S. E., Parikh, P., & Nance E. (2018). Curcumin-loaded polymeric nanoparticles for neuroprotection in neonatal rats with hypoxic-ischemic encephalopathy. *Nano Research, 11*, 5670–5688. https://doi.org/10.1007/s12274-018-2104-y

Kajon, A. E., Brown, C. C., & Spindler, K. R. (1998). Distribution of mouse adenovirus type 1 in intraperitoneally and intranasally infected adult outbred mice. *Journal of Virology*, *72*, 1219–1223.

Kalkan, H. (2021). The program cell death (apoptosis) and the therapy of cancer. In *Regulation and Dysfunction of Apoptosis*. IntechOpen.

Karmur, B. S., Philteos, J., Abbasian, A., Zacharia, B. E., Lipsman, N., Levin, V.,. Grossman, S., & Mansouri, A. (2020). Blood-brain barrier disruption in neuro-oncology: Strategies, failures, and challenges to overcome. *Frontiers in Oncology*, 10. https://doi.org/10.3389/fonc.2020.563840

Kim, K. S. (2003). Pathogenesis of bacterial meningitis: From bacteraemia to neuronal injury. *Nature Reviews Neuroscience*, *4*(5), 376–385. https://doi.org/10.1038/nrn1103

Kim, K. S. (2008). Mechanisms of microbial traversal of the blood-brain barrier. *Nature Reviews Microbiology*, *6*(8), 625–634. https://doi.org/10.1038/nrmicro1952

Kim, S. Y., Buckwalter, M., Soreq, H., Vezzani, A., & Kaufer, D. (2012). Blood–brain barrier dysfunction–induced inflammatory signaling in brain pathology and epileptogenesis. *Epilepsia*, *53*, 37–44.

Klein, R. S., Lin, E., Zhang, B., Luster, A. D., Tollett, J., Samuel, M. A., Engle, M., & Diamond, M. S. (2005). Neuronal CXCL10 directs CD8+ T-cell recruitment and control of West Nile virus encephalitis. *Journal of Virology*, *79*, 11457–11466.

Knox, E. G., Aburto, M. R., Clarke, G., Cryan, J. F., & O'Driscoll, C. M. (2022). The blood-brain barrier in aging and neurodegeneration. *Molecular Psychiatry*, *27*, 2659–2673.

Kofoed, R. H., Noseworthy, K., Wu, K., Sivadas, S., Stanek, L., Elmer, B., Hynynen, K., Shihabuddin, L. S., & Aubert, I. (2022). The engineered AAV2-HBKO promotes non-invasive gene delivery to large brain regions beyond ultrasound targeted sites. *Molecular Therapy—Methods and Clinical Development*, *27*, 167–184. https://doi.org/10.1016/j.omtm.2022.09.011

Kovacsa, Z. I., Kima, S., Jikariaa, N., Qureshia, F., Miloa, B., Lewisa, B. K., Bresler, M., Burks, S. R., & Franka, J. A. (2017). Disrupting the blood-brain barrier by focused ultrasound induces sterile inflammation. *Proceedings of the National Academy of Sciences of the United States of America*, *114*(1), E75–E84. https://doi.org/10.1073/pnas.1614777114

Krupa, P., Rehak, S., Diaz-Garcia, D., & Filip, S. (2014). Nanotechnology—new trends in the treatment of brain tumors. *Acta Medica (Hradec Kralove)*, *57*(4), 142–150.

Kumar, A., Chaudhary, R. K., Singh, R., Wang, S. Y., Hoe, Z. Y., Pan, C. T., Shiue, Y. L., Wei, D. Q., Kaushik, A. C., & Dai, X (2020). Nanotheranostic applications for detection and targeting neurodegenerative diseases. *Frontiers in Neuroscience*, *14*, 305. https://doi.org/10.3389/fnins.2020.00305

Kumar, H., Gothwal, A., Khan, I., Nakhate, K. T., Alexander, A., Ajazuddin, V., & Singh, U. G. (2017). Galactose-anchored gelatin nanoparticles for primaquine delivery and improved pharmacokinetics: A biodegradable and safe approach for effective antiplasmodial activity against p. Falciparum 3d7 and in vivo hepatocyte targeting. *Molecular Pharmaceutics*, *14*(10), 3356–3369.

Labus, J., Häckel, S., Lucka, L., & Danker, K. (2014). Interleukin-1β induces an inflammatory response and the breakdown of the endothelial cell layer in an improved human THBMEC-based in vitro blood–brain barrier model. *Journal of Neuroscience Methods*, *228*, 35–45.

Lee, R. L., & Funk, K. E. (2023). Imaging blood–brain barrier disruption in neuroinflammation and Alzheimer's disease. *Frontiers in Aging Neuroscience*, *15*, 1144036. https://doi.org/10.3389/fnagi.2023.1144036

Liddelow, S. A., Guttenplan, K. A., Clarke, L. E., Bennett, F. C., Bohlen, C. J., Schirmer, L., Bennett, M. L., Münch, A. E., Chung, W. S., & Peterson, T. C. (2017). Neurotoxic reactive astrocytes are induced by activated microglia. *Nature*, *541*, 481–487.

Lipsman, N., Lonsdale-Eccles, J. D., & Grab, D. J. (2002). Trypanosome hydrolases and the blood–brain barrier. *Trends in Parasitology*, *18*, 17–19.

Lotz, S. K., Blackhurst, B. M., Reagin, K. L., & Funk, K. E. (2021). Microbial infections are a risk factor for neurodegenerative diseases. *Frontiers in Cellular Neuroscience, 15*, 691136.

Małkiewicz, M. A., Szarmach, A., Sabisz, A., Cubała, W. J., Szurowska, E., & Winklewski, P. J. (2019). Blood-brain barrier permeability and physical exercise. *Journal of Neuroinflammation*, *16*, 1–16.

Margarita, E. N., Aleksandrova, Y. R., Sukocheva, O. A., & Klochkov, S. G. (2022). Benefits and limitations of nanomedicine treatment of brain cancers and age-dependent neurodegenerative disorders. *Seminars in Cancer Biology*, *86*(2), 805–833.

Marino, M., & Holt, M. G. (2022). AAV vector-mediated antibody delivery (A-MAD) in the central nervous system. *Frontiers in Neurology*, 13. https://doi.org/10.3389/fneur.2022.870799

Meng, Y., Bethune, A. J., Huang, Y., Lam, B., Masellis, M., Herrmann, N., Heyn, C., Aubert, I., Boutet, A., Smith, G. S., Hynynen, K., & Black, S. E. (2018). Blood-brain barrier opening in Alzheimer's disease using MR-guided focused ultrasound. *Nature Communications*, *9*(1), 2336. https://doi.org/10.1038/s41467-018-04529-6

Miyagami, M., Tsubokawa, T., Tazoe, M., & Kagawa, Y. (1990). Intra-arterial ACNU chemotherapy employing 20% mannitol osmotic blood-brain barrier disruption for malignant brain tumors. *Neurologia Medico-Chirurgica*, *30*(8), 582–590. https://doi.org/10.2176/nmc.30.582

Mo, F., Pellerino, A., Soffietti, R., & Rudà, R. (2021). Blood-brain barrier in brain tumors: Biology and clinical relevance. *International Journal of Molecular Sciences*, *22*(23). https://doi.org/10.3390/ijms222312654

Nakatsuji, H., Numata, T., Morone, N., Kaneko, S., Mori, Y., Imahori, H., & Murakami, T. (2015). Thermosensitive ion channel activation in single neuronal cells by using surface-engineered plasmonic nanoparticles. *Angewandte Chemie (International Ed. in English)*, *54*(40), 11725–11729.

Naqvi, S., Pangha, A., & Flora, S. J. S. (2020). Nanotechnology: A promising approach for delivery of neuroprotective drugs. *Frontiers in Neuroscience*, *14*. https://doi.org/10.3389/fnins.2020.00494

Ngowi, E. E., Wang, Y. Z., Qian, L., Helmy, Y. A. S. H., Anyomi, B., Li, T., Zheng, M., Jiang, E. S., Duan, S. F., Wei, J. S., Wu, D. D., & Ji, X. Y. (2021). The application of nanotechnology for the diagnosis and treatment of brain diseases and disorders. *Frontiers in Bioengineering and Biotechnology*, *9*, 629832. https://doi.org/10.3389/fbioe.2021.629832

Ozdemir-Kaynak, E., Qutub, A. A., & Yesil-Celiktas, O. (2018). Advances in glioblastoma multiforme treatment: New models for nanoparticle therapy. *Frontiers in Physiology*, *9*, 170.

Pardridge, W. M. (2005). The blood-brain barrier: Bottleneck in brain drug development. *Neurotherapeutics*, *2*(1), 3–14. https://doi.org/10.1007/bf03206638

Park, S. H., Baik, K., Jeon, S., Chang, W. S., Ye, B. S., & Chang, J. W. (2021). Extensive frontal focused ultrasound mediated blood–brain barrier opening for the treatment of Alzheimer's disease: A proof-of-concept study. *Translational Neurodegeneration*, *10*(1). https://doi.org/10.1186/s40035-021-00269-8

Patterson, C. (2018). *The State of the Art of Dementia Research: New Frontiers*. London: Alzheimer's Disease International.

Paviolo, C., & Stoddart, P. R. (2017). Gold nanoparticles for modulating neuronal behavior. *Nanomaterials*, *7*(4), 92.

Peng, F., Su, Y., Zhong, Y., Fan, C., Lee, S. T., & He, Y. (2014). Silicon nanomaterials platform for bioimaging, biosensing, and cancer therapy. *Accounts of Chemical Research*, *47*(2), 612–623.

Phares, T. W., Kean, R. B., Mikheeva, T., & Hooper, D. C. (2006). Regional differences in blood-brain barrier permeability changes and inflammation in the apathogenic clearance of virus from the central nervous system. *Journal of Immunology*, *176*, 7666–7675.

Poupot, R., & Bergozza, D. (2018). Nanoparticle-based strategies to treat neuro-inflammation. *Materials*, *11*(2), 270.

Ramos, H. J., Lanteri, M. C., Blahnik, G., Negash, A., Suthar, M. S., Brassil, M. M., Sodhi, K., Treuting, P. M., Busch, M. P., & Norris, P. J. (2012). IL-1β signaling promotes CNS-intrinsic immune control of West Nile virus infection. *PLoS Pathogens*, *8*, e1003039.

Ranaivo, H. R., Hodge, J. N., Choi, N., & Wainwright, M. S. (2012). Albumin induces upregulation of matrix metalloproteinase-9 in astrocytes via MAPK and reactive oxygen species-dependent pathways. *Journal of Neuroinflammation*, *9*, 1–12.

Reinhard, S. M., Razak, K., & Ethell, I. M. (2015). A delicate balance: Role of MMP-9 in brain development and pathophysiology of neurodevelopmental disorders. *Frontiers in Cellular Neuroscience*, *9*, 280.

Rezai, A. R., Ranjan, M., D'Haese, P. F., Haut, M. W., Carpenter, J., Najib, U., Mehta, R. I., Chazen, J. L., Zibly, Z., Yates, J. R., Hodder, S. L., & Kaplitt, M. (2020). Noninvasive hippocampal blood–brain barrier opening in Alzheimer's disease with focused ultrasound. *Proceedings of the National Academy of Sciences of the United States of America*, *117*(17), 9180–9182. https://doi.org/10.1073/pnas.2002571117

Rodriguez, A., Tatter, S. B., & Debinski, W. (2015). Neurosurgical techniques for disruption of the blood–brain barrier for glioblastoma treatment. *Pharmaceutics*, *7*(3), 175–187. https://doi.org/10.3390/pharmaceutics7030175

Roe, K., Kumar, M., Lum, S., Orillo, B., Nerurkar, V. R., & Verma, S. (2012). West Nile virus-induced disruption of the blood–brain barrier in mice is characterized by the degradation of the junctional complex proteins and increase in multiple matrix metalloproteinases. *Journal of General Virology*, *93*, 1193.

Rustenhoven, J., Jansson, D., Smyth, L. C., & Dragunow, M. (2017). Brain pericytes as mediators of neuroinflammation. *Trends in Pharmacological Sciences*, *38*, 291–304.

Růžek, D., Salát, J., Singh, S. K., & Kopecký, J. (2011). Breakdown of the blood-brain barrier during tick-borne encephalitis in mice is not dependent on CD8+ T-cells. *PLoS ONE*, *6*, e20472.

Salimi, H., & Klein, R. S. (2019). Disruption of the blood-brain barrier during neuroinflammatory and neuroinfectious diseases. In *Neuroimmune Diseases: From Cells to the Living Brain* (pp. 195–234).

Saraf, S., Gupta, A., Alexander, A., Khan, J., Jangde, M., & Saraf, S. (2015). Advancements and avenues in nano phytomedicines for better pharmacological responses. *Journal of Nanoscience and Nanotechnology*, *15*(6), 4070–4079.

Sharabi, S., Bresler, Y., Ravid, O., Shemesh, C., Atrakchi, D., Schnaider-Beeri, M., Gosselet, F., Dehouck, L., Last, D., Guez, D., Daniels, D., Mardor, Y., & Cooper, I. (2019). Transient blood-brain barrier disruption is induced by low pulsed electrical fields in vitro: An analysis of permeability and trans-endothelial electric resistivity. *Drug Delivery*, *26*(1), 459–469. https://doi.org/10.1080/10717544.2019.1571123

Sheikov, N., McDannold, N., Sharma, S., & Hynynen, K. (2008). Effect of focused ultrasound applied with an ultrasound contrast agent on the tight junctional integrity of the brain microvascular endothelium. *Ultrasound in Medicine and Biology*, *34*(7), 1093–1104. https://doi.org/10.1016/j.ultrasmedbio.2007.12.015

Shen, L., Black, E. D., Witkowski, E. D., Lencer, W. I., Guerriero, V., Schneeberger, E. E., & Turner, J. R. (2006). Myosin light chain phosphorylation regulates barrier function by remodeling tight junction structure. *Journal of Cell Science*, *119*, 2095–2106.

Silva, A. D., Aguirre-Cruz, L., Guevara, J., & Ortiz-Islas, E. (2017). Nanobiomaterials' applications in neurodegenerative diseases. *Journal of Biomaterials Applications*, *31*(7), 953–984.

Silveira, G. de B., Muller, A. P., Machado-de-Ávila, R. A., & Silveira, P. C. L. (2021). Advance in the use of gold nanoparticles in the treatment of neurodegenerative diseases: New perspectives. *Neural Regeneration Research*, *16*(12), 2425–2426. https://doi.org/10.4103/1673-5374.313040

Singh, A. K., Singh, S., Tripathi, V. K., Bissoyi, A., Garg, G., & Rizvi, S. I. (2019). Rapamycin confers neuroprotection against aging-induced oxidative stress, mitochondrial dysfunction, and neurodegeneration in old rats through activation of autophagy. *Rejuvenation Research*, *22*(1), 60–70.

Singh, M. P., Sethuraman, S. N., Miller, C., Malayer, J., & Ranjan, A. (2020). Boiling histotripsy and in-situ CD40 stimulation improve the checkpoint blockade therapy of poorly immunogenic tumors. *Theranostics*, *11*(2), 540–554. https://doi.org/10.7150/thno.49517

Sissons, J., Takata, F., Nakagawa, S., Matsumoto, J., & Dohgu, S. (2021). Blood-brain barrier dysfunction amplifies the development of neuroinflammation: Understanding of cellular events in brain microvascular endothelial cells for prevention and treatment of BBB dysfunction. *Frontiers in Cellular Neuroscience*, *15*, 661838.

Tan, X., Azad, S., & Ji, X. (2018). Hypoxic preconditioning protects SH-SY5Y cell against oxidative stress through activation of autophagy. *Cell Transplantation*, *27*(12), 1753–1762.

Trivedi, A., Noble-Haeusslein, L. J., Levine, J. M., Santucci, A. D., Reeves, T. M., & Phillips, L. L. (2019). Matrix metalloproteinase signals following neurotrauma are right on cue. *Cellular and Molecular Life Sciences*, *76*, 3141–3156.

Tsou, Y. H., Zhang, X. Q., Zhu, H., Syed, S., & Xu, X. (2017). Drug delivery to the brain across the blood-brain barrier using nanomaterials. *Small*, *13*(43).

Turner, R. J., & Sharp, F. R. (2016). Implications of MMP9 for blood brain barrier disruption and hemorrhagic transformation following ischemic stroke. *Frontiers in Cellular Neuroscience*, *10*, 56.

Viola, K. L., Sbarboro, J., Sureka, R., De, M., Bicca, M. A., Wang, J., Vasavada, S., Satpathy, S., Wu, S., Joshi, H., Velasco, P. T., MacRenaris, K., Waters, E. A., Lu, C., Phan, J., Lacor, P., Prasad, P., Dravid, V. P., & Klein, W. L. (2015). Towards non-invasive diagnostic imaging of early-stage Alzheimer's disease. *Nature Nanotechnology*, *10*(1), 91–98.

Vykhodtseva, N. I., Hynynen, K., & Damianou, C. (1995). Histologic effects of high intensity pulsed ultrasound exposure with subharmonic emission in rabbit brain in vivo. *Ultrasound in Medicine and Biology*, *21*(7), 969–979. http://dx.doi.org/10.1016/0301-5629(95)00038-S

Wang, K., Wang, H., Lou, W., Ma, L., Li, Y., Zhang, N., Wang, C., Li, F., Awais, M., & Cao, S. (2018). IP-10 promotes blood–brain barrier damage by inducing tumor necrosis factor alpha production in Japanese encephalitis. *Frontiers in Immunology*, *9*, 1148.

Wang, L., Mi, J., Song, Y., & Wang, P. (2021). Protective effect of EGb761 against Aβ1-42-induced SHSY5Y cells injury and blood-brain barrier disruption via regulating Akt/Nrf2 signaling pathway. *Tropical Journal of Pharmaceutical Research*, *20*(9), 1811–1818.

Wang, N., He, J., Pan, C., Wang, J., Ma, M., Shi, X., & Xu, Z. (2019). Resveratrol activates autophagy via the AKT/mTOR signaling pathway to improve cognitive dysfunction in rats with chronic cerebral hypoperfusion. *Frontiers in Neuroscience*, *13*, 859.

Yang, Z., Huang, C., Wu, Y., Chen, B., Zhang, W., & Zhang, J. (2019). Autophagy protects the blood-brain barrier through regulating the dynamic of claudin-5 in short-term starvation. *Frontiers in physiology*, *10*, 2.

Yang, Z., Lin, P., Chen, B., Zhang, X., Xiao, W., Wu, S., Huang, C., Feng, D., Zhang, W., & Zhang, J. (2021). Autophagy alleviates hypoxia-induced blood-brain barrier injury via regulation of CLDN5 (claudin 5). *Autophagy*, *17*(10), 3048–3067. https://doi.org/10.1080/15548627.2020.1851897

Yin, Y., Sun, G., Li, E., Kiselyov, K., & Sun, D. (2017). ER stress and impaired autophagy flux in neuronal degeneration and brain injury. *Ageing Research Reviews*, *34*, 3–14.

Yun, J. H. (2021). Interleukin-1β induces pericyte apoptosis via the NF-κB pathway in diabetic retinopathy. *Biochemical and Biophysical Research Communications*, *546*, 46–53.

Zhang, M., Viennois, E., Prasad, M., Zhang, Y., Wang, L., Zhang, Z., Han, M. K., Xiao, B., Xu, C., Srinivasan, S., & Merlin, D. (2016). Edible ginger-derived nanoparticles: A novel therapeutic approach for the prevention and treatment of inflammatory bowel disease and colitis-associated cancer. *Biomaterials*, *101*, 321–340. https://doi.org/10.1016/j.biomaterials.2016.06.018

Zhao, N., Francis, N. L., Calvelli, H. R., & Moghe, P. V. (2020). Microglia-targeting nanotherapeutics for neurodegenerative diseases. *APL Bioengineering*, *4*(3), 030902. https://doi.org/10.1063/5.001317

Zheng, M., Tao, W., Zou, Y., Farokhzad, O. C., & Shi, B. (2018). Nanotechnology-based strategies for siRNA brain delivery for disease therapy. *Trends in Biotechnology*, *36*, 562–575.

Zhu, F. D., Hu, Y. J., Yu, L., Zhou, X. G., Wu, J. M., Tang, Y., Qin, D. L., Fan, Q. Z., & Wu, A. G. (2021). Nanoparticles: A hope for the treatment of inflammation in CNS. *Frontiers in Pharmacology*, *12*, 683935. https://doi.org/10.3389/fphar.2021.683935

10 Worldwide Development and Commercialization of Nanomedicine

Hadeer N. Atalah and Esraa M. Qansuwa*

10.1 INTRODUCTION

There is a trend toward the development of smaller, faster, stronger, cheaper, and more durable products, and we are witnessing the start of a new era in drug delivery. As a result, there is an opportunity for new leadership to emerge. The impact of disruptive technologies will be more widespread and varied due to the interconnectedness of the global economy. Change is happening rapidly, necessitating investors to gain knowledge in niche markets, both domestically and internationally (Flynn & Wei, 2005). Nanotechnology has completely changed the practice of medicine by providing innovative keys for detecting, treating, and preventing diseases. Targeted therapies and improved drug delivery processes are just a few outcomes of the power of nanomedicine. However, some obstacles stand in the way of the commercialization of nanomedicine, including regulatory issues, high development costs, and limited clinical data. An overview of the prospects and difficulties facing the commercialization of nanomedicine is given in this chapter. The future of nanomedicine commercialization is also discussed, as well as the possible effects of nanomedicine on the healthcare sector.

The principal driving force behind the advancement of nanomedicine is anticipated to be the substantial financial investments directed toward the development of nanotechnology. Scientific Consulting, a Madrid-based consulting firm, has reported a significant increase in funding rates of up to 10 to 20 times the amount observed in 2004. This trend is observed in several countries, including Switzerland, Japan, Germany, the United Kingdom, Taiwan, and Singapore, which have collectively committed billions towards the development of nanotechnology (Flynn & Wei, 2005).

10.2 NANOMEDICINE DEVELOPMENT

10.2.1 Overview of Nanotechnology and Its Application in Medicine, Screening, and Diagnostics

Nanomedicine is an innovative perception in personalized medicine that solves the limitation of traditional medication (one size fits all) and improves the therapeutic results for patients (Aguado et al., 2018). Nanoparticles nowadays focus on increasing the prevalence of precision medicine, which led to personalized medicine via the creation of the precision medicine initiative (PMI) in 2015 (Collins & Varmus, 2015). This initiative targets the patients' information, such as genetic profiles, environmental exposures, or comorbidities that serve the individualization of the treatment plan. Part of the personalized medicine global plan depends on nanomedicine that could improve patient-specific therapeutic responses, representing or focusing on biomaterials and biomedical engineering innovations to overcome biological barriers and patient heterogeneity.

There have been many applications for nanomedicine in drug design and innovative diagnosis since the launch of PMI (Aguado et al., 2018). A typical example of nanotechnique screening diagnosis using graphene oxide nano flakes is in early cancer detection using a blood test (Papi et al., 2019), saving money and time compared to traditional screening tools. In addition to diagnostic screening, using nanoparticles could improve the activity of chimeric antigen receptor (CAR) T cells against cancer tumors as a crucial step in combining nanoparticles and precision medicine to advance the field of diagnosis and personalized treatment (Chen et al., 2019).

10.3 COMMERCIALIZATION OF NANOMEDICINE

In the last two decades, there has been significant advancement and expansion in the area of pharmaceutical nanotechnology. The field of nanomedicine, in particular, has garnered a lot of attention as it offers the potential to transform medical treatments by providing more effective, less harmful, and smarter therapies that can be directed to the affected body area (Halwani, 2022). During the upcoming decades, the field of nanomedicine will present a potential for a worldwide chance to revolutionize medicine and drug delivery. Several American companies have primarily concentrated on the scientific aspects of nanomedicine, neglecting its commercial applications. Consequently, there are significant challenges in introducing nanomedicine products to the market. Adopting a multidisciplinary approach to commercialization will be necessary to overcome many proposed obstacles (Flynn & Wei, 2005).

* Corresponding author: hadeeratalah@aucegypt.edu

DOI: 10.1201/9781003383376-10

Nonetheless, nanomedicine applies the advancement of science, technology, informatics, and electronics that are scaled down for developing new products. It is widely believed that nanomedicine will significantly impact almost every sector of the economy in a relatively brief period. Governments and enterprises worldwide are investing heavily in advancing nanotechnologies, and nanomedicine is already a subset of nanotechnology research and development. For instance, from 2005 to 2008, the US government authorized \$3.7 billion only for nanoscale science and engineering, while in June 2004, Europe allocated billions in funding for nanotechnologies and nanomedicine. Additionally, around 38 industrialized nations are currently pursuing biotech initiatives with different budgets. Hence, nanotechnology, including nanomedicine, is a global industry with applications worldwide (Flynn & Wei, 2005). The commercialization of initial nano-based drugs has been accomplished and positively impacts human health. The nanoparticle-based drug Doxil, an anticancer drug, was one of the first nanomedicine products introduced to the market in 1995. After it, several other nanomedicine drugs came into the light of commercialization (Saxena & Khurana, 2020). The upcoming generation of nanomedicines will undoubtedly enhance the quality of life even further once they are progressively introduced to the market. Regardless of the obstacles, there is a growing trend of commercialization in nanomedicine worldwide. The domain of nanomedicine, which boasts a prolonged developmental past, reveals a noteworthy trend. Specifically, the number of companies that have registered three or more patents about healthcare applications of liposomes nanomedicine exceeds twofold the number of companies that have openly disclosed their participation in this technology. According to Saxena and Khurana (2020), 207 companies have been identified as being actively involved in nanomedicine activities, out of which 158 are small and medium-sized enterprises (SMEs) and start-ups that primarily focus on developing nanomedicines or dedicate a significant portion of their business to them (Wagner et al., 2006).

It is predicted that nanomedicine markets will encompass present and future clinical applications. Despite the limited number of nanotechnology-based products available, the market size in 2003 was around \$500 million. In 2006, the market experienced growth to approximately \$800 million due to the emergence of novel indications. Anticipated technological advances in nanomedicine were expected to propel market expansion to \$3 billion in 2010. Additionally, the market value of nanomedicine reached \$17 billion in 2015, with nanorobots (nanobots) in biomedicine becoming more prevalent, albeit at a high cost. However, the scope of nanomedicine has expanded to encompass all healthcare applications of nanobiotechnology, including diagnostics and drugs. In 2006, the global market for healthcare applications of nanobiotechnology was valued at \$9 billion, and this was expected to grow to \$20 billion in 2010 and further to \$70 billion in 2015 (Saxena & Khurana, 2020).

A defining attribute of nanomedicine is its capacity to imbue existing cures with novel functionality, rendering them more competitive. Given the inherent challenge of quantifying the incremental value that nanomedicine confers, an established practice in business studies of nanomedicine involves gauging the economic significance of nanotechnology in an industrial sector by measuring the aggregate sales of products that incorporate nanotechnology and nanomedicine enhancements (Wagner et al., 2006). For example, Ambisome, a liposomal formulation of the fungicide Fungizone that shows reduced kidney toxicity, had total sales of \$212 million in 2004. It is estimated that the total sales of the 38 identified nanomedicine products from all sectors of nanomedicine, as defined in a study (Wagner et al., 2006), was \$6.8 billion in 2004. Considering the range of nanomedicine products in the advanced stages of development, it was anticipated that the market for such products would expand to approximately \$12 billion by 2012. Presently, drug delivery systems constitute the dominant segment of the nanomedicine market, contributing to over 75% of total sales. In the realm of in vitro diagnostics, colloidal gold in lateral flow assays is the most extensively employed nanotechnology product, utilized in swift testing for pregnancy, ovulation, HIV, and other indications. The incorporation of gold nanoparticles into these assays was introduced in the late 1980s, owing to the exceptional stability of gold conjugates essential for minimizing false positives. According to VDI Technologiezentrum GmbH in Düsseldorf, Germany, the estimated sales of in vitro diagnostic nanomedicines amounted to \$0.78 billion in 2004. Biomaterials have emerged as a prosperous commercial venture, exemplified by the successful introduction of Acticoat in 1998. Acticoat is a wound dressing comprised of silver nanoparticles, which generated sales of \$25 million in 2004, underscoring its commercial triumph. While often highlighted as significant instances of nanomedicine, it should be noted that nanotechnology-based contrast agents were a relatively small market, with an estimated sales volume of around \$12 million in 2004. Currently, all marketed contrast agents consist of superparamagnetic iron oxide nanoparticles utilized for magnetic resonance imaging (Wagner et al., 2006).

10.4 WORLDWIDE NANOMEDICINE DEVELOPMENT AND COMMERCIALIZATION

Nanomedicine encompasses various medical applications of nanotechnology. The development and commercialization of nanomedicine techniques that have the potential to improve drug administration and disease management are expected to increase the adoption of nanomedicines commercialization, leading to market growth. Furthermore, the use of nanotechnology-based drug delivery methods is becoming more prevalent, and the inadequacy of conventional therapeutics in treating diseases is driving the growth of the nanomedicine market (Nanomedicine Market Size, Share | CAGR Of 10.2%, 2023). The nanomedicine industry experienced a positive impact from the COVID-19 pandemic, as the disease led to a greater demand for nanomedicines to develop advanced therapeutics and vaccines for controlling outbreaks. Additionally, the growing interest in using nanomedicine applications to create innovative solutions for preventing, detecting, and treating other viral diseases like COVID-19 is expected to further drive industry growth (Grand View Research, 2023). In 2022, the global nanomedicine market was valued at US$139 billion, with expectations to reach US$358 billion in 2032. During the period of 2023 to 2032, this market is expected to have the highest compound annual growth rate (CAGR) of 10.2% (Grand View Research, 2023) or 11.57% (Nanomedicine Market Size, Share | CAGR Of 10.2%, 2023). The expansion of the nanomedicine market is also expected to be affected by the involvement of established pharmaceutical companies in developing new nanoscience-based therapeutics. For example, the Marble Center for Cancer Nanomedicine recently launched a fellowship program with Alloy Therapeutics in May 2022. The program's primary purpose is to encourage scientific cooperation between academia and the industry to drive innovative breakthroughs in the nanomedicine endeavor. Additionally, advances in nanotechnology and its increasing use in interventions, disease detection, etc. are anticipated further to stimulate the nanomedicine industry's growth (Grand View Research, 2023). Companies are taking advantage of the promising growth opportunities in the worldwide market due to the rising need for nanomedicines, especially after the COVID-19. In addition, the participation of major pharmaceutical firms in manufacturing cutting-edge nanotechnology-based drugs is predicted to influence the advancement of this field (Grand View Research, 2023).

In the global market, the United States leads the way in terms of commercialization efforts, with approximately 50% of all companies engaged in the development or co-development of nanomedicines being based there. By contrast, 35% of such companies are located in European countries. Furthermore, in the domain of drug delivery, only roughly 20% of European companies have successfully launched products in the market. These figures suggest that commercialization efforts in the EU nanomedicine sector are comparatively weaker, notwithstanding its considerable market potential (Wagner et al., 2006).

10.4.1 Nanomedicine Market Size and Share Analysis

The global market for nanomedicine was valued at $171,695.33 million in 2020, and from 2021 to 2030, it is predicted to increase at a CAGR of 9.2%, reaching a value of $393,046.52 million (Allied Market Research, 2022). Nanomedicine is the use of tiny particles for diagnostic, therapeutic, sensing, or actuation functions within a biological system as a nanotechnological intervention used to prevent and treat diseases. The development of novel and cutting-edge drug delivery technologies, the many advantages of nanomedicine in different healthcare applications, and the rising desire for affordable and safe therapeutic solutions all contribute to the nanomedicine market's growth (Allied Market Research, 2022). The nanomedicine's business direction underwent a significant positive shift during the COVID-19 pandemic's outbreak. As a way of effectively treating and managing the many manifestations of the virus outbreak, this was characterized by a marked increase in the demand for nanomedicines across various therapeutic fields. Remarkably, implementing nanomedicine concepts and methods in this setting created a favorable environment for creating novel products for detecting, diagnosing, and treating COVID-19 and other viral diseases (Global Market Insights, Inc., 2023). As a result of the COVID-19 pandemic, several nano-based products, including antiseptics, personal protective equipment, diagnostic tools, and nanocarrier systems for drug delivery and vaccine research, were intensively feasible in the labs. The market for nanomedicines was enhanced and positively impacted due to the increased demand for them. This urgent need for nanomedicine applications to deal with the pandemic has actively influenced the global fund for nanomedicine research. Therefore, the findings of this nanomedicine research have not only served the pandemic period but also served in other diseases, like cancer (Allied Market Research, 2022). The overall effect of the COVID-19 pandemic on the nanomedicine business was beneficial. Additionally, market participants have noticed a rise in material and resource investments, fueled by the expanding research and development activities in nanomedicines across all regions.

In 2021, the market for healthcare nanotechnology (nanomedicine) was estimated to be worth USD 172.8 billion. By 2030, it is expected to be worth USD 523.2 billion, growing at a compound annual growth rate (CAGR) of 13.14% (Verified Market Research, 2023).

The nanomedicine market is stimulated by increased nanorobotic-related research and development efforts and significant financial investments in clinical trials from the public and commercial sectors. However, the lengthy approval procedure and potential hazards of using nanomedicines, such as their influence on the environment, could hamper the market's growth. On the other hand, the rise in the out-licensing of nano-drugs and the construction of healthcare facilities in emerging nations, which gives exciting growth potential, may fuel the market growth (Allied Market Research, 2022).

10.4.2 Nanomedicine Development and Commercialization in Europe

In the European Union (EU), the oversight of market authorization for medicinal products and medical devices takes place at the EU level. The evaluation and monitoring of advanced medicinal products, which include those developed using biotechnological methods or embodying noteworthy therapeutic, scientific, or technical innovations, fall under the European Medicines Agency (EMEA) purview. All nanomedicine products accessible in the EU market have been approved in accordance with existing legislation. The European Commission is currently reviewing pertinent EU-level regulations to assess their applicability to nanotechnology and evaluate the necessity of introducing new regulatory measures (European Commission, Brussels, 2005).

The European Commission asserts that a proactive European strategy is essential to enable the EU to attain independent mastery of the opportunities presented by nanotechnology, deemed the third industrial revolution set to unfold over the next 20–30 years. A budget of €1.3 billion ($1.5 billion) has been designated for commercialization goals. In Europe, very developed and formed efforts have been exerted to consolidate the insightful outcomes of nanomedicine impacts. "Towards a European Strategy for Nanotechnology" (European Parliament, 2006) was initiated as a result of great European intentions to invest in and propagate the development of nanomedicine worldwide. It is a communication from the European Commission issued in 2004. The communication outlines a comprehensive strategy for developing and implementing nanotechnology in Europe. It aims to consolidate and coordinate research efforts in nanotechnology across Europe to promote innovation and improve the competitiveness of the European industry of nanotechnology and nanomedicine. According to a report (Wagner et al., 2008), 41 leading pharmaceutical and medical device companies currently have nanomedicine products available in the market or are engaged in development projects that incorporate nanotechnology. In the previous decade, a total of 38 medical products enhanced with nanotechnology were introduced to the market, generating an estimated EUR 5.4 billion in sales in 2004.

The communication also highlights the need for collaboration between the public and private sectors and the importance of addressing the social, ethical, and environmental concerns associated with the development and application of nanomedicine. The communication sets out a framework for developing nanomedicine in Europe, which includes research funding, regulatory measures, and promoting public awareness and education (Saxena & Khurana, 2020).

10.4.3 Nanomedicine Development and Commercialization in Developed Countries

The new trend in product development is toward smaller and more durable products, marking the start of a new era in drug delivery. There is an opportunity for new leadership to emerge in this field. The global economy's interconnectedness will widely impact the newly emerging technologies. Economy and money will play a crucial role in driving the pace of change and the need for investors to study the market both domestically and internationally (Flynn & Wei, 2005). Funding for the development of nanotechnology will be the driving force behind progress in nanomedicine. Scientific Consulting, based in Madrid, Spain, reports that funding rates have increased tenfold to twentyfold for 2004. Several countries, including Switzerland, Japan, Germany, the United Kingdom, Taiwan, and Singapore, have collectively committed billions. As funding for nanotechnology increases, the market for nanobiotechnology applications is also growing accordingly. The National Science Foundation (NSF) predicted that the market for all nanotechnologies, including nanomedicine, would reach $1 trillion by 2015. The market's influence of nanobiotechnology-based applications in the US alone was expected to be $300 billion by 2027. In 2000, $117 billion was spent annually in the United States for nanotechnological products, and it was raised to 366 billion by 2010, making the applications of nanomedicines attractive for successful investments (Flynn & Wei, 2005).

10.4.4 Nanomedicine Development and Commercialization in Developing Countries

One of the main challenges in developing countries is the need for more infrastructure and resources for research and development. Many developing countries need more funding, expertise, and equipment to conduct research in nanomedicine. This makes it difficult for them to develop their own nanomedicine products and technologies. Another challenge is the regulatory environment. Developing countries often have less-stringent regulations for approving and commercializing new drugs and medical devices. This can lead to safety concerns and hinder the adoption of new nanomedicine products. Despite these challenges, there are opportunities for developing and commercializing nanomedicine in developing countries. Many developing countries have a large population base with unmet medical needs, providing a potential market for nanomedicine products. Additionally, there is a growing interest in international collaborations and partnerships to promote the development of nanomedicine in developing countries (Wagner et al., 2008). Overall, developing and commercializing nanomedicine in developing countries is a complex issue that requires a multifaceted approach. It will require investment in research and development, improvements in infrastructure and regulatory frameworks, and collaborations between different stakeholders.

The field of nanomedicine is experiencing notable progress in countries such as China and India. Other developing nations, including Thailand, the Philippines, Chile, Argentina, and Mexico, are also actively exploring the potential of nanotechnology in health care and its commercialization. Numerous studies in this field are imperative to address the pressing medical challenges that affect these countries. Accordingly, a proposal has been put forward for an enterprise named "Addressing Global Challenges

Using Nanotechnology," which aims to hasten the implementation of nanotechnology in resolving significant sustainable development issues. This initiative is based on the model of the Foundation for the NIH/Bill and Melinda Gates Foundation's Grand Challenges in Global Health (Salamanca-Buentello et al., 2005). Many routes are being developed to pave the path for nanomedicine and nanomedicine markets.

10.5 NANOMEDICINE MARKET: SEGMENTATION ANALYSIS

The nanomedicine market displays segmentation based on modality, application, indication, and region (Figure 10.1). In terms of modality, the market is categorized into diagnostics and treatment. For application, it is divided into drug delivery, diagnostic imaging, vaccines, regenerative medicine, implants, and others. Moreover, the indication segment encloses clinical oncology, infectious diseases, clinical cardiology, orthopedics, neurology, urology, ophthalmology, immunology, and other medical indications. Geographically, the market is run across North America, Europe, Asia-Pacific, and LAMEA (Latin America, Middle East, and Africa) (Allied Market Research, 2022).

10.5.1 Nanomedicine Market: Modality Segmentation Analysis

The nanomedicine market globally is further categorized into treatment and diagnostics (Figure 10.2), where treatment is the primary contributor to the market in 2020 (Verified Market Research, 2023) and in 2022 (Global Market Insights, Inc., 2023). And it is expected to exhibit the most increased growth rate throughout the forecast period. Nanomedicine's potential mainly operates in this segment for enhanced therapeutic efficacy (Global Market Insights, Inc., 2023). Thanks to the advancements in nanotechnologies, it is revealed that this treatment segment will continue to be the major contributor to the market for the upcoming years due to the high demand of effective products for the increased incidence rate of chronic diseases (Allied Market Research, 2022). With more enhancement in nanotechnology, nanomedicine offers exceptional opportunities to improve drug delivery and targeted actions of therapeutic agents, resulting in more effective treatments. In addition, nanomedicine promotes targeted drug delivery, where therapeutic agents are hooked to nanoparticles or nanocarriers. These precisely designed nanodelivery approaches target specific cells, or tissues, enabling specific drug delivery (Reports and Data, 2022).

10.5.2 Nanomedicine Market: Indication Segmentation Analysis

Based on indication, the nanomedicine market displays various categories, including clinical oncology, infectious diseases, clinical cardiology, orthopedics, neurology, urology, ophthalmology, immunology, and others. However, the clinical oncology segment appeared as the most prominent supporter of the market's growth and is estimated to retain its dominance throughout the upcoming period. Due to the higher prevalence of oncology diseases (Verified Market Research, 2023), the oncology segment demonstrated a predominant market stake in 2020 (Allied Market Research, 2022), in 2021 (Reports and Data, 2022), and in 2022, with a share of 32.44% (Grand View Research, 2023), and is expected to own a substantial growth rate of 11.5% from 2023 to 2032 (Global Market Insights, Inc., 2023). That is because the higher the cancer incidence, the more powerful the expansion in the clinical development of cancer treatment developments and refinements in therapeutic particles and appliances. Through nanomedicine initiatives, clinical oncology includes active and passive strategies for targeting cancer (Allied Market Research, 2022). With the urgent

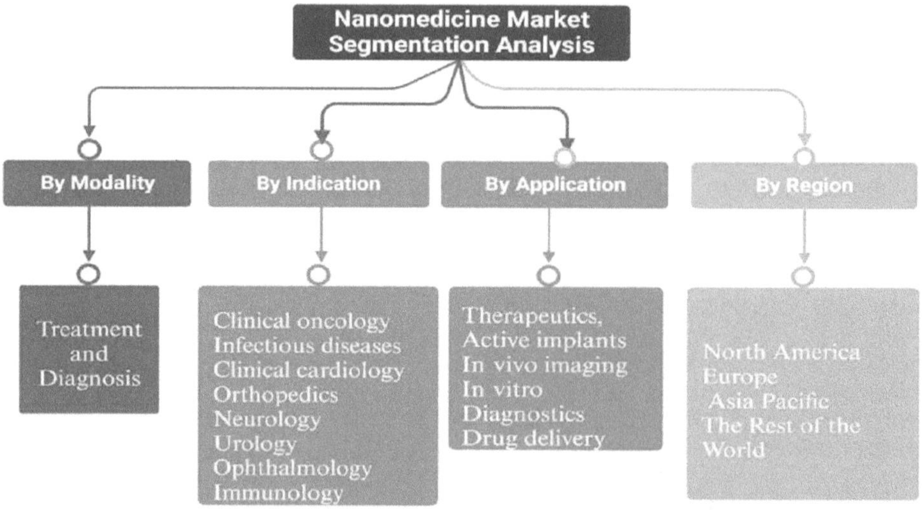

FIGURE 10.1 Nanomedicine's four main segmentation analysis and the bases of each segment.

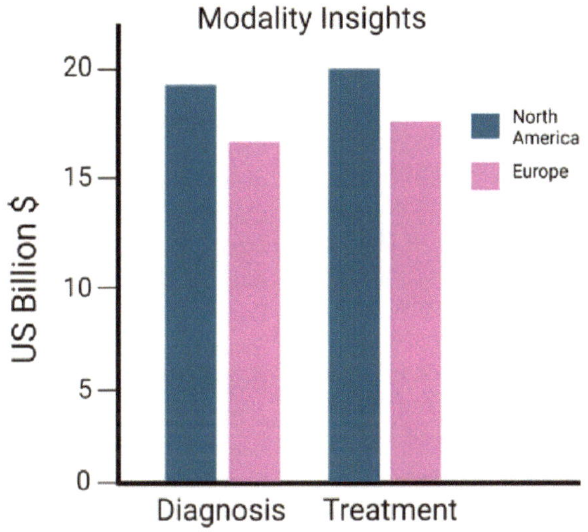

FIGURE 10.2 Modality segmentation analysis of nanomedicine market in North America and Europe.

need to develop rapid, efficient, and cost-effective medicines against infectious diseases, the infectious disease segment is predicted to display substantial growth with a rapid CAGR of 12.31% by the year 2030 (Allied Market Research, 2022).

10.5.3 Nanomedicine Market: Molecule-Type Segmentation Analysis

The nanomedicine market is classified by molecule type into nanoparticles, nanotubes, nanoshells, and nanodevices. The nanoparticles segment appeared as the market head in 2022, with a revenue share of 74%, and is anticipated to display the highest growth rate during the upcoming periods. This dominance is due to the numerous advances offered by nanoparticles (Allied Market Research, 2022). The over-consumption of metal and metal oxide particles in photodynamic therapy (PDT) for cancer treatment and infectious diseases is the reason behind the superiority of nanoparticle segments. Additionally, nanoparticles can perform various functions by forming scaffolds, proving the segment's steady growth. With a market share of 76.32% in 2022, the nanoparticles category became the market leader. This is due to the significant power of the nanoparticles segment in treating infectious diseases and cancer within their small size, high surface area, adaptable surface characteristics, imaging and drug transport capabilities, flexibility in functionalization, and biocompatibility and biodegradability (Global Market Insights, Inc., 2023). In addition, different industries develop inventive technologies using nanoparticles for targeted treatment or intervention (Grand View Research, 2023). Nanoparticles are extremely useful in targeted medication administration, imaging methods, and customized medicine due to their distinctive properties. They have the potential to significantly improve the course of therapy for a variety of illnesses and advance the science of nanomedicine.

10.5.4 Nanomedicine Market: Application Segmentation Analysis

By application, the nanomedicine market is further categorized into various applications, including therapeutics, active implants, in vivo imaging, in vitro diagnostics, and drug delivery. Among all the listed categories, the drug delivery segment has emerged as the dominant segment in the global nanomedicine market in 2022 (Nanomedicine Market Size, Share | CAGR Of 10.2%, 2023)., with a share of 34.09% (Grand View Research, 2023). That is attributed to the increasing prevalence of cancer, chronic and infectious diseases, and also the growing attention among researchers regarding the potential benefits of nanomedicine in cancer treatment (Nanomedicine Market Size, Share | CAGR Of 10.2%, 2023; Grand View Research, 2023). COVID-19 was a great trigger for the scientists to look further beyond the capabilities of nanomedicine's role in infectious diseases (Grand View Research, 2023). Additionally, nanotechnology's advantages in terms of reticule-endothelial system evasion, advantageous pharmacokinetics, and in vivo dispersion motivate its usage in drug administration. These factors play a significant role in propelling the growth of the drug delivery segment within the nanomedicine market. In March 2023, researchers from Mount Sinai Health System and Memorial Sloan Kettering Cancer Center worked together to create a breakthrough approach for delivering anticancer medications, especially to pediatric brain tumors. This represents a recent advancement in drug delivery (Grand View Research, 2023). With a predicted CAGR of 12.79% from 2023 to 2030, the therapeutics category is expected to rise fast. The use of nanomedicine in therapeutics includes the sale of pharmaceuticals, medical devices, and drug delivery methods. Additionally, it is projected that market expansion will be fueled by technological developments that allow for the creation of nanotherapeutics that can pass through biological barriers throughout the upcoming period (Grand View Research, 2023).

10.5.5 Nanomedicine Market: Regional Segmentation Analysis

The global healthcare nanotechnology (nanomedicine) market is divided into North America, Europe, Asia-Pacific, and the rest of the world based on regional analysis (Verified Market Research, 2023). The North American market accounted for the largest revenue share in 2021, per regional analysis (Verified Market Research, 2023; Reports and Data, 2022). With a roughly 50% market share, North America dominated the worldwide nanomedicine market in 2022. This dominance should persist over the course of the projection period due to the increasing number of collaborations between well-established businesses and start-ups in the field of nanomedicine (Grand View Research, 2023). The motive behind the nanomedicine market growth in the United States of America are the technological improvements and their pertinent applications in early illness diagnosis, preventative interventions, and managing acute and chronic diseases (Verified Market Research, 2023). These elements are anticipated to significantly support the nanomedicine market growth even more. For instance, according to the CDC, heart disease was a major cause of mortality in the United States in 2021 (Verified Market Research, 2023), accounting for one in every four fatalities. With 360,900 fatalities from cardiac disorders, coronary heart disease was found to be the primary cause of death; stroke came in second with 805,000 occurrences. This significant burden highlights the pressing need for nanomarket expansion in America for tissue regeneration implants, nanostructured stents, and specific therapeutic nanovectors to address these health issues (Verified Market Research, 2023). This expansion of the nanomedicine market is explained by the large amount of money invested in innovative technologies to develop the pharmaceutical and healthcare industries, such as biotechnology and artificial intelligence. The North American market's revenue growth has also been aided by the growing use of nanomedicine to treat infectious diseases and cancer. In addition, there are prospects for market expansion because of the increased R&D efforts to create novel diagnosis and treatment techniques. For example, smart dressings that include nanoparticles of growth hormones, blood-clotting agents, or sensors that can identify infections and release antibiotics are examples of wearables that use nanotechnology. These biocompatible materials used to make these nanotechnology-based bandages enable them to stay on wounds until they naturally dissolve, helping to avoid infected wounds (Reports & Data, 2022).

The nanomedicine market is expanding due in large part to increasing financing for research in America and a growing need for the prevention of serious diseases. Because of the rising interest in nanoscience and nanotechnology conferences among the general public, the Asia-Pacific area is also anticipated to develop significantly throughout the projection period (Nanomedicine Market Size, Share | CAGR Of 10.2%, 2023). North America reaps the benefits of increased R&D expenditure, government assistance, and a significant number of industrial companies pursuing nanomedicine projects. For instance, in October 2020, Advanced NanoTherapies successfully obtained $5.3 million in initial investment to expedite the development of their medication delivery platform using nanoparticles for peripheral artery disease and other uses (Grand View Research, 2023). With a CAGR of 14.07% from 2023 to 2030, the Asia-Pacific nanomedicine industry is anticipated to expand at the quickest rate. The rise in popularity of nanomedicine and the public's interest in attending conferences on nanoscience and nanotechnology are credited with this growth. For example, Beijing, China, hosted the 9th International Conference on Nanoscience and Technology, ChinaNANO in 2022. The conference's primary purpose was to exchange information about new discoveries in science, business, technology, and nanomedicine, as well as obstacles and possibilities (Grand View Research, 2023).

10.6 ARTIFICIAL INTELLIGENCE AND MACHINE LEARNING IN THE NANOMEDICINE FIELD: APPLICATIONS AND FUTURE VISION

The development of advanced nanomedicine has created a new perspective on public health and pharmaceutical products. The safe design approaches and application on living organisms are considered a challenge in the nanomedicine field associated with nano-toxicological evaluation. Artificial intelligence and machine learning would play an essential role in analyzing and interpreting a large amount of data on nano-toxicological products to better understand the massive toxic effects of nanoparticles in health care. By 2025, the global nanoparticle market is expected to reach $75.64 billion, meaning, the annual growth is approximately 13.2% between 2020 and 2025. This increase is required worldwide because of the wide range of benefits, such as nanomedicine and nanotechnology in sunscreen, cosmetics, food packaging, water filtration, medicine, and energy production (Singh et al., 2023).

We are exposed to many sources of nanoparticles, making these topics crucial to be understood on behalf of the contribution of nanoparticles in disease causes and innovative medications simultaneously. Those sources of nanoparticles might include industries, food additives, processed food, cigarette, cosmetics, packaging materials, and forest fires, contributing in harmful risks to population health, causing a punch of global disorders, such as Parkinson's, Alzheimer's, asthma, cancer, emphysema, bronchitis, arrhythmia, dermatitis, vasculitis, urticaria, Crohn's disease, hypertension, and thrombosis, etc. (Asati et al., 2021).

With the growth of nanomedicine worldwide with the increase of the volume of data, it is challenging to use traditional methods to check the massive number of toxicological components endpoints. Therefore, using machine learning models to predict or assess the toxicological levels of nanoparticles would save money and time and be accessible as a global commercial plan for the innovative use of nanomedicine (Singh et al., 2023). Understanding the mechanism of chemical actions and disease mechanisms is accessible by developing data mining approaches through advanced artificial intelligence (AI) tools (Kumar & Saha, 2022). For example, the safety of chemicals is critical in nanomedicine commercialization because the market needs machine learning modeling to check and understand the toxicity level of the newly entering chemicals that would affect people's health with adverse chemical reactions.

With the era of big data and data mining, machine learning is considered a practical approach because it plays an important role in using huge amounts of data to predict the biological, physical, toxicological, and chemical effects (Casañola-Martin & Pham, 2019). Moreover, the connection between successful applications of nanomedicine and computational approaches is always related to QSAR (quantitative–structure activity relationships) studies. These types of studies merged with statistical methods are practical choices for a computational screening for new chemicals or toxins based on a previously huge amount of data gathered. This information is used to correlate the biological effects with the kind of features describing the patterns of efficacy or the molecular descriptors gathered directly from chemical structures of the desired compounds. The defining features of nanomaterials include shape, size, composition, density, propensity to agglomerate, interaction with different molecules, and aqueous solubility that could affect the biological responses because of the properties of nanomaterials, adding novel global interest in drug discovery and design (Casañola-Martin & Pham, 2019).

There are many innovative applications merging machine learning and nanomedicine, solving the challenge of manipulating the massive amount of data generated through previous experimental research. Those applications would shape the future of nanomedicine and global collaborations. For example, using NLP (natural language processing) libraries to help in determining and specifying the nanomedicine data and non-nano of international clinical trials from other types of data, designing organized big data in clinical trials databases which facilitate the work of researchers globally (de la et al., 2014). In addition to that, there are other advanced applications focused on nanomedicine and machine learning in the prediction of biomedical responses (Jones et al., 2016), with recommendations of using the algorithms of artificial neural networks (ANN), which fit with the complexity and the significant heterogeneity of data in the nanomaterials (Tantra et al., 2015). The other remarkable application of the use of machine learning models in the prediction of biological effects are the gold nanoparticles in cell association studies (Winkler, 2016). Also, the nano-scale imaging technique and computational tomography images are used to develop machine learning models for lung tumor detectors and predictors, using classification models such as support vector machine (SVM) and K-nearest neighbor (KNN) for e-image classification and feature manipulation (Rani & Jawhar, 2019). Those models provide an accuracy of 97% in the classification of lung lesions and their diagnosis and show an interesting application for the field of medical informatics, showing interesting applications in the field of nano-informatics (Rani & Jawhar, 2019).

10.7 CHALLENGES IN THE NANOMEDICINE MARKET

Nanomedicines have the potential to make fundamental contributions and modernization of disease prevention, health care, and diagnostic processes. The global nanomedicine market growth rate is shifting from $528 billion in 2019 and $2,611 billion in 2023, with an annual increase reaching 12.6% from 2017 to 2023 (Global Opportunity Analysis and Industry Forecast from 2017 to 2023, Grand View Research, 2023). Numerous nanomedicines have been created and authorized for clinical application, with significant contributions from the academic and pharmaceutical sectors. Nonetheless, nanomedicine is still in its early stages, with little but significant accomplishments, since various obstacles are encountered in developing pharmaceutical commercialized nanomedicines. Currently, nanomedicine commercialization endeavors are primarily driven by start-ups and small- to medium-sized enterprises (SMEs). However, due to the substantial costs associated with drug and medical device development, start-ups need to work on bringing their products to the market when lacking support from larger companies in the biotech, pharmaceutical, or medical device sectors. Hence, the role of big pharma is pivotal in ensuring the success of nanomedicine as a business venture. While start-ups are actively exploring various ideas for leveraging nanotechnology to improve disease diagnosis and treatment, both start-up founders and industry R&D managers concur that major pharmaceutical companies still need to be fully committed to prioritizing nanotechnology in their business strategies. Notably, start-ups pursuing technologies without any approved products on the market face difficulties convincing pharma companies to collaborate with them or license their technology.

The commercial adoption of nanotechnology in the United States is advancing faster than in Europe, with discernible parallels to the commercialization trajectory of recombinant DNA technology observed two decades ago. Such disparities in the entrepreneurial culture and the availability of venture capital influence the pace of nanomedicine commercialization in Europe, where the availability of venture capital for start-ups is comparatively lower than in the United States (Wagner et al., 2006). The commercialization of cutting-edge medicines is impeded by healthcare markets that are regulated by cost considerations, thereby constraining the development of innovative, high-value drugs, including nanomedicines. This has prompted ongoing discourse among scientists, industry stakeholders, and regulatory bodies regarding whether new regulations are necessary to accommodate the distinctive pharmacokinetic

TABLE 10.1
Some of the challenges the commercialization of nanomedicine faces, with some possible resolutions

Commercialization challenges	Possible resolutions
The inclusion of diverse nanomaterials and nanostructures within nanomedicine commercialization increases complexity in establishing regulatory protocols and tools to ensure manufacturing and characterization, comprehensive safety and toxicology evaluation, and well-designed clinical trials (Robson et al., 2018).	Fast-track promising novel nanomedicines to clinical trials, establishment of a network of academia–industry experts and laboratories with expertise in nanomedicine characterization, preclinical studies, scale-up, and to effectively translate clinical trails (Hafner et al., 2014).
Government regulations pertaining to safety, quality assurance, patent protection, and manufacturing procedures significantly hinder the commercialization nanomedicines.	Developing clear regulatory guidelines specific for nanomedicines.
The present status questioning the safety evaluation of nanotechnology-based therapeutics is examined, highlighting the significance of risk assessment and risk mitigation in the development of nanomedicine products and markets (Robson et al., 2018).	Develop and validate an innovative and standardized applicable resolution and protocol of evaluating short-term and long-term health hazards through concluding the perceptive in vitro findings to predict chronic in vivo effects (Robson et al., 2018).
According to Robson et al. (2018), complexities associated with characterization and manufacturing, a poor understanding of pharmacokinetics and pharmacodynamics in vivo, worries about acute and chronic toxicity, and the need to take cost-effectiveness into account.	Apply interdisciplinary collaborations through encouraging scientific, engineering, and medical specialists to work together to handle the issues by combining their knowledge in manufacturing, pharmacology, toxicology, and material science.
	Investigate cutting-edge production methods that maximize nanomaterial output while upholding cost- and quality- effectiveness criteria, permitting increased accessibility and affordability.

properties of nanomedicines. One of the major challenges in the nanomedicine market are the follow-on versions and the existing regulatory evaluations for market access (Mühlebach, 2018). The nanomedicine follow-on versions authorize partially depending on the generic paradigm or small-molecule drugs that characterize the colloidal drugs and other related unknown manufacturing drug processes (Mühlebach, 2018). However, nanotechnology holds a plethora of promise for the future of industries in food, medicine, cosmetics, and health; many challenges are considered an obstacle to bringing a nanotechnological product or business to the market. For example, the technology transfer process. This technology is a crucial technological innovation in many research institutions, defined by license offices that this process is needed for institutions for product commercialization (Rambaran & Schirhagl, 2022).

According to Robson et al. (2018), clinical translation depends on achieving appropriate validation and control of nanoparticle shape. Nevertheless, achieving precise control and comprehensive characterization of nanoparticle shape is usually difficult. Regardless of whether nanomedicines are therapeutically successful or not, there might be a number of major challenges in the way of their clinical translation; some could be resolved, and some still under investigations (Table 10.1).

10.8 CONCLUSION

Nanomedicine is a fast-developing science that has the potential to revolutionize drug delivery and medicine. The translation switch of nanomedicine clinical research into practical applications, such as medication delivery, vaccine creation, and antibacterial therapy, has been made more accessible by using nanoscale materials, resulting in the thriving of the nanomedicine market. Nanomedicine is also predicted to revolutionize healthcare practices and significantly influence the worldwide market during the upcoming decades. Although countless obstacles are in the way of commercializing nanomedicine, many US business communities have paved the road in utilizing nanomedicine economically, preferring to concentrate on its scientific and clinical applications. Expensive development costs, a need for more clinical evidence, and regulatory constraints also pose significant challenges. In addition, strict regulations and exhaustive research are essential for the safe and efficient advancement of nanomedicine commercialization. Notwithstanding these obstacles, the commercialization of nanomedicine has a brighter future. Growing support for the science of nanotechnology is seen in the increased financial investments made globally in research and development. Because it allows for more precise medication administration, disease prevention, and targeted therapy, nanomedicine has great promise to transform the healthcare industry and health market.

REFERENCES

Aguado, B. A., Grim, J. C., Rosales, A. M., Watson-Capps, J. J., & Anseth, K. S. (2018). Engineering precision biomaterials for personalized medicine. Science Translational Medicine, 10, 8645.

Allied Market Research. (2022). Nanomedicine Market Size, Share, Competitive Landscape and Trend Analysis Report by Modality, Application and Indication: Global Opportunity Analysis and Industry Forecast, 2021-2030. https://www.alliedmarketresearch.com/nanomedicine-market

Asati, S., Sahu, A., & Jain, A. (2021). Nanotoxicity: The dark side of nan of ormulations. Current Nanotoxicity and Prevention (Discontinued), 1(1), 6–25. https://doi.org/10.2174/2665980801999201230095324

Casañola-Martin, G. M., & Pham-The, H. (2019). Machine learning applications in nanomedicine and nanotoxicology: An overview. International Journal of Applied Nanotechnology Research (IJANR), 4(1), 1–7. http://doi.org/10.4018/IJANR.2019010101

Chen, Q., et al. (2019). Photothermal therapy promotes tumor infiltration and antitumor activity of CAR T cells. Advanced Materials, 31, e1900192.

Collins, F. S., & Varmus, H. (2015). A new initiative on precision medicine. New England Journal of Medicine, 372, 793–795.

De La Iglesia, D., García-Remesal, M., Anguita, A., Munoz-Marmol, M., Kulikowski, C., & Maojo, V. (2014). A machine learning approach to identify clinical trials involving nanodrugs and nano devices from ClinicalTrials.gov. PLoS ONE, 9(10), e110331. http://doi.org/10.1371/journal.pone.0110331

European Parliament. (2006). Report on Nanosciences and nanotechnologies: An action plan for Europe 2005-2009 (A6-0216/2006). https://www.europarl.europa.eu/doceo/document/A-6-2006-0216_EN.html

Flynn, T., & Wei, C. (2005). The pathway to commercialization for nanomedicine. Nanomedicine: Nanotechnology, Biology and Medicine, 1(1), 47–51.

GMI Insights. (2023). Nanomedicine Market Size & Share Report, 2032. https://www.gminsights.com/industry-analysis/nanomedicine-market

Grand View Research. (2023). Nanomedicine Market Size, Share & Growth Report, 2030. https://www.grandviewresearch.com/industry-analysis/nanomedicine-market

Hafner, A., Lovrić, J., Lakoš, G. P., & Pepić, I. (2014). Nanotherapeutics in the EU: An overview on current state and future directions. International Journal of Nanomedicine, 1005–1023.

Halwani, A. A. (2022). Development of pharmaceutical nanomedicines: From the bench to the market. Pharmaceutics, 14(1), 106.

Jones, D. E., Ghandehari, H., & Facelli, J. C. (2016). A review of the applications of data mining and machine learning for the prediction of biomedical properties of nanoparticles. Computer Methods and Programs in Biomedicine, 132, 93–103. http://doi.org/10.1016/j.cmpb.2016.04.025

Kumar, R., & Saha, P. (2022). A review on artificial intelligence and machine learning to improve cancer management and drug discovery. International Journal for Research in Applied Sciences and Biotechnology, 9(3), 149–156.

Mühlebach, S. (2018). Regulatory challenges of nanomedicines and their follow-on versions: A generic or similar approach? Advanced Drug Delivery Reviews, 131, 122–131. https://doi.org/10.1016/j.addr.2018.06.024

Nanomedicine Market Size, Share | CAGR Of 10.2%. (November 2023). Market.us. https://market.us/report/nanomedicine-market/

Papi, M., Palmieri, V., Digiacomo, L., Giulimondi, F., Palchetti, S., Ciasca, G., . . . Caracciolo, G. (2019). Converting the personalized biomolecular corona of graphene oxide nanoflakes into a high-throughput diagnostic test for early cancer detection. Nanoscale, 11(32), 15339–15346.

Pelaz, B., Alexiou, C., Alvarez-Puebla, R. A., Alves, F., Andrews, A. M., Ashraf, S., . . . Parak, W. J. (2017). Diverse applications of nanomedicine. ACS Nano, 11(3), 2313–2381.

Rambaran, T. F., & Schirhagl, R. (2022). Nanotechnology from lab to industry—a look at current trends. Nanoscale Advances, 4(18), 3664–3675. https://doi.org/10.1039/d2na00439a

Rani, K. V., & Jawhar, S. J. (2019). Novel technology for lung tumor detection using nanoimage. Journal of the Institution of Electronics and Telecommunication Engineers, 1–15. https://doi.org/10.1080/03772063.2019.1565955

Robson, A. L., Dastoor, P. C., Flynn, J., Palmer, W., Martin, A., Smith, D. W., . . . Hua, S. (2018). Advantages and limitations of current imaging techniques for characterizing liposome morphology. Frontiers in Pharmacology, 9, 80. https://doi.org/10.3389/fphar.2018.00080

Reports and Data. (2022, October). Nanomedicine market size, share, trends, by type, by product, by modality, by indication, by region, forecast to 2030. https://www.reportsanddata.com/report-detail/nanomedicine-market

Salamanca-Buentello F, Persad DL, Court EB, Martin DK, Daar AS, Singer PA (2005) Nanotechnology and the Developing World. PLoS Med 2(5): e97. https://doi.org/10.1371/journal.pmed.0020097

Saxena, S. K., & Khurana, S. P. (Eds.). (2020). NanoBioMedicine (pp. 3–16). Singapore: Springer.

Singh, A. V., Varma, M., Laux, P., Choudhary, S., Datusalia, A. K., Gupta, N., . . . Nath, B. (2023). Artificial intelligence and machine learning disciplines with the potential to improve the nanotoxicology and nanomedicine fields: A comprehensive review. Archives of Toxicology, 1–17.

Tantra, R., Oksel, C., Puzyn, T., Wang, J., Robinson, K. N., Wang, X. Z., Wilkins, T., et al. (2015). Nano(Q) SAR: Challenges, pitfalls and perspectives. Nanotoxicology, 9(5), 636–642. http://doi.org/10.3109/17435390.2014.952 698

Thapa, R. K., & Kim, J. O. (2023). Nanomedicine-based commercial formulations: Current developments and future prospects. Journal of Pharmaceutical Investigation, 53(1), 19–33.

Wagner, V., Dullaart, A., Bock, A. K., & Zweck, A. (2006). The emerging nanomedicine landscape. Nature Biotechnology, 24(10), 1211–1217.

Wagner, V., Hüsing, B., Gaisser, S., & Bock, A. K. (2008). Nanomedicine: Drivers for development and possible impacts. JRC-IPTS, EUR, 23494.

Winkler, D. A. (2016). Recent advances, and unresolved issues, in the application of computational modelling to the prediction of the biological effects of nanomaterials. Toxicology and Applied Pharmacology, 299, 96–100. http://doi.org/10.1016/j.taap.2015.12.016 PMID:26723909

Verified Market Research. (2023). Healthcare Nanotechnology (Nanomedicine) Market Size & Forecast. https://www.verifiedmarketresearch.com/product/healthcare-nanotechnology-nanomedicine-market/

11 Introduction to Nanoparticles as Potential Carriers for Brain Targeting in Neurodegenerative Disorders

Adejoke Elizabeth Memudu, Michael Oluwatosin Gabriel, Victor Ebube Anadu, Olayemi Kafilat Ijomone, and Omamuyovwi Meashack Ijomone*

11.1 INTRODUCTION TO NANOPARTICLES AND NANOTECHNOLOGY

Nanotechnology involves the application of nanosized particles to effectively deliver pharmacologically active compounds (drugs or genes) to diseased areas of the human body, including the nervous system (Jagaran and Singh, 2021; Naqvi et al., 2020). Nanomedicine technology has a strong potential to effectively target factors enabling the progression of neurodegenerative diseases (ND), and giving the needed urge to combat NDs, such as AD, PD, MS, HD, nanotechnology offers a promising headway for a better life for people living with ND globally (Jagaran and Singh, 2021). The available treatments for NDs are sometimes not efficient to adequately permeate the blood–brain barrier; nevertheless, the use of nanoparticles (NPs) helps enhance brain-targeted penetration to exert a neuroprotective effect (Jagaran and Singh, 2021; Sun, 2022). This is because nanotechnology, through nanomedicine (nanoparticles), facilitates BBB penetration (Ali and Chen, 2015) mediated by immunotherapeutic strategies for ND treatment via an induce immune response (Liu et al., 2018). Nanoparticles applied in nanomedicine are the future direction for the treatment of ND (Naqvi et al., 2020; Feigin et al., 2020). Nanomedicine therapeutic usefulness for nursing many diseases is attributed to their favorable structural characteristics, such as size and surface morphology (Maney and Singh, 2019).

Nanotechnology allows deliberate engineering of nanoparticles in order to control their structural variations that suit them for different function (Chen et al., 2018a, 2018b). NP transport vehicles of 1–100 nm in size have the potential to move through significant physiological barriers/membranes, such the BBB (Venkatas and Singh, 2021). Nanotechnology has resulted in the creation of various nanoformulations with neuroprotective properties, including nerve growth factors, curcumin, and edaravone, associated with their potential to penetrate or diffuse effectively through the brain parenchyma (Cenã and Játiva, 2018; Joseph et al., 2018). This makes them suitable in brain targeting (Tereanu et al., 2018). Examples of nanoparticles are magnetic nanoparticles, nanoprobes, radiolabeled nanoparticles, gold nanoparticle, and photosensitive nanomaterials, which are effectively efficient in treating ND; however, their toxicity is a rate-limiting factor (Viola et al., 2015). Low toxicity needs to be attained in order to attain high outcome and avoid neurotoxicity or neuron damage, even though its potential in the treatment of neurological and neurodegenerative disorders is limitless (Ngowi et al., 2021).

NDs pathology includes progressive loss of neurons, which results in motor and cognitive deficits (Silveira et al., 2021; Zhu et al., 2021). ND can be triggered by neurotoxic cascade reaction which culminates in abnormal energy metabolism, aberrant tau protein accumulation, marked loss of neurons and synaptic connection (Kumar et al., 2020). ND therapeutics require optimal pharmacokinetic transport and distribution of therapeutic agents to the brain, with minimal complications; however, there is a problem with brain-targeted effective transport of the drugs (Silveira et al., 2021) associated with neuroinflammatory responses, such as activation of microglial/axoglial, secretion of pro-inflammatory mediators (Zhao et al., 2020; Zhu et al., 2021; Hosseiny et al., 2020), and the blood–brain barrier (BBB) (Chiara et al., 2018).

One measure to overcome the obstacle for successful ND treatment is to break the BBB, possible through structural modification of the BBB permeability using carrier- or receptor-mediated nanoparticles, for instance, by means of ultrasound technique or radiation therapy and use of nanomaterials. These are some methods to help drug delivery across the BBB (Naqvi et al., 2020; Poovaiah et al., 2018).

The importance of brain-targeting drug-embedded NPs is that it aids pharmacokinetics and tissue distribution of the drugs when considered with freely circulating therapeutics in the brain and spinal cord; therefore, NPs act as carriers for brain-targeted docking of nanomedicines (Hanif et al., 2020). The poor docking or traffic of appropriate drug doses to the brain tissue in NDs is a major hurdle (Ngowi

* Corresponding author: jokememudu@gmail.com or adejoke.elizabeth@edouniversity.edu.ng

DOI: 10.1201/9781003383376-11

et al., 2021) improved using" NPs in delivering brain-targeting drugs to achieve an improved outcome in ND diseases (Singh et al., 2017; Jin et al., 2020).

NPs are not only for therapeutic purposes but also for diagnosing and studying progression of NDs (Derakhshankhah et al., 2020). For instance, NPs/nanocarriers are potent diagnostic measure in AD. One of the challenges in AD progression late detection is associated with delayed clinical expression when compared to the rapid and widespread onset of its cellular pathogenesis and manifestation, resulting in marked neuronal damage and its clinical manifestations seen in memory impairment and cognitive decline when detected (Derakhshankhah et al., 2020). Fluorescent NPs have been used to demonstrate the presence of AD biomarkers for neurodegeneration/neuroinflammation, such as beta-amyloid proteins and inflammatory cytokines (Sun and Benet, 2022; Sun, 2022). Nanomedicine application in treating AD is specifically to target and alter Aβ senile plaques (Derakhshankhah et al., 2020). A report demonstrated nanoparticle docking of formulated curcumin-loaded polysorbate 80–modified cerasome (CPC) NP in MPTP-induced PD mice using ultrasound-targeted microbubble destruction of BBB, with favorable outcome in drug stability, prolonged drug delivery circulation time, and an improved penetration of the curcumin nanoformulation, which culminates in an improved cognitive behavior in comparison to the bulk curcumin (Zhang et al., 2018a).

Given the potential of NPs in therapeutic applications to NDs, here we discuss NPs, their structures and types, characteristics, and brain-targeting mechanisms, as well as applications to specific NDs.

11.2 NANOPARTICLES: STRUCTURES, TYPES, AND INVOLVEMENT IN ND THERAPEUTICS

Nanoparticles (NPs) refer to minute particles of average size in the range of 1–100 nm (Matsuno et al., 2020; Silveira et al., 2021). NPs are designed from natural compound or engineered via artificial manipulation of compounds or alloy or minerals (Hydbring and Du, 2019; Katebi et al., 2019) for medicinal purposes (Zhang et al., 2016; Li et al., 2017) associated with their potential to pass through the BBB (Hosseiny et al., 2020). The sizes of NPs influence their process of brain-specific drug trafficking; small-sized NPs are effective and rapid in releasing drug because of their surface-area-to-volume ratio, while large NPs are effective in causing a slowed release of drug to target site due to the fast breakdown of their polymer (Zhu et al., 2021). We have different types of NPs made from natural (organic) and artificial (inorganic) manipulation of compounds, such as metal and metal oxide (Zhang et al., 2018b); Jagaran and Singh, 2021), carbon-based materials (Jagaran and Singh, 2021), to extracellular vesicles, liposomal, polymeric, red blood cell membrane, proteins, polysaccharides (He et al., 2019; Liu et al., 2020a), fullerenes, nanoemulsions, solid lipid (SL), polylactide-co-glycoside (PLGA), coffee, chitosan (CS), magnetic, rare earth (RE), poly(butyl cyanoacrylate) (PBCA), PLGA, and betulinic (Ngowi et al., 2021). Organic-derived NPs, such as lipids, nanoemulsions, and polymers, and inorganic, such as gold, titanium dioxide, iron oxide, and other metal, materials (Zhu et al., 2021; Asefy et al., 2021) have different physical and chemical properties (Dong et al., 2018; Rasouli and Tabrizian, 2019; Matsuno et al., 2020). For instance, lipid nanocarriers deliver a combination of immune-modulatory agents to brain tumor cells; for example, TME-facilitated iRGD-lipid nanoparticles aid in the trafficking of CAR T cells into glioma cell lines and mouse models (Zhang et al., 2018c) via increased BBB permeability targeted to glioblastoma in mouse models (Shi et al., 2018). The design of NPs makes them carriers of therapeutics directed for brain targeting (Maney and Singh, 2019; Hosseiny et al., 2020). For instance, to target and disrupt the BBB, the surface of NPs have been designed with different kinds of biomolecules, like antibodies and peptides, designed to identify specific receptors and transporters on the BBB, as well as specific types of diseased cells such as brain tumors, and cells or cellular organelles like mitochondria, involved in the progression of oxidative stress–mediated tissue damage (Zhang et al., 2018b). According to a study by Joseph et al., (2018), curcumin-loaded PLGA-PEG nanoparticles mediate neuronal proliferation and differentiation in brain injury compared to bulk curcumin. This reversal of neurodegeneration is attributed to their small size, penetration of the BBB, and brain-focused delivery, which stabilize and regulate the release of drugs into the brain parenchyma, leading to attenuation of mediators of neuroinflammation and glial cells activation (Naqvi et al., 2020; Joseph et al., 2018), resulting in the restoration of neuronal integrity.

In NDs characterized by marked disruption of the BBB leading to uncontrolled movement of molecules, and consequently to progressive brain damage (Jagaran and Singh, 2021), in addition to challenges associated with affective and efficient brain-targeted drug delivery, NPs have demonstrated significance in the treatment of brain disorders based on their characteristic potentials as nano-drug delivery agents that target neurodegenerative processes in NDs such as Alzheimer's, Parkinson's, etc. (Borsche et al., 2020). The remodeling of NPs with polymers and ligands improves its binding affinities to the conjugated drug or gene (Aderibigbe, 2017). The noble metals gold (Au), silver (Ag), platinum (Pt), and palladium (Pd) have been commonly employed due to their favorable physiochemical, biological, and optical properties (Yaqoob et al., 2021). The physical and chemical characteristics of AuNPs can be modified in clinical application (Oladimeji et al., 2020) to treat NDs. This chapter explores the application and mechanism through which nanoparticles are applied in nanomedicine as a therapeutic strategy in neurodegenerative disorders associated with BBB disruption and brain targeting.

11.3 NANOPARTICLES CHARACTERISTIC AS CARRIERS FOR BRAIN TARGETING IN NDs

The ability of nanoparticles (NPs) to effectively transport drugs and other substances through the BBB has led to their identification as a potent therapy for neurodegenerative diseases. This is because they are soluble, biodegradable, and less toxic and can target specific regions (Rabanel et al., 2020) in the brain parenchyma. All NPs have unique optical, antibacterial, and magnetic properties (Sajid and Płotka-Wasylka, 2020). Due to their nanosize and compactness (Hasan, 2015), they exhibit properties such as stability, strength, as well as increased surface area for reactions; hence, their various applications are determined by their distinctive properties (Ealia and Saravanakumar, 2017). NPs are further categorized as soft (e.g., nanodroplets, liposomes) or hard (e.g., fullerenes) (Agashe, 2021). A characteristic feature of neurodegenerative diseases (NDs) is the progressive loss of structure and functional abilities of the brain's neurons (Santa-Cecília et al., 2019), which could result in cognitive, motor, and behavioral deficits. Nanotechnology, which utilizes engineered devices, represents an innovative and promising therapeutic approach for overcoming obstacles faced in the treatment of various NDs, by enhancing targeted drug docking and bioavailability in the brain (Fernandes et al., 2010; Teixeira et al., 2020). NPs include liposomes, micelles, polymer NPs, and inorganic NPs (Srikanth and Kessler, 2012), which carry and deliver therapeutics or imaging probes to target sites (Li et al., 2021). Earlier, the diagnosis and treatment of CNS disorders have been far from impressive, owing to the BBB's restriction on drug delivery to the brain, with an estimated 98% of small-molecule therapeutic substances failing to reach the brain (Pardridge, 2007). To overcome this, various strategies have been developed using NPs as carriers. For example, NP-based Alzheimer's disease (AD) treatment has been reported by using zwitterionic poly(carboxybetaine)–based NPs co-loaded with fingolimod, siSTAT3, and zinc oxide, henceforth named MCPZFS NPs, to improve BBB penetration and enhance brain microglia targeting. Mannose analog (4-aminophenyl α-d-mannopyranoside) was added to the NPs, resulting in improved priming of the microglial, thereby functioning to attenuate the rate of synthesis and release of pro-inflammatory mediators (such as interleukin 1-β (IL-1β), IFN-γ, IL-6, and interleukin 17A) and brain-derived neurotrophic factor in AD disease model (Liu et al., 2020a). One significant reaction in NP-based AD treatment using MCPZFS NPs is that it also helps increase Aβ transport into microglia, which enables Aβ phagocytosis and breakdown (Liu et al., 2020c). This implies that MCPZFS NP administration to AD transgenic mice results in a decline in Aβ formation, neuroinflammation, neuron damage, and loss of memory (Albert et al., 2019). Nanotechnology using the graphene quantum dots (GQDs) has the potential to inhibit α-synuclein pathogenesis and amyloidosis (Dehay et al., 2015; Aderibigbe et al., 2019), which is attributed to the BBB permeability of GQDs because of their minute size (~5–20 nm) and antibody-altered surfaces. Kim et al. (2018) reported that GQD administration in mice attenuates motor deficits, protects against loss of dopaminergic neurons and synaptic connection, inhibits neuron–neuron flow of α-synuclein, and decreases Lewy body formation in PD animal model, hence suggesting that GQDs interact with α-synuclein fibrils, preventing their development and ensuring they are degraded (Hanif et al., 2020).

11.4 MECHANISMS THROUGH WHICH NANOPARTICLES TARGET THE BRAIN

NPs can be trafficked into the brain through the oral, intranasal, or parenteral routes (Tong et al., 2020; Yang et al., 2020). Owing to their small size, they easily move across the biological barriers like the BBB to enter various sensitive organs (Bahadar et al., 2016). Recent approaches aimed at circumventing this challenge include temporary disruption of the BBB and conjugation of brain-permeable ligands to any drug of study through either intranasal delivery or direct delivery of molecules into the brain using invasive methods (Rawal et al., 2022; S Hersh et al., 2016). These methods can be classified into chemical, physical, and biological methods. The physical methods focus on paracellular transport, enhancing the transport of nanoparticles by utilizing ultrasound and magnetic fields to momentarily open the tight junctions. The chemical method involves functionalizing the targeted ligands or receptors of the NPs to increase their local concentrations in the brain endothelium, which may encourage transcytosis, while the biological method facilitates the transport of NPs across the BBB or their deposition on the brain endothelium by the use of cells (Nowak et al., 2020). The brain's endothelium cells (ECs) are connected via tight complexes and reinforced/supported by glia (e.g., astrocytes, microglia), neurons, and pericytes, which collectively form the neurovascular unit. This unit is the first hurdle that foreign substances encounter when they try to move across into the brain (Gürsoy-Özdemir and Tas, 2017). On entering the brain, NPs impede neuronal oxidative stress–mediated neurodegeneration while also destroying inflammatory molecules without affecting healthy cells (Asefy et al., 2021; Chiang et al., 2021). The nanodelivery system consists of BBB receptor ligands, peptides, as well as coupling antibodies (Gao et al., 2017; Guo et al., 2018) that enhance brain targeting with NPs, thereby facilitating the docking of therapeutic compounds into the brain tissue (Joseph et al., 2018).

A more direct approach aimed at delivering molecules like drugs to the brain includes invasive methods, such as direct intraventricular or intracerebral injections and implants (Cook et al., 2009). Intranasal delivery, another delivery route into the brain, via the olfactory region of the nasal cavity, provides direct access to the CNS via a neuro-epithelial layer (Charlton et al., 2007). Adsorptive transcytosis is an alternative, less-specific method for crossing the BBB. It generally involves the particle or molecule adhering to the cell surface (usually through an electrostatic interaction of a cationic molecule with the anionic cell surface), which then creates a vesicle to be internalized and transported through the cell (Hervé et al., 2008). Many receptors, such as the insulin receptor, nicotinic acetylcholine receptor, low-density lipoprotein receptor-related protein, transferrin (Tf) receptor, and insulin-like growth factor receptor, are overexpressed on the BBB (Guo et al., 2012; Zhang et al., 2016); due to the specificity of the interaction between receptors and ligands, the receptor-mediated transport has been the most commonly used strategy to deliver NPs to the brain through the BBB. Hu et al. (2011) used Lactoferin-NPs to deliver urocortin (a peptide medication) to the brain for PD treatment. Another technique utilizing adenosine receptors (AR) in the treatment of various CNS disorders involves modifying the tightness of the tight junctions in the brain's ECs to enhance BBB permeability; A1 and A2A, which are subtypes of AR, are shown to elevate this BBB permeability in vivo (Carman et al., 2011).

11.5 APPLICATION OR RELEVANCE OF NANOPARTICLES IN SELECTED NEURODEGENERATIVE DISEASE THERAPEUTICS

The characteristics of nanoparticles in effectively transporting agents or drugs through the blood–brain barrier provide brain-targeted therapies to specific brain regions affected by NDs.

Here we will discuss selected NDs and reports leveraging nanotechnologies for therapeutic interventions in NDs.

11.5.1 Alzheimer's Disease (AD)

AD is a brain disorder associated with ageing-linked progressive degeneration of neurons characterized by dementia and other neuropathological features. A 2016 report indicates that at least one person develops the disease every 66 seconds in the USA, with a predicted rise of up to 131 million people by 2025 (Alzheimer's Association, 2016; Prince et al., 2016). Tau accumulation is a major pathological feature associated with AD progression, as well as implicated in other NDs (Nam et al., 2020; Tagai et al., 2021). Aggregation of amyloid beta peptide is another major hypothesis in the pathogenesis of the disease, and inhibition of this peptide is of importance in the treatment and management of this disease (Aleksis et al., 2017). Dysregulation in the homeostasis of Aβ is a main characteristic of AD, and many pharmaceutical companies are trying to develop many drugs targeting Aβ, though most failed at clinical III trial (Selkoe and Hardy, 2016). The role of nanoparticles in the prevention of the aggregation of the peptides involved in the pathogenesis of AD ranges from the ability of the nanoparticles to deliver the drugs directly to the brain cells involved (Abbas, 2021). Also, as reviewed by Manel et al., the use of chitosan-based NPs has not only proved to be effective in helping delivering treatment that can pass through the BBB but can also prevent the aggregation of Aβ fibrils (Manek et al., 2020). A previous study recently shows that protein-capped cadmium sulfide and iron oxide NPs can effectively limit the polymerization and fibrillization of tau proteins (Sonawane et al., 2019). Solid lipid nanoparticles loaded with nicotinamide, an HDAC inhibitor, have been reported to significantly lower cognitive deficits linked with AD by lowering tau phosphorylation in the rat model (Vakilinezhad et al., 2018). PLGA NPs loaded with vitamin D–binding proteins were also reported to attenuate cognitive impairments in 5XFAD mice via inhibition of Aβ binding and accumulation (Jeon et al., 2019). Another study also reported that gold NPs are cytoprotective in AD rat model by promoting anti-inflammatory responses and boosting antioxidant status (De Simone et al., 2019; Derakhshankhah et al., 2020). Furthermore, gold NPs have also been shown to reduce Aβ aggregation and fibrillization as well as consequent neurotoxicity in AD experimental model (Liao et al., 2012; Moore et al., 2017; Gao et al., 2017). Molybdenum sulfide NP has also been reported to be a multifunctional therapeutic agent in amyloid-related disease, as it was reported to prevent Ca channel formation caused by Aβ fibrils (Han et al., 2017). Likewise, flavonoid-based NPs have been used to target the treatment of many neurological condition with multiple pathways (Prasanna and Upadhyay, 2021; Ul Hassan et al., 2022). Naringenin-loaded (a biflavonoid) nanoemulsion shows better protection than free naringenin against neuroblastoma cell Aβ neurotoxicity (Gaba et al., 2019) by reducing tau hyperphosphorylation and regulating amyloidogenesis (Md et al., 2018). The use of nanocleaner, which consists of the combination of rapamycin-loaded reactive oxygen–responsive PLGA core with Aβ-capturing peptide (KLVFF) and acid-cleavable DAG peptide [R@(ox-PLGA)-KcD], shows anti-AD therapeutic effect by not only reducing the deposition of Aβ but also alleviating oxidative stress damage and serving as anti-inflammatory (Lei et al., 2021). Additionally, a recent study has also developed a combination of ibuprofen-peptide NP (Ibu&FK@RNPs) consisting of polycaprolactone terminus-modified PEG with RAP peptide, tacrolimus (FK506), and ibuprofen that shows almost the same synaptic density and Morris water maze result in AD mice when compared with the control group. Their result also shows that activation of astrocytes and aggregation of Aβ were significantly lowered in the treated group than the untreated group, with additional significant changes in the Ibu&FK@RNPs than the non-RAP peptide–modified therapy (Ibu&FK@NPs) (He et al., 2022). Collectively, the aforementioned data signify that NP peptides can be used to deliver drugs targeting peptides dysregulated in AD, such as Aβ, more efficiently and effectively.

11.5.2 Prion Diseases

Prion diseases are complex, rare neurological disorders characterized by the abnormal folding or aggregation of prion proteins, transforming the cellular prion protein PrP^c to abnormal infectious isoforms rich in β-sheet PrP^{sc} with no specific treatment yet (Stöhr et al., 2008). These groups of neurological disorders are peculiar in that they can be sporadic, acquired, and/or genetically transferable. Due to the different phenotypic presentation and the way the disease mimics so many other neurological disease, including dementia, prion disease is difficult to diagnose (Baldwin and Correll, 2019). Criteria for diagnosis include progressive dementia, rapid cognitive decline, psychiatric signs, and sleep disorders (Krasnianski et al., 2014). It is a common infectious type of amyloidosis (Chesebro, 2003), the mutation in some of the prion protein genes known to cause cerebral amyloidosis that present the presence of both amyloid and tau aggregation (Ghetti et al., 2018); therefore, nanoparticles targeted at the inhibition of tau can be of great therapeutic advantage. Application of NPs in the diagnosis and delivery of drugs in prion diseases has a great impact as they have the ability to cross the BBB and have also proved to be less toxic, with the advantage of reducing oxidative damage caused by stress (Murugesan et al., 2020). Au-PEG has been reported to reduce total tau level acting in a prion fashion and has shown to improve learning ability in the animal model (Vimal et al., 2020). PEGylated nanoparticles have higher tendencies of crossing the BBB than non-PEGylated nanoparticles as they have longer circulation time in the blood and interact with cells longer whether in healthy animal or experimental scrapie-infected animal (Calvo et al., 2001a; Calvo et al., 2001b). The use of protein-capped nanoparticles has the ability to disaggregate and inhibit tau (Sonawane et al., 2019). A recent in vivo and in vitro study also shows improvement in memory impairment, suppression of neuronal cell death through inhibition of p-tau when curcumin was loaded into red blood cell membrane–coated PLGA particles (Gao et al., 2020). Zhan and his colleagues used two combined methods to study the role of carbon NPs on aggregation of PrP, and their results show that carbon NPs inhibit fibril formation through the reduction of the interpeptide interaction, therefore inhibiting the formation of β sheets. They further show that graphene has more inhibitory effect on PrP127–147 than the single-walled carbon nanotube (Zhou et al., 2018). Also, fluorinated NPs are potential inhibitors of prion proteins as they have the capacity to delay the conversion from α structure to the β structure protein (Dabur et al., 2022). Another nanoemulsion substance (nano-pomegranate seed oil [Nano-PSO]) that is rich in natural antioxidants has shown great potential in targeting mitochondrial dysfunction (Keller et al., 2019) and causing delay in the onset of prion disease; therefore, nano-PSO can be recommended in the prevention and treatment of human genetic type of prion disease (Binyamin et al., 2017; Mizrahi et al., 2014).

11.5.3 Parkinson's Disease

Parkinson's disease (PD) is a common neurodegenerative disorder with motor features, like muscular rigidity, resting tremors, and bradykinesia, and non-motor features, like cognitive, neuropsychiatric, and autonomic changes that result from the death of dopaminergic neurons in the *substantia nigra pars compacta* (SNpc), leading to a decline of dopamine in the basal nuclei (Kalia and Lang, 2015; Silva et al., 2021). Current treatment options are majorly directed at increasing dopamine levels, stimulating its receptors in the brain, mimicking its activities, and decreasing its metabolism (Baskin et al., 2021; Silva et al., 2021). The most common treatment option is levodopa (L-Dopa), a dopamine precursor that is used to control bradykinesia and improve the quality of life of patients with PD by increasing the levels of dopamine in the brain (Gandhi and Saadabadi, 2015). Despite the huge advantages of the drugs in treating PD, there is still the challenge of a limited pharmacokinetic profile and bioavailability as well as poor BBB penetration (Silva et al., 2021). Hence, there is a focus on the use of nanotechnology, which has been suggested as a strategy to enhance drug penetration across the BBB since the mid-1990s (Alonso and Couvreur, 2012). The capacity of nanocarriers to protect dopamine from being metabolized in circulation, deliver dopamine across the BBB, and promote a sustained release of dopamine has been tested both in vitro and in vivo. For instance, Pahuja et al. (2015) reported that poly(lactic-co-glycolic acid) (PLGA) nanoparticles loaded with dopamine were able to cross the BBB and were internalized by dopaminergic neurons of the substantia nigra and striatum in a rat model of PD after an intravenous administration. In this study, the treatment with dopamine-coated nanoparticle showed no adverse effect on the dopaminergic circuitry and was able to ameliorate functional deficits in the rats. Other commonly used categories of nanocarriers in PD include chitosan (Ray et al., 2018), iron oxide (magnetite, hematite, and maghemite) (Xie et al., 2018), hydrogels (Ren et al., 2017), polyvinylpyrrolidone-poly(acrylic acid) (E.R. et al., 2014), carbon nanotubes (Guo et al., 2017), carbon dots (Khan et al., 2015), etc. With the progress made in this field, some of these nanocarriers still face issues, like inadequate encapsulation and, sometimes, low drug bioavailability and distribution (García-Pardo et al., 2021). In clinical trial advances, the use of lipid nanoparticles has been most favored, and this may be due to their lipophilic, biocompatible, and biodegradable properties (Jagaran and Singh, 2022), and their physicochemical parameters make them suitable candidates as nanocarriers (Chakraborty et al., 2017). Despite the huge potential benefits of lipid nanoparticles as carriers for drug delivery, there is still a major dearth of studies exploiting it in clinical trials for the treatment of PD (Jagaran and Singh, 2022).

11.5.4 Multiple Sclerosis

Multiple sclerosis (MS) is a complex and progressive autoimmune condition characterized by chronic inflammation, axonal demyelination, and neurodegeneration (Damavandi et al., 2022; Ojha and Kumar, 2018). In this condition, there is a consequent impairment of signal transmission that produces several sensory and motor deficits, including weakness and numbness in the limb, visual impairment, etc. (Cunha et al., 2021; Zeng et al., 2020). Magnetic resonance imaging (MRI) is used for early diagnosis and monitoring of MS, which can be combined with an evaluation of cerebrospinal fluid oligoclonal bands as biomarkers based on recent diagnostic criteria (Damavandi et al., 2022; Mahmoudi et al., 2011; Zhang et al., 2021). Several treatment approaches have been used for MS, including the use of substances to control the entry of T cells into the CNS or drugs that restrict the development of lymphocytes, but they are limited by varying degrees of effectiveness or safety concerns, hence the recent interest in stem cell therapy and nanotechnology for the management of the condition (Damavandi et al., 2022). Nanoparticles made up of natural polymers, metals, and synthetic materials can be used to optimize drug stability, bioavailability, and delivery across the BBB and enhance the routes of drug administration (Nuzzo and Picone, 2021). These nanoparticle carriers have been administered intravenously, subcutaneously, and orally to deliver therapeutic agents for the treatment of MS. For example, in an animal model of MS, polymeric lactic-glycolic acid (PLGA) nanoparticles were administered subcutaneously to deliver myelin oligodendrocyte glycoprotein (MOG)35–55 and recombinant interleukin (IL)10 as a treatment strategy for MS, and the result showed beneficial outcomes in the CNS (Cappellano et al., 2014). Also, oral administration of dimethyl fumarate (DMF) lipid nanoparticles was used to treat cuprizone-induced demyelination in mice, and the DMF nanoparticle was effective in enhancing the repair of myelin sheaths (Kumar et al., 2018). Furthermore, both intravenous and subcutaneous administrations of nanoparticles loaded with peptide antigen and rapamycin were used therapeutically in the treatment of experimental autoimmune encephalomyelitis (EAE) in mice, and this delivery plan was sufficient to improve the disease condition (Maldonado et al., 2015). These studies, as well as many others that include the works of Gandomi et al. (2017) and Rittchen et al. (2015), have highlighted the huge potential of nanoparticles as drug carriers in treating MS.

11.5.5 Amyotrophic Lateral Sclerosis

In *amyotrophic lateral sclerosis (ALS)*, there is progressive damage to upper and lower motor neurons, leading to paralysis and atrophy of muscles and consequent death of patients from respiratory issues (Niccolini et al., 2022). The etiology of ALS is said to be complex, as it involves an interplay between genetic factors that implicate more than 20 genes, pathological events occurring in ageing, and environmental factors, such as lifestyle (smoking and physical activity) and exposure to injurious chemicals, metals, viruses, etc. (Masrori and Van Damme, 2020). While the pathophysiology of ALS is still not clear, some of the identified mechanisms of neuronal damage include mitochondria and oligodendrocyte dysfunction, intracellular transport deficits, oxidative stress, inflammation, nucleic acid dysfunction, proteostasis, etc. (Chia et al., 2017; Masrori and Van Damme, 2020). Also, only a few treatment options for ALS are available, including edaravone, which scavenges for free radicals, and riluzole, which inhibits glutamate excitotoxicity, both of which only mildly improve the patient's quality of life (Chia et al., 2017; Vimal et al., 2022). The efforts geared toward increasing the efficacy and spectrum of therapeutic agents against ALS involve the application of nanoparticles to package and deliver these drugs (Nabizadeh, 2022). Riluzole has been packaged in lipid nanoparticles for targeted delivery to the CNS of rats with a higher level of efficacy than free riluzole (Bondì et al., 2010). A strategy used by Yang et al. (2018) to test a putative treatment option for ALS in vitro was to use liposomes to co-deliver riluzole and verapamil, and they demonstrated that this cocktail has potential benefit for ALS therapy by reducing the efflux function of P-gp. Also, mesoporous silica nanoparticles, carbon nanotubes, poly(lactic acid)-poly(ethylene glycol) (PLA-PEG) nanoparticles, and chitosan have been evaluated for their potential for delivery of ALS drug or other symptomatic treatment drugs (Chigumbu et al., 2012; Díaz-García et al., 2022; Medina et al., 2020; Shanmukhapuvvada and Vankayalapati, 2012). Put together, these studies offer a promising approach for nanoparticles as carriers for drug delivery to treat ALS. Furthermore, since miRNA inhibition and replacement therapy represent a potential treatment approach for ALS, miRNA-based therapeutic agents must be able to pass through the BBB and require protection from degradation in the bloodstream and excretion by the renal system; hence, nanoparticles are suitable non-viral delivery options for exogenous nucleic acids (Niccolini et al., 2022). The nanoparticle carriers for miRNA delivery to cells can be lipid-based, polymer-based (polyethylenimines, PLGA, chitosan, dendrimers), or inorganic nanoparticles (gold, magnetic, and mesoporous silica) (Bai et al., 2019).

11.5.6 Huntington's Disease

Huntington's disease (HD) is an autosomal-dominant ND typified by chorea and dementia. HD results from the pathogenic expansion of the CAG trinucleotide repeat within the huntingtin gene (HTT) on chromosome 4 (Bates et al., 2015). The mutated HTT gene, mHTT, causes a misfolded protein (Huntington, htt), which, upon its accumulation in neural cells, results in gradual neuronal dysfunction and degeneration (Essa et al., 2019). Overexpression of mutant HTT triggers excessive oxidative stress and epidemic change, which disrupts cellular components, thus triggering apoptosis (Cong et al., 2019). Recent studies using *Caenorhabditis elegans* have shown that nano-selenium (Nano-Se) in vivo plays an important role in mitigating aggregation of huntingtin protein and neural degeneration, alongside scavenging for and clearing radicals. Nano-Se may therefore be a viable diet-based therapy for human HD (Cong et al., 2019). The intranasal administration of chitosan-based NPs primed with anti-HTT small interfering RNA (siRNA) to the brain is another promising therapeutic technique for the effective reduction of mutant HTT expression, as it has been shown to slow the progression of Huntington's disease. Chitosan encapsulation of siRNA functions by protecting payloads (in this case, anti-HTT) from degradation as it is being delivered to the target region (Sava et al., 2020). The use of curcumin-encapsulated solid lipid nanoparticles (C-SLNs) is a therapeutic strategy aimed at improving mitochondrial dysfunctions in 3-nitropropionc acid (3-NP)–induced HD. C-SLNs function by increasing the activity and cytochrome levels of the brain's mitochondrial complexes, thereby mitigating swelling of mitochondria, protein carbonyls, lipid peroxidation, and generation of reactive oxygen species in HD animals (Sandhir et al., 2014). The neuroprotective impacts of SLNs were further demonstrated on nasal administration of rosmarinic acid in rats, as they were shown to attenuate 3-NP-induced deficits in locomotor, and motor coordination, as well as striatal oxidative stress associated with HD animals (Bhatt et al., 2015). Another study using unique polymeric NPs made of polylactic-co-glycolic acid (PLGA) and cholesterol demonstrated that the NPs rapidly targeted neuronal cells on entering the brain. The study confirmed that systemic and recurrent injections of the hybrid-g7-Nps-chol improved endogenous cholesterol production, delayed cognitive regression, and improved motor deficits in HD animals without causing any inflammatory response, hence highlighting the advantages and safety of cholesterol delivery using advanced brain-permeable NPs for HD treatment (Birolini et al., 2021).

11.6 CONCLUDING REMARKS

The blood–brain barrier (BBB) represents a significant challenge in drug delivery to the brain in the treatment of neurodegenerative disorders, such as Alzheimer's disease, Parkinson's disease, etc. Nanoparticles are effective potential carriers for brain targeting, in the treatment of neurodegenerative disorders characterized by the progressive loss of neurons and cognitive function, due to their small size and surface modifications to permeate the BBB for targeted delivery of a diverse array of neuroprotective, anti-inflammatory, and regenerative therapies. Additionally, nanoparticles can protect fragile therapeutic molecules from degradation, hence optimizing their bioavailability and therapeutic potential while facilitating localized drug release. Nanoparticles have been engineered to carry imaging agents, enabling non-invasive monitoring of disease progression and treatment response. In conclusion, nanoparticles offer a promising approach for brain targeting treatment of neurodegenerative disorder because of their ability to bypass the BBB and deliver a wide range of localized therapeutics.

REFERENCES

Abbas, M. (2021). Potential role of nanoparticles in treating the accumulation of amyloid-beta peptide in Alzheimer's patients. *Polymers* 13. doi: 10.3390/polym13071051

Aderibigbe, B., Naki, T., Owonubi, S. (2019). Graphene-based materials for brain targeting. In: Handbook of graphene: Biomaterials. Scrivener Publishing LLC, p. 225.

Aderibigbe, B.A. (2017). Metal-based nanoparticles for the treatment of infectious diseases. *Molecules* 22:1370. doi: 10.3390/molecules22081370

Agashe, J. (2021). Advanced research in material and chemical science (Vol. I). ISBN: 978-81-951982-9-0.

Albert, M., Mairet-Coello, G., Danis, C., Lieger, S., Caillierez, R., Carrier, S., Skrobala, E., Landrieu, I., Michel, A., Schmitt, M. (2019). Prevention of tau seeding and propagation by immunotherapy with a central tau epitope antibody. *Brain* 142:1736–1750.

Aleksis, R., Oleskovs, F., Jaudzems, K., Pahnke, J., Biverstål, H. (2017). Structural studies of amyloid-β peptides: Unlocking the mechanism of aggregation and the associated toxicity. *Biochimie* 140:176–192.

Ali, I.U., Chen, X. (2015). Penetrating the blood–brain barrier: Promise of novel nanoplatforms and delivery vehicles. *ACS Nano* 9:9470–9474.

Alonso, M.J., Couvreur, P. (2012). Historical view of the design and development of nanocarriers for overcoming biological barriers. *Nanostructured Biomaterials for Overcoming Biological Barriers* 3–36. doi: 10.1039/9781849735292-00003

Alzheimer's Association. (2016). Alzheimer's disease facts and figures. *Alzheimer's Dement* 12:459–509. doi: 10.1016/j.jalz.2016.03.001

Asefy, Z., Hoseinnejhad, S., Ceferov, Z. (2021). Nanoparticles approaches in neurodegenerative diseases diagnosis and treatment. *Neurological Sciences* 42(7):2653–2660.

Bahadar, H., Maqbool, F., Niaz, K., Abdollahi, M. (2016). Toxicity of nanoparticles and an overview of current experimental models. *Iranian Biomedical Journal* 20:1.

Bai, Z., Wei, J., Yu, C., Han, X., Qin, X., Zhang, C., Liao, W., Li, L., Huang, W. (2019). Non-viral nanocarriers for intracellular delivery of microRNA therapeutics. *Journal of Materials Chemistry B* 7(8):1209–1225. doi: 10.1039/c8tb02946f

Baldwin, K.J., Correll, C.M. (2019). Prion disease. *Seminars in Neurology* 39:428–439. doi: 10.1055/s-0039-1687841

Baskin, J., Jeon, J.E., Lewis, S.J.G. (2021). Nanoparticles for drug delivery in Parkinson's disease. *Journal of Neurology* 268(5):1981–1994. doi: 10.1007/s00415-020-10291-x

Bates, G.P., Dorsey, R., Gusella, J.F., Hayden, M.R., Kay, C., Leavitt, B.R., Nance, M., Ross, C.A., Scahill, R.I., Wetzel, R. (2015). Huntington disease (primer). *Nature Reviews: Disease Primers* 1, 15005. https://doi.org/10.1038/nrdp.2015.5.

Bhatt, R., Singh, D., Prakash, A., Mishra, N. (2015). Development, characterization and nasal delivery of rosmarinic acid-loaded solid lipid nanoparticles for the effective management of Huntington's disease. *Drug Delivery* 22:931–939.

Binyamin, O., Keller, G., Frid, K., Larush, L., Magdassi, S., Gabizon, R. (2017). Continues administration of nano-PSO significantly increased survival of genetic CJD mice. *Neurobiology of Disease* 108:140–147.

Birolini, G., Valenza, M., Ottonelli, I., Passoni, A., Favagrossa, M., Duskey, J.T., Bombaci, M., Vandelli, M.A., Colombo, L., Bagnati, R. (2021). Insights into kinetics, release, and behavioral effects of brain-targeted hybrid nanoparticles for cholesterol delivery in Huntington's disease. *Journal of Controlled Release* 330:587–598.

Bondì, M.L., Craparo, E.F., Giammona, G., Drago, F. (2010). Brain-targeted solid lipid nanoparticles containing riluzole: Preparation, characterization and biodistribution. *Nanomedicine* 5(1):25–32. doi: 10.2217/nnm.09.67

Borsche, M., König, I. R., Delcambre, S., Petrucci, S., Balck, A., Brüggemann, N., Zimprich, A., Wasner, K., Pereira, S. L., Avenali, M., Deuschle, C., Badanjak, K., Ghelfi, J., Gasser, T., Kasten, M., Rosenstiel, P., Lohmann, K., Brockmann, K., Valente, E. M., Youle, R. J., . . . Klein, C. (2020) Mitochondrial damage-associated inflammation highlights biomarkers in PRKN/PINK1 parkinsonism. *Brain* 143:3041–3051. doi: 10.1093/brain/awaa246

Calvo, P., Gouritin, B., Brigger, I., Lasmezas, C., Deslys, J.P., Williams, A., Andreux, J.P., Dormont, D., Couvreur, P. (2001a). PEGylated polycyanoacrylate nanoparticles as vector for drug delivery in prion diseases. *Journal of Neuroscience Methods* 111:151–155.

Calvo, P., Gouritin, B., Chacun, H., Desmaële, D., D'Angelo, J., Noel, J.P., Georgin, D., Fattal, E., Andreux, J.P., Couvreur, P. (2001b). Long-circulating PEGylated polycyanoacrylate nanoparticles as new drug carrier for brain delivery. *Pharmaceutical Research* 18:1157–1166.

Cappellano, G., Woldetsadik, A.D., Orilieri, E., Shivakumar, Y., Rizzi, M., Carniato, F., Gigliotti, C.L., Boggio, E., Clemente, N., Comi, C., Dianzani, C., Boldorini, R., Chiocchetti, A., Renò, F., Dianzani, U. (2014). Subcutaneous inverse vaccination with PLGA particles loaded with a MOG peptide and IL-10 decreases the severity of experimental autoimmune encephalomyelitis. *Vaccine* 32(43):5681–5689. doi: 10.1016/j.vaccine.2014.08.016

Carman, A.J., Mills, J.H., Krenz, A., Kim, D.G., Bynoe, M.S. (2011). Adenosine receptor signaling modulates permeability of the blood–brain barrier. *Journal of Neuroscience* 31:13272–13280.

Cenã, V., Játiva, P. (2018). Nanoparticle crossing of blood–brain barrier: A road to new therapeutic approaches to central nervous system diseases. *Nanomedicine* 13:1513–1516. doi: 10.2217/nnm-2018-0139

Chakraborty, S., Dhakshinamurthy, G.S., Misra, S.K. (2017). Tailoring of physicochemical properties of nanocarriers for effective anti-cancer applications. *Journal of Biomedical Materials Research - Part A* 105(10):2906–2928. doi: 10.1002/jbm.a.36141

Charlton, S., Jones, N., Davis, S., Illum, L. (2007). Distribution and clearance of bioadhesive formulations from the olfactory region in man: Effect of polymer type and nasal delivery device. *European Journal of Pharmaceutical Sciences* 30:295–302.

Chatterjee, K., Roy, A., Banerjee, R., Choudhury, S., Mondal, B., Halder, S., Basu, P., Shubham, S., Dey, S., & Kumar, H. (2020). Inflammasome and α-synuclein in Parkinson's disease: A cross-sectional study. *Journal of Neuroimmunology* 338:577089. doi: 10.1016/j.jneuroim.2019.577089

Chen, W., Ouyang, J., Yi, X., Xu, Y., Niu, C., Zhang, W., Wang, L., Sheng, J., Deng, L., Liu, Y. N., & Guo, S. (2018a). Biodegradable PEG-poly(ω-pentadecalactone-co-p-dioxanone) nanoparticles for enhanced and sustained drug delivery to treat brain tumors. *Biomaterials* 178:193–203. doi: 10.1016/j.biomaterials.2018.06.024

Chen, W., Ouyang, J., Yi, X., Xu, Y., Niu, C., Zhang, W., Wang, L., Sheng, J., Deng, L., Liu, Y.N., et al. (2018b). Black phosphorus nanosheets as a neuroprotective nanomedicine for neurodegenerative disorder therapy. *Advanced Materials* 30:1703458. doi: 10.1002/adma.201703458

Chen, S., Owens, G.C., Makarenkova, H., Edelman, D.B. (2010). HDAC6 regulates mitochondrial transport in hippocampal neurons. *PLoS OnE* 5:e10848. doi: 10.1371/journal.pone.0010848

Chesebro, B. (2003). Introduction to the transmissible spongiform encephalopathies or prion diseases. *British Medical Bulletin* 66:1–20.

Chia, R., Chiò, A., Traynor, B.J. (2017). Novel genes associated with amyotrophic lateral sclerosis: Diagnostic and clinical implications. *Lancet Neurology* 17(1):94–102. http://dx.doi.org/10.1016/S1474-4422(17)30401-5%0Apapers3://publication/doi/10.1016/S1474-4422(17)30401-5

Chiang, M.C., Nicol, C.J.B., Lin, C.H., Chen, S.J., Yen, C., Huang, R.N. (2021). Nanogold induces anti-inflammation against oxidative stress induced in human neural stem cells exposed to amyloid-beta peptide. *Neurochemistry International* 145:104992.

Chiara, T., Origlia, N., Mattu, C., Accorroni, A., Chiono, V. (2018). Current limitations in the treatment of Parkinson's and Alzheimer's diseases: State-of-the-art and future perspective of polymeric carriers. *Current Medicinal Chemistry* 25:5755–5771.

Chigumbu, N., Iyuke, S., Pillay, V., Ndlovu, S. (2012). In vitro evaluation of the physicochemical effects of drug loaded carbon nanotubes on toxicity. *Journal of Nanomedicine and Nanotechnology* 3(4). doi: 10.4172/2157-7439.1000135

Cong, W., Bai, R., Li, Y.F., Wang, L., Chen, C. (2019). Selenium nanoparticles as an efficient nanomedicine for the therapy of Huntington's disease. *ACS Applied Materials & Interfaces* 11:34725–34735.

Cook, A.M., Mieure, K.D., Owen, R.D., Pesaturo, A.B., Hatton, J. (2009). Intracerebroventricular administration of drugs. *Pharmacotherapy: The Journal of Human Pharmacology and Drug Therapy* 29:832–845.

Cunha, A., Gaubert, A., Latxague, L., Dehay, B. (2021). PLGA-based nanoparticles for neuroprotective drug delivery in neurodegenerative diseases. *Pharmaceutics* 13(7). doi: 10.3390/pharmaceutics13071042

Dabur, M., Loureiro, J.A., Pereira, M.C. (2022). The current state of amyloidosis therapeutics and the potential role of fluorine in their treatment. *Biochimie* 202:123–135. doi: 10.1016/j.biochi.2022.08.003

Damavandi, A.R., Mirmosayyeb, O., Ebrahimi, N., Zalpoor, H., khalilian, P., Yahiazadeh, S., Eskandari, N., Rahdar, A., Kumar, P.S., Pandey, S. (2022). Advances in nanotechnology versus stem cell therapy for the theranostics of multiple sclerosis disease. *Applied Nanoscience (Switzerland)*. doi: 10.1007/s13204-022-02698-x

Dehay, B., Bourdenx, M., Gorry, P., Przedborski, S., Vila, M., Hunot, S., Singleton, A., Olanow, C. W., Merchant, K. M., Bezard, E., Petsko, G. A., & Meissner, W. G. (2015). Targeting α-synuclein for treatment of Parkinson's disease: Mechanistic and therapeutic considerations. *Lancet Neurology* 14:855–866.

Derakhshankhah, H., Sajadimajd, S., Jafari, S., Izadi, Z., Sarvari, S., Sharifi, M., Falahati, M., Moakedi, F., Muganda, W.C.A., Müller, M., Raoufi, M., Presley, J.F. (2020). Novel therapeutic strategies for Alzheimer's disease: Implications from cell-based therapy and nanotherapy. *Nanomedicine Nanotechnology, Biology and Medicine* 24:102149.

De Simone, A., La Pietra, V., Betari, N., Petragnani, N., Conte, M., Daniele, S., Pietrobono, D., Martini, C., Petralla, S., Casadei, R., Davani, L., Frabetti, F., Russomanno, P., Novellino, E., Montanari, S., Tumiatti, V., Ballerini, P., Sarno, F., Nebbioso, A., Altucci, L., . . . Milelli, A. (2019). Discovery of the first-in-class GSK-3β/HDAC dual inhibitor as disease-modifying agent to combat Alzheimer's disease. *ACS Medicinal Chemistry Letters* 10:469–474. doi: 10.1021/acsmedchemlett.8b00507

Díaz-García, D., Ferrer-Donato, Á., Méndez-Arriaga, J.M., Cabrera-Pinto, M., Díaz-Sánchez, M., Prashar, S., Fernandez-Martos, C.M., Gómez-Ruiz, S. (2022). Design of mesoporous silica nanoparticles for the treatment of amyotrophic lateral sclerosis (ALS) with a therapeutic cocktail based on leptin and pioglitazone. *ACS Biomaterials Science and Engineering* 8(11):4838–4849. doi: 10.1021/acsbiomaterials.2c00865

Dong, J., Liu, M., Jiang, R., Huang, H., Wan, Q., Wen, Y., Tian, J., Dai, Y., Zhang, X., Wei, Y. (2018). Synthesis and biological imaging of cross-linked fluorescent polymeric nanoparticles with aggregation-induced emission characteristics based on the combination of RAFT polymerization and the Biginelli reaction. *Journal of Colloid and Interface Science* 528:192–199. doi: 10.1016/j.jcis.2018.05.043

Ealia, S.A.M., Saravanakumar, M.A. (2017). Review on the classification, characterisation, synthesis of nanoparticles and their application. In: IOP Conference Series: Materials Science and Engineering, vol. 3. IOP Publishing, p. 032019.

Essa, M.M., Moghadas, M., Ba-Omar, T., Walid, Q.M., Guillemin, G.J., Manivasagam, T., Justin-Thenmozhi, A., Ray, B., Bhat, A., Chidambaram, S.B. (2019). Protective effects of antioxidants in Huntington's disease: An extensive review. *Neurotoxicity Research* 35:739–774.

Feigin, V.L., Vos, T., Nichols, E., Owolabi, M.O., Carroll, W.M., Dichgans, M., Deuschl, G., Parmar, P., Brainin, M., Murray, C. (2020). The global burden of neurological disorders: Translating evidence into policy. *Lancet Neurology* 19(3):255–265. doi: 10.1016/S1474-4422(19)30411-9

Fernandes, C., Soni, U., Patravale, V. (2010). Nano-interventions for neurodegenerative disorders. *Pharmacological Research* 62:166–178.

Gaba, B., Khan, T., Haider, M.F., Alam, T., Baboota, S., Parvez, S., Ali, J. (2019). Vitamin E loaded naringenin nanoemulsion via intranasal delivery for the management of oxidative stress in a 6-OHDA Parkinson's disease model. *BioMed Research International* 2382563. doi: 10.1155/2019/2382563

Gandhi, K.R., Saadabadi, A. (2015). Levodopa, l-DOPA. *Encyclopedia of Psychopharmacology* 898–899. doi: 10.1007/978-3-642-36172-2_200303

Gandomi, N., Varshochian, R., Atyabi, F., Ghahremani, M.H., Sharifzadeh, M., Amini, M., Dinarvand, R. (2017). Solid lipid nanoparticles surface modified with anti-Contactin-2 or anti-Neurofascin for brain-targeted delivery of medicines. *Pharmaceutical Development and Technology* 22(3):426–435. doi: 10.1080/10837450.2016.1226901

Gao, C., Chu, X., Gong, W., Zheng, J., Xie, X., Wang, Y., Yang, M., Li, Z., Gao, C., Yang, Y. (2020). Neuron tau-targeting biomimetic nanoparticles for curcumin delivery to delay progression of Alzheimer's disease. *Journal of Nanobiotechnology* 18:71. doi: 10.1186/s12951-020-00626-1

Gao, G., Zhang, M., Gong, D., Chen, R., Hu, X., Sun, T. (2017). The size-effect of gold nanoparticles and nanoclusters in the inhibition of amyloid-β fibrillation. *Nanoscale* 9:4107–4113.

García-Pardo, J., Novio, F., Nador, F., Cavaliere, I., Suárez-García, S., Lope-Piedrafita, S., Candiota, A.P., Romero-Gimenez, J., Rodríguez-Galván, B., Bové, J., Vila, M., Lorenzo, J., Ruiz-Molina, D. (2021). Bioinspired theranostic coordination polymer nanoparticles for intranasal dopamine replacement in Parkinson's disease. *ACS Nano* 15(5):8592–8609. doi: 10.1021/acsnano.1c00453

Ghetti, B., Piccardo, P., Zanusso, G. (2018). Dominantly inherited prion prothosphorlil amyloidoses–A modern view of Gerstmann–Sträussler–Scheinker. In: Handbook of clinical neurology, vol. 153. Elsevier, pp. 243–269.

Guo, L., Ren, J., Jiang, X. (2012). Perspectives on brain-targeting drug delivery systems. *Current Pharmaceutical Biotechnology* 13:2310–2318.

Guo, Q., You, H., Yang, X., Lin, B., Zhu, Z., Lu, Z., Li, X., Zhao, Y., Mao, L., Shen, S., Cheng, H., Zhang, J., Deng, L., Fan, J., Xi, Z., Li, R., Li, C.M. (2017). Functional single-walled carbon nanotubes "CAR" for targeting dopamine delivery into the brain of parkinsonian mice. *Nanoscale* 9(30):10832–10845. doi: 10.1039/c7nr02682j

Guo, X., Deng, G., Liu, J., Zou, P., Du, F., Liu, F., Chen, A.T., Hu, R., Li, M., Zhang, S. (2018). Thrombin-responsive, brain-targeting nanoparticles for improved stroke therapy. *ACS Nano* 12:8723–8732.

Gürsoy-Özdemir, Y., Tas, Y. (2017). Nanotechnology methods for neurological diseases and brain tumors. Academic Press.

Han, Q., Cai, S., Yang, L., Wang, X., Qi, C., Yang, R., Wang, C. (2017). Molybdenum disulfide nanoparticles as multifunctional inhibitors against Alzheimer's disease. *ACS Applied Materials & Interfaces* 9:21116–21123. doi: 10.1021/acsami.7b03816

Hanif, S., Muhammad, P., Chesworth, R., Rehman, F.U., Qian, R.J., Zheng, M., Shi, B.Y. (2020). Nanomedicine-based immunotherapy for central nervous system disorders. *Acta Pharmacologica Sinica* 41:936–953.

Hasan, S. (2015). A review on nanoparticles: Their synthesis and types. *Research Journal of Recent Sciences* 2277:2502.

He, X., Wang, X., Yang, L., Yang, Z., Yu, W., Wang, Y., Liu, R., Chen, M., Gao, H. (2022). Intelligent lesion blood–brain barrier targeting nano-missiles for Alzheimer's disease treatment by anti-neuroinflammation and neuroprotection. *Acta Pharmaceutica Sinica* 12:1987–1999.

He, W., Zou, Y., Zheng, M., Shi, B. (2019). Cell-derived biomimetic drug delivery systems for cancer therapy. *Scientia Sinica* 49:1203–1212.

Hervé, F., Ghinea, N., Scherrmann, J.M. (2008). CNS delivery via adsorptive transcytosis. *The AAPS Journal* 10:455–472.

Hosseiny, S., Reda, A., El-Sherbiny, M.I. (2020). Nano drug delivery systems targeting microglia for the treatment of brain disorder. *Journal of Nanotechnology & Advanced Materials* 8(1):31–43.

Hu, K., Shi, Y., Jiang, W., Han, J., Huang, S., Jiang, X. (2011). Lactoferrin conjugated PEG-PLGA nanoparticles for brain delivery: Preparation, characterization and efficacy in Parkinson's disease. *International Journal of Pharmaceutics* 415:273–283.

Hydbring, P., Du, J. (2019). Nanoparticle interaction with immune cells for nanoparticle-mediated (Anticancer) immunotherapy. In: Theranostic Bionanomaterials, eds. W. Cui and X. Zhao. Elsevier, pp. 55–73. doi: 10.1016/B978-0-12-815341-3.00003-1

Jagaran, K., Singh, M. (2021). Nanomedicine for neurodegenerative disorders: Focus on Alzheimer's and Parkinson's diseases. *International Journal of Molecular Sciences* 22(16):9082. doi: 10.3390/ijms22169082.

Jagaran, K., Singh, M. (2022). Lipid nanoparticles: Promising treatment approach for Parkinson's disease. *International Journal of Molecular Sciences* 23(16). doi: 10.3390/ijms23169361

Jeon, S.G., Cha, M.Y., Kim, J.I., Hwang, T.W., Kim, K.A., Kim, T.H., Song, K.C., Kim, J.J., Moon, M. (2019). Vitamin D-binding protein-loaded PLGA nanoparticles suppress Alzheimer's disease-related pathology in 5XFAD mice. *Nanomedicine* 17:297–307. doi: 10.1016/j.nano.2019.02.004

Jin, G.Z., Chakraborty, A., Lee, J.H., Knowles, J.C., Kim, H.W. (2020). Targeting with nanoparticles for the therapeutic treatment of brain diseases. *Journal of Tissue Engineering* 11. doi: 10.1177/2041731419897460

Joseph, A., Wood, T., Chen, C.C., Corry, K., Snyder, J.M., Juul, S.E., Parikh, P., Nance, E. (2018). Curcumin-loaded polymeric nanoparticles for neuroprotection in neonatal rats with hypoxic-ischemic encephalopathy. *Nano Research* 11:5670–5688.

Kalia, L.V., Lang, A.E. (2015). Parkinson's disease. *Lancet* 386(9996):896–912. doi: 10.1016/S0140-6736(14)61393-3

Katebi, S., Esmaeili, A., Ghaedi, K., Zarrabi, A. (2019). Superparamagnetic iron oxide nanoparticles combined with NGF and quercetin promote neuronal branching morphogenesis of PC12 cells. *International Journal of Nanomedicine* 14:2157–2169. doi: 10.2147/IJN.S191878

Keller, G., Binyamin, O., Frid, K., Saada, A., Gabizon, R. (2019). Mitochondrial dysfunction in preclinical genetic prion disease: A target for preventive treatment? *Neurobiology of Disease* 124:57–66.

Khan, M.S., Pandey, S., Talib, A., Bhaisare, M.L., Wu, H.F. (2015). Controlled delivery of dopamine hydrochloride using surface-modified carbon dots for neuro diseases. *Colloids and Surfaces B: Biointerfaces* 134:140–146. doi: 10.1016/j.colsurfb.2015.06.006

Kim, D., Yoo, J.M., Hwang, H., Lee, J., Lee, S.H., Yun, S.P., Park, M.J., Lee, M., Choi, S., Kwon, S.H. (2018). Graphene quantum dots prevent α-synucleinopathy in Parkinson's disease. *Nature Nanotechnology* 13:812–818.

Kumar, A., Chaudhary, R.K., Singh, R., Singh, S.P., Wang, S.Y., Hoe, Z.Y., Pan, C.T., Shiue, Y.L., Wei, D.Q., Kaushik, A.C., Dai, X. (2020). Nanotheranostic applications for detection and targeting neurodegenerative diseases. *Frontiers in Neuroscience* 14:305. doi: 10.3389/fnins.2020.00305

Kumar, P., Sharma, G., Gupta, V., Kaur, R., Thakur, K., Malik, R., Kumar, A., Kaushal, N., Raza, K. (2018). Preclinical explorative assessment of dimethyl fumarate-based biocompatible nano lipoidal carriers for the management of multiple sclerosis. *ACS Chemical Neuroscience* 9(5):1152–1158. doi: 10.1021/acschemneuro.7b00519

Krasnianski, A., Juan, P.S., Ponto, C., Bartl, M., Heinemann, U., Varges, D., Schulz-Schaeffer, W.J., Kretzschmar, H.A., Zerr, I. (2014). A proposal of new diagnostic pathway for fatal familial insomnia. *Journal of Neurology, Neurosurgery & Psychiatry* 85:654–659.

Lei, T., Yang, Z., Xia, X., Chen, Y., Yang, X., Xie, R., Tong, F., Wang, X., Gao, H. (2021). A nanocleaner specifically penetrates the blood–brain barrier at lesions to clean toxic proteins and regulate inflammation in Alzheimer's disease. *Acta Pharmaceutica Sinica B* 11:4032–4044.

Li, X., Li, W., Wang, M., Liao, Z. (2021). Magnetic nanoparticles for cancer theranostics: Advances and prospects. *Journal of Controlled Release* 335:437–448.

Li, X., Tsibouklis, J., Weng, T., Zhang, B., Yin, G., Feng, G., Cui, Y., Savina, I.N., Mikhalovska, L.I., Sandeman, S.R., Howel, C.A., Mikhalovsky, S.V. (2017). Nanocarriers for drug transport across the blood-brain barrier. *Journal of Drug Targeting* 25(1):17–28.

Liao, Y.H., Chang, Y.J., Yoshiike, Y., Chang, Y.C., Chen, Y.R. (2012). Negatively charged gold nanoparticles inhibit Alzheimer's amyloid-β fibrillization, induce fibril dissociation, and mitigate neurotoxicity. *Small* 8:3631–3639. doi: 10.1002/smll.201201068

Liu, D.Z., Cheng, Y., Cai, R.Q., Wang, B.W.W., Cui, H., Liu, M., Zhang, B.L., Mei, Q.B., Zhou, S.Y. (2018). The enhancement of siPLK1 penetration across BBB and its anti-glioblastoma activity in vivo by magnet and transferrin co-modified nanoparticle. *Nanomedicine* 14:991–1003. doi: 10.1016/j.nano.2018.01.004

Liu, H., Yang, H., Fang, Y., Li, K., Tian, L., Liu, X., Zhang, W., Tan, Y., Lai, W., Bian, L., Lin, B., Xi, Z. (2020a). Neurotoxicity and biomarkers of zinc oxide nanoparticles in main functional brain regions and dopaminergic neurons. *Science of the Total Environment* 705:135809. doi: 10.1016/j.scitotenv.2019.135809

Liu, H., Zhang, W., Fang, Y., Yang, H., Tian, L., Li, K., Lai, W., Bian, L., Lin, B., Liu, X., Xi, Z. (2020b). Neurotoxicity of aluminum oxide nanoparticles and their mechanistic role in dopaminergic neuron injury involving p53-related pathways. *Journal of Hazardous Materials* 392:122312. doi: 10.1016/j.jhazmat.2020.122312

Liu, R., Yang, J., Liu, L., Lu, Z., Shi, Z., Ji, W., Shen, J., Zhang, X. (2020c). An "Amyloid-β cleaner" for the treatment of Alzheimer's disease by normalizing microglial dysfunction. *Advanced Science* 7:1901555.

Mahmoudi, M., Sahraian, M.A., Shokrgozar, M.A., Laurent, S. (2011). Superparamagnetic iron oxide nanoparticles: Promises for diagnosis and treatment of multiple sclerosis. *ACS Chemical Neuroscience* 2(3):118–140. doi: 10.1021/cn100100e

Maldonado, R.A., LaMothe, R.A., Ferrari, J.D., Zhang, A.H., Rossi, R.J., Kolte, P.N., Griset, A.P., O'Neil, C., Altreuter, D.H., Browning, E., Johnston, L., Farokhzad, O.C., Langer, R., Scott, D.W., Von Andrian, U.H., Kishimoto, T.K. (2015). Polymeric synthetic nanoparticles for the induction of antigen-specific immunological tolerance. *Proceedings of the National Academy of Sciences of the United States of America* 112(2):E156–E165. doi: 10.1073/pnas.1408686111

Manek, E., Darvas, F., Petroianu, G.A. (2020). Use of biodegradable, chitosan-based nanoparticles in the treatment of Alzheimer's disease. *Molecules (Basel, Switzerland)* 25. doi: 10.3390/molecules25204866

Maney, V., Singh, M. (2019). The synergism of Platinum-Gold bimetallic nanoconjugates enhance 5-Fluorouracil delivery in vitro. *Pharmaceutics* 11:439. doi: 10.3390/pharmaceutics11090439

Masrori, P., Van Damme, P. (2020). Amyotrophic lateral sclerosis: A clinical review. *European Journal of Neurology* 27(10):1918–1929. doi: 10.1111/ene.14393

Matsuno, J., Kanamaru, T., Arai, K., Tanaka, R., Lee, J.H., Takahashi, R., Sakurai, K., Fujii, S. (2020). Synthesis and characterization of nanoemulsion-mediated core crosslinked nanoparticles, and in vivo pharmacokinetics depending on the structural characteristics. *Journal of Controlled Release: Official Journal of the Controlled Release Society* 324:405–412. doi: 10.1016/j.jconrel.2020.05.035

Md, S., Gan, S.Y., Haw, Y.H., Ho, C.L., Wong, S., Choudhury, H. (2018). In vitro neuroprotective effects of naringenin nanoemulsion against β-amyloid toxicity through the regulation of amyloidogenesis and tau phosphorylation. *International Journal of Biological Macromolecules* 118:1211–1219. doi: 10.1016/j.ijbiomac.2018.06.190

Medina, D.X., Chung, E.P., Teague, C.D., Bowser, R., Sirianni, R.W. (2020). Intravenously administered, retinoid activating nanoparticles increase lifespan and reduce neurodegeneration in the SOD1G93A mouse model of ALS. *Frontiers in Bioengineering and Biotechnology* 8. doi: 10.3389/fbioe.2020.00224

Mizrahi, M., Friedman-Levi, Y., Larush, L., Frid, K., Binyamin, O., Dori, D., Fainstein, N., Ovadia, H., Ben-Hur, T., Magdassi,

S. (2014). Pomegranate seed oil nanoemulsions for the prevention and treatment of neurodegenerative diseases: The case of genetic CJD. *Nanomedicine: Nanotechnology, Biology and Medicine* 10:1353–1363.

Moore, K.A., Pate, K.M., Soto-Ortega, D.D., Lohse, S., van der Munnik, N., Lim, M., Jackson, K.S., Lyles, V.D., Jones, L., Glassgow, N., Napumecheno, V.M., Mobley, S., Uline, M.J., Mahtab, R., Murphy, C.J., Moss, M.A. (2017). Influence of gold nanoparticle surface chemistry and diameter upon Alzheimer's disease amyloid-β protein aggregation. *Journal of Biological Engineering* 11:5. doi: 10.1186/s13036-017-0047-6

Murugesan, C., Manivannan, P., Gangatharan, M. (2020). Pros and cons in prion diseases abatement: Insights from nanomedicine and transmissibility patterns. *International Journal of Biological Macromolecules* 145:21–27. doi: 10.1016/j.ijbiomac.2019.12.150

Nabizadeh, F. (2022). Biomaterials in the treatment of amyotrophic lateral sclerosis. *Neurology Letters* 1(1):12–16. doi:10.52547/nl.1.1.12

Nam, E., Lee, Y.B., Moon, C., Chang, K.A. (2020). Serum tau proteins as potential biomarkers for the assessment of Alzheimer's disease progression. *International Journal of Molecular Sciences* 21:5007. doi: 10.3390/ijms21145007

Naqvi, S., Pangha, A., Flora, S.J.S. (2020). Nanotechnology: A promising approach for delivery of neuroprotective drugs. *Frontiers in Neurosc ience* 14. doi: 10.3389/fnins.2020.00494

Ngowi, E.E., Wang, Y.Z., Qian, L., Helmy, Y.A.S.H., Anyomi, B., Li, T., Zheng, M., Jiang, E.S., Duan, S.F., Wei, J.S., Wu, D.D., Ji, X.Y. (2021). The Application of nanotechnology for the diagnosis and treatment of brain diseases and disorders. *Frontiers in Bioengineering and Biotechnology* 9:629832. doi: 10.3389/fbioe.2021.629832

Niccolini, B., Palmieri, V., De Spirito, M., Papi, M. (2022). Opportunities offered by graphene nanoparticles for micrornas delivery for amyotrophic lateral sclerosis treatment. *Materials* 15(1). doi: 10.3390/ma15010126

Nowak, M., Helgeson, M.E., Mitragotri, S. (2020). Delivery of nanoparticles and macromolecules across the blood–brain barrier. *Advanced Therapeutics* 3:1900073.

Nuzzo, D., Picone, P. (2021). Multiple sclerosis: Focus on extracellular and artificial vesicles, nanoparticles as potential therapeutic approaches. *International Journal of Molecular Sciences* 22(16). doi: 10.3390/ijms22168866

Ojha, S., Kumar, B. (2018). A review on nanotechnology based innovations in diagnosis and treatment of multiple sclerosis. *Journal of Cellular Immunotherapy* 4(2):56–64. doi: 10.1016/j.jocit.2017.12.001

Oladimeji, O., Akinyelu, A., Singh, M. (2020). Co-polymer functionalised gold nanoparticles show efficient mitochondrial targeted drug delivery in cervical carcinoma cells. *Journal of Biomedical Nanotechnology* 16:853–866. doi: 10.1166/jbn.2020.2930

Pahuja, R., Seth, K., Shukla, A., Shukla, R.K., Bhatnagar, P., Chauhan, L.K.S., Saxena, P.N., Arun, J., Chaudhari, B.P., Patel, D.K., Singh, S.P., Shukla, R., Khanna, V.K., Kumar, P., Chaturvedi, R.K., Gupta, K.C. (2015). Trans-blood brain barrier delivery of dopamine-loaded nanoparticles reverses functional deficits in parkinsonian rats. *ACS Nano* 9(5):4850–4871. doi: 10.1021/nn506408v

Pardridge, W.M. (2007). Drug targeting to the brain. *Pharmaceutical Research* 24:1733–1744.

Poovaiah, N., Davoudi, Z., Peng, H., Schlichtmann, B., Mallapragada, S., Narasimhan, B., Wang, Q. (2018). Treatment of neurodegenerative disorders through the blood-brain barrier using nanocarriers. *Nanoscale* 10(36):16962–16983. doi: 10.1039/c8nr04073g

Prasanna, P., Upadhyay, A. (2021). Flavonoid-based nanomedicines in Alzheimer's disease therapeutics: Promises made, a long way to go. *ACS Pharmacology & Translational Science* 4:74–95. doi: 10.1021/acsptsci.0c00224

Prince, M., Comas-Herrera, A., Knapp, M., Guerchet, M., Karagiannidou, M. (2016). World Alzheimer report: Improving healthcare for people living with dementia: Coverage, quality and costs now and in the future. Alzheimer's Disease International (ADI), London. http://eprints.lse.ac.uk

Rabanel, J.M., Piec, P.A., Landri, S., Patten, S.A., Ramassamy, C. (2020). Transport of PEGylated-PLA nanoparticles across a blood brain barrier model, entry into neuronal cells and in vivo brain bioavailability. *Journal of Controlled Release* 328:679–695. doi: 10.1016/j.jconrel.2020.09.042

Rashed, E.R., Abd El-Rehim, H.A., El-Ghazaly, M.A. (2014). Potential efficacy of dopamine loaded-PVP/PAA nanogel in experimental models of Parkinsonism: Possible disease modifying activity. *Journal of Biomedical Materials Research – Part A*. http://www.embase.com/search/results?subaction=viewrecord&from=export&id=L53307952%0A; http://dx.doi.org/10.1002/jbm.a.35312

Rasouli, M.R., Tabrizian, M. (2019). An ultra-rapid acoustic micromixer for synthesis of organic nanoparticles. *Lab on a Chip* 19:3316–3325. doi: 10.1039/c9lc00637k

Rawal, S.U., Patel, B.M., Patel, M.M. (2022). New drug delivery systems developed for brain targeting. *Drugs* 82:749–792.

Ray, S., Sinha, P., Laha, B., Maiti, S., Bhattacharyya, U.K., Nayak, A.K. (2018). Polysorbate 80 coated crosslinked chitosan nanoparticles of ropinirole hydrochloride for brain targeting. *Journal of Drug Delivery Science and Technology* 48:21–29. doi: 10.1016/j.jddst.2018.08.016

Ren, Y., Zhao, X., Liang, X., Ma, P.X., Guo, B. (2017). Injectable hydrogel based on quaternized chitosan, gelatin and dopamine as localized drug delivery system to treat Parkinson's disease. *International Journal of Biological Macromolecules* 105:1079–1087. doi: 10.1016/j.ijbiomac.2017.07.130

Rittchen, S., Boyd, A., Burns, A., Park, J., Fahmy, T.M., Metcalfe, S., Williams, A. (2015). Myelin repair invivo is increased by targeting oligodendrocyte precursor cells with nanoparticles encapsulating leukaemia inhibitory factor (LIF). *Biomaterials* 56:78–85. doi: 10.1016/j.biomaterials.2015.03.044

Sajid, M., Płotka-Wasylka, J. (2020). Nanoparticles: Synthesis, characteristics, and applications in analytical and other sciences. *Microchemical Journal* 154:104623.

Sandhir, R., Yadav, A., Mehrotra, A., Sunkaria, A., Singh, A., Sharma, S. (2014). Curcumin nanoparticles attenuate neurochemical and neurobehavioral deficits in experimental model of Huntington's disease. *Neuromolecular Medicine* 16:106–118.

Santa-Cecília, F.V., Leite, C.A., Del-Bel, E., Raisman-Vozari, R. (2019). The neuroprotective effect of doxycycline on neurodegenerative diseases. *Neurotoxicity Research* 35:981–986.

Sava, V., Fihurka, O., Khvorovam A., Sanchez-Ramos, J. (2020). Enriched chitosan nanoparticles loaded with siRNA are effective in lowering Huntington's disease gene expression following intranasal administration. *Nanomedicine: Nanotechnology, Biology and Medicine* 24:102119.

Shanmukhapuvvada, Y., Vankayalapati, S. (2012). Design and development of riluzole loaded chitosan nanoparticles by emulsification crosslinking. *International Journal of Pharmacy and Pharmaceutical Sciences* 4(4):244–248.

Shi, Y., Jiang, Y., Cao, J., Yang, W., Zhang, J., Meng, F., Zhong, Z. (2018). Boosting RNAi therapy for orthotopic glioblastoma with nontoxic brain-targeting chimaeric polymersomes. *Journal of Controlled Release* 292:163–171. doi: 10.1016/j.jconrel.2018.10.034

Silva, S., Almeida, A.J., Vale, N. (2021). Importance of nanoparticles for the delivery of antiparkinsonian drugs. *Pharmaceutics* 13(4). doi: 10.3390/pharmaceutics13040508

Silveira, G.B., Muller, A.P., Machado-de-Ávila, R.A., Silveira, P.C.L. (2021). Advance in the use of gold nanoparticles in the treatment of neurodegenerative diseases: New perspectives. *Neural Regeneration Research* 16(12):2425–2426. doi: 10.4103/1673-5374.313040

Singh, D., Kevadiya, B., Nagpal, K., Sharma, N., Patel, S.K. (2017). The significance of nanomedicine in brain-targeted drug delivery: Crossing blood-brain barriers. *Nanomedicine Research Journal* 5(5):00132.

Selkoe, D.J., Hardy, J. (2016). The amyloid hypothesis of Alzheimer's disease at 25 years. *EMBO Molecular Medicine* 8:595–608.

Sonawane, S.K., Ahmad, A., Chinnathambi, S. (2019). Protein-capped metal nanoparticles inhibit tau aggregation in Alzheimer's disease. *ACS Omega* 4:12833–12840. doi: 10.1021/acsomega.9b01411

Srikanth, M., Kessler, J.A. (2012). Nanotechnology—novel therapeutics for CNS disorders. *Nature Reviews Neurology* 8:307–318.

Stöhr, J., Weinmann, N., Wille, H., Kaimann, T., Nagel-Steger, L., Birkmann, E., Panza, G., Prusiner, S.B., Eigen, M., Riesner, D. (2008). Mechanisms of prion protein assembly into amyloid. *Proceedings of the National Academy of Sciences* 105:2409–2414.

Sun, A., Benet, L.Z. (2020). Late-stage failures of monoclonal antibody drugs: A retrospective case study analysis. *Pharmacology* 105:145–163. doi: 10.1159/00050537

Sun, J. (2022). Neurodegerative disease. *NeurologyLive* 5(5).

Tagai, K., Ono, M., Kubota, M., Kitamura, S., Takahata, K., Seki, C., Takado, Y., Shinotoh, H., Sano, Y., Yamamoto, Y., Matsuoka, K., Takuwa, H., Shimojo, M., Takahashi, M., Kawamura, K., Kikuchi, T., Okada, M., Akiyama, H., Suzuki, H., Onaya, M., . . . Shimada, H. (2021). High-contrast in vivo imaging of tau pathologies in Alzheimer's and Non-Alzheimer's disease tauopathies. *Neuron* 109(1):42–58.e8. doi: 10.1016/j.neuron.2020.09.042

Teixeira, M.I., Lopes, C., Amaral, M.H., Costa, P. (2020). Current insights on lipid nanocarrier-assisted drug delivery in the treatment of neurodegenerative diseases. *European Journal of Pharmaceutics and Biopharmaceutics* 149:192–217.

Tereanu, D.M., Chircov, C., Grumezescu, A.M., Volceanov, A., Teleanu, R.I. (2018). Impact of nanoparticles on brain health: An up to date overview. *Journal of Clinical Medicine* 7:490. doi: 10.3390/jcm7120490

Tong, T., Wang, L., You, X., Wu, J. (2020). Nano and microscale delivery platforms for enhanced oral peptide/protein bioavailability. *Biomaterials Science* 8:5804–5823.

Ul-Hassan, S.S., Samanta, S., Dash, R., Karpiński, T.M., Habibi, E., Sadiq, A., Ahmadi, A., Bunagu, S. (2022). The neuroprotective effects of fisetin, a natural flavonoid in neurodegenerative diseases: Focus on the role of oxidative stress. *Frontiers in Pharmacology* 13.

Vakilinezhad, M.A., Amini, A., Akbari Javar, H., Baha'addini Beigi Zarandi, B.F., Montaseri, H., Dinarvand, R. (2018). Nicotinamide loaded functionalized solid lipid nanoparticles improves cognition in Alzheimer's disease animal model by reducing Tau hyperphosphorylation. *DARU Journal of Pharmaceutical Sciences* 26:165–177. doi: 10.1007/s40199-018-0221-5

Venkatas, J., Singh, M. (2021). Nanomedicine-mediated optimization of Immuno-therapeutic approaches in Cervical cancer. *Nanomedicine* 16:1311–1328. doi: 10.2217/nnm-2021-0044

Vimal, S.K., Cao, H., Dubey, A., Agrawal, L., Pathak, N., Zuo, H., Kumar, D., Bhattacharyya, S. (2022). In vivo and in silico investigations of the pegylated gold nanoparticle treatment of amyotrophic lateral sclerosis in mice. *New Journal of Chemistry* 46(25), 12252–12264. doi: 10.1039/d2nj00452f

Vimal, S.K., Zuo, H., Wang, Z., Wang, H., Long, Z., Bhattacharyya, S. (2020). Self-therapeutic nanoparticle that alters tau protein and ameliorates tauopathy toward a functional nanomedicine to tackle Alzheimer's. *Small (Weinheim an der Bergstrasse Germany)* 16:e1906861. doi: 10.1002/smll.201906861

Viola, K.L., Sbarboro, J., Sureka, R., De, M., Bicca, M.A., Wang, J., Vasavada, S., Satpathy, S., Wu, S., Joshi, H., Velasco, P.T., MacRenaris, K., Waters, E.A., Lu, C., Phan, J., Lacor, P., Prasad, P., Dravid, V.P., Klein, W.L. (2015). Towards non-invasive diagnostic imaging of early-stage Alzheimer's disease. *Nature Nanotechnology* 10(1):91–98.

Xie, W., Guo, Z., Gao, F., Gao, Q., Wang, D., Liaw, B.S., Cai, Q., Sun, X., Wang, X., Zhao, L. (2018). Shape-, size-and structure-controlled synthesis and biocompatibility of iron oxide nanoparticles for magnetic theranostics. *Theranostics* 8(12):3284–3307. doi: 10.7150/thno.25220

Yang, H., Mu, W., Wei, D., Zhang, Y., Duan, Y., Gao, J.X., Gong, X.Q., Wang, H.J., Wu X.-l., Tao, H. (2020). A novel targeted and high-efficiency nanosystem for combinational therapy for Alzheimer's disease. *Advanced Science* 7:1902906.

Yang, T., Ferrill, L., Gallant, L., McGillicuddy, S., Fernandes, T., Schields, N., Bai, S. (2018). Verapamil and riluzole cocktail liposomes overcome pharmaco resistance by inhibiting P-glycoprotein in brain endothelial and astrocyte cells: A potent approach to treat amyotrophic lateral sclerosis. *European Journal of Pharmaceutical Sciences* 120:30–39. doi: 10.1016/j.ejps.2018.04.026

Yaqoob, S.B., Adnan, R., Khan, A.M.R., Rashid, M. (2020). Gold, silver, and palladium nanoparticles: A chemical tool for biomedical applications. *Frontiers in Chemistry* 8:376. doi: 10.3389/fchem.2020.00376

Zeng, Y., Li, Z., Zhu, H., Gu, Z., Zhang, H., Luo, K. (2021). Recent advances in nanomedicines for multiple sclerosis therapy. *ACS Applied Bio Materials* 3(10):6571–6597. doi: 10.1021/acsabm.0c00953

Zhang, F., Stephan, S.B., Ene, C.I., Smith, T.T., Holland, E.C., Stephan, M.T. (2018b). Nanoparticles that reshape the tumor milieu create a therapeutic window for effective t-cell therapy in solid malignancies. *Cancer Research* 78:3718–3730.

Zhang, G., Xia, Y., Wan, F., Ma, K., Guo, X., Kou, L., Yin, S., Han, C., Liu, L., Huang, J., Xiong, N., Wang, T. (2018a). New perspectives on roles of alpha-synuclein in Parkinson's disease. *Frontiers in Aging Neuroscience* 10:370. doi: 10.3389/fnagi.2018.00370

Zhang, M., Viennois, E., Prasad, M., Zhang, Y., Wang, L., Zhang, Z., Han, M.K., Xiao, B., Xu, C., Srinivasan, S., Merlin, D. (2016). Edible ginger-derived nanoparticles: A novel therapeutic approach for the prevention and treatment of inflammatory bowel disease and colitis-associated cancer. *Biomaterials* 101:321–340. doi: 10.1016/j.biomaterials.2016.06.018

Zhang, Q., Dai, X., Zhang, H., Zeng, Y., Luo, K., Li, W. (2021). Recent advances in development of nanomedicines for multiple sclerosis diagnosis. *Biomedical Materials (Bristol)* 16(2). doi: 10.1088/1748-605X/abddf4

Zhang, W., Duan, R., Zhang, J., Cheung, W.K.C., Gao, X., Zhang, R., Zhang, Q., Wei, M., Wang, G., Zhang, Q., Mei, P.J., Chen, H.L., Kung, H., Lin, M.C., Shen, Z., Zheng, J., Zhang, L., Yao, H. (2018c). H1/pHGFK1 nanoparticles exert antitumoural and radiosensitising effects by inhibition of MET in glioblastoma. *British Journal of Cancer* 118(4):522–533. doi: 10.1038/bjc.2017.461

Zhang, T.T., Li, W., Meng, G., Wang, P., Liao, W. (2016). Strategies for transporting nanoparticles across the blood–brain barrier. *Biomaterials Science* 4:219–229.

Zhao, N., Francis, N.L., Calvelli, H.R., Moghe, P.V. (2020). Microglia-targeting nanotherapeutics for neurodegenerative diseases. *APL Bioengineering* 4(3):030902. doi: 10.1063/5.001317

Zhou, S., Zhu, Y., Yao, X., Liu, H. (2018). Carbon nanoparticles inhibit the aggregation of prion protein as revealed by experiments and atomistic simulations. *Journal of Chemical Information and Modeling* 59:1909–1918.

Zhu, F.D., Hu, Y.J., Yu, L., et al. (2021). Nanoparticles: A hope for the treatment of inflammation in CNS. *Frontiers in Pharmacology* 12:683935. doi: 10.3389/fphar.2021.683935

12 Application of Polymeric Nanoparticles in Brain Targeting

*Moshera Samy**

12.1 INTRODUCTION

The World Health Organization (WHO) estimates that approximately 20% of all humans suffer from damage to the central nervous system (CNS), such as depression, Alzheimer's disease, Parkinson's disease, epilepsy, cerebral cancer, or CNS-relevant metabolic diseases (Hartl, Adams, & Merkel, 2021). Alzheimer's disease (AD) is the most common type of dementia associated with cognitive decline, memory impairment, aphasia, and behavioral disabilities within elderly populations. The neuropathological hallmarks of AD include extracellular senile plaques consisting of amyloid-β (Aβ) deposits, intracellular neurofibrillary tangles, and cerebral atrophy. Clinical studies for many therapeutic approaches based on AD etiology treatment for the pathogenesis theories have been unsuccessful (Liu, Xie, Meng, & Kang, 2019) because of BBB's poor brain bioavailability (Kumar et al., 2017). To prevent AD, new and effective brain delivery strategies are required.

Polymeric nanoparticles (PNs) are proposed as an alternative to conventional formulations for drug administration, delivery, and targeting. The BBB is frequently a rate-limiting factor in determining the permeation of a drug into the brain. In this study, the surface-engineered long PNs were assessed for brain-specific delivery (Sharma, Badyal, & Gupta, 2015). Drug transport to the brain is restricted owing to the BBB's existence, which inhibits the administration of a range of medications into the brain. Endothelial cells with tight connections that border the cerebral capillaries make up the BBB. Because of tight connections, paracellular solute routes across the BBB are eliminated in the brain endothelium epithelium. Several efflux transport mechanisms, including p-glycoprotein and active organic acids found in brain endothelial cells, work in conjunction with tight junctions to remove undesired chemicals (Neha, Ganesh, & Preeti, 2013). Over the past ten years, both in pharmaceutical research and in the clinical context, the use of nanoparticulate pharmaceutical nanocarriers has become well-established. Through the enhanced permeability and retention (EPR) effect, drug nanocarriers are predicted to accumulate in pathological sites with damaged and leaky vasculature (tumors, inflammations, and infarcted areas) and enable targeted delivery of particular ligand-modified drugs and drug carriers into inaccessible locations (Neha et al., 2013).

12.2 POLYMER-BASED NANOCARRIER SYSTEMS

12.2.1 Polymeric Nanoparticles

Polymeric nanoparticles (PNs) are solid carriers with sizes less than 1 μm that can entrap drugs in their polymeric matrix (Samy, Abdallah, Awad, & Ayoub, 2022a). Among nanocarriers, PNs have demonstrated notable benefits for brain transport of several medicinal medicines both in vitro and in vivo (Masserini, 2013).

PNs often have substantially higher encapsulation efficacy (EE%), are more stable, and are simpler to make and scale up (Barbu, Molnàr, Tsibouklis, & Górecki, 2009; Chigumira et al., 2015). In addition to being relatively easily changed with targeting ligands to improve BBB transport or target the loading medication to particular CNS tissues, polymers can be made to be innately multipurpose. Incorporating stimuli-responsive polymeric building blocks into polymeric NPs can achieve a sustained drug release profile or permit regulated release (Domján, Manek, Geissler, & László, 2013). Natural and synthetic polymers can both be used in therapeutic purposes (Lockman, Mumper, Khan, & Allen, 2002). The ideal polymer for transport to the brain is stable in blood, biodegradable, biocompatible, nontoxic, nonthrombogenic, nonimmunogenic, and non-inflammatory. Additionally, polymeric NPs need to resist aggregation and be made in an affordable and scalable way (Gaillard, Visser, Boer, Appeldoorn, & Rip, 2014). Chitosan, PLGA, PCL, and PBCA are the four most often utilized polymers for medication transport to the brain. Diblock co-polymers, including chitosan co-polymers, PEG-PLGA, MPEG-PCL, poly(ethylene glycol)-PLA, and PBCA-chitosan, have also recently been proposed as the building blocks of nanocarriers (Andrieux & Couvreur, 2009; Hrkach et al., 2012). Additionally, they offer great pharmacokinetic control, have a high potential for surface modification by chemical transformations, and can be used to entrap and deliver a variety of medicinal compounds (Sharma et al., 2015). The creation of PNPs can be done

* Corresponding author: moshera_samy1984@yahoo.com

DOI: 10.1201/9781003383376-12

in a variety of techniques, such as ionic gelation, double emulsification, emulsification solvent diffusion, and nano-emulsification nanoprecipitation.

12.2.2 Chitosan Nanoparticles (CSNPs)

Chitosan (CS) is a naturally occurring polysaccharide. It is a by-product of the deacetylation of chitin and finds extensive usage in textiles, environmental monitoring, medicines, and tissue repairs. CS is typically recognized as harmless, biocompatible, and degradable (Samy, Abd El-Alim, Amin, & Ayoub, 2020). CS has been extensively studied for the creation and manufacturing of nanoparticles due to its favorable physicochemical and biological characteristics (Mohammed, Syeda, Wasan, & Wasan, 2017; Vozza, Danish, Byrne, Frías, & Ryan, 2018; H. Zhang, Huang, Huang, Liu, & Li, 2015). Ionic gelation has garnered a lot of attention among the several methods being researched to manufacture CSNPs since it has the benefits of being convenient, controlled, free of organic solvents, and non-toxic (Agnihotri, Mallikarjuna, & Aminabhavi, 2004). This approach relies on physical cross-linking rather than emulsifying or cross-linking chemicals, which may be hazardous. Additionally, ionic gelation reduces the chance that included medicinal substances will be damaged, especially if they are biological in nature (Berger et al., 2004). Positively charged CS chains and polyanions interact electrostatically to generate CSNPs, and sodium tripolyphosphate (STPP) is utilized as a cross-linker (Kamat, Bodas, & Paknikar, 2016; P. Zhang, Liu, Hu, Bai, & Zhang, 2016), owing to its multivalent and non-toxic properties (Samy, Abd El-Alim, et al., 2020).

Memantine hydrochloride (MEM) is the only anti-AD medication that has been used to treat mild AD disease in Europe and the United States. MEM is a well-tolerated medication that must be taken daily. However, due to low drug compliance, there is a risk that treatment will not be effective (Gothwal et al., 2019).

For instance, Ruby and Pandey (2015) mentioned the preparation of MEM-loaded CS nanoparticles via ionic gelation. As a cross-linker, tripolyphosphate anion (TPP) is used. The formulation was optimized for the particle size of nanoparticles, which was found to be 129 nm at 4 h of stirring times, 1,000 rpm stirring speed, and a 2:1 concentration ratio of chitosan and TPP. The zeta potential of the optimized formulation was found to be 54, which implies the formulation has good stability. When delivered to the brain through the olfactory nasal pathway using MEM-loaded chitosan nanoparticles, this could be a new delivery method for the treatment of Alzheimer's disease.

A non-ergoline dopamine agonist called rotigotine is reportedly very efficient in treating Parkinson's disease (PD). But because of its high first-pass metabolism, limited oral bioavailability (1%), and brief plasma half-life, oral rotigotine's therapeutic potential and use in long-term treatment have been constrained (Bi et al., 2016).

Rotigotine-loaded chitosan nanoparticles (RNPs) were synthesized by Tzeyung *et al.* (2019) via the ionic gelation method for nose-to-brain delivery. The developed RNPs exhibited a small hydrodynamic particle size (75.37 ± 3.37 nm), a small PDI (0.368 ± 0.02), a satisfactory zeta potential (25.53 ± 0.45 mV), and a very high EE% (96.08 ± 0.01). The in vitro release kinetics confirmed the sustained release of rotigotine from the formulation. A histopathological study showed that the nanoparticulate formulation produced no toxicity or structural damage to the nasal mucosa. RNPs could be promising as an efficient carrier for nose-to-brain delivery.

In another study, Bhattamisra *et al.* (2020) mentioned the preparation of rotigotine-loaded chitosan nanoparticles (RNPs) using the ionic gelation method for treating Parkinson's disease. The prepared NPs exhibited particle size zeta potential, and PDI are 75.37 ± 3.37 nm, +25.53 0.45 mV, 0.368 ± 0.02, respectively. RNPs had no cytotoxicity against SH-SY5Y human neuroblastoma cells after 24 h of exposure. The outcomes from the PD animal model demonstrate that intranasally delivered RNPs improved medication bioavailability and brain-targeting effectiveness. RNPs for nose-to-brain delivery therefore have a lot of promise to be explored as a PD therapy strategy.

Sitagliptin is a kind of dipeptidyl-peptidase (DPP)-4 inhibitor. It was observed that sitagliptin could treat AD. It has shown an anti-apoptotic effect, which was considered an anti-inflammatory agent, by affecting macrophages and microglia. Also, it cannot cross the BBB (Mozafari et al., 2022).

Recently, sitagliptin-loaded chitosan nanoparticles (SIT-CS-NPs) were synthesized by Wilson *et al.* (2021) via ionic gelation to get the specific medication into the brain after intranasal (IN) administration. The zeta potential and mean size were 20.8 mV and 188.4 ± 48.1 nm, respectively. After IN administration, the CS-NPs significantly improved SIT's ability to concentrate. Targeting SIT into the brain via the created NPs may be a secure and efficient method.

12.2.3 Chitosan (CS) Co-Polymer Nanoparticles

CS co-polymer nanoparticles play an essential role in CNS treatment in the recent years. For example, OX26, a monoclonal antibody with a high selectivity for the transferrin receptor (TfR) found in the capillary endothelial cells of the brain, is one of the most frequently utilized to route NPs to the brain (Ulbrich, Hekmatara, Herbert, & Kreuter, 2009).

Monsalve *et al.* (2015) mentioned the preparation of the monoclonal antibody OX26 loaded on PEG-g-chitosan nanoparticles (NPs) as a drug transporter system for brain drug targeting. In OX26 antibody–loaded CS-PEG, positive charge and sizes between 140 and 200 nm have been effectively attained. Furthermore, for 30 days at 4°C in a pH 6.0 buffer media, these nanoparticles demonstrated high colloidal stability. These nanoparticles were shown to be able to cross the BBB in in vivo trials (intraperitoneal administration in mice), indicating that they may be a promising drug delivery method for the CNS.

O^6-benzylguanine (BG), an inactivator of O^6-methylguanine-DNAmethyl transferase (MGMT), is used to enhance and increase the effectiveness of carmustine (BCNU) by specifically lowering the levels of MGMT in tumor cells (Qian et al., 2013).

For example, the preparation of O6-benzylguanine (BG)– and carmustine (BCNU)–loaded chitosan treated with poly(lactide-co-glycolides) nanoparticles (PLGA/CS NPs) via nanoprecipitation method for treatment of brain disease was mentioned by Liu *et al.* (2017). The diameter of the iRGD-NPs is around 127 ± 2.3 nm, and the average zeta potential is 18.5 ± 1.5 mV, with a loading efficiency of 3.28 ± 0.21% and 9.53 ± 0.19% for BCNU and BG, respectively. In comparison to other groups, iRGD-NPs showed greater cytotoxicity and cellular absorption, enhancing the effectiveness of anticancer medication delivery systems through the use of iRGD.

On the other hand, Ahmad (2017) reported the creation of chitosan-coated PLGA nanoparticles with encapsulation of rasagiline for the treatment of Parkinson's disease using double emulsification–solvent evaporation method. At various PLGA polymer compositions, the mean particle size and PDI were determined to be in the range of 63.13–402.18 nm and 0.241–0.501, respectively. When compared to the outcomes of nasal in situ gel delivery and chitosan glutamate nanoparticles, in vitro release experiments showed the produced nanoparticles to have better release profiles. These research results led to the conclusion that RSG-loaded, CS-coated PLGANPs have better brain bioavailability and are, thus, efficient therapies for Parkinson's disease. And it might be a promising method of administering intranasal medication to treat Parkinson's disease.

As a possible MIC inhibitor, catechin hydrate (CH) was selected. Natural flavan-3-ol CH has been studied for its range of health-promoting biological properties, including its anti-inflammatory, antioxidant, and anticancer actions (Lekbach et al., 2019).

In another study, CH-loaded CS–PLGA nanoparticles (NPs) were synthesized by Ahmad *et al.* (2020) via double emulsion–solvent evaporation method to increase medication brain absorption through direct CNS administration through the nose. According to SEM (Figure 12.1 a) and TEM (Figure 12.1 b) images, the obtained NPs have a smooth surface. The values of particle size, zeta potential, and PDI were observed to be 93.46 ±3.94 nm, 12.63 0.08 mV, and 0.106 ± 0.01, respectively. The obtained NPs improved the bioavailability of CH, showed a sustained release profile over 24.0 h (Figure 12.1c), prolonged retention for nose-to-brain delivery (Figure 12.1d), and avoided hepatic first-pass metabolism. The NPs produced were

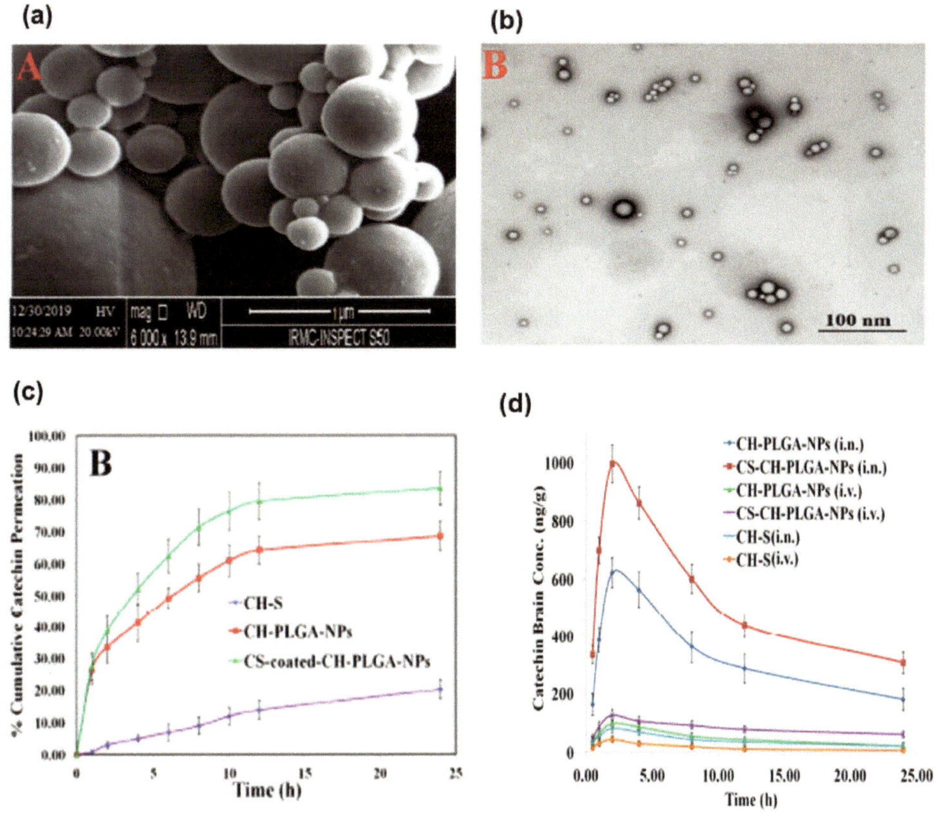

FIGURE 12.1 (a) SEM and (b) TEM images of CS-coated, CH-loaded PLGA-NPs. (c) The ex vivo permeation characteristics of produced chitosan-coated, CH-loaded PLGA-NPs and pure CH–S. (d) Pharmacokinetic profiles of CH content at various points after administration. (From Ahmad et al. (2020). Copyright 2020. Adapted from MPDI.)

demonstrated to be a unique, efficient, and secure brain-targeted delivery strategy for the treatment of epilepsy by the results in the PTZ and ICES epilepsy models.

Huperzine A (HupA) exhibits enhancements in memory and cognitive functions. It can effectively reduce oxidative injury by increasing antioxidant enzyme activity, enhancing neuronal viability, reducing the buildup of amyloid peptides, and decreasing mitochondrial dysfunction (Jiang, Jiang, Ding, & Yu, 2022).

Recently, Meng *et al.* (2018) created HupA-loaded, lactoferrin-conjugated, TMC-modified PLGA NPs using the emulsion–solvent evaporation method for treatment of Alzheimer's disease. Maleimide–TMC and TMC are displayed in Figure 12.2a. Optimized HupA Lf-TMC NPs had a particle size, zeta potential, and PDI of 153.2 ± 13.7 nm, +35.6 ± 5.2 mV, and 0.229 ± 0.078, respectively, and drug entrapment efficiency of 73.8% ± 5.7% and sustained release in vitro over a 48 h period (Figure 12.2b). HupA Lf-TMC NPs showed lower toxicity (Figure 12.2c). Lf-TMC NPs displayed a greater fluorescence intensity in the brain and a longer residence time than nontargeted NPs, according to in vivo imaging studies (Figure 12.1d). Lf-TMC NPs offer a wide range of potential applications as a nasal drug delivery carrier due to their good sustained-release action, adhesion, and targeting abilities.

12.2.4 Poly(Lactic-Co-Glycolic Acid), PLGA

PLGA is a biodegradable polymer that has been investigated for its biodegradability and biocompatibility in sustained medication delivery systems for the brain (Samy, Abdallah, Awad, & Ayoub, 2022b). It is frequently utilized in human research owing to FDA approval (Food and Drug Administration, USA). It has been utilized as a medication carrier for several different pharmaceuticals, with success for more than ten years (Lee et al., 2004). PLGA nanoparticles (PLGA NPs) are colloidal polymeric drug carriers that exhibit promise for oral drug delivery, which is by far the most popular and practical mode of administration. PLGA NPs provide a number of benefits over conventional oral

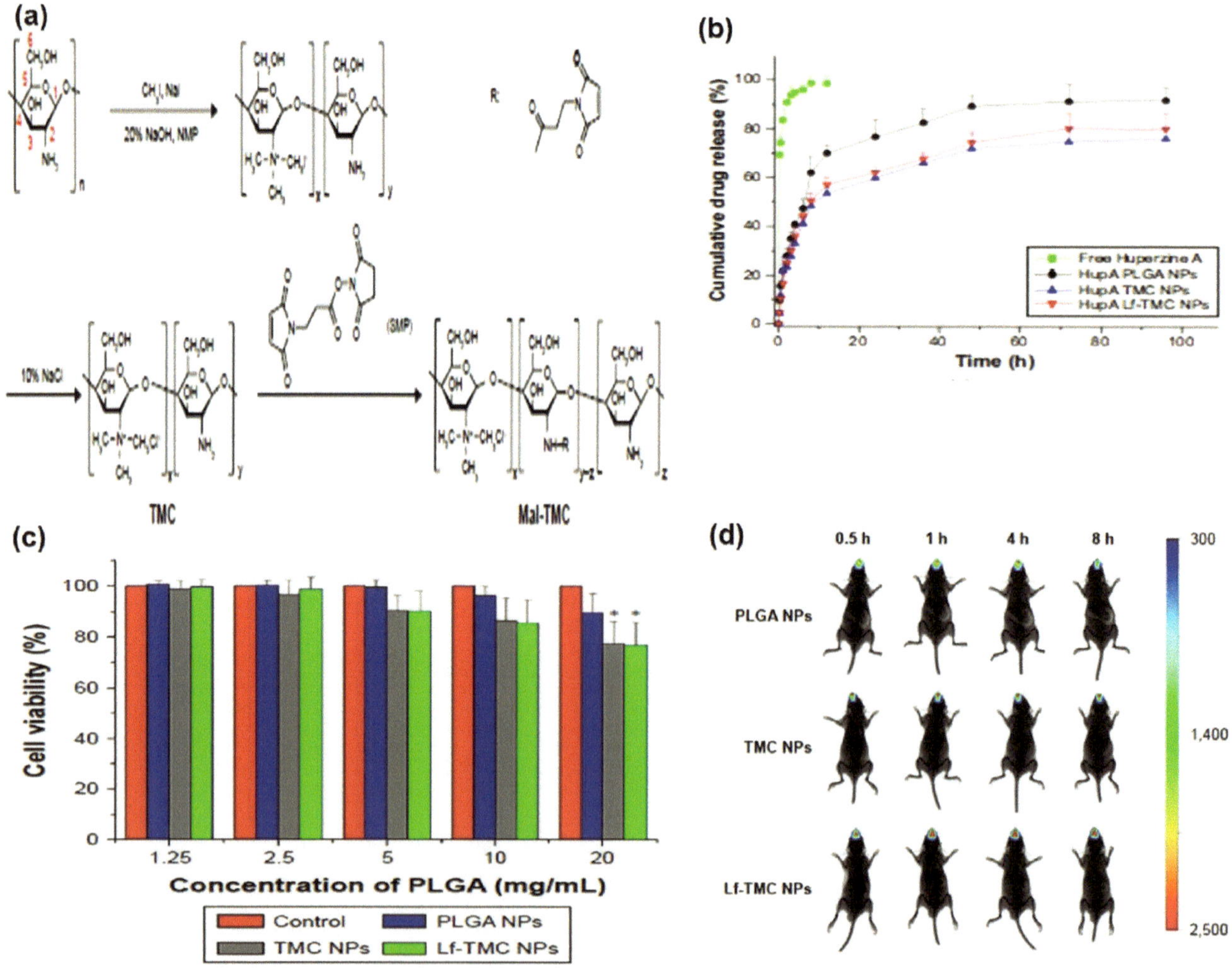

FIGURE 12.2 (a) The synthesis of TMC and Mal-TMC. (b) In vitro release study of HupA from free Huperzine A, HupA PLGA NPs, HupA TMC NPs, and HupA Lf-TMC NPs in 0.1 M, pH 7.4 PBS. (c) Cytotoxicity of free HupA, HupA-loaded NPs incubated for 24 h in 16HBE cells. (d) In vivo fluorescence image of organs of mice after IN of DiR-loaded PLGA NPs, TMC NPs, and Lf-TMC NPs. (From Meng et al. (2018). Copyright 2018. Adapted from Taylor & Francis.)

dose forms, including enhancing the oral bioavailability of poorly absorbed medications and maintaining the encapsulated chemicals in the polymer network (Samy et al., 2022b).

Loperamide (LOP) is a μ-opioid receptor agonist that is frequently available as an over-the-counter medication and is commonly used to treat diarrhea (Kang, Compton, Vaz, & Rampe, 2016).

For example, Fornaguera *et al.* (2015) used nanoemulsion tempering to create loperamide-loaded PLGA nanoparticles as effective nanocarriers for medication delivery through the BBB (Figure 12.3a). Surfactant polysorbate 80– and loperamide-loaded PLGA nanoparticles had hydrodynamic radii of 44.73 and 2.98 nm and zeta potencies of 38.70 and 7.52 mV. The obtained nanoparticles were non-cytotoxic up to a concentration of 30 mg/Ml (Figure 12.3b). Both the aqueous Tween 80 + LOP and the aqueous morphine resulted in a difference between the pre- and post-latency times (Figure 12.3c). These results will provide sufficient evidence to end up in clinical trials in the near future.

Galantamine (GAL) was chosen as a "gold standard" drug; it is a reversible, competitive AChE inhibitor used to treat neuroprotective effects, despite the fact that galantamine has been authorized for the treatment of mild to moderate symptoms (Misra, Chopra, Sinha, & Medhi, 2016).

Fornaguera *et al.* (2015) described the preparation of galantamine (GAL)–loaded PLGA nanoparticles via the nanoemulsification method as drug carrier to treat neurodegenerative diseases. The hydrodynamic radius of GAL-loaded PLGA nanoparticles was 32.73 and 0.39 nm, and the zeta potential was 24.15 and 6.74 mV. EE% more than 90% by weight and a sustained drug release profile were established, maintaining GAL for intravenous delivery, thus enabling long-term therapeutic effects. Non-cytotoxic GAL-loaded nanoparticles have been developed with the necessary characteristics to become effective in improving medication delivery devices for neurodegenerative disorders.

Loureiro *et al.* (2016) prepared PLGA NPs functionalized with two separate monoclonal antibodies, the OX26 mAb, a well-known cell surface receptor on the brain capillary endothelium, and the DE2B4, a molecule able to recognize the Aβ peptide, which have been proposed to deliver anti-amyloid peptides. The size of PLGA NPs indicates a homogeneous distribution (PDI ≤ 0.1). The spherical PLGA NPs, which were produced via nanoprecipitation, have an average diameter of about 160 nm. After optimization (drug loading, surface density), the formulation based on two antibodies, coupled with PLGANPs, is a promising approach to prevent proteolytic degradation of anti-amyloid peptides and to improve their transport through the BBB. The therapeutic peptide-transporting nanoparticles will effectively cure AD in transgenic models.

Curcumin (Cur; 5-hydroxy-1,7-bis(4-hydroxy-3-methoxyphenyl)-1,4,6-heptatrien-3-one) is a plant-derived antioxidant with two phenolic groups. It is one of the most effective natural therapeutic agents for treating AD because of its neurological benefits, for AD, Parkinson's disease,

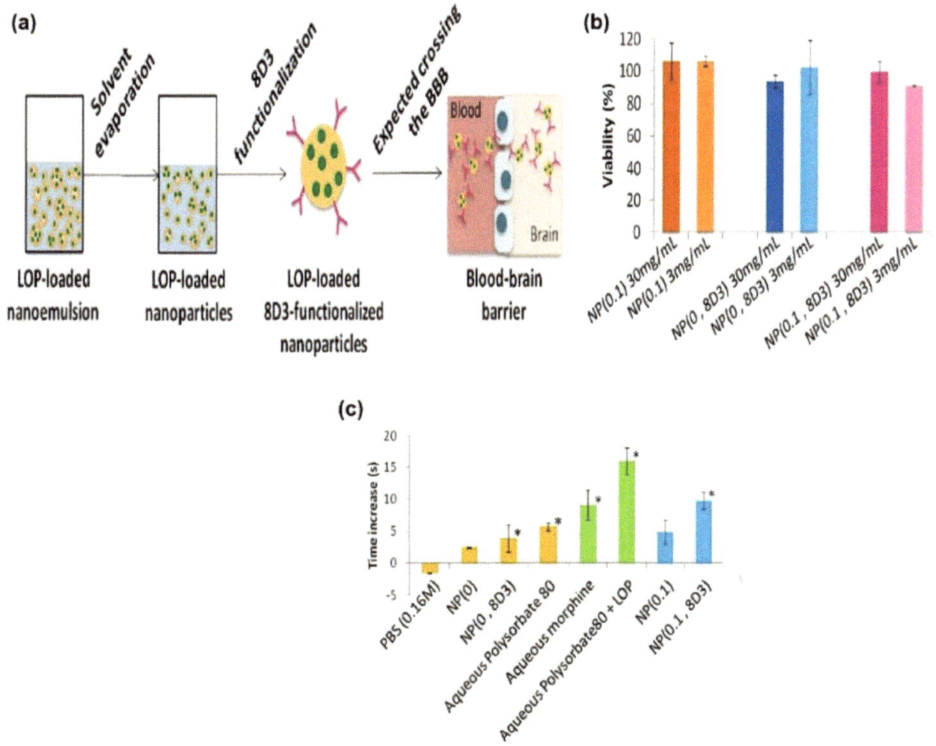

FIGURE 12.3 (a) Schematic of the preparation of LOP-loaded nanoparticles from nanoemulsion templating; (b) viability of HeLa cells for 24 h incubation at different concentrations; (c) the intervals between the post-latency time and the pre-latency time of mice of negative controls (yellow), positive controls (green), and test nanoparticles (blue). (From Fornaguera et al. (2015). Copyright 2015. Reproduced with permission from Elsevier.)

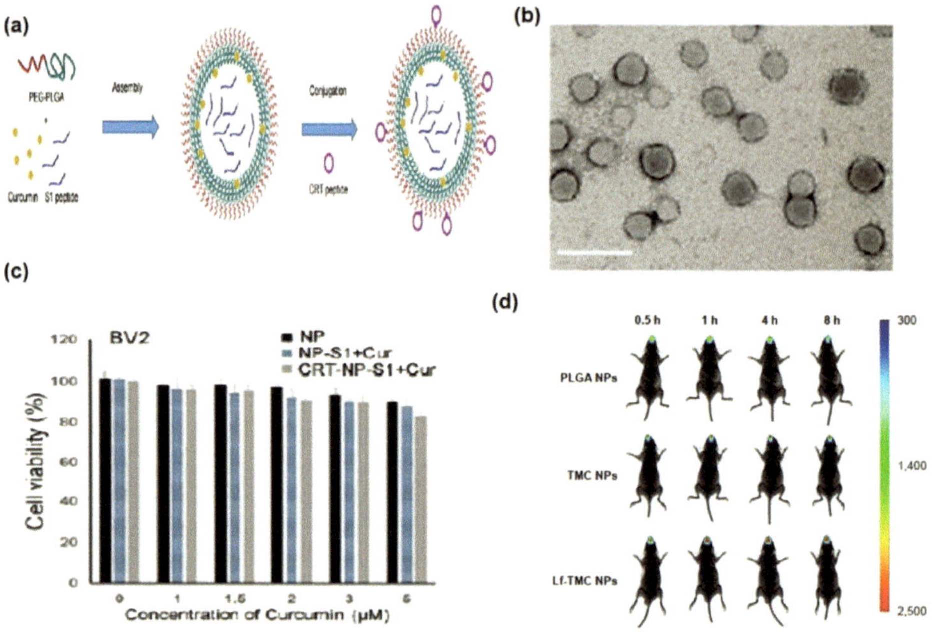

FIGURE 12.4 (a) Schematic representation of the production of PLGA. (b) TEM image of CRT-NP-S1+Cur. (c) The cytotoxicity of NP-S1+Cur conjugated with or without CRT was determined via MTT assay by adding NPs to BV2 cells. (d) Tail vein injections of ICG-labeled NP-S1+Cur and CRT-NP-S1+Cur were injected to nude mice. IVIS spectrum imaging system observed the fluorescence at various times. (From Huang et al. (2017). Copyright 2017. Adapted from Oncotarget.)

brain tumors, MS, traumatic brain injuries, ischemia, and depression (Ege, 2021).

Aβ generation inhibitor S1 (PQVGHL peptide) and curcumin-loaded PLGA nanoparticles (Figure 12.4a) were fabricated by Huang *et al.* (2017) via the double emulsion–solvent evaporation method. The average PS of spherical curcumin-loaded PLGA nanoparticles and spherical CRT-conjugated PLGA nanoparticles were 128.6 nm and 139.8 nm (Figure 12.4b), respectively, for the controlled-release formulations. Consequently, curcumin's limited brain bioavailability and prolonged release were overcome by PLGA NPs. PLGA nanoparticles did not have a significant cytotoxic effect on BV2 cells (Figure 12.4c). CRT-NPs accumulated in the mouse brain (Figure 12.4d). The best results were seen when S1 and curcumin were co-delivered via PLGA nanoparticles for the treatment of Alzheimer's (AD) disease.

Chu *et al.* (2018) described the preparation of temozolomide butyl ester (TBE)–loaded, antibody-modified PLGANPs using an emulsion–solvent evaporation technique to target glioblastoma via nose-to-brain delivery. The obtained NPs exhibited a sustained release (Figure 12.5a) and a mean PS of 145.9 ± 8.7 nm. The cellular uptake of the obtained NPs by C6 cells was dramatically increased (Figure 12.5b). The obtained NPs demonstrated high fluorescence intensity and effective accumulation of glioma tissues (Figure 12.5c), representing the targeting effect of anti-EPHA3. The apoptosis of glioma cells significantly increased in the glioma-bearing rats treated with the obtained NPs (Figure 12.5d). According to the aforementioned findings, the obtained NPs could be potentially used as a nose-to-brain drug carrier for the glioblastoma treatment.

Ramalho *et al.* (2018) prepared temozolomide (TMZ)–loaded PLGA NPs using a single emulsification solvent evaporation method for targeting the glioblastoma tumor cells. The accommodation of TMZ in the polymeric matrix resulted in an important ($p < 0.05$) rise in PS for the TMZ-PLGA NPs (181 ± 2 nm). Because BBB cells are known to overexpress this receptor, TMZ-PLGA NPs increase permeability across the BBB in vivo. As such, we propose that NPs modified with OX26 mAb for TfR could be efficiently used for dual targeting of both GBM and BBB cells. In vivo studies will be conducted in the future to assess the potential of the developed nanocarriers for the treatment of GBM.

Huo *et al.* (2019) mentioned the preparation of cur-loaded selenium nanoparticles PLGA nanospheres via an emulsion solvent evaporation method for the inhibition of amyloid β aggregation in Alzheimer's disease (AD). The produced nanoformulations have been developed for the inhibition of AD pathogenesis toxicity and Aβ aggregation. The nanospheres had PS of 160 ± 5 nm and a low PDI. The produced nanoformulations have been established as a valuable delivery system for the treatment of Alzheimer's disease.

PLGA nanoparticles were loaded with vitamin D–binding protein (DBP) using an emulsion–diffusion method by Jeon *et al.* (2019) for Alzheimer's disease (AD) pathogenesis. SEM with 226.6±44.4 nm, zeta potential of −0.144 mV, PDI of 0.039±0.013 at DBP: PLGA (1:10). DBP-PLGA nanoparticles attenuate neurodegeneration and neuroinflammation, inhibiting the activity of the obtained NPs on Aβ accumulation *in vivo* and ameliorating reasoning impairment in an animal model of AD. The obtained NPs could be a promising step in developing a therapeutic

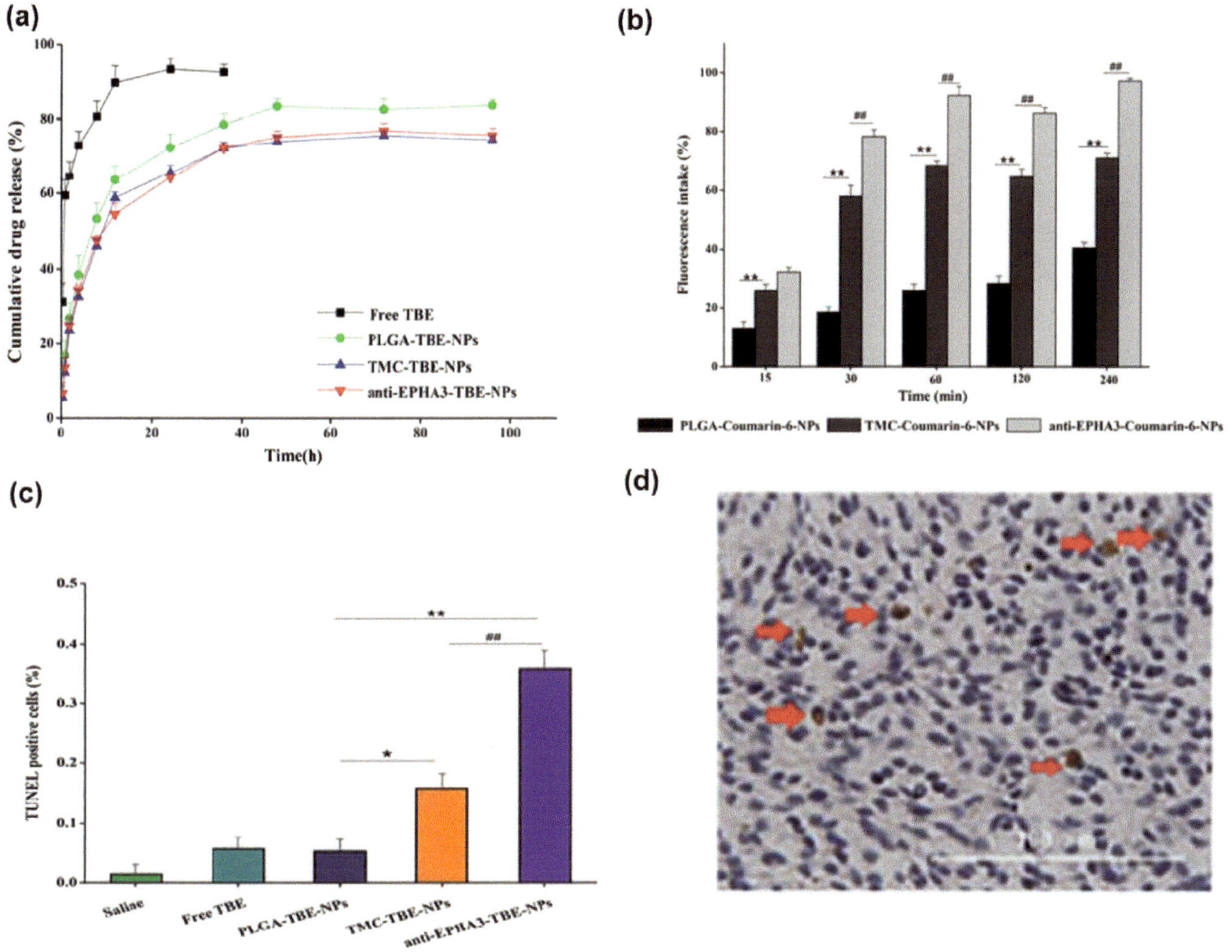

FIGURE 12.5 (a) In vitro release profiles of free TBE, T/P-TBE-NPs, P-TBE-NPs, and obtained NPs. (b) Mean proportions of coumarin-6 NPs uptake by C6 cells. (c) Apoptosis quantification in rats treated with different formulations. (d) Apoptosis of glioma cells in glioma-bearing rats treated with obtained NP. (From Chu et al. (2018). Copyright 2018. Adapted from Taylor & Francis.)

approach for the treatment of several neurodegenerative diseases made by abnormal protein aggregation.

Zhang *et al.* (2020) created Nec-1s-loaded PLGA nanoparticle conjugated with the CD47 extracellular domain via a reactive oxygen species (ROS)–responsive phenylborate ester bond exhibiting the "do not eat me" signal and the BBB-penetrating peptide CRTIGPSVC (CRT). The obtained nanoparticles were prepared by nanoprecipitation (Figure 12.6c). The obtained nanoparticles increased the halftime of obtained NPs in brain distribution and blood circulation. The resultant nanoparticles had a spherical morphology and a relatively uniform PS of 70 nm (Figure 12.1b), while the diameter of the unconjugated NPs was 72 ± 11.3 nm, with a PDI of 0.126. PLGA NPs coupled with CD47 exhibited 30% decrease in phagocytosis (Figure 12.6c) and the highest fluorescence intensity in the brain (Figure 12.6c). The obtained NPs could be promising in exploring the role of microglia in the AD pathogenesis and warrant further study for AD treatment.

Tripathi *et al.* (2021) prepared donepezil-loaded PLGA nanoparticles using the nano-precipitation method as advanced drug delivery system for Alzheimer's disease treatment. According to SEM (Figure 12.7a) and TEM (Figure 12.7b) images of donepezil-PLGA NPs, they are smooth and spherical in structure. The PS, ZP, and PDI of obtained NPs for the preferred formulation were found to be 40.8 nm (Figure 12.7c), −2.11 mV, and 0.188, respectively. The EE% was 74%, and the process yield was 78%. The obtained NPs will be beneficial when considering in vivo therapies in the future, by reducing side effects, crossing the BBB, and improving the therapeutic effect. The pharmacokinetics were consistent with the Hixson–Crowell model (Figure 12.7e).

12.2.5 PLGA–PEG Nanoparticles

PLGA–PEG NPs have attracted more attention for Parkinson's disease treatment. For example, Bi *et al.* (2016) prepared lactoferrin-PLGA-PEG nanoparticles via nanoprecipitation for Parkinson's disease treatment. The

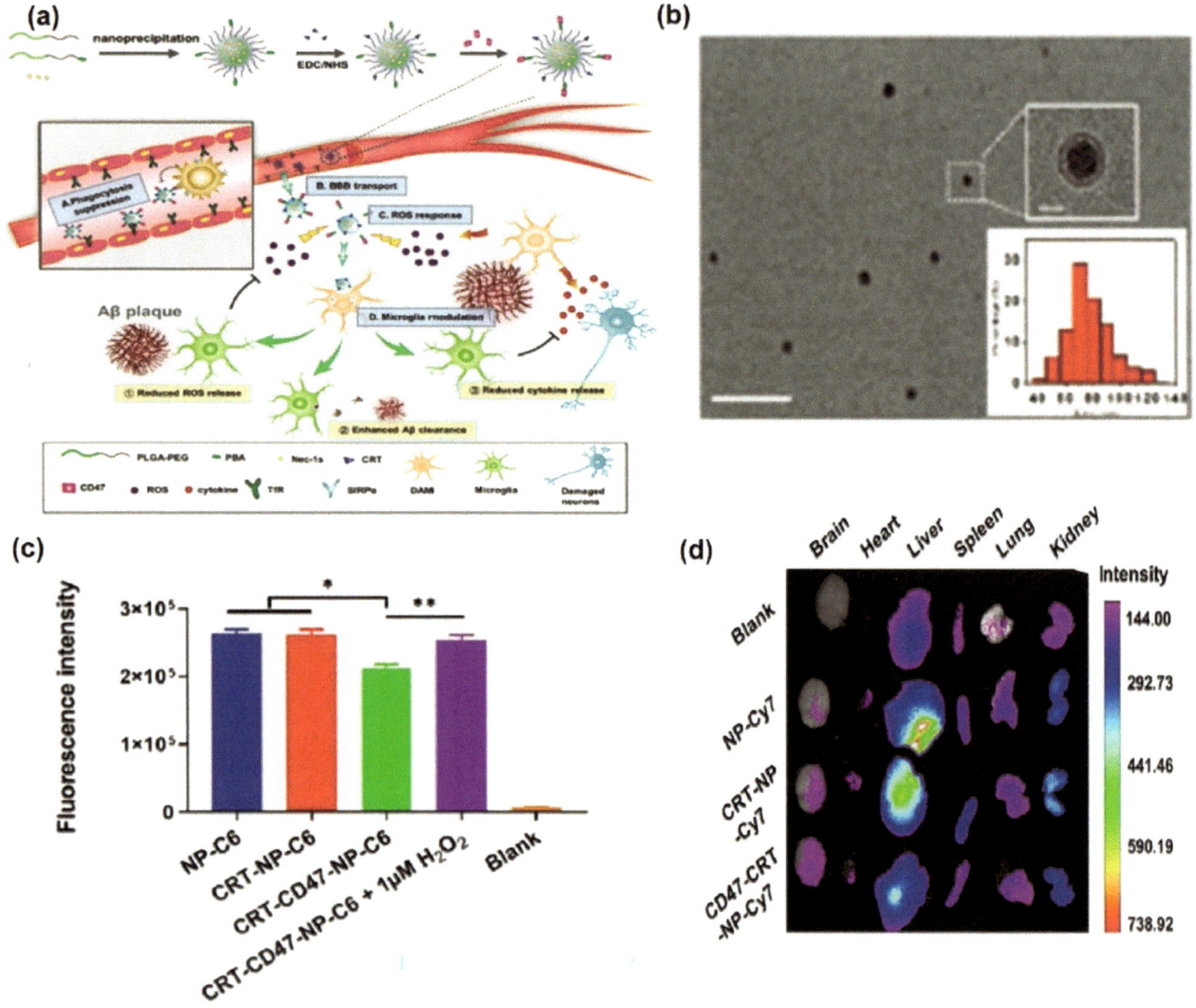

FIGURE 12.6 (a) Design of the transportation and CRT-CD47-NP-Nec-1s effect on neurons and microglia in the brains of AD. (b) TEM image of PLGA NPs. (c) Quantitative analysis of the mean fluorescence intensity in RAW 264.7 cells. (d) Fluorescence detected in different organs after blood administration of CRT-NP-Cy7, NP-Cy7, and CRT-CD47-NP-Cy7. (From L. Zhang et al. (2020). Copyright 2020. Adapted from Wiley.)

NPs showed a sphere shape with small PS of 122.0±19.3 nm, narrow size distribution of 0.194±0.023, and a negative potential of −21.28±2.15 mV. The release kinetics suggested that both Lf-NPs and NPs provided a continuous, slow release of rotigotine for 48 h. The NPs showed a spherical shape, small PS, thin size distribution, and negative zeta potential. The in vitro cytotoxicity of the unloaded and rotigotine-loaded NPs was low. Cellular uptake tests showed that obtained NPs exhibited improved cellular accumulation. The obtained NPs could be a promising, non-invasive approach for the treatment of PD.

Donepezil is a drug used for the treatment of AD. It acts as a reversible acetylcholinesterase inhibitor that increases the concentration of acetylcholine in cholinergic synapses. A major disadvantage with donepezil, however, is that although it can cross the BBB, many unwanted side effects are reported to arise in the periphery, counting hepatotoxicity, diarrhea, nausea, muscle spasms, insomnia, and cardiac arrhythmia (Topal et al., 2020).

In another study, Baysal *et al.* (2017) mentioned the preparation PLGA-b-PEG nanoparticles encapsulated with donepezil via the double emulsion technique and that it crossed the blood–brain barrier in vitro. The resultant NPs have PS of 174 ± 202 nm, a zeta potential of −20 ± 45 mV, and a PDI of 0.20. Nanoparticles demonstrated the ability to cross the BBB while also exhibiting a controlled-release profile. The obtained NP management caused a significant dose-dependent reduction in both protein expression levels and gene of IL-1b, IL-6, GM-CSF, and TNF-a. No significant variations were observed for MCP-1 and TGF-b.

On the other hand, Fan *et al.* (2018) reported the preparation of curcumin (cur)–loaded PLGA-PEG nanoparticles conjugated with B6 peptide (PLGA-PEG-B6/Cur) via solvent evaporation method (Figure 12.8a) for potential use in

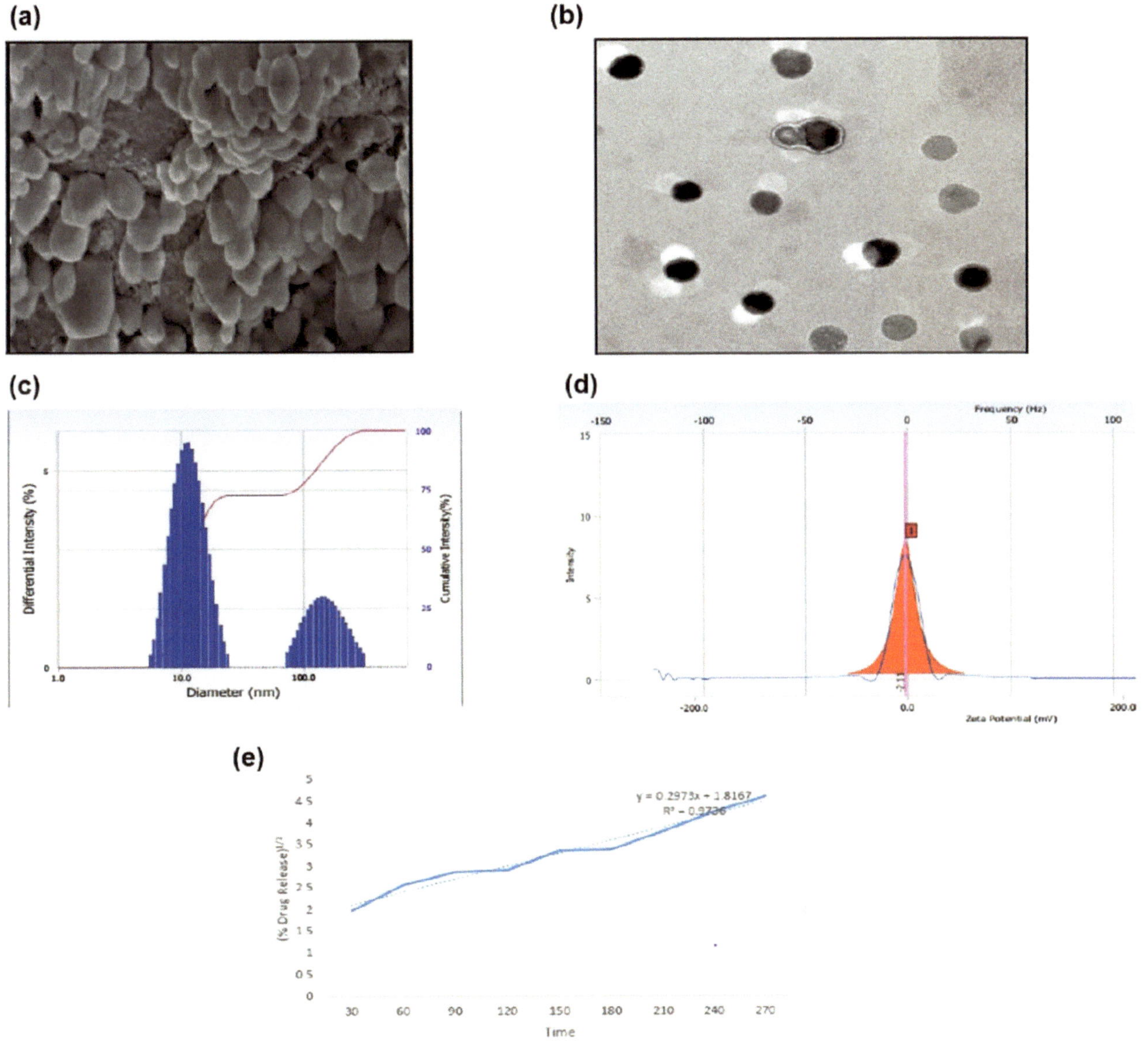

FIGURE 12.7 (a) SEM and (b) TEM images of (c) PS distribution, (b) zeta potential, and (e) plots for Hixson–Crowell model. (From Tripathi et al. (2021). Copyright 2021. Adapted from IOP Science.)

Alzheimer's disease, The results of the study suggest that the obtained nanoparticles were biocompatible, had high bioavailability, and possessed relatively low toxicity profiles (Figure 12.8b). Particle sizes of the produced nanoparticles range from 50 to 250 nm. The produced nanoparticles had a sustained release of 78.01 ± 1.55% over the course of 72 h and had a zeta potential of 3.83 ± 0.89 mV. The obtained nanoparticles shortened the escape latency compared to the APP/PS1 group (Figure 12.8c), reduced hippocampal Ab production, and decreased tau phosphorylation in APP/PS1 mice (Figure 12.8d). These findings together indicate that the obtained nanoparticles could be a promising therapeutic approach for AD treatment in the future and can significantly enhance the delivery efficiency of nanoparticles to the brain.

Recently, Senturk *et al.* (2022) loaded TMZ into magnetic poly(lactic-*co*-glycolic acid)-polyethylene glycol (PLGA-b-PEG) nanoparticles (TMZ-*m*-PNPs) via single emulsion–solvent evaporation method against glioblastoma therapy. The obtained nanoparticles displayed spherical morphology, NPs size diameter of 185 ± 45 nm, and EE% of TMZ range of 7%. In cell culture, IC_{50} values of free TMZ were found for human endothelial cells (HUVEC, ~500 μM) and human glioblastoma cells (T98G, ~350 μM) at the 72nd h. Furthermore, live/dead cell viability, and annexin V-FITC/PI assays of TMZ-*m*-PNPs, but cytotoxicity of NPs was not dose-dependent in T98G cells. The obtained nanoparticles that would provide adequate and burst release of TMZ to the tumor site without contributing to therapeutic resistance may be critical in GBM therapy.

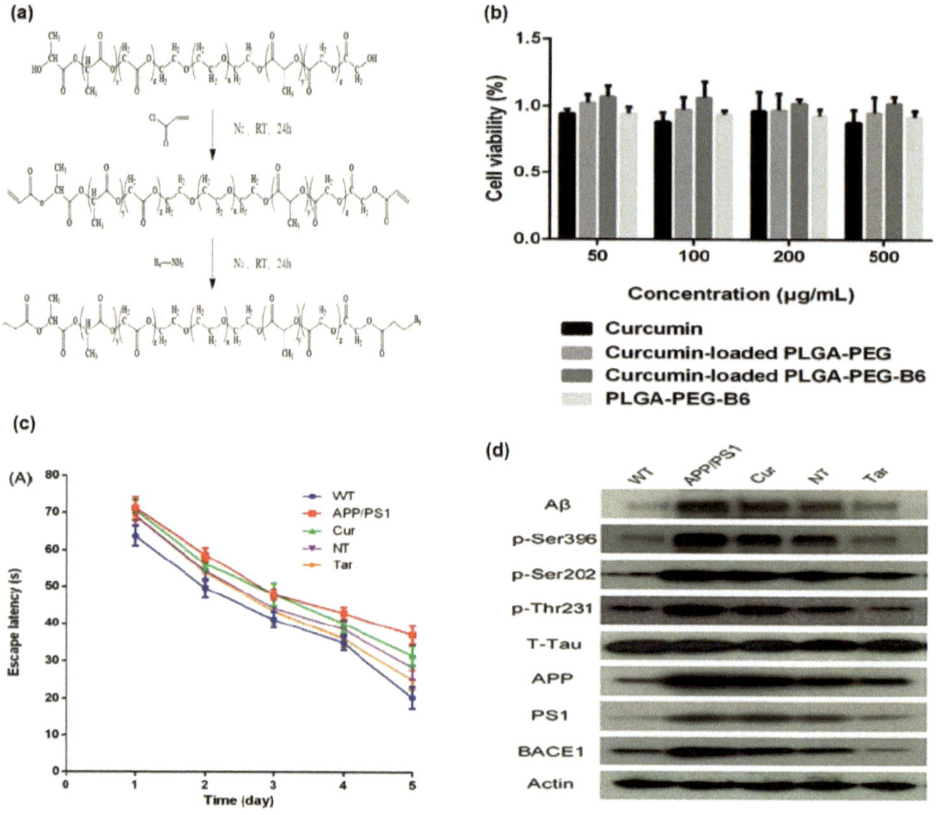

FIGURE 12.8 (a) General procedure for the synthesis of PLGA-PEG-B6 nanoparticles. (b) Results of cell viabilities by CCK-8 assay after 24 h incubation with different concentrations (50, 100, 200, and 500 mg/mL) of cur, cur-loaded PLGA-PEG, cur-loaded PLGA-PEG-B6, and PLGA-PEG-B6, respectively. (c) Comparison of escape latency of each group from 1 to 5 days in learning trials. (d) The obtained nanoparticles' effect on protein bands of groups Aβ, p-Tau (Ser202, Ser396, and Thr231), PS1, T-Tau, BACE1, APP, and actin of each group, respectively. Actin served as the internal control. (From Fan et al. (2018). Copyright 2018. Adapted from Taylor & Francis.)

12.2.6 Poly(ε-Caprolactone) Nanoparticles (PCL NPs)

PCL was selected as a valuable polymer as it is used in several medical applications, including drug delivery and tissue engineering, due to its high degree of crystallinity, hydrophobicity (Samy, Abdallah, et al., 2020), biodegradability, biocompatibility, and miscibility with a variety of other polymers and high permeability for many drugs (Sukhanova et al., 2019). PCL degrades rather slowly as compared to several other polymers (Amin, Samy, Abd El-Alim, Rabia, & Ayoub, 2018).

For intanse, Del Prado-Audelo *et al.* (2019) curcumin-loaded PCL nanoparticles via the emulsification–diffusion method for potential application in neuronal diseases. The produced nanoparticles had a mean PS of 149 ± 2.2 nm. The produced nanoparticles had EE% of 96% efficacy. Moreover, MIO-M1 and SH-SY5Y cells were used to test the cytotoxicity of nanoparticles in vitro, indicating sufficient biocompatibility. The analysis of cellular uptake used MIO-M1 cells with high internalization rates. The produced nanoparticles could be a potential alternative treatment for neuronal diseases, such as spinocerebellar ataxias and Parkinson's disease.

On other hand, Mahmoudi *et al.* (2021) memantine-loaded polycaprolactone nanocapsules via interfacial deposition of pre-formed polymers for Alzheimer's disease. Memantine was prepared using an optimized technique. The EE% of the drug in the PCL nanocapsules was higher than 80%. The results show that the produced nanoparticles had amorphous morphology with a PS within the range of 170–250 nm, PDI <0.2, and a surface negative charge of −26 mV that is stable upon 40 days of storage at 25°C, without any aggregation or decomposition. The results show that the produced nanoparticles are a promising system that could be used to treat Alzheimer's disease.

12.2.7 Methoxy Poly(Ethylene Glycol)-Co-Poly (ε-Caprolactone) (MPEG-PCL) Nanoparticles

MPEG-PCL NPs have attracted more attention for brain CNS treatment. For example, Elmowafy *et al.* (2019) methazolamide (MTZ)–loaded methoxy-poly(ethylene glycol)-b-poly(ε-caprolactone) (mPEG-PCL) polymeric micelles (PMs) using the thin-film hydration procedure. The selected formula exhibited a PS of 60 nm, EE% of 93%, and discrete spherical particles. Moreover, superior cellular, sustained release of MTZ, tissue biocompatibility, and marked anti-glaucoma efficacy, as compared to drug solution, were realized. The combined results show that the produced nanoparticles could enhanced the therapeutic benefit of ocular medication delivery based on nanotechnology.

On the other hand, Kuo *et al.* (2019), using a microemulsion–solvent evaporation method, grafted etoposide (ETO), carmustine (BCNU), and doxorubicin (DOX) into methoxy poly(ethylene glycol) (MPEG) poly(caprolactone) (PCL) nanoparticles (WFNPs). The PS was determined to be between 150 and 200 nm. The obtained nanoparticles connected to *N*-acetylglucosamine and folate receptors with the help of surface WGA and FA. The obtained nanoparticles carrying ETO, BCNU, and DOX act as dual targeting nanocarriers, and their use can be a promising approach to inhibiting GBM growth in the brain.

In another study, Zhao *et al.* (2020) discussed the preparation of ginkgolide B (GB)–loaded poly(ethylene glycol)-co-poly(ε-caprolactone) (PEG-PCL) nanoparticles using an antisolvent precipitation and stabilizer-based approach in order to enhance their capacity to accumulate in the brain and treat Parkinson's disease. The obtained nanoparticles enhanced GB brain uptakes, and GB sustained higher levels in the plasma of treated animals (Figure 12.1a). The obtained nanoparticles inhibited C6-NPs endocytosis (70.48%), followed by MβCD (68.86%), HS (38.94%), and EIPA (20.92%) (Figure 12.9a) optimized GB-NPs, which were found to have a PS of 91.26 ± 1.34 nm, a PDI of 0.17 ± 0.01, a ZP of −12.09 ± 0.97 mV, a DL capacity of 26.93%, and an EE% of 87.52%. The obtained nanoparticles enhanced cell viability (Figure 12.9c) and were sufficient to reverse these MPTP-induced effects in these PD model mice (Figure 12.9d). The obtained nanoparticles enhanced their oral bioavailability, cerebral accumulation, and bioactivity by mediating their sustained release in vivo.

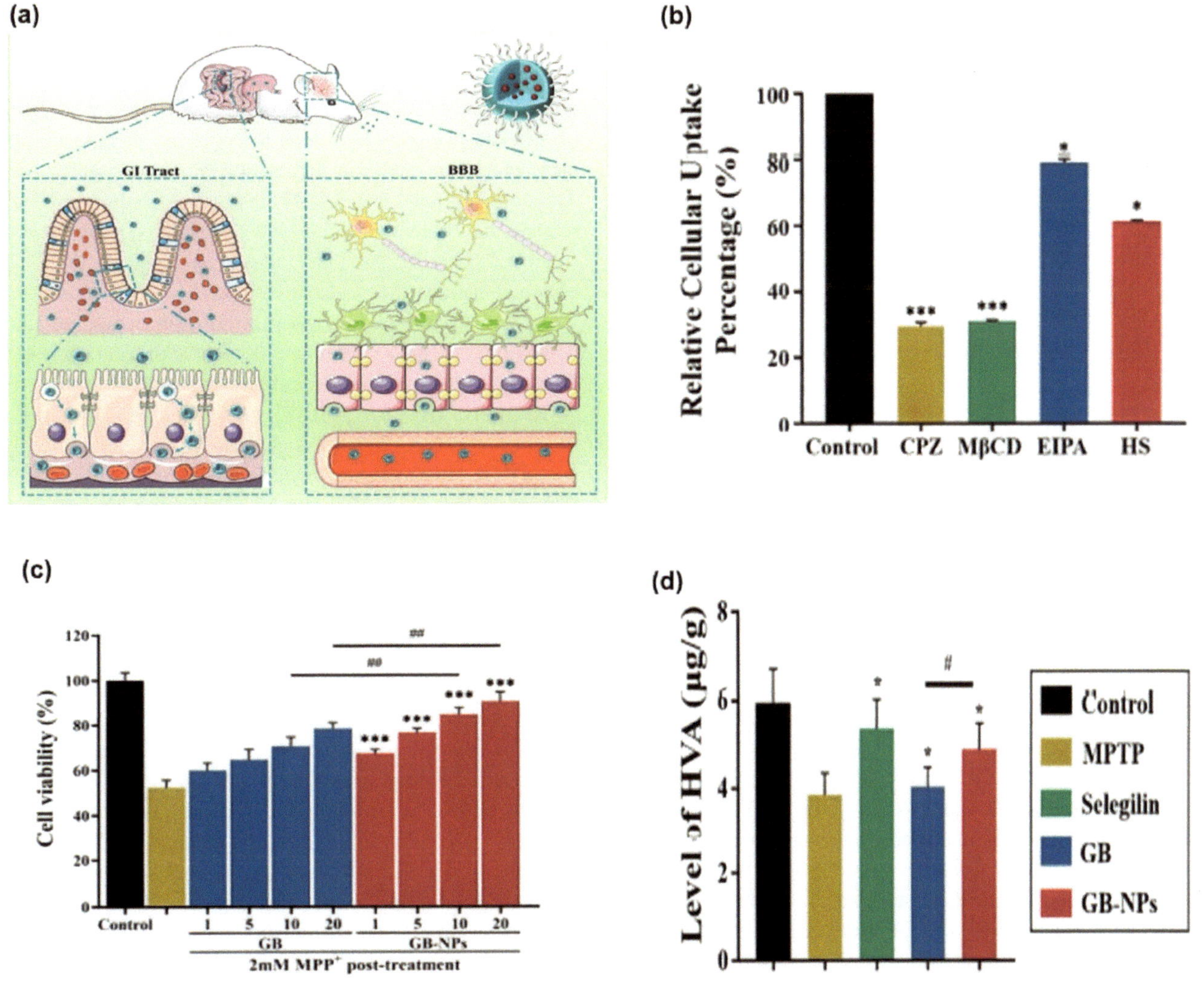

FIGURE 12.9 (a) Schematic presentation of the mechanisms influencing the absorption and accumulation of GB-NPs following oral uptake. (b) Quantification of TH⁺ neurons ($n = 5$). Striatal DA. (c) GB-NPs protected against MPP⁺-mediated cytotoxicity ($n = 3$). ***$p < 0.001$ vs the MPTP group. (d) Typical activity traces from video tracking, shown in green. (From Zhao et al. (2020). Copyright 2018. Adapted from Taylor & Francis.)

Recently, Mohamadpour *et al.* (2020) used nanoprecipitation to create rivastigmine (Riv)–loaded methoxy poly(ethylene glycol)-co-poly(ε-caprolactone) (MPEG-PCL) nanoparticles (Riv-NPs). The produced nanoparticles exhibited a PS of 98.5 ± 2.1 nm, DL of 19.2 ± 1.1 %, and sustained-release behavior of the drug. Moreover, the neuropharmacokinetic study in rats indicated that the integration plot was linear, and the brain uptake clearance of the obtained nanoparticles was significantly higher than the free drug. The results revealed that the produced nanoparticles have potential as drug carrier for drug delivery system in the brain.

12.2.8 Poly(Lactic Acid) (PLA) and Its Co-Polymers

PLA is aliphatic polyester derived from renewable sources, such as corn sugar, potato, and sugarcane (Lassalle & Ferreira, 2007). PLA is a biodegradable polymer accepted by the US Food and Drug Administration (FDA) for drug delivery, and studies have shown it to be an effective carrier for proteins, drugs, and DNA. It is used in a variety of biomedical applications, such as tissue engineering, ophthalmic implants, and especially drug delivery systems (Ramot, Haim-Zada, Domb, & Nyska, 2016). PLA is a good candidate for use as a drug carrier because of its desirable physicochemical properties, which include safety, biodegradability, biocompatibility, stability, non-cytotoxicity for many cell lines, and a lack of inherent immunogenicity (Elmowafy, Tiboni, & Soliman, 2019).

For instance, Liu *et al.* (2013) prepared coumarin, NAP, and DiR-loaded B6 modified PEG-PLA nanoparticles (B6-NP) using an emulsion/solvent evaporation method for enhanced brain delivery of a neuroprotective peptide for the treatment of Alzheimer's disease. The average particle size of the obtained nanoparticles was between 100 nm and 120 nm, with acceptably low PDI (PI <0.25). The obtained nanoparticles exhibited much higher level of brain accumulation when compared with NP. The obtained nanoparticles administration of encapsulated neuroprotective peptide NAPVSIPQ (NAP) to Alzheimer's disease mouse models showed excellent learning impairments, cholinergic disruption, and hippocampal neurons loss at a lower dose. The obtained nanoparticles might be a promising DDS for enhancing the transport of neuropeptides in the brain.

Zhang *et al.* (2014) mentioned the preparation of TGN (TGNYKALHPHNGC)– and QSH (QSHYRHISPAQVC)–loaded Mal-PEG-PLA and MePEG-PLA in a 3:7 weight ratio utilizing an emulsion/solvent evaporation technique for targeting amyloid plaques in the brains of Alzheimer's disease mice. The obtained nanoparticles had mean particle sizes of about 100 nm, and zeta potential of about −20 mV. To achieve the best targeting effect, the molar ratio of TGN/maleimide and QSH/maleimide was 1:3. With this cascade delivery technique, cytotoxicity was minimized and AD might be treated or diagnosed early.

In another study, Zhang *et al.* (2014) used an emulsion/solvent evaporation method to create TGN- and QSH-loaded H102 peptide/PEG-PLA nanoparticles (NP/H102) for the precise delivery of the sheet breaker peptide H102 to the brain lesions of AD model mice. The obtained nanoparticles had a narrow particle distribution, with mean particle sizes of about 120 nm. The zeta potential was approximately −28 mV. The loading capacities of these for the obtained nanoparticles were similar, ranging between 0.54 and 0.61%, with an EE% of approximately 57.68–65.54%. In comparison to non-modified or TGN-modified nanoparticles, this dual-functional drug delivery system efficiently raised the H102 accumulation at brain A42 localized in the hippocampus area and offered stronger neuroprotective benefits in the AD model mice. These results demonstrate that the obtained nanoparticles may be a promising carrier for peptide and protein drugs, such as H102, for entry into the CNS, followed by localization in brain AD lesions. They provide a very targeted approach to treating AD.

On the other hand, Zheng *et al.* (2017) mentioned the preparation of TGN- and QSH-loaded H102/PEG-PLA nanoparticles (NP/H102) via the emulsion/solvent evaporation technique. The mean particle size of the NP/Cou-6 particles was approximately 100 nm, which is optimal for BBB transport, and the negative zeta potential of nanoparticles was approximately −25 mV. The obtained nanoparticles targeting of amyloid plaques might lead to enhanced therapeutic efficacy, which was further evaluated in APP/PS1 transgenic mice. The obtained nanoparticles increased A-degrading enzymes, reduced tau protein phosphorylation, protected synapses, and improved spatial learning and memory in transgenic mice. The obtained nanoparticles also demonstrated good biocompatibility in subacute toxicity assays. In conclusion, the obtained nanoparticles were a useful nanodevice for the targeted distribution of biotechnology medications to treat AD.

Recently, Hu *et al.* (2017) mentioned the preparation of coumarin-6 (C6)–loaded cholesterol–polyethylene glycol (PEG) co-modified poly(ethyleneglycol)–poly(lactide) nanoparticles (CLS-PEG-PLA NPs) using an emulsion polymerization technique as a novel, biodegradable brain drug delivery system. The particle size of CLS NPs was determined to be 175.6±4.1 nm (PDI 0.26±40.017), whereas the size was increased to 185.4±4.1 nm, with PDI of 0.133±0.009. Moreover, the zeta potentials of CLS NPs and CLS-PEG NPs were of −28.5 ± 1.43 mV, and 0.66 ± 0.10 mV, respectively. The obtained nanoparticles achieved sustained release for up to 12 h (Figure 12.10a). These in vitro and in vivo results strongly imply that the obtained nanoparticles are an effective, low-toxicity drug delivery technology for targeting the brain (Figure 12.10b). The cellular absorption of medications may be greatly increased by the obtained nanoparticles, making them an effective drug delivery strategy for treating CNS diseases (Figure 12.10c).

12.2.9 Poly(n-Butylcyanoacrylate) (PBCA) and Its Co-Polymers

PBCA and its co-polymers have attracted more attention for brain CNS treatment. For example, Eslami *et al.* (2016)

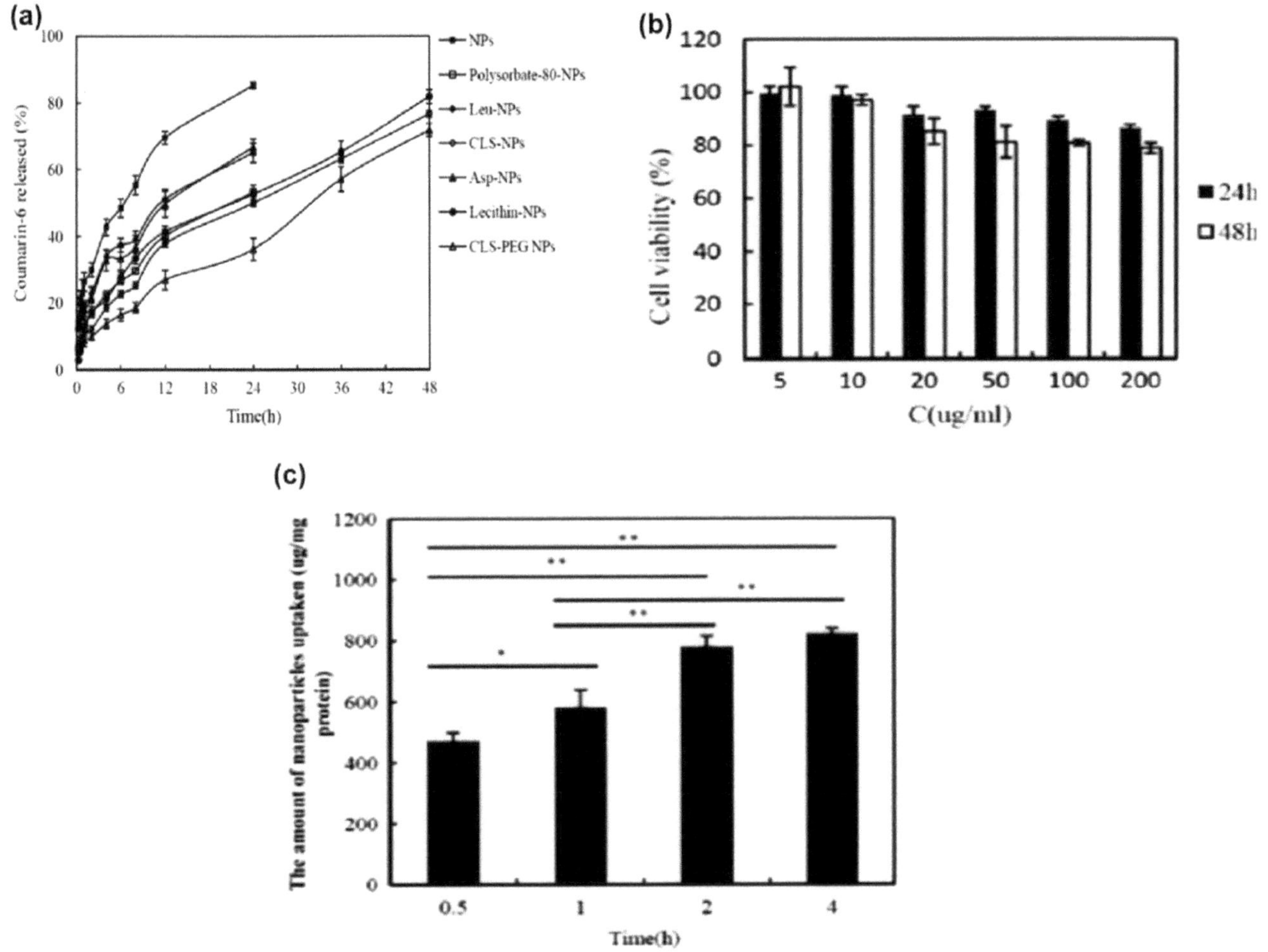

FIGURE 12.10 (a) In vitro release profiles of obtained NPs. (b) Cytotoxicity of CLS-PEG NPs encapsulated with C6 in bEnd.3 cell line at different concentrations as assessed by MTT levels (mean ± SD). And effect of (c) concentration, (a) incubation time, and (b) cellular uptake of CLS-PEG NPs. (From Hu, Yang, Liao, Li, and Zhang (2017). Copyright 2018. Adapted from Taylor & Francis.)

created the tacrine molecule with poly(n-butylcyanoacrylate) and CS as efficient carriers for CNS treatment via drug passage across the blood–brain barrier using MD simulation. According to the Flory–Huggins results, tacrine molecule exhibited higher compatibility with PBCA than CS.

On the other hand, Ćurić *et al.* (2017) used a one-step emulsion solvent evaporation method to make itraconazole-loaded PBCA NPs nanosuspension containing EE% of 50–99% with 100–150 nm spherical shape with gray color. The obtained NPs with modified surfaces provide a better way to deliver poorly soluble drugs to the desired target sites or across biological barriers, such as the blood–brain barrier. DSPE-PEG (2000) maleimide was used to modify the surface of the obtained NPs, allowing Apo E coupling.

Recently, Jahansooz *et al.* (2020) reported the preparation of dopamine (DA) loaded into poly-butyl-cyanoacrylate nanoparticles (DA-PBCA NPs) for potential drug delivery into the CNS. Particle size, zeta potential, and the PDI of the obtained NPs were 91.5 ± 1.5 nm, −7.61 ± 0.7 mV, and 0.4 ± 0.1, respectively. Moreover, the findings showed sustained drug release. The obtained NPs revealed a linearity in the Korsemeyer–Peppas model ($R^2 = 0.9985$). The produced NPs significantly increased the DA content in the brain striatum and reversed neurobehavioral deficits in PD rat models. The obtained NPs may provide a novel therapeutic approach for DA delivery in the brain for PD treatment.

12.3 CONCLUSION AND FUTURE PERSPECTIVES

This book chapter discussed chitosan, PLGA, and PBCA, as well as their co-polymers, such as chitosan co-polymers, PEG-PLGA, and methoxy poly(ethylene glycol)-co-poly(lactic-co-glycolic), which improved brain bioavailability and drug stability with sustained-release profile. Consequently, the polymeric nanoparticles described in this chapter demonstrated excellent properties. Chitosan nanoparticles and co-polymers with mucoadhesive properties were used to improve brain-targeting efficiency, drug stability, drug bioavailability, sustained release, and cytotoxicity. PLGA NPs and their co-polymers were used to improve therapeutic effect, drug delivery to the brain, permeability across the BBB, long-term therapeutic effects, fluorescence intensity in the brain, sustained real estate, low cytotoxicity, and a longer halftime compared to nontargeted NPs. PCL and its co-polymers were used to improve poly(butyl cyanoacrylate), and its co-polymers were used in parkinsonian rats to achieve high EE%, sustained drug release, and reversal of neurobehavioral deficits. Future technologies are projected to heavily rely on polymeric nanoparticles. Although significant research has been put forth, there are still few applications for polymeric nanoparticles. There are still a lot of obstacles to overcome, like the potential for large-scale production with fewer side effects. Finally, this book chapter proved that polymeric nanoparticles have become one of the most promising drug nanocarriers for improving brain bioavailability for Alzheimer's disease in brain targeting with low toxicity and will therefore be intensively investigated in the future.

TABLE 12.1
Recent studies on PNs as drug nanocarriers in brain targeting

Carrier material	Loaded drug	Method of preparation	Physicochemical characteristics	Outcome	Reference
Chitosan NPs	Memantine hydrochloride	Ionic gelation	129 nm	• Optimized formulation obtained at 2:1 concentration ratio of CS/TPP. • Good stability of NPs formulation. • Memantine-loaded chitosan nanoparticle is a potential new delivery system for treatment of Alzheimer's disease when transported via olfactory nasal pathway to the brain.	(Ruby & Pandey, 2015)
Chitosan NPs	Rotigotine	Ionic gelation	75.37 ± 3.37nm	• RNPs with small PDI (0.368 ± 0.02) and very high EE% (96.08 ± 0.01). • Sustained drug release. • Exhibits no toxicity effect. • Exhibits an efficient transport from the nose to the brain.	(Tzeyung et al., 2019)
Chitosan NPs	Rotigotine	Ionic gelation	75.37 ± 3.37 nm	• Exhibits no toxicity effect. • Intranasally administered. • Enhanced high bioavailability and targeting brain efficiency. • RNPs have significant potential to be developed as a treatment of Parkinson's disease.	(Bhattamisra et al., 2020)
Chitosan NPs	Sitagliptin	Ionic gelation	188.4 ± 48.1 nm	• Increased SIT brain concentration after IN administration. • Exhibits a safe and effective method through the nose-to-brain route.	(Wilson et al., 2021)
PEG/CS NPs	Monoclonal antibody OX26		140–200 nm	• High colloidal stability. • The obtained nanoparticles can be a promising delivery system for targeting drugs to the CNS.	(Monsalve et al., 2015)
PLGA/CS NPs	O^6-benzylguanine (BG), carmustine (BCNU)	Nanoprecipitation	127 ± 2.3 nm	• Exhibits stronger cytotoxic effect and cellular uptake. • Improved the antitumor efficacy of DDS.	(Liu et al., 2017)
PLGA/chitosan NPs	Rasagiline	Double emulsification–solvent evaporation	63.13–402.18 nm	• Improved brain bioavailability. • The obtained nanoparticles can be a promising delivery for treatment of PD.	(Ahmad, 2017)

(Continued)

TABLE 12.1 *(Continued)*

Carrier material	Loaded drug	Method of preparation	Physicochemical characteristics	Outcome	Reference
CS/PLGA NPs	Catechin hydrate (CH)	Double emulsion–solvent evaporation method	93.46 ± 3.94 nm	• Sustained release. • Higher permeation of nasal route. • The obtained NPs used as safe and effective brain-targeted delivery system for the treatment of epilepsy.	(Ahmad et al., 2020)
PLGA/N-trimethylated chitosan NPs	Huperzine A	Emulsion–solvent evaporation	153.2 ± 13.7 nm +35.6 ± 5.2 mV 73.8% ± 5.7%	• Increased cellular uptake. • Improved drug delivery to the brain. • Sustained-release effect. • The obtained NPs have good adhesion and targeting ability. • The obtained NPs have a broad application prospect as a nasal drug delivery carrier.	(Meng et al., 2018)
PLGA NPs	Loperamide (LOP)	Nanoemulsion templating using low energy	44.73 ± 2.98 nm	• Efficient nanocarriers for drug delivery across the BBB. • The obtained NPs are able to efficiently cross the BBB, with high EE%. • Active targeting moiety.	(Fornaguera, Dols-Perez, et al., 2015)
PLGA NPs	Galantamine (GAL)	Nanoemulsification	32.73 ± 0.39 nm	• Higher encapsulation efficiencies than 90 wt%. • Sustained release. • Intravenous administration. • Long-term therapeutic effects.	(Fornaguera, Feiner-Gracia, et al., 2015)
PLGA NPs	Anti-amyloidpeptides	Nanoprecipitation	160 nm	• The obtained NPs could be promising for AD treatment.	(Loureiro et al., 2016)
PLGA NPs	Curcumin	Double emulsion solvent evaporation	128.6 nm	• Overcame the drawback of low bioavailability of drug. • Sustained-release drug. • Exhibited most beneficial effect on AD treatment.	(Huang et al., 2017)
PLGA NPs	Temozolomide butyl ester (TBE)	Emulsion–solvent evaporation	145.9 ± 8.7 nm	• Sustained release. • Exhibited higher intensity of fluorescence in the brain. • The obtained NPs could be promising for glioblastoma treatment.	(Chu et al., 2018)
PLGA NPs	Curcumin (cur)	Emulsion-solvent evaporation method	160 ± 5 nm	• Inhibited toxicity and Aβ accumulation in AD treatment. • The obtained NPs promising for AD treatment.	(Huo et al., 2019)
PLGA NPs	Temozolomide (TMZ)	Single emulsion solvent evaporation	181 ± 2 nm	• Enhanced the permeability across the BBB during *in vivo* treatment. • The obtained NPs modified with a OX26 mAb could be promising for GBM and BBB cells.	(Ramalho et al., 2018)
PLGA NPs	Vitamin D-binding protein (DBP)	Emulsion–diffusion	226.6 ± 444 nm	• The obtained NPs attenuate neurodegeneration and neuroinflammation, constraining DBP activity. • PLGA on accumulation of Aβ ameliorate and *in vivo*. • Exhibited reasoning damage in an animal model of AD treatment.	(Jeon et al., 2019)
PLGA NPs	Nec-1s	Nanoprecipitation	72 ± 11.3 nm	• Increased the halftime of produced NPs in blood circulation and brain distribution. • Exhibited the highest intensity of fluorescence in the brain.	(Zhang et al., 2020)

(Continued)

TABLE 12.1 *(Continued)*
Recent studies on PNs as drug nanocarriers in brain targeting

Carrier material	Loaded drug	Method of preparation	Physicochemical characteristics	Outcome	Reference
PLGA NPs	Donepezil	Nanoprecipitation	40.8 nm	• Sustained release. • Decreased side effects, crossing the blood–brain barrier (BBB). • Enhanced therapeutic effect.	(Tripathi et al., 2021)
PEG-PLGA NPs	Lactoferrin	Nanoprecipitation	122.0 ± 19.3 nm	• Continuous release of drug. • Enhanced accumulation of cellular compared. • The obtained NPs could be promising for PD treatment. • Longer-term clinical trial treatment of PD.	(Bi et al., 2016)
PLGA -PEG NPs	Donepezil	Double emulsion	174–202 nm	• Controlled release profile. • Decreased in both protein expression and gene levels of GM-CSF, IL-1b, IL-6, and TNF-a.	(Baysal et al., 2017)
PLGA-PEG NPs	Curcumin (cur)	Solvent evaporation	50–150 nm 3.83 ± 0.89 mV	• Sustained release. • Biocompatible. • Exhibited lower toxicity effect. • Increased cellular uptake. • Exhibited higher bioavailability and good biosafety. • Improved EE%. • The obtained NPs used for AD treatment.	(Fan et al., 2018)
PLGA-b-PEG) nanoparticles	TMZ	Via single emulsion–solvent evaporation method	185 ± 45 nm	• Exhibited cytotoxicity of produced NPs was not dose-dependent in T98G cells. • Burst release of drug.	(Senturk et al., 2022)
PCL NPs	Curcumin	Emulsification–diffusion method	149 ± 2.2 nm	• Higher encapsulation efficiency of the drug more than 90%. • Adequate biocompatibility. • The obtained NPs have potential other PD treatment.	(Del Prado-Audelo et al., 2019)
PCL NPs	Memantine	Interfacial deposition of pre-formed polymer	200 nm	• Higher EE% more than 80%. • The obtained NPs are a promising system for Alzheimer's disease treatment.	(Mahmoudi et al., 2021)
mPEG-PCL NPs	Methazolamide (MTZ)	Thin-film hydration	60 nm	• Higher EE% of 93%. • Sustained release of drug. • Good biocompatibility. • The obtained NPs could potentiate the therapeutic outcome of nanotechnology ocular drug delivery.	(Elmowafy et al., 2019)
MPEG- PCL NPs	Etoposide (ETO), carmustine (BCNU), doxorubicin (DOX)	Microemulsion–solvent evaporation technique	150–200 nm	• Exhibited lower toxicity effect. • WFNPs carrying ETO, BCNU, and DOX acted as double nanocarriers targeting. • The obtained NPs could can be a promising approach to constraining brain GBM growth.	(Kuo et al., 2019)
PEG-PCL NPs	Ginkgolide B (GB)	Antisolvent precipitation and stabilizer-based	91.26 ± 1.34 nm	• Higher EE% of 87.52%. • Sustained release. • Enhanced brain uptakes of GB. • Improved bioactivity, cerebral accumulation, and bioavailability.	(Zhao et al., 2020)

(Continued)

TABLE 12.1 *(Continued)*

Carrier material	Loaded drug	Method of preparation	Physicochemical characteristics	Outcome	Reference
MPEG-PCL NPs	Rivastigmine (Riv)	Nanoprecipitation	98.5 ± 2.1 nm	• Sustained-release behavior of the drug.	(Mohamadpour et al., 2020)
PEG-PLA NPs	Coumarin, NAP, and DiR	Emulsion/solvent evaporation	100–120 nm	• Exhibited much higher accumulation in brain. • Exhibited much higher EE% to Alzheimer's disease mouse models. • Excellent amelioration at lower dose. • The obtained NPs might serve as a promising DDS for enabling neuropeptides brain delivery.	(Liu et al., 2013)
Mal-PEG-PLA NPs	TGN, QSH	Emulsion/solvent evaporation	100	• The optimal targeting effect at molar ratio 1:3 of TGN/maleimide to QSH/maleimide. • Exhibited no cytotoxicity effect. • The obtained NPs could be used for early diagnosis or treatment of AD.	(Zhang, Wan, et al., 2014)
PEG–PLA NPs/H102	TGN, QSH	Emulsion/solvent evaporation	120	• Provided better neuroprotective effects in the AD model mice. • The obtained NPs may be a promising carrier for protein and drugs peptide. • Enhanced a highly specific technique for AD treatment.	(Zhang, Zheng, et al., 2014)
PEG–PLA NPs/H102	TGN, QSH	Emulsion/solvent evaporation	100	• Enhanced therapeutic efficacy, which was further evaluated in APP/PS1 transgenic mice. • Decreased tau protein, amyloid plaques, and phosphorylation. • Increased Aβ-degrading enzymes. • Reduced protecting synapses. • Good biocompatibility. • The obtained NPs were a valuable nanodevice for delivery of biotechnological drugs to treat AD.	(Zheng et al., 2017)
CLS-PEG-PLA NPs	Coumarin-6 (C6)	Emulsion polymerization	185.4 ± 4.1	• Sustained release. • Improved the ability of C6 to pass through the BBB. • The obtained NPs exhibited low toxicity and could be a promising drug delivery system for targeting the brain.	(Hu et al., 2017)
PBCA NPs	Itraconazole	Emulsion solvent evaporation	100–150	• Exhibited higher EE%. • The obtained NPs offer a better technique to the desired target sites, for instance, the BBB.	(Ćurić et al., 2017)
PBCA NPs	Dopamine (DA)		91.5 ± 1.5	• Sustained release. • The obtained NPs revealed a linearity in the Korsemeyer–Peppas model. • Improved the drug contents in reversed neurobehavioral deficits and the striatum of brain in PD rat models. • The obtained NPs provide a novel therapeutic method for DA delivery in the brain for PD treatment.	(Jahansooz et al., 2020)
PBCA–Chitosan NPs	Tacrine	MD simulation		• Efficient carriers for dealing with CNS diseases. • Exhibited higher compatibility with PBCA than CS. • Exhibited suitable drug carriers.	(Eslami et al., 2016)

REFERENCES

Agnihotri, S. A., Mallikarjuna, N. N., & Aminabhavi, T. M. (2004). Recent advances on chitosan-based micro-and nanoparticles in drug delivery. *Journal of Controlled Release*, *100*(1), 5–28.

Ahmad, N. (2017). Rasagiline-encapsulated chitosan-coated PLGA nanoparticles targeted to the brain in the treatment of Parkinson's disease. *Journal of Liquid Chromatography & Related Technologies*, *40*(13), 677–690.

Ahmad, N., Ahmad, R., Alrasheed, R. A., Almatar, H. M. A., Al-Ramadan, A. S., Amir, M., & Sarafroz, M. (2020). Quantification and evaluations of catechin hydrate polymeric nanoparticles used in brain targeting for the treatment of epilepsy. *Pharmaceutics*, *12*(3), 203.

Amin, A., Samy, M., Abd El-Alim, S. H., Rabia, A. E. G., & Ayoub, M. M. (2018). Assessment of formulation parameters needed for successful vitamin C entrapped polycaprolactone nanoparticles. *International Journal of Polymeric Materials and Polymeric Biomaterials*, *67*(16), 942–950.

Andrieux, K., & Couvreur, P. (2009). Polyalkylcyanoacrylate nanoparticles for delivery of drugs across the blood–brain barrier. *Wiley Interdisciplinary Reviews: Nanomedicine and Nanobiotechnology*, *1*(5), 463–474.

Barbu, E., Molnàr, É., Tsibouklis, J., & Górecki, D. C. (2009). The potential for nanoparticle-based drug delivery to the brain: Overcoming the blood–brain barrier. *Expert Opinion on Drug Delivery*, *6*(6), 553–565.

Baysal, I., Ucar, G., Gultekinoglu, M., Ulubayram, K., & Yabanoglu-Ciftci, S. (2017). Donepezil loaded PLGA-b-PEG nanoparticles: Their ability to induce destabilization of amyloid fibrils and to cross blood brain barrier in vitro. *Journal of Neural Transmission*, *124*(1), 33–45.

Berger, J., Reist, M., Mayer, J. M., Felt, O., Peppas, N., & Gurny, R. (2004). Structure and interactions in covalently and ionically crosslinked chitosan hydrogels for biomedical applications. *European Journal of Pharmaceutics and Biopharmaceutics*, *57*(1), 19–34.

Bhattamisra, S. K., Shak, A. T., Xi, L. W., Safian, N. H., Choudhury, H., Lim, W. M., . . . Radhakrishnan, A. K. (2020). Nose to brain delivery of rotigotine loaded chitosan nanoparticles in human SH-SY5Y neuroblastoma cells and animal model of Parkinson's disease. *International Journal of Pharmaceutics*, *579*, 119148.

Bi, C., Wang, A., Chu, Y., Liu, S., Mu, H., Liu, W., . . . Li, Y. (2016). Intranasal delivery of rotigotine to the brain with lactoferrin-modified PEG-PLGA nanoparticles for Parkinson's disease treatment. *International Journal of Nanomedicine*, *11*, 6547.

Chigumira, W., Maposa, P., Gadaga, L. L., Dube, A., Tagwireyi, D., & Maponga, C. C. (2015). Preparation and evaluation of pralidoxime-loaded PLGA nanoparticles as potential carriers of the drug across the blood brain barrier. *Journal of Nanomaterials*, *2015*.

Chu, L., Wang, A., Ni, L., Yan, X., Song, Y., Zhao, M., . . . Wu, Z. (2018). Nose-to-brain delivery of temozolomide-loaded PLGA nanoparticles functionalized with anti-EPHA3 for glioblastoma targeting. *Drug Delivery*, *25*(1), 1634–1641.

Ćurić, A., Möschwitzer, J. P., & Fricker, G. (2017). Development and characterization of novel highly-loaded itraconazole poly (butyl cyanoacrylate) polymeric nanoparticles. *European Journal of Pharmaceutics and Biopharmaceutics*, *114*, 175–185.

Del Prado-Audelo, M., Magaña, J., Mejía-Contreras, B., Borbolla-Jiménez, F., Giraldo-Gomez, D., Piña-Barba, M., . . . Leyva-Gómez, G. (2019). In vitro cell uptake evaluation of curcumin-loaded PCL/F68 nanoparticles for potential application in neuronal diseases. *Journal of Drug Delivery Science and Technology*, *52*, 905–914.

Domján, A., Manek, E., Geissler, E., & László, K. (2013). Host–guest interactions in poly (n-isopropylacrylamide) hydrogel seen by one-and two-dimensional 1H cramps solid-state NMR spectroscopy. *Macromolecules*, *46*(8), 3118–3124.

Ege, D. (2021). Action mechanisms of curcumin in Alzheimer's disease and its brain targeted delivery. *Materials*, *14*(12), 3332.

Elmowafy, E., Gad, H., Biondo, F., Casettari, L., & Soliman, M. E. (2019). Exploring optimized methoxy poly (ethylene glycol)-block-poly (ε-caprolactone) crystalline cored micelles in anti-glaucoma pharmacotherapy. *International Journal of Pharmaceutics*, *566*, 573–584.

Elmowafy, E. M., Tiboni, M., & Soliman, M. E. (2019). Biocompatibility, biodegradation and biomedical applications of poly (lactic acid)/poly (lactic-co-glycolic acid) micro and nanoparticles. *Journal of Pharmaceutical Investigation*, *49*(4), 347–380.

Eslami, M., Nikkhah, S. J., Hashemianzadeh, S. M., & Sajadi, S. A. S. (2016). The compatibility of tacrine molecule with poly (n-butylcyanoacrylate) and chitosan as efficient carriers for drug delivery: A molecular dynamics study. *European Journal of Pharmaceutical Sciences*, *82*, 79–85.

Fan, S., Zheng, Y., Liu, X., Fang, W., Chen, X., Liao, W., . . . Ma, Q. (2018). Curcumin-loaded PLGA-PEG nanoparticles conjugated with B6 peptide for potential use in Alzheimer's disease. *Drug Delivery*, *25*(1), 1091–1102.

Fornaguera, C., Dols-Perez, A., Calderó, G., García-Celma, M., Camarasa, J., & Solans, C. (2015). PLGA nanoparticles prepared by nano-emulsion templating using low-energy methods as efficient nanocarriers for drug delivery across the blood–brain barrier. *Journal of Controlled Release*, *211*, 134–143.

Fornaguera, C., Feiner-Gracia, N., Calderó, G., García-Celma, M., & Solans, C. (2015). Galantamine-loaded PLGA nanoparticles, from nano-emulsion templating, as novel advanced drug delivery systems to treat neurodegenerative diseases. *Nanoscale*, *7*(28), 12076–12084.

Gaillard, P. J., Visser, C. C., Boer, M. d., Appeldoorn, C., & Rip, J. (2014). Blood-to-brain drug delivery using nanocarriers. In *Drug delivery to the brain* (pp. 433–454). Springer.

Gothwal, A., Kumar, H., Nakhate, K. T., Ajazuddin, Dutta, A., Borah, A., & Gupta, U. (2019). Lactoferrin coupled lower generation PAMAM dendrimers for brain targeted delivery of memantine in aluminum-chloride-induced Alzheimer's disease in mice. *Bioconjugate Chemistry*, *30*(10), 2573–2583.

Hartl, N., Adams, F., & Merkel, O. M. (2021). From adsorption to covalent bonding: Apolipoprotein E functionalization of polymeric nanoparticles for drug delivery across the blood–brain barrier. *Advanced Therapeutics*, *4*(1), 2000092.

Hrkach, J., Von Hoff, D., Ali, M. M., Andrianova, E., Auer, J., Campbell, T., . . . Horhota, A. (2012). Preclinical development and clinical translation of a PSMA-targeted docetaxel nanoparticle with a differentiated pharmacological profile. *Science Translational Medicine*, *4*(128), 128ra139–128ra139.

Hu, X., Yang, F., Liao, Y., Li, L., & Zhang, L. (2017). Cholesterol–PEG comodified poly (N-butyl) cyanoacrylate nanoparticles for brain delivery: In vitro and in vivo evaluations. *Drug Delivery*, *24*(1), 121–132.

Huang, N., Lu, S., Liu, X.-G., Zhu, J., Wang, Y.-J., & Liu, R.-T. (2017). PLGA nanoparticles modified with a BBB-penetrating peptide co-delivering Aβ generation inhibitor and curcumin attenuate memory deficits and neuropathology in Alzheimer's disease mice. *Oncotarget*, *8*(46), 81001.

Huo, X., Zhang, Y., Jin, X., Li, Y., & Zhang, L. (2019). A novel synthesis of selenium nanoparticles encapsulated PLGA nanospheres with curcumin molecules for the inhibition of amyloid β aggregation in Alzheimer's disease. *Journal of Photochemistry and Photobiology B: Biology*, *190*, 98–102.

Jahansooz, F., Hosseinzade, B. E., Zarmi, A. H., Hadi, F., Massood Hojjati, S. M., & Shahpasand, K. (2020). Dopamine-loaded poly (butyl cyanoacrylate) nanoparticles reverse behavioral deficits in Parkinson's animal models. *Therapeutic Delivery*, *11*(6), 387–399.

Jeon, S. G., Cha, M.-Y., Kim, J.-i., Hwang, T. W., Kim, K. A., Kim, T. H., . . . Moon, M. (2019). Vitamin D-binding protein-loaded PLGA nanoparticles suppress Alzheimer's disease-related pathology in 5XFAD mice. *Nanomedicine: Nanotechnology, Biology and Medicine*, *17*, 297–307.

Jiang, Y., Jiang, Y., Ding, Z., & Yu, Q. (2022). Investigation of the "nose-to-brain" pathways in intranasal Hupa nanoemulsions and evaluation of their in vivo pharmacokinetics and brain-targeting ability. *International Journal of Nanomedicine*, 3443–3456.

Kamat, V., Bodas, D., & Paknikar, K. (2016). Chitosan nanoparticles synthesis caught in action using microdroplet reactions. *Scientific Reports*, *6*(1), 1–4.

Kang, J., Compton, D. R., Vaz, R. J., & Rampe, D. (2016). Proarrhythmic mechanisms of the common anti-diarrheal medication loperamide: Revelations from the opioid abuse epidemic. *Naunyn-Schmiedeberg's Archives of Pharmacology*, *389*(10), 1133–1137.

Kumar, A., Tan, A., Wong, J., Spagnoli, J., Lam, J., Blevins, B., Thorne, L., Ashkan, K., Xie, J., & Liu, H. (2017). Nanotechnology for neuroscience: Promising approaches for diagnostics, therapeutics and brain activity mapping. *Advanced Functional Materials*, *27*, 1700489.

Kuo, Y.-C., Chang, Y.-H., & Rajesh, R. (2019). Targeted delivery of etoposide, carmustine and doxorubicin to human glioblastoma cells using methoxy poly (ethylene glycol)-poly (ε-caprolactone) nanoparticles conjugated with wheat germ agglutinin and folic acid. *Materials Science and Engineering: C*, *96*, 114–128.

Lassalle, V., & Ferreira, M. L. (2007). PLA nano-and microparticles for drug delivery: An overview of the methods of preparation. *Macromolecular Bioscience*, *7*(6), 767–783.

Lee, J. S., Chae, G. S., Kim, M. S., Cho, S. H., Lee, H. B., & Khang, G. (2004). Degradation behaviour in vitro for poly (D, L-lactide-co-glycolide) as drug carrier. *Bio-Medical Materials and Engineering*, *14*(2), 185–192.

Lekbach, Y., Dong, Y., Li, Z., Xu, D., El Abed, S., Yi, Y., . . . Wang, F. (2019). Catechin hydrate as an eco-friendly biocorrosion inhibitor for 304L stainless steel with dual-action antibacterial properties against Pseudomonas aeruginosa biofilm. *Corrosion Science*, *157*, 98–108. doi: https://doi.org/10.1016/j.corsci.2019.05.021

Liu, C., Yao, S., Li, X., Wang, F., & Jiang, Y. (2017). iRGD-mediated core-shell nanoparticles loading carmustine and O6-benzylguanine for glioma therapy. *Journal of Drug Targeting*, *25*(3), 235–246.

Liu, P.-P., Xie, Y., Meng, X.-Y., & Kang, J.-S. (2019). History and progress of hypotheses and clinical trials for Alzheimer's disease. *Signal Transduction and Targeted Therapy*, *4*(1), 1–22.

Liu, Z., Gao, X., Kang, T., Jiang, M., Miao, D., Gu, G., . . . Tu, Y. (2013). B6 peptide-modified PEG-PLA nanoparticles for enhanced brain delivery of neuroprotective peptide. *Bioconjugate Chemistry*, *24*(6), 997–1007.

Lockman, P., Mumper, R., Khan, M., & Allen, D. (2002). Nanoparticle technology for drug delivery across the blood-brain barrier. *Drug Development and Industrial Pharmacy*, *28*(1), 1–13.

Loureiro, J. A., Gomes, B., Fricker, G., Coelho, M. A., Rocha, S., & Pereira, M. C. (2016). Cellular uptake of PLGA nanoparticles targeted with anti-amyloid and anti-transferrin receptor antibodies for Alzheimer's disease treatment. *Colloids and Surfaces B: Biointerfaces*, *145*, 8–13.

Mahmoudi, M., Saeidian, H., Mirjafary, Z., & Mokhtari, J. (2021). Preparation and characterization of memantine loaded polycaprolactone nanocapsules for Alzheimer's disease. *Journal of Porous Materials*, *28*(1), 205–212.

Masserini, M. (2013). Nanoparticles for brain drug delivery. *ISRN Biochemistry*. Article ID 238428.

Meng, Q., Wang, A., Hua, H., Jiang, Y., Wang, Y., Mu, H., . . . Sun, K. (2018). Intranasal delivery of Huperzine A to the brain using lactoferrin-conjugated N-trimethylated chitosan surface-modified PLGA nanoparticles for treatment of Alzheimer's disease. *International Journal of Nanomedicine*, *13*, 705.

Misra, S., Chopra, K., Sinha, V., & Medhi, B. (2016). Galantamine-loaded solid–lipid nanoparticles for enhanced brain delivery: preparation, characterization, in vitro and in vivo evaluations. *Drug Delivery*, *23*(4), 1434–1443.

Mohamadpour, H., Azadi, A., Rostamizadeh, K., Andalib, S., Saghatchi Zanjani, M. R., & Hamidi, M. (2020). Preparation, optimization, and evaluation of methoxy poly (ethylene glycol)-co-poly (ε-caprolactone) nanoparticles loaded by rivastigmine for brain delivery. *ACS Chemical Neuroscience*, *11*(5), 783–795.

Mohammed, M. A., Syeda, J. T., Wasan, K. M., & Wasan, E. K. (2017). An overview of chitosan nanoparticles and its application in non-parenteral drug delivery. *Pharmaceutics*, *9*(4), 53.

Monsalve, Y., Tosi, G., Ruozi, B., Belletti, D., Vilella, A., Zoli, M., . . . Sierra, L. (2015). PEG-g-chitosan nanoparticles functionalized with the monoclonal antibody OX26 for brain drug targeting. *Nanomedicine*, *10*(11), 1735–1750.

Mozafari, N., Dehshahri, A., Ashrafi, H., Mohammadi-Samani, S., Shahbazi, M.-A., Heidari, R., . . . Azadi, A. (2022). Vesicles of yeast cell wall-sitagliptin to alleviate neuroinflammation in Alzheimer's disease. *Nanomedicine: Nanotechnology, Biology and Medicine*, *44*, 102575.

Neha, B., Ganesh, B., & Preeti, K. (2013). Drug delivery to the brain using polymeric nanoparticles: A review. *International Journal of Pharmaceutical and Life Sciences*, *2*(3), 107–132.

Qian, L., Zheng, J., Wang, K., Tang, Y., Zhang, X., Zhang, H., . . . Jiang, Y. (2013). Cationic core–shell nanoparticles with carmustine contained within O6-benzylguanine shell for glioma therapy. *Biomaterials*, *34*(35), 8968–8978.

Ramalho, M. J., Sevin, E., Gosselet, F., Lima, J., Coelho, M., Loureiro, J. A., & Pereira, M. (2018). Receptor-mediated PLGA nanoparticles for glioblastoma multiforme treatment. *International Journal of Pharmaceutics*, *545*(1–2), 84–92.

Ramot, Y., Haim-Zada, M., Domb, A. J., & Nyska, A. (2016). Biocompatibility and safety of PLA and its copolymers. *Advanced Drug Delivery Reviews*, *107*, 153–162.

Ruby, J. J., & Pandey, V. (2015). Chitosan nanoparticles as a nasal drug delivery for memantine hydrochloride. *International Journal of Pharmaceutical Sciences*, *7*(1), 34–37.

Samy, M., Abd El-Alim, S. H., Amin, A., & Ayoub, M. M. (2020). Formulation, characterization and in vitro release study of

5-fluorouracil loaded chitosan nanoparticles. *International Journal of Biological Macromolecules*, *156*, 783–791.

Samy, M., Abdallah, H. M., Awad, H. M., & Ayoub, M. M. (2022a). In vitro release and cytotoxicity activity of 5-fluorouracil entrapped polycaprolactone nanoparticles. *Polymer Bulletin*, *79*(8), 6645–6671.

Samy, M., Abdallah, H. M., Awad, H. M., & Ayoub, M. M. (2022b). Preparation, characterization and in vitro biological activity of 5-fluorouracil loaded onto poly (D, L-lactic-co-glycolic acid) nanoparticles. *Polymer Bulletin*, 1–23.

Samy, M., Abdallah, H. M., Ayoub, M. M., Vylegzhanina, M., Volkov, A. Y., & Sukhanova, T. (2020). Eco-friendly route for encapsulation of 5-fluorouracil into polycaprolactone nanoparticles. *Egyptian Journal of Chemistry*, *63*(1), 255–267.

Senturk, F., Cakmak, S., Gumusderelioglu, M., & Ozturk, G. G. (2022). Hydrolytic instability and low-loading levels of temozolomide to magnetic PLGA nanoparticles remain challenging against glioblastoma therapy. *Journal of Drug Delivery Science and Technology*, *68*, 103101.

Sharma, U., Badyal, P. N., & Gupta, S. (2015). Polymeric nanoparticles drug delivery to brain: A review. *International Journal of Pharmacology*, *2*(5), 60–69.

Sukhanova, T., Vylegzhanina, M., Volkov, A. Y., Gasilova, E., Kutin, A., Samy, M., . . . Ayoub, M. M. (2019). Comparative Study of polymer nanoparticles on the basis of caprolactone–polyvinyl alcohol mixtures with an encapsulated antitumor preparation by atomic force microscopy, X-Ray diffraction, and dynamic light scattering. *Technical Physics*, *64*(12), 1729–1737.

Topal, G. R., Mészáros, M., Porkoláb, G., Szecskó, A., Polgár, T. F., Siklós, L., . . . Bozkir, A. (2020). ApoE-targeting increases the transfer of solid lipid nanoparticles with donepezil cargo across a culture model of the blood–brain barrier. *Pharmaceutics*, *13*(1), 38.

Tripathi, S., Patel, B., Shukla, S., Pachouri, C., Pathak, S., & Pandey, A. (2021). *Donepezil loaded PLGA nanoparticles, from modified nano-precipitation, an advanced drug delivery system to treat Alzheimer disease*. Paper presented at the Journal of Physics: Conference Series.

Tzeyung, A. S., Md, S., Bhattamisra, S. K., Madheswaran, T., Alhakamy, N. A., Aldawsari, H. M., & Radhakrishnan, A. K. (2019). Fabrication, optimization, and evaluation of rotigotine-loaded chitosan nanoparticles for nose-to-brain delivery. *Pharmaceutics*, *11*(1), 26.

Ulbrich, K., Hekmatara, T., Herbert, E., & Kreuter, J. (2009). Transferrin-and transferrin-receptor-antibody-modified nanoparticles enable drug delivery across the blood–brain barrier (BBB). *European Journal of Pharmaceutics and Biopharmaceutics*, *71*(2), 251–256.

Vozza, G., Danish, M., Byrne, H. J., Frías, J. M., & Ryan, S. M. (2018). Application of Box-Behnken experimental design for the formulation and optimisation of selenomethionine-loaded chitosan nanoparticles coated with zein for oral delivery. *International Journal of Pharmaceutics*, *551*(1–2), 257–269.

Wilson, B., Alobaid, B. N. M., Geetha, K. M., & Jenita, J. L. (2021). Chitosan nanoparticles to enhance nasal absorption and brain targeting of sitagliptin to treat Alzheimer's disease. *Journal of Drug Delivery Science and Technology*, *61*, 102176.

Zhang, C., Wan, X., Zheng, X., Shao, X., Liu, Q., Zhang, Q., & Qian, Y. (2014). Dual-functional nanoparticles targeting amyloid plaques in the brains of Alzheimer's disease mice. *Biomaterials*, *35*(1), 456–465.

Zhang, C., Zheng, X., Wan, X., Shao, X., Liu, Q., Zhang, Z., & Zhang, Q. (2014). The potential use of H102 peptide-loaded dual-functional nanoparticles in the treatment of Alzheimer's disease. *Journal of Controlled Release*, *192*, 317–324.

Zhang, H., Huang, Q., Huang, Z., Liu, T., & Li, Y. (2015). Preparation and physicochemical properties of chitosan broadleaf holly leaf nanoparticles. *International Journal of Pharmaceutics*, *479*(1), 212–218.

Zhang, L., Liu, X.-g., Liu, D.-q., Yu, X.-l., Zhang, L.-x., Zhu, J., . . . Liu, R.-t. (2020). A conditionally releasable "do not eat me" CD47 signal facilitates microglia-targeted drug delivery for the treatment of Alzheimer's disease. *Advanced Functional Materials*, *30*(24), 1910691.

Zhang, P., Liu, X., Hu, W., Bai, Y., & Zhang, L. (2016). Preparation and evaluation of naringenin-loaded sulfobutylether-β-cyclodextrin/chitosan nanoparticles for ocular drug delivery. *Carbohydrate Polymers*, *149*, 224–230.

Zhao, Y., Xiong, S., Liu, P., Liu, W., Wang, Q., Liu, Y., . . . Wang, Q. (2020). Polymeric nanoparticles-based brain delivery with improved therapeutic efficacy of ginkgolide B in Parkinson's disease. *International Journal of Nanomedicine*, *15*, 10453.

Zheng, X., Zhang, C., Guo, Q., Wan, X., Shao, X., Liu, Q., & Zhang, Q. (2017). Dual-functional nanoparticles for precise drug delivery to Alzheimer's disease lesions: Targeting mechanisms, pharmacodynamics and safety. *International Journal of Pharmaceutics*, *525*(1), 237–248.

13 Lipid-Based Nanocarriers for Brain Delivery of Bioactives in the Treatment of Neurodegenerative Disorders

Khaled M. Kotb, Mohamed N. Sharafeldin, Hamsa Hassan,# Achouak Boutalbi, Ahmed Yahya, Haseen Fatima Niazi, Nouran A. Taha, Islam Omar, and Amgad E. Salem

13.1 INTRODUCTION

Neurodegenerative diseases affect millions of patients worldwide. The brain controls many physiological and bodily functions, so neurodegeneration affects a patient's life quality and their ability limits. Parkinson's disease (PD) and Alzheimer's disease (AD) are considered the most common neurodegenerative disorders (Lamptey et al., 2022). Despite the therapeutic advancement of neurological disorders, most of the approved medications are symptomatic treatments due to the lack of active targeting of the pathogenesis owing to the presence of the blood–brain barrier (BBB). The BBB is a physiological barrier that controls the influx of substances between the systemic circulation and the central nervous system (CNS). However, it obstructs the entrance of drugs with adequate concentration to the CNS (Khawli & Prabhu, 2013).

Nano-drug delivery is a novel approach to enhance brain targeting for diagnostic and therapeutic purposes. Several types of nanocarriers have been tested as lipid-based nanocarriers. Lipid-based nanocarriers have been widely examined owing to their enhanced bioavailability and biocompatibility (Teixeira et al., 2020; Y.-Q. Zhao et al., 2022). In 1995, the Food and Drug Administration (FDA) authorized the first liposomal product (Doxil, liposomal doxorubicin) (Xing et al., 2016). Then, it paved the way for a new area of lipid-based nanoparticles as a drug delivery system. Lipid-based nanocarriers are usually classified into unlayered lipid drug delivery systems, bilayered lipid nanocarriers, plus submicron emulsions (Witika et al., 2022).

Nguyen et al. discussed a thorough overview of lipid-based nanoparticles targeting the brain via nose-to-brain pathways for treating neurodegenerative diseases (Nguyen et al., 2021). Milan et al. conducted a systematic search for the literature and assessed the relevant data on the dermal delivery of nanosized melatonin via different platforms, including niosomes, ethosomes, and liposomes. They concluded that nanosized melatonin showed enhanced activity when compared to free melatonin (Milan et al., 2017). Moreover, a recently published book chapter reviewed the literature on other routes of administration of bioactives through lipid-based nanocarriers for managing neurodegenerative disorders effectively (Shilpi et al., 2022). Interestingly, bio-macromolecules such as antibodies and messenger RNA (mRNA) can be delivered via lipid-based nanocarriers (Zhao et al., 2022).

In this chapter, we summarize the characteristics and latest advances in different types of lipid-based nanocarriers and their applications in preclinical and clinical trials. Also, we will discuss future prospects and challenges against nano-drug delivery development and translation into clinical practice.

13.2 A PHYSIOLOGICAL BARRICADE TO CNS DRUG DELIVERY (BBB)

Although lipid-based nanocarrier forms of drugs offer several advantages over their plain counterparts, their application in brain drug delivery is frequently restricted by the BBB. The BBB is the chief overseer that governs molecule transport into and out of the CNS. It offers strict, selective permeability to the CNS using mechanisms such as passive and facilitated diffusion, along with active transport (Daneman & Prat, 2015), and only permits substances with molecular weights smaller than 400 Da to diffuse, such as oxygen, carbon dioxide, and lipophilic molecules. Protein carriers on the luminal endothelium aid in the transport of certain molecules into the CNS, such as glucose, certain amino acids, vitamins, and monocarboxylates (Pardridge, 2012; Sweeney et al., 2019). This restrictive barrier shields the brain against potentially harmful toxins and pathogens that may enter the bloodstream.

Corresponding author: hamsahassan.193@gmail.com

DOI: 10.1201/9781003383376-13

BBB presents a challenge to CNS drug development. Estimates have shown that the BBB rules out 98% of small-molecule, and nearly all large-molecule, compounds, including recombinant proteins, peptides, and antibodies (Pardridge, 2005). To overcome this, a number of invasive and non-invasive strategies have been investigated, such as lipid-based nanocarriers, disruption of the BBB by using targeted ultrasound (Kovacsa et al., 2016) or osmotic shock (Kroll & Neuwelt, 1998), exploiting endogenous transport systems such as receptor-mediated transcytosis, carrier-mediated transport, and modifying the drug properties to enhance BBB penetration (Tajes et al., 2014).

13.3 UNLAYERED LIPID DRUG DELIVERY SYSTEMS

13.3.1 Solid Lipid Nanoparticles (SLNs)

SLNs have been considered as a nanocarrier with a particle size ranging between 40 nm and 1,000 nm (Tenchov et al., 2021). SLNs have been developed as a better substitute with more advantages than their lipidic counterparts, for example, polymeric nanoparticles and liposomes. This is due to the avoidance of organic solvent use in their synthesis, which enabled SLNs to exhibit limited systemic toxicity and better biocompatibility (Mukherjee et al., 2009; Satapathy et al., 2021). Since SLNs are composed of solid lipids, this provides more stability to the formulation and protection of the active constituent against chemical degradation (Das & Chaudhury, 2011; Mukherjee et al., 2009; Müller et al., 2002; Souto & Müller, 2011).

Moreover, SLNs can be classified according to the site of the active constituent in its structure into three main types: (1) drug-loaded core, in which the active constituent is allowed to precipitate before the recrystallization of lipids; (2) drug-loaded shell, in which lipids are allowed to recrystallize and then the active constituent is added during the cooling stages; and (3) homogenous matrix, in which the active constituent becomes homogenously dispersed within the lipid matrix (Müller et al., 2002; Souto & Müller, 2011).

They can be considered promising nanocarriers meant for drug delivery purposes for several reasons, for example, (1) the capability of crossing the BBB (Satapathy et al., 2021), (2) improving stability and targeting abilities via binding to various ligands (Siddhartha et al., 2018), (3) enhanced biocompatibility through the use of physiological lipids that are generally regarded as safe instead of organic solvents (Paliwal et al., 2020; Weber et al., 2014), (4) encapsulating both lipophilic and hydrophilic active constituents (Fundarò et al., 2000; Reddy & Venkateswarlu, 2004), and (5) reduction of production costs due to using the relatively cheaper lipid raw materials as well as cheaper production methods (Müller et al., 2000; Satapathy et al., 2021).

On the contrary, a major drawback of SLNs is the possible expulsion of the active constituent in the media upon storage for a long time. This is due to polymorphic transition in the lipid matrix, leading to reducing the amorphous region in the lipid matrix (Duan et al., 2020; Esposito et al., 2019). Other drawbacks, including undesirable drug–lipid matrix interaction, poor loading capacity, and high water content, may also exist (Satapathy et al., 2021).

13.3.2 Nanostructured Lipid Carriers (NLCs)

NLCs have been evolved by Muller et al. in 1999–2000 to get over SLN limitations, and they are considered second-generation lipid carriers (Fang & Al-Suwayeh, 2013). The diameter ranges from 10 to 1,000 nm (Üner, 2016). Since the major limitations of SLNs were owed to the highly crystalline lipid matrix, the formulation of SLNs was designed to achieve a less-crystalline matrix with a solid core through altering a portion of the solid lipids with liquid lipids. Hence, the ratio between solid lipids to liquid lipids ranges between 70:30 and up to a ratio of 99.9:0.1 according to the attributes of the formulation. Thus, NLCs exhibited enhanced protection for the drug against degradation and avoided drug leakage on storage (Haider et al., 2020). Furthermore, NLCs also increased the drug loading capacity because of the voids created in the core, which can accommodate the drug, so NLCs can be an ideal nano-drug delivery system (Witika et al., 2022). Besides, their stability can be enhanced via various mechanisms, such as lyophilization, spray drying, and the addition of poloxamers or polyethylene glycol (PEG) (J. Garg et al., 2022).

Depending on their formulation parameters and lipid content, NLCs can be categorized into three major types.

13.3.2.1 Type 1 (Imperfect Type)

This type is obtained via mixing fatty acids of different chain lengths and saturation states imperfections in the crystal lattice. In this type, the amount of liquid lipids is less than the amount of solid lipids. The created voids enable the increase in drug incorporation inside the NLCs and achieve high entrapment efficiency (Müller et al., 2002).

13.3.2.2 Type 2 (Amorphous Type)

This type is formulated by using a special type of lipid, such as isopropyl myristate (Faheim et al., 2018). These blended lipids remain in the α polymorph after solidification rather than crystallization after homogenization and cooling, so their core remains amorphous. This allows the retention of the drug and minimizes its expulsion (Salvi & Pawar, 2019).

13.3.2.3 Type 3 (Multiple Type)

This type mainly depends on the concept of w/o/w emulsion and can be formulated through the hot homogenization technique via combining solid lipids with large portions of liquid lipids. Upon cooling, it results in phase separation, while oily nano-compartments are dispersed uniformly on the solid matrix. Since lipophilic drugs pose higher solubility in liquid lipids than in solid lipids, this allows for increasing the drug loading capacity of NLCs.

Furthermore, the solid lipid matrix enables controlling drug release in addition to limiting drug leakage (Haider et al., 2020; Shidhaye et al., 2008).

13.3.3 Lipid Drug Conjugates (LDC)

LDC can be defined as the drug molecule that was modified with a lipid moiety via covalent or non-covalent interaction to form a prodrug (Lambert, 2000). Generally, four types of lipids are used in LDC formulas: fatty acids, glycerides, phospholipids, and steroids (Irby et al., 2017). The desired function determines which lipid is suitable for the development of LDCs; for example, drugs undergoing extensive first-pass metabolism could benefit from glyceride-based prodrugs, as glycerides could allow transporting the drug from the route to the systemic route. Besides, glycerides may be useful for targeting the lymphatic system itself, for instance, in anticancer medications (Adhikari et al., 2017; Li et al., 2019). Modifying the active ingredient with the lipid moiety results in increasing the drug lipophilicity. Besides improving oral bioavailability and drug loading, the lipophilic prodrug can also allow controlled drug release and limit drug leakage (Banerjee & Kundu, 2018; Muchow et al., 2008).

Most LDCs are entrapped into a suitable drug delivery system, for instance, liposomes, micelles, and lipid nanoparticles, although they can form self-assembled nanoparticles without a delivery carrier (Neupane et al., 2013). The drug release from the LDC nanoparticle is dependent on two steps: (1) the release of LDC from the nanoplatforms, either by diffusion or erosion of the matrix, and (2) the bioconversion of the LDC particle into the parent drug via enzymes found in the GIT or other physiological enzymes (Date et al., 2019; Müller & Olbrich, 2004).

Although the slow drug release achieved by LDCs can be beneficial for sustained-release formulations, this will not be suitable for medical conditions requiring an immediate release of the active ingredient (Irby et al., 2017). Furthermore, the prodrug is less stable than the parent drug due to the differences in physicochemical characteristics between them; thus, there are more concerns about the polymorphic transition, chemical instability, and aqueous solubility of the LDC formulations (Jana et al., 2010).

13.3.4 Nanocapsules

Nanocapsules have either a lipid-based structure (lipid nanocapsules—LNCs) or a polymer-based structure (polymeric nanocapsules—PNCs) and can be formulated as a hybrid lipid/polymer nanocapsule (Mehandole et al., 2023). Moreover, LNCs are advanced nanocargo systems having an oily core (medium-chain triglycerides enveloped by surfactants) (Urimi et al., 2021) especially designed to avoid problems associated with liposomes and focused on achieving better encapsulation efficiency for lipophilic drugs, preventing their leakage in biological fluids, and offering good stability of up to 18 months.

Many drug molecules have limited effect on the CNS because of their incapacity to reach the BBB. LNCs pose an exceptional tendency to navigate through the BBB, delivering drug molecules efficiently to the brain. They are designed to produce optimal delivery vehicles for targeting CNS.

Formulation of LNCS involves a simple, low-energy, phase-inversion technique without the use of any organic solvents and uses FDA-approved excipients for parenteral administration, resulting in slightly negatively charged nanocapsules in the size range of 20–100 nm. The formulation can be designed to produce NCs with an aqueous core for encapsulation of hydrophilic drugs, adding diversity to them (Moura et al., 2020).

The formulation of nanocapsules can be done via emulsion diffusion, nanoprecipitation, layer-by-layer assembly, emulsion coacervation, double emulsification, and polymer coating. Moreover, they can be given to the patients parenterally, orally, and topically, in addition to ocular and nasal routes. Among various applications of nanocapsules, gene delivery, peptide delivery, cell targeting, organ targeting, and theragnostic applications have garnered the most attention from researchers (Mehandole et al., 2023).

13.3.5 Micelles

Micelles can be termed as tiny clusters of molecules (amphiphilic in nature) at a certain concentration, so-called critical micelle concentration (CMC). The micelles contain a hydrophilic head and a hydrophobic tail generally, and they self-assemble so the hydrophilic heads point outward toward a solvent and all the hydrophobic tails join inward, making a center of the micelle (Recent et al., 2023). With this configuration, the hydrophobic tails require less energy to dissolve, and the micelle is more stable in solution (Ghezzi et al., 2021).

The amount of surfactant in the solution, the type of surfactant, and the characteristics of the solvent are only a few of the variables that impact the size and form of micelles. Individual surfactant molecules are distributed in solution at low surfactant concentrations. However, as the solution achieves CMC, surfactant molecules self-assemble into aggregates, and micelles begin to form above this concentration (Krawczyk, 2023). Polymeric micelles have a high-molecular weight, low CMC, and can be preserved for a longer time than their counterparts made of low-molecular-weight surfactants. Their size ranges between 10 and 100 nm.

Polymeric micelles are modified by biocompatible ligands for active, targeted drug delivery. Sugar molecules, peptides, antibodies, and proteins are often employed in ligands. The targeted response is achieved via certain natural stimuli, like pH, temperature, and particular enzymes. The disintegration of micelles releases actives/drugs at the target sites. Additionally, external factors like light, ultrasound, and magnetic fields can cause micelles to release drugs (Ghezzi et al., 2021).

Micelles are advantageous in a range of applications due to their distinctive features. They increase bioavailability, solubilize hydrophobic compounds, and shield them from deterioration. Additionally, micelles can be engineered to target particular tissues or cells and are employed as delivery systems for a variety of substances, for instance, drugs, peptides, and proteins (Gong et al., 2012).

13.3.6 Self-Nanoemulsifying Drug Delivery

Despite being one of the safest routes, the oral route of drug administration depicts some kinetic and stability issues. Researchers have suggested that self-nanoemulsifying drug delivery systems (SNEDDSs) may enhance oral bioavailability, especially those with solubility issues. This technology has addressed several concerns with oral route medications, including insufficient bioavailability, undesirable toxicity, poor absorption, degradation, and solubility problems at the site of action (Baloch et al., 2019; Nair et al., 2022; Parveen et al., 2023).

They involve an oily phase (comprising medium-long-chain triglycerides), surfactants, drug components, and co-surfactants/co-solvents/solubilizers which, when agitated gently upon interaction with the aqueous phase, form a nanoemulsion having droplets sized around 20–200 nm (followed by dilution), demonstrating better and consistent dissolution rates and bioavailability. Moreover, SNEDDSs are comparatively stable and easy to formulate on any scale of production with an agenda to obtain a pre-concentrated solution, avoiding the dissolution step (Ashfaq et al., 2022). Despite the numerous benefits, there are some challenges, such as drug loading capacity, precipitation chances, and safety risks of oral administration of surfactants and other excipients (Buya et al., 2020).

The development of commercial SEDDSs formulas, for example, Norvir® (ritonavir), Fortavase® (saquinavir), and Neoral® (cyclosporine), increased formulators' interest in the use of SNEDDs in other applications to improve their bioavailability and solubility. Recently, a study demonstrated the new roles of SNEDDSs as solid supersaturated, mucus-permeating, and targeted systems to resolve issues with their conventional versions, showcasing more applications in various treatments (Buya et al., 2020).

13.4 BILAYERED LIPID NANOCARRIERS

13.4.1 Liposomes

Liposomes, which are spherical lipid vesicles with particle sizes typically ranging from 50 to 500 nm, have garnered extensive attention as the predominant nanocarriers in targeted drug delivery systems. These nanocarriers are formed by emulsifying natural or synthetic lipids in an aqueous medium and are composed of one or more lipid bilayers (Jha et al., 2016). Upon introduction of an amphiphilic lipid to water or other hydrophilic liquids, nanocarriers of 50 to 500 nm are formed very instantly (Peng et al., 2020). This system enables hydrophilic drug molecules to be encapsulated by simply dissolving the drug in the liquid used to make the nanocarriers. Moreover, drugs that are hydrophobic or amphiphilic can be enclosed by adding them directly to the lipid solution before the nanocarriers are formed, resulting in a layer of drug molecules between the lipid bilayer (Sun et al., 2014).

Liposome nanoemulsions stand out as the most commonly utilized nanoparticles in the field of nanomedicine. This preference is attributed to their biocompatibility, stability, straightforward synthesis process, excellent drug-loading capability, enhanced bioavailability, and the use of safe excipients within these formulations (Nsairat et al., 2022). According to their size, they are classified into small liposomes (20–100 nm), large liposomes (>100 nm), and giant vesicles (>10 μm) (Vemuri & Rhodes, 1995). Furthermore, they can be categorized based on the number of bilayers to uni-, oligo-, and multilamellar (Larivi6re et al., 1991). Also, liposomes are classified by the phospholipid charge as neutral, anionic, or cationic (Storm & Crommelin, 1998). According to the function, liposomes may be adopted for conventional, ligand-targeted, long-release, and triggered-release applications (Sharma & Sharma, 1997). In addition, there are further classifications based on composition and mode of drug delivery (Friend et al., 1996; Karanth & Murthy, 2010; Lee, 2020; Zelphati & Szoka, 1996).

Currently, a wide range of liposome preparation techniques are accessible, each with unique characteristics and advantages. Novel liposomal formulations can be simply and efficiently created if appropriate methodologies and related novel techniques are applied.

13.4.1.1 Reverse-Phase Evaporation

It is commonly used as a substitute for thin-film hydration in various applications (Pradhan et al., n.d.). Initially, the lipids are dissolved in an organic solvent and then mixed with an aqueous buffer containing the hydrophilic drug. Subsequently, the organic solvent is evaporated using a low-pressure rotary evaporator, leading the lipid vesicles to disperse in the aqueous solution (Gu et al., 2020). This approach works well for molecules with large molecular weight; while the method of using organic solvents and sonication can be effective in creating lipid vesicles, it is essential to be cautious, especially when dealing with therapeutic peptides. These peptides may be susceptible to denaturation or structural changes under such conditions, potentially affecting their therapeutic efficacy. (Szoka, 1980).

13.4.1.2 Thin-Film Hydration Method (Bangham Method)

Using a round-bottom flask, all lipids, as well as hydrophobic drugs, are dissolved in an adequate organic solvent. Afterward, the thin film is hydrated by adding an aqueous buffer solution at temperatures higher than the lipid's transition temperature (Tm). The hydration solution can include a hydrophilic drug that will be encapsulated within the aqueous core of the liposomes (Pradhan et al., n.d.).

13.4.1.3 Supercritical Fluid

Instead of employing organic solvents, this approach used a supercritical fluid (i.e., carbon dioxide—CO_2) to dissolve lipids. An aqueous phase is continuously fed into a cell containing the supercritical lipid solution using a high-performance liquid pump, allowing the dissolved phospholipids to phase transition. After eliminating CO_2, liposomes will develop as a result of a sudden reduction in pressure. This approach resulted in fivefold greater encapsulation efficiency. Even when employing ecologically safe and inexpensive carbon dioxide, this technology suffers from high costs, limited yield, and specialized infrastructure (William et al., 2020).

13.4.1.4 Solvent Injection Method

An aqueous phase was immediately injected using an organic solvent that dissolves the lipids and hydrophobic active ingredients. This technique produced predominantly liposomal formulations with higher polydispersity indices (Jaafar-Maalej et al., 2011). Furthermore, prolonged exposure to high temperatures and chemical solvents may impair drug and lipid stability (Maherani et al., 2011).

13.4.1.5 Dehydration–Rehydration Method

After drying, the buffer containing tiny unilamellar vesicles is removed and then rehydrated with the aqueous solution containing the drug that will be integrated. This results in the dispersion of solid lipids in tiny subdivided forms, and the vesicles are subsequently rehydrated. Liposomes generated using this method are typically oligolamellar vesicles (Jangde & Singh, 2016).

Following drug release from liposomes, liposome-associated drugs (DL) are absorbed from the site of injection into the central compartment as the DL package, or as a free drug (D) or drug-free liposomes (L). The DL, D, or L are either removed from the central compartment or disseminated into the tissue compartment, where they are destroyed subsequently (Peschka et al., 1998). Liposomes interact with the cell membrane in a variety of forms, including selective (receptor-mediated) or nonselective endocytosis (Kelly et al., 2011), local fusion (adhesion), phagocytosis (Caracciolo et al., 2005), and cell membrane absorption (Bozzuto & Molinari, 2015). Besides, a multitude of parameters modifies liposome–cell interactions, including composition, liposome size, surface charge (Nejat Duzguneş et al., 1999), presence of targeting ligands on the liposome surface, and biological environment (Kono et al., n.d.).

13.4.2 Pharmacosomes

Pharmacosomes, as new drug delivery systems, refer to colloidal dispersions of drugs covalently bonded to lipids. These entities can exist as ultrafine vesicular, micellar, or hexagonal aggregates which depend on the chemical nature of the drug–lipid complex (Semalty et al., 2009). The criteria for the formation of vesicular pharmacosomes involve both surface and bulk interactions between lipids and the drug. These interactions play a significant role in determining the vesicular structure of the pharmacosomes. Any substance having an active hydrogen atom, such as -COOH, -OH, -NH_2, etc., can be esterified to the lipid, either with or without a spacer chain, ultimately leading to the formation of an amphiphilic molecule (Pathak & Manchanda, 2020).

There are three components involved in this delivery method:

- *Drugs*. Any drug containing an active hydrogen atom (-COOH, -OH, -NH_2, etc.) can be esterified to lipid, with or without a spacer chain, to form amphiphilic complexes. These amphiphilic complexes (pharmacosomes) play a crucial role in facilitating the transfer of drugs across cell membrane, tissue, or cell wall transfer in the organism (Patil & Jadhav, 2014).
- *Lipids*. Phospholipids are the essential molecular components of cell membranes. Phosphoglycerides and sphingolipids are the two most frequently encountered kinds of phospholipids. Phosphatidylcholine is the most common phospholipid (ResearchGate, n.d.).
- *Solvents*. During the production of pharmacosomes, a high-purity, volatile organic solvent of analytical quality with intermediate polarity is employed. This solvent is chosen based on the polarity of the medication and the phospholipids. The phospholipids and the medication are then dissolved in the chosen solvent to form the pharmacosome complex (Pandita & Sharma, 2013).

Initially, a drug–lipid conjugate is required for the production of pharmacosomes. To do this, the drug's salt form is converted into an acidic form, exposing the functional hydrogen atom and creating a complex. The drug's aqueous solution is acidified following extraction with chloroform and recrystallized. Then, the phospholipid content is quantified and dissolved in an organic solvent before being evaporated under a vacuum at a suitable temperature. After being placed in a desiccator overnight, the complex is recovered as a dry residue (Pandita & Sharma, 2013).

Pharmacosomes have a higher stability profile and a longer shelf life. They can improve drug absorption and distribution. Pharmacosomes can increase penetration rates by increasing membrane fluidity (Ali, 2017). Additionally, the transition temperature of vesicles, whether in the form of vesicles and micelles, significantly affects their interaction with biomembranes; in consequence, they increase drug transport across the membrane. Several studies have utilized this strategy effectively to enhance the therapeutic performance of numerous drugs, such as pindolol diglyceride, amoxicillin, taxol, cytarabine, dermatansulfate, bupranolol hydrochloride, and others, as mentioned in the following table (Table 13.1) (Pandita & Sharma, 2013;

Pharmaceutical Dosage Forms, 2020; Semalty et al., 2009; Talegaonkar et al., 2006; Taskintuna et al., 1997).

Protein PL complexes (insulin and salmon calcitonin) have been effectively created and have been shown to increase bioavailability (Gu et al., 2020). Besides, Peng-Fei Yue et al. created and characterized geniposide-pharmacosomes. They discovered that it can increase physiologically active ingredient absorption and penetration (Pandita & Sharma, 2013; Venkatesh et al., 2014). Pharmacosomes were created for a variety of poorly soluble nonsteroidal anti-inflammatory medicines, such as aceclofenac, diclofenac, aspirin, and fenoprofen. These studies demonstrate that pharmacosomes can improve medication solubility and penetration. *In vitro*, percutaneous absorption of fenoprofen employing a flow-through diffusion cell improved drug permeation over the skin.

In another study, Yi-Guang et al. (2005) developed acyclovir pharmacosomes and then discovered that plasma proteins in blood absorbed these pharmacosomes and interfered with erythrocyte connections, resulting in a diminished hemolytic response (Yi-Guang et al., 2005).

Finally, pharmacosomes are an emerging, promising carrier system in the pharmaceutical industry. Even with the limitations of being fused and aggregated, they remain an important tool for targeted and sustained drug delivery. Further, drug destiny and biological activity may be altered when spacer groups and connections improve. However, further efforts are needed to investigate the mechanism of action and non-bilayer phases. Eventually, pharmacosomes promise to enhance drug distribution in the case of both natural and synthesized active components (Veena & Savitha, 2019). Thereby, pharmacosomes have superior advantages over liposomes and niosomes. The main features and problems associated with each vesicular system can be included in Table 13.2 (Sharma et al., 2013).

13.4.3 Ethosomes and Transferosomes

Ethosomes and transfersomes vesicles exert their action via the skin (i.e., transdermal), owing to their structure and flexibility. The primary mode of transportation for liposomes is through hair, while transfersomes and niosomes are mainly transported in the pores between keratinocytes (Bhasin et al., 2018). Ethosomes and transfersomes differ mostly in their structure. Transfersomes comprise phospholipids plus an edge activator, such as a surfactant. Instead, ethosomes contain phospholipids, ethanol, and water. The ethanol component evaporates when the solution is spread on the skin without covering the area, while surface-active molecules such as surfactants remain on top (Sudhakar et al., 2021).

TABLE 13.1
Drug effects after incorporation in pharmacosomes

Drug	The effect after incorporation in pharmacosomes
Pindolol diglyceride	Three- to fivefold high plasma concentration Lower renal clearance
Cytarabine	Improved biological activity
Taxol	Enhanced biological activity
Amoxicillin	Enhanced cytoprotection as well as treatment of *H. pylori* infections in male rats
Dermatan sulfate	Improved biological activity
Bupranolol hydrochloride	Enhanced effect on intraocular pressure Enhance lymph transport

TABLE 13.2
Comparison between conventional vesicular systems and pharmacosomes

Vesicular system	Features	Problems
Liposomes	Microscopic vesicle (25–100 µm) of one or more lipid bilayers, separated by water or aqueous buffer compartment	Expensive to prepare, degradation by oxidation, sedimentation, leaching of the drug, lack of purity of natural phospholipids.
Niosome	Non-ionic surfactant vesicle	Leaching the drug, time-consuming, and insufficient stability.
Transferosome	Versatile and well-suited for delivering both low- and high-molecular-weight drugs and also for lipophilic as well hydrophilic drugs.	Expensive, oxidative degradation, lack of purity of natural phospholipids.
Pharmacosomes	Prepared easily, high entrapment efficiency, more stable, more efficient	Pharmacosomes can only encapsulate water-insoluble pharmaceuticals in very tiny hydropic patches inside membrane bilayers. Pharmacosomes suffer fusion and aggregation during storage, as well as chemical hydrolysis.

Ethosomes are less abrasive to the skin and more stable at 4°C. Variations in the amounts of ethanol and lipids in formulations have a great impact on how well medications are delivered by ethosomes (Chauhan et al., 2022). Interestingly, ethosomes and transfersomes can be used in transdermal drug delivery of a wide range of medications. For instance, ethosomes have been attempted to deliver diclofenac, ibuprofen, and lidocaine. Additionally, transfersomes have been utilized to administer insulin, corticosteroids, and anticancer medications (Pierre & Dos Santos Miranda Costa, 2011).

Ethosomes and transfersomes are different from liposomes and niosomes in terms of their structure and mode of operation. Niosomes constitute non-ionic surfactants, while liposomes contain phospholipids and cholesterol. Due to their structural flexibility, ethosomes and transfersomes penetrate the skin more deeply than do liposomes and niosomes (Abd El-Alim et al., 2019).

13.4.4 Niosomes

Niosomes are vesicular structures comparable to liposomes. They can carry both amphiphilic and lipophilic medications. Moreover, niosomes are formed by hydrating synthetic non-ionic surfactants, either with or without the addition of lipids, such as cholesterol. Niosomes are vesicular structures made up of a non-ionic surfactant bilayer (Muzzalupo & Tavano, 2015). Non-ionic surfactants self-assemble to construct the bilayer in an aqueous medium. Niosomes are tiny lamellar (unilamellar or multilamellar) structures that are spherical in shape. Niosomes are created using ionic compounds, while vesicles are formed using non-ionic surfactants. The hydrophobic head or hydrocarbon segments of the non-ionic surfactants are aligned in a way that contact with the aqueous media is reduced, yet the polar heads are aligned toward the aqueous bulk (media) (Momekova et al., 2021).

Niosomes exhibit plasticity and flexibility in their structural characterization and are considered biodegradable, biocompatible, and non-immunogenic. Controlled discharge has been used to treat cancers and other health conditions. Niosomes can encapsulate both hydrophilic and lipophilic molecules, potentially increasing the duration of action of the drug. Drug encapsulation in vesicular systems may prolong their presence in systemic circulation, enhance penetration into target tissues, and possibly reduce toxicity if targeted uptake can be achieved (Muzzalupo & Tavano, 2015).

Niosomes shield medicinal components from biological enzymes and acids while enhancing drug stability. They can increase the oral bioavailability of medicines that are not well absorbed, by delaying clearance from the bloodstream. Regarding the encapsulation and distribution of hydrophilic medicines, niosomes are thought to be reliable and efficient nanocarriers. Liposomes fulfill the requirements for drug protection, controlled release of the active ingredient, and targeted delivery. They are useful for achieving these goals (Cheng et al., 2016).

However, niosome transdermal administration encounters some difficulties. Drug penetration is prevented by the stratum corneum, the outermost layer of the skin. Niosome transdermal penetration can be impacted by the size and charge of the niosomes. Niosome stability may be impacted by the skin's environment, which is characterized by a low pH, high temperature, and high enzymatic activity. The environment on the skin may also have an impact on how rapidly drugs are released from niosomes. Nevertheless, the penetration of niosomes into the skin can be fostered by the application of gelling agents (Yasamineh et al., 2022).

13.4.5 Cubosomes

Unlike the single bilayered vesicle, known as liposomes, cubosomes are a type of multilamellar bilayered vesicle, as they are formed of distinct amphiphilic lipid components in particular ratios. These systems are composed of curvaceous, bicontinuous, bilayered lipid membranes in a porous, lattice-like 3D system that are divided channels that can be filled by a variety of bioactive components, such as pharmacological molecules (Dawoud et al., 2020; Karami & Hamidi, 2016). Cubosomes feature pores ranging in size from 5 to 10 nm and a size of 10–500 nm. Consequently, they can encapsulate a wide range of pharmacological compounds, such as hydrophilic, hydrophobic (Garg et al., 2007), and amphipathic molecules (Karami & Hamidi, 2016).

Additionally, unlike lamellar liposomes, the nanostructure's central core is evenly rich in a lipid matrix. They also have up to three times the lipophilic load compared to liposomes. However, just about 60% of their outer surface comes into contact with water. Therefore, cubosomes are ideal for hydrophilic drug targeting (Azmi et al., 2015).

Beyond the structural standpoint, cubosomes possess key properties that enable effective and targeted drug delivery. These properties include lipid biodegradability (Bei et al., 2010; Rizwan et al., 2007), biocompatibility, and bioadhesives. In addition, Cubosomes exhibit a bicontinuous, cubic liquid-crystalline phase that can retain its structure over an extended duration, even when exposed to an excess amount of water (Avantika et al., 2020).

Interestingly, cubosomes are classified as generally regarded as safe. They are also approved and marketed in the United Kingdom as a non-parenteral pharmaceutical, and they are on the US Food and Drug Administration's list of inactive substance guidelines. Furthermore, they can be adapted to create selective and regulated drug delivery systems on a small scale with a simple production procedure (Garg et al., 2007). The hallmark of cubosomes is their capacity to encapsulate hydrophilic, hydrophobic, and amphiphilic molecules. In addition, many proteins can be readily entrapped in their water channels, which enhances the delivery and provides a superior stability profile towards enzymatic breakdown (Rajabi & Mousa, 2016).

Another key aspect of cubosomes is the ability to modify the composition and structure of their bilayer membrane.

This can be achieved by introducing lipids with different acyl groups to alter the charges in the aqueous networks, which in turn affects the interaction between the bilayer membrane and guest molecules. The diameters of the water channels within the host cubic phase can be modified to regulate the effectiveness of transport, and the electrostatic interaction between the channel surfaces and the guest components can be carefully adjusted to increase flexibility (Clogston et al., 2005; Negrini et al., 2014, 2015). Furthermore, the diameter of the aqueous channels can be changed by adding certain additives, such as octylglucoside surfactant, which has been shown to increase the diameter of hydrophilic channels (Angelov et al., 2003). Some cubosome systems also incorporate symmetrical lipid-based liquid crystal phases that can respond to pH changes, making them useful for controlled drug delivery (Negrini et al., 2015).

It is also worth noting that cubosomes are a promising approach to drug delivery via the nasal route because of their special qualities. Specifically, their mucoadhesive capabilities, and their components that increase penetration, make them well-suited to address the numerous challenges of nasal drug delivery, such as fast mucociliary clearance of the formulation, restricted reach of the nasal epithelium, and enzymatic degradation. Additionally, cubosomes protect the therapeutic compound from enzymatic hydrolysis by integrating it into their core, which allows for immediate contact with enzymes. This protection ensures the therapeutic compound remains stable and effective, making cubosomes an ideal option for nasal drug delivery (Ali et al., 2017; Crowe et al., 2018; Kwon & Kim, 2010; Mittal et al., 2014).

Cubosomes offer three methods of drug loading: incorporation within the lipid layer, attachment to the lipid bilayer, or confinement inside the aqueous platforms of the cubic phase.

In the first method, drugs are incorporated into the lipid layer of cubosomes during their formation. This can be achieved through co-lyophilization of the active compounds with the bilayer films or by introducing the payload to molten lipids. This method is known as pre-loading and offers a higher encapsulation efficiency since drugs can fuse into the liquid-crystalline framework of cubosomes.

In the second method, drugs are anchored to the lipid bilayer of cubosomes after their formation. This is known as post-loading and involves adsorbing drugs onto premodeled cubosomes. Before the medication is integrated into cubosomes, the post-incorporated particles go through a pre-heat processing step. Additionally, the final product produced by sonication often shows a lower percentage of vesicles.

In the third method, drugs are localized inside aqueous platforms within cubic phases of cubosomes after their formation. This method involves loading cubosomes after incubation and is another form of post-loading.

Cubosomes have exhibited the potential for controlled drug release through customization. This means that researchers can modify the surface of cubosomes to regulate the pace and timing of drug release. For example, they might attach targeting molecules to the surface of cubosomes to direct them to specific cells or tissues in the body, or they might add stimuli-responsive groups that cause the cubosomes to release their contents in response to changes in pH, temperature, or other environmental factors.

Moreover, factors such as size, morphology, and intrinsic physical/chemical properties can affect drug release from cubosomes. For example, smaller cubosomes may release drugs more quickly than larger ones because their surface-area-to-volume ratio is larger. Similarly, cubosomes with a more porous structure may release drugs more quickly than those with a denser structure. Other factors that could affect drug release include the composition of the lipid bilayer (which can affect how easily drugs diffuse through it), as well as external factors, such as temperature and pH (which can affect how stable cubosomes are and how easily they break down) (Abourehab et al., 2022).

13.5 LIPID-BASED EMULSIONS

The lipid-based emulsion system, which falls under the category of emulsion technology, specifically micro- and nanoemulsions, has become an attractive alternative for improving solubility of lipophilic compounds. This section is concerned with highlighting the subtle differences between both systems in terms of structure and characteristics, as well as the advantages compared to traditional emulsions, and the drawbacks of both systems.

13.5.1 Microemulsion

A type of lipid-based emulsion system known as a microemulsion is one in which microscopic droplets of one liquid are dispersed in another liquid and then stabilized by a film of a surfactant at the interface between the two liquids. The droplet size in a microemulsion usually has a range of 10–100 nm (Table 13.3) (Gupta et al., 2016). Oil-in-water (o/w), water-in-oil (w/o), and bicontinuous microemulsions are the three structural subcategories of microemulsions.

In o/w microemulsions, the continuous aqueous phase is where oil droplets are disseminated, whereas the opposite can be seen in w/o microemulsions. A bicontinuous microemulsion may form in systems with equivalent quantities of water and oil. When the surfactant concentration is increased enough to produce a network that crosses both phases, the bicontinuous structure forms. Surfactant molecules are organized in such a way that they can form a continual interface between the oil and water phases, forming a three-dimensional system of nodes and channels (Talegaonkar et al., 2008).

The microemulsion's interior structure is critical for phase diffusivity, which impacts drug release kinetics. Microemulsions can display a variety of structural types, including lamellar structures, anisotropic crystalline hexagonal or cubic phases, and regular emulsions. The presence of these different structures is dependent on the ratio of the components used to create the microemulsion (Talegaonkar et al., 2008).

Furthermore, surfactant film flexibility is an important factor for determining the structure and behavior of microemulsions. If the surfactant film is too rigid, it will be unable to form the necessary interconnections between both the oil and water phases required for the bicontinuous structure to form properly (Talegaonkar et al., 2008). Since drug delivery systems must be safe and biocompatible, special consideration should be given to the components used and how much of each should be used. However, creating microemulsions requires a lot of surfactants, which might not be desired (Talegaonkar et al., 2008).

13.5.2 Nanoemulsion

Nanoemulsions are a form of colloidal dispersion system that combines two immiscible liquids with the help of emulsifying agents, such as surfactants and co-surfactants. These emulsifiers play an important role in nanoemulsion stability by reducing the surface tension between both the oil and water phases. The resulting nanosized droplets, which range from 20 to 500 nm (Table 13.3) (Gupta et al., 2016), have high surface area and free energy, making them a useful transport system for a variety of applications (Sabjan et al., 2019).

Following are the types of nanoemulsion systems (Sabjan et al., 2019):

1. Oil-in-water (o/w) nanoemulsions, where oil droplets are scattered in the continuous aqueous phase.
2. Nanoemulsions of water-in-oil (w/o), in which the oil phase is continuously mixed with water droplets.
3. Bicontinuous nanoemulsions, which include small pockets of water and oil dispersed throughout the structure.
4. Multiple emulsions, which are categorized as either water-in-oil-in-water (w/o/w) or oil-in-water-in-oil (o/w/o) based on how their dispersed and continuous phases are arranged.

13.6 DIFFERENCES BETWEEN MICROEMULSIONS AND NANOEMULSIONS

Emulsions, such as micro- and nanoemulsions, are composed of droplets of small dimensions, below 1 μm, that is, on the nanometer scale. However, their application in pharmaceutics is determined by their physiochemical properties (Müller et al., 2002; Rao & McClements, 2011).

13.6.1 Droplet Size

Microemulsions have a particle size between 10 and 100 nm (Gupta et al., 2016), while nanoemulsions have a particle size between 20 and 500 nm (Gupta et al., 2016). This indicates that microemulsions have smaller droplets compared to nanoemulsions (Dhaval et al., 2022).

13.6.2 Preparation Procedure

Another substantial difference is the process of preparation. Microemulsion formation is a spontaneous process that does not require external energy. It typically involves mixing the oil, water, and surfactant using a low-energy approach, like vortex mixing or gentle stirring (Table 13.1) (Dhaval et al., 2022).

Nanoemulsions, on the other hand, cannot be spontaneously formed and are mostly prepared by energy-driven methods, which means that energy has to be supplied externally to break the internal phase into the nanosized particles (Dhaval et al., 2022).

13.6.3 Stability

Microemulsions exhibit thermodynamical stability under specific conditions (e.g., compositions and storage temperatures) (Table 13.3) as this system exhibits a lower free energy state when the colloidal droplets are dispersed in water compared to the state when they are simply separate phases. Hence, when the surfactant, oil, and water components are combined, microemulsions are expected to form spontaneously, and then they should remain stable indefinitely, provided the initial conditions remain unaltered (McClements, 2012).

As for nanoemulsions, they are thermodynamically unstable, as the system exhibits a higher free energy state when the colloidal droplets are dispersed in water compared to the state when they are simply separate phases, which means that a nanoemulsion will always break down when given sufficient time (McClements, 2012). Additional differences between the two systems are highlighted in Table 13.3.

13.6.4 Advantages over Traditional Emulsions

Microemulsions and nanoemulsions have substantial advantages compared to traditional emulsions. Some of the key advantages of both systems include (Sabjan et al., 2019; Talegaonkar et al., 2008):

1. *Increased Surface Area.* Both systems have a higher surface area than traditional emulsions, indicating that they are effective transport systems for various applications, such as drug delivery.
2. *Improved Bioavailability.* The small droplet size of both systems enhances the solubilization of poorly soluble drugs, which leads to improved therapeutic efficacy.
3. *Enhanced Stability.* Microemulsions and nanoemulsions are thermodynamically stable and do not exhibit issues associated with traditional

emulsions, which leads to prolonged shelf life and improved product quality.

(NB: It is interesting to note that nanoemulsions are relatively stable compared to traditional emulsions. In other words, regarding stability, all three systems can be ordered from the most stable to the least stable as follows: microemulsions, nanoemulsions, and regular emulsions.)

4. *Versatile Formulation Options.* Both microemulsions and nanoemulsions can be formulated as foams, creams, liquids, sprays, and other forms, depending on the application requirements.

Following are their limitations (Sabjan et al., 2019; Talegaonkar et al., 2008):

1. *Limited-Drug Loading Capacity.* Due to their high surface area and small droplet size, which can limit the amount of drug that can be encapsulated within the system.
2. *Potential Toxicity.* While they are generally regarded as safe, the incorporation of surfactants and co-surfactants in their formulation carries a potential for toxicity or adverse effects.
3. *Manufacturing Complexity.* The preparation of nanoemulsions can be more complex than the preparation of other drug delivery systems due to the need for precise control over surfactant and co-surfactant concentrations, as well as the use of high-energy emulsification methods.
4. *Stability Issues.* In the case of nanoemulsions, they are thermodynamically unstable, so they are susceptible to physical and chemical instability over time due to factors such as droplet aggregation or phase separation.
5. *Limited Understanding of Physicochemical Behavior.* Despite the extensive research, there is still a limited understanding of their physicochemical behavior, particularly *in vivo.* This can make it difficult to predict their behavior and efficacy in different biological environments.
6. *Challenges in Targeting Specific Tissues.* Although both systems have been investigated as potential tools for targeted drug delivery, there are still challenges associated with achieving effective targeting due to barriers, such as biological membranes and tissue-specific physiological conditions.

13.7 APPLICATION OF LIPID NANOCARRIERS FOR NEURODEGENERATIVE DISEASES (NDGs)

13.7.1 Alzheimer's Disease

AD and PD are among the most globally prevalent neurodegenerative disorders (Collaborators, 2019). *Alzheimer's dementia* refers to a gradually progressive clinical syndrome which affects mainly adults older than 65 years. Approximately 60–80% of dementia cases are attributable to AD, which is defined by a memory and cognition deterioration with subsequent impairment in daily activities, disorientation, performance, communication, and function ("2021 Alzheimer's Disease Facts and Figures," 2021; Atri, 2019). The progressive decline in cognition seen in AD is a result of the combined mutual existence of two types of unusual structures: extracellular amyloid plaques and intraneuronal neurofibrillary tangles, which are both composed of densely packed and highly insoluble filaments. The fundamental constituents of these structures consist of amyloid-β (Aβ) peptides, which form the plaques, and the tau protein, which makes up the tangles. The Aβ peptides are derived from the amyloid precursor protein, a molecule that spans the membrane, while tau is a protein associated with microtubules, predominantly located in neuronal axons (Bloom, 2014).

Despite extensive research and numerous clinical trials, there is currently no effective curative treatment for the disease (Cummings et al., 2019; Scheltens et al., 2016). Drug delivery across the BBB with lipid-based nanocarriers is a promising approach, which makes them good candidates for AD treatment. This is attributable to their desirable properties, including, but not limited to, their biodegradable nature, non-toxicity, and non-immunogenicity (Chen & Liu, 2012). They also possess an affinity and binding to endothelial cells and can provide sustained drug delivery through adsorption into blood capillaries. The surface properties of these molecules can also be modified by binding to certain cell-penetrating peptides and functional proteins (Arora et al., 2021; Kristensson & Olsson, 1971).

Few drugs that target Aβ and tau proteins are commercially available. The degradation process is slowed down by these drugs, which also enhance cognitive impairments. The first-line medications for the treatment of mild to moderate and severe cases include acetylcholinesterase (AChE) inhibitors, such as rivastigmine (RS), galantamine (GH), and donepezil (DPL) (Twohig & Nielsen, 2019). Memantine, an N-methyl D-aspartate (NMDA) antagonist, is employed to treat moderate to severe cases of AD, where it reduces neurotoxicity and consequent neuronal death by lowering excess stimulation of the glutaminergic pathway (Patnaik, 2015; Robinson & Keating, 2006). The United States Food and Drug Administration (FDA) approved the monoclonal antibody known as aducanumab in 2021 to cure or manage the symptoms of certain diseases, overcoming the limitations of earlier, older treatments. Aducanumab acts by lowering the Aβ load. Aducanumab has since been regarded as a novel agent, and research is ongoing to determine its clinical effectiveness, side effects, and cost–benefit ratio (Briggs et al., 2016).

Clinical studies are also being conducted on a number of different biologics, including therapeutic antibodies for Aβ and τ proteins, nerve growth factors (neurotrophin), and therapeutic antibodies for the triggering receptor expressed on myeloid cells 2 (TREM2). The lack of clinical significance led approximately 40% of these studies to discontinue, and only 12% entered phase III clinical trials. The fundamental reason behind their therapeutic failure is these drugs' inability to pass the BBB (Therapeutics | ALZFORUM, n.d.).

DPL, a hydrophilic small molecule for the treatment of AD, is a drug with restricted permeability to the BBB, thus requires repeated dosing and consequent significant cholinergic side effects. To minimize this, DPL-loaded SLNs were designed for administration intranasally to the brain. Glyceryl monostearate (GMS) was employed as the lipid, and a mixture of Tween® 80 and poloxamer 188 (1:1) was employed as the surfactant in the preparation of the system using the solvent emulsification diffusion technique. Male albino Wistar rats were used for the in vivo research of comparative brain-targeting delivery. DPL-SLN was administered intranasally, and DPL solution was administered intravenously. Intranasal administration resulted in increased brain delivery and a much higher concentration of the drug in the brain. According to stability experiments, DP-SLN remained stable for six months at 4 ± 2°C and 25 ± 2°C/60 ± 5% RH, being a promising candidate for commercial applications as a drug delivery system (Yasir et al., 2018).

Curcumin and DPL co-encapsulated nanolipid carriers (NLCs) were developed using the microemulsion technique, given intravenously, and compared to intravenous delivery of the free medication. NLC concentration in the brain was found to be greater in the *in vivo* pharmacokinetic experiments as opposed to the free drug given intravenously. Also, NLC used on Wistar rats in this study yielded a significantly higher acetylcholine level and reduced oxidative stress, demonstrating enhanced learning and memory relative to those receiving the free drug (Sood et al., 2013).

Campisi et al. (2022) studied curcumin SLNs (CUR-SLNs), which were synthesized using the surfactant Lutrol F68® and the lipid Compritol 888 ATO in a solvent evaporation method. In this study, wild-type and CRND8 transgenic mice were categorized into four groups and administered either SLNs or SLNs-CUR intraperitoneally trice weekly for three consecutive weeks. Control groups were treated similarly with saline solution. A step-down test was carried out to examine memory performance. There were no significant distinctions between the groups' training results during the step-down inhibitory avoidance test. However, in the 24 h retention test, the step-down latencies of SLN transgenic mice were significantly reduced compared to those of the wild-type groups. Treatment with SLNs-CUR significantly improved the performance of transgenic mice, reaching levels similar to those displayed by WT-SLN-CUR and wild-type SLNs mice. According to this study, systemic administration of SLNs-CUR in transgenic mice improved their cognitive performance and memory function. Additionally, systemic treatment of transgenic mice utilizing SLNs-CUR increased TG2-L expression to levels comparable with the ones seen in wild-type SLNs-CUR-administered mice. Along with that, it decreased TG2-S expression to levels comparable to those in mice treated with SLNs-CUR wild type. The study found that treatment with SLNs-CUR in transgenic mice increased Bcl-2 expression, as well as a decrease in caspase 3 cleavage. These findings imply that SLNs-CUR is capable of reducing the production of TG2's alternative splice variant, which is involved in cell death and apoptosis. Moreover, it suggests that the TG2 conformation change from "open state" during injury to a "close state," which participates in encouraging neural repair, might be accomplished by SLNs-CUR therapy. GH, another AChE inhibitor, has the same drawbacks as DPL. Microemulsification technique was used to prepare GH-SLN, where the lipid was glyceryl behenate (Compritol®), the surfactant was Pluronic® f-127, and the co-surfactant was Tween® 80. When compared to pure GH, which released 81% of its drug within the first hour, GH-SLN had a 24 h drug release rate of 99%, according to *in vitro* testing, while oral administration of the drug increased the drug's bioavailability by two folds. On the contrary, the Morris water maze test used in the in vivo pharmacodynamics testing demonstrated the efficacy of the SLN strategy by demonstrating a notable decrease in the mean escape latency in groups treated with either GH pure or GH-SLN compared to cognitive deficit group induced by isoproterenol. The GH-SLN also showed good stability at 25 ± 2°C and 4 ± 2°C storage conditions (Misra et al., 2016).

The least bioavailable (40%) with the shortest half-life (2 h) among the AChE inhibitors is RS. This justifies the negative treatment outcomes seen with RS. Extensive research has been conducted on RS lipid nanoparticles to improve BBB penetration (Birks et al., 2015). For instance, RS liposomes coupled with a cell-penetrating peptide, administered intranasally, yielded CNS cerebral tissues with significantly increased drug concentrations, compared to normal liposomes and free drug (Yang et al., 2013). Furthermore, RS liposomes were studied in various animal models, administered subcutaneously (Ismail et al., 2013), orally, and intraperitoneally (Mutlu et al., 2011). The results showed an increased drug brain penetration, with higher AChE inhibition compared to the free drug.

Another study tested intranasal administration of RS-loaded SLNs synthesized using polysorbate 80, 90% methanol, and glyceryl monostearate, where the *in vitro* release of RS from RS solution was about 95% within 6 h, while RS-SLNs released the drug at a slower rate over 24 h. The initial release of RS from the RS solution was higher than from SLNs. This is due to the drug's adsorbed state on the surface of RS-SLN nanoparticles, which caused the drug to be released quickly at first prior to sustained release. An *ex vivo* diffusion study was performed employing goat nasal mucosa as a biological membrane to simulate an *in vivo* barrier. The results from the in vitro investigations were merely preliminary because the cellulose acetate membranes utilized in them were artificially generated. In comparison to RS solution, RS-SLN demonstrated higher flux and diffusion coefficients, indicating a better capacity for lipid nanoparticles to penetrate the lipoidal membrane and, in turn, a further rise in the drug's solubilization in the membrane. Additionally, the surfactants boosted permeability across the nasal mucosa and served as penetration enhancers, whereas the hydrophilic properties of RS solution prevent it from readily crossing the lipoidal membrane. The formulation showed high safety for nasal administration, as shown by nasal histopathology testing done on goat nasal mucosa. The formulation had a pH of 5.8, which is within the range of the cilia in the human nasal membrane (5–6.5), and there was no obvious evidence of damage to the nasal membrane. RS-SLN intranasally resulted in higher AUC and a longer half-life in the brain, showing greater drug targeting compared to RS solution given intravenously and intranasally. T_{max} for the brain was shorter than that for plasma, due to direct nose-to-brain transfer after injection intranasally. The brain/blood ratios for RS-SLN were consistently higher, indicating that SLN's lipophilic properties may have facilitated higher drug penetration into the brain. The stability study showed that RS-SLN was stable at 4 ± 2°C and 25 ± 2°C/60 ± 5% RH, but it was unstable at 40 ± 2°C/75 ± 5% RH, where it showed aggregation and heterogeneity, possibly due to humidity and high temperature (Arora et al., 2022).

In Ismail et al.'s (2013) study, two methods were used—the heating method and the lipid layer hydration—to formulate liposomes loaded with different molar RS concentrations. An acute-toxicity study was performed to compare the safety of the SC formulation. Six groups of a total of 30 male albino rats were used, with three groups receiving RS solution and three groups receiving RS-liposomes. The rats were administered different dose levels of the solution and formulation (1, 10, and 20 mg/kg), and LD50 was determined for both formulations over 24 h. The liposome formulation showed low LD50 in comparison with that of the solution. Moreover, the liposome formulation decreased the toxicity of the drug contrasting the free drug solution, which might be due to different pharmacokinetic and/or tissue distribution profile from the liposome formulation compared to the solution, when given by the same route of administration. Treatment with RS solution and RS liposome was tested in an aluminum chloride–induced model for AD, where they significantly reduced plasma levels of C-reactive protein (CRP), homocysteine (Hcy), asymmetric dimethylarginine (ADMA). RS liposome normalized AChE and Na^+/K^+-ATPase activities. Additionally, the RS solution reduced AChE activity and enhanced Na^+/K^+-ATPase activity, while RS liposome showed a comparable effect on biochemical parameters as the solution (Ismail et al., 2013).

Rivastigmine hydrogen tartrate (RHT)–loaded SLNs were also constructed by the microemulsion technique using Tween 80 as a surfactant and tocopherol succinate as the lipid. The release of RHT-SLNs *in vitro* was tested and showed more consistent release over time compared to RHT alone in the water. The results suggest that encapsulating the hydrophilic RS in tocopherol succinate SLNs using the microemulsion technique is a potential strategy to enhance rivastigmine's brain uptake. However, the researchers did not study the efficacy of the SLNs on any outcome models (Malekpour-Galogahi et al., 2018).

Quercetin, a naturally present flavonoid in many fruits and vegetables, has a protective role against oxidative stress, with a subsequent increase in neuronal survival, as well as inducing neuronal differentiation and maintaining cellular balance (Dok-Go et al., 2003; Heo & Lee, 2004; Sagara et al., 2004; Zhang et al., 2011). However, quercetin has poor water solubility and is rapidly degraded in the body, making it challenging to use it in AD management (Almeida et al., 2018; Guo & Bruno, 2015). RVG29 is a peptide with 29 amino acids that is produced from the glycoprotein of the rabies virus. It can interact with nicotinic acetylcholine receptors (nAChR) and thus has been used for brain delivery of several drugs (Hua et al., 2018; Kumar et al., 2007; Lafon, 2005; Liu et al., 2009; Oswald et al., 2017; You et al., 2018).

Brain endothelial cell line (hCMEC/D3) was used to evaluate the safety and efficacy of RVG29 peptide–functionalized NLC (composed of cetyl palmitate and miglitol-812) and SLNs (composed of cetyl palmitate). The nanoparticles had an entrapment efficacy of over 80% and could capture up to 90% of the quercetin. There was no cytotoxicity in the cell line, as shown by the LDH assay. RVG29 nanoparticles increased the *in vitro* BBB permeability by 1.5 fold after 4 h incubation compared to non-functionalized nanoparticles. The thioflavin T binding assay also showed that the nanosystem was capable of inhibiting Aβ aggregation (Pinheiro et al., 2020b).

The same research group conducted another study to assess the efficacy of transferrin-functionalized lipid nanoparticles of quercetin in the same cell line. Around 80–90% of the quercetin was effectively entrapped. After incubating with the hCMEC/D3 cell line for 4 h, the greatest concentration of nanoparticles did not exhibit cytotoxicity in the LDH cytotoxicity assays. Results of permeability studies across hCMEC/D3 monolayers showed that NLC had better permeability than SLN, both for non-functionalized and functionalized nanoparticles. However, functionalization with transferrin did not improve permeability, possibly due to receptor saturation. Unloaded nanoparticles inherently act as nuclei for aggregation, as shown in the aβ1–42 fibrillation testing. Comparatively to the control sample with A1–42 alone, the quercetin-loaded nanoparticles, particularly the transferrin-functionalized NLC, were able to reduce peptide aggregation, reverse the aggregation impact of unloaded nanoparticles, and suppress fibril formation (Pinheiro et al., 2020a).

Preclinical results suggest that nicotinamide adenine dinucleotide (NAD^+) is a neuroprotective and bioenergetics stimulant agent to treat AD. Primary outcomes of preclinical studies revealed that learning and memory are improved by NAD+. Secondary end points also showed that NAD^+ had neuroprotective effects in different AD models. The hypothesized neuroprotective mechanisms included improving mitochondrial function and reduction of oxidative stress, inflammation, and apoptosis (Hosseini et al., 2021).

Recently, Yulug et al. (2023) demonstrated that nicotinamide riboside, along with other metabolic activators (L-serine, N-acetyl-L-cysteine, and L-carnitine tartrate), has been shown to improve clinical parameters and cognitive functions. These improvements are likely due to enhanced phenomics, metabolomics, proteomics, and imaging analysis in patients receiving the supplementation.

One study assessed the safety and efficacy of nicotinamide-loaded SLN, augmented with phosphatidylserine, polysorbate 80, or phosphatidic acid. The functionalized SLNs demonstrated significantly greater biodistribution to the brain compared to the nonfunctionalized SLNs, with phosphatidylserine being more effective than phosphatidic acid. Histopathology, biochemical tests, spatial learning, and memory testing via Morris water maze test showed the efficacy of intraperitoneally administered phosphatidylserine-functionalized SLNs in protecting the neuronal cells, lowering tau hyperphosphorylation and improving the cognition in an adult male Sprague–Dawley rat model of Alzheimer's disease (Vakilinezhad et al., 2018).

One trial tested another cell-penetrating peptide, RVG-9R, obtained from rabies virus glycoprotein. The investigators prepared a complex of RVG-9R and siRNA to protect the oligonucleotide and enhance receptor-mediated endocytosis inside neurons, and then the complex was encapsulated inside SLNs to facilitate delivery across olfactory epithelium and penetration through respiratory epithelium mucosal surface to increase penetration to reach the trigeminal nerve endings. The SLNs were coated with chitosan to open tight junctions and increase paracellular transport, while simultaneously increasing mucoadhesiveness to the system. The study did not assess either the safety or efficacy of the SLNs, but it provided a new approach for the enhancement of siRNA cellular transport from SLNs using Caco-2 as a model of epithelial-like phenotypes (Rassu et al., 2017).

Memantine (MEM), N-methyl-d-aspartate (NMDA) receptor antagonist, inhibitor, is used for the management of patients with moderate to severe AD (Winblad et al., 2007). The safety and stability of Lipoyl (LA)–MEM as a co-drug were tested after encapsulation in SNLs. Good stability of nanoparticles was seen after incubation at pH 1.2 for 2 h, in the absence or presence of pepsin, where there was no significant detection of aggregation by dynamic light scattering. However, the stability of particles after incubation in intestinal simulated fluids for 2 h at pH 6.8, in the absence and presence of pancreatin differed, which was hypothesized to be due to hydrophobic interactions between the particles and the co-drug released in alkaline PH. It was also hypothesized that the coating was not enough to protect the particles from those interactions. The co-drug release from SLNs in both simulated gastric and intestinal fluids showed that there was a small release burst of about 20% in the first hour and after 2 to 4 h. The researchers suggest this is a beneficial feature since about 80% of the co-drug was still entrapped and available to reach the CNS. The trial also evaluated the cytotoxicity of different SLN formulations on cell viability and oxidative stress in cultured peripheral human whole blood cells and mouse N2a neuroblastoma cells. MTT and LDH assays showed no significant changes in cell viability for most SLN concentrations (Laserra et al., 2015).

Andrade et al. (2023) showed that caffeic acid (CA) liposomes were produced by various techniques, with their surface modified with transferrin for improved targeting ability. The liposomes showed high stability over the storage period at 4°C and remained stable for 2 months. The release pattern was studied *in vitro*, where it showed an initial burst release, followed by a slow and regulated release. CA is poorly soluble in water; therefore, it is anticipated to be kept in the liposomes' phospholipid bilayer rather than released to the aqueous medium. The release pattern allows for the minimal release of CA into the circulation, and as the liposomes degrade when reaching the brain, the remaining CA will be released, allowing it to exert its therapeutic effect. The liposomes were shown to delay Aβ1–42 monomers aggregation into smaller aggregates; hence, they reduce the number of formed fibrils. Additionally, this nanosystem showed a mature fibrils disaggregation rate of approximately 15%.

Resveratrol (Resv), a naturally occurring flavonoid mainly present in grape seed and skin, had been shown to protect neurons against oxidative damage and to prevent apoptotic neuronal death and thus provide neuroprotective effects in neurological diseases (Markus & Morris, 2008; Neves et al., 2012). In less than 2 h, Resv is rapidly metabolized into glucuronic acid and sulfate phenolic group conjugates by liver and intestinal epithelial cells after intravenous injection. This results in low bioavailability, with limited biological and pharmacological benefits. Additionally, it poses poor water solubility; moreover, it is chemically unstable, as exposure to high temperatures, pH changes, UV light, and particular enzymes can degrade it through isomerization (Anekonda, 2006; Neves et al., 2012).

The extracts of grape seeds and skins were found to contain other polyphenols that make an aggregation of Aβ1–42 inhibitory effect more pronounced than Resv alone, so Loureiro et al. (2017) studied the stability and BBB penetration efficacy of SLNs loaded with grape seed and skin extracts, functionalized with OX26 monoclonal antibody, which binds to cells expressing transferrin receptors as BBB cells (Jefferies et al., 1984). The stability of loaded and unloaded SLNs was studied over a period of two months at room temperature. SLNs showed no significant variations in the size and zeta potential, indicating their stability during this period. An *in vivo* BBB model showed receptors and transporters, the low paracellular passage of low-molecular-weight molecules. This model consisted of endothelial cells derived from hematopoietic umbilical cord blood cells seeded on a Transwell filter with brain pericytes. Findings demonstrated the decoration of OX-26 antibody-loaded SLNs had an increased uptake and transcytosis by

BBB cells compared to another antibody (LB 509) or nonfunctionalized SLN. However, the efficacy of the loaded and functionalized SLNs was not assessed in relation to any AD outcomes (Loureiro et al., 2017).

Epoetin (EPO), a hematopoietic cytokine, has neuroprotective effects in several disorders, including, but not limited to, neurodegenerative disorders, such as AD, through regulation of neurogenesis and maintaining survival of neuronal cells (Cevik et al., 2017). Studies have illustrated that very low levels of BBB penetration, because of its large molecular weight (30.4 kD), hinder transport through lipid-mediated transport (Noguchi et al., 2007).

Dara et al. (2019) studied the stability and efficacy of erythropoietin-loaded SLNs (EPO-SLNs) in ameliorating memory deficits in a rat model of AD. Researchers prepared SLNs by $w_1/o/w_2$ double-emulsion method, where glycerin monostearate was used as a lipid, Span 80/Span 60 as an internal-phase surfactant, dichloromethane as a solvent, and Tween 80 as external-phase surfactant. The researchers used a rat model of AD induced by intracerebral injection of Aβ oligomer. The Aβ deposition showed a diminished neuronal density as a cell death marker in the AD model, and EPO treatment was able to hinder this effect with EPO-SLNs able to potentiate this function. The researchers evaluated the efficacy EPO-SLNs through behavioral tests, such as the Morris water maze, which assessed spatial memory, where rats injected with Aβ had longer escape latency and traveled a longer distance, indicating poor memory and learning performance. Treatment with EPO, particularly with EPO-SLN formulation, improved memory function in the Aβ-injected rats significantly. EPO treatment significantly decreased the levels of ADP/ATP, lipid peroxidation, and reactive oxygen species (ROS) induced by Aβ. The EPO-SLN was also found to be more effective than the native EPO. In addition, EPO treatment also increased the thiol level in the hippocampus, with SLNs being more efficient.

Piperine is a naturally occurring alkaloid with a polyene bond system which offers a potent antioxidant effect and provides tertiary nitrogen similar to acetylcholine, inhibiting AChE or enhancing the brain's cholinergic system. First-pass metabolism and pH-mediated metabolism to piperidine and photoisomerization make it difficult to administer piperine to the brain (Bhat & Chandrasekhara, 1986; Hashimoto et al., 1996).

AD model was performed in albino rats by ibotenic acid–stimulated lesions of basalis magnocellular. Then, the emulsification–solvent diffusion technique was used to prepare piperine SLNs with polysorbate 80 coating. Forced swimming tests, superoxide dismutase, AChE enzymatic assays, and histopathology of the brain cortex were conducted to evaluate the therapeutic effects of piperine SLNs. The results showed that piperine SLNs reduced oxidative stress, evidenced by reduced superoxide dismutase activity; increased the acetylcholine level through reduction of AChE activity; reduced immobility in a force swimming test; and even showed superior effects to 5 mg/kg DPL. Histopathologic examination showed reduced brain plaques and tangles (Yusuf et al., 2013).

Sesamol, a major component of sesame seed oil derived from *Sesamum indicum* L., has shown strong antioxidant properties, where it has been found to suppress lipid peroxidation induced by UV radiation and Fe^{3+}/ascorbate in the rat brain. Additionally, sesamol acts as a scavenger of hydroxyl and lipid peroxyl radicals and can reduce radiation-induced deoxyribose degradation (Joshi et al., 2005; Parihar et al., 2006; Prasad et al., 2005; Uchida et al., 1996). Polysorbate 80, soy lecithin, water, and lipid were used to prepare SLNs and sesamol. Sesamol has limited bioavailability and a low ability to penetrate the brain, so it was incorporated into the aqueous phase containing polysorbate 80. Male Wistar rats were administered intracerebroventricular streptozotocin to produce cognitive deficits similar to those seen in Alzheimer's disease. The rats were divided into ten groups. Group 1 served as the control and received a vehicle injection. Group 2 received an intracerebroventricular injection of streptozotocin (ICV-STZ) as a positive control. Groups 3 to 5 were ICV-STZ-treated rats administered different doses of sesamol. Group 6 received unloaded SLNs. Groups 7 to 9 were ICV-STZ-treated rats administered SLNs loaded with different doses of sesamol. Group 10 received rivastigmine as a standard treatment. The treatment was administered for 21 days, starting before the ICV-STZ injection. The antioxidant profile in ICV-STZ-treated rats showed reduced enzyme activity (catalase, superoxide dismutase, and reduced glutathione) and increased lipid peroxidation and nitrosative stress levels. Treatment with sesamol and sesamol-SLNs improved antioxidant enzyme activity, reduced lipid peroxidation, and inhibited nitrosative stress. The effects were dose-dependent, with significant differences observed between treatment groups and the ICV-STZ group at various doses. Notably, the effects were dose-dependent, with sesamol-SLN showing significant differences at the highest dose. Sesamol and sesamol-SLNs also decreased serum tumor necrosis factor-alpha levels in ICV-STZ rats. Treatment with sesamol and sesamol-SLNs reduced AChE activity in the brains of ICV-STZ-treated rats and improved memory performance in the Morris water maze and elevated plus maze tests. However, there was no significant effect on locomotor activity. These findings suggest the potential of sesamol and sesamol-SLNs as therapeutic interventions for memory impairment associated with ICV-STZ-induced neurodegeneration (Sachdeva et al., 2015).

Similarly, Polysorbate 80, soy lecithin, water, and lipid were used to produce SLNs, and curcumin was added to the aqueous phase. Young male Lacca mice (8–12 weeks old) were used to produce an AD animal model by administration of aluminum chloride ($AlCl_3$) orally to mice for 12 weeks. Memory lapses were observed at 12 weeks, confirming AD-like symptoms. Following 12 weeks of $AlCl_3$ treatment, mice were orally treated with free curcumin or curcumin-loaded SLNs at different doses, blank SLNs, or rivastigmine for 6 weeks. Mice underwent

Morris water maze tests and biochemical estimations, including AChE measurements. Histopathological studies were also conducted on preserved brain samples. The mice exhibited impaired spatial navigation and increased escape latency, indicating AD-like symptoms. Treatment with curcumin-SLNs and free curcumin for an additional six weeks significantly improved spatial learning abilities, with curcumin-SLNs showing superior effects compared to free curcumin. Elevated AChE activity was observed in the $AlCl_3$-treated mice and was returned to normal by SLNs, while free curcumin showed limited efficacy. $AlCl_3$-induced oxidative stress and lipid peroxidation were attenuated by curcumin-SLNs and glutathione levels, and the activity of antioxidant enzymes was restored. Histological analysis revealed that curcumin-SLNs and rivastigmine restored brain microstructural elements, including neuronal integrity and reduced edema. Overall, curcumin-SLNs exhibited promising therapeutic effects in alleviating cognitive impairment and neurodegenerative changes associated with AD in this animal model (Kakkar & Kaur, 2011).

The homogenization–sonication technique was adopted to create DPL-loaded SLNs. The lipid phase consisted of heated lipid dynasty 116 and the active substance DPL HCl, while the water phase involved dissolved Tween 80 in phosphate-buffered saline. The DPL-loaded SLNs were then functionalized with ApoE-targeting ligand. The researchers studied the transfer of DPL using SLNs across different brain cell lines. The researchers found that the ApoE-targeted SLNs were significantly taken up more than the non-targeted SLNs by both rat and human brain endothelial cells, as well as by a human differentiated neuronal cell line. The researchers also found that the ApoE-targeted SLNs were able to penetrate the BBB more effectively than the non-targeted SLNs. These results suggest that ApoE-targeted SLNs could be a promising, new approach for delivering drugs and other therapies to the brain (Topal et al., 2020).

Ferulic acid is an active constituent of plant origin with antioxidant and anti-inflammatory activity. It can also remove ROS and scavenge free radicals (Li et al., 2021). Empty and ferulic acid–loaded SLNs were prepared via a warm oil-in-water microemulsion using Compritol 888 ATO as the lipid matrix, Epikuron 200 as the surfactant, and taurocholate sodium salt as co-surfactant. The researchers investigated the effects of free or loaded ferulic acid into SLNs against the toxicity of rAb42 oligomers on LAN5 neuroblastoma cells. The results showed that treatment with rAb42 oligomers caused morphological changes and reduced cell viability. However, when FA, either in its free form or loaded into SLNs, showed recovery in cell morphology and improved cell viability, the FA-loaded SLNs were particularly more effective in protecting against rAb42-induced toxicity. Additionally, FA, in both forms, restored the mitochondrial membrane potential and inhibited intracellular ROS generation induced by rAb42 treatment. The FA-loaded SLNs showed a total recovery of rAb42-induced toxicity and complete inhibition of ROS formation. Furthermore, FA and FA-loaded SLNs reduced the release of cytochrome c from mitochondria and inhibited apoptosis induced by rAb42 oligomers. The FA-loaded SLNs also downregulated the activation of caspase 9 and caspase 3, which are involved in apoptotic cell death. The study revealed that FA reduced the expression of peroxiredoxin 3 (Prx3), an antioxidant protein, and reduced ERK1/2 pathway activation induced by rAb42 oligomers. Overall, the results suggest that FA, particularly when loaded into SLNs, has protective effects against rAb42 oligomer toxicity, including restoring cellular morphology, improving cell viability, maintaining the mitochondrial function, inhibiting ROS generation, reducing apoptosis, and modulating signaling pathways associated with neuronal death (Picone et al., 2009).

α-Bisabolol is an unsaturated monocyclic sesquiterpene alcohol that is extracted from chamomile, with therapeutic properties against oxidative stress, inflammation, infections, neurodegenerative diseases, cancers, and metabolic disorders. The mechanism of these effects is hypothesized to be due to the reduction of ROS, enhancing antioxidant enzymes, suppressing of apoptosis-related factors, modulating inflammatory pathways, inhibiting microbial growth, and improving cognitive function (Ramazani et al., 2022). Hence, α-Bisabolol-loaded SLNs were prepared, where soy lecithin, Tween 80, and water were mixed to form the aqueous phase. The lipid phase contained cholesterol and α-bisabolol. The researchers found that α-Bisabolol exhibited antioxidant properties by reducing the production of ROS and reactive nitrogen species induced by the Aβ25–35 peptide. Furthermore, α-Bisabolol mitigated Aβ-induced loss of mitochondrial membrane and prevented lipid and protein oxidation. Also, α-Bisabolol inhibited the activity of β-secretase, an enzyme responsible for processing amyloid precursor protein, which plays a role in Aβ peptide formation and AChE. Additionally, α-Bisabolol exhibited anti-apoptotic effects by preventing Aβ-induced apoptosis in neuronal cells, as evidenced by reduced caspase 3 activity, increased expression of Bcl-2, and reduced expression of Bax proteins. Overall, the conclusion suggests α-Bisabolol poses potential therapeutic effects against AD by targeting various pathological mechanisms associated with the disease (Sathya et al., 2020).

13.7.2 Parkinson's Disease (PD)

Parkinson's disease (PD) is a deteriorating neurodegenerative disorder characterized by rigidity, bradykinesia, resting tremor (4–6 Hz), and loss of postural reflexes (Kalia & Lang, 2015). PD ranks as the second most common neurodegenerative disorder, following Alzheimer's disease, and demonstrates a higher prevalence and incidence in older adults. The reported prevalence rates of 3% and 1% are observed in patients aged >80 years and >65 years, respectively. Additionally, the incidence of PD is reported as 21 per 100,000 person-years in the United States, while its prevalence tends to increase with age (Homayoun, 2018).

The pathophysiology of PD involves an intertwining of genetic and environmental contributors that result in neuronal dysfunction, degeneration, and death. The accumulation of aggregated α-synuclein in Lewy bodies is a hallmark of the disease, and there is increasing evidence for the involvement of mitochondria dysfunction, oxidative stress, and impaired proteostasis. The toxic oligomers of α-synuclein can impair various cellular processes, including mitochondrial function, autophagy, and calcium homeostasis, resulting in neuronal death. Furthermore, the activation of immune cells and the subsequent neuroinflammation are believed to be involved in the pathogenesis of PD. Early changes can occur in the striatum years before the diagnosis, and loss of nigrostriatal axon terminals occurs (Poewe et al., 2017).

Management of PD typically involves medications to treat motor symptoms. The primary medication in use is levodopa, which is transformed into dopamine (DA) in the brain to replace the DA that is lost in PD. However, long-term use can lead to motor complications, such as dyskinesias. Other medications, such as DA agonists, monoamine oxidase type B (MAO-B) inhibitors, and catechol-O-methyltransferase (COMT) inhibitors, can be used alone or in combination with levodopa to manage symptoms and extend the therapeutic effect of levodopa. Additionally, medications can be used to manage non-motor symptoms, such as antidepressants for the accompanying depressive symptoms, antipsychotics for psychosis, and cholinesterase inhibitors for dementia. Non-pharmacological interventions, such as physical and occupational therapy, speech therapy, and psychological support, can also help manage symptoms. Treatment should be individualized based on the patient's symptoms, disease stage, and comorbidities. Regular monitoring and adjustment of medications are necessary to manage symptoms effectively and minimize side effects. Surgical interventions, such as deep brain stimulation, can also be considered in refractory cases or those with motor complications (Poewe et al., 2017).

Ropinirole (RP), a novel non-ergoline D2/D3 DA receptor agonist, exhibits specific binding to D2 receptors in the striatum and substantia nigra, displaying a selectivity profile similar to that of DA. Its proposed mechanism of antiparkinsonian effects involves the augmentation of striatal neuronal firing rates via selective activation of D2 DA receptors. However, RP has low systemic availability after oral intake (50%) because of the extensive exposure to hepatic first-pass metabolism, and it has an elimination half-life of 5.8 h, with a T_{max} of less than 1 h. Presently, RP is commercially available in conventional tablet dosage forms ranging from 3 to 9 mg, which fails in achieving adequate oral bioavailability. Intravenous (IV) administration of RP is highly irritating and, therefore, is contraindicated (Pahwa et al., 2004; Shill & Stacy, 2009).

Therefore, RP-SLNs and RP-nanolipid carriers (NLCs) were prepared, and hydrogel formulations of RP-SLNs and RP-NLCs were developed using 1% w/v concentration of Carbopol 934 as a gelling agent. To achieve a suitable pH for topical application, triethanolamine (0.3% w/v) was used for neutralizing the pH of the hydrogel formulations. Additionally, a conventional RP hydrogel formulation was prepared using the same drug loading concentration. Haloperidol-induced PD model was performed in male albino Wistar rats. Motor impairment was confirmed through preliminary tests after administering haloperidol. The development of PD was further validated using the standard bar and rotarod tests (Dudhipala & Gorre, 2020). Dudhipala and Gorre (2020) evaluated the effects of RP-SLNs and RP-NLCs and corresponding hydrogel formulations on various parameters associated with PD. The results showed that PD induction led to a significant increase in lipid peroxidation levels, as indicated by elevated thiobarbituric acid reactive substances (TBARS) concentration in the striatum. However, treatment with RP-SLN and RP-NLC formulations effectively restored TBARS levels, with RP-NLC hydrogel exhibiting the highest restoration. Similarly, glutathione levels, which were depleted in the PD group, were restored by the different formulations, with RP-NLC hydrogel showing the highest restoration. The activity of catalase, an important enzyme involved in antioxidant defense, was significantly decreased in the PD group, but the formulations restored its activity, with RP-NLC-C demonstrating the highest restoration. DA levels, which were depleted in the PD group, were also restored to varying degrees by the formulations, with RP-NLC hydrogel showing the highest restoration (Dudhipala & Gorre, 2020).

Moreover, Pardeshi et al. (2013) used the emulsification–solvent diffusion technique for the preparation of SLNs (Pardeshi et al., 2013). In this study, the materials used included RP HCl, Dynasan 114, Pluronic F-68, glucose, sucrose, lactose, mannitol, stearyl amine, Phospholipon® 90G, and benzyl alcohol. The experimental methodology used a two-factor three-level factorial approach to assess the impact of Pluronic F-68 concentration and stearyl amine concentration on particle size, zeta potential, and entrapment efficiency. The preparation of ROPI-loaded SLNs involved the emulsification–solvent diffusion technique with modifications. *In vitro* drug release and permeation studies were conducted, and the stability of the SLN formulation was evaluated. Non-invasive *in vivo* pharmacodynamic investigations were performed to assess the anti-tremor activity in mice. The histopathological examination indicated no severe damage to the nasal mucosa after treatment with the SLN formulation. The drug release from the SLNs followed a sustained pattern. Overall, the SLN formulation demonstrated potential for nasal drug delivery. This study sought to assess the lyophilized SLNs for nasal drug delivery. Histopathological examination of sheep nasal mucosa treated with the SLN formulation showed no severe signs of damage, indicating that the integrity of the nasal mucosa remained intact. *In vitro* drug release studies demonstrated sustained release patterns, attributable to the drug loading and entrapment within the cavities of the SLN formulation. The release data exhibited the best fit with a zero-order kinetic model, showing a linear, time-dependent drug release. *Ex vivo* drug permeation studies across sheep nasal mucosa revealed a permeation rate of approximately 70% after 8 h. Stability studies indicated no significant change in mean particle size after three months of storage, although a slight reduction in drug encapsulation efficiency was observed. *In vivo* pharmacodynamic studies in mice showed that the therapeutic activity of the lyophilized SLN formulation was comparable to that of the marketed oral formulation, with a significant reduction in tremors. Overall, the findings suggested the potential of the lyophilized SLN formulation for nasal drug delivery with improved therapeutic efficacy.

Piribedil (PBD) is a non-ergot DA agonist that has been used in the treatment of PD. It has been in clinical practice since 1969 and is commonly available in the form of 50 mg tablets. However, the clinical application of PBD has been significantly constrained by various challenges, such as its extremely low oral bioavailability (less than 10%) caused by extensive first-pass metabolism in the liver, as well as its short biological half-life, requiring high daily doses of 3–5 tablets (Deleu et al., 2002; Millan, 2010).

Palmitic acid was selected as the lipid and polyvinyl alcohol as the stabilizer for the development of PBD-SLNs. The SLNs were prepared using a combination of hot homogenization and ultrasonication techniques. Stability studies were conducted to evaluate the long-term stability of both PBD-SLNs (stored at room temperature) and PBD-SLNs gel (stored under refrigeration at 2–8°C) for 60 days. No significant changes were observed in any of the measured parameters between the freshly prepared samples and the samples stored under the mentioned conditions for the entire 60-day period. To delay mucociliary clearance after nasal administration in male Wistar rats, the SLNs were loaded in thermoresponsive methyl cellulose in situ gel. The *in vivo* pharmacokinetic studies demonstrated that PBD-SLNs gel significantly increased the brain exposure of PBD by a fourfold increase in the AUC compared to plain intranasal suspension of PBD. Additionally, PBD-SLN gel reduced the plasma C_{max} of PBD by 2.3-fold compared to PBD suspension. In contrast, PBD suspension showed limited direct uptake from the nose to the brain, with direct transport percentage values below zero. However, the optimized PBD-SLN-ISG demonstrated a DTP value of 27%, indicating efficient and direct nose-to-brain uptake of PBD (Uppuluri et al., 2021).

Curcumin, a constituent derived from *Curcuma longa*, is a dietary polyphenol recognized for its diverse range of beneficial properties, including neuroprotection, antioxidant effects, inflammation-reducing properties, and potential antitumor capabilities. The neuroprotective effects of curcumin can be attributed to its capacity to reduce inflammation and oxidative stress. Several scientific investigations were dedicated to the development of delivery systems incorporating curcumin for the treatment of neurodegenerative disorders, specifically PD. The clinical applications of curcumin are hindered by its limited bioavailability, which can be linked to factors such as low solubility, poor absorption, rapid metabolism, and excretion. These factors collectively contribute to the challenges associated with the utilization of the full potential of curcumin in clinical settings (Babazadeh et al., 2016; Bollimpelli et al., 2016; Don et al., 2021; Meng et al., 2021; Wang et al., 2010).

Wang et al. (2010) prepared and optimized curcumin SLNs via a microemulsion technique using glyceryl monostearate as the lipid component, stabilized by Tween 80 in distilled water. The mechanical properties, drug release, permeation studies, histology, and neuroprotective activity of curcumin-loaded microneedles were tested. The mechanical properties of the microneedles were assessed using Parafilm M® film, demonstrating uniformity and limited permeation through successive layers. *In vitro* release studies revealed a burst release followed by sustained release, attributed to the dissolution of microneedle tips. *Ex vivo* permeation studies on goat skin showed significant drug permeation through the microchannels created by the microneedles. In addition, histological examination confirmed the successful delivery of curcumin into the dermis, indicating the potential for transdermal drug delivery. Also, the microneedles exhibited improved motor coordination and balance in neuroprotective activity assessments, as demonstrated by decreased bradykinesia and improved performance in the pole and rotarod tests. Additionally, skin irritation tests reported no signs of irritation or sensitivity, indicating the safety of the microneedle patch. These findings collectively support the potential of curcumin-loaded microneedles for transdermal drug delivery and their neuroprotective effects (Wang et al., 2010).

Bacopa monnieri, commonly referred to as Brahmi, is a medicinal plant that falls under the Scrophulariaceae family. It has gained significant recognition as an herbal remedy for treating many neurological disorders and enhancing memory and cognitive function. Several studies have explored the pharmacological properties of *B. monnieri*, with its therapeutic effects primarily attributed to specific saponins called "bacosides." However, traditional drug delivery methods face challenges in effectively targeting the CNS. The utilization of drug-loaded nanocarriers, such as SLNs, has shown great potential in facilitating CNS targeting and resolving the barriers that hinder the delivery of medications across the BBB (Abbott & Friedman, 2012; Bandiwadekar et al., 2021; Giacoppo et al., 2015; Roodenrys et al., 2001).

Moreover, citicoline (CIT) is a choline donor and plays a role in the biosynthesis of phosphatidylcholine. When administered orally or parenterally, CIT breaks down into cytidine and choline, providing these components to the body. As a nutritional supplement, CIT is a provider of choline and cytidine and has shown potential neuroprotective and cognitive-enhancing properties. It has been used in addition to levodopa for PD treatment. While the exact mode of action remains unclear, CIT has been found to enhance the effects of levodopa, leading to improvements in motor and cognitive symptoms in patients with PD. One limitation of CIT is its hydrophilic nature, which prevents it from crossing the BBB (Barrachina et al., 2003; Cheong et al., 2019; Conant & Schauss, 2004; Eid et al., 2022; Fresta et al., 2011; Que & Jamora, 2021; Savitt & Jankovic, 2019).

Margari et al. (2022) formulated SLNs using Gelucire® 50/13, CIT sodium salt, polysorbate 85 (Tween® 85), acetic acid, potassium bromide, porcine stomach mucin, fluorescein isothiocyanate (FITC), and PBS (Margari et al., 2022). The human neuroblastoma cell line SH-SY5Y was cultivated in Dulbecco's Modified Eagle's Medium (DMEM) low-glucose medium supplemented with fetal bovine serum, penicillin, and streptomycin. For the experiments, the cells were treated with 6-hydroxyDA (6-OHDA) and either free CIT or CIT-SLNs. The quantitative determination of CIT in CIT-SLNs was performed using spectrophotometric analysis. The CIT-SLNs and FITC-SLNs were prepared using the melt emulsification technique. Biological evaluations were made to assess the effectiveness of CIT-SLNs in delivering a fluorescent dye and rescuing cell viability in SH-SY5Y cells exposed to the neurotoxic compound 6-OHDA. Confocal images showed that the FITC dye released from SLNs was diffusely localized in the cytoplasm of the cells but not in the nuclei. The viability of SH-SY5Y cells treated with 6-OHDA was significantly decreased, but pre-treatment with CIT-SLNs containing 50 μM CIT significantly increased cell survival compared to cells treated with 6-OHDA alone. Plain SLNs and free CIT did not have a significant effect on cell viability. Moreover, the presence of CIT-SLNs reduced cell retraction, swelling, and the appearance of cellular debris induced by 6-OHDA. Additionally, CIT-SLNs showed a reduction in 6-OHDA-induced DNA fragmentation and chromatin condensation, indicating a potential protective effect. These findings suggest that CIT-SLNs can deliver CIT and enhance cell viability while mitigating the toxic effects of 6-OHDA, highlighting their potential therapeutic value.

Bromocriptine (BK), a DA agonist that was introduced to the market over 30 years ago, has a slow onset of action (1–2 h) and a long half-life (3–5 h) (Bunjes et al., 2001). This characteristic likely contributes to its lower tendency to induce dyskinesia compared to levodopa (Pearce et al., 1998). In fact, the stabilization of plasma levodopa levels or using a longer-acting DA agonist has been proved to reduce dyskinesia (Obeso et al., 1994, 2000).

Esposito et al. (2008) prepared SLNs using various lipidic mixtures, such as pure tribehenin, pure tristearin, pure monostearin, combinations of tristearin/monostearin, tristearin/tricaprin, and tribehenin/tricaprin. The lipidic mixture was melted and dispersed in an aqueous poloxamer 188 solutions. The resulting emulsion was subjected to either homogenization or ultrasonication and then cooled to room temperature. In the case of BK-containing dispersions, the drug was added to the lipidic mixture before dispersion. The encapsulation yield of BK in different lipid mixtures for the preparation of NLCs was evaluated. The highest encapsulation yield (84%) was achieved with a tristearin-tricaprin (2:1) mixture, indicating that the homogeneous lipid nano-compartments of NLCs promote BK encapsulation. *In vitro* release kinetics demonstrated that BK release from the NLCs was slower compared to that of free BK, with the slowest release observed for carriers produced by the tristearin–tricaprin mixture. Additionally, *in vivo* tests conducted on 6-OHDA hemilesioned rats showed that both free and encapsulated BK reduced immobility time and attenuated akinesia compared to vehicle-treated animals. However, encapsulated BK exhibited a more rapid and prolonged onset of action compared to free BK, which was slower and short-lasting. Overall, these findings highlight the potential of NLCs as a promising maneuver for controlled release and enhanced therapeutic effects of BK.

Geraniol (GER), a monoterpene found in essential oils of many aromatic plants, shows various pharmacological properties, including neuroprotection, anti-inflammatory, and antioxidant effects (Rekha & Selvakumar, 2014). Also, Rekha and Selvakumar showed that oral administration of GER enhances the preservation of dopaminergic neurons. The proposed mechanisms involve increased production of antioxidant enzymes and neurotrophic factors and decreased expression of apoptotic markers (Rekha et al., 2013). Furthermore, several studies in PD animal models have shown positive effects of GER (Rekha et al., 2013; Rekha & Selvakumar, 2014; Siddique et al., 2016). Thus, with this favorable combination of properties and low toxicity (Lapczynski et al., 2008), GER can be regarded as a viable therapeutic choice for PD. However, GER's short half-life is a limitation for long-term therapies via conventional administration routes (Pavan et al., 2018; Rekha & Selvakumar, 2014).

Ursodeoxycholic acid (UDCA), a secondary bile acid commonly prescribed for liver disease, has been clinically used to treat primary biliary cirrhosis for over 30 years, with extensively studied clinical pharmacokinetics. In addition, it has recently been proven to be effective in maintaining mitochondrial function in individuals with PD. Currently, UDCA is undergoing a phase II clinical trial specifically for PD patients (Hofmann, 2009; Mortiboys et al., 2013, 2015; Roda et al., 1991). It had been shown that UDCA interacts with the glucocorticoid receptor, leading to increased phosphorylation of the Akt protein. This interaction protects against mitochondria-dependent programmed cell death (Chun & Low, 2012). Furthermore, Parry et al. (2010) showed the safety and tolerability of UDCA at doses up to 50 mg/kg per day in patients with motor neuron disease (Parry et al., 2010).

De Oliveira Junior et al. (2020) successfully synthesized a conjugate of GER and UDCA (GER-UDCA) to utilize its combined pharmacological properties as a novel therapeutic strategy for treating PD. However, the resulting compound had low water solubility in water and showed a sticky and highly viscous physical appearance. To address these issues, the researchers prepared lipid- and polymer-based nanoparticles as delivery systems for GER-UDCA. The researchers prepared nanoparticles loaded with GER and GER-UDCA using different methods. SLNs were prepared using an emulsification/evaporation solvent method, where compritol ATO 888, Span 85, and GER-UDCA were dissolved in chloroform and then emulsified with an aqueous phase containing Tween 80 and taurocholate sodium salt. The resulting emulsion was sonicated, and the solvent was evaporated. The SLNs were purified by dialysis and freeze-dried. Then, poly(lactic-co-glycolic acid) (PLGA) nanoparticles were obtained using a nanoprecipitation method. PLGA and GER-UDCA were dissolved in a solvent mixture and added to an aqueous phase containing pluronic-127. The formulation was stirred to remove the organic solvents, and the nanoparticles were purified by centrifugation and freeze-dried. Additionally, the possibility of encapsulating GER alone in liquid form in the nanoparticles was investigated. GER-loaded SLNs and NPs were prepared using a similar protocol. In this study, the researchers investigated the pharmacokinetics and potential uptake of GER-UDCA in the CNS using male Sprague–Dawley rats. The rats were fasted for 24 h and anesthetized during the experiment. The researchers administered GER-UDCA through a femoral IV infusion and nasally to evaluate its effects. They found that a 500 μg IV infusion of GER-UDCA resulted in a maximum concentration of 13.1 μg/mL in whole blood, which decreased over time. However, GER was not detected in the blood or cerebrospinal fluid (CSF) within 90 min after infusion. A higher dose of 2.5 mg of GER-UDCA intravenous infusion showed a concentration of 73.5 μg/mL at the end of infusion, with a half-life of 21.6 min. When GER was administered nasally at a dose of 4 mg/kg, the CSF concentration was 4.3 μg/mL at 30 min and 2.16 μg/mL at 60 min. At 90 min, there was a significant increase to 122 μg/mL in the CSF. Nasal administration of GER-UDCA-loaded SLNs at a dose of 2 mg/kg resulted in detectable concentrations of GER-UDCA in the CSF. The concentrations ranged from 4.65 μg/mL at 30 min to 1.1 μg/mL at 150 min. Notably, no GER or GER-UDCA was detected in the bloodstream after nasal administration. Histopathological examination of the nasal mucosa showed no damage after the administration of PBS 7.4 or GER-UDCA-SLNs. However, separation of the pseudostratified columnar epithelium from the basement membrane was observed with isopropyl alcohol and free GER administration. This study found that nasal administration of GER-UDCA-loaded SLNs showed increased dissolution rate, and permeation to the CSF without mucosal irritation, suggesting a promising non-invasive approach for central effects in PD treatment.

The treatment of DA depletion in PD is challenging due to the failure of exogenous DA to cross the BBB by itself. As a result, current "DA replacement therapy" primarily involves the administration of DA prodrugs or DA receptor agonists that can penetrate the BBB. L-Dopa remains the preferred treatment for PD, but its long-term use is accompanied by severe side effects, driving researchers to explore alternative strategies for reducing PD symptoms. Intranasal administration of DA has been inspected in several studies using techniques such as autoradiography with tritium DA, microdialysis experiments, and examination of DA metabolism and nasal mucosal transport (Dahlin et al., 2001; de Souza Silva et al., 2008), which showed that the absorption of the neurotransmitter through nasal administration is lower compared to its uptake in the brain. This implies that the associated systemic side effects are likely to be reduced and potentially be less significant. Therefore, these findings suggest that intranasal administration of DA could be utilized as a "DA replacement strategy therapy" for PD and serve as a substitute to the current oral levodopa treatment.

Cometa et al. (2020) used Gelucire® 50/13, DA hydrochloride (DA), Tween 85, acetic acid (HAc), and glycol chitosan (GCS) as materials. They prepared DA-SLN and GCS-DA-SLN using a method called melt emulsification. They melted Gelucire® 50/13 and heated a solution of HAc with Tween 85 separately. Then they combined DA with the HAc solution and added it to the melted Gelucire 50/13 to create an emulsion. After cooling, the nanosuspension was centrifuged, and the resulting pellet was resuspended in water. For GCS-DA-SLN, they used an aqueous phase containing GCS. Control SLNs were prepared without DA and GCS or without DA but with GCS. The SLNs were freeze-dried for further analysis. *In vitro* release of DA from DA-SLN and DA-GCS-SLN was investigated for potential intranasal administration. Both SLN formulations showed sustained release of DA over 48 h, with DA-SLN exhibiting an initial burst effect, followed by slower release, while DA-GCS-SLN demonstrated continuous and sustained release. The release profiles suggest that DA is localized on the particles' surface, contributing to the initial burst release. Intact DA was detected in the release medium for up to 48 h, but autoxidation processes were observed thereafter. Lipophilic carriers, such as liposomes and SLN, showed low DA release compared to hydrophilic delivery systems. The sustained release of DA from SLN holds promise for clinical treatments, offering a more controlled and prolonged delivery compared to current approaches. Additionally, DA-GCS-SLN protected autoxidation, which was not observed with chitosan nanoparticles.

In addition to the aforementioned drawbacks of DA therapy, DA can undergo automatic oxidation reaction under neutral or alkaline conditions, facilitated by molecular oxygen. This process involves the formation of aminochrome and the subsequent synthesis of polymer compounds, including neuromelanins, through the action of ROS. These reactions are thought to have a significant influence on the development of neurodegenerative conditions, including PD (Li et al., 2020; Trapani et al., 2017, 2018).

Trapani et al. (2021) used various materials for liposomes and SLNs formulation: DA hydrochloride, cholesterol, glycol chitosan, Fluorescein 5, 6-FITC, 6-Coumarin (6-COUM), and carboxyl ester hydrolase, soybean phosphatidyl choline (PC) and phosphatidyl glycerol (PG), Gelucire® 50/13, and hydroxyethyl cellulose (HEC). Chitosan–glutathione conjugate (CS-GSH) was prepared, and a thin-film hydration method was employed for the formulation of the liposomes, whereas melt homogenization method was carried out for the SLNs formulation. The liposomes were either uncoated or coated with CS-GSH, and fluorescent variants were also prepared. The SLNs were formulated with Gelucire® 50/13 and could be loaded with dopamine and/or GCS. In the study, different formulations of drug delivery systems were evaluated for their cytotoxicity and uptake by olfactory ensheathing cells (OECs). SLN formulations, namely, SLN1 and SLN2, were prepared using Gelucire® 50/13 as the lipid base. SLN1 was DA-loaded without any coating, while SLN2 was DA-loaded and coated with a chitosan–glutathione conjugate (CS-GSH). Both SLN formulations showed no cytotoxicity to OECs at various concentrations. Liposomal formulations, Lip1 and Lip2, were also investigated. Lip1 were uncoated DA-loaded liposomes, while Lip2 were DA-loaded liposomes coated with CS-GSH. Both liposomal formulations exhibited low cytotoxicity to OECs. Furthermore, uptake studies demonstrated that SLN2 and Lip2 were internalized by a significant percentage of OECs, indicating their potential for intranasal drug delivery.

Apomorphine (APO), a dopamine-receptor agonist, is commonly used in PD therapy (Gómez-Jeria & Ojeda-Vergara, 1997). However, its oral bioavailability is less than 2% due to hepatic metabolism, necessitating subcutaneous administration either through separated injections or as a sustained infusion (Campbell et al., 1980). This mode of delivery has drawbacks, including injection site reactions and the need for multiple injections (Manson et al., 2001). Sublingual administration with APL-130277 shows promise as an alternative, but finding a substitute for subcutaneous infusion remains challenging (Navia et al., 2018). Buccal mucosa, which bypasses liver metabolism, is under investigation as a potential administration site (Gershkovich et al., 2009). Experiments using freshly excised porcine buccal mucosa were conducted to assess permeability (Brandl & Bauer-Brandl, 2019; Rathbone et al., 2015). Pig models are suitable for buccal permeability research due to their similarity to humans and practical considerations (Nielsen & Rassing, 2000).

Itin et al. (2020) studied the permeability of APO through the buccal mucosa, with particular emphasis on the effect of nano-liposphere particles (NLF). The research revealed that APO permeates the buccal mucosa through passive transcellular diffusion. In order to modify the permeability, different techniques were explored, including pH alteration and trying different organic solvents. Notably, the introduction of NLF resulted in delayed delivery of APO due to its storage in the lipidic core of the particles. Simulations were run to determine the clinical relevance of buccal delivery of APO, showing its potential as a prolonged-release and convenient alternative to parenteral administration. However, the development of suitable delivery devices that consider the microenvironment and prevent washout by saliva flow remains crucial.

In PD patients, the expression levels of brain-derived neurotrophic factor (BDNF) and basic fibroblast growth factor (bFGF) are reduced in the remaining dopaminergic neurons in the substantia nigra pars compacta (Howells et al., 2000; Parain et al., 1999; Tooyama et al., 1993, 1994). Depletion of BDNF specifically from the midbrain–hindbrain region results in a decrease in the expression of tyrosine hydroxylase (TH), an enzyme involved in dopamine synthesis (Baquet et al., 2005). Studies using neurotoxin-induced degeneration models have shown that bFGF plays a protective role in dopamine neurons. Higher levels of dopamine neuron death were observed in mice lacking bFGF, while increased preservation of dopamine neurons was seen in mice overexpressing bFGF (Timmer et al., 2007).

The use of neurotrophic factors (NTFs) as a therapeutic option in PD has gained attention (Rangasamy et al., 2010). BDNF can cross the BBB through a high-capacity, saturable transport system (Pan et al., 1998). However, bFGF is unstable in solution and has a short half-life, which hinders its ability to cross the BBB in sufficient quantities for pharmacological efficacy in the central nervous system (CNS) (Deguchi et al., 2000; Edelman et al., 1993; Whalen et al., 1989). Therefore, the limited ability of bFGF to penetrate the BBB restricts its therapeutic potential in CNS disorders.

Zhao et al. (2014) prepared a water-in-water emulsion, and a freeze-drying technique was employed to prepare gelatin nanoparticles (GNs) loaded with bFGF. The resulting bFGF-GNs and gelatin nanoparticle liposomes (bFGF-GNLs) exhibited an amount of 2 mg/mL of bFGF. The GNs were formulated by dispersing gelatin in an aqueous phase, followed by the addition of bFGF and subsequent emulsification using high-shear homogenization. The emulsion was then freeze-dried to obtain stable nanoparticles with a mean particle size of 200 nm. To establish a hemiparkinsonian rat model, rats were injected with 6-OHDA in the right-side striatum, while sham animals received vehicle injections. The behavioral evaluation involved assessing apomorphine-induced rotations, which were recorded in an opaque cylinder. The rats showing rotational behavior of more than seven contralateral turns per minute were selected for further experiments. For intranasal administration of bFGF-GNLs, deeply anesthetized rats were positioned supine, with their head and neck maintained in a horizontal position. Alternating nostrils were used for drug administration, with the unmedicated nostril obstructed to enhance the absorption. The distribution of bFGF in the rat brain was examined by obtaining samples from the olfactory bulb, pallium, striatum, and hippocampus, followed by Western blot analysis. Additionally, nasal mucosae were collected to assess potential toxicity using immunohistochemistry after 14 days of daily administration. To evaluate bFGF's impact, rats were divided into different treatment groups and examined for rotational behavior and changes in DA levels in the striatum. Dopamine levels were estimated using high-performance liquid chromatography coupled with electrochemical detection. Immunohistochemical staining with anti-tyrosine hydroxylase antibodies was carried out to assess dopaminergic neuron loss. Furthermore, the regional distribution of exogenous bFGF in the brain was investigated by analyzing protein levels in various brain regions. Results showed that intranasal administration of bFGF-GNLs led to increased bFGF distribution in the olfactory bulb and striatum compared to the control group. Moreover, treatment with bFGF-GNLs resulted in elevated DA, dihydroxyphenylacetic acid, and homovanillic acid levels in the striatum, indicating enhanced dopaminergic neurotransmission. Immunohistochemical analysis revealed improved retention of dopaminergic neurons in the substantia nigra in the bFGF-GNLs/IN group compared to the control group. Importantly, no mucosal toxicity or damage was observed in the nasal mucosa following two weeks of daily treatments. Taken together, intranasal delivery of bFGF-GNLs shows promise in improving dopaminergic neuron survival and reducing motor deficits in PD, with a well-received and safe delivery method.

13.7.3 Amyotrophic Lateral Sclerosis

Amyotrophic lateral sclerosis (ALS), also referred to as Lou Gehrig's disease, is a degenerative neurological condition that affects upper and lower motor neurons, which results in loss of muscle control and eventual paralysis (Orsini et al., 2015).

There are many published studies describing glutamate's role in ALS progression. Glutamate is the main chemical responsible for nerve cell excitation in the neurotransmission process. When there is an excess of glutamate in the synaptic cleft, it can cause N-methyl-D-aspartate (NMDA) receptors to become over-stimulated and eventually results in nerve cell death. This event is known as glutamate-induced excitotoxicity (Blasco et al., 2014; Rizzo et al., 2014; Zarei et al., 2015). Riluzole, a drug that shows anti-glutamatergic activities, has been investigated as a treatment for ALS. Bondì et al. (2010) formulated riluzole into SLNs and tested it on Sprague–Dawley rat models (Bondì et al., 2010). They found that the formulation demonstrated 1.3 higher efficacy in crossing the BBB and targeting the central nervous system compared to the free riluzole. They also found that riluzole-containing SLNs decreased drug distribution in non-target organs (kidneys, heart, spleen, lungs, and liver) compared to the free riluzole. Another approach to maintain glutamate homeostasis is by upregulating the expression of glutamate transporter 1 (GLT-1). GLT-1 is a type of transporter protein responsible for clearing excess glutamate from the synaptic cleft, ensuring the prevention of glutamate excitotoxicity. Several reports have shown that ceftriaxone, a beta-lactam antibiotic, has GLT-1 upregulation capability (Bajrektarevic & Nistri, 2017; Hu et al., 2015; Wilkie et al., 2021). Recently, ceftriaxone has been encapsulated into SLNs by Kumar et al. (2016), and it exhibited a 24 h sustained ceftriaxone release while it followed a burst release (Kumar et al., 2016).

13.7.4 Multiple Sclerosis (MS)

Multiple sclerosis is a progressive, neurologically disabling, persistent disorder where the white matter of the CNS becomes demyelinated. It is a multifaceted immune-mediated process that is impacted by environmental and genetic factors. In most cases, the early phases of the disease are marked by temporary episodes of neurological dysfunction that remain for a few days or weeks. Over time, irreversible cognitive deficits develop. However, a minority of patients exhibit a progressive disease trajectory from the outset. At the pathological level, multiple sclerosis (MS) is typified by the presence of demyelinated lesions in the brain and spinal cord, which can be linked to neuroaxonal injury (Filippi et al., 2018). Teriflunomide (TFM) is a disease-modifying, remyelinating drug used in the management of relapsing multiple sclerosis cases. TFM inhibits dihydroorotate dehydrogenase, an enzyme required for de novo pyrimidine synthesis, resulting in decreased proliferation of activated T and B, which helps alleviate the symptoms of multiple sclerosis (Alroughani et al., 2022). Recently, TFM was incorporated into intranasal NLCs by Gadhave and Kokare (2019). When compared to the control, their formulation achieved rapid remyelination in treated male Wistar rats, and there were no significant alterations in liver biomarkers or subacute toxicity (Gadhave & Kokare, 2019).

Dimethyl fumarate (DMF) is another treatment option that can shift the immune profile of relapsing-remitting MS patients (Mills et al., 2018). Kumar et al. (2017) have encapsulated DMF in NLCs. They found that the systemic availability after oral intake was increased by 4.09-fold, and brain distribution was significantly improved over the free drug, which can facilitate MS management by reducing the dosing frequency (Kumar et al., 2017).

Baclofen, a muscle relaxant that is used to reduce muscle spasticity associated with multiple sclerosis, was formulated in NLCs by Ghasemian et al. (2017). Plasma and brain drug concentrations were compared to an aqueous baclofen solution. They demonstrated that the baclofen-containing NLC increased the drug's half-life in plasma and brain by up to 10- and 1.5-fold, respectively, in comparison with the aqueous counterpart (Ghasemian et al., 2017).

13.7.5 Huntington's Disease

Huntington's disease (HD) is a hereditary neurodegenerative condition of the CNS that is marked by chorea, dance-like involuntary movements caused by excess dopamine in the brain (Chen et al., 2013), behavioral and mental problems, and dementia. The Caucasian population has a prevalence of 1/10,000–1/20,000. The average age of symptoms development ranges from 30 to 50 years. In some cases, it may appear before 20 years of age, including behavioral issues and learning difficulties. Chorea is a characteristic symptom, which slowly affects all muscles. All psychomotor processes are drastically hindered. Patients exhibit psychiatric symptoms as well as cognitive impairment. HD is an inherited autosomal dominant disorder. As the condition progresses, patients become completely dependent throughout daily life, necessitating full-time care (Roos, 2010). There are many factors contributing to HD progression, including oxidative damage, gene transcription defect, neuroinflammation, and excitotoxicity (Jimenez-Sanchez et al., 2017).

Considering the association between the oxidative stress and the pathogenesis of HD, rosmarinic acid (RA), an antioxidant (Kelsey et al., 2010), was formulated by Bhatt et al. (2015) as SLNs and was administered to 3-nitro propionic acid (3NP)–treated male Wistar rats through intranasal route. The drug delivery via the nasal route achieved a significant effect when compared to the intravenous route. In addition, RA-containing SLNs treatment alleviated the oxidative stress impairments in locomotor activity and behavioral abnormalities in the treated rats (Bhatt et al., 2015).

Thymoquinone is an active compound obtained from the volatile oil of *Nigella sativa* seeds; it has antioxidant, anti-inflammatory, and immunomodulatory abilities. Recently, Ramachandran and Thangarajan (2018) incorporated thymoquinone into SLNs. HD male Wistar rat models were utilized in the study. One cohort received treatment with thymoquinone SLNs, and the other cohort was administered thymoquinone suspension. The SLN-treated group exhibited suppression of microglial activation and NMDA receptor desensitization. In addition, better improvement of motor functions compared to the thymoquinone suspension–treated group (Ramachandran & Thangarajan, 2018).

Tetrabenazine (TBZ), a chorea-alleviating drug that selectively inhibits monoamine transporter-2, leading to dopamine reduction in pre-synaptic vesicles (Marshall, 2006), was incorporated by Arora et al. (2020) into a nanoemulsion form and was administrated to Wistar rats intranasally. The formulation showed increased TBZ brain concentration, and *ex vivo* drug permeation increased by 1.68-fold, while the aqueous form showed poor solubility and was susceptible to the first-pass effect (Arora et al., 2020).

13.8 NOSE-TO-BRAIN DRUG DELIVERY

Alternative methods for conquering the BBB have been proposed by researchers, including intrathecal injection and intracerebroventricular injection. The intracerebroventricular injection is a painful, invasive method that skilled practitioners can only use under very specific circumstances. It appears to be almost unusable (Jiang et al., 2011). Additionally, utilizing these alternate routes as the intracerebroventricular or intracerebral medication delivery route targeting the brain poses some challenges because injection solutions and the drug itself must meet several requirements, such as the pH, volume, diluents, and preservatives (Cook et al., 2009). Although intrathecal injection is increasingly popular, particularly in oncology, its use in neurology is still somewhat limited (Lee & Minko, 2021).

The intranasal (IN) delivery method has gained significant interest as a possible drug delivery pathway to the brain because each of the aforementioned procedures is invasive and costly. Intranasal administration has typically been used to treat hay fever or allergies by promoting local effects. However, due to its beneficial features, such as non-invasiveness, high patient adherence, and convenience of administration, it has been used for systemic distribution. It has demonstrated beneficial impacts in smoking cessation, migraine management, and the flu vaccine, among other effects, so its use has been increasing in the medical field (Pires et al., 2009). Numerous studies have revealed that the intranasal route can successfully reach the brain with either small or large molecules. The olfactory mucosa is in direct touch with the brain and is the only area that the BBB does not shield. Additionally, therapeutic molecules' accumulation in vital organs (such as the spleen, liver, and kidneys) could be prohibited, thus minimizing systemic adverse effects (Dhuria et al., 2010; Lee & Minko, 2021; Tian et al., 2019).

Some labs in preclinical trials have shown that many nervous system ailments can be effectively treated with intranasal medication administration. Jiang et al (2011) showed that insulin-like growth factor I (IGF-I) intranasal delivery enhanced neurological function in rats with ischemic stroke (Jiang et al., 2011). There have also been encouraging outcomes for the therapeutic approaches using IN delivery of drugs in treating epilepsy (Florence et al., 2011), tumors (Shingaki et al., 2010), Alzheimer's disease (Reger et al., 2008), and ALS (Martinez et al., 2008).

Recently, in a nose-to-brain clinical trial, Alzheimer's disease (AD) patients showed statistical improvement (Bourganis et al., 2018). This all led to more trust in the IN route to efficiently transport medications to CNS, thus subsequently enhancing patient outcomes in neurological and neurodegenerative diseases.

Nanocarrier-based formulations are considered a potential approach for improving the effectiveness of nose-to-brain delivery via increasing nasal absorption of pharmaceuticals. Drug delivery through the nasal mucosa may be aided by lipid- and polymer-based formulations. This occurs while preventing the enzymatic breakdown of their payload and guaranteeing prolonged mucosal surface retention, which boosts drug concentration at the desired location (Kammona & Kiparissides, 2012; Ozsoy et al., 2009). Additionally, the design of the nanocarriers can be adjusted in terms of excipient selection, physicochemical characteristics (zeta potential, particle size distribution), biocompatibility, biodegradability, drug loading, and temporal and spatial controlled drug release, among others (Adair et al., 2010).

13.9 CHALLENGES AGAINST THE CLINICAL IMPLEMENTATION AND FUTURE OUTLOOKS

Nanotechnology shows potential use in pharmaceutical delivery, improving drug-targeting CNS and overcoming physiological barriers (i.e., BBB). Lipid-based nanoformulas demonstrated increased therapeutic effectiveness and bioavailability for treating neurodegenerative disorders, such as

TABLE 13.3
Subtle differences between nanoemulsions and microemulsions

Feature	Nanoemulsion	Microemulsion	References
Droplet size	20 to 500 nm	10 to 100 nm	(Gupta et al., 2016)
Appearance	Transparent to translucent, milky appearance	Transparent and isotropic	(McClements, 2012)
Stability	Thermodynamically unstable, but kinetically stable	Thermodynamically stable	(Anton & Vandamme, 2011)
Preparation procedure	High-energy methods (ultrasonication, high-pressure homogenization) or low-energy methods (phase inversion temperature, spontaneous emulsification)	Spontaneous or with low-energy input formation via simple mixing under specific conditions (e.g., temperature, surfactant concentration)	(Tadros, 2013)
Surfactant type	Either hydrophilic or lipophilic surfactants	Preferentially uses lipophilic surfactants	(Tadros, 2013)
Surfactant ratio/concentration	Lower surfactant concentration	Higher surfactant concentration	(McClements, 2012)
Interfacial tension	Reduced, but not to the same extent as microemulsions	Extremely low	(Anton & Vandamme, 2011; McClements, 2012)
Storage requirements	May require refrigeration or other precautions to maintain stability	Less-stringent storage conditions, as the system is thermodynamically stable	(Anton & Vandamme, 2011; McClements, 2012; Solans & Solé, 2012)

Parkinson's diseases and Alzheimer's, in respect to the corresponding free drug. Despite the aforementioned benefits, there is no authorized CNS-targeting product for marketing to date. High production costs and unclear regulations may attribute to the difficulty of translation into clinical trials and clinical practice. Moreover, lipid-based nanosystems may exhibit safety concerns in clinical trials, owing to the complexity of human bodies when compared to *in vitro* and animal models. The impact of ageing on the dynamics of BBB and disruption of BBB after certain neurological conditions (e.g., multiple sclerosis) should be considered while developing lipid-based nanoparticles for adequate drug delivery, taking advantage of structural damages in neurological diseases (Al-Ahmady, 2018). Safety and toxicological studies are needed to assess the pharmacokinetics of lipid-based nanocarriers with and without drugs, including metabolism and elimination from the brain and systemic circulation, in addition to biocompatibility on the long run (Naqvi et al., 2020). Currently, the majority of lipid-based nanosystems occur in laboratories. Large-scale production, stability, and sterility are considered big challenges against translation into clinical practice. For instance, lipid-based nanocarriers may be prone to damage under some sterilization conditions (Toh & Chiu, 2013). Concerning stability, lipid oxidation, embedding material escape, and other issues may happen during storage, which, in turn, may lower the efficacy or lead to fatal adverse effects. On top of that, lipid moiety selection enhances biological distribution and efficacy; thus, searching for novel and appropriate lipid material is crucial for developing functionalized lipid-based nanosystems (Álvarez-Benedicto et al., 2022). Real-time monitoring techniques must be established in order to predict the destiny of lipid-based nanoparticles in vivo, thus helping in guiding the design of effective lipid-based nanosystems, and this will be reflected in the therapeutic efficacy.

REFERENCES

(2021). 2021 Alzheimer's disease facts and figures. *Alzheimers Dement*, *17*(3), 327–406. https://doi.org/10.1002/alz.12328

Abbott, N. J., & Friedman, A. (2012). Overview and introduction: The blood–brain barrier in health and disease. *Epilepsia*, *53*, 1–6. https://doi.org/10.1111/J.1528-1167.2012.03696.X

Abd El-Alim, S. H., Kassem, A. A., Basha, M., & Salama, A. (2019). Comparative study of liposomes, ethosomes and transfersomes as carriers for enhancing the transdermal delivery of diflunisal: In vitro and in vivo evaluation. *International Journal of Pharmaceutics*, *563*, 293–303. https://doi.org/10.1016/j.ijpharm.2019.04.001

Aburehab, M. A. S., Ansari, M. J., Singh, A., Hassan, A., Abdelgawad, M. A., Shrivastav, P., Abualsoud, B. M., Amaral, L. S., & Pramanik, S. (2022). Cubosomes as an emerging platform for drug delivery: A review of the state of the art. *Journal of Materials Chemistry B*, *10*(15), 2781–2819. https://doi.org/10.1039/d2tb00031h

Adair, J. H., Parette, M. P., Altınoğlu, E. İ., & Kester, M. (2010). Nanoparticulate alternatives for drug delivery. *ACS Nano*, *4*(9), 4967–4970. https://doi.org/10.1021/nn102324e

Adhikari, P., Pal, P., Das, A. K., Ray, S., Bhattacharjee, A., & Mazumder, B. (2017). Nano lipid-drug conjugate: An integrated review. *International Journal of Pharmaceutics*, *529*(1–2), 629–641. www.sciencedirect.com/science/article/abs/pii/S0378517317306385

Al-Ahmady, Z. S. (2018). Selective drug delivery approaches to lesioned brain through blood brain barrier disruption. *Expert Opinion on Drug Delivery*, *15*(4), 335–349. https://doi.org/10.1080/17425247.2018.1444601

Ali, G. A. A. (2017). Pharmacosomes: An updated review. *Universal Journal of Pharmaceutical Research*, *2*(1), 33–36. https://doi.org/10.22270/ujpr.v2i1.RW1

Ali, M. A., Kataoka, N., Ranneh, A. H., Iwao, Y., Noguchi, S., Oka, T., & Itai, S. (2017). Enhancing the solubility and oral bioavailability of poorly water-soluble drugs using monoolein cubosomes. *Chemical and Pharmaceutical Bulletin*, *65*(1), 42–48. https://doi.org/10.1248/CPB.C16-00513

Almeida, A. F., Borge, G. I. A., Piskula, M., Tudose, A., Tudoreanu, L., Valentova, K., Williamson, G., & Santos, C. N. (2018). Bioavailability of quercetin in humans with a focus on interindividual variation. *Comprehensive Reviews in Food Science and Food Safety*, *17*(3), 714–731. https://doi.org/10.1111/1541-4337.12342

Alroughani, R., Inshasi, J., Al Khawajah, M., Ahmed, S. F., Al Malik, Y., Alkhabouri, J., Shatila, A., Aljarallah, S., Cupler, E. J., Qureshi, S. Al, Thakre, M., Elhasin, H., Ezzat, A., & Roushdy, S. (2022). Real-world effectiveness and safety profile of teriflunomide in the management of multiple sclerosis in the Gulf Cooperation Council countries: An expert consensus narrative review. *Multiple Sclerosis Journal—Experimental, Translational and Clinical*, *8*(1). https://doi.org/10.1177/20552173221077185/ASSET/IMAGES/LARGE/10.1177_20552173221077185-FIG1.JPEG

Álvarez-Benedicto, E., Farbiak, L., Márquez Ramírez, M., Wang, X., Johnson, L. T., Mian, O., Guerrero, E. D., & Siegwart, D. J. (2022). Optimization of phospholipid chemistry for improved lipid nanoparticle (LNP) delivery of messenger RNA (mRNA). *Biomaterials Science*. https://pubs.rsc.org/en/content/articlelanding/2022/bm/d1bm01454d

Andrade, S., Pereira, M. C., & Loureiro, J. A. (2023). Caffeic acid loaded into engineered lipid nanoparticles for Alzheimer's disease therapy. *Colloids and Surfaces B: Biointerfaces*, *225*, 113270. https://doi.org/10.1016/j.colsurfb.2023.113270

Anekonda, T. S. (2006). Resveratrol—a boon for treating Alzheimer's disease? *Brain Research Reviews*, *52*(2), 316–326. https://doi.org/10.1016/j.brainresrev.2006.04.004

Angelov, B., Angelova, A., Ollivon, M., Bourgaux, C., & Campitelli, A. (2003). Diamond-type lipid cubic phase with large water channels. *Journal of the American Chemical Society*, *125*(24), 7188–7189. https://doi.org/10.1021/JA034578V/SUPPL_FILE/JA034578VSI20030423_123607.PDF

Anton, N., & Vandamme, T. F. (2011). Nano-emulsions and micro-emulsions: Clarifications of the critical differences. *Pharmaceutical Research*, *28*(5), 978–985. https://doi.org/10.1007/s11095-010-0309-1

Arora, A., Kumar, S., Ali, J., & Baboota, S. (2020). Intranasal delivery of tetrabenazine nanoemulsion via olfactory region for better treatment of hyperkinetic movement associated with Huntington's disease: Pharmacokinetic and brain delivery study. *Chemistry and Physics of Lipids*, *230*, 104917. https://doi.org/10.1016/J.CHEMPHYSLIP.2020.104917

Arora, D., Bhatt, S., Kumar, M., Verma, R., Taneja, Y., Kaushal, N., Tiwari, A., Tiwari, V., Alexiou, A., Albogami, S., Alotaibi, S. S., Mittal, V., Singla, R. K., Kaushik, D., & Batiha, G. E. (2022). QbD-based rivastigmine tartrate-loaded solid lipid nanoparticles for enhanced intranasal delivery to the brain for Alzheimer's therapeutics. *Frontiers in Aging Neuroscience*, *14*, 960246. https://doi.org/10.3389/fnagi.2022.960246

Arora, S., Layek, B., & Singh, J. (2021). Design and validation of liposomal ApoE2 gene delivery system to evade blood-brain barrier for effective treatment of Alzheimer's disease. *Molecular Pharmaceutics*, *18*(2), 714–725. https://doi.org/10.1021/acs.molpharmaceut.0c00461

Ashfaq, M., Shah, S., Rasul, A., Hanif, M., Khan, H. U., Khames, A., Abdelgawad, M. A., Ghoneim, M. M., Ali, M. Y., Abourehab, M. A. S., Maheen, S., Iqbal, O., Abbas, G., & El Sisi, A. M. (2022). Enhancement of the solubility and bioavailability of pitavastatin through a Self-Nanoemulsifying Drug Delivery System (SNEDDS). *Pharmaceutics*, *14*(3). https://doi.org/10.3390/PHARMACEUTICS14030482

Atri, A. (2019). The Alzheimer's disease clinical spectrum: Diagnosis and management. *Medical Clinics of North America*, *103*(2), 263–293. https://doi.org/10.1016/j.mcna.2018.10.009

Avantika, D., Raj, S. D., Vinay, P., Singh, A. M., Pravin, K., & Dhadwal, A. (2020). Cubosomes: A novel carrier for transdermal drug delivery. *Journal of Drug Delivery and Therapeutics*, *10*(1), 123–130. https://doi.org/10.22270/JDDT.V10I1.3814

Azmi, I. D. M., Moghimi, S. M., & Yaghmur, A. (2015). Cubosomes and hexosomes as versatile platforms for drug delivery. *Therapeutic Delivery*, *6*(12), 1347–1364. https://doi.org/10.4155/tde.15.81

Babazadeh, A., Ghanbarzadeh, B., & Hamishehkar, H. (2016). Novel nanostructured lipid carriers as a promising food grade delivery system for rutin. *Journal of Functional Foods*, *26*, 167–175. https://doi.org/10.1016/J.JFF.2016.07.017

Bajrektarevic, D., & Nistri, A. (2017). Ceftriaxone-mediated upregulation of the glutamate transporter GLT-1 contrasts neurotoxicity evoked by kainate in rat organotypic spinal cord cultures. *NeuroToxicology*, *60*, 34–41. https://doi.org/10.1016/J.NEURO.2017.02.013

Baloch, J., Sohail, M. F., Sarwar, H. S., Kiani, M. H., Khan, G. M., Jahan, S., Rafay, M., Chaudhry, M. T., Yasinzai, M., & Shahnaz, G. (2019). Self-Nanoemulsifying Drug Delivery System (SNEDDS) for improved oral bioavailability of chlorpromazine: In Vitro and in Vivo evaluation. *Medicina (Kaunas, Lithuania)*, *55*(5). https://doi.org/10.3390/MEDICINA55050210

Bandiwadekar, A., Jose, J., Khayatkashani, M., Habtemariam, S., Reza, H., Kashani, K., & Nabavi, S. M. (2021). Emerging novel approaches for the enhanced delivery of natural products for the management of neurodegenerative diseases. *Journal of Molecular Neuroscience*, *72*(3), 653–676. https://doi.org/10.1007/S12031-021-01922-7

Banerjee, S., & Kundu, A. (2018). Lipid-drug conjugates: A potential nanocarrier system for oral drug delivery applications. *DARU Journal of Pharmaceutical Sciences*, *26*, 65–75. www.ncbi.nlm.nih.gov/pmc/articles/PMC6154489/

Baquet, Z. C., Bickford, P. C., & Jones, K. R. (2005). Brain-derived neurotrophic factor is required for the establishment of the proper number of dopaminergic neurons in the substantia nigra pars compacta. *Journal of Neuroscience*, *25*(26), 6251–6259. https://doi.org/10.1523/JNEUROSCI.4601-04.2005

Barrachina, M., Domínguez, I., Ambrosio, S., Secades, J., Lozano, R., & Ferrer, I. (2003). Neuroprotective effect of citicoline in 6-hydroxydopamine-lesioned rats and in 6-hydroxydopamine-treated SH-SY5Y human neuroblastoma cells. *Journal of the Neurological Sciences*, *215*(1–2), 105–110. https://doi.org/10.1016/S0022-510X(03)00204-1

Bei, D., Meng, J., & Youan, B. B. C. (2010). Engineering nanomedicines for improved melanoma therapy: Progress and promises. *Nanomedicine*, *5*(9), 1385–1399. https://doi.org/10.2217/NNM.10.117

Bhasin, B., Patel, S. P., & Road, V. L. M. (2018). An overview of transfersomal drug delivery. *International Journal of Pharmaceutical Sciences and Research*, *9*(6), 2175–2184. https://doi.org/10.13040/IJPSR.0975-8232.9(6).2175-84

Bhat, B. G., & Chandrasekhara, N. (1986). Studies on the metabolism of piperine: Absorption, tissue distribution and excretion of urinary conjugates in rats. *Toxicology*, *40*(1), 83–92. https://doi.org/10.1016/0300-483x(86)90048-x

Bhatt, R., Singh, D., Prakash, A., & Mishra, N. (2015). Development, characterization and nasal delivery of rosmarinic acid-loaded solid lipid nanoparticles for the effective management of Huntingtons disease. *Drug Delivery*, *22*(7), 931–939. https://doi.org/10.3109/10717544.2014.880860

Birks, J. S., Chong, L. Y., & Grimley Evans, J. (2015). Rivastigmine for Alzheimer's disease. *Cochrane Database of Systematic Reviews*, *9*(9), CD001191. https://doi.org/10.1002/14651858.CD001191.pub4

Blasco, H., Mavel, S., Corcia, P., & Gordon, P. H. (2014). The glutamate hypothesis in ALS: Pathophysiology and drug development. *Current Medicinal Chemistry*, *21*(31), 3551–3575. https://doi.org/10.2174/0929867321666140916120118

Bloom, G. S. (2014). Amyloid-beta and tau: The trigger and bullet in Alzheimer disease pathogenesis. *JAMA Neurology*, *71*(4), 505–508. https://doi.org/10.1001/jamaneurol.2013.5847

Bollimpelli, V. S., Kumar, P., Kumari, S., & Kondapi, A. K. (2016). Neuroprotective effect of curcumin-loaded lactoferrin nano particles against rotenone induced neurotoxicity. *Neurochemistry International*, *95*, 37–45. https://doi.org/10.1016/J.NEUINT.2016.01.006

Bondì, M. L., Craparo, E. F., Giammona, G., & Drago, F. (2010). Brain-targeted solid lipid nanoparticles containing riluzole: Preparation, characterization and biodistribution. *Nanomedicine*, *5*(1), 25–32. https://doi.org/10.2217/nnm.09.67

Bourganis, V., Kammona, O., Alexopoulos, A., & Kiparissides, C. (2018). Recent advances in carrier mediated nose-to-brain delivery of pharmaceutics. *European Journal of Pharmaceutics and Biopharmaceutics: Official Journal of Arbeitsgemeinschaft Fur Pharmazeutische Verfahrenstechnik e.V*, *128*, 337–362. https://doi.org/10.1016/j.ejpb.2018.05.009

Bozzuto, G., & Molinari, A. (2015). Liposomes as nanomedical devices. *International Journal of Nanomedicine*, *10*, 975–999. https://doi.org/10.2147/IJN.S68861

Brandl, M., & Bauer-Brandl, A. (2019). Oromucosal drug delivery: Trends in in-vitro biopharmaceutical assessment of new chemical entities and formulations. *European Journal of Pharmaceutical Sciences*, *128*, 112–117. https://doi.org/10.1016/j.ejps.2018.11.031

Briggs, R., Kennelly, S. P., & O'Neill, D. (2016). Drug treatments in Alzheimer's disease. *Clinical Medicine (London, England)*, *16*(3), 247–253. https://doi.org/10.7861/clinmedicine.16-3-247

Bunjes, H., Drechsler, M., Koch, M. H. J., & Westesen, K. (2001). Incorporation of the model drug ubidecarenone into solid lipid nanoparticles. *Pharmaceutical Research*, *18*(3), 287–293. https://doi.org/10.1023/A:1011042627714/METRICS

Buya, A. B., Beloqui, A., Memvanga, P. B., & Préat, V. (2020). Self-nano-emulsifying drug-delivery systems: From the development to the current applications and challenges in oral drug delivery. *Pharmaceutics*, *12*(12), 1–52. https://doi.org/10.3390/PHARMACEUTICS12121194

Campbell, A., Kula, N. S., Jeppsson, B., & Baldessarini, R. J. (1980). Oral bioavailability of apomorphine in the rat with a portacaval venous anastomosis. *European Journal of Pharmacology*, *67*(1), 139–142. https://doi.org/10.1016/0014-2999(80)90022-9

Campisi, A., Sposito, G., Pellitteri, R., Santonocito, D., Bisicchia, J., Raciti, G., Russo, C., Nardiello, P., Pignatello, R., Casamenti, F., & Puglia, C. (2022). Effect of unloaded and curcumin-loaded solid lipid nanoparticles on tissue transglutaminase isoforms expression levels in an experimental model of Alzheimer's disease. *Antioxidants (Basel)*, *11*(10). https://doi.org/10.3390/antiox11101863

Caracciolo, G., Pozzi, D., Caminiti, R., & Amenitsch, H. (2005). Lipid mixing upon deoxyribonucleic acid-induced liposomes fusion investigated by synchrotron small-angle x-ray scattering. *Applied Physics Letters*, *87*(13), 1–3. https://doi.org/10.1063/1.2058202

Cevik, B., Solmaz, V., Yigitturk, G., Cavusoglu, T., Peker, G., & Erbas, O. (2017). Neuroprotective effects of erythropoietin on Alzheimer's dementia model in rats. *Advances in Clinical and Experimental Medicine*, *26*(1), 23–29. https://doi.org/10.17219/acem/61044

Chauhan, N., Vasava, P., Khan, S. L., Siddiqui, F. A., Islam, F., Chopra, H., & Emran, T. B. (2022). Ethosomes: A novel drug carrier. *Annals of Medicine and Surgery*, *82*, 8–11. https://doi.org/10.1016/j.amsu.2022.104595

Chen, J. Y., Wang, E. A., Cepeda, C., & Levine, M. S. (2013). Dopamine imbalance in Huntington's disease: A mechanism for the lack of behavioral flexibility. *Frontiers in Neuroscience*, *7*, 114. https://doi.org/10.3389/FNINS.2013.00114/BIBTEX

Chen, Y., & Liu, L. (2012). Modern methods for delivery of drugs across the blood-brain barrier. *Advanced Drug Delivery Reviews*, *64*(7), 640–665. https://doi.org/10.1016/j.addr.2011.11.010

Cheng, Y.-L., Lee, C.-Y., Huang, Y.-L., Buckner, C. A., Lafrenie, R. M., Dénommée, J. A., Caswell, J. M., Want, D. A., Gan, G. G., Leong, Y. C., Bee, P. C., Chin, E., Teh, A. K. H., Picco, S., Villegas, L., Tonelli, F., Merlo, M., Rigau, J., Diaz, D., . . . Mathijssen, R. H. J. (2016). We are IntechOpen, the world's leading publisher of open access books built by scientists, for scientists TOP 1%. *Intech*, *11*, 13. www.intechopen.com/books/advanced-biometric-technologies/liveness-detection-in-biometrics

Cheong, S. L., Federico, S., Spalluto, G., Klotz, K. N., & Pastorin, G. (2019). The current status of pharmacotherapy for the treatment of Parkinson's disease: Transition from single-target to multitarget therapy. *Drug Discovery Today*, *24*(9), 1769–1783. https://doi.org/10.1016/j.drudis.2019.05.003

Chun, H. S., & Low, W. C. (2012). Ursodeoxycholic acid suppresses mitochondria-dependent programmed cell death induced by sodium nitroprusside in SH-SY5Y cells. *Toxicology*, *292*(2–3), 105–112. https://doi.org/10.1016/J.TOX.2011.11.020

Clogston, J., Craciun, G., Hart, D. J., & Caffrey, M. (2005). Controlling release from the lipidic cubic phase by selective alkylation. *Journal of Controlled Release*, *102*(2), 441–461. https://doi.org/10.1016/J.JCONREL.2004.10.007

Collaborators, G. B. D. N. (2019). Global, regional, and national burden of neurological disorders, 1990–2016: A systematic analysis for the Global Burden of Disease Study 2016. *Lancet Neurology*, *18*(5), 459–480. https://doi.org/10.1016/S1474-4422(18)30499-X

Cometa, S., Bonifacio, M. A., Trapani, G., Di Gioia, S., Dazzi, L., De Giglio, E., & Trapani, A. (2020). In vitro investigations on dopamine loaded solid lipid nanoparticles. *Journal of Pharmaceutical and Biomedical Analysis*, *185*, 113257. https://doi.org/10.1016/j.jpba.2020.113257

Conant, R., & Schauss, A. (2004). Therapeutic applications of citicoline for stroke and cognitive dysfunction in the elderly: A review of the literature. *Alternative Medicine Review : A Journal of Clinical Therapeutic*, *9*(1), 17–31.

Cook, A. M., Mieure, K. D., Owen, R. D., Pesaturo, A. B., & Hatton, J. (2009). Intracerebroventricular administration of drugs. *Pharmacotherapy*, *29*(7), 832–845. https://doi.org/10.1592/phco.29.7.832

Crowe, T. P., Greenlee, M. H. W., Kanthasamy, A. G., & Hsu, W. H. (2018). Mechanism of intranasal drug delivery directly to the brain. *Life Sciences*, *195*, 44–52. https://doi.org/10.1016/j.lfs.2017.12.025

Cummings, J., Lee, G., Ritter, A., Sabbagh, M., & Zhong, K. (2019). Alzheimer's disease drug development pipeline: 2019. *Alzheimer's and Dementia (New York, N. Y.)*, *5*, 272–293. https://doi.org/10.1016/j.trci.2019.05.008

Dahlin, M., Jansson, B., & Bjork, E. (2001). Levels of dopamine in blood and brain following nasal administration to rats. *European Journal of Pharmaceutical Sciences*, *14*(1), 75–80. https://doi.org/10.1016/s0928-0987(01)00151-8

Daneman, R., & Prat, A. (2015). The blood-brain barrier. *Cold Spring Harbor Perspectives in Biology*, *7*(1). https://doi.org/10.1101/CSHPERSPECT.A020412

Dara, T., Vatanara, A., Sharifzadeh, M., Khani, S., Vakilinezhad, M. A., Vakhshiteh, F., Nabi Meybodi, M., Sadegh Malvajerd, S., Hassani, S., & Mosaddegh, M. H. (2019). Improvement of memory deficits in the rat model of Alzheimer's disease by erythropoietin-loaded solid lipid nanoparticles. *Neurobiology of Learning and Memory*, *166*, 107082. https://doi.org/10.1016/j.nlm.2019.107082

Das, S., & Chaudhury, A. (2011). Recent advances in lipid nanoparticle formulations with solid matrix for oral drug delivery. *AAPS PharmSciTech*, *12*, 62–76. www.ncbi.nlm.nih.gov/pmc/articles/PMC3066374/

Date, T., Paul, K., Singh, N., & Jain, S. (2019). Drug–lipid conjugates for enhanced oral drug delivery. *AAPS PharmSciTech*, *20*, 1–11. https://link.springer.com/article/10.1208/s12249-018-1272-0

Dawoud, M., Abourehab, M. A. S., & Abdou, R. (2020). Monoolein cubic nanoparticles as novel carriers for docetaxel. *Journal of Drug Delivery Science and Technology*, *56*, 101501. https://doi.org/10.1016/J.JDDST.2020.101501

Deguchi, Y., Naito, T., Yuge, T., Furukawa, A., Yamada, S., Pardridge, W. M., & Kimura, R. (2000). Blood-brain barrier transport of 125I-labeled basic fibroblast growth factor. *Pharmaceutical Research*, *17*(1), 63–69. https://doi.org/10.1023/a:1007570509232

Deleu, D., Northway, M. G., & Hanssens, Y. (2002). Clinical pharmacokinetic and pharmacodynamic properties of drugs used in the treatment of Parkinson's disease. *Clinical Pharmacokinetics*, *41*(4), 261–309. https://doi.org/10.2165/00003088-200241040-00003

de Oliveira, E. R., Jr., Truzzi, E., Ferraro, L., Fogagnolo, M., Pavan, B., Beggiato, S., Rustichelli, C., Maretti, E., Lima, E. M., Leo, E., & Dalpiaz, A. (2020). Nasal administration of nanoencapsulated geraniol/ursodeoxycholic acid conjugate: Towards a new approach for the management of Parkinson's disease. *Journal of Controlled Release*, *321*, 540–552. https://doi.org/10.1016/j.jconrel.2020.02.033

de Souza Silva, M. A., Topic, B., Huston, J. P., & Mattern, C. (2008). Intranasal dopamine application increases dopaminergic activity in the neostriatum and nucleus accumbens and enhances motor activity in the open field. *Synapse*, *62*(3), 176–184. https://doi.org/10.1002/syn.20484

Dhaval, M., Vaghela, P., Patel, K., Sojitra, K., Patel, M., Patel, S., Dudhat, K., Shah, S., Manek, R., & Parmar, R. (2022). Lipid-based emulsion drug delivery systems—a comprehensive review. *Drug Delivery and Translational Research*, *12*(7), 1616–1639. https://doi.org/10.1007/s13346-021-01071-9

Dhuria, S. V., Hanson, L. R., & Frey, W. H. (2010). Intranasal delivery to the central nervous system: Mechanisms and experimental considerations. *Journal of Pharmaceutical Sciences*, *99*(4), 1654–1673. https://doi.org/10.1002/jps.21924

Dok-Go, H., Lee, K. H., Kim, H. J., Lee, E. H., Lee, J., Song, Y. S., Lee, Y. H., Jin, C., Lee, Y. S., & Cho, J. (2003). Neuroprotective effects of antioxidative flavonoids, quercetin, (+)-dihydroquercetin and quercetin 3-methyl ether, isolated from Opuntia ficus-indica var. saboten. *Brain Research*, *965*(1–2), 130–136. https://doi.org/10.1016/s0006-8993(02)04150-1

Don, T. M., Chang, W. J., Jheng, P. R., Huang, Y. C., & Chuang, E. Y. (2021). Curcumin-laden dual-targeting fucoidan/chitosan nanocarriers for inhibiting brain inflammation via intranasal delivery. *International Journal of Biological Macromolecules*, *181*, 835–846. https://doi.org/10.1016/J.IJBIOMAC.2021.04.045

Duan, Y., Dhar, A., Patel, C., Khimani, M., Neogi, S., Sharma, P., Kumar, N. S., & Vekariya, R. L. (2020). A brief review on solid lipid nanoparticles: Part and parcel of contemporary drug delivery systems. *RSC Advances*, *10*(45), 26777–26791. https://pubs.rsc.org/en/error/pageloaderror

Dudhipala, N., & Gorre, T. (2020). Neuroprotective effect of ropinirole lipid nanoparticles enriched hydrogel for Parkinson's disease: In Vitro, Ex Vivo, pharmacokinetic and pharmacodynamic evaluation. *Pharmaceutics*, *12*(5). https://doi.org/10.3390/pharmaceutics12050448

Duzgunes, N., & Nir, S. (1999). Mechanisms and kinetics of liposome-cell interactions. *Advanced Drug Delivery Reviews*, *40*, 3–18. www.elsevier.com/locate/drugdeliv

Edelman, E. R., Nugent, M. A., & Karnovsky, M. J. (1993). Perivascular and intravenous administration of basic fibroblast growth factor: Vascular and solid organ deposition. *Proceedings of the National Academy of Sciences of the United States of America*, *90*(4), 1513–1517. https://doi.org/10.1073/pnas.90.4.1513

Eid, H. M., Ali, A. A., Ali, A. M. A., Eissa, E. M., Hassan, R. M., El-Ela, F. I. A., & Hassan, A. H. (2022). Potential use of tailored citicoline chitosan-coated liposomes for effective wound healing in diabetic rat model. *International Journal of Nanomedicine*, *17*, 555–575. https://doi.org/10.2147/IJN.S342504

Esposito, E., Fantin, M., Marti, M., Drechsler, M., Paccamiccio, L., Mariani, P., Sivieri, E., Lain, F., Menegatti, E., Morari, M., & Cortesi, R. (2008). Solid lipid nanoparticles as delivery systems for bromocriptine. *Pharmaceutical Research*, *25*(7), 1521–1530. https://doi.org/10.1007/S11095-007-9514-Y/FIGURES/5

Esposito, E., Sguizzato, M., Drechsler, M., Mariani, P., Carducci, F., Nastruzzi, C., Valacchi, G., & Cortesi, R. (2019). Lipid nanostructures for antioxidant delivery: A comparative preformulation study. *Beilstein Journal of Nanotechnology*, *10*(1), 1789–1801. www.beilstein-journals.org/bjnano/articles/10/174

Faheim, S. H., Gardouh, A. R., Nouh, A. T., & Ghorab, M. M. (2018). Review article on nanoemulsions and nanostructured lipid carriers. *Records of Pharmaceutical and Biomedical Sciences*, *2*(2), 23–31. https://rpbs.journals.ekb.eg/article_23370_e0566d5dffccb87159178e93904c80a8.pdf

Fang, C. L., Al-Suwayeh, A. S., & Fang, J. Y. (2013). Nanostructured Lipid Carriers (NLCs) for drug delivery and targeting. *Recent Patents on Nanotechnology*, *7*, 41–55.

Filippi, M., Bar-Or, A., Piehl, F., Preziosa, P., Solari, A., Vukusic, S., & Rocca, M. A. (2018). Multiple sclerosis.

Nature Reviews Disease Primers, *4*(1), 1–27. https://doi.org/10.1038/s41572-018-0041-4

Florence, K., Manisha, L., Kumar, B. A., Ankur, K., Kumar, M. A., & Ambikanandan, M. (2011). Intranasal clobazam delivery in the treatment of status epilepticus. *Journal of Pharmaceutical Sciences*, *100*(2), 692–703. https://doi.org/10.1002/jps.22307

Fresta, M., Puglisi, G., Di Giacomo, C., & Russo, A. (2011). Liposomes as In-vivo carriers for citicoline: Effects on rat cerebral post-ischaemic reperfusion. *Journal of Pharmacy and Pharmacology*, *46*(12), 974–981. https://doi.org/10.1111/J.2042-7158.1994.TB03252.X

Friend, D. S., Papahadjopoulos, D., & Debs, R. J. (1996). Endocytosis and intracellular processing accompanying transfection mediated by cationic liposomes. *Biochimica et Biophysica Acta*, *1278*, 41–50.

Fundarò, A., Cavalli, R., Bargoni, A., Vighetto, D., Zara, G. P., & Gasco, M. R. (2000). Non-stealth and stealth solid lipid nanoparticles (SLN) carrying doxorubicin: Pharmacokinetics and tissue distribution after iv administration to rats. *Pharmacological Research*, *42*(4), 337–343. www.sciencedirect.com/science/article/abs/pii/S1043661800906959

Gadhave, D. G., & Kokare, C. R. (2019). Nanostructured lipid carriers engineered for intranasal delivery of teriflunomide in multiple sclerosis: Optimization and in vivo studies. *Drug Development and Industrial Pharmacy*, *45*(5), 839–851. https://doi.org/10.1080/03639045.2019.1576724

Garg, G., Saraf, S., & Saraf, S. (2007). Cubosomes: An overview. *Biological & Pharmaceutical Bulletin*, *30*(2), 350–353. https://doi.org/10.1248/BPB.30.350

Garg, J., Pathania, K., Sah, S. P., & Pawar, S. V. (2022). Nanostructured lipid carriers: A promising drug carrier for targeting brain tumours. *Future Journal of Pharmaceutical Sciences*, *8*(1), 25. https://link.springer.com/article/10.1186/s43094-022-00414-8

Gershkovich, P., Itin, C., Yacovan, A., Amselem, S., & Hoffman, A. (2009). Effect of abdominal surgery on the intestinal absorption of lipophilic drugs: Possible role of the lymphatic transport. *Translational Research*, *153*(6), 296–300. https://doi.org/10.1016/j.trsl.2009.02.008

Ghasemian, E., Vatanara, A., Navidi, N., & Rouini, M. R. (2017). Brain delivery of baclofen as a hydrophilic drug by nanolipid carriers: Characteristics and pharmacokinetics evaluation. *Journal of Drug Delivery Science and Technology*, *37*, 67. https://doi.org/10.1016/j.jddst.2016.06.012

Ghezzi, M., Pescina, S., Padula, C., Santi, P., Del Favero, E., Cantù, L., & Nicoli, S. (2021). Polymeric micelles in drug delivery: An insight of the techniques for their characterization and assessment in biorelevant conditions. *Journal of Controlled Release: Official Journal of the Controlled Release Society*, *332*, 312–336. https://doi.org/10.1016/J.JCONREL.2021.02.031

Giacoppo, S., Galuppo, M., Montaut, S., Iori, R., Rollin, P., Bramanti, P., & Mazzon, E. (2015). An overview on neuroprotective effects of isothiocyanates for the treatment of neurodegenerative diseases. *Fitoterapia*, *106*, 12–21. https://doi.org/10.1016/J.FITOTE.2015.08.001

Gómez-Jeria, J. S., & Ojeda-Vergara, M. (1997). Electrostatic medium effects and formal quantum structure–activity relationships in apomorphines interacting with D1 and D2 dopamine receptors. *International Journal of Quantum Chemistry*, *61*(6), 997–1002. https://doi.org/10.1002/(SICI)1097-461X(1997)61:6<997::AID-QUA12>3.0.CO;2-Z

Gong, J., Chen, M., Zheng, Y., Wang, S., & Wang, Y. (2012). Polymeric micelles drug delivery system in oncology. *Journal of Controlled Release: Official Journal of the Controlled Release Society*, *159*(3), 312–323. https://doi.org/10.1016/J.JCONREL.2011.12.012

Gu, Z., Da Silva, C. G., van der Maaden, K., Ossendorp, F., & Cruz, L. J. (2020). Liposome-based drug delivery systems in cancer immunotherapy. *Pharmaceutics*, *12*(11), 1–25. https://doi.org/10.3390/PHARMACEUTICS12111054

Guo, Y., & Bruno, R. S. (2015). Endogenous and exogenous mediators of quercetin bioavailability. *Journal of Nutritional Biochemistry*, *26*(3), 201–210. https://doi.org/10.1016/j.jnutbio.2014.10.008

Gupta, A., Eral, H. B., Hatton, T. A., & Doyle, P. S. (2016). Nanoemulsions: Formation, properties and applications. *Soft Matter*, *12*(11), 2826–2841. https://doi.org/10.1039/c5sm02958a

Haider, M., Abdin, S. M., Kamal, L., & Orive, G. (2020). Nanostructured lipid carriers for delivery of chemotherapeutics: A review. *Pharmaceutics*, *12*(3), 288. www.mdpi.com/1999-4923/12/3/288/pdf

Hashimoto, K., Yaoi, T., Koshiba, H., Yoshida, T., Maoka, T., Fujiwara, Y., Yamamoto, Y., & Mori, K. (1996). Photochemical isomerization of piperine, a pungent constituent in pepper. *Food Science and Technology International, Tokyo*, *2*, 24–29.

Heo, H. J., & Lee, C. Y. (2004). Protective effects of quercetin and vitamin C against oxidative stress-induced neurodegeneration. *Journal of Agricultural and Food Chemistry*, *52*(25), 7514–7517. https://doi.org/10.1021/jf049243r

Hofmann, A. F. (2009). Pharmacology of ursodeoxycholic acid, an enterohepatic drug. *Scandinavian Journal of Gastroenterology*, *29*(S204), 15. https://doi.org/10.3109/00365529409103618

Homayoun, H. (2018). Parkinson disease. *Annals of Internal Medicine*, *169*(5), ITC33–ITC47. https://doi.org/10.7326/AITC201809040

Hosseini, L., Mahmoudi, J., Pashazadeh, F., Salehi-Pourmehr, H., & Sadigh-Eteghad, S. (2021). Protective effects of nicotinamide adenine dinucleotide and related precursors in Alzheimer's disease: A systematic review of preclinical studies. *Journal of Molecular Neuroscience*, *71*(7), 1425–1435. https://doi.org/10.1007/s12031-021-01842-6

Howells, D. W., Porritt, M. J., Wong, J. Y., Batchelor, P. E., Kalnins, R., Hughes, A. J., & Donnan, G. A. (2000). Reduced BDNF mRNA expression in the Parkinson's disease substantia nigra. *Experimental Neurology*, *166*(1), 127–135. https://doi.org/10.1006/exnr.2000.7483

Hu, Y. Y., Xu, J., Zhang, M., Wang, D., Li, L., & Li, W. B. (2015). Ceftriaxone modulates uptake activity of glial glutamate transporter-1 against global brain ischemia in rats. *Journal of Neurochemistry*, *132*(2), 194–205. https://doi.org/10.1111/JNC.12958

Hua, H., Zhang, X., Mu, H., Meng, Q., Jiang, Y., Wang, Y., Lu, X., Wang, A., Liu, S., Zhang, Y., Wan, Z., & Sun, K. (2018). RVG29-modified docetaxel-loaded nanoparticles for brain-targeted glioma therapy. *International Journal of Pharmaceutics*, *543*(1–2), 179–189. https://doi.org/10.1016/j.ijpharm.2018.03.028

Irby, D., Du, C., & Li, F. (2017). Lipid–drug conjugate for enhancing drug delivery. *Molecular Pharmaceutics*, *14*(5), 1325–1338. www.ncbi.nlm.nih.gov/pmc/articles/PMC5477224/

Ismail, M. F., Elmeshad, A. N., & Salem, N. A. (2013). Potential therapeutic effect of nanobased formulation of rivastigmine on rat model of Alzheimer's disease. *International Journal of Nanomedicine*, *8*, 393–406. https://doi.org/10.2147/IJN.S39232

Itin, C., Komargodski, R., Domb, A. J., & Hoffman, A. (2020). Controlled delivery of apomorphine through buccal mucosa, towards a noninvasive administration method in Parkinson's disease: A preclinical mechanistic study. *Journal of Pharmaceutical Sciences*, *109*(9), 2729–2734. https://doi.org/10.1016/j.xphs.2020.05.017

Jaafar-Maalej, C., Charcosset, C., & Fessi, H. (2011). A new method for liposome preparation using a membrane contactor. *Journal of Liposome Research*, *21*(3), 213–220. https://doi.org/10.3109/08982104.2010.517537

Jana, S., Mandlekar, S., & Marathe, P. (2010). Prodrug design to improve pharmacokinetic and drug delivery properties: Challenges to the discovery scientists. *Current Medicinal Chemistry*, *17*(32), 3874–3908. www.ingentaconnect.com/content/ben/cmc/2010/00000017/00000032/art00006

Jangde, R., & Singh, D. (2016). Preparation and optimization of quercetin-loaded liposomes for wound healing, using response surface methodology. *Artificial Cells, Nanomedicine and Biotechnology*, *44*(2), 635–641. https://doi.org/10.3109/21691401.2014.975238

Jefferies, W. A., Brandon, M. R., Hunt, S. V, Williams, A. F., Gatter, K. C., & Mason, D. Y. (1984). Transferrin receptor on endothelium of brain capillaries. *Nature*, *312*(5990), 162–163. https://doi.org/10.1038/312162a0

Jha, S., Sharma, P. K., & Malviya, R. (2016). Liposomal drug delivery system for cancer therapy: Advancement and patents. *Recent Patents on Drug Delivery & Formulation*, *10*(3), 177–183. https://doi.org/10.2174/1872211310666161004155757

Jiang, Y., Wei, N., Lu, T., Zhu, J., Xu, G., & Liu, X. (2011). Intranasal brain-derived neurotrophic factor protects brain from ischemic insult via modulating local inflammation in rats. *Neuroscience*, *172*, 398–405. https://doi.org/10.1016/j.neuroscience.2010.10.054

Jiang, Y., Zhu, J., Xu, G., & Liu, X. (2011). Intranasal delivery of stem cells to the brain. *Expert Opinion on Drug Delivery*, *8*(5), 623–632. https://doi.org/10.1517/17425247.2011.566267

Jimenez-Sanchez, M., Licitra, F., Underwood, B. R., & Rubinsztein, D. C. (2017). Huntington's disease: Mechanisms of pathogenesis and therapeutic strategies. *Cold Spring Harbor Perspectives in Medicine*, *7*(7), a024240. https://doi.org/10.1101/CSHPERSPECT.A024240

Joshi, R., Kumar, M. S., Satyamoorthy, K., Unnikrisnan, M. K., & Mukherjee, T. (2005). Free radical reactions and antioxidant activities of sesamol: Pulse radiolytic and biochemical studies. *Journal of Agricultural and Food Chemistry*, *53*(7), 2696–2703. https://doi.org/10.1021/jf0489769

Kakkar, V., & Kaur, I. P. (2011). Evaluating potential of curcumin loaded solid lipid nanoparticles in aluminium induced behavioural, biochemical and histopathological alterations in mice brain. *Food and Chemical Toxicology*, *49*(11), 2906–2913. https://doi.org/10.1016/j.fct.2011.08.006

Kalia, L. V., & Lang, A. E. (2015). Parkinson's disease. *Lancet*, *386*(9996), 896–912. https://doi.org/10.1016/S0140-6736(14)61393-3

Kammona, O., & Kiparissides, C. (2012). Recent advances in nanocarrier-based mucosal delivery of biomolecules. *Journal of Controlled Release*, *161*(3), 781–794. https://doi.org/10.1016/j.jconrel.2012.05.040

Karami, Z., & Hamidi, M. (2016). Cubosomes: Remarkable drug delivery potential. *Drug Discovery Today*, *21*(5), 789–801. https://doi.org/10.1016/J.DRUDIS.2016.01.004

Karanth, H., & Murthy, R. S. R. (2010). pH-sensitive liposomes-principle and application in cancer therapy. *Journal of Pharmacy and Pharmacology*, *59*(4), 469–483. https://doi.org/10.1211/jpp.59.4.0001

Kelly, C., Jefferies, C., & Cryan, S.-A. (2011). Targeted liposomal drug delivery to monocytes and macrophages. *Journal of Drug Delivery*, *2011*, 1–11. https://doi.org/10.1155/2011/727241

Kelsey, N. A., Wilkins, H. M., & Linseman, D. A. (2010). Nutraceutical antioxidants as novel neuroprotective agents. *Molecules*, *15*(11), 7792–7814. https://doi.org/10.3390/MOLECULES15117792

Khawli, L. A., & Prabhu, S. (2013). Drug delivery across the blood–brain barrier. *Molecular Pharmaceutics*, *10*(5), 1471–1472. https://doi.org/10.1021/MP400170B

Kono, K., Nakai, R., Morimoto, K., & Takagishi, T. (n.d.). *Temperature-dependent interaction of thermo-sensitive polymer-modi¢ed liposomes with CV1 cells. FEBS Letters*, *456*(2), 306–310. https://doi.org/10.1016/S0014-5793(99)00975-8

Kovacsa, Z. I., Kima, S., Jikariaa, N., Qureshia, F., Miloa, B., Lewisa, B. K., Breslera, M., Burksa, S. R., & Franka, J. A. (2016). Disrupting the blood–brain barrier by focused ultrasound induces sterile inflammation. *Proceedings of the National Academy of Sciences*, *114*(1), E75–E84. https://doi.org/10.1073/PNAS.1614777114

Krawczyk, J. (2023). Thermodynamic studies of the micellar properties of a surfactant used for membrane protein solubilization and stabilization. *Sustainability (Switzerland)*, *15*(8), 6618. https://doi.org/10.3390/SU15086618/S1

Kristensson, K., & Olsson, Y. (1971). Uptake of exogenous proteins in mouse olfactory cells. *Acta Neuropathologica*, *19*(2), 145–154. https://doi.org/10.1007/BF00688493

Kroll, R. A., & Neuwelt, E. A. (1998). Outwitting the blood-brain barrier for therapeutic purposes: Osmotic opening and other means. *Neurosurgery*, *42*(5), 1083–1100. https://doi.org/10.1097/00006123-199805000-00082

Kumar, P., Sharma, G., Kumar, R., Malik, R., Singh, B., Katare, O. P., & Raza, K. (2017). Enhanced brain delivery of dimethyl fumarate employing tocopherol-acetate-based nanolipidic carriers: Evidence from pharmacokinetic, biodistribution, and cellular uptake studies. *ACS Chemical Neuroscience*, *8*(4), 860–865. https://doi.org/10.1021/ACSCHEMNEURO.6B00428/SUPPL_FILE/CN6B00428_SI_001.PDF

Kumar, P., Wu, H., McBride, J. L., Jung, K. E., Kim, M. H., Davidson, B. L., Lee, S. K., Shankar, P., & Manjunath, N. (2007). Transvascular delivery of small interfering RNA to the central nervous system. *Nature*, *448*(7149), 39–43. https://doi.org/10.1038/nature05901

Kumar, S., Bhanjana, G., Kumar, A., Taneja, K., Dilbaghi, N., & Kim, K. H. (2016). Synthesis and optimization of ceftriaxone-loaded solid lipid nanocarriers. *Chemistry and Physics of Lipids*, *200*, 126–132. https://doi.org/10.1016/J.CHEMPHYSLIP.2016.09.002

Kwon, T. K., & Kim, J. C. (2010). Preparation and In Vitro skin permeation of cubosomes containing hinokitiol. *Journal of Dispersion Science and Technology*, *31*(7), 1004–1009. https://doi.org/10.1080/01932690903224862

Lafon, M. (2005). Rabies virus receptors. *Journal of NeuroVirology*, *11*(1), 82–87. https://doi.org/10.1080/13550280590900427

Lambert, D. M. (2000). Rationale and applications of lipids as prodrug carriers. *European Journal of Pharmaceutical Sciences*, *11*, S15–S27. https://citeseerx.ist.psu.edu/document?repid=rep1&type=pdf&doi=bc0fed3d22d98954651de32063b66853df568727

Lamptey, R. N. L., Chaulagain, B., Trivedi, R., Gothwal, A., Layek, B., & Singh, J. (2022). A review of the common neurodegenerative disorders: Current therapeutic approaches and the potential role of nanotherapeutics. *International Journal of Molecular Sciences*, *23*(3). https://doi.org/10.3390/IJMS23031851

Lapczynski, A., Bhatia, S. P., Foxenberg, R. J., Letizia, C. S., & Api, A. M. (2008). Fragrance material review on geraniol. *Food and Chemical Toxicology*, *46*(11), S160–S170. https://doi.org/10.1016/J.FCT.2008.06.048

Larivire, B., Soda, M. E., Soucyy, Y., Tr6panier, G., Paquin, P., & Vuillemard, J. C. (1991). Microfluidized liposomes for the acceleration of cheese ripening. *International Journal of Food Science & Technology*, *1*, 111–124.

Laserra, S., Basit, A., Sozio, P., Marinelli, L., Fornasari, E., Cacciatore, I., Ciulla, M., Turkez, H., Geyikoglu, F., & Di Stefano, A. (2015). Solid lipid nanoparticles loaded with lipoyl-memantine codrug: Preparation and characterization. *International Journal of Pharmaceutics*, *485*(1–2), 183–191. https://doi.org/10.1016/j.ijpharm.2015.03.001

Lee, D., & Minko, T. (2021). Nanotherapeutics for nose-to-brain drug delivery: An approach to bypass the blood brain barrier. *Pharmaceutics*, *13*(12). https://doi.org/10.3390/pharmaceutics13122049

Lee, M. K. (2020). Liposomes for enhanced bioavailability of water-insoluble drugs: In vivo evidence and recent approaches. *Pharmaceutics*, *12*(3). https://doi.org/10.3390/pharmaceutics12030264

Li, A., Goodfriend, C., Sokol, J., & Kruse-Jarres, R. (2019). Patterns and predictors of emicizumab adherence in people with hemophilia. *Blood*, *134*(Supplement_1), 2178–2178. https://doi.org/10.1182/BLOOD-2019-128083

Li, D., Rui, Y. X., Guo, S. D., Luan, F., Liu, R., & Zeng, N. (2021). Ferulic acid: A review of its pharmacology, pharmacokinetics and derivatives. *Life Sciences*, *284*, 119921. https://doi.org/10.1016/j.lfs.2021.119921

Li, H., Yang, P., Knight, W., Guo, Y., Perlmutter, J. S., Benzinger, T. L. S., Morris, J. C., & Xu, J. (2020). The interactions of dopamine and oxidative damage in the striatum of patients with neurodegenerative diseases. *Journal of Neurochemistry*, *152*(2), 235–251. https://doi.org/10.1111/jnc.14898

Liu, Y., Huang, R., Han, L., Ke, W., Shao, K., Ye, L., Lou, J., & Jiang, C. (2009). Brain-targeting gene delivery and cellular internalization mechanisms for modified rabies virus glycoprotein RVG29 nanoparticles. *Biomaterials*, *30*(25), 4195–4202. https://doi.org/10.1016/j.biomaterials.2009.02.051

Loureiro, J. A., Andrade, S., Duarte, A., Neves, A. R., Queiroz, J. F., Nunes, C., Sevin, E., Fenart, L., Gosselet, F., Coelho, M. A. N., Pereira, M. C., & Latruffe, N. (2017). Resveratrol and grape extract-loaded solid lipid nanoparticles for the treatment of Alzheimer's disease. *Molecules*, *22*(2), 277. https://doi.org/10.3390/MOLECULES22020277

Maherani, B., Arab-Tehrany, E., Mozafari, M. R., Gaiani, C., & Linder, M. (2011). Liposomes: A review of manufacturing techniques and targeting strategies. *Current Nanoscience*, *7*, 436–452.

Malekpour-Galogahi, F., Hatamian-Zarmi, A., Ganji, F., Ebrahimi-Hosseinzadeh, B., Nojoki, F., Sahraeian, R., & Mokhtari-Hosseini, Z. B. (2018). Preparation and optimization of rivastigmine-loaded tocopherol succinate-based solid lipid nanoparticles. *Journal of Liposome Research*, *28*(3), 226–235. https://doi.org/10.1080/08982104.2017.1349143

Manson, A. J., Hanagasi, H., Turner, K., Patsalos, P. N., Carey, P., Ratnaraj, N., & Lees, A. J. (2001). Intravenous apomorphine therapy in Parkinson's disease: Clinical and pharmacokinetic observations. *Brain*, *124*(Pt 2), 331–340. https://doi.org/10.1093/brain/124.2.331

Margari, A., Monteduro, A. G., Rizzato, S., Capobianco, L., Crestini, A., Rivabene, R., Piscopo, P., D'Onofrio, M., Manzini, V., Trapani, G., Quarta, A., Maruccio, G., Ventra, C., Lieto, L., & Trapani, A. (2022). The encapsulation of citicoline within solid lipid nanoparticles enhances its capability to counteract the 6-hydroxydopamine-induced cytotoxicity in human neuroblastoma SH-SY5Y cells. *Pharmaceutics*, *14*(9), 1827. https://doi.org/10.3390/PHARMACEUTICS14091827/S1

Markus, M. A., & Morris, B. J. (2008). Resveratrol in prevention and treatment of common clinical conditions of aging. *Clinical Interventions in Aging*, *3*(2), 331–339. www.ncbi.nlm.nih.gov/pubmed/18686754

Marshall, F. J. (2006). Tetrabenazine as antichorea therapy in Huntington disease. *Neurology*, *66*(3), 366–372. https://doi.org/10.1212/01.WNL.0000198586.85250.13

Martinez, J. A., Francis, G. J., Liu, W. Q., Pradzinsky, N., Fine, J., Wilson, M., Hanson, L. R., Frey, W. H., Zochodne, D., Gordon, T., & Toth, C. (2008). Intranasal delivery of insulin and a nitric oxide synthase inhibitor in an experimental model of amyotrophic lateral sclerosis. *Neuroscience*, *157*(4), 908–925. https://doi.org/10.1016/j.neuroscience.2008.08.073

McClements, D. J. (2012). Nanoemulsions versus microemulsions: Terminology, differences, and similarities. *Soft Matter*, *8*(6), 1719–1729. https://doi.org/10.1039/c2sm06903b

Mehandole, A., Walke, N., Mahajan, S., Aalhate, M., Maji, I., Gupta, U., Mehra, N. K., & Singh, P. K. (2023). Core-shell type lipidic and polymeric nanocapsules: The transformative multifaceted delivery systems. *AAPS PharmSciTech*, *24*(1). https://doi.org/10.1208/S12249-023-02504-Z

Meng, R., Wu, Z., Xie, Q. T., Cheng, J. S., & Zhang, B. (2021). Preparation and characterization of zein/carboxymethyl dextrin nanoparticles to encapsulate curcumin: Physicochemical stability, antioxidant activity and controlled release properties. *Food Chemistry*, *340*, 127893. https://doi.org/10.1016/J.FOODCHEM.2020.127893

Milan, A. S., Calpena Campmany, A. C., & Naveros, B. C. (2017). Antioxidant nanoplatforms for dermal delivery: Melatonin. *Current Drug Metabolism*, *18*(5), 437–453. https://doi.org/10.2174/1389200218666170222145908

Millan, M. J. (2010). From the cell to the clinic: A comparative review of the partial D(2)/D(3)receptor agonist and alpha2-adrenoceptor antagonist, piribedil, in the treatment of Parkinson's disease. *Pharmacology & Therapeutics*, *128*(2), 229–273. https://doi.org/10.1016/j.pharmthera.2010.06.002

Mills, E. A., Ogrodnik, M. A., Plave, A., & Mao-Draayer, Y. (2018). Emerging understanding of the mechanism of action for dimethyl fumarate in the treatment of multiple sclerosis. *Frontiers in Neurology*, *9*, 5. https://doi.org/10.3389/FNEUR.2018.00005/BIBTEX

Misra, S., Chopra, K., Sinha, V. R., & Medhi, B. (2016). Galantamine-loaded solid-lipid nanoparticles for enhanced brain delivery: Preparation, characterization, in vitro and in vivo evaluations. *Drug Delivery*, *23*(4), 1434–1443. https://doi.org/10.3109/10717544.2015.1089956

Mittal, D., Ali, A., Md, S., Baboota, S., Sahni, J. K., & Ali, J. (2014). Insights into direct nose to brain delivery: Current status and future perspective. *Drug Delivery*, *21*(2), 75–86. https://doi.org/10.3109/10717544.2013.838713

Momekova, D. B., Gugleva, V. E., & Petrov, P. D. (2021). Nanoarchitectonics of multifunctional niosomes for advanced drug delivery. *ACS Omega*, *6*(49), 33265–33273. https://doi.org/10.1021/acsomega.1c05083

Mortiboys, H., Aasly, J., & Bandmann, O. (2013). Ursocholanic acid rescues mitochondrial function in common forms of familial Parkinson's disease. *Brain*, *136*(10), 3038–3050. https://doi.org/10.1093/BRAIN/AWT224

Mortiboys, H., Furmston, R., Bronstad, G., Aasly, J., Elliott, C., & Bandmann, O. (2015). UDCA exerts beneficial effect on mitochondrial dysfunction in LRRK2G2019S carriers and in vivo. *Neurology*, *85*(10), 846–852. https://doi.org/10.1212/WNL.0000000000001905

Moura, R. P., Pacheco, C., Pêgo, A. P., des Rieux, A., & Sarmento, B. (2020). Lipid nanocapsules to enhance drug bioavailability to the central nervous system. *Journal of Controlled Release: Official Journal of the Controlled Release Society*, *322*, 390–400. https://doi.org/10.1016/J.JCONREL.2020.03.042

Muchow, M., Maincent, P., & Müller, R. H. (2008). Lipid nanoparticles with a solid matrix (SLN®, NLC®, LDC®) for oral drug delivery. *Drug Development and Industrial Pharmacy*, *34*(12), 1394–1405. www.tandfonline.com/doi/abs/10.1080/03639040802130061

Mukherjee, S., Ray, S., & Thakur, R. S. (2009). Solid lipid nanoparticles: A modern formulation approach in drug delivery system. *Indian Journal of Pharmaceutical Sciences*, *71*(4), 349. www.ncbi.nlm.nih.gov/pmc/articles/PMC2865805/

Müller, R. H., Mäder, K., & Gohla, S. (2000). Solid lipid nanoparticles (SLN) for controlled drug delivery–a review of the state of the art. *European Journal of Pharmaceutics and Biopharmaceutics*, *50*(1), 161–177. http://kinampark.com/PL/files/Muller 2000%2C Solid lipid nanoparticles %28SLN%29 for controlled drug delivery.pdf

Müller, R. H., & Olbrich, C. (2004). *Lipid matrix-drug conjugates particle for controlled release of active ingredient*. https://patentimages.storage.googleapis.com/cc/c9/4b/19e3eb4f4d019a/US6770299.pdf

Müller, R. H., Radtke, M., & Wissing, S. A. (2002). Solid lipid nanoparticles (SLN) and nanostructured lipid carriers (NLC) in cosmetic and dermatological preparations. *Advanced Drug Delivery Reviews*, *54*, S131–S155. https://doi.org/10.1016/S0169-409X(02)00118-7

Mutlu, N. B., Degim, Z., Yilmaz, S., Essiz, D., & Nacar, A. (2011). New perspective for the treatment of Alzheimer diseases: Liposomal rivastigmine formulations. *Drug Development and Industrial Pharmacy*, *37*(7), 775–789. https://doi.org/10.3109/03639045.2010.541262

Muzzalupo, R., & Tavano, L. (2015). Niosomal drug delivery for transdermal targeting: Recent advances. *Research and Reports in Transdermal Drug Delivery*, *23*. https://doi.org/10.2147/rrtd.s64773

Nair, A. B., Singh, B., Shah, J., Jacob, S., Aldhubiab, B., Sreeharsha, N., Morsy, M. A., Venugopala, K. N., Attimarad, M., & Shinu, P. (2022). Formulation and evaluation of self-nanoemulsifying drug delivery system derived tablet containing sertraline. *Pharmaceutics*, *14*(2), 336. https://doi.org/10.3390/PHARMACEUTICS14020336

Naqvi, S., Panghal, A., & Flora, S. J. S. (2020). Nanotechnology: A promising approach for delivery of neuroprotective drugs. *Frontiers in Neuroscience*, *14*, 494. https://doi.org/10.3389/FNINS.2020.00494/BIBTEX

Navia, B., Factor, S., Pahwa, R., Hauser, R., Worden, M., Bhargava, P., Vakili, G., & Blum, D. (2018). Efficacy and safety of sublingual apomorphine film (APL-130277) for the treatment of off episodes in patients with Parkinson's disease: Results from a double-blind, placebo-controlled trial, 3. *MDS Abstracts*. Retrieved May 3, 2024, from https://www.mdsabstracts.org/abstract/safety-of-sublingual-apomorphine-film-apl-130277-for-the-treatment-of-off-episodes-in-patients-with-parkinsons-disease-results-from-a-phase-3-double-blind-placebo-controlled-trial/

Negrini, R., Fong, W. K., Boyd, B. J., & Mezzenga, R. (2015). pH-responsive lyotropic liquid crystals and their potential therapeutic role in cancer treatment. *Chemical Communications*, *51*(30), 6671–6674. https://doi.org/10.1039/C4CC10274F

Negrini, R., Sánchez-Ferrer, A., & Mezzenga, R. (2014). Influence of electrostatic interactions on the release of charged molecules from lipid cubic phases. *Langmuir*, *30*(15), 4280–4288. https://doi.org/10.1021/LA5008439/SUPPL_FILE/LA5008439_SI_001.PDF

Neupane, Y. R., Sabir, M. D., Ahmad, N., Ali, M., & Kohli, K. (2013). Lipid drug conjugate nanoparticle as a novel lipid nanocarrier for the oral delivery of decitabine: Ex vivo gut permeation studies. *Nanotechnology*, *24*(41), 415102. www.academia.edu/download/32220392/paper.pdf

Neves, A. R., Lucio, M., Lima, J. L., & Reis, S. (2012). Resveratrol in medicinal chemistry: A critical review of its pharmacokinetics, drug-delivery, and membrane interactions. *Current Medicinal Chemistry*, *19*(11), 1663–1681. https://doi.org/10.2174/092986712799945085

Nguyen, T. T., Nguyen, T. T. D., Tran, N. M. A., & Van Vo, G. (2021). Lipid-based nanocarriers via nose-to-brain pathway for central nervous system disorders. *Neurochemical Research*, *47*(3), 552–573. https://doi.org/10.1007/S11064-021-03488-7

Nielsen, H. M., & Rassing, M. R. (2000). TR146 cells grown on filters as a model of human buccal epithelium: IV. Permeability of water, mannitol, testosterone and beta-adrenoceptor antagonists. Comparison to human, monkey and porcine buccal mucosa. *International Journal of Pharmaceutics*, *194*(2), 155–167. https://doi.org/10.1016/s0378-5173(99)00368-3

Noguchi, C. T., Asavaritikrai, P., Teng, R., & Jia, Y. (2007). Role of erythropoietin in the brain. *Critical Reviews in Oncology/Hematology*, *64*(2), 159–171. https://doi.org/10.1016/j.critrevonc.2007.03.001

Nsairat, H., Khater, D., Sayed, U., Odeh, F., Al Bawab, A., & Alshaer, W. (2022). Liposomes: Structure, composition, types, and clinical applications. *Heliyon*, *8*(5), e09394. https://doi.org/10.1016/J.HELIYON.2022.E09394

Obeso, J. A., Grandas, F., Herrero, M. T., & Horowski, R. (1994). Review. *European Journal of Neuroscience*, *6*(6), 889–897. https://doi.org/10.1111/J.1460-9568.1994.TB00584.X

Obeso, J. A., Olanow, W., Nutt, J. G., & Givup, G. (2000). Introduction levodopa motor complications in Parkinson's disease. *Trends in Neurosciences*, 2–7.

Orsini, M., Oliveira, A. B., Nascimento, O. J. M., Reis, C. H. M., Leite, M. A. A., de Souza, J. A., Pupe, C., de Souza, O. G., Bastos, V. H., de Freitas, M. R. G., Teixeira, S., Bruno, C., Davidovich, E., & Smidt, B. (2015). Amyotrophic lateral sclerosis: New perpectives and update. *Neurology International*, *7*(2), 39–47). https://doi.org/10.4081/ni.2015.5885

Oswald, M., Geissler, S., & Goepferich, A. (2017). Targeting the central nervous system (CNS): A review of rabies virus-targeting strategies. *Molecular Pharmaceutics*, *14*(7), 2177–2196. https://doi.org/10.1021/acs.molpharmaceut.7b00158

Ozsoy, Y., Gungor, S., & Cevher, E. (2009). Nasal delivery of high molecular weight drugs. *Molecules*, *14*(9), 3754–3779. https://doi.org/10.3390/molecules14093754

Pahwa, R., Lyons, K. E., & Hauser, R. A. (2004). Ropinirole therapy for Parkinson's disease. *Expert Review of Neurotherapeutics*, *4*(4), 581–588. https://doi.org/10.1586/14737175.4.4.581

Paliwal, R., Paliwal, S. R., Kenwat, R., Kurmi, B. Das, & Sahu, M. K. (2020). Solid lipid nanoparticles: A review on recent perspectives and patents. *Expert Opinion on Therapeutic Patents*, *30*(3), 179–194. www.tandfonline.com/doi/abs/10.1080/13543776.2020.1720649

Pan, W., Banks, W. A., Fasold, M. B., Bluth, J., & Kastin, A. J. (1998). Transport of brain-derived neurotrophic factor across the blood–brain barrier. *Neuropharmacology*, *37*(12), 1553–1561.

Pandita, A., & Sharma, P. (2013). Pharmacosomes: An emerging novel vesicular drug delivery system for poorly soluble synthetic and herbal drugs. *ISRN Pharmaceutics*, *2013*, 1–10. https://doi.org/10.1155/2013/348186

Parain, K., Murer, M. G., Yan, Q., Faucheux, B., Agid, Y., Hirsch, E., & Raisman-Vozari, R. (1999). Reduced expression of brain-derived neurotrophic factor protein in Parkinson's disease substantia nigra. *Neuroreport*, *10*(3), 557–561. https://doi.org/10.1097/00001756-199902250-00021

Pardeshi, C. V., Rajput, P. V., Belgamwar, V. S., Tekade, A. R., & Surana, S. J. (2013). Novel surface modified solid lipid nanoparticles as intranasal carriers for ropinirole hydrochloride: Application of factorial design approach. *Drug Delivery*, *20*(1), 47–56. https://doi.org/10.3109/10717544.2012.752421

Pardridge, W. M. (2005). The blood-brain barrier: Bottleneck in brain drug development. *NeuroRx: The Journal of the American Society for Experimental NeuroTherapeutics*, *2*(1), 3–14. https://doi.org/10.1602/NEURORX.2.1.3

Pardridge, W. M. (2012). Drug transport across the blood-brain barrier. *Journal of Cerebral Blood Flow and Metabolism: Official Journal of the International Society of Cerebral Blood Flow and Metabolism*, *32*(11), 1959–1972. https://doi.org/10.1038/JCBFM.2012.126

Parihar, V. K., Prabhakar, K. R., Veerapur, V. P., Kumar, M. S., Reddy, Y. R., Joshi, R., Unnikrishnan, M. K., & Rao, C. M. (2006). Effect of sesamol on radiation-induced cytotoxicity in Swiss albino mice. *Mutation Research*, *611*(1–2), 9–16. https://doi.org/10.1016/j.mrgentox.2006.06.037

Parry, G. J., Rodrigues, C. M. P., Aranha, M. M., Hilbert, S. J., Davey, C., Kelkar, P., Low, W. C., & Steer, C. J. (2010). Safety, tolerability, and cerebrospinal fluid penetration of ursodeoxycholic acid in patients with amyotrophic lateral sclerosis. *Clinical Neuropharmacology*, *33*(1), 17–21. https://doi.org/10.1097/WNF.0B013E3181C47569

Parveen, N., Sheikh, A., Abourehab, M. A. S., Karwasra, R., Singh, S., & Kesharwani, P. (2023). Self-nanoemulsifying drug delivery system for pancreatic cancer. *European Polymer Journal*, 111993. https://doi.org/10.1016/J.EURPOLYMJ.2023.111993

Pathak, D., & Manchanda, S. (2020). Pharmacosomes: A novel vesicular drug delivery system. *IJISET-International Journal of Innovative Science, Engineering & Technology*, *7*. www.ijiset.com

Patil, Y. P., & Jadhav, S. (2014). Novel methods for liposome preparation. *Chemistry and Physics of Lipids*, *177*, 8–18). https://doi.org/10.1016/j.chemphyslip.2013.10.011

Patnaik, N. (2015). Cure for Alzheimer's disease. *World Journal of Neuroscience*, *5*(5), 328.

Pavan, B., Dalpiaz, A., Marani, L., Beggiato, S., Ferraro, L., Canistro, D., Paolini, M., Vivarelli, F., Valerii, M. C., Comparone, A., De Fazio, L. D., & Spisni, E. (2018). Geraniol pharmacokinetics, bioavailability and its multiple effects on the liver antioxidant and xenobiotic-metabolizing enzymes. *Frontiers in Pharmacology*, *9*, 332229. https://doi.org/10.3389/FPHAR.2018.00018/BIBTEX

Pearce, R. K. B., Banerji, T., Jenner, P., & Marsden, C. D. (1998). De novo administration of ropinirole and bromocriptine induces less dyskinesia than L-dopa in the MPTP-treated marmoset. *Movement Disorders*, *13*(2), 234–241. https://doi.org/10.1002/MDS.870130207

Peng, Y., Bariwal, J., Kumar, V., Tan, C., & Mahato, R. I. (2020). Organic nanocarriers for delivery and targeting of therapeutic agents for cancer treatment. *Advanced Therapeutics*, *3*(2). https://doi.org/10.1002/adtp.201900136

Peschka, R., Dennehy, C., & Szoka, F. C. (1998). A simple in vitro model to study the release kinetics of liposome encapsulated material. *Journal of Controlled Release*, *56*, 41–51.

Pharmaceutical Dosage Forms. (2020). *Pharmaceutical dosage forms*. https://doi.org/10.1201/9781003067368/PHARMACEUTICAL-DOSAGE-FORMS-HERBERT-LIEBERMAN-GILBERT-BANKER-MARTIN-RIEGER

Picone, P., Bondi, M. L., Montana, G., Bruno, A., Pitarresi, G., Giammona, G., & Di Carlo, M. (2009). Ferulic acid inhibits oxidative stress and cell death induced by Ab oligomers: Improved delivery by solid lipid nanoparticles. *Free Radical Research*, *43*(11), 1133–1145. https://doi.org/10.1080/10715760903214454

Pierre, M. B. R., & Dos Santos Miranda Costa, I. (2011). Liposomal systems as drug delivery vehicles for dermal and transdermal applications. *Archives of Dermatological Research*, *303*(9), 607–621. https://doi.org/10.1007/s00403-011-1166-4

Pinheiro, R. G. R., Granja, A., Loureiro, J. A., Pereira, M. C., Pinheiro, M., Neves, A. R., & Reis, S. (2020a). Quercetin lipid nanoparticles functionalized with transferrin for Alzheimer's disease. *European Journal of Pharmaceutical Sciences*, *148*, 105314. https://doi.org/10.1016/j.ejps.2020.105314

Pinheiro, R. G. R., Granja, A., Loureiro, J. A., Pereira, M. C., Pinheiro, M., Neves, A. R., & Reis, S. (2020b). RVG29-functionalized lipid nanoparticles for quercetin brain delivery and Alzheimer's disease. *Pharmaceutical Research*, *37*(7), 139. https://doi.org/10.1007/s11095-020-02865-1

Pires, A., Fortuna, A., Alves, G., & Falcão, A. (2009). Intranasal drug delivery: How, why and what for? *Journal of Pharmacy & Pharmaceutical Sciences: A Publication of the Canadian Society for Pharmaceutical Sciences, Societe Canadienne Des Sciences Pharmaceutiques*, *12*(3), 288–311. https://doi.org/10.18433/j3nc79

Poewe, W., Seppi, K., Tanner, C. M., Halliday, G. M., Brundin, P., Volkmann, J., Schrag, A. E., & Lang, A. E. (2017). Parkinson disease. *Nature Reviews Disease Primers*, *3*(1), 1–21. https://doi.org/10.1038/nrdp.2017.13

Pradhan, B., Kumar, N., Saha, S., & Roy, A. (n.d.). *Liposome: Method of preparation, advantages, evaluation and its application*. www.japtronline.com

Prasad, N. R., Mahesh, T., Menon, V. P., Jeevanram, R. K., & Pugalendi, K. V. (2005). Photoprotective effect of sesamol on UVB-radiation induced oxidative stress in human blood lymphocytes in vitro. *Environmental Toxicology and Pharmacology*, *20*(1), 1–5. https://doi.org/10.1016/j.etap.2004.09.009

Que, D. L. S., & Jamora, R. D. G. (2021). Citicoline as adjuvant therapy in Parkinson's disease: A systematic review. *Clinical Therapeutics*, *43*(1), e19–e31. https://doi.org/10.1016/j.clinthera.2020.11.009

Rajabi, M., & Mousa, S. A. (2016). Lipid nanoparticles and their application in nanomedicine. *Current Pharmaceutical Biotechnology*, *17*(8), 662–672. https://doi.org/10.2174/1389201017666160415155457

Ramachandran, S., & Thangarajan, S. (2018). Thymoquinone loaded solid lipid nanoparticles counteracts 3-nitropropionic acid induced motor impairments and neuroinflammation in rat model of Huntington's disease. *Metabolic Brain Disease*, *33*(5), 1459–1470. https://doi.org/10.1007/S11011-018-0252-0/FIGURES/7

Ramazani, E., Akaberi, M., Emami, S. A., & Tayarani-Najaran, Z. (2022). Pharmacological and biological effects of alpha-bisabolol: An updated review of the molecular mechanisms. *Life Sciences*, *304*, 120728. https://doi.org/10.1016/j.lfs.2022.120728

Rangasamy, S. B., Soderstrom, K., Bakay, R. A., & Kordower, J. H. (2010). Neurotrophic factor therapy for Parkinson's disease. *Progress in Brain Research*, *184*, 237–264. https://doi.org/10.1016/S0079-6123(10)84013-0

Rao, J., & McClements, D. J. (2011). Formation of flavor oil microemulsions, nanoemulsions and emulsions: Influence of composition and preparation method. *Journal of Agricultural and Food Chemistry*, *59*(9), 5026–5035. https://doi.org/10.1021/JF200094M/ASSET/IMAGES/MEDIUM/JF-2011-00094M_0001.GIF

Rassu, G., Soddu, E., Posadino, A. M., Pintus, G., Sarmento, B., Giunchedi, P., & Gavini, E. (2017). Nose-to-brain delivery of BACE1 siRNA loaded in solid lipid nanoparticles for Alzheimer's therapy. *Colloids and Surfaces B: Biointerfaces*, *152*, 296–301. https://doi.org/10.1016/j.colsurfb.2017.01.031

Rathbone, M. J., Senel, S., & Pather, I. (2015). *Oral mucosal drug delivery and therapy*. Springer.

Recent, A., Mateus, A., Rosa Dias, J., Marisa Fernandes Patrício, T., Mustafai, A., Zubair, M., Hussain, A., & Ullah, A. (2023). Recent progress in proteins-based micelles as drug delivery carriers. *Polymers*, *15*(4), 836. https://doi.org/10.3390/POLYM15040836

Reddy, J. S., & Venkateswarlu, V. (2004). Novel delivery systems for drug targeting to the brain. *Drugs Future*, *29*(1), 63–83. https://hero.epa.gov/hero/index.cfm/reference/details/reference_id/69406

Reger, M. A., Watson, G. S., Green, P. S., Wilkinson, C. W., Baker, L. D., Cholerton, B., Fishel, M. A., Plymate, S. R., Breitner, J. C. S., DeGroodt, W., Mehta, P., & Craft, S. (2008). Intranasal insulin improves cognition and modulates beta-amyloid in early AD. *Neurology*, *70*(6), 440–448. https://doi.org/10.1212/01.WNL.0000265401.62434.36

Rekha, K. R., & Selvakumar, G. P. (2014). Gene expression regulation of Bcl2, Bax and cytochrome-C by geraniol on chronic MPTP/probenecid induced C57BL/6 mice model of Parkinson's disease. *Chemico-Biological Interactions*, *217*, 57–66. https://doi.org/10.1016/J.CBI.2014.04.010

Rekha, K. R., Selvakumar, G. P., Santha, K., & Inmozhi Sivakamasundari, R. (2013). Geraniol attenuates α-synuclein expression and neuromuscular impairment through increase dopamine content in MPTP intoxicated mice by dose dependent manner. *Biochemical and Biophysical Research Communications*, *440*(4), 664–670. https://doi.org/10.1016/J.BBRC.2013.09.122

Rekha, K. R., Selvakumar, G. P., Sethupathy, S., Santha, K., & Sivakamasundari, R. I. (2013). Geraniol ameliorates the motor behavior and neurotrophic factors inadequacy in MPTP-induced mice model of Parkinson's disease. *Journal of Molecular Neuroscience*, *51*(3), 851–862. https://doi.org/10.1007/S12031-013-0074-9/FIGURES/7

ResearchGate. (n.d.). www.researchgate.net/search.Search.html?query=Pharmacosomes%3A+Opening+new+doors+for+drug+delivery&type=publication

Rizwan, S. B., Dong, Y. D., Boyd, B. J., Rades, T., & Hook, S. (2007). Characterisation of bicontinuous cubic liquid crystalline systems of phytantriol and water using cryo field emission scanning electron microscopy (cryo FESEM). *Micron*, *38*(5), 478–485. https://doi.org/10.1016/J.MICRON.2006.08.003

Rizzo, F., Riboldi, G., Salani, S., Nizzardo, M., Simone, C., Corti, S., & Hedlund, E. (2014). Cellular therapy to target neuroinflammation in amyotrophic lateral sclerosis. *Cellular and Molecular Life Sciences*, *71*(6), 999–1015. https://doi.org/10.1007/s00018-013-1480-4

Robinson, D. M., & Keating, G. M. (2006). Memantine: A review of its use in Alzheimer's disease. *Drugs*, *66*(11), 1515–1534. https://doi.org/10.2165/00003495-200666110-00015

Roda, A., Hrelia, P., Paolini, M., Calzolari, M., Grigolo, B., Simoni, P., Aldini, R., & Cantelli Forti, G. (1991). Pharmacokinetics of ursodeoxycholic acid in rat. *Pharmacological Research*, *23*(4), 327–335. https://doi.org/10.1016/1043-6618(91)90048-3

Roodenrys, S., Booth, D., Bulzomi, S., Phipps, A., Micallef, C., & Smoker, J. (2001). Chronic effects of brahmi (bacopa monnieri) on human memory. *Neuropsychopharmacology*, *27*(2), 279–281. https://doi.org/10.1016/s0893-133x(01)00419-5

Roos, R. A. C. (2010). Huntington's disease: A clinical review. *Orphanet Journal of Rare Diseases*, *5*(1), 1–8. https://doi.org/10.1186/1750-1172-5-40/TABLES/5

Sabjan, K. B., Munawar, S. M., Rajendiran, D., Vinoji, S. K., & Kasinathan, K. (2019). Nanoemulsion as oral drug delivery—a review. *Current Drug Research Reviews*, *12*(1), 4–15. https://doi.org/10.2174/2589977511666191024173508

Sachdeva, A. K., Misra, S., Pal Kaur, I., & Chopra, K. (2015). Neuroprotective potential of sesamol and its loaded solid lipid nanoparticles in ICV-STZ-induced cognitive deficits: Behavioral and biochemical evidence. *European Journal of Pharmacology*, *747*, 132–140. https://doi.org/10.1016/j.ejphar.2014.11.014

Sagara, Y., Vanhnasy, J., & Maher, P. (2004). Induction of PC12 cell differentiation by flavonoids is dependent upon extracellular signal-regulated kinase activation. *Journal of Neurochemistry*, *90*(5), 1144–1155. https://doi.org/10.1111/j.1471-4159.2004.02563.x

Salvi, V. R., & Pawar, P. (2019). Nanostructured lipid carriers (NLC) system: A novel drug targeting carrier. *Journal of Drug Delivery Science and Technology*, *51*, 255–267. www.sciencedirect.com/science/article/abs/pii/S1773224718314473

Satapathy, M. K., Yen, T.-L., Jan, J.-S., Tang, R.-D., Wang, J.-Y., Taliyan, R., & Yang, C.-H. (2021). Solid lipid nanoparticles (SLNs): An advanced drug delivery system targeting brain through BBB. *Pharmaceutics*, *13*(8), 1183. www.mdpi.com/1999-4923/13/8/1183/pdf

Sathya, S., Shanmuganathan, B., & Devi, K. P. (2020). Deciphering the anti-apoptotic potential of alpha-bisabolol loaded solid lipid nanoparticles against Abeta induced neurotoxicity in Neuro-2a cells. *Colloids and Surfaces B: Biointerfaces*, *190*, 110948. https://doi.org/10.1016/j.colsurfb.2020.110948

Savitt, D., & Jankovic, J. (2019). Targeting α-synuclein in Parkinson's disease: Progress towards the development of disease-modifying therapeutics. *Drugs*, *79*(8), 797–810. https://doi.org/10.1007/S40265-019-01104-1

Scheltens, P., Blennow, K., Breteler, M. M., de Strooper, B., Frisoni, G. B., Salloway, S., & Van der Flier, W. M. (2016). Alzheimer's disease. *Lancet*, *388*(10043), 505–517. https://doi.org/10.1016/S0140-6736(15)01124-1

Semalty, A., Semalty, M., Rawat, B. S., Singh, D., & Rawat, M. S. M. (2009). Pharmacosomes: The lipid-based new drug delivery system. *Expert Opinion on Drug Delivery*, *6*(6), 599–612. https://doi.org/10.1517/17425240902967607

Sharma, A., & Sharma, U. S. (1997). Review Liposomes in drug delivery: Progress and limitations. *International Journal of Pharmaceutics*, *154*, 140.

Sharma, P. H., Powar, P. V., & Sharma, S. S. (2013). Pharmacosomes: A novel drug delivery system. *The Pharma Innovation Journal*, *3*(10), 94–100. www.thepharmajournal.com

Shidhaye, S. S., Vaidya, R., Sutar, S., Patwardhan, A., & Kadam, V. J. (2008). Solid lipid nanoparticles and nanostructured lipid carriers—innovative generations of solid lipid carriers. *Current Drug Delivery*, *5*(4), 324–331. http://kinampark.com/PL/files/Shidhaye 2008%2C Solid lipid nanoparticles and nanostructured lipid carriers-innovative generations of solid lipid carriers.pdf

Shill, H. A., & Stacy, M. (2009). Update on ropinirole in the treatment of Parkinson's disease. *Neuropsychiatric Disease and Treatment*, *5*, 33–36. www.ncbi.nlm.nih.gov/pubmed/19557097

Shilpi, S., Dixit, S., Jain, A., Jain, S. K., Saraogi, G., & Yadav, A. K. (2022). Lipid nanocarrier-based drug delivery for the treatment of brain-related disorders. *Nanomedical Drug Delivery for Neurodegenerative Diseases*, 55–65. https://doi.org/10.1016/B978-0-323-85544-0.00014-9

Shingaki, T., Inoue, D., Furubayashi, T., Sakane, T., Katsumi, H., Yamamoto, A., & Yamashita, S. (2010). Transnasal delivery of methotrexate to brain tumors in rats: A new strategy for brain tumor chemotherapy. *Molecular Pharmaceutics*, *7*(5), 1561–1568. https://doi.org/10.1021/mp900275s

Siddhartha, V. T., Pindiprolu, S. K. S. S., Chintamaneni, P. K., Tummala, S., & Nandha Kumar, S. (2018). RAGE receptor targeted bioconjuguate lipid nanoparticles of diallyl disulfide for improved apoptotic activity in triple negative breast cancer: In vitro studies. *Artificial Cells, Nanomedicine, and Biotechnology*, *46*(2), 387–397. www.tandfonline.com/doi/full/10.1080/21691401.2017.1313267

Siddique, Y. H., Naz, F., Jyoti, S., Ali, F., Fatima, A., Rahul, & Khanam, S. (2016). Protective effect of geraniol on the transgenic drosophila model of Parkinson's disease. *Environmental Toxicology and Pharmacology*, *43*, 225–231. https://doi.org/10.1016/J.ETAP.2016.03.018

Solans, C., & Solé, I. (2012). Nano-emulsions: Formation by low-energy methods. *Current Opinion in Colloid & Interface Science*, *17*(5), 246–254. https://doi.org/10.1016/J.COCIS.2012.07.003

Sood, S., Jain, K., & Gowthamarajan, K. (2013). P1–382: Curcumin-donepezil–loaded nanostructured lipid carriers for intranasal delivery in an Alzheimer's disease model. *Alzheimer's & Dementia*, *9*, P299–P299.

Souto, E. B., & Müller, R. H. (2011). Solid lipid nanoparticles and nanostructured lipid carriers lipid nanoparticles for medicals and pharmaceuticals. In *Encyclopedia of Nanoscience and Nanotechnology* (Vol. 23, pp. 313–328). American Scientific Publishers SV.

Storm, G., & Crommelin, D. J. A. (1998). Liposomes: Quo vadis? *Pharmaceutical Science & Technology Today*, *1*(1), 19–31. https://doi.org/10.1016/S1461-5347(98)00007-8

Sudhakar, K., Fuloria, S., Subramaniyan, V., Sathasivam, K. V., Azad, A. K., Swain, S. S., Sekar, M., Karupiah, S., Porwal, O., Sahoo, A., Meenakshi, D. U., Sharma, V. K., Jain, S., Charyulu, R. N., & Fuloria, N. K. (2021). Ultraflexible liposome nanocargo as a dermal and transdermal drug delivery system. *Nanomaterials*, *11*(10). https://doi.org/10.3390/nano11102557

Sun, T., Zhang, Y. S., Pang, B., Hyun, D. C., Yang, M., & Xia, Y. (2014). Engineered nanoparticles for drug delivery in cancer therapy. *Angewandte Chemie—International Edition*, *53*(46), 12320–12364. https://doi.org/10.1002/anie.201403036

Sweeney, M. D., Zhao, Z., Montagne, A., Nelson, A. R., & Zlokovic, B. V. (2019). Blood-brain barrier: From physiology to disease and back. *Physiological Reviews*, *99*(1), 21–78. https://doi.org/10.1152/PHYSREV.00050.2017

Szoka, F. (1980). Comparative properties and methods of preparation of lipid vesicles (LIPOSOMES)l. *Annual Review of Biophysics and Bioengineering*, *9*, 467–508. www.annualreviews.org

Tadros, T. F. (2013). Emulsion formation, stability, and rheology. *Emulsion Formation and Stability*, 1–75. https://doi.org/10.1002/9783527647941.CH1

Tajes, M., Ramos-Fernández, E., Weng-Jiang, X., Bosch-Morató, M., Guivernau, B., Eraso-Pichot, A., Salvador, B., Fernàndez-Busquets, X., Roquer, J., & Muñoz, F. J. (2014). The blood-brain barrier: Structure, function and therapeutic approaches to cross it. *Molecular Membrane Biology*, *31*(5), 152–167. https://doi.org/10.3109/09687688.2014.937468

Talegaonkar, S., Azeem, A., Ahmad, F. J., Khar, R. K., Pathan, S. A., & Khan, Z. I. (2008). Microemulsions: A novel approach to enhanced drug delivery. *Recent Patents on Drug Delivery & Formulation*, *2*, 238–257.

Talegaonkar, S., Mishra, P. R., Khar, R. K., & Biju, S. S. (2006). Vesicular systems: An overview. *Indian Journal of Pharmaceutical Sciences*, *68*(2), 141. https://doi.org/10.4103/0250-474X.25707

Taskintuna, I., Banker, A. S., Floresaguilar, M., Bergeronlynn, G., Aldern, K. A., Hostetler, K. Y., & Freeman, W. R. (1997). Evaluation of a novel lipid prodrug for intraocular drug delivery. *Retina*, *17*(1), 57–64. https://doi.org/10.1097/00006982-199701000-00011

Teixeira, M. I., Lopes, C. M., Amaral, M. H., & Costa, P. C. (2020). Current insights on lipid nanocarrier-assisted drug delivery in the treatment of neurodegenerative diseases. *European Journal of Pharmaceutics and Biopharmaceutics: Official Journal of Arbeitsgemeinschaft Fur Pharmazeutische Verfahrenstechnik e.V*, *149*, 192–217. https://doi.org/10.1016/J.EJPB.2020.01.005

Tenchov, R., Bird, R., Curtze, A. E., & Zhou, Q. (2021). Lipid nanoparticles— from liposomes to mRNA vaccine delivery, a landscape of research diversity and advancement. *ACS Nano*, *15*(11), 16982–17015. https://pubs.acs.org/doi/pdf/10.1021/acsnano.1c04996

Therapeutics | ALZFORUM. (n.d.). Retrieved June 12, 2023, from www.alzforum.org/therapeutics

Tian, L., Shang, Y., Chen, R., Bai, R., Chen, C., Inthavong, K., & Tu, J. (2019). Correlation of regional deposition dosage for inhaled nanoparticles in human and rat olfactory. *Particle and Fibre Toxicology*, *16*(1), 6. https://doi.org/10.1186/s12989-019-0290-8

Timmer, M., Cesnulevicius, K., Winkler, C., Kolb, J., Lipokatic-Takacs, E., Jungnickel, J., & Grothe, C. (2007). Fibroblast growth factor (FGF)-2 and FGF receptor 3 are required for the development of the substantia nigra, and FGF-2 plays a crucial role for the rescue of dopaminergic neurons after 6-hydroxydopamine lesion. *Journal of Neuroscience*, *27*(3), 459–471. https://doi.org/10.1523/JNEUROSCI.4493-06.2007

Toh, M. R., & Chiu, G. N. C. (2013). Review. *Asian Journal of Pharmaceutical Sciences*, *2*(8), 88–95. https://doi.org/10.1016/J.AJPS.2013.07.011

Tooyama, I., Kawamata, T., Walker, D., Yamada, T., Hanai, K., Kimura, H., Iwane, M., Igarashi, K., McGeer, E. G., & McGeer, P. L. (1993). Loss of basic fibroblast growth factor in substantia nigra neurons in Parkinson's disease. *Neurology*, *43*(2), 372–376. https://doi.org/10.1212/wnl.43.2.372

Tooyama, I., McGeer, E. G., Kawamata, T., Kimura, H., & McGeer, P. L. (1994). Retention of basic fibroblast growth factor immunoreactivity in dopaminergic neurons of the substantia nigra during normal aging in humans contrasts with loss in Parkinson's disease. *Brain Research*, *656*(1), 165–168. https://doi.org/10.1016/0006-8993(94)91378-1

Topal, G. R., Meszaros, M., Porkolab, G., Szecsko, A., Polgar, T. F., Siklos, L., Deli, M. A., Veszelka, S., & Bozkir, A. (2020). ApoE-targeting increases the transfer of solid lipid nanoparticles with donepezil cargo across a culture model of the blood-brain barrier. *Pharmaceutics*, *13*(1). https://doi.org/10.3390/pharmaceutics13010038

Trapani, A., De Giglio, E., Cometa, S., Bonifacio, M. A., Dazzi, L., Di Gioia, S., Hossain, M. N., Pellitteri, R., Antimisiaris, S. G., & Conese, M. (2021). Dopamine-loaded lipid based nanocarriers for intranasal administration of the neurotransmitter: A comparative study. *European Journal of Pharmaceutics and Biopharmaceutics*, *167*, 189–200. https://doi.org/10.1016/j.ejpb.2021.07.015

Trapani, A., Mandracchia, D., Tripodo, G., Cometa, S., Cellamare, S., De Giglio, E., Klepetsanis, P., & Antimisiaris, S. G. (2018). Protection of dopamine towards autoxidation reaction by encapsulation into non-coated- or chitosan- or thiolated chitosan-coated-liposomes. *Colloids and Surfaces B: Biointerfaces*, *170*, 11–19. https://doi.org/10.1016/j.colsurfb.2018.05.049

Trapani, A., Tricarico, D., Mele, A., Maqoud, F., Mandracchia, D., Vitale, P., Capriati, V., Trapani, G., Dimiccoli, V., Tolomeo, A., & Scilimati, A. (2017). A novel injectable formulation of 6-fluoro-l-DOPA imaging agent for diagnosis of neuroendocrine tumors and Parkinson's disease. *International Journal of Pharmaceutics*, *519*(1–2), 304–313. https://doi.org/10.1016/j.ijpharm.2017.01.038

Twohig, D., & Nielsen, H. M. (2019). Alpha-synuclein in the pathophysiology of Alzheimer's disease. *Molecular Neurodegeneration*, *14*(1), 23. https://doi.org/10.1186/s13024-019-0320-x

Uchida, M., Nakajin, S., Toyoshima, S., & Shinoda, M. (1996). Antioxidative effect of sesamol and related compounds on lipid peroxidation. *Biological and Pharmaceutical Bulletin*, *19*(4), 623–626. https://doi.org/10.1248/bpb.19.623

Üner, M. (2016). Characterization and imaging of solid lipid nanoparticles and nanostructured lipid carriers. In *Handbook of nanoparticles* (pp. 117–141). Springer. https://link.springer.com/referenceworkentry/10.1007/978-3-319-15338-4_3

Uppuluri, C. T., Ravi, P. R., & Dalvi, A. V. (2021). Design, optimization and pharmacokinetic evaluation of piribedil loaded solid lipid nanoparticles dispersed in nasal in situ gelling system for effective management of Parkinson's disease. *International Journal of Pharmaceutics*, *606*, 120881. https://doi.org/10.1016/J.IJPHARM.2021.120881

Urimi, D., Widenbring, R., Pérez García, R. O., Gedda, L., Edwards, K., Loftsson, T., & Schipper, N. (2021). Formulation development and upscaling of lipid nanocapsules as a drug delivery system for a novel cyclic GMP analogue intended for retinal drug delivery. *International Journal of Pharmaceutics*, *602*. https://doi.org/10.1016/J.IJPHARM.2021.120640

Vakilinezhad, M. A., Amini, A., Akbari Javar, H., Baha'addini Beigi Zarandi, B. F., Montaseri, H., & Dinarvand, R. (2018). Nicotinamide loaded functionalized solid lipid nanoparticles improves cognition in Alzheimer's disease animal model by reducing Tau hyperphosphorylation. *Daru*, *26*(2), 165–177. https://doi.org/10.1007/s40199-018-0221-5

Veena, K. R., & Savitha, S. K. (2019). Pharmacosomes: A novel strategy for controlled drug delivery. *Journal of Pharmaceutical Sciences and Research*, *11*.

Vemuri, S., & Rhodes, C. T. (1995). Preparation and characterization of liposomes as therapeutic delivery systems: A review. *Pharmaceutics Acta Helvetiae*, *70*, 95–111.

Venkatesh, D. N., Kalyani, K., Tulasi, K., Priyanka, V. S., Ali, S. A., & Kumar, S. S. (2014). Pharmacosomes: A potential vesicular drug delivery system. *International Journal of Pharmaceutical Sciences and Drug Research*, *6*(2), 90–94. www.ijpsdr.com/index.php/ijpsdr/article/view/309

Wang, M. S., Boddapati, S., Emadi, S., & Sierks, M. R. (2010). Curcumin reduces α-synuclein induced cytotoxicity in Parkinson's disease cell model. *BMC Neuroscience*, *11*(1), 1–10. https://doi.org/10.1186/1471-2202-11-57/FIGURES/4

Weber, S., Zimmer, A., & Pardeike, J. (2014). Solid lipid nanoparticles (SLN) and nanostructured lipid carriers (NLC) for pulmonary application: A review of the state of the art. *European Journal of Pharmaceutics and Biopharmaceutics*, *86*(1), 7–22. www.sciencedirect.com/science/article/abs/pii/S0939641113002920

Whalen, G. F., Shing, Y., & Folkman, J. (1989). The fate of intravenously administered bFGF and the effect of heparin. *Growth Factors*, *1*(2), 157–164. https://doi.org/10.3109/08977198909029125

Wilkie, C. M., Barron, J. C., Brymer, K. J., Barnes, J. R., Nafar, F., & Parsons, M. P. (2021). The effect of GLT-1 upregulation on extracellular glutamate dynamics. *Frontiers in Cellular Neuroscience*, *15*, 87. https://doi.org/10.3389/FNCEL.2021.661412/BIBTEX

Winblad, B., Jones, R. W., Wirth, Y., Stoffler, A., & Mobius, H. J. (2007). Memantine in moderate to severe Alzheimer's disease: A meta-analysis of randomised clinical trials. *Dementia and Geriatric Cognitive Disorders*, *24*(1), 20–27. https://doi.org/10.1159/000102568

William, B., Noémie, P., Brigitte, E., & Géraldine, P. (2020). Supercritical fluid methods: An alternative to conventional methods to prepare liposomes. In *Chemical Engineering Journal* (Vol. 383). Elsevier B.V. https://doi.org/10.1016/j.cej.2019.123106

Witika, B. A., Poka, M. S., Demana, P. H., Matafwali, S. K., Melamane, S., Khamanga, S. M. M., & Makoni, P. A. (2022). Lipid-based nanocarriers for neurological disorders: A review of the state-of-the-art and therapeutic success to date. *Pharmaceutics*, *14*(4), 1–44. https://doi.org/10.3390/pharmaceutics14040836

Xing, H., Hwang, K., & Lu, Y. (2016). Recent developments of liposomes as nanocarriers for theranostic applications. *Theranostics*, *6*(9), 1336. https://doi.org/10.7150/THNO.15464

Yang, Z. Z., Zhang, Y. Q., Wang, Z. Z., Wu, K., Lou, J. N., & Qi, X. R. (2013). Enhanced brain distribution and pharmacodynamics of rivastigmine by liposomes following intranasal administration. *International Journal of Pharmaceutics*, *452*(1–2), 344–354. https://doi.org/10.1016/j.ijpharm.2013.05.009

Yasamineh, S., Yasamineh, P., Ghafouri Kalajahi, H., Gholizadeh, O., Yekanipour, Z., Afkhami, H., Eslami, M., Hossein Kheirkhah, A., Taghizadeh, M., Yazdani, Y., & Dadashpour, M. (2022). A state-of-the-art review on the recent advances of niosomes as a targeted drug delivery system. *International Journal of Pharmaceutics*, *624*, 121878. https://doi.org/10.1016/j.ijpharm.2022.121878

Yasir, M., Sara, U. V. S., Chauhan, I., Gaur, P. K., Singh, A. P., Puri, D., & Ameeduzzafar. (2018). Solid lipid nanoparticles for nose to brain delivery of donepezil: Formulation, optimization by box–behnken design, in vitro and in vivo evaluation. *Artificial Cells, Nanomedicine, and Biotechnology*, *46*(8), 1838–1851.

Yi-Guang, J., Ai, P., Li, M., & Xin-Pu, H. (2005). Preparation and properties of acyclovir pharmacosomes. *Chinese Journal of Pharmaceuticals*, 36.

You, L., Wang, J., Liu, T., Zhang, Y., Han, X., Wang, T., Guo, S., Dong, T., Xu, J., Anderson, G. J., Liu, Q., Chang, Y. Z., Lou, X., & Nie, G. (2018). Targeted brain delivery of rabies virus glycoprotein 29-modified deferoxamine-loaded nanoparticles reverses functional deficits in parkinsonian mice. *ACS Nano*, *12*(5), 4123–4139. https://doi.org/10.1021/acsnano.7b08172

Yulug, B., Altay, O., Li, X., Hanoglu, L., Cankaya, S., Lam, S., Velioglu, H. A., Yang, H., Coskun, E., Idil, E., Nogaylar, R., Ozsimsek, A., Bayram, C., Bolat, I., Oner, S., Tozlu, O. O., Arslan, M. E., Hacimuftuoglu, A., Yildirim, S., Arif, M., Shoaie, S., Zhang, C., Nielsen, J., Turkez, H., Boren, J., Uhlen, M., & Mardinoglu, A. (2023). Combined metabolic activators improve cognitive functions in Alzheimer's disease patients: A randomised, double-blinded, placebo-controlled phase-II trial. *Translational Neurodegeneration*, *12*(1), 4. https://doi.org/10.1186/s40035-023-00336-2

Yusuf, M., Khan, M., Khan, R. A., & Ahmed, B. (2013). Preparation, characterization, in vivo and biochemical evaluation of brain targeted Piperine solid lipid nanoparticles in an experimentally induced Alzheimer's disease model. *Journal of Drug Targeting*, *21*(3), 300–311. https://doi.org/10.3109/1061186X.2012.747529

Zarei, S., Carr, K., Reiley, L., Diaz, K., Guerra, O., Altamirano, P. F., Pagani, W., Lodin, D., Orozco, G., & Chinea, A. (2015). A comprehensive review of amyotrophic lateral sclerosis. *Surgical Neurology International*, *6*(1). https://doi.org/10.4103/2152-7806.169561

Zelphati, O., & Szoka, F. C. (1996). Mechanism of oligonucleotide release from cationic liposomes. *Proceedings of the National Academy of Sciences of the United States of America*, *93*, 11493–11498.

Zhang, Z. J., Cheang, L. C., Wang, M. W., & Lee, S. M. (2011). Quercetin exerts a neuroprotective effect through inhibition of the iNOS/NO system and pro-inflammation gene expression in PC12 cells and in zebrafish. *International Journal of Molecular Medicine*, *27*(2), 195–203. https://doi.org/10.3892/ijmm.2010.571

Zhao, Y.-Q., Li, L.-J., Zhou, E.-F., Wang, J.-Y., Wang, Y., Guo, L.-M., & Zhang, X.-X. (2022). Lipid-based nanocarrier systems for drug delivery: Advances and applications. *Pharmaceutical Fronts*, *4*(2), e43–e60. https://doi.org/10.1055/s-0042-1751036

Zhao, Y. Z., Li, X., Lu, C. T., Lin, M., Chen, L. J., Xiang, Q., Zhang, M., Jin, R. R., Jiang, X., Shen, X. T., Li, X. K., & Cai, J. (2014). Gelatin nanostructured lipid carriers-mediated intranasal delivery of basic fibroblast growth factor enhances functional recovery in hemiparkinsonian rats. *Nanomedicine*, *10*(4), 755–764. https://doi.org/10.1016/j.nano.2013.10.009

14 Autophagy and Extracellular Nanovesicles

Communication and Their Therapeutic Application in Neurodegenerative Diseases

Mohamed Abdelkader and Noha M. AboQuella*

14.1 INTRODUCTION

Autophagy is a self-regulatory process that aims to maintain a healthy intracellular homeostasis by systematically eliminating misfolded proteins and/or ageing-damaged organelles through lysosome-mediated degradation (Levy et al., 2017). It is a recycling and housekeeping mechanism that is highly conserved in all eukaryotes. In eukaryotes, cells primarily use two distinct proteolytic systems to maintain cellular homeostasis, the ubiquitin–proteasome system (UPS) and lysosome-based autophagy. Whereas UPS identifies only short-lived ubiquitinated protein substrates, lysosome-based autophagy can degrade cytoplasmic constituents, extracellular matrix, plasma membrane (PM) proteins, and even whole organelles. Autophagy also conserves cellular energy by catabolizing proteins, lipids (lipophagy), carbohydrates (glycophagy), and iron (ferritinophagy) (Kaur & Debnath, 2015). Autophagy plays an important role in cell survival and maintenance under physiological and/or pathological conditions by providing a signal transduction machinery to regulate cell proliferation, differentiation, and replicative senescence (Kroemer, 2015; Rockel & Kapoor, 2016). Although it is usually a nonselective process that occurs in response to starvation (Takáts et al., 2019), basal autophagy has been shown to selectively occur under nutrient-rich conditions, such as mitochondrial autophagy (mitophagy), peroxisome autophagy, aggregate autophagy (aggrephagy), etc. (Martens & Behrends, 2020).

Autophagy can be classified according to the path taken by the cargo to lysosomes into three morphologically distinct categories: microautophagy, chaperone-mediated autophagy (CMA), and macroautophagy (Cao et al., 2021). However, all three finally culminate by the cargo being delivered and degraded by the lysosomes (Figure 14.1).

14.2 MICROAUTOPHAGY

Microautophagy describes the process of direct uptake of cytoplasmic contents through the invagination of the lysosomal membrane (Ghosh et al., 2016). A very early *in vitro* study revealed that isolated rat liver lysosomes had a microphagocytic activity toward Percol particles, and this activity was increased with the increase in incubation period and temperature (Marzella et al., 1980). Another study showed that endosomal microautophagy provides a transfer mechanism for cytosolic bodies into late endosomes/multivesicular bodies (Sahu et al., 2011). One more revealed a potential role for rapamycin in activating mammalian microautophagy like it does with macroautophagy (Sato et al., 2019). After all, there is still so much to know about microautophagy that remains hidden until now.

14.3 CHAPERONE-MEDIATED AUTOPHAGY (CMA)

Unlike microautophagy and macroautophagy, which are essentially non-specific toward bulk cytosolic bodies, chaperone-mediated autophagy (CMA) describes a highly specific mechanism in which only intracellular proteins (soluble) bound to molecular protein tags, the chaperones, are transported across the lysosomal membrane and digested by lysosomal enzymes (Oshima et al., 2019). CMA is now accepted to modulate several intracellular mechanisms through timely degradation of a wide range of specific cellular proteins, including, among others, calcium-binding proteins, lipid-binding proteins, some glycolytic enzymes, transcription factors inhibitors, protease subunits, and other proteins involved in synaptic vesicle trafficking and neurotransmitter release (α-synuclein and tau). This way, CMA is believed to maintain intracellular protein quality control and provide a solid defensive mechanism against proteotoxicity (Arias & Cuervo, 2011). Besides, any disruption of CMA by ageing or other factors can be correlated to the development of ageing-associated neurodegenerative diseases and cancers (Kaushik & Cuervo, 2018).

The basis of CMA's selectivity lies in the fact that not all cytosolic proteins can be degraded. Interestingly, it was

* Corresponding author: m.kasem3193@aucegypt.edu

DOI: 10.1201/9781003383376-14

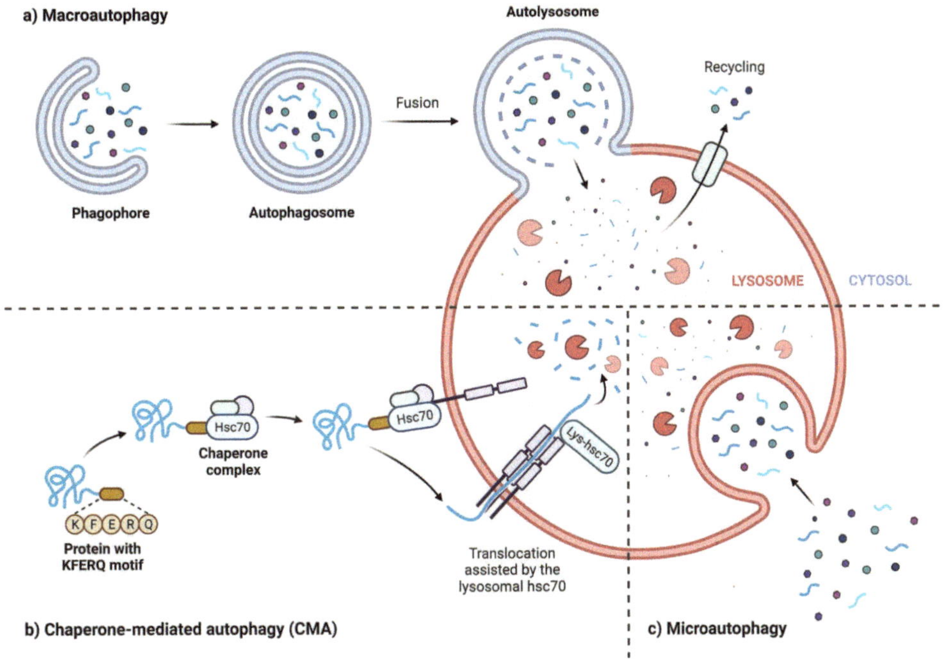

FIGURE 14.1 Different types of autophagy process. (Created with BioRender.com.)

evident that the pentapeptide-targeting motif KFERQ is necessary for the protein substrate to be identified by the CMA-targeting chaperone HSC70 and degraded by lysosomal enzymes (Dice, 1990). This 11-amino acid sequence is estimated to exist in at least 30% of cytosolic proteins. Besides, approximately 40% of the mammalian proteome have a canonical KFERQ-like motif (Ikami et al., 2022). However, the residues that form the motif are considered the main determinants of substrate affinity to the HSC70 chaperone.

The motif contains one or two positively charged K and R residues, one or two hydrophobic residues F, L, I, or V, and one negatively charged E or D residue. Intriguingly, some mechanisms can provide a potential reserve of CMA substrates, even if they lack the targeting motif in their structure. For example, some substrate proteins can generate the targeting motif through post-translational modifications. In addition, phosphorylation of S, T, or Y in a targeting motif containing hydrophobic and positively charged canonical residues and missing the negatively charged one can prime the protein, making it amenable for degradation by the CMA (Kaushik & Cuervo, 2016; Quintavalle et al., 2014).

14.4 MACROAUTOPHAGY

Macroautophagy depicts the process through which an endoplasmic reticulum–derived membrane expands around the unwanted material and forms cytosolic vesicles that finally fuse with late endosomes/lysosomes and transfer their contents for degradation. It essentially passes through five phases: autophagy initiation; isolation membrane extension and nucleation; autophagosome formation and maturation; autophagosome–lysosomal fusion and cargo transfer; and lysosomal regeneration through the lysosomal reformation cycle (Al-Bari & Xu, 2020). Apparently, macroautophagy differs from microautophagy and CMA in the way that the cargo in vesicles, termed *autophagosomes*, is sequestered by membrane expansion and does not bud from an already-existing organelle (Stavoe & Holzbaur, 2019). Unlike yeasts, autophagosome formation in mammals begins at multiple sites in the cytoplasm rather than at a single phagosome assembly site (PAS). To date, the source of the sequestering vesicle membrane has not been strictly identified. However, the endoplasmic reticulum (ER) theory gains the most weight in this controversy. As per several studies, omegasomes of the ER may function as initiation sites in mammals (Park et al., 2020). Other studies suggest different organelles, such as Golgi complex, plasma membrane, and mitochondria, as possible sources for the sequestering vesicle isolation membrane (Mizushima et al., 2011; Ravikumar et al., 2010; Takahashi et al., 2011).

Initiation begins with the formation of an isolation membrane through a sophisticated mechanism and series of reactions. Following initiation, the membrane gradually expands, forming the initial double-membrane compartment, termed a *phagophore*. The isolation membrane bends itself around intracellular-damaging material, misfolded proteins, ageing-damaged organelles, and other unwanted components and turns into a cargo-containing spherical component termed an *autophagosome*. Different autophagosomes vary in sizes from 0.2 to 0.5 μm in baker's yeast, *Saccharomyces cerevisiae*, to 0.5 to 1.5 μm in mammals based on the size of their cargo contents (Backues et al., 2014; Mizushima et al., 2002). The autophagosome then docks and fuses with a late endosome/lysosome and forms an autolysosome. Within the autolysosome, the unwanted cargo contents are degraded by resident lysosomal enzymes (such as proteases, lipases, glycases, and nucleases) and the acidic lumen (Al-Bari & Xu, 2020). The yield of degradation is then exported back into the cytoplasm through lysosomal permeases for reuse in cellular processes and energy production (He, 2022).

14.5 GLYMPHATIC SYSTEM

The term *glymphatic system*, a glial–lymphatic system, which was first introduced by Jeffrey Iliff, describes a large-scale waste clearance system like the peripheral lymphoid system, but specifically within the brain (Iliff et al., 2012). Several studies have demonstrated that the glymphatic system has a crucial role in regulating the dispersion of various substances in the brain. This includes not only its primary function of clearing waste but also its involvement in the distribution and transportation of essential nutrients, like glucose (Lundgaard et al., 2015). Due to its essential role, the glymphatic system is thought to be associated with the development of different disorders, such as neurodegenerative diseases (Rasmussen et al., 2018) (e.g., Alzheimer's disease and Parkinson's disease).

The brain possesses an intricate network of blood vessels, with most lying in the brain parenchyma, being enveloped by astrocytes end feet (Nagelhus & Ottersen, 2013). The glymphatic system acts mainly through the cerebrospinal fluid filling the space between these astrocytic end feet and the vascular wall. The water channel known as aquaporin-4 (AQP4) covers around half of the surface area of astrocyte processes, especially those attached to blood vessels (Nielsen et al., 1997). CSF enters the brain's stroma after passing through the subarachnoid space, facilitated by AQP4 in the area around the artery. This influx pushes interstitial fluid to transport solutes to the perivenous space, after which it exits the brain and enters the cervical lymphatic system. In this process, astrocytes, particularly through the PVS, act as the main players, and this transport system motored by the AQP4 channel is recognized as the glymphatic system (Murtha et al., 2014).

In murine brains, the bulk movement of cerebrospinal fluid (CSF) in the perivascular space is, in part, impacted by the pulsations generated by arterial pressure waves during the cardiac systolic phase (Iliff et al., 2013). Through using advanced magnetic resonance encephalography, similar pulsations and vasomotor changes were found traversing the human brain and probably contributing to the glymphatic flow (Kiviniemi et al., 2016). Conversely, within the ventricular compartments, CSF flow appears to be primarily affected by respiration, which is also believed to play a vital role in facilitating the outflow of interstitial fluid (ISF) along perivenous spaces and white matter tracts (Dreha-Kulaczewski et al., 2015).

The production of CSF contributes to its entry into the brain's parenchyma, and experiments in rodents have shown that inhibiting CSF production using acetazolamide significantly impairs CSF clearance (Lundgaard et al., 2017). Furthermore, the posture of the body can influence this clearance process; in sleeping mice, CSF clearance works much better when the head is positioned laterally, unlike the prone position, where CSF flow is directed more toward the spinal cord (Lee et al., 2015).

14.6 MECHANISTIC REGULATION OF AUTOPHAGY MACHINERY

Several studies have examined the mechanisms of autophagy regulation under physiological conditions in mammals in addition to their potential contribution to the pathogenesis of various health outcomes. *Autophagy* is a complex process, and many upstream signaling pathways are involved in autophagy regulation under physiological conditions. However, the most widely accepted pathways at present are mainly the mTOR and PI3K pathways. Other direct or indirect signaling pathways also play roles in the regulation of autophagy.

14.7 MTOR-DEPENDENT PATHWAYS

mTOR is a critical regulatory component of autophagy. It is mainly composed of two signaling complexes, termed mTORC1 and mTORC2, both of which are functionally overlapping with each other. mTORC1 is usually considered the principal regulator of autophagy in normal cells and tumor cells. mTORC1 has five core components: raptor (a regulatory-associated protein of mTOR), mTOR, mLST8 (mammalian lethal with Sec13 protein 8, also known as GßL), and the two inhibitory subunits, PRAS40 (proline-rich AKT substrate 40, kDa) and DEPTOR (DEP domain–containing mTOR-interacting protein) (Saxton & Sabatini, 2017). Under stress conditions, mTORC1 is inhibited to allow for ULK1 complex activation and autophagy progress. However, through pathways mediated by Akt and MAPK signaling, mTORC1 can lead to inhibition of autophagy. As is known, the AMPK and p53 signaling pathways negatively regulate mTOR pathways to promote autophagy by interacting with the ULK1 complex, enhancing its activity and promoting autophagy. AMPK is strictly ATP dependent and is activated under starvation conditions to induce phosphorylation at serines 317, 467, 555, 574, 637 and 777 of ULK1 (Kim et al., 2011; Stamenkovic et al., 2019). mTORC1, in turn, inactivates the ULK1 complex by binding to serine 757 of ULK1, leading to ULK1 phosphorylation and inhibition of the AMPK-ULK1 interaction, and hence autophagy (Hosokawa et al., 2009).

Another signaling pathway for inhibiting autophagy through mTOR is mediated through the PI3K lipid kinase complex. PI3K complex plays a key role in the regulation of autophagy (Strappazzon et al., 2019). In response to insulin availability, RTK autophosphorylation induces phosphorylation of insulin receptor substrates (IRSs), which end by initiating activity of PI3K complex. PI3K pathway, in turn, increases phosphatidylinositol (3,4,5)-trisphosphate (PIP3) (Lane et al., 2017). PIP3 recruits PDK1 (PI-dependent kinase 1), which, through a phosphorylated-dependent activation of AKT, induces suppression of autophagy (Ziemba et al., 2013). Meanwhile, hyperactivation of the phosphatidylinositol (PI) phosphatase and tensin homolog (PTEN) promotes autophagy through inhibiting the PI3K/Akt signaling pathway (Saxton & Sabatini, 2017).

The important function of AMBRA1 in autophagy has been discussed extensively in recent research. AMBRA1 is an autophagy-related gene, which is a key factor regulating autophagy. AMBRA1 promotes autophagy by binding to Becn1 and combining with Vps34/PI3KC3, which are key regulators of autophagosome formation. This combination further facilitates the formation of the PI3K complex III, which essentially orchestrates autophagy, as mentioned earlier. AMBRA1 has also been shown to regulate autophagy and apoptosis at the same time *in vitro* in embryonic stem cells and fibroblasts. Interestingly, AMBRA1 promoted autophagy and suppressed apoptosis at the same time. Moreover, AMBRA1 was found to dynamically bind Beclin2 in mitochondria to promote the Becn1-dependent program of autophagy.

Besides its direct function as a negative regulator of autophagy, mTORC1 also regulates signaling via the mTORC2 pathway (Hsu et al., 2011). Unlike mTORC1, mTORC2 is not sensitive to starvation. mTORC2 also plays a role in regulating autophagy through the FOXO3 (fork box protein O3) transcription factor pathway. mTORC2 enhances AKT activity through direct phosphorylation at Ser473 (Guertin et al., 2006). Then, through AKT-mediated phosphorylation, phosphorylated FOXO3 binds to the 14–3–3 protein and thus inhibits the activation of autophagy gene transcription (Saxton & Sabatini, 2017). Moreover, FOXO1 inhibits mTORC1 activity by inducing the activity of Sestrin3, the negative regulator of mTORC1, thereby balancing AKT and mTORC1 activity under stress conditions to maintain cellular energy homeostasis (Chen et al., 2010).

14.8 MTOR-INDEPENDENT PATHWAYS

Several mTOR-independent pathways have been studied as additional regulators of autophagy in mammals. The Ca2+ signaling pathway is considered one of the most puzzling due to its stimulatory and inhibitory effects on the autophagy process. Under normal physiological conditions, IP3-mediated Ca2 release signaling arises from the ER to the mitochondria in order to promote mitochondrial functions and meet cellular energy demands (Leipnitz et al., 2018). This signal leads to an increased overall Ca load in the cytoplasm as well as in other organelles, including lysosomes, due to membrane contact sites (Raffaello et al., 2016). However, under stressful conditions, cytosolic Ca overload can lead to autophagy, apparently due to a lack of energy supply. This explains the mechanism by which the cell switches its Ca2 signaling mode from autophagy to pro-autophagy. Several channels have been proposed to channel Ca from lysosomes; however, bipore segment channels and transient receptor (TRP) channels are the main ones. TRPs have been proposed, specifically, to mediate the regulation of autophagy via its subfamily channel, TRPML1 (mucolipin1) (Rosato et al., 2019).

Other mTOR-independent pathways have been proposed to regulate autophagy, including receptor tyrosine kinases, HMGB1, Pk, Prk, casein kinases, Class3K Ras-MAPK, p53, PTEN, and ER stress (Ji et al., 2018; Min et al., 2018). In addition, recent research has shown that some pro-apoptotic signals, such as TNF, TRAIL, and FADD, can promote autophagy (Campbell et al., 2018). After all, more research is still needed to relate the autophagy regulatory mechanisms to different disease states.

14.9 OTHER PATHWAYS RELATED TO MITOPHAGY

The mitochondria serve as the main source for cellular energy controlling metabolism and physiology (Galluzzi et al., 2012). To preserve cellular metabolism and homeostasis, it is critical to remove any mitochondrion via mitophagy, a control process for mitochondria (Kim et al., 2007). Mitophagy is very selective when compared to autophagy as it only removes the malfunctions of mitochondria using different mechanisms, either by a ubiquitin-dependent or ubiquitin-independent manner (Ashrafi & Schwarz, 2013). Studies have revealed that mitochondrial damage accumulation is one of the main reasons involved in many diseases, such as neurodegenerative diseases (NDs), including Parkinson's disease (PD), Alzheimer's disease (AD), Huntington's disease (HD), and amyotrophic lateral sclerosis (ALS) (Liu et al., 2019; Palikara et al., 2018). There is a wide variety of cell types that are affected by mitochondrial damage, but the most common type known to be susceptible to mitochondrial dysfunction is the neuron because of its substantial amount of energy requirements (Martinez-Vicente, 2017).

There are different signaling cascades that orchestrate mitophagy. The first pathway that regulates mitochondrial degradation in a ubiquitin-dependent manner is called PINK/Parkin pathway–mediated mitophagy, also known as canonical mitophagy. This pathway reacts to severe cellular conditions that damage the mitochondria. The second type of mitochondrial degradation pathway is known as non-canonical mitophagy in a ubiquitin-independent manner. This pathway consists of receptors that are expressed on both the inner and outer mitochondrial membrane (Khaminets et al., 2016).

14.10 UBIQUITIN-DEPENDENT MITOPHAGY (CANONICAL PATHWAY)

Parkin (PARK2) is an E3 ubiquitin–protein ligase with mutations associated with the development of Parkinson's disease. PINK1 (phosphatase and tensin homolog (PTEN)–induced putative kinase 1) is a serine/threonine kinase that preserves mitochondria (Shimizu et al., 2000; Whitworth & Pallanck, 2009). These two proteins are the main key players in the canonical mitophagy pathway as they control the ubiquitin-dependent mitophagy pathway (Pickles et al., 2018). PINK1 attaches to the translocase of the outer membrane (TOM) complex and then transfers from the cytoplasm to the inner mitochondrial membrane (IMM), where it is cleaved by different proteases, leading to the production of truncated protein (Lazarou et al., 2012; Sekine & Youle, 2018). Ultimately, it is liberated into the cytoplasm, at which it is degraded by the ubiquitin–proteasome system. This cellular mechanism reduces PINK1 levels in mitochondria to function normally (Yamano & Youle, 2013). Upon mitochondrial damage, depolarization of the membrane potential takes place, and thereby the PINK1 cannot be transferred to the IMM; instead, it aggregates on the outer mitochondrial membrane (OMM). This aggregation results in PINK1 autophosphorylation at both Ser 228 and Ser 402, which, in turn, promotes the activation of the PINK1 kinase. Following this, Parkin is brought to the mitochondrial surface and commences its E3 ubiquitin ligase function through PINK1-induced phosphorylation at Ser 65. Intriguingly, even if Parkin carries a mutation at Ser 65, it can still be recruited because PINK1 phosphorylates ubiquitin at Ser 65, thus also activating Parkin's E3 ligase activity (Harper et al., 2018; Kane et al., 2014). When both PINK1 and Parkin get activated, they trigger the formation of phosphorylated poly-Ub chains attached to Lys 63 and Lys 27 to OMM proteins. Subsequently, these poly-Ub chains will be identified by autophagic receptors, including P62/SQSTM1 and optineurin (OPTN), which bind to the autophagosomal light chain 3 (LC3) protein. Through the LIR motif, LC3 then attaches to the isolation membrane to form an autophagosome. Finally, the autophagosome integrates with the lysosome to form an autolysosome, which, in turn, destroys the malfunctioning mitochondria (Geisler et al., 2010; Koyano et al., 2014). Then, mitophagy, facilitated by the PINK1/Parkin pathway, serves to remove damaged mitochondria, leading to reactive oxygen species (ROS) production, thus preventing oxidative stress and preserving energy hemostasis (Xu et al., 2020).

14.11 UBIQUITIN-INDEPENDENT MITOPHAGY (NON-CANONICAL PATHWAY)

In contrast to the ubiquitin-dependent method, there are some mitophagy receptors that facilitate mitophagy directly by engaging in protein–protein interactions. These mitophagy receptor proteins, such as FUNDC1, NIX, and BNIP3, among others, are found to be expressed on the OMM. They interact with Atg8 family proteins by binding the LIR motif and play a role in the inclusion of mitochondria by autophagosomes, thus promoting mitophagy (Stolz et al., 2014).

In mammalian cells, specific OMM proteins like NIX (also known as BNIP3L), BNIP3, and FUNDC1 have been recognized as mitophagy receptors that are ubiquitin-independent (Cen et al., 2021). These receptors play a role in initiating mitophagy in response to different situations, including hypoxia, stress, and cellular differentiation. NIX and BNIP3, which are members of the Bcl2-family protein group, share a 56% similarity in their amino acid sequences (Matsushima et al., 1998). BNIP3, as a mitophagy receptor, possesses an LIR motif and operates in a manner like NIX. Notably, the interaction between BNIP3 and LC3 prompts mitophagy and the removal of unwanted endoplasmic reticulum (ER) (Zhu et al., 2013). Interestingly, BNIP3 is rendered essential in apoptosis, indicating a cross-regulation between mitophagy and apoptosis. In addition to these proteins, accumulating research suggests the involvement of other Bcl2-family proteins in the regulation of mitophagy (Hollville et al., 2014). Bcl2, recognized as an anti-apoptotic protein, exerts a negative regulatory influence on mitophagy by directly interacting with Parkin, thereby inhibiting its function. Another essential anti-apoptotic protein, MCL-1, has recently been identified as a novel mitophagy receptor.

14.12 AUTOPHAGY OF NEURODEGENERATIVE DISEASE

The proper function of cell autophagy is essential for the preservation of neuronal health, which, in turn, is dispensable for the function of the central nervous system (CNS) (Diab et al., 2023). Postmitotic neuron stability relies significantly on basal autophagy in comparison with other cell types because the aggregation of misfolded proteins and defected organelles cannot be thinned out during cell cycles. In addition, neurons have very specific structures that enable them to communicate, build networks, and work synchronously, such as axons, dendrites, and synapses (Son et al., 2012). These structures need proteins, organelles, and autophagosomes to travel long distances from the soma. Thus, autophagy as the primary intracellular mechanism responsible for breaking down the aggregated proteins and the defective organelles is implicated in the pathological changes of most NDs (Guo et al., 2018).

14.13 AUTOPHAGY IN SYNAPSE DEVELOPMENT

Neuronal autophagy is critical for synapse development and connections. A study reported that the change in the autophagy process rate either via increase or decrease has a great effect on the size of the synapse. For example, in the neuromuscular junction (NMJ) of *Drosophila*, neuronal autophagy has been found to enhance synaptic development (Shen & Ganetzky, 2009). Highly expressed Atg1 (autophagy-related gene), a serine/threonine kinase and a vital controller of autophagy, is involved in a substantial increase in autophagy levels and, consequently, an improvement in synaptic growth. Conversely, mutations in autophagy-related genes result in a decrease in autophagy activity, thereby reducing the size of synapses.

Moreover, one of the main significant inhibitory factors of synaptic development in the neuromuscular junction is the ubiquitin–proteasome pathway system (UPS) (Wan et al., 2000). Indeed, synapse development is regulated by the convergence of neuronal autophagy and the UPS, which jointly control Hiw, an E3 ubiquitin ligase. Interestingly, Atg1 is thought to be implicated in synaptic structure control through the reduction of MAPK/ERK activity (Wairkar et al., 2009). In the Atg1 mutated mouse model, a reduction in synaptic density is caused due to a significant increase in ERK activity, indicating that activated ERK negatively affects the formation of synapses. As autophagy is triggered by various developmental and environmental signals, neuronal autophagy might have a significant function in promoting synapse growth and plasticity, which are required for learning and memory processes (Son et al., 2012).

14.14 ALZHEIMER'S DISEASE AUTOPHAGY

Alzheimer's disease (AD) is the most recognized type of dementia in humans, generally associated with highly characteristic intra- and extracellular protein aggregates (MAPT/tau tangles, beta-amyloid peptide (Aβ) plaques, respectively). These aggregates are implicated in several deficits, including neuronal death and cognitive decline. The link between autophagy and AD arises from the observation of the increased size of autophagic compartments in the brains of individuals diagnosed with AD (Nixon et al., 2005). Multilayer proteomics analysis of the brain has indicated that there is an accumulation of the autophagic substrate SQSTM1/p62 (sequestosome1) in AD, indicative of disrupted autophagic flow (Bai et al., 2020), which is corresponding to the same substrate in experimental models of AD (Yu et al., 2005). Reinforcing this theory, functional autophagy is critical to break down both soluble and aggregated variants of MAPT/tau (Silva et al., 2020). Lysosomal membrane damage induced by MAPT/tau oligomers initiates a coordinated program involving LGALS3 (galectin 3), which results in the activation of autophagy (Jia et al., 2020). In mice experiments, genetic inhibition of SQSTM1/p62 drove hyperphosphorylated MAPT/tau to accumulate and, consequently, neurodegeneration to develop (Ramesh et al., 2008). Interestingly, aggregation of MAPT/tau tangles hinders autophagosome retrograde transport via axons since it interferes with the dynein–DCTN (dynactin) complex. This eventually leads to the accumulation of MAPT/tau-containing autophagic vesicles (Butzlaff et al., 2015).

Importantly, the NFE2L2/NRF2 (nuclear factor, erythroid-derived 2, like 2)–dependent transcription of the autophagy regulator CALCOCO2/NDP52 (calcium-binding and coiled-coil domain 2) has an indispensable role in facilitating the degradation of MAPT/tau due to exposure to oxidative stress (Jo et al., 2014). Additionally, SQSTM1/p62 is a target gene for NFE2L2/NRF2 (Jain et al., 2010) that has been shown to be involved in the degradation of aggregated MAPT/tau (Xu et al., 2019). Recently, the malfunction of the endosomal-sorting complex, called the retromer, has been linked to several NDs, including Alzheimer's disease (AD). Low expression levels of the retromer proteins and variants of the core retromer component VPS35 (vacuolar protein sorting 35) are coupled with neurodegenerative diseases, most notably coinciding with brain MAPT/tau aggregation (Carosi et al., 2021; Seaman, 2021). Recent studies revealed that the autophagy–lysosomal axis is critical for the clearance of aggregated MAPT/tau, and drainage of VPS35 blocks autophagy, while high expression levels have the counter effect (Carosi et al., 2021). Therefore, the interaction between the retromer and autophagy pathways could have an essential role in the protection against various neurodegenerative disorders. This is achieved by ensuring the timely removal of harmful protein aggregates as they form, thereby preventing their accumulation and potential detrimental effects (Klionsky et al., 2021).

Other factors that influence autophagy can impact the levels of Aβ protein within cells. For example, NRBF2 (nuclear receptor–binding factor 2, a component of the PtdIns3K complex I) interacts with APP (Yang et al., 2017) and promotes its clearance through the lysosomal pathway. This is evident from the observation that a reduction in NRBF2 leads to an accumulation of intracellular APP levels in cells and Aβ accumulation in AD mouse models (Lachance et al., 2019); however, overexpression of NRBF2 results in low levels of Aβ and enhances mouse memory.

Recently, the correlation between the activation of autophagy in the microglial compartment and AD has been demonstrated. Crucially, the elimination of the gene encoding for TREM2 (triggering receptor expressed on myeloid cells 2), a surface receptor important for microglial activation in inflammation and neurodegeneration, leads to the improper buildup of autophagosomes and disrupts the normal clustering of microglia around plaques. This effect is linked to the dysregulated activation of MTORC1, which, in turn, triggers metabolic abnormalities in microglial cells (Ulland et al., 2017).

Furthermore, non-canonical functions of the ATG in microglia play a role in mitigating the harmful effects linked to the accumulation of Aβ plaques in the 5XFAD mouse model. Significantly, the generic removal of Atg5 or Rubcn from myeloid cells leads to increased Aβ plaque formation and abnormal production of inflammatory cytokines. This, in turn, accelerates neuronal deterioration and cognitive decline. Mechanistically, ATG5 and RUBCN participate in the process of MAP1LC3/LC3 (microtubule-associated protein 1 light chain 3) conjugation to endosomal membranes containing Aβ. These membranes are positively marked by RAB5 and clathrin. This process, known as LC3-associated endocytosis (LANDO), seems to facilitate the recycling of potential Aβ receptors (e.g., TLR4, TREM2) from internalized endosomes back to the plasma membrane of microglial cells (Heckmann et al., 2019). Although it is not yet fully understood whether LANDO is implicated in Aβ receptor degradation, its activation plays a critical role in decreasing Aβ accumulation and restraining neuroinflammation in AD. In a similar context, when LANDO is impaired in aged mice through the deletion of the wild-type domain of ATG16L1 (which is not required for canonical autophagy), it worsens the neuroinflammatory characteristics linked to AD-like symptoms (Heckmann et al., 2020).

In addition, chaperone-mediated autophagy (CMA) participates in neuronal MAPT/tau degeneration under normal conditions. However, other MAPT/tau mutations and post-translational modifications, including acetylation, inhibit CMA-mediated MAPT/tau degradation by inhibiting normal CMA functioning (Caballero et al., 2018). Moreover, the CMA blockade resulted in redirecting some of the pathogenic forms of MAPT/tau to the endosomal microautophagy. Since both pathways involve the use of the same chaperone, HSPA8, it facilitates the release of MAPT/tau variants outside the neurons via helping fusion of late endosomes to the plasma membrane (Caballero et al., 2021). This process contributes to the propagation of MAPT/tau. Recently, a reduction in neuronal CMA activity has proven to exist in the brains of AD patients (Bourdenx et al., 2021).

14.15 PARKINSON'S DISEASE AUTOPHAGY

Parkinson's disease (PD) is one of the prominent neurodegenerative disorders that is pathologically characterized by the loss of dopaminergic neurons in the substantia nigra (SN) and the accumulation of Lewy bodies that are composed of SNCA/α-synuclein SNCA/α-synuclein aggregates besides other polyubiquitinated proteins. The main symptoms of PD are motor and autonomic dysfunction and, in some cases, cognitive and psychological deficits. Initial findings indicated that both chaperone-mediated autophagy (CMA) and macroautophagy play roles in the degradation of SNCA/α-synuclein (Cuervo et al., 2004; Webb et al., 2003).

Overexpression of wild-type SNCA/α-synuclein, mutations, and unfavorable post-translational modifications, including the formation of dopamine adducts, are toxic to the CMA via prohibiting multimerization of LAMP2A and subsequent lysosomal internalization of cargo proteins (Martinez-Vicente et al., 2008). Recent studies have shown that selective autophagy plays a crucial role in clearing SNCA/α-synuclein released by neurons. This process involves the autophagy receptor SQSTM1/p62 in microglia, providing protection to dopaminergic neurons (Choi et al., 2020a). In correspondence with these findings, the induction of autophagy reduces the buildup of SNCA/α-synuclein (Nakamura et al., 2019).

In contrast, unregulated expression of wild-type or mutated forms of SNCA/α-synuclein interrupts the autophagic flux and interferes with TFEB-mediated lysosomal biogenesis. This interference occurs by preventing the nuclear translocation of TFEB (Decressac et al., 2013). High levels of SNCA/α-synuclein influence the intracellular localization of ATG9 through RAB1A(RAB1A, member RAS oncogene family) and, hence, disturb autophagy dynamics in the overexpressing SNCA/α-synuclein genetically modified mice brains (Winslow et al., 2010). One of the most common genetic risk factors associated with PD is the GBA (glucosylceramidase beta) loss-of-function gene mutation that has been reported to hinder the autophagic flux and, as a result, lead to the aggregation of SNCA/α-synuclein (Murphy et al., 2014). Additionally, autosomal-dominant mutation impacting VPS35 diminishes autophagy via modifying ATG9 localization (Zavodszky et al., 2014).

One of the main causes of familial PD is the dominant LRRK2 (leucine-rich repeat kinase 2) mutation expression, which has been linked with dysfunctional autophagy (Ramonet et al., 2011). While autophagy upregulation results in the removal of protein aggregates, it also increases the risk of degeneration of the axodendritic arbor. This susceptibility to autophagy-induced degeneration has been reported in cultured dopaminergic, sympathetic, and cortical neurons, as well as in the axons of dopaminergic neurons in living organisms, as demonstrated by Atg7 knockdown or knockout (H. C. Cheng et al., 2011; Plowey et al., 2008). In addition, the expression of dominant-negative ULK1, the expression of a mutant form of LC3, which mocks lack of autophagy, has been observed to provide protection against the shrinking of dendrites caused by disease-related LRRK2 mutations and the MPP+-induced Parkinson's disease (Ramonet et al., 2011).

The relationship between the autosomal recessive forms of PD and mutations impacting the mitophagy regulators PINK1 (PTEN-induced putative kinase 1) and PRKN/PARK2 (Parkin RBR E3 ubiquitin protein ligase) has been demonstrated (Matsuda et al., 2010; Valente et al., 2004). Mouse models designed to track mitophagy reveal that dopaminergic neurons exhibit increased levels of baseline mitophagy (McWilliams et al., 2018). PINK1 and PRKN deficiencies are not involved in most of the defects under normal conditions; however, knockout PRKN mice models have shown defective striatal neural plasticity (Kitada et al., 2009). Importantly, when mitophagy is impaired due to the removal of PRKN (Pickrell et al., 2015) or PINK1 (Gautier et al., 2008), mice become more susceptible to oxidative stress, and the situation worsens when combined with mitochondrial dysfunction (observed in mitochondrial DNA mutator–PRKN/Parkin–KO mice), resulting in increased neural damage (Pickrell et al., 2015). In neurons, there are alternative pathways for mitophagy, and research involving the abolishing of PINK1 or PRKN in mouse and fly mitophagy biosensor models suggests that neither protein is essential for maintaining normal levels of basal brain mitophagy (Chu et al., 2013; McWilliams et al., 2018). Furthermore, when these mice with PRKN mutations are crossed with STING1/STING-deficient mice, serological markers of inflammation, also observed in individuals with PRKN mutations, are reduced, resulting in the reversal of neuronal degeneration (Sliter et al., 2018).

14.16 HUNTINGTON'S DISEASE AUTOPHAGY

Huntington's disease is the most common type of polyglutamine disease, which is a debilitating autosomal dominant neurodegenerative disease. HD is defined by the presence of a CAG trinucleotide repeat mutation in the first exon of the huntingtin (HTT) gene. This repeatedly causes polyglutamine (polyQ) expansions within the huntingtin protein, resulting in the formation of pathogenic protein aggregates (Imarisio et al., 2008; Jimenez-Sanchez et al., 2017). Aggregated autophagosomes have been reported in many HD models (Martinez-Vicente et al., 2010); however, the generation of the autophagosome is not affected by the pathology of HD. In models of HD, the reduction of huntingtin levels results in the abnormal accumulation of autophagosomes containing engulfed mitochondria, indicating a malfunction in the degradation of cellular cargo (Zheng et al., 2010).

Furthermore, there are dysregulated interactions between autophagy and the development of Huntington's disease (HD). One specific genetic variation in the Atg7 gene is associated with an early-onset HD (Metzger et al., 2010). Beclin1, another autophagy-related protein, has been revealed to decrease the levels of the mutant huntingtin mRNA as individuals age (Shibata et al., 2006). In cellular and animal models of HD, there is a malfunction in the process of loading cargo into autophagosomes, resulting in impaired degradation of autophagic proteins, despite an increase in the number of autophagic vesicles.

The autophagy-selective substrate p62/SQSTM1 is a critical marker for monitoring autophagic flux, particularly in the machinery responsible for transporting substrates to autophagosomes. This deficiency in the machinery is widespread in HD models (Martinez-Vicente et al., 2010). Furthermore, the increased activity of casein kinase 2 (CK2), which phosphorylates p62/SQSRM1, has been found to reduce the formation of large inclusions containing mutant huntingtin (Matsumoto et al., 2011).

Upon the comparison of mutant HTT with non-mutant HTT, it appears that they have specific roles in coordinating with autophagy. For example, wild-type HTT can interact with p62, thereby improving its function in the autophagy process. Additionally, it can bind with ULK1 to stimulate autophagy. Moreover, the protein Atg11 shares a similar structure to HTT and plays a role in the generation of autophagosomes. Conversely, when the dynein protein is knocked out, there is an observed elevation in the levels of autophagosomes and malfunction in autolysosomes. This condition is accompanied by an augmented aggregation of mutant huntingtin, the protein associated with HD (Ravikumar et al., 2005).

14.17 THERAPEUTIC APPLICATIONS OF AUTOPHAGY NANOVESICLES IN NEURODEGENERATIVE DISEASES

It is widely accepted that autophagy and extracellular vesicles (EVs) share common characteristics in terms of their origin and influence on each other. Interestingly, EV secretion and transport show a close relationship with autophagosome formation. Since autophagosomes have the ability not only to fuse with lysosomes but also to merge into the multivesicular bodies (MVBs), known precursors of EVs, to form amphiphiles that finally fuse with lysosomes and shed the inner material of intraluminal vesicles (ILVs) for hydrolysis (Xu et al., 2018; You et al., 2017), herefore, accelerated intracellular autophagy could significantly interfere with EV secretion and transport due to the simultaneous consumption of MVBs in the autophagy process (Fader & Colombo, 2005; Fader et al., 2008). On the other hand, eukaryotic and bacterial EVs (BEVs) may be able to enhance autophagy in various NDs. These EVs may play important physiological functions in NDs as essential regulators of microglia activity, promoters of demyelination, deregulators of synaptic activity, and enhancers of autophagy (Gallart-Palau et al., 2020; Scesa et al., 2016). For example, EVs isolated from the conditioned medium of umbilical cord MSCs (ucMSCs-CM) exposed to BV2 microglia showed altered levels of autophagy-related proteins (LC3, Beclin-1, and p62), enhanced phagocytic function of BV2 cells, and decreased cell death induced by Aβ25–35 (Xu et al., 2018). Moreover, microglial α-synuclein-containing EVs had the potential to trigger α-synuclein aggregation in recipient neurons and to impair autophagy flux by upregulating PELI1, resulting in the degradation of LAMP2 in activated microglia. Meanwhile, depleting microglia slowed down the protein aggregation and transmission (Guo et al., 2020). These results indicate that the regulating autophagy and inhibiting microglia delivery of α-synuclein via EVs are likely to serve as a therapeutic target for PD (Xia et al., 2019).

In addition, BEVs have shown promising results as therapeutic drug delivery agents in animal models. The results show that BEVs of mesenchymal stem cell origin can provide active siRNA that targets tau protein in the brain and reduces its phosphorylation and clustering in a mouse model of AD (Panaro et al., 2020; Reza-Zaldivar et al., 2019). Also, BEVs have proved to protect neurons in vitro by preventing the accumulation of unfolded proteins, promoting autophagy, and improving lysosomal function (Wu et al., 2020; Zhang et al., 2022). Furthermore, recent preclinical studies in murine models of NDs demonstrated the ability of BEVs to modulate the immune response, reduce neuroinflammation, and enhance cognitive function. Thus, they have emerged as a good method for treating diseases that destroy neurons (Choi et al., 2020b). However, extensive research is still needed to further detail the mechanisms beyond its action and validate its safety before it can be used on a large clinical scale.

REFERENCES

Al-Bari, M. A. A., & Xu, P. (2020). Molecular regulation of autophagy machinery by mTOR-dependent and -independent pathways. *Annals of the New York Academy of Sciences, 1467*(1), 3–20. https://doi.org/10.1111/NYAS.14305

Arias, E., & Cuervo, A. M. (2011). Chaperone-mediated autophagy in protein quality control. *Current Opinion in Cell Biology*, *23*(2), 184–189. https://doi.org/10.1016/J.CEB.2010.10.009

Ashrafi, G., & Schwarz, T. L. (2013). The pathways of mitophagy for quality control and clearance of mitochondria. *Cell Death Differ*, *20*(1), 31–42. https://doi.org/10.1038/cdd.2012.81

Backues, S. K., Chen, D., Ruan, J., Xie, Z., & Klionsky, D. J. (2014). Estimating the size and number of autophagic bodies by electron microscopy. *Autophagy*, *10*(1), 155–164. https://doi.org/10.4161/AUTO.26856/SUPPL_FILE/KAUP_A_10926856_SM0001.ZIP

Bai, B., Wang, X., Li, Y., Chen, P. C., Yu, K., Dey, K. K., . . . Peng, J. (2020). Deep multilayer brain proteomics identifies molecular networks in Alzheimer's disease progression. *Neuron*, *106*(4), 700. https://doi.org/10.1016/j.neuron.2020.04.031

Bourdenx, M., Martín-Segura, A., Scrivo, A., Rodriguez-Navarro, J. A., Kaushik, S., Tasset, I., . . . Cuervo, A. M. (2021). Chaperone-mediated autophagy prevents collapse of the neuronal metastable proteome. *Cell*, *184*(10), 2696–2714. e2625. https://doi.org/10.1016/j.cell.2021.03.048

Butzlaff, M., Hannan, S. B., Karsten, P., Lenz, S., Ng, J., Voßfeldt, H., . . . Voigt, A. (2015). Impaired retrograde transport by the Dynein/Dynactin complex contributes to tau-induced toxicity. *Human Molecular Genetics*, *24*(13), 3623–3637. https://doi.org/10.1093/hmg/ddv107

Caballero, B., Bourdenx, M., Luengo, E., Diaz, A., Sohn, P. D., Chen, X., . . . Cuervo, A. M. (2021). Acetylated tau inhibits chaperone-mediated autophagy and promotes tau pathology propagation in mice. *Nature Communications*, *12*(1), 2238. https://doi.org/10.1038/s41467-021-22501-9

Caballero, B., Wang, Y., Diaz, A., Tasset, I., Juste, Y. R., Stiller, B., . . . Cuervo, A. M. (2018). Interplay of pathogenic forms of human tau with different autophagic pathways. *Aging Cell*, *17*(1). https://doi.org/10.1111/acel.12692

Campbell, G. R., Bruckman, R. S., Chu, Y. L., Trout, R. N., & Spector, S. A. (2018). SMAC mimetics induce autophagy-dependent apoptosis of HIV-1-infected resting memory CD4+ T cells. *Cell Host and Microbe*, *24*(5), 689–702.e7. https://doi.org/10.1016/j.chom.2018.09.007

Cao, W., Li, J., Yang, K., & Cao, D. (2021). An overview of autophagy: Mechanism, regulation and research progress. *Bulletin Du Cancer*, *108*(3), 304–322. https://doi.org/10.1016/J.BULCAN.2020.11.004

Carosi, J. M., Denton, D., Kumar, S., & Sargeant, T. J. (2021). Retromer dysfunction at the nexus of tauopathies. *Cell Death Differ*, *28*(3), 884–899. https://doi.org/10.1038/s41418-020-00727-2

Carosi, J. M., Hein, L. K., van den Hurk, M., Adams, R., Milky, B., Singh, S., . . . Sargeant, T. J. (2021). Retromer regulates the lysosomal clearance of MAPT/tau. *Autophagy*, *17*(9), 2217–2237. https://doi.org/10.1080/15548627.2020.1821545

Cen, X., Chen, Y., Xu, X., Wu, R., He, F., Zhao, Q., . . . Xia, H. (2020). Pharmacological targeting of MCL-1 promotes mitophagy and improves disease pathologies in an Alzheimer's disease mouse model. *Nature Communications*, *11*(1), 5731. https://doi.org/10.1038/s41467-020-19547-6

Cen, X., Zhang, M., Zhou, M., Ye, L., & Xia, H. (2021). Mitophagy regulates neurodegenerative diseases. *Cells*, *10*(8). https://doi.org/10.3390/cells10081876

Chen, C. C., Jeon, S. M., Bhaskar, P. T., Nogueira, V., Sundararajan, D., Tonic, I., . . . Hay, N. (2010). FoxOs inhibit mTORC1 and activate AKT by inducing the expression of sestrin3 and rictor. *Developmental Cell*, *18*(4), 592–604. https://doi.org/10.1016/j.devcel.2010.03.008

Cheng, H. C., Kim, S. R., Oo, T. F., Kareva, T., Yarygina, O., Rzhetskaya, M., . . . Burke, R. E. (2011). Akt suppresses retrograde degeneration of dopaminergic axons by inhibition of macroautophagy. *Journal of Neuroscience*, *31*(6), 2125–2135. https://doi.org/10.1523/jneurosci.5519-10.2011

Choi, I., Zhang, Y., Seegobin, S. P., Pruvost, M., Wang, Q., Purtell, K., . . . Yue, Z. (2020a). Microglia clear neuron-released α-synuclein via selective autophagy and prevent neurodegeneration. *Nature Communications*, *11*(1), 1386. https://doi.org/10.1038/s41467-020-15119-w

Choi, J. H., Moon, C. M., Shin, T. S., Kim, E. K., McDowell, A., Jo, M. K., . . . Kim, Y. K. (2020b). Lactobacillus paracasei-derived extracellular vesicles attenuate the intestinal inflammatory response by augmenting the endoplasmic reticulum stress pathway. *Experimental & Molecular Medicine*, *52*(3), 423–437. https://doi.org/10.1038/s12276-019-0359-3

Chu, C. T., Ji, J., Dagda, R. K., Jiang, J. F., Tyurina, Y. Y., Kapralov, A. A., . . . Kagan, V. E. (2013). Cardiolipin externalization to the outer mitochondrial membrane acts as an elimination signal for mitophagy in neuronal cells. *Nature Cell Biology*, *15*(10), 1197–1205. https://doi.org/10.1038/ncb2837

Cuervo, A. M., Stefanis, L., Fredenburg, R., Lansbury, P. T., & Sulzer, D. (2004). Impaired degradation of mutant alpha-synuclein by chaperone-mediated autophagy. *Science (New York, N.Y.)*, *305*(5688), 1292–1295. https://doi.org/10.1126/science.1101738

Decressac, M., Mattsson, B., Weikop, P., Lundblad, M., Jakobsson, J., & Björklund, A. (2013). TFEB-mediated autophagy rescues midbrain dopamine neurons from α-synuclein toxicity. *Proceedings of the National Academy of Sciences of the United States of America*, *110*(19), E1817–1826. https://doi.org/10.1073/pnas.1305623110

Diab, R., Pilotto, F., & Saxena, S. (2023). Autophagy and neurodegeneration: Unraveling the role of C9ORF72 in the regulation of autophagy and its relationship to ALS-FTD pathology. *Frontiers in Cellular Neuroscience*, *17*, 1086895. https://doi.org/10.3389/fncel.2023.1086895

Dreha-Kulaczewski, S., Joseph, A. A., Merboldt, K. D., Ludwig, H. C., Gärtner, J., & Frahm, J. (2015). Inspiration is the major regulator of human CSF flow. *Journal of Neuroscience*, *35*(6), 2485–2491. https://doi.org/10.1523/jneurosci.3246-14.2015

Fader, C. M., & Colombo, M. I. (2005). Multivesicular bodies and autophagy in erythrocyte maturation. *Autophagy*, *2*(2), 122–125. https://doi.org/10.4161/AUTO.2.2.2350

Fader, C. M., Sánchez, D., Furlán, M., & Colombo, M. I. (2008). Induction of autophagy promotes fusion of multivesicular bodies with autophagic vacuoles in K562 cells. *Traffic*, *9*(2), 230–250. https://doi.org/10.1111/J.1600-0854.2007.00677.X

Fred Dice, J. (1990). Peptide sequences that target cytosolic proteins for lysosomal proteolysis. *Trends in Biochemical Sciences*, *15*(8), 305–309. https://doi.org/10.1016/0968-0004(90)90019-8

Gallart-Palau, X., Guo, X., Serra, A., & Sze, S. K. (2020). Alzheimer's disease progression characterized by alterations in the molecular profiles and biogenesis of brain extracellular vesicles. *Alzheimer's Research and Therapy*, *12*(1), 1–15. https://doi.org/10.1186/S13195-020-00623-4/TABLES/2

Galluzzi, L., Kepp, O., Trojel-Hansen, C., & Kroemer, G. (2012). Mitochondrial control of cellular life, stress, and death. *Circulation Research*, *111*(9), 1198–1207. https://doi.org/10.1161/circresaha.112.268946

Gautier, C. A., Kitada, T., & Shen, J. (2008). Loss of PINK1 causes mitochondrial functional defects and increased sensitivity to oxidative stress. *Proceedings of the National Academy of Sciences of the United States of America*, *105*(32), 11364–11369. https://doi.org/10.1073/pnas.0802076105

Geisler, S., Holmström, K. M., Skujat, D., Fiesel, F. C., Rothfuss, O. C., Kahle, P. J., & Springer, W. (2010). PINK1/Parkin-mediated mitophagy is dependent on VDAC1 and p62/SQSTM1. *Nature Cell Biology*, *12*(2), 119–131. https://doi.org/10.1038/ncb2012

Ghosh, A. K., Mau, T., O'Brien, M., Garg, S., & Yung, R. (2016). Impaired autophagy activity is linked to elevated ER-stress and inflammation in aging adipose tissue. *Aging*, *8*(10), 2525–2537. https://doi.org/10.18632/AGING.101083

Guertin, D. A., Stevens, D. M., Thoreen, C. C., Burds, A. A., Kalaany, N. Y., Moffat, J., . . . Sabatini, D. M. (2006). Ablation in mice of the mTORC components raptor, rictor, or mLST8 reveals that mTORC2 is required for signaling to Akt-FOXO and PKCα, but not S6K1. *Developmental Cell*, *11*(6), 859–871. https://doi.org/10.1016/j.devcel.2006.10.007

Guo, F., Liu, X., Cai, H., & Le, W. (2018). Autophagy in neurodegenerative diseases: Pathogenesis and therapy. *Brain Pathology*, *28*(1), 3–13. https://doi.org/10.1111/bpa.12545

Guo, M., Wang, J., Zhao, Y., Feng, Y., Han, S., Dong, Q., . . . Tieu, K. (2020). Microglial exosomes facilitate α-synuclein transmission in Parkinson's disease. *Brain*, *143*(5), 1476–1497. https://doi.org/10.1093/BRAIN/AWAA090

Harper, J. W., Ordureau, A., & Heo, J. M. (2018). Building and decoding ubiquitin chains for mitophagy. *Nature Reviews Molecular Cell Biology*, *19*(2), 93–108. https://doi.org/10.1038/nrm.2017.129

He, C. (2022). Balancing nutrient and energy demand and supply via autophagy. *Current Biology*, *32*(12), R684–R696. https://doi.org/10.1016/J.CUB.2022.04.071

Heckmann, B. L., Teubner, B. J. W., Boada-Romero, E., Tummers, B., Guy, C., Fitzgerald, P., . . . Green, D. R. (2020). Noncanonical function of an autophagy protein prevents spontaneous Alzheimer's disease. *Science Advances*, *6*(33), eabb9036. https://doi.org/10.1126/sciadv.abb9036

Heckmann, B. L., Teubner, B. J. W., Tummers, B., Boada-Romero, E., Harris, L., Yang, M., . . . Green, D. R. (2019). LC3-associated endocytosis facilitates β-amyloid clearance and mitigates neurodegeneration in murine Alzheimer's disease. *Cell*, *178*(3), 536–551.e514. https://doi.org/10.1016/j.cell.2019.05.056

Hollville, E., Carroll, R. G., Cullen, S. P., & Martin, S. J. (2014). Bcl-2 family proteins participate in mitochondrial quality control by regulating Parkin/PINK1-dependent mitophagy. *Molecular Cell*, *55*(3), 451–466. https://doi.org/10.1016/j.molcel.2014.06.001

Hosokawa, N., Hara, T., Kaizuka, T., Kishi, C., Takamura, A., Miura, Y., . . . Mizushima, N. (2009). Nutrient-dependent mTORCl association with the ULK1-Atg13-FIP200 complex required for autophagy. *Molecular Biology of the Cell*, *20*(7), 1981–1991. https://doi.org/10.1091/MBC.E08-12-1248/ASSET/IMAGES/LARGE/ZMK0070990130007.JPEG

Hsu, P. P., Kang, S. A., Rameseder, J., Zhang, Y., Ottina, K. A., Lim, D., . . . Sabatini, D. M. (2011). The mTOR-regulated phosphoproteome reveals a mechanism of mTORC1-mediated inhibition of growth factor signaling. *Science*, *332*(6035), 1317–1322. https://doi.org/10.1126/SCIENCE.1199498/SUPPL_FILE/HSU.SOM.PDF

Ikami, Y., Terasawa, K., Sakamoto, K., Ohtake, K., Harada, H., Watabe, T., . . . Hara-Yokoyama, M. (2022). The two-domain architecture of LAMP2A regulates its interaction with Hsc70. *Experimental Cell Research*, *411*(1), 112986. https://doi.org/10.1016/J.YEXCR.2021.112986

Iliff, J. J., Wang, M., Liao, Y., Plogg, B. A., Peng, W., Gundersen, G. A., . . . Nedergaard, M. (2012). A paravascular pathway facilitates CSF flow through the brain parenchyma and the clearance of interstitial solutes, including amyloid β. *Science Translational Medicine*, *4*(147), 147ra111. https://doi.org/10.1126/scitranslmed.3003748

Iliff, J. J., Wang, M., Zeppenfeld, D. M., Venkataraman, A., Plog, B. A., Liao, Y., . . . Nedergaard, M. (2013). Cerebral arterial pulsation drives paravascular CSF-interstitial fluid exchange in the murine brain. *Journal of Neuroscience*, *33*(46), 18190–18199. https://doi.org/10.1523/jneurosci.1592-13.2013

Imarisio, S., Carmichael, J., Korolchuk, V., Chen, C. W., Saiki, S., Rose, C., . . . Rubinsztein, D. C. (2008). Huntington's disease: From pathology and genetics to potential therapies. *Biochemical Journal*, *412*(2), 191–209. https://doi.org/10.1042/bj20071619

Jain, A., Lamark, T., Sjøttem, E., Larsen, K. B., Awuh, J. A., Øvervatn, A., . . . Johansen, T. (2010). p62/SQSTM1 is a target gene for transcription factor NRF2 and creates a positive feedback loop by inducing antioxidant response element-driven gene transcription. *Journal of Biological Chemistry*, *285*(29), 22576–22591. https://doi.org/10.1074/jbc.M110.118976

Ji, T., Zhang, C., Ma, L., Wang, Q., Zou, L., Meng, K., . . . Jiao, J. (2018). TRPC6-Mediated Ca2+ signaling is required for hypoxia-induced autophagy in human podocytes. *Cellular Physiology and Biochemistry*, *48*(4), 1782–1792. https://doi.org/10.1159/000492351

Jia, J., Claude-Taupin, A., Gu, Y., Choi, S. W., Peters, R., Bissa, B., . . . Deretic, V. (2020). Galectin-3 coordinates a cellular system for lysosomal repair and removal. *Developmental Cell*, *52*(1), 69–87.e68. https://doi.org/10.1016/j.devcel.2019.10.025

Jimenez-Sanchez, M., Licitra, F., Underwood, B. R., & Rubinsztein, D. C. (2017). Huntington's disease: Mechanisms of pathogenesis and therapeutic strategies. *Cold Spring Harbor Perspectives in Medicine*, *7*(7). https://doi.org/10.1101/cshperspect.a024240

Jo, C., Gundemir, S., Pritchard, S., Jin, Y. N., Rahman, I., & Johnson, G. V. (2014). Nrf2 reduces levels of phosphorylated tau protein by inducing autophagy adaptor protein NDP52. *Nature Communications*, *5*, 3496. https://doi.org/10.1038/ncomms4496

Kane, L. A., Lazarou, M., Fogel, A. I., Li, Y., Yamano, K., Sarraf, S. A., . . . Youle, R. J. (2014). PINK1 phosphorylates ubiquitin to activate Parkin E3 ubiquitin ligase activity. *Journal of Cell Biology*, *205*(2), 143–153. https://doi.org/10.1083/jcb.201402104

Kaur, J., & Debnath, J. (2015). Autophagy at the crossroads of catabolism and anabolism. *Nature Reviews Molecular Cell Biology*, *16*(8), 461–472. https://doi.org/10.1038/nrm4024

Kaushik, S., & Cuervo, A. M. (2016). AMPK-dependent phosphorylation of lipid droplet protein PLIN2 triggers its degradation by CMA. *Autophagy*, *12*(2), 432–438. https://doi.org/10.1080/15548627.2015.1124226

Kaushik, S., & Cuervo, A. M. (2018). The coming of age of chaperone-mediated autophagy. *Nature Reviews Molecular Cell Biology*, *19*(6), 365–381. https://doi.org/10.1038/s41580-018-0001-6

Khaminets, A., Behl, C., & Dikic, I. (2016). Ubiquitin-dependent and independent signals in selective autophagy. *Trends in Cell Biology*, *26*(1), 6–16. https://doi.org/10.1016/j.tcb.2015.08.010

Kim, I., Rodriguez-Enriquez, S., & Lemasters, J. J. (2007). Selective degradation of mitochondria by mitophagy. *Archives of Biochemistry and Biophysics*, *462*(2), 245–253. https://doi.org/10.1016/j.abb.2007.03.034

Kim, J., Kundu, M., Viollet, B., & Guan, K. L. (2011). AMPK and mTOR regulate autophagy through direct phosphorylation of Ulk1. *Nature Cell Biology*, *13*(2), 132–141. https://doi.org/10.1038/ncb2152

Kitada, T., Pisani, A., Karouani, M., Haburcak, M., Martella, G., Tscherter, A., . . . Shen, J. (2009). Impaired dopamine release and synaptic plasticity in the striatum of Parkin-/- mice. *Journal of Neurochemistry*, *110*(2), 613–621. https://doi.org/10.1111/j.1471-4159.2009.06152.x

Kiviniemi, V., Wang, X., Korhonen, V., Keinänen, T., Tuovinen, T., Autio, J., . . . Nedergaard, M. (2016). Ultra-fast magnetic resonance encephalography of physiological brain activity—Glymphatic pulsation mechanisms? *Journal of Cerebral Blood Flow & Metabolism*, *36*(6), 1033–1045. https://doi.org/10.1177/0271678x15622047

Klionsky, D. J., Petroni, G., Amaravadi, R. K., Baehrecke, E. H., Ballabio, A., Boya, P., . . . Pietrocola, F. (2021). Autophagy in major human diseases. *EMBO Journal*, *40*(19), e108863. https://doi.org/10.15252/embj.2021108863

Koyano, F., Okatsu, K., Kosako, H., Tamura, Y., Go, E., Kimura, M., . . . Matsuda, N. (2014). Ubiquitin is phosphorylated by PINK1 to activate parkin. *Nature*, *510*(7503), 162–166. https://doi.org/10.1038/nature13392

Kroemer, G. (2015). Autophagy: A druggable process that is deregulated in aging and human disease. *Journal of Clinical Investigation*, *125*(1), 1–4. https://doi.org/10.1172/JCI78652

Lachance, V., Wang, Q., Sweet, E., Choi, I., Cai, C. Z., Zhuang, X. X., . . . Yue, Z. (2019). Autophagy protein NRBF2 has reduced expression in Alzheimer's brains and modulates memory and amyloid-beta homeostasis in mice. *Molecular Neurodegeneration*, *14*(1), 43. https://doi.org/10.1186/s13024-019-0342-4

Lane, J. D., Korolchuk, V. I., Murray, J. T., Rabanal-Ruiz, Y., Otten, E. G., & Korolchuk, V. I. (2017). mTORC1 as the main gateway to autophagy. *Essays in Biochemistry*, *61*(6), 565–584. https://doi.org/10.1042/EBC20170027

Lazarou, M., Jin, S. M., Kane, L. A., & Youle, R. J. (2012). Role of PINK1 binding to the TOM complex and alternate intracellular membranes in recruitment and activation of the E3 ligase Parkin. *Developmental Cell*, *22*(2), 320–333. https://doi.org/10.1016/j.devcel.2011.12.014

Lee, H., Xie, L., Yu, M., Kang, H., Feng, T., Deane, R., . . . Benveniste, H. (2015). The effect of body posture on brain glymphatic transport. *Journal of Neuroscience*, *35*(31), 11034–11044. https://doi.org/10.1523/jneurosci.1625-15.2015

Leipnitz, G., Mohsen, A. W., Karunanidhi, A., Seminotti, B., Roginskaya, V. Y., Markantone, D. M., . . . Vockley, J. (2018). Evaluation of mitochondrial bioenergetics, dynamics, endoplasmic reticulum-mitochondria crosstalk, and reactive oxygen species in fibroblasts from patients with complex I deficiency. *Scientific Reports*, *8*(1), 1–14. https://doi.org/10.1038/s41598-018-19543-3

Levy, J. M. M., Towers, C. G., & Thorburn, A. (2017). Targeting autophagy in cancer. *Nature Reviews Cancer*, *17*(9), 528–542. https://doi.org/10.1038/nrc.2017.53

Liu, J., Liu, W., Li, R., & Yang, H. (2019). Mitophagy in Parkinson's disease: From pathogenesis to treatment. *Cells*, *8*(7). https://doi.org/10.3390/cells8070712

Lundgaard, I., Li, B., Xie, L., Kang, H., Sanggaard, S., Haswell, J. D., . . . Nedergaard, M. (2015). Direct neuronal glucose uptake heralds activity-dependent increases in cerebral metabolism. *Nature Communications*, *6*, 6807. https://doi.org/10.1038/ncomms7807

Lundgaard, I., Lu, M. L., Yang, E., Peng, W., Mestre, H., Hitomi, E., . . . Nedergaard, M. (2017). Glymphatic clearance controls state-dependent changes in brain lactate concentration. *Journal of Cerebral Blood Flow & Metabolism*, *37*(6), 2112–2124. https://doi.org/10.1177/0271678x16661202

Martens, S., & Behrends, C. (2020). Molecular mechanisms of selective autophagy. *Journal of Molecular Biology*, *432*(1), 1–2. https://doi.org/10.1016/J.JMB.2019.11.010

Martinez-Vicente, M. (2017). Neuronal mitophagy in neurodegenerative diseases. *Frontiers in Molecular Neuroscience*, *10*, 64. https://doi.org/10.3389/fnmol.2017.00064

Martinez-Vicente, M., Talloczy, Z., Kaushik, S., Massey, A. C., Mazzulli, J., Mosharov, E. V., . . . Cuervo, A. M. (2008). Dopamine-modified alpha-synuclein blocks chaperone-mediated autophagy. *Journal of Clinical Investigation*, *118*(2), 777–788. https://doi.org/10.1172/jci32806

Martinez-Vicente, M., Talloczy, Z., Wong, E., Tang, G., Koga, H., Kaushik, S., . . . Cuervo, A. M. (2010). Cargo recognition failure is responsible for inefficient autophagy in Huntington's disease. *Nature Neuroscience*, *13*(5), 567–576. https://doi.org/10.1038/nn.2528

Marzella, L., Ahlberg, J., & Glaumann, H. (1980). In vitro uptake of particles by lysosomes. *Experimental Cell Research*, *129*(2), 460–466. https://doi.org/10.1016/0014-4827(80)90515-7

Matsuda, N., Sato, S., Shiba, K., Okatsu, K., Saisho, K., Gautier, C. A., . . . Tanaka, K. (2010). PINK1 stabilized by mitochondrial depolarization recruits Parkin to damaged mitochondria and activates latent Parkin for mitophagy. *Journal of Cell Biology*, *189*(2), 211–221. https://doi.org/10.1083/jcb.200910140

Matsumoto, G., Wada, K., Okuno, M., Kurosawa, M., & Nukina, N. (2011). Serine 403 phosphorylation of p62/SQSTM1 regulates selective autophagic clearance of ubiquitinated proteins. *Molecular Cell*, *44*(2), 279–289. https://doi.org/10.1016/j.molcel.2011.07.039

Matsushima, M., Fujiwara, T., Takahashi, E., Minaguchi, T., Eguchi, Y., Tsujimoto, Y., . . . Nakamura, Y. (1998). Isolation, mapping, and functional analysis of a novel human cDNA (BNIP3L) encoding a protein homologous to human NIP3. *Genes Chromosomes Cancer*, *21*(3), 230–235.

McWilliams, T. G., Prescott, A. R., Montava-Garriga, L., Ball, G., Singh, F., Barini, E., . . . Ganley, I. G. (2018). Basal mitophagy occurs independently of PINK1 in mouse tissues of high metabolic demand. *Cell Metabolism*, *27*(2), 439–449.e435. https://doi.org/10.1016/j.cmet.2017.12.008

Metzger, S., Saukko, M., Van Che, H., Tong, L., Puder, Y., Riess, O., & Nguyen, H. P. (2010). Age at onset in Huntington's disease is modified by the autophagy pathway: Implication of the V471A polymorphism in Atg7. *Human Genetics*, *128*(4), 453–459. https://doi.org/10.1007/s00439-010-0873-9

Min, Y., Kim, M. J., Lee, S., Chun, E., & Lee, K. Y. (2018). Inhibition of TRAF6 ubiquitin-ligase activity by PRDX1 leads to inhibition of NFKB activation and autophagy activation. *Autophagy*, *14*(8), 1347–1358. https://doi.org/10.1080/15548627.2018.1474995/SUPPL_FILE/KAUP_A_1474995_SM5858.DOCX

Mizushima, N., Ohsumi, Y., & Yoshimori, T. (2002). Autophagosome formation in mammalian cells. *Cell Structure and Function*, *27*(6), 421–429. https://doi.org/10.1247/CSF.27.421

Mizushima, N., Yoshimori, T., & Ohsumi, Y. (2011). The role of Atg proteins in autophagosome formation. *Annual Review of Cell and Developmental Biology*, *27*, 107–132. https://doi.org/10.1146/ANNUREV-CELLBIO-092910-154005

Murphy, K. E., Gysbers, A. M., Abbott, S. K., Tayebi, N., Kim, W. S., Sidransky, E., . . . Halliday, G. M. (2014). Reduced glucocerebrosidase is associated with increased α-synuclein in sporadic Parkinson's disease. *Brain*, *137*(Pt 3), 834–848. https://doi.org/10.1093/brain/awt367

Murtha, L. A., Yang, Q., Parsons, M. W., Levi, C. R., Beard, D. J., Spratt, N. J., & McLeod, D. D. (2014). Cerebrospinal fluid is drained primarily via the spinal canal and olfactory route in young and aged spontaneously hypertensive rats. *Fluids Barriers CNS*, *11*, 12. https://doi.org/10.1186/2045-8118-11-12

Nagelhus, E. A., & Ottersen, O. P. (2013). Physiological roles of aquaporin-4 in brain. *Physiological Reviews*, *93*(4), 1543–1562. https://doi.org/10.1152/physrev.00011.2013

Nakamura, S., Oba, M., Suzuki, M., Takahashi, A., Yamamuro, T., Fujiwara, M., . . . Yoshimori, T. (2019). Suppression of autophagic activity by Rubicon is a signature of aging. *Nature Communications*, *10*(1), 847. https://doi.org/10.1038/s41467-019-08729-6

Nielsen, S., Nagelhus, E. A., Amiry-Moghaddam, M., Bourque, C., Agre, P., & Ottersen, O. P. (1997). Specialized membrane domains for water transport in glial cells: High-resolution immunogold cytochemistry of aquaporin-4 in rat brain. *Journal of Neuroscience*, *17*(1), 171–180. https://doi.org/10.1523/jneurosci.17-01-00171.1997

Nixon, R. A., Wegiel, J., Kumar, A., Yu, W. H., Peterhoff, C., Cataldo, A., & Cuervo, A. M. (2005). Extensive involvement of autophagy in Alzheimer disease: An immuno-electron microscopy study. *Journal of Neuropathology & Experimental Neurology*, *64*(2), 113–122. https://doi.org/10.1093/jnen/64.2.113

Oshima, M., Seki, T., Kurauchi, Y., Hisatsune, A., & Katsuki, H. (2019). Reciprocal regulation of chaperone-mediated autophagy/microautophagy and exosome release. *Biological and Pharmaceutical Bulletin*, *42*(8), 1394–1401. https://doi.org/10.1248/BPB.B19-00316

Palikaras, K., Lionaki, E., & Tavernarakis, N. (2018). Mechanisms of mitophagy in cellular homeostasis, physiology and pathology. *Nature Cell Biology*, *20*(9), 1013–1022. https://doi.org/10.1038/s41556-018-0176-2

Panaro, M. A., Benameur, T., & Porro, C. (2020). Extracellular vesicles miRNA cargo for microglia polarization in traumatic brain injury. *Biomolecules*, *10*(6). https://doi.org/10.3390/biom10060901

Park, S., Zuber, C., & Roth, J. (2020). Selective autophagy of cytosolic protein aggregates involves ribosome-free rough endoplasmic reticulum. *Histochemistry and Cell Biology*, *153*(2), 89–99. https://doi.org/10.1007/S00418-019-01829-W/FIGURES/7

Pickles, S., Vigié, P., & Youle, R. J. (2018). Mitophagy and quality control mechanisms in mitochondrial maintenance. *Current Biology*, *28*(4), R170–R185. https://doi.org/10.1016/j.cub.2018.01.004

Pickrell, A. M., Huang, C. H., Kennedy, S. R., Ordureau, A., Sideris, D. P., Hoekstra, J. G., . . . Youle, R. J. (2015). Endogenous parkin preserves dopaminergic substantia nigral neurons following mitochondrial DNA mutagenic stress. *Neuron*, *87*(2), 371–381. https://doi.org/10.1016/j.neuron.2015.06.034

Plowey, E. D., Cherra, S. J., 3rd, Liu, Y. J., & Chu, C. T. (2008). Role of autophagy in G2019S-LRRK2-associated neurite shortening in differentiated SH-SY5Y cells. *Journal of Neurochemistry*, *105*(3), 1048–1056. https://doi.org/10.1111/j.1471-4159.2008.05217.x

Quintavalle, C., di Costanzo, S., Zanca, C., Tasset, I., Fraldi, A., Incoronato, M., . . . Condorelli, G. (2014). Phosphorylation-regulated degradation of the tumor-suppressor form of PED by chaperone-mediated autophagy in lung cancer cells. *Journal of Cellular Physiology*, *229*(10), 1359–1368. https://doi.org/10.1002/JCP.24569

Raffaello, A., Mammucari, C., Gherardi, G., & Rizzuto, R. (2016). Calcium at the center of cell signaling: Interplay between endoplasmic reticulum, mitochondria, and lysosomes. *Trends in Biochemical Sciences*, *41*(12), 1035–1049. https://doi.org/10.1016/J.TIBS.2016.09.001

Ramesh Babu, J., Lamar Seibenhener, M., Peng, J., Strom, A. L., Kemppainen, R., Cox, N., . . . Wooten, M. W. (2008). Genetic inactivation of p62 leads to accumulation of hyperphosphorylated tau and neurodegeneration. *Journal of Neurochemistry*, *106*(1), 107–120. https://doi.org/10.1111/j.1471-4159.2008.05340.x

Ramonet, D., Daher, J. P., Lin, B. M., Stafa, K., Kim, J., Banerjee, R., . . . Moore, D. J. (2011). Dopaminergic neuronal loss, reduced neurite complexity and autophagic abnormalities in transgenic mice expressing G2019S mutant LRRK2. *PLoS One*, *6*(4), e18568. https://doi.org/10.1371/journal.pone.0018568

Rasmussen, M. K., Mestre, H., & Nedergaard, M. (2018). The glymphatic pathway in neurological disorders. *Lancet Neurology*, *17*(11), 1016–1024. https://doi.org/10.1016/s1474-4422(18)30318-1

Ravikumar, B., Acevedo-Arozena, A., Imarisio, S., Berger, Z., Vacher, C., O'Kane, C. J., . . . Rubinsztein, D. C. (2005). Dynein mutations impair autophagic clearance of aggregate-prone proteins. *Nature Genetics*, *37*(7), 771–776. https://doi.org/10.1038/ng1591

Ravikumar, B., Moreau, K., Jahreiss, L., Puri, C., & Rubinsztein, D. C. (2010). Plasma membrane contributes to the formation of pre-autophagosomal structures. *Nature Cell Biology*, *12*(8), 747–757. https://doi.org/10.1038/ncb2078

Reza-Zaldivar, E. E., Hernández-Sapiéns, M. A., Gutiérrez-Mercado, Y. K., Sandoval-Ávila, S., Gomez-Pinedo, U., Márquez-Aguirre, A. L., . . . Canales-Aguirre, A. A. (2019). Mesenchymal stem cell-derived exosomes promote neurogenesis and cognitive function recovery in a mouse model of Alzheimer's disease. *Neural Regeneration Research*, *14*(9). https://journals.lww.com/nrronline/Fulltext/2019/14090/Mesenchymal_stem_cell_derived_exosomes_promote.29.aspx

Rockel, J. S., & Kapoor, M. (2016). Autophagy: Controlling cell fate in rheumatic diseases. *Nature Reviews Rheumatology*, *12*(9), 517–531. https://doi.org/10.1038/nrrheum.2016.92

Sahu, R., Kaushik, S., Clement, C. C., Cannizzo, E. S., Scharf, B., Follenzi, A., . . . Santambrogio, L. (2011). Microautophagy of cytosolic proteins by late endosomes. *Developmental Cell*, *20*(1), 131–139. https://doi.org/10.1016/j.devcel.2010.12.003

Sato, M., Seki, T., Konno, A., Hirai, H., Kurauchi, Y., Hisatsune, A., & Katsuki, H. (2019). Rapamycin activates mammalian microautophagy. *Journal of Pharmacological Sciences*, *140*(2), 201–204. https://doi.org/10.1016/J.JPHS.2019.05.007

Saxton, R. A., & Sabatini, D. M. (2017). mTOR signaling in growth, metabolism, and disease. *Cell*, *168*(6), 960–976. https://doi.org/10.1016/J.CELL.2017.02.004

Scesa, G., Moyano, A. L., Bongarzone, E. R., & Givogri, M. I. (2016). Port-to-port delivery: Mobilization of toxic sphingolipids via extracellular vesicles. *Journal of Neuroscience Research*, *94*(11), 1333–1340. https://doi.org/10.1002/JNR.23798

Scotto Rosato, A., Montefusco, S., Soldati, C., di Paola, S., Capuozzo, A., Monfregola, J., . . . Medina, D. L. (2019). TRPML1 links lysosomal calcium to autophagosome biogenesis through the activation of the CaMKKβ/VPS34 pathway. *Nature Communications*, *10*(1), 1–16. https://doi.org/10.1038/s41467-019-13572-w

Seaman, M. N. J. (2021). The retromer complex: From genesis to revelations. *Trends in Biochemical Sciences*, *46*(7), 608–620. https://doi.org/10.1016/j.tibs.2020.12.009

Sekine, S., & Youle, R. J. (2018). PINK1 import regulation; a fine system to convey mitochondrial stress to the cytosol. *BMC Biology*, *16*(1), 2. https://doi.org/10.1186/s12915-017-0470-7

Shen, W., & Ganetzky, B. (2009). Autophagy promotes synapse development in drosophila. *Journal of Cell Biology*, *187*(1), 71–79. https://doi.org/10.1083/jcb.200907109

Shibata, M., Lu, T., Furuya, T., Degterev, A., Mizushima, N., Yoshimori, T., . . . Yuan, J. (2006). Regulation of intracellular accumulation of mutant huntingtin by Beclin 1. *Journal of Biological Chemistry*, *281*(20), 14474–14485. https://doi.org/10.1074/jbc.M600364200

Shimizu, N., Asakawa, S., Minoshima, S., Kitada, T., Hattori, N., Matsumine, H., . . . Mizuno, Y. (2000). PARKIN as a pathogenic gene for autosomal recessive juvenile Parkinsonism. *Journal of Neural Transmission. Supplementa*, (58), 19–30. https://doi.org/10.1007/978-3-7091-6284-2_2

Silva, M. C., Nandi, G. A., Tentarelli, S., Gurrell, I. K., Jamier, T., Lucente, D., . . . Haggarty, S. J. (2020). Prolonged tau clearance and stress vulnerability rescue by pharmacological activation of autophagy in tauopathy neurons. *Nature Communications*, *11*(1), 3258. https://doi.org/10.1038/s41467-020-16984-1

Sliter, D. A., Martinez, J., Hao, L., Chen, X., Sun, N., Fischer, T. D., . . . Youle, R. J. (2018). Parkin and PINK1 mitigate STING-induced inflammation. *Nature*, *561*(7722), 258–262. https://doi.org/10.1038/s41586-018-0448-9

Son, J. H., Shim, J. H., Kim, K. H., Ha, J. Y., & Han, J. Y. (2012). Neuronal autophagy and neurodegenerative diseases. *Experimental & Molecular Medicine*, *44*(2), 89–98. https://doi.org/10.3858/emm.2012.44.2.031

Stamenkovic, M., Janjetovic, K., Paunovic, V., Ciric, D., Kravic-Stevovic, T., & Trajkovic, V. (2019). Comparative analysis of cell death mechanisms induced by lysosomal autophagy inhibitors. *European Journal of Pharmacology*, *859*, 172540. https://doi.org/10.1016/J.EJPHAR.2019.172540

Stavoe, A. K. H., & Holzbaur, E. L. F. (2019). Neuronal autophagy declines substantially with age and is rescued by overexpression of WIPI2. *Autophagy*, *16*(2), 371–372. https://doi.org/10.1080/15548627.2019.1695401

Stolz, A., Ernst, A., & Dikic, I. (2014). Cargo recognition and trafficking in selective autophagy. *Nature Cell Biology*, *16*(6), 495–501. https://doi.org/10.1038/ncb2979

Strappazzon, F., di Rita, A., Peschiaroli, A., Leoncini, P. P., Locatelli, F., Melino, G., & Cecconi, F. (2019). HUWE1 controls MCL1 stability to unleash AMBRA1-induced mitophagy. *Cell Death & Differentiation*, *27*(4), 1155–1168. https://doi.org/10.1038/s41418-019-0404-8

Takahashi, Y., Meyerkord, C. L., Hori, T., Runkle, K., Fox, T. E., Kester, M., . . . Wang, H. G. (2011). Bif-1 regulates Atg9 trafficking by mediating the fission of Golgi membranes during autophagy. *Autophagy*, *7*(1), 61–73. https://doi.org/10.4161/AUTO.7.1.14015/SUPPL_FILE/KAUP_A_10914015_SM0001.ZIP

Takáts, S., Tóth, S., Szenci, G., & Juhász, G. (2019). Investigating non-selective autophagy in drosophila. *Methods in Molecular Biology*, *1880*, 589–600. https://doi.org/10.1007/978-1-4939-8873-0_38/FIGURES/5

Ulland, T. K., Song, W. M., Huang, S. C., Ulrich, J. D., Sergushichev, A., Beatty, W. L., . . . Colonna, M. (2017). TREM2 maintains microglial metabolic fitness in Alzheimer's disease. *Cell*, *170*(4), 649–663.e613. https://doi.org/10.1016/j.cell.2017.07.023

Valente, E. M., Abou-Sleiman, P. M., Caputo, V., Muqit, M. M., Harvey, K., Gispert, S., . . . Wood, N. W. (2004). Hereditary early-onset Parkinson's disease caused by mutations in PINK1. *Science (New York, N.Y.)*, *304*(5674), 1158–1160. https://doi.org/10.1126/science.1096284

Wairkar, Y. P., Toda, H., Mochizuki, H., Furukubo-Tokunaga, K., Tomoda, T., & Diantonio, A. (2009). Unc-51 controls active zone density and protein composition by downregulating ERK signaling. *Journal of Neuroscience*, *29*(2), 517–528. https://doi.org/10.1523/jneurosci.3848-08.2009

Wan, H. I., DiAntonio, A., Fetter, R. D., Bergstrom, K., Strauss, R., & Goodman, C. S. (2000). Highwire regulates synaptic growth in Drosophila. *Neuron*, *26*(2), 313–329. https://doi.org/10.1016/s0896-6273(00)81166-6

Webb, J. L., Ravikumar, B., Atkins, J., Skepper, J. N., & Rubinsztein, D. C. (2003). Alpha-synuclein is degraded by both autophagy and the proteasome. *Journal of Biological Chemistry*, *278*(27), 25009–25013. https://doi.org/10.1074/jbc.M300227200

Whitworth, A. J., & Pallanck, L. J. (2009). The PINK1/Parkin pathway: A mitochondrial quality control system? *Journal of Bioenergetics and Biomembranes*, *41*(6), 499–503. https://doi.org/10.1007/s10863-009-9253-3

Winslow, A. R., Chen, C. W., Corrochano, S., Acevedo-Arozena, A., Gordon, D. E., Peden, A. A., . . . Rubinsztein, D. C. (2010). α-Synuclein impairs macroautophagy: Implications for Parkinson's disease. *Journal of Cell Biology*, *190*(6), 1023–1037. https://doi.org/10.1083/jcb.201003122

Wu, S., Shen, Y., Zhang, S., Xiao, Y., & Shi, S. (2020). Salmonella interacts with autophagy to offense or defense. *Frontiers in Microbiology*, *11*, 515222. https://doi.org/10.3389/FMICB.2020.00721/BIBTEX

Xia, Y., Zhang, G., Han, C., Ma, K., Guo, X., Wan, F., . . . Wang, T. (2019). Microglia as modulators of exosomal alpha-synuclein transmission. *Cell Death & Disease*, *10*(3), 1–15. https://doi.org/10.1038/s41419-019-1404-9

Xu, Y., Shen, J., & Ran, Z. (2020). Emerging views of mitophagy in immunity and autoimmune diseases. *Autophagy*, *16*(1), 3–17. https://doi.org/10.1080/15548627.2019.1603547

Xu, Y., Zhang, S., & Zheng, H. (2019). The cargo receptor SQSTM1 ameliorates neurofibrillary tangle pathology and spreading through selective targeting of pathological MAPT (microtubule associated protein tau). *Autophagy*, *15*(4), 583–598. https://doi.org/10.1080/15548627.2018.1532258

Xu, Z., Nan, W., Zhang, X., Sun, Y., Yang, J., Lu, K., . . . Lin, J. (2018). Umbilical cord mesenchymal stem cells conditioned medium promotes Aβ25–35 phagocytosis by modulating autophagy and Aβ-degrading enzymes in BV2 cells. *Journal of Molecular Neuroscience*, *65*(2), 222–233. https://doi.org/10.1007/S12031-018-1075-5/FIGURES/7

Yamano, K., & Youle, R. J. (2013). PINK1 is degraded through the N-end rule pathway. *Autophagy*, *9*(11), 1758–1769. https://doi.org/10.4161/auto.24633

Yang, C., Cai, C. Z., Song, J. X., Tan, J. Q., Durairajan, S. S. K., Iyaswamy, A., . . . Lu, J. H. (2017). NRBF2 is involved in the autophagic degradation process of APP-CTFs in Alzheimer disease models. *Autophagy*, *13*(12), 2028–2040. https://doi.org/10.1080/15548627.2017.1379633

You, L., Mao, L., Wei, J., Jin, S., Yang, C., Liu, H., . . . Qian, W. (2017). The crosstalk between autophagic and endo-/exosomal pathways in antigen processing for MHC presentation in anticancer T cell immune responses. *Journal of Hematology & Oncology*, *10*(1), 1–9. https://doi.org/10.1186/S13045-017-0534-8

Yu, W. H., Cuervo, A. M., Kumar, A., Peterhoff, C. M., Schmidt, S. D., Lee, J. H., . . . Nixon, R. A. (2005). Macroautophagy–a novel Beta-amyloid peptide-generating pathway activated in Alzheimer's disease. *Journal of Cell Biology*, *171*(1), 87–98. https://doi.org/10.1083/jcb.200505082

Zavodszky, E., Seaman, M. N., Moreau, K., Jimenez-Sanchez, M., Breusegem, S. Y., Harbour, M. E., & Rubinsztein, D. C. (2014). Mutation in VPS35 associated with Parkinson's disease impairs WASH complex association and inhibits autophagy. *Nature Communications*, *5*, 3828. https://doi.org/10.1038/ncomms4828

Zhang, W., Xu, C., Sun, J., Shen, H. M., Wang, J., & Yang, C. (2022). Impairment of the autophagy–lysosomal pathway in Alzheimer's diseases: Pathogenic mechanisms and therapeutic potential. *Acta Pharmaceutica Sinica B*, *12*(3), 1019–1040. https://doi.org/10.1016/J.APSB.2022.01.008

Zheng, S., Clabough, E. B., Sarkar, S., Futter, M., Rubinsztein, D. C., & Zeitlin, S. O. (2010). Deletion of the huntingtin polyglutamine stretch enhances neuronal autophagy and longevity in mice. *PLoS Genetics*, *6*(2), e1000838. https://doi.org/10.1371/journal.pgen.1000838

Zhu, Y., Massen, S., Terenzio, M., Lang, V., Chen-Lindner, S., Eils, R., . . . Brady, N. R. (2013). Modulation of serines 17 and 24 in the LC3-interacting region of Bnip3 determines pro-survival mitophagy versus apoptosis. *Journal of Biological Chemistry*, *288*(2), 1099–1113. https://doi.org/10.1074/jbc.M112.399345

Ziemba, B. P., Pilling, C., Calleja, V., Larijani, B., & Falke, J. J. (2013). The PH domain of phosphoinositide-dependent kinase-1 exhibits a novel, phospho-regulated monomer-dimer equilibrium with important implications for kinase domain activation: Single-molecule and ensemble studies. *Biochemistry*, *52*(28), 4820–4829. https://doi.org/10.1021/BI400488F/ASSET/IMAGES/MEDIUM/BI-2013-00488F_0005.GIF

15 Mesoporous Silica Nanoparticles as Promising Carriers for Brain Delivery

*Sudarshan Naidu Chilamakuri, Sivakumar Manickam, Rajukumar Malayandi, Ravichandiran Velayutham, and Subramanian Natesan**

15.1 INTRODUCTION

15.1.1 Exploration of Mesoporous Silica Nanoparticles (MSNs)

Over the past decade, nanotechnology has seen a significant rise in its application for drug delivery in treating various diseases. This growth is attributed to the unique properties of nanoparticles (NPs) (Li et al., 2019). Utilizing NPs with distinct properties enhances drug delivery precision, promising targeted transportation of diverse therapeutic agents to specific body locations (Yetisgin et al., 2020). Porous nanoparticles, notably mesoporous silica nanoparticles (MSNs), have recently gained considerable attention due to their role as storage reservoirs for biological and pharmaceutical molecules. In drug delivery and imaging applications, MSNs are highly valued for their high porosity and surface property modulation through a process known as multifunctionalization (Mitran et al., 2019). Creating solid porous materials with specific ordered structures is achievable using various materials, but silica is a common and preferred choice. The International Union of Pure and Applied Chemistry (IUPAC) classifies materials based on their pore size. Although microporous materials such as zeolites demonstrate exceptional performance in catalysis, their application in drug delivery encounters obstacles due to their extremely tiny pore dimensions (0.74 nm), making it challenging to effectively encapsulate drugs (Khodaverdi et al., 2015). To address the limitation posed by the pore size zeolites, Mobil Company in the USA developed Mobil Composition of Matter No. 41 (MCM-41) mesoporous silica in 1992 (Kresge et al., 1992). MCM-41 mesoporous silica was employed in drug delivery as a matrix material for delivering ibuprofen, marking significant progress in the field (Vallet-Regi et al., 2000). Since then, extensive research has been conducted to enhance the synthesis and surface functionalization of nano-silica for biomedical applications (Lin et al., 2012). Smaller silica derivatives are also widely used for delivering peptides, proteins, and gene sequences (Ahmed et al., 2022). The initial modification of MSNs for the controlled release of neurotransmitters and drug molecules involved stimulus-responsive functionalization using 2-(propyl disulfanyl) ethylamine (Shen et al., 2017). Scientists have also explored MSN modifications with dual diagnostic and therapeutic effects for controlled drug release and targeting (Shen et al., 2017; Wang et al., 2010).

Synthesized mesoporous silica is commonly employed for imaging purposes, often by including imaging agents like fluorophores for fluorescence imaging and supermagnetic NPs (Kim et al., 2005; Wu et al., 2008; Lin et al., 2005). Ongoing research has been dedicated to studying the *in vivo* biodistribution, therapeutic outcomes, and potential toxicity of mesoporous silica over an extended period (Huang et al., 2011; Kim et al., 2008).

Addressing the challenge of delivering biomolecules such as peptides and amino acids, which are typically challenging to retain within MSNs, can be achieved by altering the surface reactivity. This is achieved by introducing an amino functional group, which changes the surface charge of MSNs from negative to positive. This modification aids in retaining these cargo molecules and, in addition to surface alterations, enables the controlled release of drugs and biomolecules at specific target sites to treat various diseases (Vallet-Regí, 2022).

15.1.2 Hallmark of MSNs

MSNs are equipped with silanol groups, which are silicic acid species that are biocompatible and non-toxic when present in appropriate amounts (Tarn et al., 2013). Including pores in MSNs for cargo molecule delivery protects the cargo from enzymatic degradation (Niculescu, 2020). These pores are adjustable, with sizes ranging from 2 to 6 nm and a uniform distribution, greatly facilitating drug loading and absorption. Compared to non-porous silica NPs, MSNs exhibit higher biodegradability as they increase exposure to biological fluids, leading to greater dispersion of the MSNs for degradation (Yamada et al., 2012). MSNs also possess remarkable durability, with a solid structure that resists degradation from acids, bases, alkalis, heat, and hydrolysis. This durability makes them an attractive alternative to other drug carriers based on polymers.

MSNs possess both an outer and an inner surface, increasing the potential to selectively attach drug molecules to the interior

* Corresponding author: snatesan@niperkolkata.ac.in

DOI: 10.1201/9781003383376-15

and/or exterior of MSNs. The versatility can be harnessed for a wide range of therapeutic applications. Furthermore, the cost-effective production of MSNs makes them an economically viable and attractive choice for utilization and fabrication (Pednekar et al., 2017). In addressing the influence of drug molecules, whether hydrophilic or hydrophobic, surface functionalization is necessary. By forming chemical bonds with other materials, such as stimuli-responsive, luminescent, or capping, MSNs can acquire intelligent and multifunctional properties, expanding their versatility (Bharti et al., 2015).

15.2 CLASSIFICATION OF SILICA

MSNs are a class of nanomaterials characterized by their extensive surface area and a uniform network of mesopores, which are pores with diameters ranging from 2 to 50 nm integrated into their structure. These mesopores confer various applications to MSNs, including drug delivery, catalysis, and sensing. Researchers have been dedicated to the development of MSNs for several decades. Figure 15.2 shows a general timeline highlighting key advancements in various types of MSNs.

15.3 SYNTHESIS OF MSNS

MSNs are generated by carefully assembling silica precursors, followed by removing a template, resulting in a porous structure. Over the years, various approaches to MSN synthesis have been developed, each offering unique advantages and disadvantages. The primary methods employed in synthesizing MSNs include the sol-gel process, the soft and hard template method, and hydrothermal treatment.

15.3.1 Sol–Gel Synthesis

The sol–gel method involves a solution phase that contains colloidal particles generated under either alkaline or acidic pH conditions. This is achieved through condensation and hydrolysis reactions. Subsequently, the gel phase forms due to the condensation of these colloidal particles, leading to the development of a 3D structure with cross-linking siloxane bonds (Corbalan et al., 2011). The sol–gel synthesis process is represented using Equations 15.1 to 15.5..

$$\equiv Si - OH + OH - Si \equiv \leftrightarrow \equiv Si - O - Si \equiv + HOH \quad (15.1)$$

$$\equiv Si - OH + RO - Si \equiv \leftrightarrow \equiv Si - O - Si \equiv + ROH \quad (15.2)$$

where R = alkyl group (methyl, ethyl, propyl, etc.).

$$\text{Hydrolysis: } Si\,(OC_2H_5)_4 + H_2O \leftrightarrow Si\,(OC_2H_3)_3\,OH + C_2H_5OH \quad (15.3)$$

Water condensation:

$$Si - OH + OH - Si \equiv \leftrightarrow \equiv Si - O - Si \equiv + HOH \quad (15.4)$$

Alcohol condensation:

$$Si - O - C_2H_5 + OH - Si \leftrightarrow Si - O - Si + C_2H_5OH \quad (15.5)$$

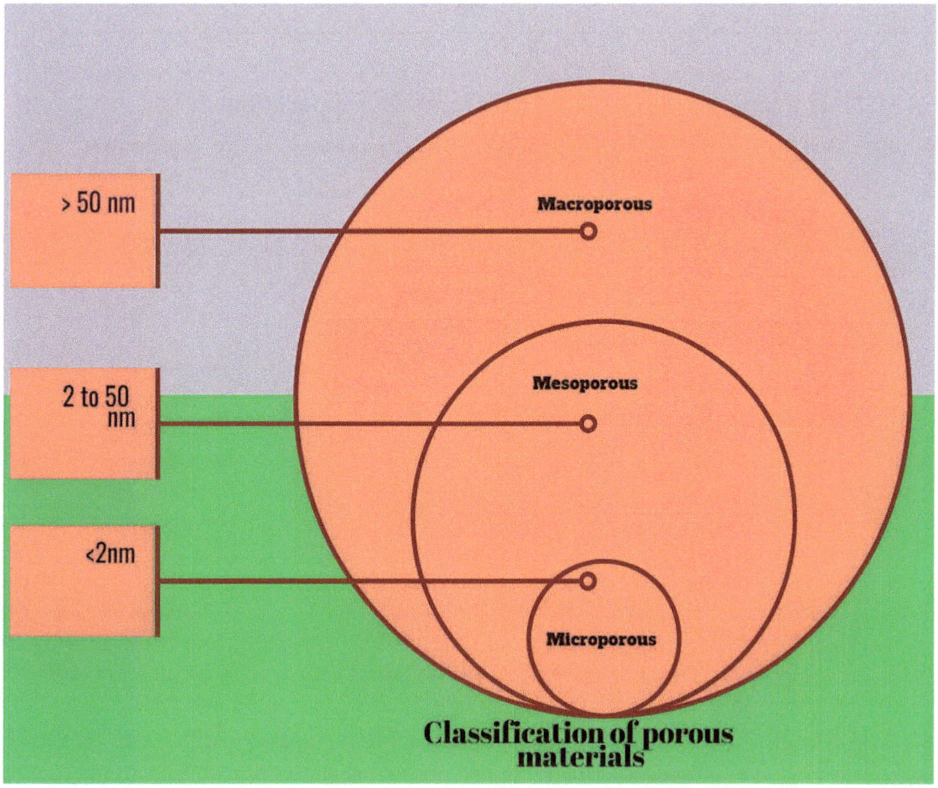

FIGURE 15.1 Classification of porous materials.

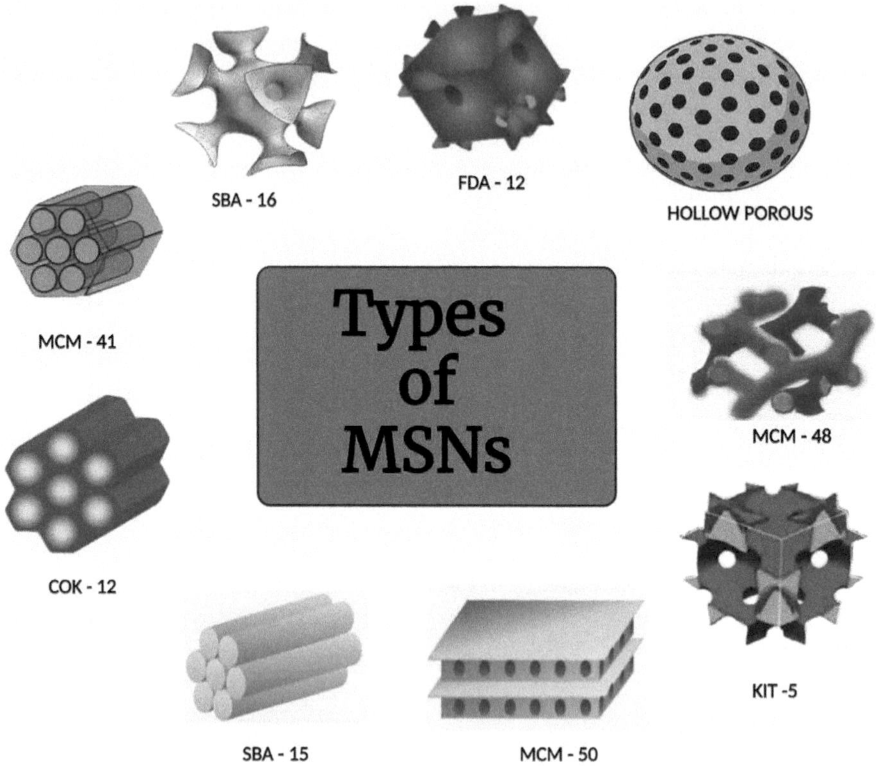

FIGURE 15.2 Types of MSNs. MCM – Mobile Crystalline Matter: SBA – Santa Barbara Amorphous: KIT – Korea Advanced Institute of Science and Technology: FDU – Fairleigh Dickinson University: COK – Centre for Research Chemistry and Catalysis Mobil Crystalline.

TABLE 15.1
Classification of MSNs and their pore size, surfactants used, and applications

MSN family	MSN types	Pore size (nm)	Pore symmetry	Surfactants used	Applications	References
M41S	MCM-41	1.5–8	2D hexagonal	Cetyltrimethylammonium chloride	Drug delivery	Kumar et al., 2001
	MCM-48	2–5	3D cubic	Hexadecyltrimethylammonium bromide	Drug delivery	Wang, 2009; Izquierdo-Barba et al., 2005
	MCM-50	2–5	Lamellar	Cetyltrimethylammonium bromide	Drug delivery	Wang & Li, 2006
SBA	SBA-11	2.1–3.6	3D cubic	Polyethylene oxide (Brij 56)	Drug delivery	Kim et al., 2015
	SBA-12	3.1	3D hexagonal	Pluronic F127	Drug and protein delivery	Mayoral et al., 2013
	SBA-15	5–9	2D hexagonal	Pluronic P123	Drug delivery	Albayati et al., 2019
	SBA-16	5–15	Cubic cage	Pluronic F127	Drug delivery	Andrade et al., 2012
KIT	KIT-5	1–9.3	Cubic	Pluronic F-127	Drug delivery	Javad Kalbasi & Zirakbash, 2015
COK	Cok-12	6–9	Hexagonal P6m	Pluronic P123		Colmenares et al., 2018
FDU	FDU-1	10–26 nm	Hexagonal prism	Pluronic F-127	Adsorbents and drug delivery	Ghaforinejad et al., 2021
FSM	FSM-16	2–50 nm	Hexagonal channels	Quaternary ammonium surfactant	Catalytic activity and adsorbent	Tozuka et al., 2005
TUD	TUD-1	2.5–25 nm	Foam-type	Free of surfactant	Catalytic activity and drug delivery	Heikkila et al., 2007
HMM	HMM-33	4–15 nm	Spherical	Cetyltrimethylammonium bromide (CTAB)	Catalytic activity and adsorbent	Nandiyanto et al., 2009

Several modifications to Stober's method, instrumental in the early synthesis of MSNs, have led to monodisperse and highly ordered nanosized MSNs. One such modification involves using MCM-41 formed by cationic surfactants as a template, resulting in MSNs with a spherical shape and properties similar to Stober's hexagonal MCM-41 (Grün et al., 1997). This method offers advantages, such as reduced use of excipients, simplicity in production, and cost-effectiveness (Bharti et al., 2015). Adjustments in reaction conditions, such as changes in the pH of the reaction mixture, temperature, surfactant concentration, and the source of silica, can yield MSNs with different sizes and shapes (Narayan et al., 2018). Additionally, biogenic MSNs, derived from natural compounds found in wheat and rice husks, can be synthesized using the sol–gel method to deliver substances like doxorubicin (Porrang et al., 2021).

15.3.2 Soft or Hard Templating

To increase drug loading capacity for drug delivery, researchers have synthesized hollow silica mesoporous structures with larger pore sizes using either the soft or hard templating method. In the soft templating approach, vesicular structures like micelles and microemulsions serve as soft templates to generate pores, which are subsequently removed by heating to obtain pure MSNs (Djayanti et al., 2023). While the soft template method is a direct and effective way to produce NPs, it has certain limitations that may restrict its applicability. One concern is that heating may not eliminate all the vesicular structures, resulting in a non-uniform distribution of NP sizes and mesoporous structures.

To address the challenges associated with the soft template approach, hard templates composed of silica colloids, polymer lattices, or metal oxides have been employed to obtain MSNs with a narrower particle size distribution (Ghaferi et al., 2020). The hard templating or nano-casting process involves using capillary forces to impregnate the mesopores of the template material, such as silica, with precursor chemicals. Subsequently, the template material is chemically or thermally removed, resulting in a duplicate framework with an equivalent pore size distribution and structure mirroring the mesoporous silica (Farjadian et al., 2019).

Studies by Saputra et al. (2022) have yielded promising results, using the hard templating method to produce curcumin-loaded hollow mesoporous silica particles with high drug loading. The hard template formation was validated via X-ray diffraction (XRD) and transmission electron microscopy (TEM) analysis, employing sonic co-precipitation with L-serine.

15.3.3 Hydrothermal Technique

Innovative methods for synthesizing nanoparticles remain an ongoing pursuit in materials research. The development of the hydrothermal method can be attributed to Roderick Murchison, a renowned British geologist. Murchison's groundbreaking insight recognized that minerals could be formed through the cooling of magmas by hot water solutions, laying the foundation for the hydrothermal method.

The generation of mesoporous silica NPs is achieved through hydrothermal processes, which involve a reaction conducted within a sealed container at elevated pressure and temperature. In the hydrothermal process, the initial template is removed, and subsequently, the mixture is heated in an autoclave lined with Teflon at a specific temperature. The hydrothermal method yields MSNs that exhibit greater regularity and hydrothermal stability, two of its advantages. The elevated pressure and temperature conditions foster stronger bonding between the silica framework and any functional groups on the surface of the MSN, creating well-ordered structures (Siddiqui et al., 2022).

15.4 CHARACTERIZATION OF MSNs

15.4.1 Physisorption

A common approach to gaining a deeper understanding of the characteristics of porous materials is gas sorption analysis. In physisorption measurements, nitrogen gas is typically the standard adsorbate used. Gas sorption allows for calculating specific surface area, pore size distribution (PSD), and material pore volume. These parameters play a vital role in assessing the suitability and functionality of porous materials in various applications, such as catalysis and gas separation.

15.4.1.1 BET Method

The Brunauer-Emmet-Teller (BET) method is widely employed for assessing the surface area of mesoporous silica nanoparticles (MSNs) and similar porous materials. This approach is based on the Langmuir theory, with slight modifications for multilayer adsorption.

15.4.2 X-Ray Diffraction

X-ray diffraction (XRD) is a widely employed technique for investigating structural symmetry and determining the distances between adjacent pores. The intensity of X-rays scattered during this process is plotted as a function of the angle 2θ. Applying Bragg's law to the peaks observed in the XRD pattern makes it possible to determine the spacing between the crystal lattices. It is important to note that XRD peak intensities are strongly influenced by the mesopore diameter and unit cell size, making XRD a valuable tool for analyzing mesoporous materials.

$$n\lambda = 2d \sin\theta \tag{15.6}$$

15.4.3 Optical Microscopes

15.4.3.1 Scanning Electron Microscopy (SEM) and Transmission Electron Microscopy (TEM)

Electron microscopy is a valuable tool for investigating various aspects of NPs, including their shape, external

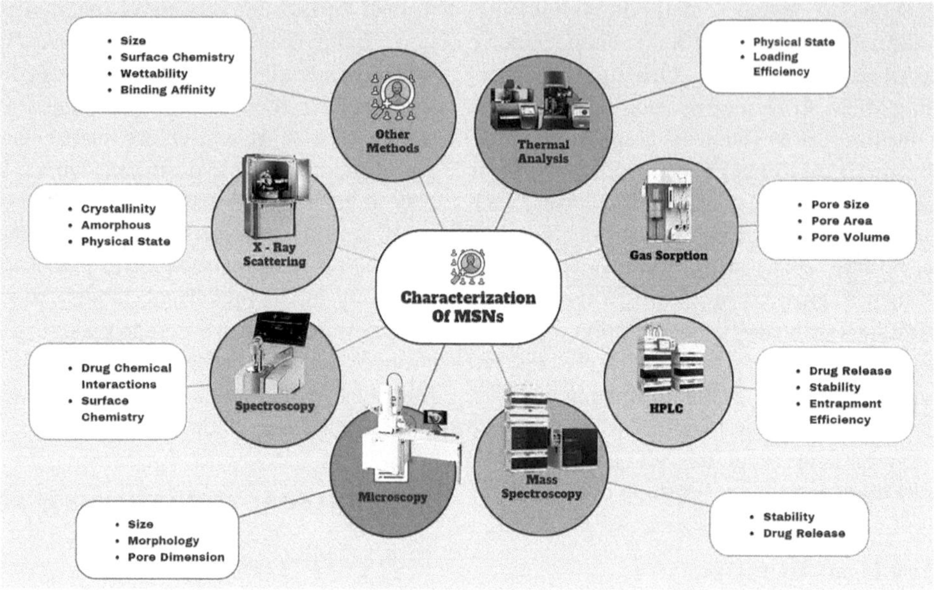

FIGURE 15.3 Different types of characterization techniques for MSNs.

diameter, surface area, and pore architecture. This technique provides rich structural information at the nanoscale, making it well-suited for studying MSNs and other nanostructured materials. Microscopy finds multiple applications, such as imaging API/MSNs systems; however, it does come with certain limitations. Some potential issues to consider include drying or vacuum conditions during sample preparation, which can lead to problematic and potentially destructive artifacts. Furthermore, to obtain an accurate assessment of the properties of the entire sample, it is necessary to characterize a significant number of individual particles.

15.4.4 Thermal Analysis (DSC and TGA)

DSC (differential scanning calorimetry) analysis is a valuable technique to determine whether a crystalline drug is present within a sample. DSC measures the depression in the melting point of a crystalline drug and can be utilized to quantify and identify the amounts of the drug enclosed within the pores of a material. It is important to note that melting point depression is only observable if the drug contained within the pores is crystalline.

Thermal gravimetric analysis (TGA) is a useful method for determining the total drug load in a sample. This technique quantifies the overall quantity of drug present by monitoring changes in weight as the drug undergoes decomposition and desorption at varying temperatures. It is important to note that quantifying the drug fraction loaded within the mesopores of mesoporous materials is a complex process, and it cannot be achieved solely by measuring the total drug content in the sample. Alongside other techniques like XRD, DSC, and high-performance liquid chromatography (HPLC), TGA is a significant approach for assessing the amount of drug loading (Zhang et al., 2010).

15.5 MSN-BASED SMART DRUG DELIVERY AGENTS

The effectiveness of active targeting in drug delivery can be influenced by an NP's ability to surmount biological barriers and its circulation time within the bloodstream. Various characteristics, including surface composition, surface charge, size, and shape, govern the cellular and tissue uptake and accumulation of nanoparticles. MSNs hold promise for drug delivery applications due to their substantial surface area, biocompatibility, and ease of functionalization. The substantial surface area of MSNs allows for the effective inclusion of drugs, even at high concentrations. However, there is a risk of premature drug release before reaching the intended target. To address this issue, using MSN pores aids in drug loading, and certain molecules can close these pores. Interestingly, MSNs can enhance their carrier efficiency by selectively opening their pores in response to targeted stimulation. This feature can be advantageous for achieving precise drug delivery.

15.5.1 Passive Targeting

In solid malignancies, NPs play a role in passive targeting. Passive targeting involves the natural accumulation of drug carriers in solid tumors that possess permeable vasculature and limited lymphatic drainage, allowing for the retention of these carriers. A deeper understanding of tumor pathology is crucial for comprehending the enhanced permeability and retention (EPR) effect. Solid tumors, characterized by their rapid growth, require an abundant supply of nutrients and oxygen, which leads to angiogenesis and the formation of new blood vessels. These newly formed blood vessels often have irregular, discontinuous endothelial linings and disorganized structures, creating gaps through which nanoparticles can infiltrate the tumor. This phenomenon increases permeability and contributes to the EPR effect (Narum et al., 2020).

To effectively use MSNs as passive drug delivery agents, it is essential to control their size, shape, and surface characteristics. These particles must be within a 50–300 nm size range to enter cancer cells through the leaky vasculature, ensuring they are larger than the kidneys can filter out but still manageable. However, it is worth noting that while particle size is important, it may not be the sole determinant of achieving an effective EPR effect, considering the wide range of methods and materials used to generate MSNs (Martínez-Carmona et al., 2015). Other factors, such as surface properties and shape, can also play a role in passive targeting.

15.5.2 Active Targeting

Active targeting is a method employed to overcome the challenges associated with passive targeting. It involves using strategies that specifically target receptors on particular cells. This is achieved by functionalizing the surface of MSNs with a ligand to bind with these receptors. Various ligands have been utilized for targeting and have been functionalized on MSNs, including antibodies, aptamers, transferrin, folic acid, and hyaluronic acid. The applications of these ligands are further discussed in the subsequent sections (Aquib et al., 2019).

15.5.3 Stimulus-Responsive

MSNs have evolved from passive targeting and active targeting to the current stimulus-responsive systems in their role as drug delivery agents. Literature suggests that successful cargo capture in stimulus-responsive systems often involves a combination of passive and active targeting mechanisms with the inclusion of a gate component. Both passive and active targeting approaches have drawbacks, including the risk of premature cargo release during circulation. To address this issue, a gate mechanism is employed to seal the pores of MSNs. These gates become accessible when the particles reach predetermined areas in response to specific stimuli. There are two primary categories of stimuli: internal stimuli, which encompass factors like pH, redox conditions, and enzymes, and external stimuli, including light and temperature (Mohamed Isa et al., 2021).

15.6 CHALLENGES ASSOCIATED WITH REACHING THE BRAIN

Several challenges exist when it comes to delivering drugs to the brain. Barriers such as the blood–brain barrier (BBB), arachnoid membrane, blood–cerebrospinal fluid barrier, ependymal barrier, and others create obstacles that make it difficult for drug formulations to reach the intended brain target (Figure 15.4). The complex structure of the brain also contributes to these challenges. As a result of these barriers, it can be challenging to ensure that medications reach the specific brain regions where they need to be effective or that molecules reach their intended brain destinations (Pandit et al., 2020). Understanding the challenges associated with drug transport to the brain is crucial for developing successful treatments for neurological diseases. Photodynamic and photothermal therapies offer ways to bypass the BBB and treat various brain disorders, although they come with challenges, such as photosensitization and potential skin tissue damage (Tang et al., 2021). In recent years, NPs, particularly multifunctional smart NPs, have shown promising potential in drug delivery by providing a non-invasive means to cross the BBB for treating CNS diseases.

15.6.1 Blood–Brain Barrier (BBB)

Neurons are highly sensitive to variations in chemical composition and require a constant supply of oxygen and nutrients to function properly. To ensure this supply, neurons are typically located within a distance of 8–20 mm from capillaries, which are crucial for maintaining brain homeostasis. The blood–brain barrier (BBB) is a specialized structural and biochemical barrier that has evolved in the brain's blood vessels to fulfil these vital functions. It regulates the brain's intake of nutrients and acts as a protective shield against external influences. The brain's arteries predominantly form this barrier, exhibiting distinctive features such as tight junctions (TJs) and elevated concentrations of metabolic enzymes and efflux transport proteins like P-glycoprotein (P-gp) and multidrug resistance proteins (MRPs), enhancing its complexity and functionality (Wanat, 2020).

BBB isolates the brain from the diverse array of cells in the body, facilitating communication between neurons while impeding the movement of molecules, thus slowing it down (Tajes et al., 2014). The BBB presents a significant challenge that must be overcome for drugs to access the brain. Because the BBB is predominantly semi-permeable, it acts as a barrier preventing drugs intended for treating CNS diseases from entering the brain (Burgess & Hynynen, 2014). This barrier, created by tight junctions in the epithelial and endothelial capillaries that lack fenestrations, imposes high electrical resistance and hinders the paracellular passage of drugs into the brain (Nguyen et al., 2021).

15.6.2 Arachnoid Membrane

The meninges include the avascular arachnoid membrane, comprising three layers: the outer dura, the arachnoid, and the inner pia mater. This membrane envelops the brain and spinal cord, forming the arachnoid barrier, a protective shield that safeguards these vital organs from external influences. The arachnoid barrier is a partition between the cerebrospinal fluid (CSF) located in the subarachnoid space and the extracellular fluids in the blood vessels of the superficial dura layers, consisting of fenestrated capillaries. The subarachnoid space is between the arachnoid and the pia mater (Mastorakos & McGavern, 2019).

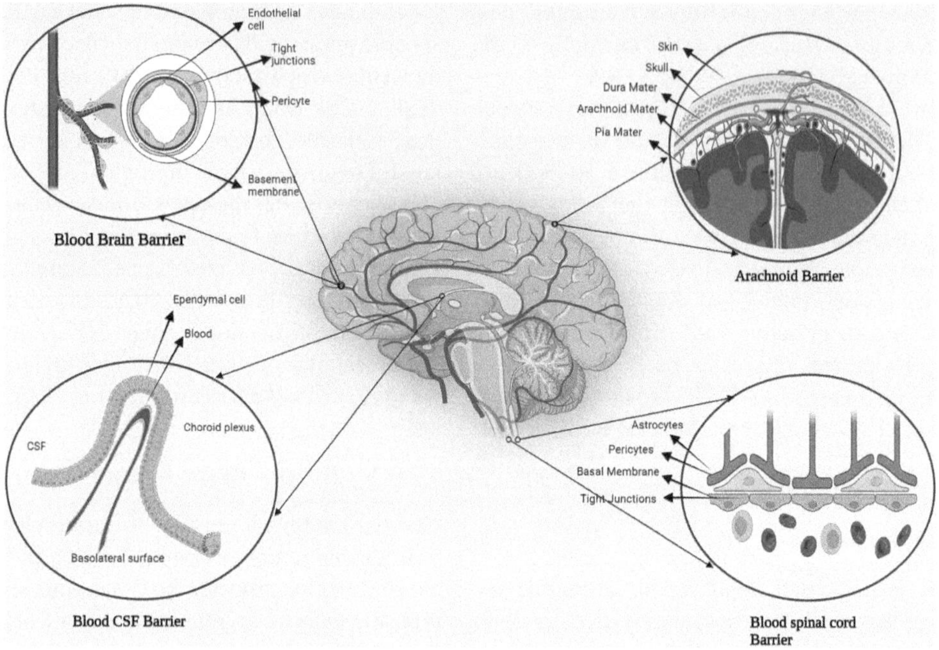

FIGURE 15.4 Barriers to drug delivery to the brain.

15.6.3 Blood–CSF Barrier

The choroid plexus (CP) is a complex and highly vascularized tissue located within the cerebral ventricles of the brain. It plays a pivotal role in CSF production and is a crucial component of the blood–CSF barrier. The choroidal blood vessels, which are fenestrated, lack tight junctions that typically protect the CNS. However, they are surrounded by a ciliated epithelial cell layer expressing tight junctions. Choroid cells have a distinct advantage over the endothelial cells of the BBB when transporting large molecules, such as proteins, from the blood into the CSF, as they possess a substantial number of pinocytotic vesicles. Maintaining the structural integrity of the choroid plexus is essential to keep the blood–CSF barrier permeable and to ensure the brain's normal functioning.

Recent research conducted both in laboratory settings and clinical environments has highlighted the critical role played by the CP in developing various CNS disorders. The CP serves as a common entry point for brain infections and is also implicated in the pathogenesis of neurodegenerative and autoimmune diseases, such as multiple sclerosis and Alzheimer's disease. Additionally, the CP is crucial in restoring brain homeostasis following disruptions caused by injuries or strokes (Solár et al., 2020).

15.6.4 The Circumventricular Organs

Several circumventricular organs (CVOs) in the human brain include the pineal gland, subfornical organ, area postrema, and median eminence. These organs are located in the brain region that connects the third and fourth ventricles. They are essential in regulating homeostasis and releasing hormones into the circulatory system. The blood vessels within the CVOs are not highly efficient at preventing the diffusion of substances because they lack particularly tight connections. Tanycytes are ependymal cells found in the circumventricular organs (CVOs). These cells are responsible for monitoring the CSF within the brain. Tanycytes come into direct contact with the CSF through their apical processes extending into the CSF. Additionally, they have distal processes that interact with fenestrated capillaries. Fenestrated capillaries are blood vessels with small openings that regulate the movement of various substances between the blood and CSF. Tanycytes play a crucial role in maintaining a consistent environment within the brain by controlling the movement of molecules between different fluid compartments. The vasculature surrounding these organs is similar to the BBB in the peripheral nervous system (Verheggen et al., 2020).

15.6.5 Ependyma

An ependymal barrier between the ventricular CSF and the brain parenchyma simplifies the passage of chemicals between these two compartments within the brain. However, this process is disrupted when the ependymal barrier is compromised, as in cases of encephalopathy, when there is a breach in the brain's protective covering. Mature ependymal cells can take on various forms, such as polygonal, cuboidal, or columnar arrangements, depending on their location within the ventricles. These adult ependymal cells exhibit distinct morphological and functional characteristics and are polarized. "Ependyma" refers to the simple, ciliated epithelium that borders the cerebral ventricles in the CNS. While it lacks specialization in CVO and the choroid plexus, ependymal cells establish partial contact between the brain parenchyma and the CSF through gap junctions. This interface allows communication between the CSF and the brain parenchyma, with ependymal cells responsible for creating this connection. Importantly, this interface does not act as a barrier to the diffusion of molecules, making it accessible even to large proteins (Jiménez et al., 2014; Saunders et al., 2016).

15.7 MSN AS DRUG CARRIERS TO TARGET THE BRAIN

The exceptional combination of physicochemical properties renders MSNs a highly favorable choice for drug delivery. Despite their compact size, the high surface-area-to-volume ratio enables them to transport many drugs efficiently. Furthermore, MSNs can be designed to be biocompatible with biological systems, biodegradable, and capable of targeting specific areas by functionalizing their surface. The BBB presents a formidable challenge to effective drug delivery to the brain, as it restricts the passage of chemicals and medications. However, it has been demonstrated that NPs like MSNs can overcome the BBB due to their small size and surface characteristics. MSNs possess a distinctive mesoporous structure that enables them to load and transport drugs in a controlled manner. The pore size of MSNs can be precisely customized to accommodate different types of medications, and the particles can be functionalized with targeting molecules, such as antibodies, peptides, or aptamers, to enhance specificity and efficacy.

Compared to other NP delivery systems, using MSNs as drug carriers for brain targeting offers several advantages. They are biocompatible, are biodegradable, and have a low toxicity profile, making them an attractive choice for the pharmaceutical industry. Moreover, they can be readily synthesized in large quantities, offering a cost-effective solution for mass production. Overall, using MSNs as drug carriers for brain targeting holds significant promise for treating neurological conditions. Their ability to cross the BBB, combined with their unique mesoporous structure, positions them as a valuable option for drug delivery. Investigations into the characteristics of MSNs and their applications in medication administration are essential for developing effective treatments for neurological disorders (Geng et al., 2008).

15.7.1 MSN-Based Therapy in Alzheimer's Disease

Neurological conditions such as Alzheimer's disease (AD), Parkinson's disease (PD), glioma, and multiple sclerosis (MS) can lead to brain damage. Numerous other neurological conditions can also result in brain damage. Neuroinflammation is a defining feature of Alzheimer's disease, the predominant form of neurodegeneration, which eventually leads to the formation of amyloid plaques and neurofibrillary tangles. A reduction in the levels of the neurotransmitter acetylcholine is another hallmark of the disease. Alzheimer's disease can have a genetic basis or occur randomly. It presents a wide range of symptoms, including, but not limited to, depressive episodes, agitation, memory loss, mild impairment in essential functions, and cognitive decline (Kocahan & Doğan, 2017; Petrovic et al., 2007).

There is a connection between the accumulation of extracellular amyloid-beta protein (A), a hallmark of the prevalent type of cognitive decline, and elevated levels of certain trace metal ions (Fe^{2+}, Cu^{2+}, and Zn^{2+}). Another characteristic of dementia is the presence of hyperphosphorylated tau protein, cell and synapse loss, the formation of tangled neurofibrils, and the growth of reticular structures. Some individuals may experience excitotoxicity, a condition where N-methyl-D-aspartate (NMDA) receptor activation gradually decreases. The aggregation of amyloid-beta is widely considered the primary pathophysiological process underlying Alzheimer's disease development, with metal ions playing a critical role in this pathology. Aberrant interactions between amyloids (A42) and metal ions such as Fe^{2+}, Cu^{2+}, or Zn^{2+} can lead to neurotoxicity. These interactions result in peptide aggregation, oxidation, and the release of reactive oxygen species (ROS) (Strozyk et al., 2009).

Alzheimer's disease (AD) is characterized by the accumulation of amyloid-beta (A42) peptides outside of cells, and inhibiting the aggregation of these peptides is crucial for AD therapy and prevention. Recent studies suggest that chiral inhibitors, which selectively target the discordant region (from His 13 to Lys 16, HHQK) of A42 peptides, demonstrate enantioselectivity in preventing A42 aggregation. A chiral amide gel–directed synthetic technique generated these chiral inhibitors to produce mesoporous silica ($mSiO_2$) nanospheres with molecular-scale chirality in the silicate skeletons. The interaction between micelles and chiral amide gels led to the formation of chiral silica sols with a specific molecular arrangement.

This study holds promise in addressing the challenges associated with Alzheimer's disease by targeting A42 peptide aggregation more effectively, and it demonstrates the potential of chiral nanospheres for drug delivery to the brain. However, additional investigations and clinical trials are necessary to assess the safety and effectiveness of these formulations in treating Alzheimer's disease. The generation of large molecularly chiral dendritic mesoporous silica nanospheres, achieved through modular self-assembly, has demonstrated substantial circular dichroism (CD) activity. These nanospheres possess large mesopores (10.1 nm), a significant pore volume (1.8 $cm^3\ g^{-1}$), and a large surface area (525 $m^2\ g^{-1}$). The chirality of chiral amide gels was effectively transferred to composite micelles and subsequently to independently self-assembled asymmetric silica polymeric frameworks. This breakthrough has significant implications for designing nanomaterials with molecular chirality arrangements, which could find applications in visual and medicinal research. Developing these chiral nanospheres with large mesopores and CD activity opens up new possibilities for various fields, including drug delivery, optics, and materials science. Further exploration of their potential applications and properties will be essential to harnessing their benefits fully (Xu et al., 2023).

Curcumin, a natural compound, has shown promise in treating Alzheimer's disease due to its various beneficial properties. However, its therapeutic potential is limited due to challenges like low water solubility, extensive metabolism, poor bioavailability, and instability in the body (Liu et al., 2016). Researchers have explored innovative methods to address these limitations to enhance curcumin's delivery and effectiveness. One such approach involves encapsulating curcumin within MSNs embedded in a thermoresponsive hydrogel. This strategy offers several advantages, including improved drug bioavailability, protection from mucociliary clearance processes, and ease of administration through the nasal route.

Studies have conducted *in vitro* and *in vivo* experiments to characterize the MSN-loaded drug hydrogel. *In vivo* experiments involved the use of a streptozocin-induced Alzheimer's disease model in mice, which showed promising results in reversing cognitive decline. This suggests the potential use of this approach for treating Alzheimer's disease. The development of drug delivery systems like MSNs-loaded hydrogels offers new possibilities for improving the therapeutic effectiveness of compounds like curcumin in addressing neurological conditions like Alzheimer's disease. Further research and clinical investigations will be necessary to validate these findings and explore their clinical applicability (Taebnia et al., 2016).

The development of dual-targeted magnetic MSNs is an exciting advancement in the potential treatment of Alzheimer's disease. These nanoparticles, HA-MMSN-1F12, are designed to target and eliminate Aβ plaques, which play an important role in developing Alzheimer's disease. HA-MMSN-1F12 is designed with dual targeting capabilities, utilizing surface-coupled Aβ42-targeting antibody 1F12 and the CD44-targeting ligand hyaluronic acid (HA). HA is particularly interesting because it targets the CD44 receptor on endothelial cells, offering a potential advantage over using anti-TfR antibodies for drug delivery to the brain. This approach can address immunogenicity and specificity associated with traditional antibodies.

In experimental studies involving APP/PS1 mice, HA-MMSN-1F12 was administered intravenously, leading to successful BBB crossing and depolymerizing brain Aβ plaques into soluble Aβ species. These results are promising, suggesting the potential efficiency of this MSN-based technique for treating Alzheimer's disease. This study showcases the innovative strategies being developed to target and treat Alzheimer's disease, opening up new possibilities for effective therapies. Further investigations and clinical studies will be necessary to validate these findings and explore their application in human patients (Liu et al., 2022).

Rivastigmine hydrogen tartrate (RT) has been explored for treating Alzheimer's disease (AD) by temporarily inhibiting the metabolism of acetylcholinesterase and butyrylcholinesterase in the brain. However, this compound faces significant challenges, including its limited ability to penetrate the brain, with a short half-life of 1.5 h due to hepatic first-pass metabolism. These factors result in poor therapeutic efficacy, necessitating frequent dosing. Researchers have turned to nanocarriers to address these challenges, making mesoporous silica nanoparticles (MSNs) a prominent choice. In a specific study, they developed P1-MSN, which are MSNs loaded with RT and featuring larger pores. The nanoparticles were further modified using succinic anhydride (S-P1-MSN) and 3-aminopropyltriethoxysilane (APTES) (AP-CO-P1-MSN) through grafting and co-condensation methods. These modifications aimed to enhance drug loading and control its release. Results from this study showed that AP-CO-P1-MSN-RT exhibited sustained drug release in simulated stomach and bodily fluids. Importantly, these nanoparticles demonstrated efficient uptake by cells and accumulation in various cellular compartments, as evidenced by transmission electron microscopy (TEM) (Karimzadeh et al., 2017).

Berberine (BBR) is an isoquinoline alkaloid that safeguards the brain by blocking the Akt/ERK1/2 survival/apoptotic signaling pathway, GSK3β, and caspase 3 activity, as well as the Jun amino-terminal kinase (JNK) pathway. Strategies involving lipid-coated MSNs were employed to enhance its effectiveness while minimizing hepatic metabolism. These plasma coatings on MSNs were designed to slow drug release, preventing loaded medications from entering the physiological buffer. Unlike unmodified silica MSNs, the lipid-coated MSNs demonstrated a significantly extended half-life and greater biocompatibility. Both *in vitro* and *in vivo* assessments were conducted, revealing a significant increase in acetylcholinesterase (AChE) inhibition (p <0.05), reduced amyloid fibrillation, and lower levels of malondialdehyde (MDA) (p <0.05). Furthermore, the area under the curve for the obtained MSNs was eight times larger than that of pure BBR. These findings suggest that using lipid-coated MSNs as a drug delivery system to the brain is promising (Singh et al., 2021).

The studies revealed that anti-Aβ single-chain variable fragments (scFvs) with a molecular weight of 27 kDa, specifically the 1E8 scFvs, exhibited strong binding and selectivity for the Aβ peptide, a substance commonly used as a food additive. MSNs were engineered with larger pores to facilitate the delivery of these scFvs to the brain. Utilizing scFvs as nanodepletors, it was possible to hinder the aggregation of Aβ, reduce the accumulation of Aβ aggregates, and extend the lifespan of cells. When these nanodepletors were stereotaxically injected into the hippocampus and cortex of transgenic mice carrying five familial Alzheimer's disease mutations (5xFAD animals), they resulted in a 30% reduction in plaque formation. Furthermore, after administering an Aβ nanodepletor, there was a 20% decrease in the presence of A42 in the brain. *In vitro*/*in vivo* (IV/IV) studies revealed that the Aβ nanodepletors effectively reduced the pool of Aβ peptides and prevented amyloid aggregation by binding to scFvs anchored on the porous silica substrate. In contrast, non-porous silica NPs displayed an eightfold lower capacity for loading scFvs than the porous nanostructures with exceptionally large pores loaded with Aβ nanodepletors. These nanodepletors also reversed neurotoxic effects and exhibited strong binding to Aβ peptides (Jung et al., 2020).

15.7.2 MSN-Based Therapy in Parkinson's Disease

Parkinson's disease (PD) is a neurodegenerative condition characterized by an unexplained disruption of the motor system and cognitive decline, presenting considerable physical and financial challenges to the elderly population. What distinguishes Parkinson's disease from other neurodegenerative conditions is the progressive decline of neuronal cells in the substantia nigra (SN). The limited understanding regarding the mechanisms initiating and advancing Parkinson's disease has resulted in a low success rate for existing monotherapy. It has led to a low success rate for current monotherapy, which targets specific signaling pathways. Evidence links Lewy bodies, primarily composed of misfolded alpha-synuclein (α-Syn), to neuronal cell death. Abnormal aggregation of α-Syn monomers results in the formation of insoluble α-Syn fibrils, leading to elevated levels of reactive oxygen species (ROS) and impaired mitochondrial function. Furthermore, the death of dopaminergic neurons and the advancement of Parkinson's disease are both attributed to aggregated α-syn forms interacting with neuronal cell membranes, creating pore-like structures. This disrupts the normal cell function by depolarizing the membrane potential (Lei et al., 2023).

Unfortunately, there are currently only a limited number of drugs available for treating Parkinson's disease. One of the challenges in developing effective treatments is the difficulty of delivering anti-Parkinsonian drugs to the brain and CNS because of the BBB's extremely low permeability. While the BBB is essential for protecting the CNS from external threats, it also poses a significant obstacle to drug delivery efforts. Various methods have been proposed to increase the permeability of the BBB, with receptor-mediated transcytosis in combination with nanotechnology being the most commonly used approach. However, a major issue with this approach is that specialized targeting ligands with high binding affinity tend to remain attached to receptors after recognition. This can make it challenging for drugs to escape from lysosomes, limiting their release into the brain parenchyma. Therefore, there is a pressing need to develop innovative techniques for effectively transporting therapeutic drugs across the BBB to treat Parkinson's disease while preserving the normal functioning of the CNS.

Utilizing mesoporous silica–encapsulated gold nanorods (MSNs-AuNRs) offers a promising strategy to enhance BBB permeability. This approach can reliably generate an efficient photothermal effect when exposed to second near-infrared (NIR-II) irradiation. Researchers loaded MSNs-AuNRs with the commonly used anti-Parkinson's disease drug quercetin (QCT), resulting in the generation of MSNs-AuNRs@QCT. Following NIR-II irradiation of the BBB, a substantial increase in the transfer of QCT across the BBB was observed. These findings suggest that MSNs-AuNRs may enhance BBB permeability by leveraging the photothermal effect, as demonstrated in an *in vitro* model. Pharmacokinetic studies on rats further confirmed these results, indicating a significant enhancement in brain absorption associated with the combination of MSNs-AuNRs and NIR-II irradiation. *In vitro* toxicity assessments were carried out to evaluate cytotoxicity, and these experiments involved the use of rats to test the hypotheses. In a mouse model of Parkinson's disease induced by 1-methyl-4-phenyl-1,2,3,6-tetrahydropyridine, it was demonstrated that intravenous injection of MSNs-AuNRs@QCT followed by NIR-II irradiation exhibited a remarkable effect in reducing neurological dysfunction in these animals. This was evident through a significant improvement in the mice's condition. When these components are combined, they hold great potential as a nanoplatform for delivering neuroprotective drugs in treating Parkinson's disease (Liu et al., 2020).

Pramipexole, an agonist for the D3-dopamine receptor, has been successfully incorporated into MCM-41 mesoporous silica particles. This innovative approach has led to the generation of drug delivery systems sensitive to pH changes. Nanoparticles were coated with chitosan and/or sodium alginate to control the release of pramipexole. Various characterization techniques were employed to analyze the pramipexole-loaded NPs. Post-coating with chitosan and sodium alginate polymers significantly altered the physicochemical properties of MCM-41 particles loaded with pramipexole.

The study's findings indicate that the combined use of both polymers led to a substantial reduction in the initial rapid drug release phenomenon by approximately 50% at pH levels of 1.2 and 6.8. In contrast to uncoated MCM-41 particles, which released pramipexole entirely within 15 min, the double-coated MCM-41 particles required 300 min to achieve the same effect. The study demonstrated that using MCM-41 particles loaded with pramipexole was more effective in mitigating oxidative damage induced by H_2O_2 in human neuroblastoma SH-SY5Y cells, which is a promising outcome. *In vitro*, cytotoxicity assessments revealed that neither free pramipexole nor empty nanoparticles significantly affected cell viability compared to the untreated control group. Within the concentration range of 0.21–10.57 μM, drug-loaded MCM-41 did not significantly increase cytotoxicity. However, exposure to 21.13 μM of pramipexole-loaded MCM-41 led to a slight decrease in the viability of SH-SY5Y cells, particularly when coated with alginate and chitosan/alginate. The results of this study indicate that loading the drug into MCM-41 and applying a coating of chitosan/sodium alginate could serve as an efficient strategy for drug delivery, ensuring controlled release and providing neuronal protection against oxidative damage, particularly in individuals diagnosed with Parkinson's disease (Tzankov et al., 2019).

The generation of amorphous nanostructured silica–dopamine materials with aggregated particles and varying particle sizes is dependent on dopamine concentration within the material. This unique structure is formed by establishing weak hydrogen bonds between dopamine and the Si-OH groups present in silica. *In vitro* studies of dopamine release profiles reveal two distinct stages: an initial 24 h period of rapid and sustained release, followed by a subsequent phase with a constant and slower rate of dopamine release. One of the significant challenges in medication is achieving a consistent and continuous supply of L-DOPA to the brain to alleviate symptoms. To address this challenge, engineered MSNs were developed using the concept of drug structure–directing agents (DSDAs) possessing inherent therapeutic activity. These MSNs were designed for the targeted delivery of L-DOPA, which was modified through amidation with fatty acids to obtain anionic surfactants capable of forming micelles as templates for assembling inorganic precursors into the silica framework. This template-based construction method yielded MSNs with adjustable sizes ranging from 100 nm to 1 μm, featuring various shapes, including spherical structures with either solid cores encompassing radial mesopores and porous shells or hollow shells with large internal void cavities, as well as elongated structures with extensive hollow regions covered by mesoporous shells. This innovative approach, employing DSDAs for creating drug nanocarriers, offers an alternative to the conventional process involving surfactant removal and subsequent drug-loading steps in producing MSNs (Morales et al., 2021).

A variety of MSNs, including MCM-41, MCM-48, SBA-15, PHTS, and MCF, was synthesized and extensively explored for drug delivery purposes. The mesoporous structures were created using diverse synthetic techniques. To characterize these MSNs, SEM and BET analyses were employed. The research focused on evaluating the loading and release capabilities of the L-DOPA drug, a well-known treatment for Parkinson's disease, using the selected MSNs with varying pore sizes and morphologies. The L-DOPA drug loading and release patterns were analyzed using UV-Vis spectroscopy. Among the mesoporous silica materials studied, SBA-15 was identified as the most effective vehicle for L-DOPA drug delivery. This approach holds promise for treating Parkinson's disease (Swar et al., 2019).

The administration of the dopamine agonist apomorphine-induced rotational asymmetry in hemiparkinsonian rats was studied. However, this effect was restored by intrastriatal implants of silica–dopamine. Apomorphine increased rotational behavior in rats without implants but did not produce the same effect in animals with empty silica reservoirs. Rats implanted with silica–dopamine nanomaterial exhibited a 57% reduction in apomorphine-induced rotations. Notably, these rats did not display any signs of dyskinesias at any point, indicating that dopamine was released slowly and consistently. In contrast, orofacial dyskinesias were observed as a result of sporadic or intermittent systemic injection of L-DOPA. In summary, nanostructured silica is compatible with cerebral tissue and holds potential as a mechanism for controlled dopamine release in Parkinson's disease patients (Lopez, 2010).

15.7.3 MSN-Based Therapy in Brain Tumors

Brain tumors account for the highest proportion of malignant tumors within the CNS, ranking as the 13th major contributor to cancer-related fatalities on a global scale, as reported by the Global Cancer Observation (Sung et al., 2021). Among CNS cancers, glioblastoma (GBM), originating from astrocytes, stands out as the most prevalent form. Gliomas, a category of malignant CNS tumors, constitute 49% of such cases, making them the most common type of CNS cancer. Despite extensive research efforts over the years, the prognosis for individuals diagnosed with GBM remains grim. GBM patients typically have a median survival of only 14 months, and the five-year survival rate ranges from 5 to 10%, even when standard-of-care surgery is followed by prompt initiation of radiation and chemotherapy treatments (Tang et al., 2021).

Similarly, the five-year survival rate for individuals with any primary malignant brain tumor is a mere 20%. Effective treatments for brain cancer face numerous challenges. Tumors often recur after surgical removal, and they may develop in locations where complete resection is either difficult or risky. Consequently, alternative or combination therapy approaches are essential for managing such conditions. Moreover, most drugs struggle to penetrate both the BBB and the blood–brain tumor barrier at levels sufficient to halt tumor progression. While stereotactic radiosurgery is highly effective for many types of brain tumors that can be precisely located using MRI or other neuroimaging techniques, its effectiveness diminishes for recurrent cancers, like metastatic brain tumors and GBM. This is because stereotactic radiosurgery is primarily designed to remove tumors at their origin (Caraway et al., 2022).

A multicomponent NP, consisting of an iron oxide core and a mesoporous silica shell, has demonstrated the capability to effectively transport drugs, breach the BBB, and infiltrate the cells of glioma tumors (Turan et al., 2019). The drug molecules trapped within the pores of the silica shell can be released when the NP is exposed to periodic low-power radiofrequency (RF) fields induced by the mechanical vibrations of the NP. To precisely target the areas of the brain tumor near blood vessels and exhibit altered endothelium, ligands specifically binding to fibronectin, a perivascular biomarker overexpressed in glioblastomas, were employed. When RF activates the NPs, it becomes considerably easier for the therapeutic agent to traverse the blood–brain tumor interface, leading to rapid drug release within the NPs. This approach was further validated by examining the accumulation of NPs targeting fibronectin in the CNS-1 rodent glioma model.

In a study aimed at targeting glioblastoma (GB), ultra-small, large-pore silica NPs (USLP) were utilized to load the drug temozolomide (TMZ). To minimize drug efflux, these NPs were surface-modified with PEGylation. Furthermore, they were decorated with lactoferrin, a targeting protein, to facilitate efficient uptake across the BBB and into glioblastoma. The research findings indicate that the use of USLP leads to an enhancement in the permeability of the BBB in *in vitro* experiments. This was observed using a Transwell model mimicking the tight junctions of endothelial cells, with increased permeation achieved using PEGylated particles. Data from the *in vitro* Transwell blood–brain barrier model loaded with TMZ suggests that USLP formulations can potentially reduce the efflux ratio of TMZ significantly. Furthermore, the results suggest that USLP formulations are more effective in inducing apoptosis in glioblastoma cells when compared to pure TMZ. A preclinical mouse study was conducted to provide proof of concept for the efficacy of USLP particles. Overall, the data indicates that this novel drug delivery system demonstrates efficiency in extravasation from the bloodstream and penetration of the BBB, thus showing potential to enhance the effectiveness of TMZ in treating glioblastoma (Janjua et al., 2023).

The field of biomedicine is currently exploring injectable hydrogels as a promising method for precisely delivering pharmaceuticals. Additionally, a compelling need exists to investigate composite materials that leverage the potential benefits of combining nanotechnology with hydrogels. In a study by Erthal et al. (2023), a new formulation, GlioGel, was developed, characterized, and evaluated. This formulation comprises an injectable hydrogel containing unbound temozolomide and components that respond to external stimuli. This study uses responsive MSNs loaded with paclitaxel for targeted glioblastoma multiforme (GBM) treatment following surgical intervention. The MSNs were synthesized and modified with polyethylene glycol (PEG), serving as a molecular gate that responds to redox stimuli. This modification involved attaching PEG to the nanoparticle surface through a disulfide bond. This enabled the targeted and controlled release of the chemotherapeutic agent from the mesoporous structure within the intracellular environment. Various analytical techniques, including Zetasizer, XRD, FTIR, and SEM, were employed to characterize the MSN-PTX-PEG, and the hydrogel was characterized using cryo-SEM and a rheometer. The study assesses the effectiveness of a localized therapeutic strategy for treating glioblastoma multiforme (GBM), both *in vitro* and in a preclinical GBM mouse model after tumor resection. The PEGylated MSN component of GlioGel was specifically designed to facilitate responsive drug release within the intracellular environment, marked by a decrease in oxidative potential. Additionally, PEGylated MSNs demonstrated superior penetration capabilities in GBM spheroids compared to non-PEGylated nanoparticles. Furthermore, incorporating paclitaxel (PTX) into the conventional TMZ therapy showed efficacy in inhibiting spheroid growth throughout the study period. After resection surgery, the GlioGel formulation was administered into the brains of mice near the resection cavity walls. The innovative formulation effectively slowed down tumor regrowth *in vivo*, increased the survival rate of mice with U-87 tumors, and improved their overall well-being (Erthal et al., 2023).

One potential approach for precisely delivering pharmaceuticals to the brain is receptor-mediated endocytosis. It has been identified that several receptors and transporters in the brain hold potential as therapeutic targets for brain tumors. Some of these include the endothelial growth factor receptor (EGFR), the TfR1 receptor, the low-density lipoprotein receptor (LDLR), the insulin receptor (IR), and the glucose transporter (GLU). The researchers demonstrated the mechanism of receptor-mediated endocytosis and employed a template-based strategy to generate mesoporous nanoplatforms with redox control and dual targeting capabilities. This study's ultimate goal is to enhance the effectiveness of glioma treatment by precisely delivering therapeutic agents to the target site.

Brain capillary endothelial cells, which are a significant part of the BBB, as well as malignant brain cancer cells, express elevated levels of the transferrin receptor (TfR). Utilizing nanomedicine delivery techniques with transferrin (Tf) can facilitate the crossing of the BBB and target glial cells in the brain. This approach treats gliomas expressing TfR and nucleolin by combining Tf and aptamer (Apt) NPs. By employing this method, specific tumor types can be precisely targeted to achieve high specificity for gliomas. The targeted delivery of NPs relies on the permeability and retention effects, in contrast to conventional NPs that rely on passive targeting for accumulation at specific cancer sites. As exemplified by RBT@MRN-SS-Tf/Apt, the active targeting of receptors and ligands can bypass the BBB effectively and selectively target gliomas.

Disulfide-mediated redox-responsive mechanisms have made it easier to implement drug delivery strategies that can be activated on demand. Furthermore, loading rose bengal tetra (RBT) onto MRN-SS-Tf/Apt produces ROS when the system is exposed to 808 nm laser radiation, leading to apoptosis in neoplastic cells. This approach presents a potentially effective strategy for applying photodynamic therapy to treat gliomas. On the one hand, it has been demonstrated that RBT@MRN-SSTf/Apt has a significant inhibitory effect on U87 3D tumor cell spheres. On the other hand, RBT loaded onto MRN-SS-Tf/Apt can thoroughly infiltrate 3D cancer cell spheres. These findings suggest that RBT@MRN-SS-Tf/Apt can accommodate the invasive growth characteristics of glioma and significantly enhance its treatment efficacy. The data from the study indicate that the median survival rate of nude mice increased when they were administered RBT@MRN-SS-Tf/Apt. Moreover, an even higher survival rate was observed in the group treated with a combination of RBT@MRN-SS-Tf/Apt and PDT, highlighting the considerable potential of combination therapy as the optimal therapeutic approach. This work, in conjunction with photodynamic treatment, introduces unique concepts and perspectives that can be applied in the design of multi-targeted nanoscale drug delivery platforms. Additionally, photodynamic therapy is anticipated to find applications in other cancer drug delivery systems shortly (Zhu et al., 2018).

In a study by Zhu et al., receptor-mediated endocytosis was utilized to traverse the BBB. They developed ANG-LP-MSN-PTX, a lipid-coated mesoporous silica NP modified with angiopep-2, to facilitate the delivery of paclitaxel (PTX) for treating brain cancer. This transport pathway is mediated by a low-density lipoprotein receptor–related protein 1 (LRP1), which is overexpressed in BBB and glioma cells. The ANG-LP-MSN-PTX possessed several notable features, including uniform size, a substantial drug loading capacity of 11.08%, and an extended-release profile. *In vitro* experiments demonstrated that ANG-LP-MSN-PTX was significantly more effective in targeting than PTX alone. Additionally, ANG-LP-MSN-PTX induced a greater degree of apoptosis in C6 cells. Pharmacokinetic studies, along with blood and intracerebral microdialysis assessments, showed that ANG-LP-MSN-PTX (20.6%) exhibited superior targeting effectiveness compared to LP-MSN-PTX (10.6%). Rats treated with ANG-LP-MSN-PTX for intracranial C6 gliomas showed significantly higher survival rates compared to those treated with conventional chemotherapy. The combination of ANG-LP-MSN-PTX has shown promising results as a targeted delivery approach in treating gliomas (Zhu et al., 2021).

Islas et al. demonstrated the amalgamation, characterization, and evaluation of mesoporous silica NPs ($MNPSiO_2$) functionalized with folic acid (FA) and loaded with cisplatin (Cis-Pt) to target and eliminate glioblastoma cancer cells. They produced $MNPSiO_2$ particles using the Stober method, with hexadecyltrimethylammonium bromide as a template, and removed the templating agent by calcination at 550°C. Folic acid was immobilized onto the silica NPs through a carbodiimide reaction. The physicochemical properties of $MNPSiO_2$ were extensively analyzed using various techniques. The release of Cis-Pt was modeled theoretically to understand its behavior. The cytotoxicity of $MNPSiO_2$ was assessed through experiments on LN 18 cells, a human glioblastoma multiforme cell line. The study successfully synthesized homogeneous NPs with well-defined properties, consistent porosity, and even distribution. The functionalization of mesoporous nanoparticles was confirmed through spectroscopic analysis. $MNPSiO_2$ exhibited a substantial surface area and porosity. The Korsmeyer–Peppas kinetic model provided a good fit for the system, while the Higuchi model showed superior fitting performance. $MNPSiO_2$-FA particles were biocompatible and enhanced cisplatin's cytotoxic effects (Ortiz-Islas et al., 2021).

The study demonstrated that silica nanostructures serve as effective carriers for the controlled release of cisplatin. The process involves the following steps: Mesoporous silica nanoparticles were obtained using 1-hexadecyltrimethylammonium bromide as a templating agent. These nanoparticles were then loaded with cisplatin. Then, the sol–gel silica was rapidly synthesized by using an excess of acetic acid in the hydrolysis–condensation reactions of tetraethylorthosilicate. Cisplatin was added to the reaction mixture simultaneously. *In vitro* experiments were conducted to determine the amount of cisplatin released using artificial CSF as the medium. Cytotoxicity experiments were carried out using the C6 cancer cell line to assess the level of toxicity exhibited by each silica nanostructure. The spectroscopic analysis indicated that cisplatin was efficiently stabilized within two silica nanostructures. Mesoporous silica NPs presented a significantly higher surface area than did their non-mesoporous counterparts, which had relatively lower surface areas. This difference in surface area was quite noticeable. Cisplatin was observed to be released at a significantly higher rate from the mesoporous silica channels compared to the interior of aggregated silica nanoparticles. The cytotoxicity of cisplatin against cancer cells was established, both for free cisplatin and cisplatin released from silica nanostructures. In contrast, neither of the two silica constructs without medication exhibited any signs of toxicity.

Shahein et al. detailed using MSNs as carriers for the natural compound thymoquinone (TQ), aiming to target glioma cancer cells through novel core–shell nanoformulations. These nanoformulations comprise MSNs loaded with TQ and enveloped by a bilayer shell of two distinct polymers. Significantly, these nanoformulations exhibit a pronounced and specific cytotoxic impact on glioma cells while minimally affecting healthy brain cells. The pH-dependent release of TQ plays a crucial role, with gradual release observed under acidic conditions (pH 5.5 and 6.8), but not under pH 7.4. In terms of molecular targeting, the nanoformulations outperformed free TQ. The toxic elements within the shell proved effective against cancer cells. The anticancer mechanisms induced by TQ involve apoptosis and cell cycle arrest at the G2/M phase. TQ, recognized as an anticancer agent, demonstrated superior efficacy in this strategy compared to free TQ. The *in vitro* optimization of TQ formulations highlights the innovative nature of current drug delivery strategies. However, the study underscores the necessity for *in vivo* testing and further development, emphasizing the potential significance of these findings (Shahein et al., 2019).

15.8 MSN ELIMINATION

MSNs are effective tools in solid drug delivery systems by effectively controlling premature drug release. However, the challenge lies in the potential difficulty of disintegrating NP formulations, posing a challenge to their elimination from the human body. Typically, positively charged MSNs are absorbed by hepatic cells, subsequently following the hepatobiliary excretion pathway and ultimately being excreted in fecal matter. This cycle continues until the MSNs are no longer present in the body. Conversely, negatively charged MSNs may accumulate in the liver's Kupffer cells without metabolization, potentially leading to hepatotoxicity. In physiological conditions, various types of MSNs can dissolve into soluble silica, which can be eliminated in urine as silicic acid or oligomeric silica compounds. This dissolution is facilitated by the high solubility of silica in water, allowing MSNs to transform into soluble forms, enabling their safe removal from the body (Mendiratta et al., 2019; Chen et al., 2008).

15.9 FUTURE PERSPECTIVES AND CONCLUSION

Biological barriers, including mucosal, cellular, gastrointestinal, blood–tumor, and blood–brain barriers, pose significant challenges to drug transport, with their underlying biological processes remaining poorly understood despite decades of exploring MSNs for various applications. Understanding these processes is crucial for determining the fate of MSNs linked to their physicochemical features. Researchers are strategically developing MSNs to address these challenges and enhance pharmaceutical delivery efficacy. These strategies improve mucosal membrane adhesion and permeability, enhance cellular uptake and transport, facilitate endosome escape, and optimize drug delivery to specific targets. Clinical studies have demonstrated the therapeutic potential of mesoporous silica in dental implants, showcasing a decrease in bacterial count at the implant abutment junction and an increase in implant rigidity when utilizing MSN (Mirzaali et al., 2023). Hollow MSNs, characterized by low toxicity, biodegradability, and minimal organ interference, based on *in vivo* investigations, are gaining attention in biomedical applications. While they offer promising uses in treating various diseases, including Alzheimer's, Parkinson's, and gliomas, potential risks need thorough exploration. More research is necessary to fully understand the significance and safety of nanostructured materials, ensuring the mitigation of potential hazards. The book chapter focused on using mesoporous silica particles as drug delivery systems for neurological disorders, like Alzheimer's, Parkinson's, and gliomas. The unique properties of MSNs, such as a large surface-area-to-volume ratio, adaptable pore sizes, and successful crossing of the blood–brain barrier, make them particularly promising. Nanoscavenging properties of MSNs contribute to their potential in treating neurological disorders by penetrating glioma cancer cells and selectively binding and eliminating harmful compounds in the brain, such as amyloid beta peptides in Alzheimer's disease or α-synuclein in Parkinson's disease, through tailored surface functionalization on MSNs (Kong et al., 2017).

REFERENCES

Ahmed, H., Gomte, S. S., Prathyusha, E. A. P., Agrawal, M., & Alexander, A. (2022, October). Biomedical applications of mesoporous silica nanoparticles as a drug delivery carrier. *Journal of Drug Delivery Science and Technology*, *76*, 103729. https://doi.org/10.1016/j.jddst.2022.103729

Albayati, T. M., Salih, I. K., & Alazzawi, H. F. (2019, October). Synthesis and characterization of a modified surface of SBA-15 mesoporous silica for a chloramphenicol drug delivery system. *Heliyon*, *5*(10), e02539. https://doi.org/10.1016/j.heliyon.2019.e02539

Andrade, G. F., Soares, D. C. F., Almeida, R. K. D. S., & Sousa, E. M. B. (2012). Mesoporous Silica SBA-16 functionalized with alkoxysilane groups: Preparation, characterization, and release profile study. *Journal of Nanomaterials*, 1–10. https://doi.org/10.1155/2012/816496

Aquib, M., Farooq, M. A., Banerjee, P., Akhtar, F., Filli, M. S., Boakye-Yiadom, K. O., Kesse, S., Raza, F., Maviah, M. B. J., Mavlyanova, R., & Wang, B. (2019, August 19). Targeted and stimuli–responsive mesoporous silica nanoparticles for drug delivery and theranostic use. *Journal of Biomedical Materials Research Part A*, *107*(12), 2643–2666. https://doi.org/10.1002/jbm.a.36770

Bharti, C., Gulati, N., Nagaich, U., & Pal, A. (2015). Mesoporous silica nanoparticles in target drug delivery system: A review. *International Journal of Pharmaceutical Investigation*, *5*(3), 124. https://doi.org/10.4103/2230-973x.160844

Burgess, A., & Hynynen, K. (2014, March 20). Drug delivery across the blood–brain barrier using focused ultrasound. *Expert Opinion on Drug Delivery*, *11*(5), 711–721. https://doi.org/10.1517/17425247.2014.897693

Caraway, C. A., Gaitsch, H., Wicks, E. E., Kalluri, A., Kunadi, N., & Tyler, B. M. (2022, July 21). Polymeric nanoparticles in brain cancer therapy: A review of current approaches. *Polymers*, *14*(14), 2963. https://doi.org/10.3390/polym14142963

Chen, Z., Chen, H., Meng, H., Xing, G., Gao, X., Sun, B., Shi, X., Yuan, H., Zhang, C., & Liu, R. (2008, August 1). Biodistribution and metabolic paths of silica coated CdSeS quantum dots. *Toxicology and Applied Pharmacology*, *230*(3), 364–371. https://doi.org/10.1016/j.taap.2008.03.022

Colmenares, M. G., Simon, U., Schmidt, F., Dey, S., Schmidt, J., Thomas, A., & Gurlo, A. (2018, September). Tailoring of ordered mesoporous silica COK-12: Room temperature synthesis of mesocellular foam and multilamellar vesicles. *Microporous and Mesoporous Materials*, *267*, 142–149. https://doi.org/10.1016/j.micromeso.2018.03.015

Corbalan, J. J., Medina, C., Jacoby, A., Malinski, T., & Radomski, M. W. (2011). *Amorphous silica nanoparticles aggregate human platelets: potential implications for vascular homeostasis*. https://doi.org/10.2147/IJN.S28293

Djayanti, K., Maharjan, P., Cho, K. H., Jeong, S., Kim, M. S., Shin, M. C., & Min, K. A. (2023, March 28). Mesoporous silica nanoparticles as a potential nanoplatform: Therapeutic applications and considerations. *International Journal of Molecular Sciences*, *24*(7), 6349. https://doi.org/10.3390/ijms24076349

Erthal, L. C. S., Shi, Y., Sweeney, K. J., Gobbo, O. L., & Ruiz-Hernandez, E. (2023, March 29). Nanocomposite formulation for a sustained release of free drug and drug-loaded responsive nanoparticles: An approach for a local therapy of glioblastoma multiforme. *Scientific Reports*, *13*(1). https://doi.org/10.1038/s41598-023-32257-5

Farjadian, F., Roointan, A., Mohammadi-Samani, S., & Hosseini, M. (2019, March). Mesoporous silica nanoparticles: Synthesis, pharmaceutical applications, biodistribution, and biosafety assessment. *Chemical Engineering Journal*, *359*, 684–705. https://doi.org/10.1016/j.cej.2018.11.156

Geng, J., Yu, H., Ren, J., & Qu, X. (2008, November). Rapid label-free detection of metal-induced Alzheimer's amyloid β peptide aggregation by electrochemical method.

Electrochemistry Communications, *10*(11), 1797–1800. https://doi.org/10.1016/j.elecom.2008.09.020

Ghaferi, M., Koohi Moftakhari Esfahani, M., Raza, A., Al Harthi, S., Ebrahimi Shahmabadi, H., & Alavi, S. E. (2020, September 8). Mesoporous silica nanoparticles: synthesis methods and their therapeutic use-recent advances. *Journal of Drug Targeting*, *29*(2), 131–154. https://doi.org/10.1080/1061186x.2020.1812614

Ghaforinejad, H., Mazaheri, H., Hassani Joshaghani, A., & Marjani, A. (2021, January 22). Study on novel modified large mesoporous silica FDU-12/polymer matrix nanocomposites for adsorption of Pb(II). *PLoS One*, *16*(1), e0245583. https://doi.org/10.1371/journal.pone.0245583

Grün, M., Lauer, I., & Unger, K. K. (1997, March). The synthesis of micrometer- and submicrometer-size spheres of ordered mesoporous oxide MCM-41. *Advanced Materials*, *9*(3), 254–257. https://doi.org/10.1002/adma.19970090317

Heikkila, T., Salonen, J., Tuura, J., Hamdy, M., Mul, G., Kumar, N., Salmi, T., Murzin, D., Laitinen, L., & Kaukonen, A. (2007, February 22). Mesoporous silica material TUD-1 as a drug delivery system. *International Journal of Pharmaceutics*, *331*(1), 133–138. https://doi.org/10.1016/j.ijpharm.2006.09.019

Huang, X., Li, L., Liu, T., Hao, N., Liu, H., Chen, D., & Tang, F. (2011, June 8). The shape effect of mesoporous silica nanoparticles on biodistribution, clearance, and biocompatibility in Vivo. *ACS Nano*, *5*(7), 5390–5399. https://doi.org/10.1021/nn200365a

Izquierdo-Barba, I., Martinez, A., Doadrio, A. L., Pérez-Pariente, J., & Vallet-Regí, M. (2005, December). Release evaluation of drugs from ordered three-dimensional silica structures. *European Journal of Pharmaceutical Sciences*, *26*(5), 365–373. https://doi.org/10.1016/j.ejps.2005.06.009

Janjua, T. I., Cao, Y., Ahmed-Cox, A., Raza, A., Moniruzzaman, M., Akhter, D. T., Fletcher, N. L., Kavallaris, M., Thurecht, K. J., & Popat, A. (2023, May). Efficient delivery of temozolomide using ultrasmall large-pore silica nanoparticles for glioblastoma. *Journal of Controlled Release*, *357*, 161–174. https://doi.org/10.1016/j.jconrel.2023.03.040

Javad Kalbasi, R., & Zirakbash, A. (2015). Synthesis, characterization and drug release studies of poly(2-hydroxyethyl methacrylate)/KIT-5 nanocomposite as an innovative organic–inorganic hybrid carrier system. *RSC Advances*, *5*(16), 12463–12471. https://doi.org/10.1039/c4ra13930e

Jiménez, A. M., Domínguez-Pinos, M. D., Guerra, M. E., Fernández-Llebrez, P., & Pérez-Fígares, J. M. (2014). Structure and function of the ependymal barrier and diseases associated with ependyma disruption. *Tissue barriers*, 2(1), e28426. https;//doi.org/10.4161/tisb.28426

Jung, H., Chung, Y. J., Wilton, R., Lee, C. H., Lee, B. I., Lim, J., Lee, H., Choi, J., Kang, H., Lee, B., Rozhkova, E. A., Park, C. B., & Lee, J. (2020, February 17). Silica nanodepletors: Targeting and clearing Alzheimer's β-amyloid plaques. *Advanced Functional Materials*, *30*(15), 1910475. https://doi.org/10.1002/adfm.201910475

Karimzadeh, M., Rashidi, L., & Ganji, F. (2017, January 8). Mesoporous silica nanoparticles for efficient rivastigmine hydrogen tartrate delivery into SY5Y cells. *Drug Development and Industrial Pharmacy*, *43*(4), 628–636. https://doi.org/10.1080/03639045.2016.1275668

Khodaverdi, E., Soleimani, H. A., Mohammadpour, F., & Hadizadeh, F. (2016, March 21). Synthetic zeolites as controlled-release delivery systems for anti-inflammatory drugs. *Chemical Biology & Drug Design*, *87*(6), 849–857. https://doi.org/10.1111/cbdd.12716

Kim, J., Kim, H., Lee, N., Kim, T., Kim, H., Yu, T., Song, I., Moon, W., & Hyeon, T. (2008, October 20). Multifunctional uniform nanoparticles composed of a magnetite nanocrystal core and a mesoporous silica shell for magnetic resonance and fluorescence imaging and for drug delivery. *Angewandte Chemie International Edition*, *47*(44), 8438–8441. https://doi.org/10.1002/anie.200802469

Kim, J., Lee, J. E., Lee, J., Yu, J. H., Kim, B. C., An, K., Hwang, Y., Shin, C. H., Park, J. G., Kim, J., & Hyeon, T. (2005, December 31). Magnetic fluorescent delivery vehicle using uniform mesoporous silica spheres embedded with monodisperse magnetic and semiconductor nanocrystals. *Journal of the American Chemical Society*, *128*(3), 688–689. https://doi.org/10.1021/ja0565875

Kim, K. S., Park, M., Kim, T. W., Kim, J. E., Papoulis, D., Komarneni, S., & Choi, J. (2015, June 30). Adsorbate-dependent uptake behavior of topographically bi-functionalized ordered mesoporous silica materials. *Journal of Porous Materials*, *22*(5), 1297–1303. https://doi.org/10.1007/s10934-015-0008-8

Kocahan, S., & Doğan, Z. (2017). Mechanisms of Alzheimer's disease pathogenesis and prevention: The brain, neural pathology, N-methyl-D-aspartate receptors, Tau protein and other risk factors. *Clinical Psychopharmacology and Neuroscience*, *15*(1), 1–8. https://doi.org/10.9758/cpn.2017.15.1.1

Kong, M., Tang, J., Qiao, Q., Wu, T., Qi, Y., Tan, S., Gao, X., & Zhang, Z. (2017). Biodegradable hollow mesoporous silica nanoparticles for regulating tumor microenvironment and enhancing antitumor efficiency. *Theranostics*, *7*(13), 3276–3292. https://doi.org/10.7150/thno.19987

Kresge, C. T., Leonowicz, M. E., Roth, W. J., Vartuli, J. C., & Beck, J. S. (1992, October). Ordered mesoporous molecular sieves synthesized by a liquid-crystal template mechanism. *Nature*, *359*(6397), 710–712. https://doi.org/10.1038/359710a0

Kumar, D., Schumacher, K., du Fresne von Hohenesche, C., Grün, M., & Unger, K. (2001, August). MCM-41, MCM-48 and related mesoporous adsorbents: Their synthesis and characterisation. *Colloids and Surfaces A: Physicochemical and Engineering Aspects*, *187–188*, 109–116. https://doi.org/10.1016/s0927-7757(01)00638-0

Lei, L., Tu, Q., Zhang, X., Xiang, S., Xiao, B., Zhai, S., Yu, H., Tang, L., Guo, B., Chen, X., & Zhang, C. (2023, April). Stimulus-responsive curcumin-based polydopamine nanoparticles for targeting Parkinson's disease by modulating α-synuclein aggregation and reactive oxygen species. *Chemical Engineering Journal*, *461*, 141606. https://doi.org/10.1016/j.cej.2023.141606

Li, T., Shi, S., Goel, S., Shen, X., Xie, X., Chen, Z., Zhang, H., Li, S., Qin, X., Yang, H., Wu, C., & Liu, Y. (2019, April). Recent advancements in mesoporous silica nanoparticles towards therapeutic applications for cancer. *Acta Biomaterialia*, *89*, 1–13. https://doi.org/10.1016/j.actbio.2019.02.031

Lin, Y. S., Hurley, K. R., & Haynes, C. L. (2012, January 18). Critical considerations in the biomedical use of mesoporous silica nanoparticles. *Journal of Physical Chemistry Letters*, *3*(3), 364–374. https://doi.org/10.1021/jz2013837

Liu, N., Liang, X., Yang, C., Hu, S., Luo, Q., & Luo, H. (2022). Dual-targeted magnetic mesoporous silica nanoparticles reduce brain amyloid-β burden via depolymerization and intestinal metabolism. *Theranostics*, *12*(15), 6646–6664. https://doi.org/10.7150/thno.76574

Liu, W., Zhai, Y., Heng, X., Che, F. Y., Chen, W., Sun, D., & Zhai, G. (2016, March 17). Oral bioavailability of curcumin: Problems and advancements. *Journal of Drug Targeting*, *24*(8), 694–702. https://doi.org/10.3109/1061186x.2016.1157883

Liu, Y., Zhu, D., Luo, J., Chen, X., Gao, L., Liu, W., & Chen, T. (2020, August 19). NIR-II-activated yolk–shell nanostructures as an intelligent platform for Parkinsonian therapy. *ACS Applied Bio Materials*, *3*(10), 6876–6887. https://doi.org/10.1021/acsabm.0c00794

Lopez, T. (2010, December). Treatment of Parkinson's disease: Nanostructured sol– gel silica– dopamine reservoirs for controlled drug release in the central nervous system. *International Journal of Nanomedicine*, *19*. https://doi.org/10.2147/ijn.s13223

Martínez-Carmona, M., Colilla, M., & Vallet-Regí, M. (2015, November 6). Smart mesoporous nanomaterials for antitumor therapy. *Nanomaterials*, *5*(4), 1906–1937. https://doi.org/10.3390/nano5041906

Mastorakos, P., & McGavern, D. (2019, July 5). The anatomy and immunology of vasculature in the central nervous system. *Science Immunology*, *4*(37). https://doi.org/10.1126/sciimmunol.aav0492

Mayoral, A., Blanco, R. M., & Diaz, I. (2013, June). Location of enzyme in lipase-SBA-12 hybrid biocatalyst. *Journal of Molecular Catalysis B: Enzymatic*, *90*, 23–25. https://doi.org/10.1016/j.molcatb.2013.01.012

Mendiratta, S., Hussein, M., Nasser, H. A., & Ali, A. A. A. (2019, August 15). Multidisciplinary role of mesoporous silica nanoparticles in brain regeneration and cancers: From crossing the blood–brain barrier to treatment. *Particle & Particle Systems Characterization*, *36*(9), 1900195. https://doi.org/10.1002/ppsc.201900195

Mirzaali, F., Dizaj, S. M., Shahi, S., Memar, M. Y., & Parnia, F. (2023, April). The antibacterial effect of tetracycline-loaded mesoporous silica nanoparticles in the gingival fluid at implant-abutment junction: A randomized clinical trial study. *Pharmaceutical Nanotechnology*, *11*(2), 208–216. https://doi.org/10.2174/2211738511666230106151403

Mitran, R. A., Deaconu, M., Matei, C., & Berger, D. (2019). Mesoporous silica as carrier for drug-delivery systems. *Nanocarriers for Drug Delivery*, 351–374. https://doi.org/10.1016/b978-0-12-814033-8.00011-4

Mohamed Isa, E. D., Ahmad, H., Abdul Rahman, M. B., & Gill, M. R. (2021, January 24). Progress in mesoporous silica nanoparticles as drug delivery agents for cancer treatment. *Pharmaceutics*, *13*(2), 152. https://doi.org/10.3390/pharmaceutics13020152

Morales, V., McConnell, J., Pérez-Garnes, M., Almendro, N., Sanz, R., & García-Muñoz, R. A. (2021). L-Dopa release from mesoporous silica nanoparticles engineered through the concept of drug-structure-directing agents for Parkinson's disease. *Journal of Materials Chemistry B*, *9*(20), 4178–4189. https://doi.org/10.1039/d1tb00481f

Nandiyanto, A. B. D., Kim, S. G., Iskandar, F., & Okuyama, K. (2009, April). Synthesis of spherical mesoporous silica nanoparticles with nanometer-size controllable pores and outer diameters. *Microporous and Mesoporous Materials*, *120*(3), 447–453. https://doi.org/10.1016/j.micromeso.2008.12.019

Nguyen, T. T., Dung Nguyen, T. T., Vo, T. K., Tran, N. M. A., Nguyen, M. K., Van Vo, T., & Van Vo, G. (2021, November). Nanotechnology-based drug delivery for central nervous system disorders. *Biomedicine & Pharmacotherapy*, *143*, 112117. https://doi.org/10.1016/j.biopha.2021.112117

Niculescu, V. C. (2020, February 19). Mesoporous silica nanoparticles for bio-applications. *Frontiers in Materials*, *7*. https://doi.org/10.3389/fmats.2020.00036

Narayan, R., Nayak, U., Raichur, A., & Garg, S. (2018, August 6). Mesoporous silica nanoparticles: A comprehensive review on synthesis and recent advances. *Pharmaceutics*, *10*(3), 118. https://doi.org/10.3390/pharmaceutics10030118

Narum, S. M., Le, T., Le, D. P., Lee, J. C., Donahue, N. D., Yang, W., & Wilhelm, S. (2020). Passive targeting in nanomedicine: Fundamental concepts, body interactions, and clinical potential. *Nanoparticles for Biomedical Applications*, 37–53. https://doi.org/10.1016/b978-0-12-816662-8.00004-7

Ortiz-Islas, E., Sosa-Arróniz, A., Manríquez-Ramírez, M. E., Rodríguez-Pérez, C. E., Tzompantzi, F., & Padilla, J. M. (2021, January 1). Mesoporous silica nanoparticles functionalized with folic acid for targeted release Cis-Pt to glioblastoma cells. *Reviews on Advanced Materials Science*, *60*(1), 25–37. https://doi.org/10.1515/rams-2021-0009

Pandit, R., Chen, L., & Götz, J. (2020). The blood-brain barrier: Physiology and strategies for drug delivery. *Advanced Drug Delivery Reviews*, 165–166, 1–14. https://doi.org/10.1016/j.addr.2019.11.009

Pednekar, P. P., Godiyal, S. C., Jadhav, K. R., & Kadam, V. J. (2017). Mesoporous silica nanoparticles: A promising multifunctional drug delivery system. *Nanostructures for Cancer Therapy*, 593–621. https://doi.org/10.1016/b978-0-323-46144-3.00023-4

Petrovic, M., Hurt, C., Collins, D., Burns, A., Camus, V., Liperoti, R., Marriott, A., Nobili, F., Robert, P., Tsolaki, M., Vellas, B., Verhey, F., & Byrne, E. (2007, December). Clustering of behavioural and psychological symptoms in dementia (bpsd): A European Alzheimer's disease consortium (eadc) study. *Acta Clinica Belgica*, *62*(6), 426–432. https://doi.org/10.1179/acb.2007.062

Porrang, S., Rahemi, N., Davaran, S., Mahdavi, M., & Hassanzadeh, B. (2021, August). Preparation and in-vitro evaluation of mesoporous biogenic silica nanoparticles obtained from rice and wheat husk as a biocompatible carrier for anti-cancer drug delivery. *European Journal of Pharmaceutical Sciences*, *163*, 105866. https://doi.org/10.1016/j.ejps.2021.105866

Saputra, O. A., Wibowo, F. R., & Lestari, W. W. (2022, September). High storage capacity of curcumin loaded onto hollow mesoporous silica nanoparticles prepared via improved hard-templating method optimized by Taguchi DoE. *Engineering Science and Technology, an International Journal*, *33*, 101070. https://doi.org/10.1016/j.jestch.2021.10.002

Saunders, N. R., Habgood, M. D., Møllgård, K., & Dziegielewska, K. M. (2016). The biological significance of brain barrier mechanisms: Help or hindrance in drug delivery to the central nervous system? *F1000Research*, *5*, 313. https://doi.org/10.12688/f1000research.7378.1

Shahein, S. A., Aboul-Enein, A. M., Higazy, I. M., Abou-Elella, F., Lojkowski, W., Ahmed, E. R., Mousa, S. A., & AbouAitah, K. (2019, July). Targeted anticancer potential against glioma cells of thymoquinone delivered by mesoporous silica core-shell nanoformulations with pH-dependent release. *International Journal of Nanomedicine*, *14*, 5503–5526. https://doi.org/10.2147/ijn.s206899

Shen, S. C., Dong, Y. C., Letchmanan, K., & Ng, W. K. (2017). Mesoporous materials and technologies for development of oral medicine. *Nanostructures for Oral Medicine*, 699–749. https://doi.org/10.1016/b978-0-323-47720-8.00024-9

Siddiqui, B., Rehman, A., Haq, I. U., Al-Dossary, A. A., Elaissari, A., & Ahmed, N. (2022, December). Exploiting recent trends for the synthesis and surface functionalization of mesoporous silica nanoparticles towards biomedical applications. *International Journal of Pharmaceutics: X*, *4*, 100116. https://doi.org/10.1016/j.ijpx.2022.100116

Singh, A. K., Singh, S. S., Rathore, A. S., Singh, S. P., Mishra, G., Awasthi, R., Mishra, S. K., Gautam, V., & Singh, S. K. (2021, July 23). Lipid-coated MCM-41 mesoporous silica nanoparticles loaded with berberine improved inhibition of acetylcholine esterase and amyloid formation. *ACS*

Biomaterials Science & Engineering, *7*(8), 3737–3753. https://doi.org/10.1021/acsbiomaterials.1c00514

Solár, P., Zamani, A., Kubíčková, L., Dubový, P., & Joukal, M. (2020, May 6). Choroid plexus and the blood–cerebrospinal fluid barrier in disease. *Fluids and Barriers of the CNS*, *17*(1). https://doi.org/10.1186/s12987-020-00196-2

Strozyk, D., Launer, L. J., Adlard, P. A., Cherny, R. A., Tsatsanis, A., Volitakis, I., Blennow, K., Petrovitch, H., White, L. R., & Bush, A. I. (2009, July). Zinc and copper modulate Alzheimer Aβ levels in human cerebrospinal fluid. *Neurobiology of Aging*, *30*(7), 1069–1077. https://doi.org/10.1016/j.neurobiolaging.2007.10.012

Sung, H., Ferlay, J., Siegel, R. L., Laversanne, M., Soerjomataram, I., Jemal, A., & Bray, F. (2021, February 4). Global cancer statistics 2020: GLOBOCAN estimates of incidence and mortality worldwide for 36 cancers in 185 countries. *CA: A Cancer Journal for Clinicians*, *71*(3), 209–249. https://doi.org/10.3322/caac.21660

Swar, S., Máková, V., & Stibor, I. (2019, September 30). Effectiveness of diverse mesoporous silica nanoparticles as potent vehicles for the drug L-DOPA. *Materials*, *12*(19), 3202. https://doi.org/10.3390/ma12193202

Taebnia, N., Morshedi, D., Yaghmaei, S., Aliakbari, F., Rahimi, F., & Arpanaei, A. (2016, December 6). Curcumin-loaded amine-functionalized mesoporous silica nanoparticles inhibit α-synuclein fibrillation and reduce its cytotoxicity-associated effects. *Langmuir*, *32*(50), 13394–13402. https://doi.org/10.1021/acs.langmuir.6b02935

Tajes, M., Ramos-Fernández, E., Weng-Jiang, X., Bosch-Morató, M., Guivernau, B., Eraso-Pichot, A., Salvador, B., Fernàndez-Busquets, X., Roquer, J., & Muñoz, F. J. (2014, July 21). The blood-brain barrier: Structure, function and therapeutic approaches to cross it. *Molecular Membrane Biology*, *31*(5), 152–167. https://doi.org/10.3109/09687688.2014.937468

Tang, L., Feng, Y., Gao, S., Mu, Q., & Liu, C. (2021, November 25). Nanotherapeutics overcoming the blood-brain barrier for glioblastoma treatment. *Frontiers in Pharmacology*, *12*. https://doi.org/10.3389/fphar.2021.786700

Tarn, D., Ashley, C. E., Xue, M., Carnes, E. C., Zink, J. I., & Brinker, C. J. (2013, February 6). Mesoporous silica nanoparticle nanocarriers: Biofunctionality and biocompatibility. *Accounts of Chemical Research*, *46*(3), 792–801. https://doi.org/10.1021/ar3000986

Tozuka, Y., Wongmekiat, A., Kimura, K., Moribe, K., Yamamura, S., & Yamamoto, K. (2005). Effect of pore size of FSM-16 on the entrapment of flurbiprofen in mesoporous structures. *Chemical and Pharmaceutical Bulletin*, *53*(8), 974–977. https://doi.org/10.1248/cpb.53.974

Turan, O., Bielecki, P., Perera, V., Lorkowski, M., Covarrubias, G., Tong, K., Yun, A., Rahmy, A., Ouyang, T., Raghunathan, S., Gopalakrishnan, R., Griswold, M. A., Ghaghada, K. B., Peiris, P. M., & Karathanasis, E. (2019). Delivery of drugs into brain tumors using multicomponent silica nanoparticles. *Nanoscale*, *11*(24), 11910–11921. https://doi.org/10.1039/c9nr02876e

Tzankov, B., Tzankova, V., Aluani, D., Yordanov, Y., Spassova, I., Kovacheva, D., Avramova, K., Valoti, M., & Yoncheva, K. (2019, June). Development of MCM-41 mesoporous silica nanoparticles as a platform for pramipexole delivery. *Journal of Drug Delivery Science and Technology*, *51*, 26–35. https://doi.org/10.1016/j.jddst.2019.02.008

Vallet-Regí, M. (2022, January). Our contributions to applications of mesoporous silica nanoparticles. *Acta Biomaterialia*, *137*, 44–52. https://doi.org/10.1016/j.actbio.2021.10.011

Vallet-Regi, M., Rámila, A., del Real, R. P., & Pérez-Pariente, J. (2000, December 29). A new property of MCM-41: Drug delivery system. *Chemistry of Materials*, *13*(2), 308–311. https://doi.org/10.1021/cm0011559

Verheggen, I. C. M., de Jong, J. J. A., van Boxtel, M. P. J., Postma, A. A., Verhey, F. R. J., Jansen, J. F. A., & Backes, W. H. (2020, October 28). Permeability of the windows of the brain: feasibility of dynamic contrast-enhanced MRI of the circumventricular organs. *Fluids and Barriers of the CNS*, *17*(1). https://doi.org/10.1186/s12987-020-00228-x

Wanat, K. (2020, March 5). Biological barriers, and the influence of protein binding on the passage of drugs across them. *Molecular Biology Reports*, *47*(4), 3221–3231. https://doi.org/10.1007/s11033-020-05361-2

Wang, L. S., Wu, L. C., Lu, S. Y., Chang, L. L., Teng, I. T., Yang, C. M., & Ho, J. A. A. (2010, July 12). Biofunctionalized phospholipid-capped mesoporous silica nanoshuttles for targeted drug delivery: Improved water suspensibility and decreased nonspecific protein binding. *ACS Nano*, *4*(8), 4371–4379. https://doi.org/10.1021/nn901376h

Wang, S. (2009, January). Ordered mesoporous materials for drug delivery. *Microporous and Mesoporous Materials*, *117*(1–2), 1–9. https://doi.org/10.1016/j.micromeso.2008.07.002

Wang, S., & Li, H. (2006, December). Structure directed reversible adsorption of organic dye on mesoporous silica in aqueous solution. *Microporous and Mesoporous Materials*, *97*(1–3), 21–26. https://doi.org/10.1016/j.micromeso.2006.08.005

Wu, S. H., Lin, Y. S., Hung, Y., Chou, Y. H., Hsu, Y. H., Chang, C., & Mou, C. Y. (2008, January 4). Multifunctional mesoporous silica nanoparticles for intracellular labeling and animal magnetic resonance imaging studies. *ChemBioChem*, *9*(1), 53–57. https://doi.org/10.1002/cbic.200700509

Xu, L., Guo, M., Hung, C. T., Shi, X. L., Yuan, Y., Zhang, X., Jin, R. H., Li, W., Dong, Q., & Zhao, D. (2023, April 1). Chiral skeletons of mesoporous silica nanospheres to mitigate Alzheimer's β-amyloid aggregation. *Journal of the American Chemical Society*, *145*(14), 7810–7819. https://doi.org/10.1021/jacs.2c12214

Yamada, H., Urata, C., Aoyama, Y., Osada, S., Yamauchi, Y., & Kuroda, K. (2012, April 6). Preparation of colloidal mesoporous silica nanoparticles with different diameters and their unique degradation behavior in static aqueous systems. *Chemistry of Materials*, *24*(8), 1462–1471. https://doi.org/10.1021/cm3001688

Yetisgin, A. A., Cetinel, S., Zuvin, M., Kosar, A., & Kutlu, O. (2020, May 8). Therapeutic nanoparticles and their targeted delivery applications. *Molecules*, *25*(9), 2193. https://doi.org/10.3390/molecules25092193

Zhang, Y., Zhi, Z., Jiang, T., Zhang, J., Wang, Z., & Wang, S. (2010, August). Spherical mesoporous silica nanoparticles for loading and release of the poorly water-soluble drug telmisartan. *Journal of Controlled Release*, *145*(3), 257–263. https://doi.org/10.1016/j.jconrel.2010.04.02

Zhu, J., Zhang, Y., Chen, X., Zhang, Y., Zhang, K., Zheng, H., Wei, Y., Zheng, H., Zhu, J., Wu, F., Piao, J. G., Zhu, Z., & Li, F. (2021, January). Angiopep-2 modified lipid-coated mesoporous silica nanoparticles for glioma targeting therapy overcoming BBB. *Biochemical and Biophysical Research Communications*, *534*, 902–907. https://doi.org/10.1016/j.bbrc.2020.10.076

Zhu, X., Zhou, H., Liu, Y., Wen, Y., Wei, C., Yu, Q., & Liu, J. (2018, December). Transferrin/aptamer conjugated mesoporous ruthenium nanosystem for redox-controlled and targeted chemo-photodynamic therapy of glioma. *Acta Biomaterialia*, *82*, 143–157. https://doi.org/10.1016/j.actbio.2018.10.012

16 Potential of Dendrimers as Nanocarriers for Brain Drug Delivery

*Arab Iram Saba Gaus Mohammed, Victor Hmingthansanga, Sivakumar Manickam, V. Ravichandiran, and Subramanian Natesan**

16.1 INTRODUCTION

Brain illnesses are a growing medical concern because, as the world's population ages, the number of patients suffering from them will increase and have a significant social impact due to their severe morbidity or mortality (1). Brain diseases such as Alzheimer's, Parkinson's, and gliomas are the most prevalent (2). Numerous drugs have been approved and are currently used clinically to treat this disease (3). However, drugs administered through any route for targeting brain diseases frequently fail to reach the target site due to the blood–brain barrier (BBB), a known biological hindrance (1). BBB is an essential part of the brain that helps maintain brain health. The BBB shields the brain neurons from toxic and damaging substances found in blood and prevents many therapeutic drugs from penetrating and entering the brain (4). BBB comprises neurons, astrocytes, pericytes, and the brain capillaries endothelial cells that actively pump nutrients in and out of the brain. The BBB only allows low-molecular-weight (400–500 Da) and tiny lipophilic molecules to enter the brain, making it difficult for large, lipid-insoluble drugs to cross (5). The brain is impervious to more than 98% of small-molecule medications and about 100% of large-molecule medications, such as recombinant protein and monoclonal antibodies (6). The ability to administer drugs through the neurovascular unit (NVU) interfaces is among the most promising strategies for effective brain targeting, since nearly every neuron is linked to a micro-vessel (7, 8). In recent years, the investigation has been navigated toward exploring and developing new and easy ways of achieving and making remarkable injected agents through investigating numerous nanoparticles (NPs), leading to a specific understanding of mechanisms of NP absorption in the brain (9). Dendrimers, among these accomplishments, showed excellent promising non-invasive therapy. Dendrimers and dendritic nanomaterials, an emerging class of synthesized and multifunctional polymers with an architecture resembling a tree or dendron, have drawn much attention primarily because of their distinctive molecular architectures, multifunctional properties, and ease of surface functionalization with multiple functional agents (10). Since Donald A. Tomalia authored the synthesis and analysis of a neoteric class of poly(amidoamine) (PAMAM) polymers in 1984 and referred to these hyperbranched polymers as dendrimers, these nanoscopic molecules have undergone extensive research for potential therapeutic applications (11). Different dendrimers have been created, each with a unique structure and purpose. It is noteworthy that Tomalia's important contribution launched a brand-new area of study involving nanotechnological methods. In contrast to most polymers, many dendrimers are considered well-defined, discrete, adjustable macromolecules with great molecular uniformity and monodispersed. Dendritic polymers have been viewed as attractive delivery systems for particular drugs because of their abundance of interior voids and surface functions (12). Examples of intrinsic dendrimeric medicinal value, anti-prion (11, 13) antitoxin (14), as well as anti-amyloidogenic properties (15), have all been put forth. Several correlated biological properties, such as CNS bioavailability and selectivity to specific cells, can be improved by enclosing the drug in a dendritic molecular framework. In this sense, dendrimers represent the ideal drug delivery system to the brain. In this chapter, the anatomy of the BBB, the transport route across the BBB, dendrimers and their synthesis, dendrimers acting via transporters, toxicities, the role of various dendrimer-based nanocarriers in treating brain illnesses, and lastly, the future perspectives are discussed.

16.2 ANATOMY OF THE BBB

The human brain is protected against pathogens, viruses, and other outside stimuli by several anatomical barriers collectively called the BBB. Moreover, these barriers help govern the flow of vital proteins, ions, nutrients, and metabolites in and out of the brain and help maintain the homeostasis of the brain (7). The entire barrier system creates an anatomical border to selectively isolate the brain from the circulatory system (7). Anatomically, the BBB comprises a variety of cells, including nerve cells, pericytes, astrocytes, and brain capillaries endothelial cells (BCECs). The tight junction of BCECs only allows tiny lipophilic molecules to penetrate. Moreover, the tight junctions do not allow active molecules

* Corresponding author: snatesan@niperkolkata.ac.in

DOI: 10.1201/9781003383376-16

to transport paracellularly across the BBB, shield the brain against pathogenic and antigenic attacks, and increase transendothelial electrical resistance, allowing small ions to pass across the BBB (16). The BBB also serves as an enzymatic barrier and contains a variety of efflux transporters, including P-glycoprotein (Pgp), ATP-binding cassette, and multidrug resistance–related protein, which efflux out the unwanted substances from the brain into the blood and prevent neurotoxic substances from entering the brain (17). The brain's endothelial cells luminal and abluminal membrane surfaces possess certain enzymes that can metabolize drugs and nutrients, such as glutamyl transpeptidase, aromatic acid decarboxylase, and alkaline phosphatase. Many mitochondria and minimal pinocytotic activity in BCECs further restrict drug molecule transport (17). Thus, the BBB only permits extremely hydrophobic molecules with molecular weights of 400 to 500 Da or less to cross the BCECs passively. In contrast, more than 98% of low-molecular-weight molecules and almost 100% of high-molecular-weight molecules cannot passively cross the BBB (17). Although the BBB does not completely block all molecules from entering the brain, it allows the movement of vital endogenous nutrients to guarantee the brain's proper operation. Various saturable transporters are also present at the BBB to guarantee efficient transportation of endogenous chemicals and nutrients into the brain (18). The human brain has the highest blood flow of any organ in the body; it contains 100 billion blood capillaries that stretch 400 km (19). The BBB also keeps the brain's ionic composition stable for proper signal conduction. The coats of astrocytes and pericytes endothelial cells represent the BBB's structural integrity (20) (Figure 16.1).

Pericytes give the blood vessels structural stiffness, whereas astrocytes protect basal endothelial layers and connect to neurons. Current studies have concentrated on producing soluble developmental signals to preserve barrier rigidity from the covering layer of astrocytes and pericytes (21, 22). Due to the absence of small apertures, this tight junction only permits extremely tiny molecules and gaseous materials to flow through it, which hinders any transendothelial passage of soluble substances from the blood's circulation to the brain tissue or the other way around. The integrity and activity of the barrier can be directly impacted by the absence of any one BBB component. Contrarily, the BBB's high electrical resistance (2,000 cm^2) compared to other vascular tissues (33 cm^2) further prevents drugs from diffusing into brain cells through paracellular pathways.

16.3 TRANSPORT ROUTE ACROSS THE BBB

Due to physiological barriers like the BBB, most pharmacological interventions cannot enter the brain. Although treating CNS illnesses has always been difficult, knowing the characteristics and structure of the BBB helps scientists and researchers create drug delivery methods that are specifically tailored for the brain. By overcoming physiological barriers and studying several mechanisms, such as receptor-mediated endocytosis, absorptive endocytosis, paracellular and transcellular transport, and transportation proteins, it may be possible to deliver medications to the brain specifically to target the diseased cells. Another approach that researchers have looked into is normally modifying impermeable pharmaceuticals to mimic the architecture of

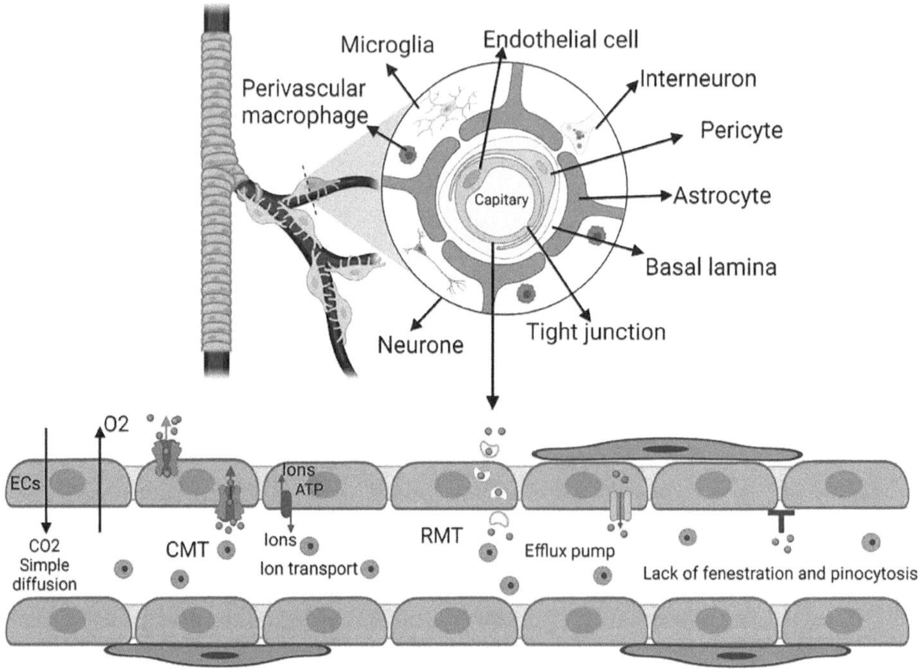

FIGURE 16.1 Structure of the BBB.

endogenous compounds like glucose. Dendrimers and their physiochemical and biological properties are used for targeted drug delivery; various nanomaterials and nanotechnological systems, such as dendrimers, carbon nanotubes, quantum dots, polymeric, as well as metallic nanoparticles, nanocrystals, nanoemulsions, and lipid-based nanoparticles, have been tested (16). Similarly, several nanocarriers have been investigated for targeted delivery to the brain via passing through the BBB to achieve therapeutic concentrations of drugs in the brain (16). Cell-penetrating peptides, such as Tat-derived peptides, Syn-B vectors, or cationic proteins, such as hexamethylene diamine, have also been investigated for targeted delivery to the brain. This is accomplished by adhesion and absorbing cationic particles on the luminal surface of the brain's capillary endothelial cells, followed by luminal exocytosis. Receptor-mediated transcytosis, a more focused technique, is based on finding a single receptor exclusively expressed by the BBB, such as transferrin, insulin, insulin-like growth factor, low-population-density lipoproteins, etc. Scientists have studied a variety of nanomaterials with and without ligand conjugation, as well as individualized medicine systems, using three-dimensional printing nanotechnology for drug delivery to the brain (23).

16.4 DENDRIMERS AND THEIR SYNTHESIS

The "cascade molecules," a new type of extremely branched molecules, were first described by Vögtle et al. toward the end of 1970 and gave rise to dendrimers (24). Denkewalter, Tomalia, Newkome, Frechet, and others later expanded the complexity of these systems, causing the formation of larger structures and coining the word "dendrimers" for them (25). The word *dendrimer*, which alludes to the distinctive arrangement of its branched building elements, is derived from the Greek words *dendron* ("tree" or "branch") and *meros* ("part"). A bi- or multifunctional central point, building units covalently connected to the core and arranged in layers called "generations" (G), and several end functional units on their perimeter are the components of dendrimers. Due to their step-by-step manufacturing process, dendrimers are related to molecular chemistry. Because of the structure, composed of repeating branching units comprised of monomers, they are linked to polymer science/chemistry. Divergent or convergent approaches are used to synthesize them (26). In the divergent approach, monomers are added first from a multivalent core to a surface. In contrast, in the convergent method, dendrimers are synthesized from the surface to a core, generating a wedge-shaped unit (inward–out method). On the other hand, two fundamental steps make up the convergent synthesis approach. While the branches are joined to a common core in the following phase, the monomers are linked together in the first step to create the outer layers with a focal point. Consequently, it generates a three-dimensional topology that is heavily branching. Due to the large number of repeating reactions required for this procedure, the yield of a final dendrimeric composition is low. Moreover, this synthesis procedure has fewer adverse effects, allowing the management of the end product's molecular weight and functional group positions. Eventually, it is discovered to be favorable to purify the produced dendrimer utilizing the convergent technique (25).

As an alternative, the branch monomers are introduced to a functional unit of a core via the divergent technique, where at least two protective branching sites initially react with the activated core. More monomers must be added to the reaction mixture in order for the reaction to proceed. First-generation dendrimers are created by removing the branching monomers' protective spots. Second-generation dendrimers are then obtained by adding another branching monomer, and so on. Dendrimers of the necessary generations are produced by repeatedly adding monomers, and as a result, the size of dendrimers likewise grows. Then, using modified chromatographic techniques, the resulting dendrimer of interest is separated from the unreacted core or monomers (27). So it might be conceivable to add a new substituent to the dendrimer's outermost layer or any ligand to a dendrimer in favor of targeting a particular receptor (28) (Figure 16.2).

Although the concurrent technique is an "inward" or "descending" mode of synthesis, where its reaction begins from the peripheral and moves toward the core, the divergent approach starts first from the core. Then, it builds up from generation to generation to a periphery (29). PAMAM and PPI dendrimers are among the dendrimers created using the divergent technique (12, 30). Poly(benzylether), created by Frechet et al., and poly-(aryl ethynyl), generated by Moore, are two dendrimers obtained via the convergent approaches (31).

Other methods of synthesizing dendrimers include hypercores, branching monomer growth, double exponential growth, lego chemistry, and click chemistry. This technique aims to quicken the dendrimer's synthesis process. By using fewer stages in this technology and chemically attaching monomers to the oligomeric forms of the active attachment group in the hyper core, high-generation dendrimers having greater manufacturing yield are attainable (32). Lego chemistry has been developed to reduce the lengthy synthesis time and associated costs of synthesizing dendrimers. The price and time for synthesizing are decreased by using this method. The technique produces an amplified end surface group of 48 to 250 in a single step by reacting the strongly functionalized core using branched monomers. Also, the fast purification of the finished dendrimeric product is made possible by using a minimal volume of reacting solvent (33). Dendrimers can be produced more quickly using this approach, which is another synthesis method. The primary benefit of this synthesis method is the outstanding yield and high purity of final dendrimers (33). Implementing convergent and divergent growth from a similar starting material using a single process represents the most recent innovation in the production of dendrimers. The two process products are combined in this quick-growing method to create a trimer, which can produce greater generations of dendrimers in less time (34).

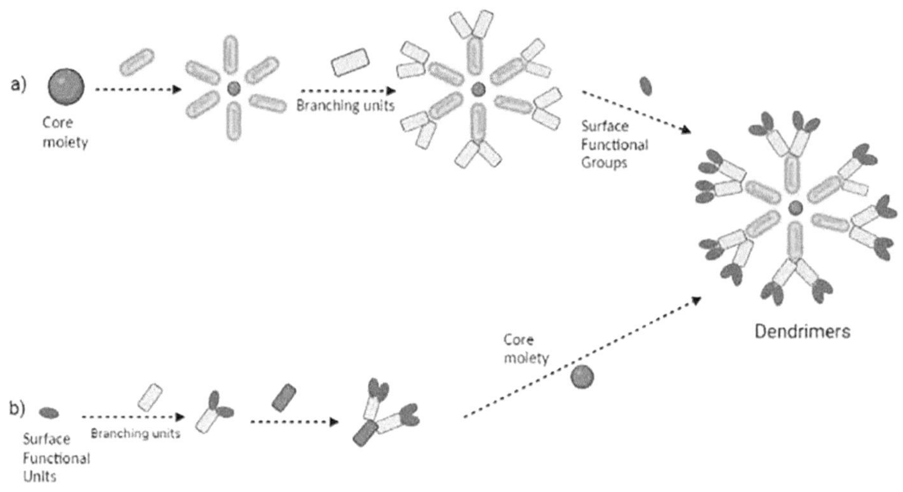

FIGURE 16.2 Synthesis of dendrimers: (a) divergent method, (b) convergent method.

16.5 DENDRIMERS AND THEIR DERIVATIVES

16.5.1 Polyamidoamine Dendrimers

PAMAM dendrimer derivatives are a type of nanocarrier composed of a core and multiple branches. They are synthesized by a step-growth polymerization process and are characterized by their highly branched spherical structure. PAMAM dendrimer derivatives have a wide range of applications in drug delivery, gene therapy, and nanomedicine. They can be used to encapsulate drugs, proteins, and other molecules and can be tailored. The term "polyamidoamine dendrimers" (PAMAM) is typically used to describe dendrimers with a core (the initial core was an ethylenediamine core), branches that protrude out from an inner point and create cavity walls, with an amine moiety on the periphery. Numerous PAMAM dendrimers have various surface amine moieties simultaneously (35) (Figure 16.3).

One well-investigated dendrimer, PAMAM dendrimer, has peripheral functional groups and inside cavities that can be further changed to enclose cargos for therapeutic applications. As dendrimer creator Donald Tomalia suggested, dendrimers might be used as nano-agents to combat cancer, bacteria, and viruses (36). The cytotoxicity of PAMAM, particularly those ended with positive charge groups, is a problem, and it depends on both concentration and production *in vitro*. PAMAM dendrimers of the G5 or lower generations are considered harmless (37). Modifying the surface with PEG has been done to overcome this restriction. PAMAM has the broadest range of uses compared to other dendrimers in drug delivery.

16.5.2 Polypropylene Imine

The first dendrimer known to be commercially employed for drug delivery is polypropylene imine (PPI). It comprises repeating propyleneimine units with a diaminobutane core. The second most common abbreviation for polypropyleneimine is DAB-dendrimer, since diaminobutane (DAB) is a fundamental component of PPI (35) (Figure 16.4). PPI's aminoterminal groups offer appropriate water solubility. PPI dendrimers can then increase the hydrophobic substances confined in the hydrophobic inner cavities of PPI by making them more soluble in water. However, PPI's positive charge surface frequently weakens cell membranes and results in cell lysis. Another issue is that PPI has a lower drug loading than PAMAM (38).

Compared to PPI, the PPI/drug compound is less stable. PEGylation and acetylation are two examples of surface group modifications (39). Acetylation is more favored because of its strong effectiveness and high penetration ability. Moreover, the PEG chain's steric hindrance can affect how well medicinal molecules interact with surface functional groups (39).

16.5.3 Poly-L-Lysine

Dendrimers made of lysine residues are referred to as poly-L-lysine (PLL) dendrimers (also known as dendri-grafted poly-L-lysine or DGL) (Figure 16.5). These dendrimers benefit from better biocompatibility, reduced cytotoxicity, simple enzymatic breakdown, and consequent excretion of lower-molecular-weight metabolites (40). Also, it is highly tempting to have the ability to add stimulus responsiveness when needed by including particular amino acid sequences in PLL dendrimers (41). Gene delivery has considerably used PLL dendrimers and their variants (41). Compared to PLL dendrimers, the larger generation of PLL displays superior gene transfection, while typical linear poly(lysine) demonstrates poor gene transfection efficacy and higher cytotoxicity (41).

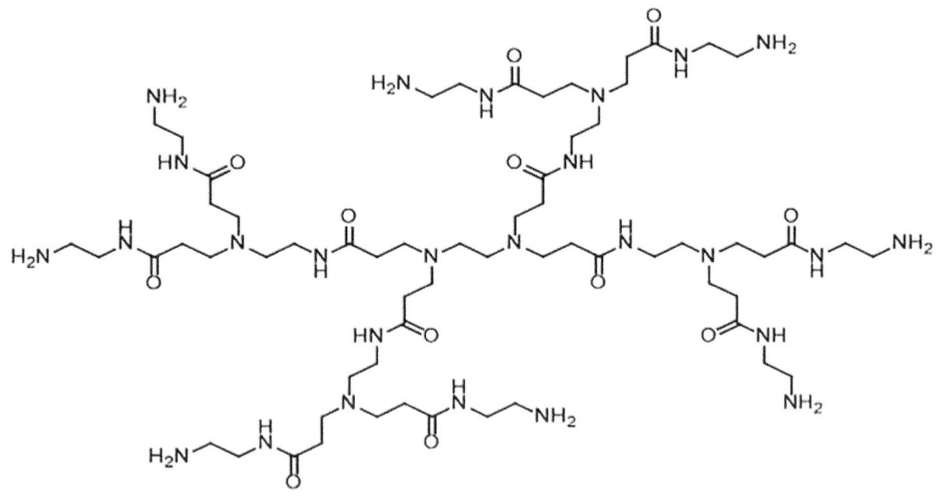

FIGURE 16.3 Partial structure of PAMAM dendrimers.

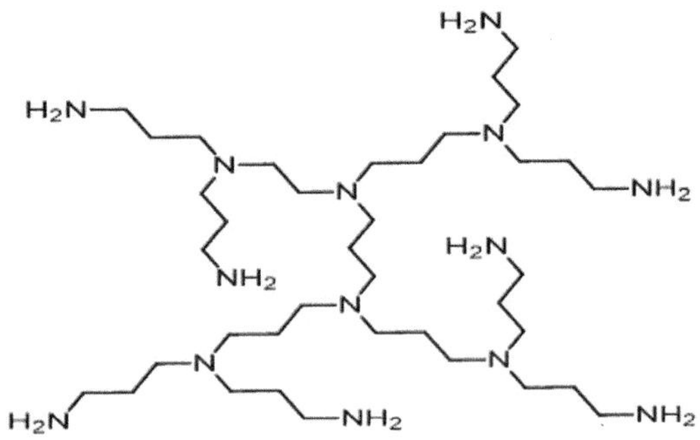

FIGURE 16.4 Partial structure of PPI dendrimer.

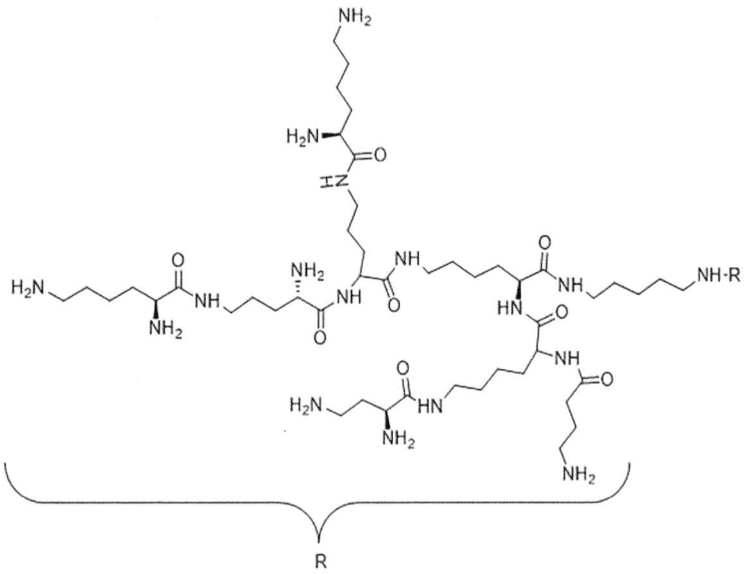

FIGURE 16.5 Partial structure of PLL dendrimer.

16.5.4 Carbosilane

Many carbosilane dendritic macromolecules with hydrophobic scaffolds and exceptional thermal stability were also developed through the extensive application of silicon chemistry (42). Because nucleophilic compounds can easily access electrophilic silicon (Si^+), silicon chemistry is frequently used to generate dendrimers (43) (Figure 16.6). Importantly, the low polarity and high energy of the C-Si bond, giving it a high degree of hydrophobicity, are related to the supremacy of carbosilane dendrimers (44). Although it has such a hydrophobic core structure, carbosilane dendrimers can be surface-functionalized with polar moieties to transform them into hydrophilic molecules. Reactive compounds such as Si–H, Si–Cl, Si–CH=CH_2, or Si–CH2CH=CH_2, which can introduce fascinating chemical, natural, and organometallic substituents, are finding greater use in the pharmaceutical industry (35).

16.5.5 Polyphosphorhydrazone (PPH) Dendrimers

Polyphosphorhydrazone (PPH) dendrimers, made specifically for drug delivery systems, fall into the category of cationic dendrimers and are particularly important. PPH dendrimers, or dendrimers with phosphorus atoms at every branching point, are unique because of their simple production and various chemistry, typically attributed to reactive terminal groups like aldehydes or P(S)Cl_2 groups (45) (Figure 16.7). Such dendrimers have the potential to have a hydrophilic surface and a hydrophobic backbone, enabling them to interact with cell membranes and efficiently enter into the cells (45, 46). According to recent investigations, PPH dendrimers hold great promise for biological applications. Such dendritic materials have been shown to affect tau protein and amyloid peptide aggregation in neurodegenerative disorders (47). Moreover, gene treatment targeting HIV infection (48) and the ability of PPH dendrimers to deliver anticancer siRNAs to specific sites (49) have been reported.

16.5.6 Janus

Recent amphiphilic dendrimers for pharmaceutical applications are "Janus" dendrimers, which are derived from Greek and also known as bow-tie dendrimers or diblock dendrimers and co-dendrimers "surface-block" dendrimers and distinguished by two dendrimeric wedges with distinct terminal groups. The ancient deity of gates and doors, known as "Janus," is typically pictured as having two heads facing opposite directions (50). Janus dendrimers' broken symmetry offers the chance to create complex self-assembled substances and introduces novel properties that are now impossible for homogeneous or symmetrical dendrimers. Therefore, Janus dendrimers, whose characteristics are closely related to the two distinct peripheral surface functional groups, are highly desired. Due to their bio-functional nature and thermal characteristics, these Janus dendrimers attract interest, notably in self-assembling and medicinal applications (51).

16.5.7 Linear–Dendritic Block Co-Polymers

A unique amphiphilic block co-polymer known as linear–dendritic block co-polymer was created by developing dendrimers (LDBC). They are made up of a dendrimer covalently bonded to linear chains. Most LDBCs require ligating hydrophilic linear chains to hydrophobic dendrons (52). The sturdy core–shell structure created by the amphiphilic co-polymers in aqueous solution has exceptional characteristics that set them apart from traditional micelles,

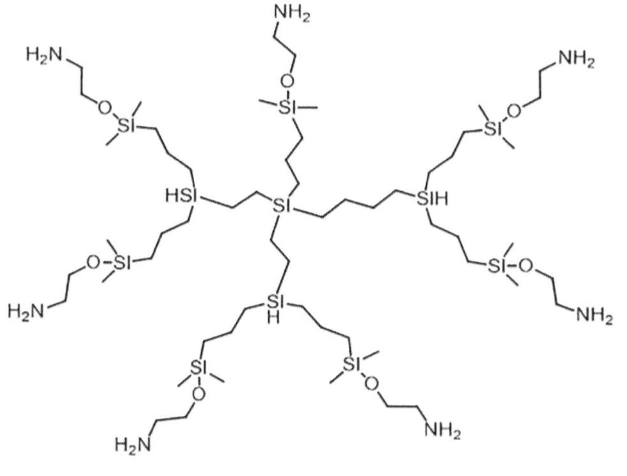

FIGURE 16.6 Partial structure of carbosilane dendrimer.

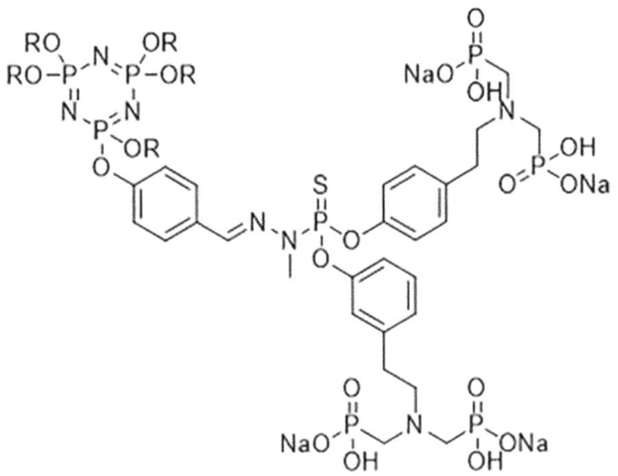

FIGURE 16.7 Partial structure of PPH dendrimer.

such as a reduced critical micelle concentration (CMC) or increased stability in the blood (53). In terms of design, co-polymers can be divided into six primary groups: linear–hyperbranched biopolymers, side chain functional and dendronized linear–dendritic co-polymers, multi-arm star co-polymers, and AB diblock linear–dendritic co-polymers, co-polymers and polymers with linear dendrimer grafts which use the linear A block or dendritic B block as building blocks (53).

16.5.8 Other Types

In addition to the LDBC dendrimers indicated earlier, which are members of the hybrid dendrimer group, numerous varieties of dendrimers with clearly defined nanoscopic sizes and plenty of active terminal groups are now readily available (54) (Figure 16.8).

Core–shell (tecto) dendrimers, a polymeric architectural style with highly organized arrangement achieved through the controlled covalent connection of dendrimer building blocks, are among these dendrimers (55). The functionalization of GATG dendrimers and poly(ethyleneglycol) (PEG)-dendritic block co-polymers with a range of ligands of medicinal importance, such as anionic and cationic groups, carbohydrates, peptides, or anti-ageing drugs, is made possible by the presence of terminal azides in these compounds (54). Another dendrimer type is PETIM (polypropyl ether imine), which expands three-dimensionally and contains tertiary amines (55). Polyether copolyester (PEPE) has low toxicity with higher BBB permeability among all dendrimers. PAMAM–organosilicon (PAMAMOS) dendrimers, which consist of hydrophilic as well as nucleophilic PAMAM interiors but also hydrophobic organosilicon (OS) exteriors, are another type (55, 56).

16.6 DENDRIMERS ACTING VIA TRANSPORTERS

CNS disorders like Alzheimer's disease, Huntington's disease, Parkinson's disease, and brain cancer currently account for 11% of all diseases worldwide. Due to their lack of ability to pass the BBB, a crucial controlling factor for drug availability to the brain that serves as an opening gateway for different nutrients while shielding it from possibly toxic compounds, there are still very few drugs on the market that have been accepted for the therapy of CNS disorders (57, 58). Physical, transport, and metabolic barriers are all combined to provide the BBB's barrier function (59). The low and specific penetration of BBB can be linked to its distinct biological properties: endothelial cells lack fenestrations, vesicular transport, and pinocytosis (60); the physical barrier created by the existence of tight junctions between adjacent endothelial cells (61); the transport barriers brought on by the expression of various transporters, such as GLUT-1, LAT1, transferrin receptors (TfRs), insulin receptors, LRP1, LRP2, and the ATP family of efflux transporters. Only specific chemicals are transported to or from the brain by these transporters and receptors, which are produced on the capillary endothelium of the BBB (62) (Figure 16.9). Although having a crucial part to play in how the CNS develops into a complex, interconnected network, the multitasking BBB creates a significant obstacle for therapies that transfer medicines as well as nucleic acids to a brain to treat CNS illnesses. Using the unique transport mechanisms exhibited upon that BBB for such delivery of medicines to the brain can, however, solve this problem.

Based on this delivery method, numerous dendrimers have been created and preclinically tested. Due to their many benefits, including monodispersity, non-immunogenicity,

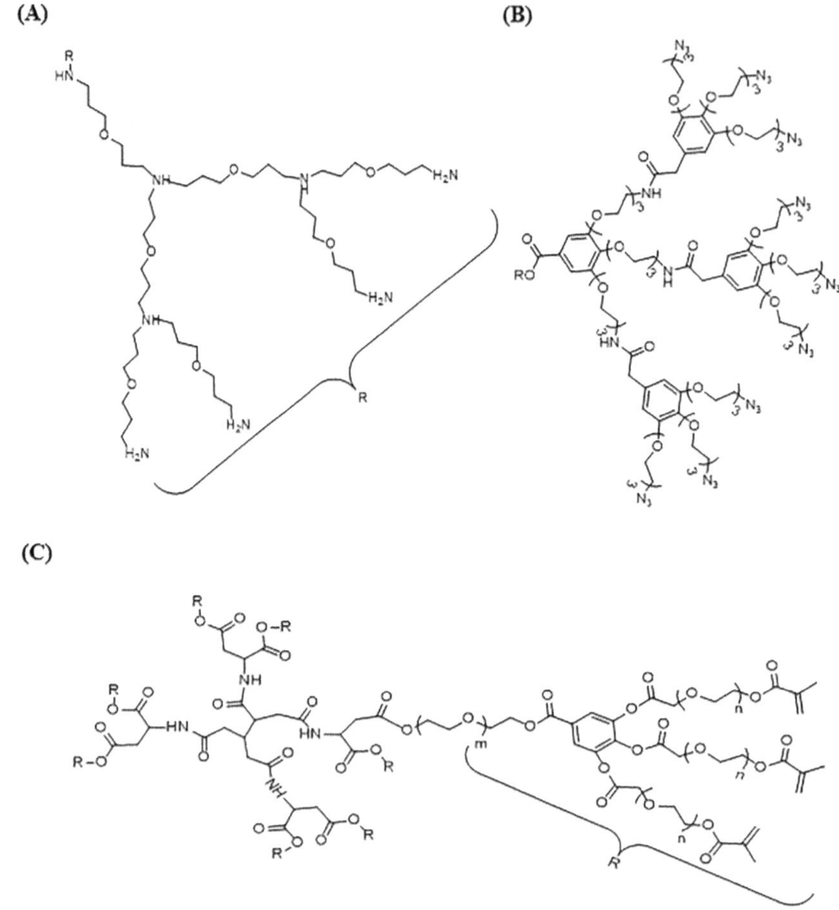

FIGURE 16.8 Partial structure of (A) PETIM, (B) GATG, (C) PEPE dendrimer.

penetrability through biological membranes, improved drug stability, and maintenance of drug concentrations in the therapeutically good range, they stand out among other nanomaterials as being particularly pledging in this application (63).

16.6.1 TfR

Of all the body's tissues, the brain capillary endothelium, which forms the BBB, exhibits the highest level of TfRs expression (26). TfRs are responsible for the transport and distribution of iron throughout the body via the iron-binding glycoprotein transferrin. TfRs have been thoroughly studied over the past 20 years as a specific focus for drug/gene delivery systems to deliver therapies to the brain (64). Dendrimers have been identified as intriguing possibilities for brain delivery due to their distinct physicochemical characteristics. The most investigated possibility for TfR-targeted medicinal delivery to the brain is the PAMAM dendrimer. Huang et al. conjugated Tf to PAMAM (G5) using a dual-purpose PEG spacer, and then it was complexed with plasmid DNA expressing GFP (65). Following intravenous delivery to mice, PAMAM-PEG-Tf/DNA could cross the BBB, as evidenced by a systemic distribution study using 125I-labeled dendrimers. Qualitative examination revealed GFP expression in several brain regions, including the fourth ventricle, cortical layer, hippocampus, caudate putamen, and substantia nigra. It was approximately two times greater when comparing the gene expression to PAMAM/DNA and PAMAM-PEG/DNA. This study shows that TfRs can serve as a potential target for brain drug delivery.

16.6.2 LRP1 and LRP2

The BBB expresses the multifunctional scavenger receptors LRP1 and LRP2. They can attach to various molecules, such as proteinases, proteinase inhibitor complexes, and lipoproteins rich in lipase, and then cause those molecules to cross the BBB (66). The Tf family of proteins includes lactoferrin (Lf), a single-chain, cationic iron-binding protein. Lf can cross the BBB by attaching to low-density lipoprotein receptor–related protein (LRP) receptors. Non-viral brain-targeted gene delivery was engineered by Huang et al., which was made up of PAMAM (G5) coupled to Lf via a PEG spacer and then complexed to a plasmid DNA expressing GFP. Results of their study suggest that GFP

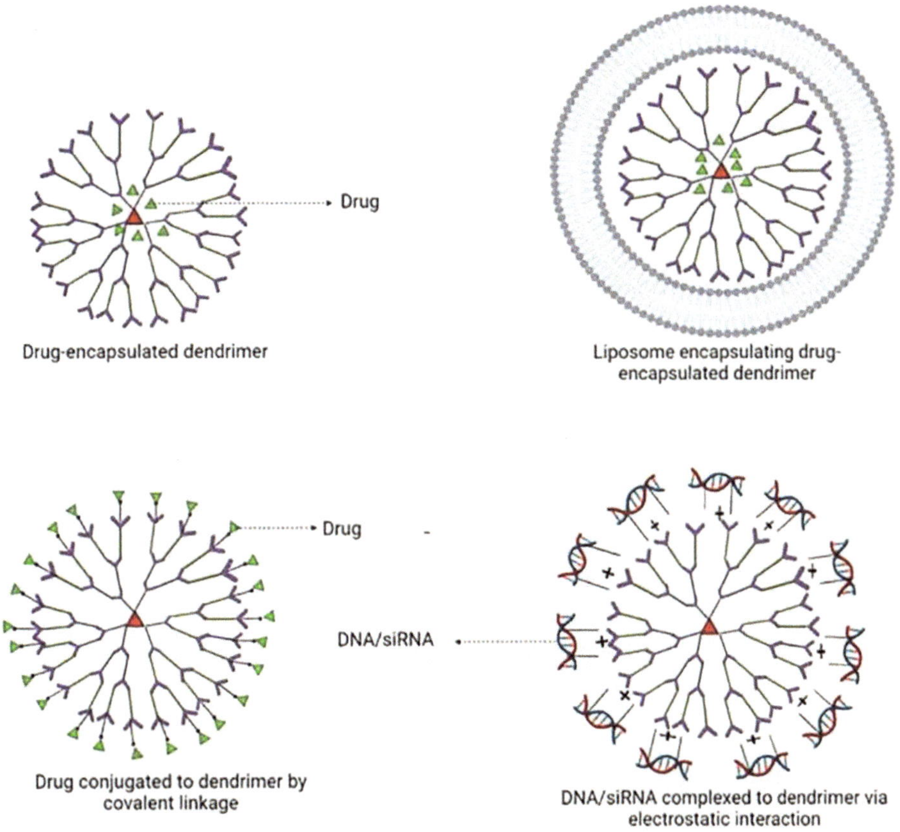

FIGURE 16.9 Nucleic acid/drug–dendrimer delivery system.

expression was seen in the cortical layer, hippocampus, caudate putamen, substantia nigra, and fourth ventricle of mice after intravenous treatment of PAMAM-PEG-Lf/DNA (67). The expression of this gene was 5.2 times higher than that of PAMAM-PEG/DNA and PAMAM/DNA. Also, the authors have shown that the increased uptake of the PAMAM-PEG-Lf/DNA in the cerebral capillary endothelial cells was caused by adsorptive-mediated and LRP receptor–mediated pathways (68). Aprotinin's Kunitz domains are the sources of peptides known as angiopeps, which are excellent brain-targeting ligands. Via LRP1-mediated transcytosis, they are transferred across the BBB (69). Angiopep-2 was coupled to PAMAM via a particular reaction with the bifunctional PEG derivative's terminal N-hydroxysuccinimidyl, and it was used for gene brain delivery. The brain absorption of DNA was up to 8.4 times greater with PAMAM-PEG-angiopep cross-linked compared to PAMAM/DNA after intravenous injection in mice. The cerebral layer, caudate putamen, hippocampus, and substantia nigra all showed signs of gene expression. Adsorptive endocytosis and LRP-mediated endocytosis both played a role in the cellular uptake of PAMAM-PEG-angiopep (70). In another work, the neuroprotective impact of angiopep was assessed in a Parkinson's disease model by conjugating it with the dendrigraft poly-L-lysine (DGL; G3) through a PEG linkage and complexing it with a plasmid DNA expressing hGDNF. Compared to the unmodified DGL, the DGL-PEG-angiopep/hGDNF showed more cellular DNA absorption or gene expression within brain cells, enhancing the locomotor performance and healing of dopaminergic neurons (71).

16.6.3 GLUT-1

The BBB has a high expression of the glucose transporter GLUT-1, which transports the brain's required amounts of glucose for normal function (72). Based on this, Dhanikula et al. created polyether-copolyester dendrimers (G2) packed mostly with up to 20.8% w/w methotrexate (MTX) for treating gliomas. Polyether-copolyester dendrimers (G2) were conjugated to D-glucosamine through carbamate linkages. Since GLUT-1 is also clearly expressed in brain tumors, the polyether-copolyester dendrimers loaded with MTX transport were improved compared to free drugs. In addition, the system allowed the accumulation of significant portions of MTX inside the central necrotic areas of the vascularized brain tumor spheroids (73).

16.6.4 Leptin

In response to food consumption, adipocytes release leptin, a 146-amino acid polypeptide that controls hunger and prevents weight gain. The choroid plexus and the luminal surface of the cerebral micro-vessels both contain leptin receptors (74). Although endogenous leptin's saturation of the leptin receptors occurs at relatively low concentrations (0.15–5 ng/mL), exogenous leptin cannot be employed as a targeted ligand for brain absorption. However, crucial regions for leptin-specific receptors on the BBB are present in peptide fragments generated from leptin. In particular, leptin 30 (sequence 61–90) is absorbed by the brain and exhibits comparable brain concentrations to its parent molecule, leptin (75). A dendrigraft poly-L-lysine (DGL) was coupled to leptin 30 through a PEG spacer and tested for gene delivery to the brain based on this outcome. Results suggest that gene expression was higher after intravenous treatment of DGL-PEG-leptin 30 applied in combination with a plasmid DNA than after DGL-PEG/DNA and DGL/DNA. Moreover, it was primarily visualized in the regions of the substantia nigra, caudate putamen, cortical layer, and hippocampus (76).

16.7 CYTOTOXICITY OF DENDRIMERS

The amount and type of surface functional groups greatly influence dendrimer cytotoxicity. Anionic or neutral dendrimers have minimal toxicity, whereas cationic dendrimers frequently demonstrate severe toxicity. In contrast, grafted carbosilane-poly(ethylene oxide) (CSi-PEO) dendrimers and other dendrimers terminated with neutral and anionic groups seem to be much less toxic (77). For instance, PAMAM or PPI dendrimers possessing principal terminal amines are characterized by concentration- and generation-dependent toxicity (78, 79). As a result, the cationic dendrimer's cytotoxicity is reduced when its surface is modified with negatively charged and neutral moieties (79). This section examines current studies on the cytotoxicity using dendrimers in light of future clinical uses. Felczak et al. investigated the cytotoxic and antibacterial properties of PPI dendrimers. They tested the antibacterial activity of the original PPI (G4) dendrimers against the Gram-positive bacteria *Staphylococcus aureus* or *Staphylococcus epidermidis*, the Gram-negative bacteria *Escherichia coli* or *Pseudomonas aeruginosa*, as well as the yeast *Candida albicans*. Cytotoxicity was examined on B14, rat liver cell lines, human liver hepatocellular carcinoma (HepG2), N2a, or all tested dendrimers (BRL-3A). The results showed that PPI dendrimers that had not been modified and PPI dendrimers that had been modified by 25% maltose exhibited antibacterial activity or a unique selectivity toward *S. aureus* at concentrations of 1 M while being safe for eukaryotic cell lines (80). Another kind of dendritic substance are metal-based dendrimer complexes, or metallodendrimers. It was discovered that polyamide metallodendrimers have antibacterial activity against *Salmonella typhi*, *Bacillus subtilis*, and Gram-positive *S. aureus* (Gram-negative). The MFC cell line was also susceptible to the antitumor effects of metallodendrimers. Pt(II)-containing metallodendrimers were found to be more effective than Pd(II)-containing metallodendrimers in the induction of MFC-7 mortality, and both showed lesser cytotoxicity than cisplatin (standard medication) (81). The effects of a PAMAM dendrimer coupled with the anticancer drug camptothecin (PAMAM-CPT) on individual colorectal carcinoma cells were studied by Thiagarajan et al. (HCT-116). The compound was stable in phosphate-buffered saline (PBS) or growth media (with 10% FBS) at physiological pH (7.4) with negligible drug release at 48 h of 4% and 6%, respectively. PAMAM-CPT prevented HCT-116 cells from proliferating and caused cell-cycle arrest, including up to 68% of cells remaining inside the gap 2 (G2) phase. The PAMAM-CPT combination was, therefore, effective *in vitro* at suppressing the proliferation and causing nuclear fragmentation of colorectal cancer cells. Testing this conjugate's cytotoxicity against normal cells might be helpful (82). However, it has been demonstrated that surface changes of dendrimers improve their capacity to penetrate cells, which helps explain why the drug–dendrimer complex inhibits cancer cell proliferation. Jia et al. added acryloyloxyethyl phosphorylcholine (APC), a chemical having biomimetic characteristics, to PAMAM dendrimers G5. The anticancer medication adriamycin (ADR) was added to the hydrophobic core of the improved dendrimer (G5-PC), and the G5-PC demonstrated sustained release behavior for ADR. In the malignant HepG2 cell line, cell survival and morphology tests showed that G5-PC had substantially lower toxicity than the unaltered PAMAM dendrimer. Moreover, it was demonstrated that cancer cells internalized the drug-loaded G5-PC, significantly reducing the cells' vitality (83). In a different investigation, fluorouracil (5-FU) was loaded onto a PEGylated PAMAM dendrimer from G3 and evaluated *in vitro* on the MCF-7 cell line. According to the findings, the PAMAM dendrimer was more cytotoxic than its PEGylated equivalent. PEG-modified PAMAM dendrimers, however, showed antiproliferative action against cancer cell lines when loaded with 5-FU. A gradual release profile of a drug was another advantage of encapsulating fluorouracil within PEGylated dendrimer. Moreover, *in vivo* research showed that the tested compound significantly reduced the size of tumors produced by MCF-7 cancerous cells (84). In conclusion, surface properties and pendant group of the dendrimers which is used for various purposes regulate the toxicity of the dendrimers.

16.8 APPLICATION OF DENDRIMERS FOR TREATING BRAIN DISEASES

Neurological illnesses are major healthcare concerns with complex etiologies and challenging treatment (85). Due to their prominence as the primary cause of impairment and mortality worldwide, neurological illnesses are a significant reason for worry (86). While the incidence of causality is

16.8%, the prevalence of these conditions is 10.2% (87). The coronavirus has been linked to multiple reports of neurological symptoms (88). More focus must be placed on the early detection and treatment of neurological illnesses due to the rise in their prevalence (89). Dendrimers have been used to deliver various drugs to treat neurological illnesses like brain cancer (90), inflammation in the brain (91), brain damage (92), and Alzheimer's disease (47). The outcome of this application will be discussed thoroughly in the next sections.

16.8.1 Use of Dendrimers in Brain Damage

Because of the absence of public awareness, brain damage is a serious problem that could result in a lifelong handicap (15). The etiology of traumatic brain injury (TBI) is a complicated process that is brought on by both fundamental and secondary injuries. Acute or chronic subsequent pathological processes brought on by TBI can last from days to weeks, including inflammatory, chemical, and molecular cascades that result in additional cerebral damage (93). Some secondary pathogenic processes include blood–brain barrier disruptions, seizures, sleep issues, brain edema, neurodegenerative illnesses, neuroendocrine dysregulation, or psychiatric ailments (94). Concussion, contusion, subdural and epidural hemorrhages, cerebral edema, subarachnoid hemorrhage, or pneumocephalus are all mild TBI (80). Some moderate TBI somatic symptoms are nausea, vertigo, headaches, impaired vision, auditory disturbances, and exhaustion. In contrast, emotional and behavioral issues include disinhibition or emotional lability, and cognitive complaints involve memory and executive functioning (18). TBI can result from falls, sports, explosion-related injuries, and car accidents (95). Although TBI is a frequent and significant problem, there are not many therapeutic alternatives. The phenomena of microglial activation occur in TBI. Microglia become activated quickly after damage, and this activation might extend for months, according to some human investigations (96). When brain cells are damaged, microglia aid in removing cellular debris and offer neuroprotection (97). Nevertheless, pro-inflammatory cytokines like TNF and IL-1 are produced when microglia are activated for a long time. These inflammatory mediators impair tryptophan metabolism and encourage macrophage growth in the wounded tissue. Hence, dendrimers to target microglia and stop pro-inflammatory responses may be a viable way to treat early or late-stage TBI. With their capacity to penetrate compromised BBB and improved absorption by brain parenchymal cells, G4 or G6 PAMAM dendrimers have successfully targeted microglia or key inflamed cells in wounded brain tissues (90). Dendrimer phagocytosis was shown to be more active than usual in activated microglia. Hence, dendrimers can be a great delivery system for an anti-inflammatory medication to reach the active microglia and astrocytes in the wounded brain and facilitate cellular uptake. In TBI-induced *in vitro* or *in vivo* models, the bioactive natural substance sinomenine has demonstrated a strong anti-inflammatory effect (86). Dendrimers can target activated microglia with the aid of terminal hydroxyl groups. As a result, the alkyne in the study underwent partial alteration of hydroxyl groups (by about 15 numbers). A covalent enzyme-sensitive bond that was linked to the membrane of a hydroxyl PAMAM dendrimer via sinomenine. The sinomenine conjugation improved the absorption of dendrimers by murine macrophages, which was an intriguing discovery in this investigation. Sinomenine conjugation results in nearly 1.8-fold greater uptake than non-drug-attached dendrimers, according to confocal microscopy or flow cytometry analysis using Cy5 (fluorescence marker)–labeled sinomenine conjugated dendrimers. This could occur due to the hydrophobicity that sinomenine imparts to conjugated dendrimers, which may assist them in interacting favorably with the lipid bilayer on the cell membrane (98, 99). Sinomenine-loaded dendrimers were used in the study to decrease TNF cytokine secretion in LPS-treated TNF secretory murine macrophages in a dose-dependent manner. Dendrimers that contained sinomenine significantly inhibited NF-kB translocation. A similar kind of dose-dependent suppression of active microglia and associated inflammatory mediators was observed in the *in vivo* study using a rabbit model of TBI. For instance, it has been reported that in rabbit kits, sinomenine-conjugated dendrimers drastically reduced the expression of TNF-α and increased the production of the anti-inflammatory cytokine IL 10 (71). In the brain damage model, dendrimers have been demonstrated to target macrophages and microglia based on ligands. Microglia, macrophages, peripheral dendritic cells, or astrocytes in the brain all contain the endocytic receptor mannose receptor (CD 206) (100). Mannose receptors are abundantly expressed in inflammatory states (101). In a neonatal rabbit brain damage model, the biodistribution of mannose-conjugated G4 dendrimers with 64 end groups has been investigated (102). By conjugating mannose as a ligand, Cy5 as a fluorescent dye, and an anti-inflammatory medication, the researchers have created tri-functional dendrimers with hydroxyl, acetylene, or amine functional groups. Nevertheless, some hydroxyl groups must be free for dendrimers to retain their innate capacity for brain transport. An *in vitro* study utilizing LPS-treated RAW 264.7 murine macrophages revealed increased absorption of mannose-conjugated dendrimers. A higher uptake suggests that the anti-inflammatory drug may be more effective. This study suggests a viable strategy to enhance the internalization and uptake of anti-inflammatory therapeutics in inflammatory cells. The strategy must, however, be backed up by further comparable investigations. Many illness models have shown the effectiveness of cell-based therapies. Transplanting mesenchymal stem cells (MSCs) into patients with acute TBI can effectively modulate the inflammatory response by changing the blood levels of pro- and anti-inflammatory chemokines and cytokines (102)

16.8.2 Use of Dendrimers in Glioblastoma

Brain tumor is the most lethal type of cancer among all malignancies (103). Brain tumors can either be primary brain tumors that start in the brain or metastatic/secondary brain tumors that start in other body parts and spread to the brain (104). Based on the tumor grade and cell source, the heterogeneous nature of the brain tumor is categorized (97). Brain tumors can be benign or cancerous. Glioblastoma accounts for 14.6% of all malignant brain tumors, while meningiomas account for 37.6% of all non-malignant brain tumors (71). Frequent migraines, mood swings, speech abnormalities, difficulties concentrating, seizures or memory loss, gait problems, and a lower quality of life are among the typical symptoms (105). Traditional methods for treating brain cancer involve surgically removing the tumur, radiation therapy, and systemic chemotherapy. The majority of systemic chemotherapy, however, has a very narrow therapeutic index, poor water solubility, low bioavailability, and restricted tumor cell drug uptake (106). Nanotechnology-based carriers now employ dendrimers to identify and treat specific tumors to circumvent the conventional technique's problems. Only nanoparticles smaller than 30 nm can penetrate gliomas, which are solid brain tumors, and achieve tumor dispersion (107). In glioblastoma or glioma, dendrimers are used as nanocarriers. Their small particle size facilitates passage through the blood–tumor barrier's microscopic pores into tumor cells. Some targeted ligands linked with dendrimers increase tumor cells' penetration and retention effects. Hence, inside glioma cells, the cellular uptake of nanoparticles is amplified. The conjugated ligand may also improve the persistence of the nanoparticles inside tumor cells. Due to the greater availability of their 128 surface amino groups, G5 PAMAM dendrimers are desirable for ligand conjugation. The tumor-homing peptide CREKA was coupled with G5 PAMAM dendrimers for better nanoparticle anchoring in tumor cells. PEGylated PAMAM dendrimers were employed in this study (73). Dendrimer glycation lessens toxicity and extends the circulation duration within the body (108). First, the scientists generated fluorescein isothiocyanate isomer I (FITC-I), a pigment attached to PAMAM dendrimers. Then, they attached another near-infrared dye (IR783-NHS) to the dendrimers for imaging. CREKA's PEGylation and surface modification followed the last phase. The *in vivo* biodistribution or glioma-targeting effectiveness in the U87 orthotopic glioma-producing mice model was assessed using fluorescent imaging. Dendrimer–extracellular fibrin interaction is a factor in the improved distribution within glioma tissues. This theory was backed by an *in vitro* investigation on fibrin clotting, where it was found that dendrimers and fibrin formed aggregates more readily. Dendrimers are immobilized inside glioma tissue, in which they can connect with the protein integrin on the cell surface, leading to higher accumulation. Nanoparticle immobilization improves internalization by cells. In the study by Zhao et al., the increased fluorescence intensity in the CREKA-conjugated dendrimers treating group over the control or untargeted group revealed prolonged and in glioma tissues (73).

16.8.3 Use of Dendrimers in Alzheimer's Disease (AD)

Alzheimer's is the leading cause of mortality and disability among the elderly. The three primary features of AD are the accumulation of amyloid surrounding dying neurons and the deterioration of neurofibrillary tangles, or cerebrovascular angiopathies. As the clumps of beta-amyloid plaque harm neurons, they develop between neurons and impede cell activity (80). The breakdown of neurofibrillary tangles is caused by tau protein. The tau protein attaches to and keeps the microtubules stable in a healthy brain. Nevertheless, tau undergoes a post-translational change in the AD brain that causes it to aggregate inside the neurons. This prevents the neurons from communicating synaptically (109). Memory loss is the earliest indication of AD, followed by loss of cortical functions, including emotional regulation and neuropsychiatric symptoms, like agitation, depression, and apathy, which cause psychological and behavioral abnormalities (110). The two main treatment modalities are those that support illness progression and those that impede its pathways (111). Various delivery approaches, including dendrimers, have been employed to treat AD. Agents proved anti-amyloidogenic have been loaded or conjugated to dendrimers and have been shown to prevent the etiology and development of AD preclinically (112). G4 poly(propylene imine) (PPI) dendrimers with the use of a histidine–maltose shell have been created to protect the synapse and memory in AD conditions (113). First, histidine was added to the G4 PPI dendrimers to create G4His, and later, maltose was added to this G4His to create G4HisMal PBS (5 g per day to 7 L) was employed as a control. The developed dendrimer conjugation was administered intranasally to 5 transgenic and 6 WT mice. After the mice received therapy with G4HisMal, memory assessment in the V-maze tests was conducted, significantly improving memory conditions. *In vitro* or *in vivo*, G4HisMal reduced the synthesis of fibrils, and long-term drug administration prevents the animals from having memory loss. Moreover, following the G4HisMal therapy in AD, the level of synaptic markers, such as PSD95, synaptophysin, or debris, remains the same. This study implies that synapse protection against soluble A neurotoxicity may be the mechanism underlying memory management in transgenic mice. A plaque quantification could not indicate a decrease in plaque burden inside the neocortex of treated mice. Still, compared to the A plaques found in the cortex of control mice, a different A plaque shape was visible in the cerebral cortex of treated animals. Compared to the control group, mice from the G4HisMal group had significantly improved memory in the two-object recognition test. The examination of the plaque area as just a ratio of fibrillar A (OC positive) to the total A deposits, based on the amyloid plaque morphology, showed a considerable augmentation of fibrillar contents in amyloid plaques following G4HisMal therapy (6E10 positive). According to the study's findings, persistent treatment of G4HisMal for three months allowed it to pass the blood–brain barrier and lessen memory deficits in mice by protecting synapses against soluble A. Several researchers have proposed distinct dendrimer processes that work to prevent the development and spread of amyloid in the central nervous system. Dendrimers prevent the development of sheet amyloid fibrils. Because amyloid accumulation in AD is directly related to polymeric amyloid fibrils, which are often rich in sheet structure (114). According to the findings, amyloid beta aggressively attaches to sialic acid residues upon that cell surface, producing neurotoxicity. As a result, sialic acid–attached dendrimers can efficiently inhibit amyloid adherence on the cell membrane and hence reduce neurotoxicity by removing sialic acid residue and mimicking cell surfaces. Therefore, 32(G3) and 64(G4) terminal group sialic acid–functionalized PAMAM dendrimers were synthesized and examined for their activity. A physiological connection of the anomeric hydroxyl group and a non-physiological attachment seen between carboxylic acid units of the sialic acid or dendrimers terminals were used to conjugate sialic acid to dendrimers. The sialic acid–attached dendrimers significantly decreased the neurotoxicity assessed in experiments on SH-SY5Y neuroblastoma cells. Moreover, according to *in vitro* results, sialic acid–linked dendrimers bind to lysozymes and aid in producing amyloid fibrils, which assist in reducing neurotoxicity. Benseny-Cases et al. examined the effects of PAMAM dendrimers (G3, 32 terminal groups), PPI dendrimers (polypropylene imine, G4-PPI), or dendrimers with 96 cationic end groups (G4–96) on the inhibition of amyloid fibril formation (115). They discovered that the interference with amyloid fibril production changes depending on the dendrimer-peptide ratio. Low dendrimer-peptide ratios increased fibril formation, but high dendrimer-peptide ratios slowed down polymerization and fibril production. According to one theory on the pathophysiology of AD, neurofibrillary entanglement is caused by microtubule-associated phosphorylated tau protein (MAP). Abnormal hyperphosphorylation results in the loss of tau function to support stable microtubule assembly, which is the primary cause of tau dysfunction in AD (116). When hyperphosphorylated tau cannot connect to microtubules, it disrupts them and kills neurons. The lack of equilibrium in the two enzymatic activities, tau kinase or tau phosphatase, causes hyperphosphorylation. Preventing tau abnormalities has recently emerged as an important AD treatment approach. A study revealed that tau aggregation and amyloid fibrillation are inhibited by cationic phosphorus dendrimers having 48 end groups (G3, dendrimers III) or 96 terminal groups (G4, dendrimers II) (47). The thioflavin T fluorescence indicator and this assay were used to investigate the aggregation of A 1–28 and MAP-tau. Dendrimer interference with fibrillation was demonstrated to be concentration-dependent. Dendrimers damaged the fibrils at high concentrations but increased fibrillation frequency at low quantities. At the molar ratio of 1:1.5, the cationic dendrimers successfully blocked the generation of MAP-tau amyloid. The effect was similar to untreated groups at low dendrimer doses. According to the transmission electron micrograph, both G3 and G4 cationic dendrimers encourage amorphous tau aggregation more than filamentous formations, which ultimately results in disruption. This study demonstrated the potential of targeting MAP-tau accumulation in AD with cationic phosphorous dendrimers.

16.8.4 Use of Dendrimers in Prion Disease

Like AD, prion disease also produces amyloid fibrils and is caused by infectious particles that contain a pathogenic isoform of a prion protein, which is generally found in cell membranes. The most prevalent human prion disease is Creutzfeldt–Jakob disease (CJD) (117, 118). A different or abnormal isoform of PrP triggers the transformation of PrPC to nascent PrPSc, which results in neurodegeneration in the brain in this condition (119). The clinical symptoms of a prion disease of the CJD type include rapidly developing dementia and cerebellar dysfunction involving muscular incoordination or visual, speech, and gait abnormalities. Dementia is the most frequent symptom, accompanied by spontaneous or induced myoclonia. Behavioral abnormalities are observed throughout the illness, like agitation, confusion, and depression, as well as reflexes, tremors, spasticity, stiffness, or extrapyramidal dysfunction symptoms. When the illness worsens, most patients experience akinetic mutism (they stop responding to external stimuli) (47). Using modified PPI dendrimers containing neutral carbohydrate units comprising maltose (mPPI) and maltotriose to replace positive charged amino surface groups, Fischer et al. studied the anti-prion action of dendrimers (m3PPI). However, surface group modification significantly reduced the unmodified PPI's anti-prion activity (50% by maltose). This finding does not prove that positive surface groups are necessary for dendrimer action. Three days of exposure to different dendrimer concentrations were given to the ScN2a cells, and the protease-resistant PrPsc (PrP res) levels were assessed on an immunoblot following PK degradation of cellular lysates. PPI and mPPI dendrimers were similarly effective in reducing protease-resistant PrP Sc levels in ScN2a cells or pre-existing aggregate in homogenates from the infected brain in a dose- and time-dependent manner. While generation 5 dendrimers from PPI and mPPI reduced PrPres level in ScN2a cells comparably well, generation 3 dendrimers performed differently. In contrast to mPPI-G3, PPI-G3 did not successfully lower the PrPres level in ScN2a cells. This is due to the mPPI's hydrogen bonding interactions with the prion protein PrP 185–208, which were discovered to be generationally dependent with rising generation numbers. It was found that phosphorus dendrimers significantly reduced PrPSc levels in the spleens of scrapie-infected mice (114). Heparin, an endogenous protein, was found to be involved in the creation of fibrils and, based on its concentration, could attenuate prion illness. Electrostatic interactions allowed polyamidoamine, PPI, or phosphorous dendrimers to bind with heparin (114). Dendrimers and heparin's respective cationic or anionic charges cause these interactions, which halt the development of amyloid fibrils. Phosphorous G4 dendrimers were created, and their inhibitory action upon that amyloid fibrillation procedure of PrP 185–208 prion peptide was reported in a few earlier research studies (120, 121). It lowered the fibril concentration and decelerated the nucleation activity, thereby reducing the aggregation. Phosphorus G3 dendrimers and amyloid proteins share a region similar to the PrP 185–208 prion protein, leading to the formation of a compound. In mouse neuroblastoma cells, these nanocomplexes decreased amyloid accumulation and neurodegeneration.

16.8.5 Use of Dendrimers in Parkinson's Disease

Parkinson's disease (PD) is a neurological condition that primarily affects those over 65 globally who are elderly (122). The etiology of PD involves serotonergic or noradrenergic neurons as well as dopaminergic neurons. The main features of PD are impairments of somatic nerve systems, such as stiffness, bradykinesia, postural imbalance, gait dysfunction, or tremor (86), as well as impairments of non-motor functions, including dementia, hyposmia, or gastrointestinal changes. Lewy bodies (LB) and Lewy neurites (LN), which are the accumulation of α-synuclein, are filamentous, cytoplasmic inclusions that accumulate as a result of Parkinson's disease (PD). Neuronal death is caused by the phosphorylation, fibrillization, and subsequent production of LB in α-synuclein (123). The selective decline of dopaminergic neurons throughout a substantia nigra pars compacta, which results in dopamine depletion inside the striatum, is a contributing factor in the development of Parkinson's disease's movement disorder. In contrast, non-motor symptoms are brought on by more generalized neurodegeneration that affects the cortex and various brain stem regions. Preparations of levodopa are employed in several of the current therapies. Levodopa use for an extended period produces detrimental side effects, even though these drugs can aid Parkinson's disease patients with motor symptoms. Deep brain stimulation (DBS), which particularly aids in managing motor activities (124), and therapies that lower the synthesis of synuclein production (125) are two further treatments. One method for treating Parkinson's disease (PD) that targets drug delivery to the brain are dendrimers. Rekas et al. investigated the morphological aspects of α-synuclein accumulation in PAMAM dendrimers. α-synuclein was incubated with G4, G5, and G6 dendrimers at a PAMAM concentration to produce amyloid fibrils. The fluorescent dye known as thioflavin T (TFT) typically attaches to proteins found only in stacked cross-sheet structures, which causes fluorescence emission to be enhanced. The interaction of TFT using PAMAM dendrimer revealed a decrease in fluorescence, which may have been caused by the α-synuclein fibrils clumping together. While PAMAM G4 and G5 effectively decreased α-synuclein fibrillation, G6 was roughly eight times more effective. The TEM supported the TFT binding tests by showing a more pronounced inhibitory effect on fibrillation with increasing PAMAM dendrimer concentration and generation number. According to this study, dendrimers have been demonstrated to reduce fibrillation and remove pre-existing PrPSc fibrils (125). To prevent the buildup of α-synuclein(α-syn) in PD, Lorente et al. used cationic carbosilane dendrimers with various topologies, including spherical, dendritic wedge, and bow-tie. In this investigation, they assessed changes in zeta potential to determine the impact of the system architecture on interactions between α-syn and the dendritic system. Due to the prevalence of negative charges at pH 7.4, the α-syn zeta potential was approximately 21 mV. However, as dendritic system concentrations increased, the zeta potential turned positive, until 20 mV was attained. Nanoconjugates made with α-syn are anticipated to have positive charge values independent of the system topology since dendritic structures are cationic. Due to their inhibitory efficacy at low concentrations, bow-tie dendrimer structures were chosen for additional interaction experiments. To examine the impact of cationic carbosilane dendrimers on the survivability of PD patients' dopamine neurons in culture, DAns derived from iPSC lines originating from PD patients having the G2019S mutation, the LRRK2 gene (LRRK2-PD), demonstrating aberrant accumulation of α-syn were utilized. Positively charged carbosilane dendrimers (bow-tie mechanism) were added to DAn cultures at two concentrations (1 and 3 M). Compared to cultures treated with vehicle alone, the greatest dendrimer concentration caused a 60% drop in DAn counts. Dendrimers were also not hazardous to DAns when applied at a concentration of 1 M for 48 h, at least in cell viability. After 48 h of treatment with 1 M of cationic carbosilane dendrimers, the percentage of abnormally accumulating α-syn in DAns from PD patients was diminished to levels that were not statistically different from all those observed in control DAns. Analysis of the ability of dendrimers to inhibit the intracellular buildup of α-syn aggregates revealed that LRKK2-PD DAns had a considerably larger percentage of α-syn puncta than control DAns (51.8 8.7% vs. 19.4 4.7%). These results show that the DAns of PD patients directly absorb bow-tie-shaped cationic carbosilane dendrimers and interact with α-syn aggregates, most likely aiding in their solubilization into tiny aggregates that are more easily eliminated (125).

16.9 FUTURE PERSPECTIVES

Due to their unique regulated globular structure, dendrimers are nanoscale materials emerging in numerous biomedical therapeutic applications due to their nanosize, multivalency, and low polydispersity. Dendritic systems have already passed the clinical evaluation as contrast-enhancer MRI agents (Gadomer®-17), anti-microbicides (Vivagel®), as well as drug carriers for solid tumors therapy (DEPTM docetaxel), in addition to the diagnosis and *in vitro* dendrimer-based products that are currently on the market. Nevertheless, despite these documented uses, dendrimer use inside the CNS remains in its infant stages. No reports of commercially available products or ongoing clinical studies using dendrimers to treat CNS diseases exist. The capacity of therapeutic bioactive to penetrate the BBB and enter the brain with an appropriate bioavailability is one of the biggest obstacles in treating CNS diseases (57). Many studies have shown that dendrimers can pass through this barrier and enter the brain following systemic delivery because of their controlled nanosize. Moreover, dendrimers have demonstrated the ability to function as standalone therapeutic as well as theranostic compounds and as vectors for the delivery and protection of bioactive to the CNS. However, since the appropriate *in vivo* dendrimer properties are rare and the safety profile has not yet been thoroughly established, their progress toward clinical evaluation may be hampered. Non-specific dendrimer toxicity is strongly correlated with its size, production, concentration, and internal chemical composition, but the type of terminal surface groups mostly impacts it. Unaltered cationic dendrimers are typically more cytotoxic than their anionic and neutral counterparts (86). Similarly, due to their higher cationic surface charge, unmodified dendrimers in higher concentrations and generations have a more pronounced cytotoxic and hemolytic impact. Nonetheless, there continues to be a discrepancy in the literature concerning how the toxicity profile of a particular dendrimer is affected by its production. In particular, some publications have demonstrated that PAMAM and PPI dendrimers of low generations are not harmful *in vivo* (91). The same dendrimers have been linked to a toxic profile by others (82). This highlights the necessity of adjusting dendrimer composition, production, dosage, and mode of administration and evaluating the safety and efficacy at various stages following administration. According to previous research, unmodified cationic dendrimers and other cationic macromolecules strongly interact with negatively charged cell membranes. This interaction may weaken the membrane's integrity and allow intracellular substances to leak out, which can cause cell death and toxicity (79). Moreover, when given systemically, unmodified dendrimers' cationic terminal groups may react with RBCs, which could cause hemolysis. Similar to how they affect other hematological variables, these cationic dendrimers can also affect WBC and RBC numbers and morphologies, hemoglobin content, hematocrit, and mean corpuscular hemoglobin (MCH) values (109). Therefore, neutral or anionic dendrimers and/or surface engineering (e.g., the well-known PEGylation and carbohydrate coating) prevent strong electrostatic interactions of dendrimers or bioactive/dendrimer systems with cellular membranes, a step that must be carefully considered to minimize non-specific toxicity. For instance, Greish et al. contrasted the toxicity profiles of the amine- and hydroxyl-terminated PAMAM dendrimers. Unlike amine-terminated PAMAM, neutral hydroxyl– or anionic carboxyl–terminated PAMAM dendrimers displayed a decrease in their cytotoxic profile and no indication of producing protein opsonization or blood hemolysis (86). Several researchers confirmed the same results. Overall, these results show that surface engineering techniques and the use of neutral as well as anionic dendrimers are crucial tools for improving biocompatibility and lowering the overall immunogenicity of all these systems. Moreover, optimizing the size of dendritic networks is another tactic that can be used to control their biocompatibility. As previously mentioned, the diameter of dendrimers and dendrimer/bioactive complexes can be precisely adjusted according to the intended application to enhance, for instance, the EPR effect or quick *in vivo* clearance. Dendrimers' surface chemistry and size mostly influence their distribution in living things. Larger dendrimers are more slowly removed from circulation and are more effectively absorbed by macrophages, leading to greater reticuloendothelial system (RES) clearance. Smaller dendrimers have short circulation times and are quickly cleared by renal filtration (94). Furthermore, cationic dendrimers have a higher clearance rate from circulation than their anionic counterparts due to the possibility of stronger interactions with the vasculature of strongly perfused organs. In addition to reducing charge-dependent toxicity, the surface above engineering of shielding agents as well as targeting ligands may also aim to increase solubility in water as well as circulation time (particularly when systems amended with anti-biofouling polymers, such as PEG, are explored), while minimizing off-target effects by improving the structures targeting potential as well as payload efficacy. For instance, Zhu and colleagues showed that, in comparison to their real equivalents, enhancing the extent of PEGylation through PAMAM dendrimers could, indeed, decrease liver and spleen accumulation as well as general toxicity, prolong the time spent in circulation, and improve the accumulation in malignant cells of a PEGylated dendrimers conjugates (19). Kojima et al. confirmed dendrimer PEGylation's effect on its biodistribution or circulation time. In addition, several promising anti-biofouling/non-fouling polymers, including zwitterionic polymers, poly(2-oxazoline)s, poly(peptoids), and poly(carbonates), have recently come to light as potential PEG substitutes. Although some of those polymers have already been linked to dendritic frameworks in a few studies (126), more research is anticipated to investigate the beneficial features of these alternate polymers whenever attached to dendrimers. Similarly, Kesharwani et al. contrasted three distinct ligand-conjugated PPI dendrimers with their basic (unmodified) PPI counterparts in terms of their capability for targeting tumors and toxicity profiles (34). They found that the receptor-conjugated PPI dendrimers had improved tumor-targeting capacity but dramatically lessened hemolytic toxicity compared to the plain PPI dendrimers. The BBB penetration capability of dendrimers in CNS applications can be viewed as a two-edged sword. While the increased bioavailability for therapeutic bioactive inside the CNS may result from the improved BBB permeability, it can also enhance the dangers of overexposing target cells to dendrimers and therapeutic bioactive, leading to CNS toxicity. Its toxicity to cells is a characteristic that non-biodegradable dendrimer use is frequently linked to. The formation of reactive oxygen species (ROS) and reduced metabolic activity may result from non-biodegradable dendrimers because of their potential for long-term accumulation inside lysosomes or cytoplasm (97). Consequently, using biodegradable dendrimers is strongly advised in a biomedical context to avoid dendrimer bioaccumulation–associated toxicity (90). Under physiological settings, such dendrimers should break down into smaller pieces that can be removed or expelled through metabolic processes. Dendrimers have undergone extensive review in recent years, but significant work needs to be done to transform these nanoconstructs quickly into therapeutically relevant CNS nanomedicines. It is now obvious that dendrimers' potential benefits are directly related to their physicochemical characteristics, which should be adjusted to the intended therapeutic use. However, this review shows the limited diversity of dendrimer groups that received preclinical investigation. Most dendrimers employed for CNS applications belonged to just three primary families: PAMAM, PPI, or PLL dendrimers. Further breakthroughs in creating highly biocompatible and biodegradable dendrimers, novel linker technologies, and novel targeting agents are eagerly anticipated to improve the bioavailability of dendrimer-based systems at the target tissues. The sector will also benefit from work on creating effective multigram synthesis to meet pharmaceutical industry demand and green synthetic pathways. In addition to affecting the environment, reducing the usage and production of hazardous compounds would also help advance biocompatibility and purifying techniques. So more research will be required to comprehend further the interaction between dendrimers and cells, bodies, and their *in vivo* fate to create such improved systems. When examining the behavior of nanosystems at the whole-body scale, the resolution or lack of sensitivity of presently available bioimaging technologies must be considered (86). The field focuses only on assessing overall biodistribution/accumulation in important organs, time to clearance, systemic inflammatory response, and related toxicity. The ability to investigate several delivery routes while thinking about the CNS brings up new possibilities for enhancing the effectiveness and, when appropriate, the treatment target. However, the etiology and specifics of many neurological illnesses are still poorly understood. Furthermore, the field will need to be continuously fed from advancements in the basic sciences of neuroscience or neurology. In conclusion, due to this system's uniqueness and utmost significate, successfully introducing dendrimers as potent weapons in treating CNS illnesses would unquestionably demand a multi- or interdisciplinary effort. In summary, dendrimers can be employed in the CNS for variations, including the delivery of active therapeutic molecules; such a surprising capability reveals their enormous promise as precise theragnostic multifunctional agents, which we anticipate will eventually expand other unstudied CNS illnesses.

16.10 CONCLUSION

The invention of PAMAM dendrimer, in particular, has attracted much attention and paved the way for delivering medicines in a spatially, temporally, and dosage-controlled manner for brain targeting. Dendrimers are a fascinating nanovector for treating and diagnosing brain diseases as a promising nanomedicine integrating all three components: functionalization, targeting, or imaging. They have considerable potential as protean DDS for tiny molecules, genes, oligonucleotides, proteins, as well as peptides. There is optimism for new therapeutic alternatives shortly due to several promising new nanoparticles being developed, even though some dendrimer-based approaches have failed when clinically translated. For neurological disorders, the non-invasive drug administration method, improved therapeutic effectiveness, and reduced adverse effects would contribute to nanosystems' preponderance in penetrating the BBB. Using esoteric tools for brain imaging and manipulation will help us better understand how the human brain generates thought and action. Additionally, the huge potential of dendrimer-based nanomaterials will undoubtedly encourage the development of new dependable ways to deliver various drugs to the brain precisely.

REFERENCES

1. Moscariello, P., Ng, D. Y., Jansen, M., Weil, T., Luhmann, H. J., & Hedrich, J. (2018). Brain delivery of multifunctional dendrimer protein bioconjugates. *Advanced Science*, *5*(5), 1700897.
2. Srikanth, M., & Kessler, J. A. (2012). Nanotechnology-novel therapeutics for CNS disorders. *Nature Reviews Neurology*, *8*(6), 307–318.
3. Wegener, G., & Rujescu, D. (2013). The current development of CNS drug research. *International Journal of Neuropsychopharmacology*, *16*(7), 1687–1693.
4. Saeedi, M., Eslamifar, M., Khezri, K., & Dizaj, S. M. (2019). Applications of nanotechnology in drug delivery to the central nervous system. *Biomedicine & Pharmacotherapy*, *111*, 666–675.
5. Sharma, G., Sharma, A. R., Lee, S. S., Bhattacharya, M., Nam, J. S., & Chakraborty, C. (2019). Advances in nanocarriers enabled brain targeted drug delivery across blood brain barrier. *International Journal of Pharmaceutics*, *559*, 360–372.
6. Yu, S., Xu, X., Feng, J., Liu, M., & Hu, K. (2019). Chitosan and chitosan coating nanoparticles for the treatment of brain disease. *International Journal of Pharmaceutics*, *560*, 282–293.
7. Banks, W. A. (2016). From blood–brain barrier to blood–brain interface: New opportunities for CNS drug delivery. *Nature Reviews Drug Discovery*, *15*(4), 275–292.
8. Begley, D. J., & Brightman, M. W. (2003). Structural and functional aspects of the blood-brain barrier. *Peptide Transport and Delivery into the Central Nervous System*, 39–78.
9. Poldrack, R. A., & Farah, M. J. (2015). Progress and challenges in probing the human brain. *Nature*, *526*(7573), 371–379.
10. Palma-Florez, S., Lopez-Canosa, A., Moralez-Zavala, F., Castaño, O., Kogan, M. J., Samitier, J., & Mir, M. (2023).

BBB-on-a-chip with integrated micro-TEER for permeability evaluation of multi-functionalized gold nanorods against Alzheimer's disease. *Journal of Nanobiotechnology*, *21*(1), 1–17.

11. Aulenta, F., Hayes, W., & Rannard, S. (2003). Dendrimers: A new class of nanoscopic containers and delivery devices. *European Polymer Journal*, *39*(9), 1741–1771.
12. Akiyama, H., Miyashita, K., Hari, Y., Obika, S., & Imanishi, T. (2013). Synthesis of novel polyesteramine dendrimers by divergent and convergent methods. *Tetrahedron*, *69*(33), 6810–6820.
13. Huang, D., & Wu, D. (2018). Biodegradable dendrimers for drug delivery. *Materials Science and Engineering: C*, *90*, 713–727.
14. Lim, Y. B., Mays, C. E., Kim, Y., Titlow, W. B., & Ryou, C. (2010). The inhibition of prions through blocking prion conversion by permanently charged branched polyamines of low cytotoxicity. *Biomaterials*, *31*(8), 2025–2033.
15. McCarthy, J. M., Rasines Moreno, B., Filippini, D., Komber, H., Maly, M., Cernescu, M., & Rogers, M. S. (2013). Influence of surface groups on poly (propylene imine) dendrimers antiprion activity. *Biomacromolecules*, *14*(1), 27–37.
16. Alexander, A., Agrawal, M., Uddin, A., Siddique, S., Shehata, A. M., Shaker, M. A., & Shaker, M. A. (2019). Recent expansions of novel strategies towards the drug targeting into the brain. *International Journal of Nanomedicine*, 5895–5909.
17. Koehn, L. M. (2020). ABC efflux transporters at blood-central nervous system barriers and their implications for treating spinal cord disorders. *Neural Regeneration Research*, *15*(7), 1235.
18. Polinder, S., Cnossen, M. C., Real, R. G., Covic, A., Gorbunova, A., Voormolen, D. C., & Von Steinbuechel, N. (2018). A multidimensional approach to post-concussion symptoms in mild traumatic brain injury. *Frontiers in Neurology*, *9*, 1113.
19. Van Tellingen, O., Yetkin-Arik, B., De Gooijer, M. C., Wesseling, P., Wurdinger, T., & De Vries, H. E. (2015). Overcoming the blood–brain tumor barrier for effective glioblastoma treatment. *Drug Resistance Updates*, *19*, 1–12.
20. Obermeier, B., Daneman, R., & Ransohoff, R. M. (2013). Development, maintenance and disruption of the blood-brain barrier. *Nature Medicine*, *19*(12), 1584–1596.
21. Amjad, M. W., Amin, M. C. I. M., Katas, H., Butt, A. M., Kesharwani, P., & Iyer, A. K. (2015). In vivo antitumor activity of folate-conjugated cholic acid-polyethylenimine micelles for the codelivery of doxorubicin and siRNA to colorectal adenocarcinomas. *Molecular Pharmaceutics*, *12*(12), 4247–4258.
22. Mizee, M. R., & de Vries, H. E. (2013). Blood-brain barrier regulation: Environmental cues controlling the onset of barrier properties. *Tissue Barriers*, *1*(5), 1660–71.
23. Ahmad, J., Gautam, A., Komath, S., Bano, M., Garg, A., & Jain, K. (2019). Topical nano-emulgel for skin disorders: Formulation approach and characterization. *Recent Patents on Anti-infective Drug Discovery*, *14*(1), 36–48.
24. Buhleier, E. (1978). "Cascade and nonskid-chain-like" syntheses of molecular cavity topologies. *Synthesis*, 1978(02), 155–158.
25. Watkins, D. M., Sayed-Sweet, Y., Klimash, J. W., Turro, N. J., & Tomalia, D. A. (1997). Dendrimers with hydrophobic cores and the formation of supramolecular dendrimer– surfactant assemblies. *Langmuir*, *13*(12), 3136–3141.
26. Fréchet, J. M., Hawker, C. J., Gitsov, I., & Leon, J. W. (1996). Dendrimers and hyperbranched polymers: Two families of three-dimensional macromolecules with similar but clearly distinct properties. *Journal of Macromolecular Science, Part A: Pure and Applied Chemistry*, *33*(10), 1399–1425.
27. Kaur, D., Jain, K., Mehra, N. K., Kesharwani, P., & Jain, N. K. (2016). A review on comparative study of PPI and PAMAM dendrimers. *Journal of Nanoparticle Research*, *18*, 1–14.
28. Gorain, B., Tekade, M., Kesharwani, P., Iyer, A. K., Kalia, K., & Tekade, R. K. (2017). The use of nanoscaffolds and dendrimers in tissue engineering. *Drug Discovery Today*, *22*(4), 652–664.
29. Sohail, I., Bhatti, I. A., Ashar, A., Sarim, F. M., Naveed, R., Yasir, M., Iqbal, M., & Nazir, A. (2020). Polyamidoamine (PAMAM) dendrimers synthesis, characterization and adsorptive removal of nickel ions from aqueous solution. *Journal of Materials Research and Technology*, *9*(1), 498–506.
30. Mintzer, M. A., & Grinstaff, M. W. (2011). Biomedical applications of dendrimers: A tutorial. *Chemical Society Reviews*, *40*(1), 173–190.
31. Lee, J. W., Kim, J. H., Kim, B. K., Kim, J. H., Shin, W. S., & Jin, S. H. (2006). Convergent synthesis of PAMAM dendrimers using click chemistry of azide-functionalized PAMAM dendrons. *Tetrahedron*, *62*(39), 9193–9200.
32. Kesharwani, P., Banerjee, S., Gupta, U., Amin, M. C. I. M., Padhye, S., Sarkar, F. H., & Iyer, A. K. (2015). PAMAM dendrimers as promising nanocarriers for RNAi therapeutics. *Materials Today*, *18*(10), 565–572.
33. Svenson, S., & Tomalia, D. A. (2012). Dendrimers in biomedical applications—reflections on the field. *Advanced Drug Delivery Reviews*, *64*, 102–115.
34. Kesharwani, P., Tekade, R. K., Gajbhiye, V., Jain, K., & Jain, N. K. (2011). Cancer targeting potential of some ligand-anchored poly (propylene imine) dendrimers: A comparison. *Nanomedicine: Nanotechnology, Biology and Medicine*, *7*(3), 295–304.
35. Pedziwiatr-Werbicka, E., Milowska, K., Dzmitruk, V., Ionov, M., Shcharbin, D., & Bryszewska, M. (2019). Dendrimers and hyperbranched structures for biomedical applications. *European Polymer Journal*, *119*, 61–73.
36. Tomalia, D. A., Christensen, J. B., & Boas, U. (2012). *Dendrimers, dendrons, and dendritic polymers: Discovery, applications, and the future*. Cambridge University Press.
37. Beezer, A. E., King, A. S. H., Martin, I. K., Mitchel, J. C., Twyman, L. J., & Wain, C. F. (2003). Dendrimers as potential drug carriers: encapsulation of acidic hydrophobes within water soluble PAMAM derivatives. *Tetrahedron*, *59*(22), 3873–3880.
38. Richter-Egger, D. L., Tesfai, A., & Tucker, S. A. (2001). Spectroscopic investigations of poly (propyleneimine) dendrimers using the solvatochromic probe phenol blue and comparisons to poly (amidoamine) dendrimers. *Analytical Chemistry*, *73*(23), 5743–5751.
39. Fant, K., Esbjorner, E. K., Jenkins, A., Grossel, M. C., Lincoln, P., & Norden, B. (2010). Effects of PEGylation and acetylation of PAMAM dendrimers on DNA binding, cytotoxicity and in vitro transfection efficiency. *Molecular Pharmaceutics*, *7*(5), 1734–1746.
40. Ryan, G. M., Kaminskas, L. M., Bulitta, J. B., McIntosh, M. P., Owen, D. J., & Porter, C. J. (2013). PEGylated polylysine dendrimers increase lymphatic exposure to doxorubicin when compared to PEGylated liposomal and

solution formulations of doxorubicin. *Journal of Controlled Release*, *172*(1), 128–136.
41. Byrne, M., Victory, D., Hibbitts, A., Lanigan, M., Heise, A., & Cryan, S. A. (2013). Molecular weight and architectural dependence of well-defined star-shaped poly (lysine) as a gene delivery vector. *Biomaterials Science*, *1*(12), 1223–1234.
42. Uchida, H., Kabe, Y., Yoshino, K., Kawamata, A., Tsumuraya, T., & Masamune, S. (1990). General strategy for the systematic synthesis of oligosiloxanes: silicone dendrimers. *Journal of the American Chemical Society*, *112*(19), 7077–7079.
43. Zhou, L. L., & Roovers, J. (1993). Synthesis of novel carbosilane dendritic macromolecules. *Macromolecules*, *26*(5), 963–968.
44. Sepulveda-Crespo, D., Gomez, R., De La Mata, F. J., Jimenez, J. L., & Muñoz-Fernández, M. Á. (2015). Polyanionic carbosilane dendrimer-conjugated antiviral drugs as efficient microbicides: Recent trends and developments in HIV treatment/therapy. *Nanomedicine: Nanotechnology, Biology and Medicine*, *11*(6), 1481–1498.
45. Caminade, A. M., & Majoral, J. P. (2013). Positively charged phosphorus dendrimers: an overview of their properties. *New Journal of Chemistry*, *37*(11), 3358–3373.
46. Taghavi Pourianazar, N., Mutlu, P., & Gunduz, U. (2014). Bioapplications of poly (amidoamine)(PAMAM) dendrimers in nanomedicine. *Journal of Nanoparticle Research*, *16*, 1–38.
47. Wasiak, T., Marcinkowska, M., Pieszynski, I., Zablocka, M., Caminade, A. M., Majoral, J. P., & Klajnert-Maculewicz, B. (2015). Cationic phosphorus dendrimers and therapy for Alzheimer's disease. *New Journal of Chemistry*, *39*(6), 4852–4859.
48. Briz, V., Serramia, M. J., Madrid, R., Hameau, A., Caminade, A. M., Majoral, J. P., & Munoz-Fernandez, M. A. (2012). Validation of a generation 4 phosphorus-containing polycationic dendrimer for gene delivery against HIV-1. *Current Medicinal Chemistry*, *19*(29), 5044–5051.
49. Shcharbin, D., Shcharbina, N., Pedziwiatr-Werbicka, E., de la Mata, J., Gomez-Ramirez, R., Mignani, S., Kulchitsky, V. A., Munoz-Fernandez, M. A., Caminade, A. M., Majoral, J. P., & Bryszewska, M. (2018). Phosphorus dendrimers as vectors for gene therapy in cancer. In *Phosphorus dendrimers in biology and nanomedicine* (pp. 227–244). Jenny Stanford Publishing.
50. Caminade, A. M., Laurent, R., Delavaux-Nicot, B., & Majoral, J. P. (2012). "Janus" dendrimers: Syntheses and properties. *New Journal of Chemistry*, *36*(2), 217–226.
51. Tuuttila, T., Lahtinen, M., Kuuloja, N., Huuskonen, J., & Rissanen, K. (2010). Synthesis and thermal behavior of Janus dendrimers: part 1. *Thermochimica Acta*, *497*(1–2), 101–108.
52. Kalva, N., Aswal, V. K., & Ambade, A. V. (2014). Effect of the branching pattern of hydrophobic dendrons on the core structure of linear-dendritic copolymer micelles. *Macromolecular Chemistry and Physics*, *215*(15), 1456–1465.
53. Huang, D., Wang, Y., Yang, F., Shen, H., Weng, Z., & Wu, D. (2017). Charge-reversible and pH-responsive biodegradable micelles and vesicles from linear-dendritic supramolecular amphiphiles for anticancer drug delivery. *Polymer Chemistry*, *8*(43), 6675–6687.
54. Fan, X., Zhao, Y., Xu, W., & Li, L. (2016). Linear–dendritic block copolymer for drug and gene delivery. *Materials Science and Engineering: C*, *62*, 943–959.
55. Sadler, K., & Tam, J. P. (2002). Peptide dendrimers: Applications and synthesis. *Reviews in Molecular Biotechnology*, *90*(3–4), 195–229.
56. Boas, U., Sontjens, S. H., Jensen, K. J., Christensen, J. B., & Meijer, E. W. (2002). New dendrimer–peptide host–guest complexes: Towards dendrimers as peptide carriers. *ChemBioChem*, *3*(5), 433–439.
57. Pardridge, W. M. (2005). The blood-brain barrier: Bottleneck in brain drug development. *NeuroRx*, *2*, 3–14.
58. Pardridge, W. M. (2007). Blood–brain barrier delivery. *Drug Discovery Today*, *12*(1–2), 54–61.
59. Abbott, N. J., & Romero, I. A. (1996). Transporting therapeutics across the blood-brain barrier. *Molecular Medicine Today*, *2*(3), 106–113.
60. Stewart, P. A. (2000). Endothelial vesicles in the blood–brain barrier: Are they related to permeability? *Cellular and Molecular Neurobiology*, *20*, 149–163.
61. Persidsky, Y., Ramirez, S. H., Haorah, J., & Kanmogne, G. D. (2006). Blood–brain barrier: Structural components and function under physiologic and pathologic conditions. *Journal of Neuroimmune Pharmacology*, *1*, 223–236.
62. Abbott, N. J., Ronnback, L., & Hansson, E. (2006). Astrocyte-endothelial interactions at the blood–brain barrier. *Nature Reviews Neuroscience*, *7*(1), 41–53.
63. Liu, M., & Fréchet, J. M. (1999). Designing dendrimers for drug delivery. *Pharmaceutical Science & Technology Today*, *2*(10), 393–401.
64. Dufes, C., Al Robaian, M., & Somani, S. (2013). Transferrin and the transferrin receptor for the targeted delivery of therapeutic agents to the brain and cancer cells. *Therapeutic Delivery*, *4*(5), 629–640.
65. Huang, R. Q., Qu, Y. H., Ke, W. L., Zhu, J. H., Pei, Y. Y., & Jiang, C. (2007). Efficient gene delivery targeted to the brain using a transferrin-conjugated polyethyleneglycol-modified polyamidoamine dendrimer. *FASEB Journal*, *21*(4), 1117–1125.
66. Herz, J., & Strickland, D. K. (2001). LRP: A multifunctional scavenger and signaling receptor. *Journal of Clinical Investigation*, *108*(6), 779–784.
67. Huang, R., Ke, W., Liu, Y., Jiang, C., & Pei, Y. (2008). The use of lactoferrin as a ligand for targeting the polyamidoamine-based gene delivery system to the brain. *Biomaterials*, *29*(2), 238–246.
68. Huang, R., Ke, W., Han, L., Liu, Y., Shao, K., Ye, L., & Pei, Y. (2009). Brain-targeting mechanisms of lactoferrin-modified DNA-loaded nanoparticles. *Journal of Cerebral Blood Flow & Metabolism*, *29*(12), 1914–1923.
69. Demeule, M., Currie, J. C., Bertrand, Y., Ché, C., Nguyen, T., Régina, A., & Béliveau, R. (2008). Involvement of the low-density lipoprotein receptor-related protein in the transcytosis of the brain delivery vector Angiopep-2. *Journal of Neurochemistry*, *106*(4), 1534–1544.
70. Ke, W., Shao, K., Huang, R., Han, L., Liu, Y., Li, J., & Jiang, C. (2009). Gene delivery targeted to the brain using an angiopep-conjugated polyethyleneglycol-modified polyamidoamine dendrimer. *Biomaterials*, *30*(36), 6976–6985.
71. Huang, R., Ma, H., Guo, Y., Liu, S., Kuang, Y., Shao, K., Li, J., Liu, Y., Han, L., Huang, S., & An, S. (2013). Angiopep-conjugated nanoparticles for targeted long-term gene therapy of Parkinson's disease. *Pharmaceutical Research*, *30*, 2549–2559.
72. Pardridge, W. M., Boado, R. J., & Farrell, C. R. (1990). Brain-type glucose transporter (GLUT-1) is selectively localized to the blood-brain barrier: Studies with quantitative western

blotting and in situ hybridization. *Journal of Biological Chemistry*, *265*(29), 18035–18040.
73. Dhanikula, R. S., Argaw, A., Bouchard, J. F., & Hildgen, P. (2008). Methotrexate loaded polyether-copolyester dendrimers for the treatment of gliomas: Enhanced efficacy and intratumoral transport capability. *Molecular Pharmaceutics*, *5*(1), 105–116.
74. Golden, P. L., Maccagnan, T. J., & Pardridge, W. M. (1997). Human blood-brain barrier leptin receptor: binding and endocytosis in isolated human brain microvessels. *Journal of Clinical Investigation*, *99*(1), 14–18.
75. Barrett, G. L., Trieu, J., & Naim, T. (2009). The identification of leptin-derived peptides that are taken up by the brain. *Regulatory Peptides*, *155*(1–3), 55–61.
76. Liu, Y., Li, J., Shao, K., Huang, R., Ye, L., Lou, J., & Jiang, C. (2010). A leptin derived 30-amino-acid peptide modified pegylated poly-L-lysine dendrigraft for brain targeted gene delivery. *Biomaterials*, *31*(19), 5246–5257.
77. Ciolkowski, M., Petersen, J. F., Ficker, M., Janaszewska, A., Christensen, J. B., Klajnert, B., & Bryszewska, M. (2012). Surface modification of PAMAM dendrimer improves its biocompatibility. *Nanomedicine: Nanotechnology, Biology and Medicine*, *8*(6), 815–817.
78. Jevprasesphant, R., Penny, J., Jalal, R., Attwood, D., McKeown, N. B., & D'Emanuele, A. (2003). The influence of surface modification on the cytotoxicity of PAMAM dendrimers. *International Journal of Pharmaceutics*, *252*(1–2), 263–266.
79. Wiwattanapatapee, R., Carreño-Gómez, B., Malik, N., & Duncan, R. (2000). Anionic PAMAM dendrimers rapidly cross adult rat intestine in vitro: A potential oral delivery system? *Pharmaceutical Research*, *17*, 991–998.
80. Felczak, A., Wrońska, N., Janaszewska, A., Klajnert, B., Bryszewska, M., Appelhans, D., . . . Lisowska, K. (2012). Antimicrobial activity of poly (propylene imine) dendrimers. *New Journal of Chemistry*, *36*(11), 2215–2222.
81. Ahamad, T., Mapolie, S. F., & Alshehri, S. M. (2012). Synthesis and characterization of polyamide metallodendrimers and their anti-bacterial and anti-tumor activities. *Medicinal Chemistry Research*, *21*, 2023–2031.
82. Thiagarajan, G., Ray, A., Malugin, A., & Ghandehari, H. (2010). PAMAM-camptothecin conjugate inhibits proliferation and induces nuclear fragmentation in colorectal carcinoma cells. *Pharmaceutical Research*, *27*, 2307–2316.
83. Jia, L., Xu, J. P., Wang, H., & Ji, J. (2011). Polyamidoamine dendrimers surface-engineered with biomimetic phosphorylcholine as potential drug delivery carriers. *Colloids and Surfaces B: Biointerfaces*, *84*(1), 49–54.
84. Ly, T. U., Tran, N. Q., Hoang, T. K. D., Phan, K. N., Truong, H. N., & Nguyen, C. K. (2013). Pegylated dendrimer and its effect in fluorouracil loading and release for enhancing antitumor activity. *Journal of Biomedical Nanotechnology*, *9*(2), 213–220.
85. Williams, R. J., Mohanakumar, K. P., & Beart, P. M. (2016). Neuro-nutraceuticals: Further insights into their promise for brain health. *Neurochemistry International*, *95*, 1–3.
86. Feigin, V. L., Nichols, E., Alam, T., Bannick, M. S., Beghi, E., Blake, N., Culpepper, W. J., Dorsey, E. R., Elbaz, A., Ellenbogen, R. G., & Fisher, J. L. (2019). Global, regional, and national burden of neurological disorders, 1990–2016: A systematic analysis for the global burden of disease study 2016. *Lancet Neurology*, *18*(5), 459–480.
87. Gautam, R., & Sharma, M. (2020). Prevalence and diagnosis of neurological disorders using different deep learning techniques: A meta-analysis. *Journal of Medical Systems*, *44*(2), 49.
88. Needham, E. J., Chou, S. H. Y., Coles, A. J., & Menon, D. K. (2020). Neurological implications of COVID-19 infections. *Neurocritical Care*, *32*, 667–671.
89. Aarli, J. A., Abramsky, O., Chin, J. H., Vora, N. (2014). The global burden of neurologic diseases. *Neurology*, *83*, 349–351.
90. Liaw, K., Zhang, F., Mangraviti, A., Kannan, S., Tyler, B., & Kannan, R. M. (2020). Dendrimer size effects on the selective brain tumor targeting in orthotopic tumor models upon systemic administration. *Bioengineering & Translational Medicine*, *5*(2), e10160.
91. Yellepeddi, V. K., Mohammadpour, R., Kambhampati, S. P., Sayre, C., Mishra, M. K., Kannan, R. M., & Ghandehari, H. (2018). Pediatric oral formulation of dendrimer-N-acetyl-l-cysteine conjugates for the treatment of neuroinflammation. *International Journal of Pharmaceutics*, *545*(1–2), 113–116.
92. Urandur, S., Banala, V. T., Shukla, R. P., Gautam, S., Marwaha, D., Rai, N., Sharma, M., Sharma, S., Ramarao, P., & Mishra, P. R. (2020). Theranostic lyotropic liquid crystalline nanostructures for selective breast cancer imaging and therapy. *Acta Biomaterialia*, *113*, 522–540.
93. Galgano, M., Toshkezi, G., Qiu, X., Russell, T., Chin, L., & Zhao, L. R. (2017). Traumatic brain injury: Current treatment strategies and future endeavors. *Cell Transplantation*, *26*(7), 1118–1130.
94. Bramlett, H. M., & Dietrich, W. D. (2015). Long-term consequences of traumatic brain injury: Current status of potential mechanisms of injury and neurological outcomes. *Journal of Neurotrauma*, *32*(23), 1834–1848.
95. Harvey, L. A., & Close, J. C. (2012). Traumatic brain injury in older adults: Characteristics, causes and consequences. *Injury*, *43*(11), 1821–1826.
96. Johnson, V. E., Stewart, J. E., Begbie, F. D., Trojanowski, J. Q., Smith, D. H., & Stewart, W. (2013). Inflammation and white matter degeneration persist for years after a single traumatic brain injury. *Brain*, *136*(1), 28–42.
97. Loane, D. J., & Kumar, A. (2016). Microglia in the TBI brain: The good, the bad, and the dysregulated. *Experimental Neurology*, *275*, 316–327.
98. Paulo, C. S., das Neves, R. P., & Ferreira, L. S. (2011). Nanoparticles for intracellular-targeted drug delivery. *Nanotechnology*, *22*(49), 494002.
99. Chandra, S., Dietrich, S., Lang, H., & Bahadur, D. (2011). Dendrimer–doxorubicin conjugate for enhanced therapeutic effects for cancer. *Journal of Materials Chemistry*, *21*(15), 5729–5737.
100. Marzolo, M. P., Von Bernhardi, R., Bu, G., & Inestrosa, N. C. (2000). Expression of α2-macroglobulin receptor/low density lipoprotein receptor-related protein (LRP) in rat microglial cells. *Journal of Neuroscience Research*, *60*(3), 401–411.
101. Varin, A., & Gordon, S. (2009). Alternative activation of macrophages: Immune function and cellular biology. *Immunobiology*, *214*(7), 630–641.
102. Sharma, A., Porterfield, J. E., Smith, E., Sharma, R., Kannan, S., & Kannan, R. M. (2018). Effect of mannose targeting of hydroxyl PAMAM dendrimers on cellular and organ biodistribution in a neonatal brain injury model. *Journal of Controlled Release*, *283*, 175–189.
103. Siegel, R. L., Miller, K. D., & Jemal, A. (2018). Cancer statistics, 2018. *CA: A Cancer Journal for Clinicians*, *68*(1), 7–30.

104. Preusser, M., & Marosi, C. (2017). Advances in brain tumour classification and therapy. *Nature Reviews Neurology, 13*(2), 71–72.
105. Ostrom, Q. T., Cioffi, G., Gittleman, H., Patil, N., Waite, K., Kruchko, C., & Barnholtz-Sloan, J. S. (2019). CBTRUS statistical report: Primary brain and other central nervous system tumors diagnosed in the United States in 2012–2016. *Neuro-Oncology, 21*(Supplement-5), v1-v100.
106. Gallego, P. L. J., & Hildebrand, J. (2014). Brain metastases. *Handbook of Clinical Neurology, 121*, 1143–57.
107. Kesharwani, P., & Iyer, A. K. (2015). Recent advances in dendrimer-based nanovectors for tumor-targeted drug and gene delivery. *Drug Discovery Today, 20*(5), 536–547.
108. Zhao, J., Zhang, B., Shen, S., Chen, J., Zhang, Q., Jiang, X., & Pang, Z. (2015). CREKA peptide-conjugated dendrimer nanoparticles for glioblastoma multiforme delivery. *Journal of Colloid and Interface Science, 450*, 396–403.
109. Gorantla, N. V., & Chinnathambi, S. (2018). Tau protein squired by molecular chaperones during Alzheimer's disease. *Journal of Molecular Neuroscience, 66*, 356–368.
110. Joe, E., & Ringman, J. M. (2019). Cognitive symptoms of Alzheimer's disease: Clinical management and prevention. *BMJ, 367.*
111. Jan, A. T., Azam, M., Rahman, S., Almigeiti, A. M., Choi, D. H., Lee, E. J., Haq, Q. M. R., & Choi, I. (2017). Perspective insights into disease progression, diagnostics, and therapeutic approaches in Alzheimer's disease: A judicious update. *Frontiers in Aging Neuroscience, 9*, 356.
112. Aso, E., Martinsson, I., Appelhans, D., Effenberg, C., Benseny-Cases, N., Cladera, J., Gouras, G., Ferrer, I., & Klementieva, O. (2019). Poly (propylene imine) dendrimers with histidine-maltose shell as novel type of nanoparticles for synapse and memory protection. *Nanomedicine: Nanotechnology, Biology and Medicine, 17*, 198–209.
113. Arotiba, O., Owino, J., Songa, E., Hendricks, N., Waryo, T., Jahed, N., Baker, P., & Iwuoha, E. (2008). An electrochemical DNA biosensor developed on a nanocomposite platform of gold and poly (propyleneimine) dendrimer. *Sensors, 8*(11), 6791–6809.
114. Klajnert, B., Cortijo-Arellano, M., Cladera, J., Majoral, J. P., Caminade, A. M., & Bryszewska, M. (2007). Influence of phosphorus dendrimers on the aggregation of the prion peptide PrP 185–208. *Biochemical and Biophysical Research Communications, 364*(1), 20–25.
115. Benseny-Cases, N., Klementieva, O., & Cladera, J. (2012). Dendrimers antiamyloidogenic potential in neurodegenerative diseases. *New Journal of Chemistry, 36*(2), 211–216.
116. Iqbal, K., Alonso, A. D. C., Chen, S., Chohan, M. O., El-Akkad, E., Gong, C. X., Khatoon, S., Li, B., Liu, F., Rahman, A., & Tanimukai, H. (2005). Tau pathology in Alzheimer disease and other tauopathies. *Biochimica et Biophysica Acta (BBA)-Molecular Basis of Disease, 1739*(2–3), 198–210.
117. Walker, L. C., & Jucker, M. (2015). Neurodegenerative diseases: Expanding the prion concept. *Annual Review of Neuroscience, 38*, 87–103.
118. Johnson, R. T. (2005). Prion diseases. *Lancet Neurology, 4*(10), 635–642.
119. Imran, M., & Mahmood, S. (2011). An overview of human prion diseases. *Virology Journal, 8*(1), 1–9.
120. Raza, C., & Anjum, R. (2019). Parkinson's disease: Mechanisms, translational models and management strategies. *Life Sciences, 226*, 77–90.
121. Buddhala, C., Loftin, S. K., Kuley, B. M., Cairns, N. J., Campbell, M. C., Perlmutter, J. S., & Kotzbauer, P. T. (2015). Dopaminergic, serotonergic, and noradrenergic deficits in Parkinson disease. *Annals of Clinical and Translational Neurology, 2*(10), 949–959.
122. Chaudhuri, K. R., & Martinez-Martin, P. (2008). Quantitation of non-motor symptoms in Parkinson's disease. *European Journal of Neurology, 15*, 2–8.
123. Kalia, S. K., Sankar, T., & Lozano, A. M. (2013). Deep brain stimulation for Parkinson's disease and other movement disorders. *Current Opinion in Neurology, 26*(4), 374–380.
124. Rekas, A., Lo, V., Gadd, G. E., Cappai, R., & Yun, S. I. (2009). PAMAM dendrimers as potential agents against fibrillation of α-synuclein, a Parkinson's disease-related protein. *Macromolecular Bioscience, 9*(3), 230–238.
125. Ferrer-Lorente, R., Lozano-Cruz, T., Fernández-Carasa, I., Miłowska, K., de la Mata, F. J., Bryszewska, M., Consiglio, A., Ortega, P., Gómez, R., & Raya, A. 2021). Cationic carbosilane dendrimers prevent abnormal α-synuclein accumulation in Parkinson's disease patient-specific dopamine neurons. *Biomacromolecules, 22*(11), 4582–4591.
126. Kojima, C., Regino, C., Umeda, Y., Kobayashi, H., & Kono, K. (2010). Influence of dendrimer generation and polyethylene glycol length on the biodistribution of PEGylated dendrimers. *International Journal of Pharmaceutics, 383*(1–2), 293–296.

17 Surface-Active Ligands for Brain Targeting

Mariam Zewail and Heba Abd-El-Azim*

17.1 INTRODUCTION

Brain-related disorders are regarded as high-risk disorders due to the complexity of drug permeation through the blood–brain barrier (BBB) to the brain parenchyma. Only a small number of low-molecular-mass drugs can cross the BBB. The BBB restricts the flow of chemicals as a natural defense mechanism that prevents the entry of poisons and microbes. The tight connections between capillary endothelial cells regulate the permeability of the BBB. Polyglycoprotein (P-gp) and breast cancer resistance protein are two of the main barriers to drug transport through the BBB, and these efflux pumps play an important role in the efflux of drugs acting on various CNS system. A lot of receptors are overexpressed on the BBB, such as transferrin receptors (TF), lactoferrin (LF), low-density receptors, growth receptors, choline receptors, insulin receptors, and folic acid receptors, among others. Targeting the aforementioned receptors is considered one of the effective strategies for brain targeting. To achieve effective active brain targeting through these receptors, nanocarriers are considered an ideal approach, as they enable drug administration without anesthesia or hospitalization. Also, their small size enables them to penetrate the small capillaries of the brain and be absorbed by cells for non-invasive active targeting, protecting the drug from chemical and enzymatic degradation [1]. In this chapter, we introduce these ligands and discuss the main results of studies carried out with functionalized lipid nanosystems for the treatment of brain diseases.

17.2 TRANSFERRIN

Transferrin (TF) is a monomeric glycoprotein with a molecular weight of approximately 80 kDa. the main function of TF is to transport iron throughout the body, and each TF molecule binds up to two iron atoms [2]. Jefferies et al. [3] reported that the TF receptor is more expressed in brain capillary endothelial cells than in other tissues, enabling their targeting using TF-modified nanocarriers [4].

Liposomes are one of the most extensively studies nanocarriers for brain targeting. They are surrounded by one or more lipid bilayers surrounding an aqueous core. Since only specific sizes can pass through the BBB in neurotherapy, small vesicles (<100 nm) are often preferred. However, previous studies have illustrated that liposomes of 100–140 nm have pros, such as a prolonged circulation half-life and the absence of plasma proteins [5].

Kong et al. [6] prepared ostiole-loaded TF-modified liposomes (TF-Ost-Lip). *In vitro* results demonstrated that the prepared TF-modified liposomes increased the intracellular uptake of hCMEC/D3 cells and APP-SH-SY5Y cells, and hence, the drug concentration through the BBB was elevated. Also, TF-Ost-Lip had a protective effect on APP-SH-SY5Y cells. Pharmacokinetics studies and Ost distribution studies in brain tissue illustrated that TF-Ost-Lip could prolong the circulation time in mice and increase Ost brain accumulation, and TF-Ost-Lip could alleviate Alzheimer's disease symptoms [6].

Pinheiro et al. [7] prepared TF-functionalized, quercetin-loaded SLN and NLC for AD. They reported that the inclusion of TF did not affect the colloidal characteristics and the stability of SLNs and NLCs. Cytotoxicity and permeability *in vitro* assays were conducted on hCMEC/D3 cells and revealed that the prepared lipid-based nanocarriers had no cytotoxic effects even at the highest quercetin concentration (30 μM). Compared to non-functionalized nanocarriers, TF-SLN and NLC did not show an increase in permeability capacity, which may be due to the saturation of TF. However, TF-functionalized quercetin SLNs and NLCs, to a greater extent, could inhibit amyloid-beta (Aβ) fibrillation and aggregation, suggesting their application for AD management [7].

Lopalco et al. [8] encapsulated dopamine hydrochloride in TF-modified liposomes to reduce side effects associated with the long-term administration of dopamine hydrochloride in the management of Parkinson's disease. TF-functionalized liposomes significantly increased transport across hCMEC/D3 cell monolayers, with the value for paracellular permeability being fivefold higher than that of unfunctionalized liposomes (4.97 ± 0.41 × 103 cm/min *vs.* 0.92 ± 0.24 × 103 cm/min, respectively) [8].

Kuo et al. [9] prepared LF-modified neuron growth factor (NGF)–loaded liposomes. The developed liposomes could deliver NGF through the blood–brain barrier (BBB) and improve the vitality of neuron-like SK-N-MC cells with deposited β-amyloid peptide (Aβ) and could effectively manage Aβ AD-induced neurotoxicity [9].

* Corresponding author: mariamzewail@gmail.com

DOI: 10.1201/9781003383376-17

Dual-functionalized PEGylated immunoliposomes loaded with epigallocatechin-3-gallate were prepared by Loureiro et al. [10] to control Parkinson's disease (PD) symptoms. The prepared liposomes were functionalized by anti-transferrin receptor OX26 and anti-α-synuclein LB509 antibodies. The binding ability of the dual-functionalized immunoliposomes to TR and α-synuclein protein was evaluated by ELISA. The dual-functionalized liposomes demonstrated a sustained-release profile; about 40% of the drug was released within 44 h. The cellular uptake of the dual-functionated immunoliposomes increased by twofold in cultured brain endothelial cells hCMEC/D3 compared with unfunctionalized liposomes [10].

Empty α-M and TF-modified liposomes were prepared by thin-film hydration; TF was covalently attached to α-M liposomes via a PEG linker, and the formulated liposomes were also characterized and evaluated for brain uptake and pharmacokinetic behavior. The prepared had an appropriate particle size range that was less than 200 nm. Meanwhile, the zeta potential was below 25 mV, and the PDI was 0.2, reflecting the stability of the formulation during the period of the study. It was well observed by the authors that this strategy significantly increased drug incorporation efficiency by up to 90%.

Furthermore, *in vitro* drug release studies showed a prolonged extended-release profile. Brain targeting has been proven by *in vitro* uptake studies carried out using advanced BBB models (b End 3 cells and astrocytes), which gave more precise results. The results confirmed that TF-α-M20-liposomes more efficiently crossed the BBB via FT receptor–mediated transcytosis in comparison with free α-M solution. The brain uptake and pharmacokinetic studies were conducted by SD21 rat groups. Drug concentrations were measured in the brain and peripheral organs over time.

TF-α-M-liposomes showed higher brain accumulation in comparison with α-M solution and α-M-liposomes. In addition, the TF-functionalized preparations increased the half-life, AUC, and mean residence time. Thus, this approach suggests that TF-functionalized liposomes could effectively deliver the drug to the brain and facilitate drug transfer through the BBB without altering the integrity of the drug [11].

Kuo et al. [12] prepared RMP-723/TF-modified liposomes to deliver NGF to the brain for AD management. This study was done to ensure the capability of the delivery system to suppress apoptosis of SK-N-MC cells, as well as the ability to target the brain. Liposomal systems were formulated with a mixture of lipid (PC24 and DPPC), cholesterol, DSPEPEG-2000-CA25, and DSPE-PEG-200026, followed by NGF encapsulation and cross-linking.

PEGylation increases the stability of liposomes in the biological environment by shielding them from enzymatic breakdown and extends the circulation time of liposomes, facilitating their transfer to the BBB. Furthermore, the addition of cholesterol serves to provide a stable configuration to the liposomes. RMP-7 was also included in the formulation as it has a specific binding capacity to the bradykinin receptor of the BBB; RMP-7 has a longer half-life and is considered a safer therapeutic agent. Meanwhile, TF also exhibits selective binding to TF on brain endothelial cells. These tactics act synergistically and enable efficient drug targeting via transcytosis. Further evaluation of the prepared liposomes showed that as the concentration of RMP-7 and the amount of targeting ligand increased, particle size increased. *In vitro* studies were conducted with HBMECs and SK-N-MC cells to assess cell lining permeability and viability. *In vitro* assays were performed using HBMECs and SK-N-MC cells (mimicking neuronal phenotype). The findings confirmed that RMP-7/TF/NGF-liposomes can cross the BBB and reduce cell apoptosis–induced neurotoxicity in SK-N-MC cells [12].

Narendra et al. [13] prepared TF-conjugated theranostic liposomes loaded with docetaxel (DXL) for glioma management and diagnosis. The liposomes were prepared by the solvent-injection method. The liposomal particle size was nearly 200 nm, with a spherical shape and high entrapment efficiency that reached 75.93%, and demonstrated a sustained-release profile. Results of IC_{50} values revealed that the prepared TF-modified liposomes had a higher cytotoxicity on C6 glioma cells compared with unmodified liposomes [13].

Li et al. [14] formulated element and cabazitaxel co-loaded liposomes with active functionalized surfaces by TF and RG2, which are cell membrane proteins present on glioma cells, to prepare active targeted biomimetic liposomes that can overcome BBB and target glioma [14].

Temozolomide and bromodomain inhibitor JQ1 were loaded in TF-functionalized liposomes by Lam et al. [15]. TF-functionalized liposomes could accumulate in brain micro-vessels and diffuse through the BBB in glioma-bearing mice, along with their ability to inhibit tumor development and signaling. This eventually increased DNA damage and apoptosis. Also, it resulted in reducing tumor burden by 1.5- to 2-fold, with a corresponding elevation in the survival of mice treated with functionalized liposomes [15].

Another type of nanocarriers that can be functionalized to target TF receptors are polymeric nanoparticles, which vary in their composition, for example, single polymers, co-polymers, protein-based, lipid–polymer hybrids, metal–polymer hybrids, etc. Each of these nanocarriers can be optimized to formulate nanocarriers of different sizes, different patterns, and different degradation profiles, besides the ability to modify their surface to target different ligands or receptors [2].

Recently, clonidine (CLD)–loaded TF-conjugated polylactic co-glycolic acid (PLGA) nanoparticles (NPs) for targeted nose-to-brain delivery were formulated by Sharma et al. [16]. The prepared NPs demonstrated a sustained-release profile that extended for 72 h. TF-conjugated NPs enhanced cell uptake with no significant cytotoxicity signs in N2a cells. *In vivo* studies confirmed that TF-conjugated NPs increased the brain concentration of CLD and improved the behavior of mice [16].

In addition to liposomes and polymeric NPs, nanoemulsion preparations were reported in the active targeting of brain ligands. They can be administered by different routes of administration, including nose-to-brain delivery. Mucoadhesive polymers can be included in nanoemulsions' composition to increase their contact time with nasal mucosa [17].

Darunavir (DRV), an antiretroviral drug, was formulated to control HIV encephalopathy as a nanoemulsion functionalized with TF. The particle size of DRV-loaded nanoemulsion droplets was 130 nm, and *in vitro* cytotoxicity studies in 293 T and hCMEC/D3 cells showed that TF-modified nanoemulsion had lower cytotoxicity compared with free DRV. Also, DRV-loaded nanoemulsion had higher permeability through the BBB hCMEC/D3 model. Biodistribution studies of brain tissues dissected from mice were conducted along with *ex vivo* imaging. Results also revealed that TF-modified nanoemulsions had a superior effect compared with free drugs [18].

Muntoni et al. [19] encapsulated methotrexate in the insulin of TF-functionalized SLNs functionalized by using a maleimide–thiol reaction. The functionalized liposomes had good permeation through the hCMEC/D3 cell line (a human BBB *in vitro* model). Brain accumulation was enhanced by intravenous administration in Wistar rats, suggesting the potential of functionalized liposomes to deliver nanocarriers across the BBB for glioma treatment [19].

Several studies reported the use of functionalized niosomes for brain targeting using some ligands, such as glucose-targeted, folic acid, and TF-targeted niosomes. They were utilized to deliver various pharmaceutical agents, such as vasoactive intestinal peptide (VIP), neuropeptide DynB, and doxorubicin [20].

17.3 MELANOTRANSFERRIN

Melanotransferrin (p97) is a GPI-anchored iron-binding protein that is highly expressed in melanomas and BBB endothelial cells [21]. SLNs loaded by etoposide were formulated by a cross-linking reaction between SLNs and a p97 antibody. P97-modified SLNs showed less cytotoxic effect to HBMECs and human astrocytes; on the other hand, they improved the drug uptake and growth inhibition effect on U87 cells [22].

p97 was covalently conjugated to paclitaxel or adriamycin [23]. Brain accumulation of the p97–drug complex was tenfold higher compared with their corresponding free drugs. Moreover, adriamycin's pharmacological effects were preserved, and significantly, the dissemination of intracranial gliomas and breast tumors was suppressed compared with non-targeted adriamycin. According to several reports, the melanotransferrin receptor is considered the most preferred target for brain-targeted drug delivery [18, 21].

Etoposide-loaded SLNs were modified by conjugation with melanotransferrin antibody (MA) to target glioblastoma multiforme through human brain microvascular endothelial cells (HBMECs) and U87MG cells targeting. The results revealed that increasing the molecular weight ratio of tripalmitin weight percentage in lipids increased the particle size and viability of U87MG cells. On the contrary, encapsulation and MA conjugation efficiency were decreased. Also, the permeability coefficient for etoposide via the BBB was reduced. Upon increasing MA concentration on the particle surface, the atomic ratio of nitrogen to phosphorus and permeability coefficient for propidium iodide and etoposide across the BBB were elevated. Meanwhile they decreased MA conjugation efficiency, transendothelial electrical resistance, and U87MG cells viability. Also, MA on etoposide SLNs activated melanotransferrin-mediated transcytosis and increased the growth-inhibitory efficacy of U87MG cells. MA-conjugated etoposide-SLNs are considered a potential effective nanocarrier for glioblastoma management [22].

Doxorubicin-loaded SLNs bound to aprotinin (Apr) and melanotransferrin antibody (Anti-MTf) were developed for glioblastoma multiforme (GBM) treatment. They were functionalized nanocarriers that had a suitable particle that enhanced their permeability across the BBB, and as a result, they inhibited the growth of GBM [24].

17.4 LACTOFERRIN

Lactoferrin (LT) is an iron-binding glycoprotein with immunomodulatory and antibacterial effects that help in preventing infections. LF receptors are present in the brain microvasculature and neurons [2], BBB endothelial cells, and the surface of glioblastoma multiforme (GM) cells [25].

Kuo et al. [9] prepared LF-modified neuron growth factor (NGF)–loaded liposomes. The developed liposomes could deliver NGF via the BBB and improve the viability of neuron-like SK-N-MC cells with deposited β-amyloid peptide (Aβ) and could effectively manage Aβ AD-induced neurotoxicity [9].

Chen et al. [26] formulated doxorubicin-loaded LF-modified pro cationic liposomes. In BCECs and C6 glioma cells, doxorubicin-loaded LF-modified liposomes had higher cell uptake and were superior in inhibiting the proliferation of C6 cells in comparison to doxorubicin solution and unmodified liposomes. In *in vivo* studies, LF liposomes demonstrated superior anti-glioma activity and showed a significant effect in prolonging the mean survival time in an intracranial tumor rat model with lower toxicity compared with doxorubicin solution and unmodified liposomes [26].

Singh et al. [27] formulated docetaxel-loaded LF-functionalized SLNs and reported that LF-modified carriers possessed good biocompatibility and stability and had better brain-targeting potential compared with unmodified SLNs, suggesting their use for targeted CNS treatments.

Kuo et al. [28] developed LF-modified carmustine and tamoxifen dual-loaded SLNs for the treatment of glioma. The incorporation of tamoxifen was useful to overcome multidrug resistance linked to efflux transporters like P-glycoprotein. LF and tamoxifen enhanced the dissolution profile of carmustine SLNs and prevented its rapid release. Permeation of the dual-loaded LF SLN increased by tenfold through a monolayer of BCECs and human astrocytes than LF-unmodified SLNs. Tamoxifen successfully inhibited the activity of the efflux pump, and that was evidenced by fluorescent staining images that demonstrated the intracellular signal for calcein-AM. Furthermore, MTT assays showed that this formulation had no effect on the viability of human brain capillary endothelial cells (HBCE); meanwhile, it inhibited the growth of U87 glioma cells [28].

Nimodipine (NMD) is a calcium antagonist used for the prophylaxis and management of cerebral ischemic stroke. Zhao et al. [29] formulated NMD-loaded LF-NLCs. The uptake of NMD-loaded LF-NLC was higher than that of plain NLCs or PEGylated NLCs, and internalization via LT receptors endocytosis was observed. To assess the neuroprotective potential, PC12 cells were co-cultured with NLCs and sodium nitroprusside (SNP), an inorganic compound that induces neuronal cell injury, and LF-NLCs showed higher cell viability and lower apoptosis rate in comparison to the control group and plain NLCs. This is due to the LF portion of the NLCs. This was directly related to the increased intracellular drug accumulation as a result of the efficient uptake of NLCs facilitated by the LF portion. Furthermore, *in vivo* biodistribution studies using BALB/c nude mice demonstrated that these functionalized NLCs successfully transported the therapeutic drug to brain tissue, indicating its usefulness [29].

Meng et al. [30] prepared curcumin (CUR)–loaded NLCs for AD management (LF-NLCs-CUR). LF was adsorbed on the NLCs' surface as a result of the electrostatic interaction with carboxylated polyethene glycol 100 monostearate. The prepared formulation was able to penetrate BCECs and release CUR. *Ex vivo* studies demonstrated that LF-NLCs-CUR passed the BBB and increased the brain absorption by 2,79-fold in the brain of an AD rat model in comparison to unmodified NLCs. Histopathological examination of the hippocampus after staining revealed that the group treated with LF-NLCs-CUR showed less neuronal damage. Also, plasma levels of malondialdehyde were significantly decreased, assuring the superiority of LF-NLCs in reducing oxidative damage, relieving pathological impairment, and suppressing AD dissemination [30].

Recently, riluzole-encapsulated LF-functionalized NLCs were produced by ultrasonication. The particle size of the prepared NLCs was about 220 nm, and they were formulated from GRAS excipients. LF-functionalized NLCs were biocompatible, as evidenced by MTT assay in two cell lines: hCMEC/D3 monolayer and mouse motor neuron-like hybrid cells (NSC-34 cell line). No significant decrease in cell viability was observed upon using NLCs at riluzole concentrations up to 10 μM, suggesting their promising application for brain-targeted delivery of riluzole [18].

Karami et al. [31] reported that indinavir-loaded LF-modified and non-modified nanoemulsions were able to increase drug concentration in the brain by 4.1- and 1.6-fold, respectively, in comparison to the free drug [31].

Jiang et al. [32] reported that LF-modified huperzine A, which is a specific acetylcholinesterase inhibitor used to AD management, revealed the superiority of intranasal administration of LF-modified huperzine A nanoemulsions in brain delivery, and that was confirmed by pharmacokinetic studies [32].

17.5 RABIES VIRUS GLYCOPROTEIN (RVG)

Rabies is a high-neurotropism virus. The *N*-acetylcholine receptor on neuronal membranes and the rabies virus glycoprotein (RVG) are specifically conjugated. These receptors are responsible for the entry of the virus to the brain. A modified RVG peptide with 9 arginine residues that binds siRNA was utilized as a vector for transvascular delivery [21].

Peptides derived from RVG are utilized to improve brain-targeted delivery through binding with the acetylcholine receptor or the γ-aminobutyric acid (GABA) receptor on neuronal cells [33].

Zhao et al. [34] functionalized liposomes encapsulating curcumin with RDP, which is an RVG-derived peptide, and utilized the prepared liposomes for the management of mouse glioma model. Their findings revealed that modified liposomes improved the delivery of curcumin to the CNS and prolonged mice survival time to reach ten days. Also, in human glioma cells (U251MG), RDP stimulated the specific uptake of the liposomes via nAchR-mediated endocytosis, and this, in turn, reduced cell proliferation by promoting apoptosis and arresting the S phase of the cell cycle [34].

Likewise, RDP-functionalized liposomes loaded with polypyridyl ruthenium complex decreased the inflammation and microbial activity in cryptococcal encephalitis in mice, with no toxic signs [2].

Arora et al. [35] formulated dual-functionalized liposomes with RVG and mannose to deliver the ApoE2 gene for AD management. Dual-functionalized liposomes demonstrated greater expression of ApoE2 in bEnd.3 cells, primary neurons, and astrocytes in comparison to single modified or unmodified liposomes. These results were further ascertained by the *in vivo* study in C57BL/6 mice that demonstrated that ApoE2 transfection was increased and the prepared liposomes were biocompatible, and this was confirmed by the absence of dead cells or inflammatory cells' infiltration in tissues dissected from treated animals [35].

RVG-functionalized anionic liposomes could effectively deliver siRNA and induce α-synuclein gene silencing in mouse primary cortical and hippocampal cells, reducing the levels of PD's characteristic protein. These anionic liposomes have proven long circulation time and superior effectiveness [36].

Conceição et al. [37] prepared siRNA-loaded liposomes conjugated with RVG-9r. The prepared liposomes efficiently silenced the mutant ataxin-3 in Neuro-2a cells and in two distinct mouse models of Machado–Joseph disease

(MJD). Also, MJD-related neuropathology and abnormal motor behavior were decreased in the event that the proposed system was used to treat spinocerebellar ataxia [37].

Pinheiro et al. [38] prepared quercetin-loaded RVG-29 conjugated SLNs and NLCs. The prepared nanocarriers enhanced the permeation through a hCMEC/D3 BBB model by 1.5-fold. In addition, they suppressed Aβ aggregation in the thioflavin T binding assay. Results demonstrated the neuroprotective potential of these lipid nanoparticles in AD management [38].

Rassu et al. [39] prepared chitosan-coated and uncoated SLNs for nose-to-brain targeting for AD. RVG-9R was complexed with BACE1 siRNA and loaded into SLNs. RVG-9R was incorporated to enhance the transcellular pathway within the neuronal cells. Chitosan coating significantly affected the particle size and zeta potential of SLNs. It also aided in the protection of siRNA. Mucoadhesiveness of chitosan-coated SLNs increased their residence time in the nasal mucosa. This might be explained by the interaction of negatively charged mucous and positively charged chitosan. The cellular transport of siRNA was studied using Caco-2 as a model of epithelial-like phenotypes. Results revealed that chitosan-coated SLNs demonstrated superior permeation compared with uncoated SLNs and free siRNA [39].

RVG surface-conjugated polymeric nanoparticles encapsulating micro-RNA (miR)-124 were prepared to alleviate PD neuroinflammation. The miR-NPs decreased the levels of pro-inflammatory cytokines and increased the levels of neuroprotective molecules by targeting the MEKK3 and nuclear factor kappa light chain enhancer of activated B cell (NF-Kb) pathways. The delivery of miR-NPs decreased apoptosis and showed potential in inhibiting pro-inflammatory signaling and enhancing neuroprotection in PD [40].

You et al. [41] prepared deferoxamine (DFO)–loaded polymeric NPs functionalized with RVG. Particle size and zeta potential of the prepared nanocarriers were 168.8 nm ± 1.9 and −27.40 ± 0.71, respectively, and about 75% of DFO was released within 24 h. This research demonstrated that the mechanism of NPs' uptake was receptor-mediated endocytosis. NPs have proven a remarkable reduction in oxidative stress, the content of iron, and an improvement in the neurobehavioral disturbances, with no toxic effects on the brain and the rest of mouse organs. These findings are promising for the brain delivery of DFO for PD management [41].

17.6 DISIALOGANGLIOSIDE (GD2) ANTIGEN

The disialoganglioside GD2 antigen is found in high concentrations in brain tumor cells. The high presence of GD2 in neuroblastoma tumor cells compared with normal tissues suggests the possibility of using GD2 antibodies to target neuroblastoma [2].

CPT-11 and panobinostat were dual-loaded liposomes bound with anti-GD2 antibody developed to synergistically target glioma cells. Targetability was confirmed using the U87MG human glioma cell line and its drug-resistant variant (U87DR). Results demonstrated that the dual-loaded functionalized liposomes could effectively bind specifically to GD 2.

In vitro cytotoxicity assay confirmed the synergistic chemotherapeutic effect of CPT-11 and panobinostat-loaded liposomes. Confocal microscopy and flow cytometry results revealed that drug uptake increased by 6-fold in liposomes and 1.8-fold in targeted liposomes. This may be attributed to the enhanced intracellular trafficking for targeted liposomes with significantly high cytotoxicity toward U87DR cells *in vitro*. Apoptosis rate was elevated by 1.6-fold when compared to non-functionalized liposomes.

Using a xenograft nude mouse model subcutaneously implanted with U87DR cells, Cy5.5-labeled liposomes were used in *in vivo* imaging system studies that showed a similar *in vivo* distribution profile; on the other hand, accumulation rate increased by 5.1-fold in the functionalized liposomes. Targeted liposomes were shown to significantly improve the anticancer effects against U87DR tumors that were manifested by a marked decrease in tumor size and increase the chance of survival in experimental animals. Furthermore, cell proliferation markers levels decreased; on the other hand, levels of apoptosis marker proteins increased [42].

17.7 CELL-PENETRATING PEPTIDES

Cell-penetrating peptides (CPPs) are short cationic sequences of amino acids that are capable of crossing cell membranes and translocating into the intracellular space in accordance with the "model" represented by the HIV trans-activator of transcription (Tat) protein.

CPPs are non-cell-type-specific and capable of transporting various cargos (small molecules, proteins, nucleic acids, nanocarriers) across different biological barriers, such as the intestinal wall, BBB, and skin. The aforementioned properties suggest the utilization of CPPs as an effective ligand for brain targeting through the nasal mucosa [43].

Tat protein was conjugated to PEGylated cholesterol and utilized to create liposomes, resulting in higher brain distribution and higher chances for animals to survive, indicating the importance of PEGylation and long circulation for nanocarrier accumulation in the brain [44].

Gartziandia et al. [45] conjugated CPPs (Tat or Penetratin) to the surface of polymer and lipid nanoparticles and tested the potential of CPPs to improve the ability of nanocarriers to permeate through an *in vitro* primary rat olfactory cell monolayer model [45].

Also, PEG–PLA nanoparticles functionalized with low-molecular-weight protamine showed greater accumulation in 16HBE14o human bronchial epithelial cells compared with the unfunctionalized nanoparticles. Upon the loading of functionalized nanoparticles with the fluorescence dye coumarin-6 and intranasal administration in rats, the fluorescence intensity increased by more than twofold in various brain structures in comparison with the unfunctionalized nanoparticles [46].

In addition, coumarin-loaded methoxy poly(ethylene glycol)-poly(ε-caprolactone) (mPEG–PCL) amphiphilic block co-polymers micelles conjugated with Tat had higher accumulation in C6 rat glioma cells in rats with glioma model compared with unmodified micelles. No statistically significant differences between conjugated and non-conjugated micelles were noted in fluorescence brain distribution 1 h after nasal administration. On the other hand, after 4 h, the fluorescence signal for unconjugated micelles was reduced; meanwhile, it increased in animals treated with Tat-conjugated micelles. This may be attributed to their higher intracellular penetration [47].

The aforementioned micelles were dual-loaded with camptothecin and small interfering RNA (siRNA). The former was included to treat brain tumors, while the latter to suppress the genes related to CNS disorders pathogenesis. Camptothecin-loaded Tat-mPEG–PCL micelles demonstrated greater cytotoxicity in rat C6 glioma cells *in vitro* compared with unmodified micelles. Results revealed modified micelles. For siRNA, nucleic acids were condensed with Tat-mPEG–PCL to obtain 50–100 nm polyplexes. Brain distribution studies were conducted after encapsulating Alexa–dextran in the micelles and intranasal administration. Results demonstrated that Tat-modified micelles increased brain accumulation compared with the controls (IV Tat-mPEG–PCL micelles or IN mPEG–PCL micelles). These findings revealed that Tat enhanced the permeation of micelles through the nasal mucosa and improved their uptake by the olfactory and trigeminal nerves [48].

Kanazawa et al. [49] formulated CPP-modified micelles composed of polyethylene glycol–polycaprolactone (PEG–PCL) co-polymers bound to Tat for intranasal delivery. These micelles were loaded by either siRNA or dextran (Mw: 10,000 Da) as a model siRNA by utilizing MPEG–PCL–Tat. The intranasal route enhanced the brain delivery of dextran compared with the intravenous route. Furthermore, results revealed that MPEG–PCL–Tat enhanced dextran delivery to the brain as a result of high permeation of olfactory mucosa [49].

In spite of the translocation abilities of CPPs, they possess a few disadvantages, like endosome sequestration, poor subcellular localization and efficient transport to the cytoplasm [46].

Zhao et al. [50] conjugated fibrin-binding peptide (CREKA) to PEGylated polyamidoamine (PAMAM) dendrimer, where PAMAM was utilized as a result of their smaller size and better permeation into tumors. CREKA was utilized in glioblastoma multiforme (GBM) targeting, as it had greater entrapment within the tumor. Furthermore, fluorescence imaging in nude mice *in vivo*, brain imaging, ex vivo, and frozen slices were carried out. Fluorescence imaging demonstrated that the CREKA-modified PAMAM exhibited greater accumulation and deeper penetration in GBM tissue compared with unmodified micelles. These findings suggest that the CREKA-modified PAMAM penetrated deeper into the GBM tissue and improved its retention effects, which is considered a promising approach for brain tumor targeting [50].

Chlorotoxin-anchored PEGylated liposomes showed enhanced uptake by glioma and endothelial cell lines, resulting in significantly reduced brain tumor volumes compared to nontargeted liposomes and free DOX, indicating the importance of targeted delivery for effective tumor growth inhibition [51].

The presentation of solute carriers (SLC) in the brain is high and shows a unique pattern at the BBB. Veszelka et al. [52] prepared triple-functionalized nanocarriers by using three targeting ligands—ascorbic acid, leucine, and glutathione—on nanoparticles, and their results showed that these triple-functionalized NPs increased the internalization of encapsulated albumin in cultured primary rat brain endothelial cells. Furthermore, they reported that the triple-functionalized NPs were able to deliver albumin to midbrain organoids after overcoming the BBB model. The cellular uptake was temperature- and energy-dependent based on metabolic inhibition, and cellular uptake was attributed to an endocytotic mechanism. The uptake of the drug-loaded triple-targeted nanoparticles was enhanced after altering the negative zeta potential of endothelial cells by treatment with a cationic lipid or after enzymatic cleavage of sugar chains. Fusion of the prepared NPs with endothelial cell membranes also resulted in increased fluidity of cell membranes [52].

17.8 APOLIPOPROTEIN (ApoE)

Apolipoprotein E peptide (ApoE, [LRKLRKRLL]2C) has a specific affinity to low-density lipoprotein receptor members (LDLRs) and enhances BBB permeability for targeted brain delivery for many diseases, including glioblastoma [53]. Wagner et al. [54] reported that the uptake of ApoE-modified nanoparticles occurred by an active endocytotic uptake mechanism and demonstrated the involvement of LDLR family members in the uptake of the ApoE functionalized nanoparticles into the brain capillary endothelial cells [54].

Michaelis et al. [55] prepared loperamide-loaded ApoE-functionalized human serum albumin NPs. ApoE was chemically conjugated by linkers to loperamide-loaded human serum albumin nanoparticles. After intravenous injection into ICR mice, this formulation induced antinociceptive effects in the tail-flick test. On the contrary, these effects were not induced by NPs bound to an apolipoprotein E mutant that does not recognize lipoprotein receptors. These findings suggest that apolipoprotein E bound to the NPs' surface facilitates drug transport across the BBB, possibly after interacting with lipoprotein receptors on the endothelial cell membranes of capillaries [55].

Tamaru et al. [56] reported the ability of nucleic acid (RNA or DNA)–loaded ApoE-functionalized liposomes containing proton-ionizable aminolipids (YSK-MEND). In mouse neuron-derived cell lines (Neuro-2a and CAD), ApoE modification significantly enhanced the cellular uptake and function of YSK-MEND encapsulating short interfering RNA or plasmid DNA. In addition, transgene expression increased compared to YSK-MEND without ApoE modification upon intracerebroventricular injection of ApoE-modified YSK-MEND. Furthermore, expression of fluorescently labeled ApoE/YSK-MEND and plasmid DNA-derived mCherry (fluorescent protein) was observed, indicating that these liposomes may be useful for transduction and expression of transgenes in neural stem and progenitor cells [56].

Neves et al. [57] prepared ApoE-functionalized SLNs and carried full *in vitro* characterization, including *in vitro* cytotoxicity study, that was assessed by MTT and LDH assays in the human cerebral microvascular endothelial cells (hCMEC/D3) cell line, a model of BBB. Their findings demonstrated that no cytotoxic effects were noted for up to 1.5 mg ml^{-1} after incubation for 4 h. hCMEC/D3 monolayers were assessed for brain permeability in a Transwell device and showed a 1.5-fold increase in barrier permeability for modified SLNs in comparison with unmodified SLNs [57].

Also, Topal et al. [58] functionalized donepezil-loaded SLNs with ApoE and evaluated their efficiency in AZ management. Cytotoxicity of donepezil SLNs was evaluated by metabolic assay and impedance-based cell analysis. ApoE-functionalized SLNs enhanced drug uptake in cultured brain endothelial cells and neurons. Moreover, the permeability across the BBB was also increased for ApoE-modified SLNs [58].

Zhang et al. [59] prepared paclitaxel (PTX)–loaded nanomicelles for glioma treatment. The prepared micelles were functionalized with amyloid β-protein (Aβ)-CN peptide, which can bind specifically to the lipid-binding domain of ApoE and forms an ApoE-enriched protein corona that surrounds Aβ-CN-PMs (ApoE/PTX/Aβ-CN-PMs) *in vivo*. The receptor-binding domain of the ApoE then binds to low-density lipoprotein receptor (LDLr) and LDLr-related protein 1 receptor (LRP1r) present at the BBB and in glioma, ensuring effective brain targeting. The aforementioned micelles were successfully formulated by thin-film hydration technique and had suitable particle size and positive surface charge. Nanomicelles modified with ApoE corona showed enhanced BBB permeability, enhanced brain distribution, anti-glioma effects, and higher median survival in mice in comparison to those treated with unmodified micelles [59].

Kato et al. [60] developed PEGylated liposomes modified with apolipoprotein E mimetic tandem dimer peptide (ApoEdp) that was conjugated with serine–glycine (SG)5 spacer (ApoEdp-SG-lipid). In mice, the prepared modified liposomes increased brain accumulation by 3.9-fold compared with unmodified PEGylated liposomes. Also, ApoEdp-functionalized PEGylated liposomes were able to reach and concentrate beyond the BBB into the brain parenchyma, as shown by three-dimensional imaging. These findings suggested that ApoEdp-SG-lipid-modified liposomes are considered a promising approach for effective brain targeting and increasing BBB permeability [60].

17.9 VITAMINS

17.9.1 Folate (FA)

Folate, or folic acid, is a B-complex vitamin that is essential for various physiological functions, such as *de novo* nucleotide synthesis, nervous system development during pregnancy, protein metabolism, *etc.* FA deficiency is associated with neurological disorders like AD and PD. Folate receptors are markedly overexpressed in BBB and glioma cells in case of neurological disorders, suggesting the application of FA-functionalized nanocarriers to enhance brain delivery. One of the reported FA-conjugated nanocarriers is liposomes. Gao et al. [61] used DSPE-PEG2000-FA to formulate dual-functionalized FA and TF-liposomes encapsulating doxorubicin (DOX) to treat mice with the C6 glioma cell model. The bifunctional liposomes demonstrated specific and targeted delivery of DOX to glioma cells, compared with normal brain tissue. Furthermore, the bifunctionalized liposomes demonstrated less toxicity and enhanced pharmacological effects compared with free DOX and unfunctionalized DOX-loaded liposomes. This was confirmed by the reduction in tumor volume, survival time prolongation, and increased apoptotic tumor cell death activity [61].

Li et al. [62] prepared dual liposomes with FA and cell-penetrating peptide dNP2-modified surface and loaded them with PTX. The liposomes were formulated using a thin-film hydration technique utilizing two formerly modified phospholipids: DSPE-PEG-dNP2 and hydrazone-based acid-cleavable FA-Hz-PEG-DSPE. dNP2 enhanced BBB penetration in a bEnd.3-cell model. Moreover, FA was cleaved at pH 6.8 (typical pH of human brain tumor microenvironment) and exposed to the dNP2 peptide; as a result, liposomal uptake and PTX cytotoxicity were enhanced [62].

Upon intravenous administration in orthotopic glioma-bearing mice, the dual-functionalized liposomes demonstrated higher tumor accumulation and, in turn, better pharmacological effects compared with non-cleavable FA-TF-liposomes and liposomes with single functionalization [62].

SLNs were also functionalized by FA to target glioma. Etoposide was nanoencapsulated in SLNs grafted with FA and p-aminophenyl-α-D-manno-pyranoside (a mannose derivate). The prepared functionalized SLNs showed enhanced brain delivery with greater gliomal inhibitory effect compared with unfunctionalized SLNs [63].

17.9.2 Thiamine (TH)

Thiamine or vitamin B1 is a water-soluble nutrient that is mainly found in the mitochondria and acts as a catalyst for various enzymes. TH transporters are present in the BBB to facilitate its transport to the brain [2, 21].

Lockman et al. [64] formulated radiolabeled TH-functionalized SLNs by using emulsifying wax and Brij 78. TH was bound to a distearoylphosphatidylethanolamine

(DSPE) anchor by a PEG spacer. Kinetic modeling of the SLNs was conducted after *in situ* rat brain perfusion. Results revealed that TH-SLNs enhanced cell uptake compared to uncoated SLNs. However, BBB-specific targeting was wasted following systemic administration, with no marked differences in the radioactivity levels of the brain between TH-SLNs and the control. The difference between intravitreal and systemic administration may be a result of their competition for BBB binding sites between endogenous TH and TH-functionalized SLNs or insufficient density of TH coating of the SLNs produced, which inhibits brain absorption [64].

17.10 MANNOSE

Mannose is one of the transmembrane glucose transporter proteins (GLUTs) found in the BBB that have a pivotal role in the passage of metabolic substrates, such as mannose, glucose, and other sugars to the brain. Consequently, mannose or its derivatives are used to facilitate drug passage to the brain through the GLUTs on the BBB [18].

Lewicky et al. [65] formulated a mannosylated liposomal system for targeted delivery of dynantin to the CNS. In order to increase the stability of the nanodispersion, cholesterol was included in the formulation. The activity of the peptide on its receptor was assessed by the *in vitro* PRESTO-TANGO assay system. Moreover, the *in vivo* distribution studies in mice demonstrated that the mannosylated glycol liposomes enhanced the delivery of dynactin to the brain and the potential use of mannosylated liposomes for specific brain delivery [65].

Kuo et al. [66] reported that p-aminophenyl-α-D-mannopyranoside- and ApoE-functionalized liposomes surface could effectively deliver neuron growth factor through the BBB besides increasing degenerated neuron survival in an AD *in vitro* model [66].

Other studies reported that encapsulation of curcumin in liposomes and DTX in SLNs and nanocarriers' conjugation to p-aminophenyl-α-D-manno-pyranoside improved the targetability and the antitumoral effects in both U87 and C6 glioblastoma cells and glioma mice model [18].

17.11 GLUTATHIONE (GSH)

GSH is an endogenous tripeptide and antioxidant with a wide presentation in different body cells. GSH is actively transported via the BBB, through a sodium-dependent GSH transporter that is highly present in the BBB epithelium [5]. GSH was investigated as a promising molecule to aid carrier-mediated transcytosis of lipid nanocarriers to the brain. GSH PEGylated liposomes (G-Technology®) have shown effectiveness in promoting drug delivery to the brain while maintaining safety [2].

Rip et al. [67] reported that carboxifluorescein-loaded GSH-coated PEGylated liposomes enhanced its fluorescence intensity in the brain by fourfold compared with uncoated liposomes [67].

Maussang et al. [68] reported that the ribavirin uptake rate via the BBB, both *in vitro* and *in vivo*, is directly linked to the amount of GSH ligand bound to the liposomal surface. Also, they reported that nanocarriers' composition can affect GSH brain targeting efficiency [68].

Kuo et al. [69] formulated GSH-coated liposomes encapsulating ceftriaxone, FK506, and nilotinib. The coated liposomes were biocompatible and enhanced drug transport through the BBB in a co-culture model comprised of HBMECs, human astrocytes, and human brain vascular pericytes. Liposomes were superior to free drugs in rescuing injured dopaminergic neurons by more effectively downregulating the expression of apoptotic factors and dopamine markers [69].

DTX-loaded GSH-coated PEG-PLGA NPs were formulated by Grover et al. [70] to target DTX to brain tumors [70]. *In vitro* results demonstrated that GSH-coated NPs provided a sustained release profile for ten days with an initial burst effect of 18.49% ± 5.21%. MTT assays in C6 and RG2 cells showed that DTX-loaded GSH-coated NPs decreased cell viability compared with untreated cells, uncoated NPs, and DTX suspension. The permeability of the drug was assessed using the *in vitro* BBB (Transwell™) model. GSH-coated NPs demonstrated a time-dependent release of DTX in the first 16 h, followed by a constant release profile for 48 h, proposing the capability of NPs to cross the BBB [70].

17.12 GLUCOSE

Glycosyl derivatives of cholesterol were synthesized and used in the formulation of liposome to effectively target (glucose transporters) GLUTs on the BBB. Coumarin-6 was used as fluorescent probe. The results revealed that glucose-mediated brain-targeting liposome containing coumarin-6 was less than that of conventional liposome and was less cytotoxic compared with conventional liposomes to brain capillary endothelial cells (BCECs) in an *in vitro* model of the BBB model that was made by co-culturing of BCECs and rat astrocytes (ACs) to evaluate the transendothelial ability of liposomes to cross the BBB.

The results showed that the transendothelial ability increased depending on the amount of glycosylated cholesterol derivatives in the liposomes following intravenous administration of liposomes (LIP), control liposomes (CLP), and GLP-4, resulting in the coumarin-6AUC0-t value of GLP-4 being 2.85-fold higher than that of LIP and 3.33-fold higher than that of CLP. The Cmax value of CLP-4 was 1.43-fold of LIP and 3.10-fold of CLP. Both pharmacokinetics and distribution were studied in mice, indicating that this novel brain-targeted drug delivery system is expected to effectively target brain tissue.

Also, the results revealed that glucose-mediated brain-targeting liposomes loaded with coumarin-6 were less cytotoxic to brain capillary endothelial cells (BCECs) than conventional liposomes; an *in vitro* BBB model was established by co-culturing BCECs and rat astrocytes (ACs), testing the ability of transendothelial migration through the BBB. These findings demonstrated that the transendothelial ability increased depending on the amount of glycosylated cholesterol derivatives in the liposomes. Intravenous administration of LIP, control liposomes (CLP), and GLP-4 resulted in the coumarin-6AUC0-t value of GLP-4 being 2.85-fold higher than that of LIP and 3.33-fold higher than that of CLP. The Cmax value of CLP-4 was 1.43-fold of LIP and 3.10-fold of CLP. Both pharmacokinetics and distribution have been evaluated in mice and suggested that this new brain-targeted drug delivery system can be effectively used for brain delivery [71].

17.13 INSULIN

Insulin transport from the bloodstream to the brain occurs by a transcytosis mechanism via the insulin receptor located at the vascular endothelial cell surface. The human anti-insulin receptor monoclonal antibody (HIRMAb) has been shown to block TNFα effects in the inflammatory brain region, as well as the potential for the same endogenous ligands and HIRMAb-modified drugs, which may reduce efficacy and change receptor activity [21].

17.14 LEPTIN

Leptin is a low-molecular-weight hormone that has a major role in regulating appetite, body weight, and energy. On the luminal side of the BBB capillaries, leptin receptors (OBRa and OBRc) are overexpressed. To signal satiety levels, leptin is transferred to the brain stem via receptor-mediated transcytosis [12].

Liposomes functionalized with a leptin-derived peptide (Lep70–89) possessed higher permeation in mouse brain–derived endothelial cells (MBEC4) than unmodified PEGylated liposomes. Micropinocytosis was primarily used to internalize the Lep70–89 liposomes, which may suppress lysosomal degradation [72].

Kuo et al. [73] prepared leptin-functionalized liposomes and loaded them epigallocatechin gallate (EGCG) and RSV and used them in PD management. Immunofluorescence experiments showed that Leptin/RSV-EGCG-liposomes had better uptake in *in vitro* BBB model of HBMECs and SH-SY5Y cells compared with unfunctionalized liposomes. Furthermore, the nanocarriers enhanced the neuroprotective effects of RSV and EGCGs by decreasing apoptosis and downregulating α-synuclein expression, meanwhile increasing the expression of tyrosine hydroxylase and dopamine transporters [73].

Astragaloside IV– and nestin-1-loaded liposomes were functionalized with leptin and wheat germ agglutinin (WGA) to ensure effective brain targeting for a variety of PD pathogenesis, as the loaded drugs possess antioxidant, anti-apoptotic, and neuroprotective effects [74]. The formulated functionalized liposomes were able to decrease the expression of phosphorylated tau and α-synuclein protein and improve tyrosine hydroxylase and dopamine transporter expression. Also, they could trigger anti-apoptotic pathways in SH-SY5Y cells treated with the neurotoxin 1-methyl-4-phenylpyridinium (MPP+), halting neuron degeneration [74].

17.15 CARDIOLIPIN

Cardiolipin is a phospholipid that is present in high amounts in the innermost membrane in the mitochondria. It is essential for mitochondrial energy metabolism. It has a positive surface charge, which aids in its adsorption to the BBB. Cardiolipin-functionalized liposomes can perform several functions as they have a strong preference for A fibrils in the brain. Also, they can control plasma Aβ levels, and this may, in turn, decrease tau phosphorylation and plaque accumulation in AD.

In SK-N-MC human neuroblastoma cells, liposomes that are composed of cholesterol, 1,2-dipalmitoyl-glycero-3-phosphocholine (DPPC), and DSPE-PEG (2000) were prepared and bound with wheat germ agglutinin and cardiolipin. The prepared conjugated liposomes increased the neuronal survival and reduced tau phosphorylation. Also, liposomes' pharmacological effects were evaluated *in vivo* in an AD rat model, and results demonstrated that Aβ plaque deposition, lipid peroxidation, and acetylcholinesterase enzyme activity in the hippocampus of the rats were significantly decreased. As a result, the viability of neuronal cell and cholinergic function were improved [75].

In addition, liposomes were TAT-grafted with cardiolipin and phosphatidic acid and loaded with anti-apoptotic chemicals like rosmarinic acid, quercetin, curcumin, and NGF. The use of the aforementioned liposomes resulted in decreasing Aβ deposition and hindering tau-hyperphosphorylated neurodegeneration [76].

17.16 OTHER LIGANDS

17.16.1 Fas Ligand Antibody

Lu et al. [77] prepared butylphthalide (dl-NBP)–loaded Fas-conjugated PEGylated lipid nanoparticles that can cross the BBB to target brain ischemia, which was confirmed by fluorescent analysis in mice. Also, the confocal immunohistochemical study revealed that prepared nanoparticles had preferential accumulation on OX42 positive microglia cells in the ischemic region of the mice model. Furthermore, dl-NBP-loaded nanocarriers significantly improved traumatic brain injury and neurological deficit after ischemia, with significantly reduced dosages compared to unfunctionalized nanocarriers [77].

17.16.2 CD133 mAb

Immunoliposomes loaded with temozolomide (TMZ) were functionalized by angiopep-2 (An2) and anti-CD133 monoclonal antibody (CD133 mAb) to dual-target glioblastoma stem cells (GSCs). Particle size, zeta potential, and entrapment efficiency of the dual-targeting immunoliposomes were 203.4 nm, −1.6 mV, and 99.2%, respectively.

The *in vitro* cytotoxicity against U87MG GSCs, apoptosis, and antimigration ability of the prepared dual-targeted formulation was significantly increased in comparison with that of free TMZ and non-targeted TMZ liposomes.

In vivo study has shown that intravenous administration of dual-targeted liposomes to mice with brain tumors significantly reduced tumor size. In addition, dual-targeted liposomes increased the chance of survival of mice with carcinoma (211.2% in ILS and 49.2 days in MST) compared to free TMZ (0% in ILS and 23.3 days in MST). These data demonstrated that binding of both An2 peptide and CD133 mAb to TMZ-loaded liposome is highly efficient in delivering TMZ to GSCs *via* the BBB and suggest that dual-targeted liposomes could be used as a therapeutic tool for GBM [78].

17.16.3 Lectins

Lectins are plant-derived proteins or glycoproteins that can identify and conjugate with high specificity to glycolipid and proteins sequences that exist on various cell surfaces. Therefore, they are considered as potential targeting ligands for drug delivery since 1988 and were initially used to enhance absorption throughout the gastrointestinal tract [79].

Immunogenicity is one of the main obstacles facing the application of lectins in drug delivery. Since odorranalectin (OL) is the smallest peptide with lectin-like activity, it was suggested as a promising targeting approach with decreased immunological reactions. Urocortin, which is a peptide that can restore the nigrostriatal function in PD, was encapsulated in OL-conjugated PEG–PLG NPs. Results of *in vivo* study in rats with semi-PD rats revealed the prepared NPs showed better ability to improve dopaminergic lesion symptoms compared with unmodified NPs. This may be attributed to a partial but insufficient improvement of monoamine neurotransmitter levels [80].

In conclusion, surface-active ligands are considered a promising effective approach for delivering different therapeutic agents to the brain to manage different neurological disorders by circumventing the BBB.

REFERENCES

1. Rajwar, T.K., et al., Receptor attacking ligands for active targeting to brain, in *Nanomedicine-based approaches for the treatment of dementia*. Elsevier, 2023, pp. 57–70.
2. Teixeira, M.I., et al., *Surface-modified lipid nanocarriers for crossing the blood-brain barrier (BBB): A current overview of active targeting in brain diseases*. Colloids and Surfaces B: Biointerfaces, 2022: p. 112999.
3. Jefferies, W.A., et al., *Transferrin receptor on endothelium of brain capillaries*. Nature, 1984. **312**: pp. 162–163.
4. Yoo, J., et al., *Active targeting strategies using biological ligands for nanoparticle drug delivery systems*. Cancers, 2019. **11**(5): p. 640.
5. Ross, C., et al., *Liposome delivery systems for the treatment of Alzheimer's disease*. International Journal of Nanomedicine, 2018: pp. 8507–8522.
6. Kong, L., et al., *Transferrin-modified osthole PEGylated liposomes travel the blood-brain barrier and mitigate Alzheimer's disease-related pathology in APP/PS-1 mice*. International Journal of Nanomedicine, 2020: pp. 2841–2858.
7. Pinheiro, R., et al., *Quercetin lipid nanoparticles functionalized with transferrin for Alzheimer's disease*. European Journal of Pharmaceutical Sciences, 2020. **148**: p. 105314.
8. Lopalco, A., et al., *Transferrin functionalized liposomes loading dopamine HCl: Development and permeability studies across an in vitro model of human blood–brain barrier*. Nanomaterials, 2018. **8**(3): p. 178.
9. Kuo, Y.-C. and C.-T. Wang, *Protection of SK-N-MC cells against β-amyloid peptide-induced degeneration using neuron growth factor-loaded liposomes with surface lactoferrin*. Biomaterials, 2014. **35**(22): pp. 5954–5964.
10. Loureiro, J.A., et al., *Immunoliposomes doubly targeted to transferrin receptor and to α-synuclein*. Future Science OA, 2015. **1**(4).
11. Agrawal, M., et al., *Recent advancements in liposomes targeting strategies to cross blood-brain barrier (BBB) for the treatment of Alzheimer's disease*. Journal of Controlled Release, 2017. **260**: pp. 61–77.
12. Kuo, Y.-C. and P.-R. Chou, *Neuroprotection against degeneration of SK-N-MC cells using neuron growth factor-encapsulated liposomes with surface cereport and transferrin*. Journal of Pharmaceutical Sciences, 2014. **103**(8): pp. 2484–2497.
13. Narendra, et al., *Formulation and in vitro evaluation of upconversion nanoparticle-loaded liposomes for brain cancer*. Therapeutic Delivery, 2020. **11**(9): pp. 557–571.
14. Li, J., et al., *Active targeting of orthotopic glioma using biomimetic liposomes co-loaded elemene and cabazitaxel modified by transferritin*. Journal of Nanobiotechnology, 2021. **19**(1): pp. 1–19.
15. Lam, F.C., et al., *Enhanced efficacy of combined temozolomide and bromodomain inhibitor therapy for gliomas using targeted nanoparticles*. Nature Communications, 2018. **9**(1): p. 1991.
16. Sharma, S., A. Tyagi, and S. Dang, *Nose to brain delivery of transferrin conjugated PLGA nanoparticles for clonidine*. International Journal of Biological Macromolecules, 2023: p. 126471.
17. Bonferoni, M.C., et al., *Nanoemulsions for "nose-to-brain" drug delivery*. Pharmaceutics, 2019. **11**(2): p. 84.
18. Teixeira, M.I., et al., *Formulation, characterization, and cytotoxicity evaluation of lactoferrin functionalized lipid nanoparticles for riluzole delivery to the brain*. Pharmaceutics, 2022. **14**(1): p. 185.
19. Muntoni, E., et al., *Methotrexate-loaded solid lipid nanoparticles: Protein functionalization to improve brain biodistribution*. Pharmaceutics, 2019. **11**(2): p. 65.
20. Gharbavi, M., et al., *Niosome: A promising nanocarrier for natural drug delivery through blood-brain barrier*. Advances in Pharmacological and Pharmaceutical Sciences, 2018. **2018**.
21. Georgieva, J.V., D. Hoekstra, and I.S. Zuhorn, *Smuggling drugs into the brain: An overview of ligands targeting transcytosis for drug delivery across the blood–brain barrier*. Pharmaceutics, 2014. **6**(4): pp. 557–583.
22. Kuo, Y.C. and I.W. Chao, *Conjugation of melanotransferrin antibody on solid lipid nanoparticles for mediating brain*

cancer malignancy. Biotechnology progress, 2016. **32**(2): pp. 480–490.
23. Karkan, D., et al., *A unique carrier for delivery of therapeutic compounds beyond the blood-brain barrier*. PloS One, 2008. **3**(6): p. e2469.
24. Kuo, Y.-C. and I.-H. Lee, *Delivery of doxorubicin to glioblastoma multiforme in vitro using solid lipid nanoparticles with surface aprotinin and melanotransferrin antibody for enhanced chemotherapy*. Journal of the Taiwan Institute of Chemical Engineers, 2016. **61**: pp. 32–45.
25. Jnaidi, R., A.J. Almeida, and L.M. Gonçalves, *Solid lipid nanoparticles and nanostructured lipid carriers as smart drug delivery systems in the treatment of glioblastoma multiforme*. Pharmaceutics, 2020. **12**(9): p. 860.
26. Chen, H., et al., *Lactoferrin modified doxorubicin-loaded procationic liposomes for the treatment of gliomas*. European Journal of Pharmaceutical Sciences, 2011. **44**(1–2): pp. 164–173.
27. Singh, I., et al., *Lactoferrin bioconjugated solid lipid nanoparticles: A new drug delivery system for potential brain targeting*. Journal of Drug Targeting, 2016. **24**(3): pp. 212–223.
28. Kuo, Y.-C. and S.-J. Cheng, *Brain targeted delivery of carmustine using solid lipid nanoparticles modified with tamoxifen and lactoferrin for antitumor proliferation*. International Journal of Pharmaceutics, 2016. **499**(1–2): pp. 10–19.
29. Zhao, C., et al., *Design of lactoferrin modified lipid nanocarriers for efficient brain-targeted delivery of nimodipine*. Materials Science and Engineering: C, 2018. **92**: pp. 1031–1040.
30. Meng, F., et al., *A novel LDL-mimic nanocarrier for the targeted delivery of curcumin into the brain to treat Alzheimer's disease*. Colloids and Surfaces B: Biointerfaces, 2015. **134**: pp. 88–97.
31. Karami, Z., et al., *Neuropharmacokinetic evaluation of lactoferrin-treated indinavir-loaded nanoemulsions: Remarkable brain delivery enhancement*. Drug Development and Industrial Pharmacy, 2019. **45**(5): pp. 736–744.
32. Jiang, Y., et al., *The optimization design of lactoferrin loaded HupA nanoemulsion for targeted drug transport via intranasal route*. International Journal of Nanomedicine, 2019: pp. 9217–9234.
33. Wang, Q., et al., *Application progress of RVG peptides to facilitate the delivery of therapeutic agents into the central nervous system*. RSC Advances, 2021. **11**(15): pp. 8505–8515.
34. Zhao, M., et al., *Targeted therapy of intracranial glioma model mice with curcumin nanoliposomes*. International Journal of Nanomedicine, 2018: pp. 1601–1610.
35. Arora, S., B. Layek, and J. Singh, *Design and validation of liposomal ApoE2 gene delivery system to evade blood–brain barrier for effective treatment of Alzheimer's disease*. Molecular Pharmaceutics, 2020. **18**(2): pp. 714–725.
36. Schlich, M., et al., *Anionic liposomes for small interfering ribonucleic acid (siRNA) delivery to primary neuronal cells: Evaluation of alpha-synuclein knockdown efficacy*. Nano Research, 2017. **10**: pp. 3496–3508.
37. Conceição, M., et al., *Intravenous administration of brain-targeted stable nucleic acid lipid particles alleviates Machado–Joseph disease neurological phenotype*. Biomaterials, 2016. **82**: pp. 124–137.
38. Pinheiro, R., et al., *RVG29-functionalized lipid nanoparticles for quercetin brain delivery and Alzheimer's disease*. Pharmaceutical Research, 2020. **37**: pp. 1–12.
39. Rassu, G., et al., *Nose-to-brain delivery of BACE1 siRNA loaded in solid lipid nanoparticles for Alzheimer's therapy*. Colloids and Surfaces B: Biointerfaces, 2017. **152**: pp. 296–301.
40. Gan, L., et al., *Rabies virus glycoprotein (RVG29)-linked microRNA-124-loaded polymeric nanoparticles inhibit neuroinflammation in a Parkinson's disease model*. International Journal of Pharmaceutics, 2019. **567**: p. 118449.
41. You, L., et al., *Targeted brain delivery of rabies virus glycoprotein 29-modified deferoxamine-loaded nanoparticles reverses functional deficits in parkinsonian mice*. Acs Nano, 2018. **12**(5): p. 4123–4139.
42. Jose, G., et al., *Co-delivery of CPT-11 and panobinostat with anti-GD2 antibody conjugated immunoliposomes for targeted combination chemotherapy*. Cancers, 2020. **12**(11): p. 3211.
43. Xie, J., et al., *Cell-penetrating peptides in diagnosis and treatment of human diseases: From preclinical research to clinical application*. Frontiers in Pharmacology, 2020. **11**: p. 697.
44. Gajbhiye, K., et al., *PEGylated nanocarriers: A promising tool for targeted delivery to the brain*. Colloids and Surfaces B: Biointerfaces, 2020. **187**: p. 110770.
45. Gartziandia, O., et al., *Nanoparticle transport across in vitro olfactory cell monolayers*. International Journal of Pharmaceutics, 2016. **499**(1–2): pp. 81–89.
46. Sonvico, F., et al., *Surface-modified nanocarriers for nose-to-brain delivery: From bioadhesion to targeting*. Pharmaceutics, 2018. **10**(1): p. 34.
47. Kanazawa, T., et al., *Cell-penetrating peptide-modified block copolymer micelles promote direct brain delivery via intranasal administration*. Pharmaceutical Research, 2011. **28**: pp. 2130–2139.
48. Taki, H., et al., *Intranasal delivery of camptothecin-loaded tat-modified nanomicells for treatment of intracranial brain tumors*. Pharmaceuticals, 2012. **5**(10): pp. 1092–1102.
49. Kanazawa, T., et al., *Delivery of siRNA to the brain using a combination of nose-to-brain delivery and cell-penetrating peptide-modified nano-micelles*. Biomaterials, 2013. **34**(36): pp. 9220–9226.
50. Zhao, J., et al., *CREKA peptide-conjugated dendrimer nanoparticles for glioblastoma multiforme delivery*. Journal of Colloid and Interface Science, 2015. **450**: pp. 396–403.
51. Lyons, S.A., J. O'Neal, and H. Sontheimer, *Chlorotoxin, a scorpion-derived peptide, specifically binds to gliomas and tumors of neuroectodermal origin*. Glia, 2002. **39**(2): pp. 162–173.
52. Veszelka, S., et al., *A triple combination of targeting ligands increases the penetration of nanoparticles across a blood-brain barrier culture model*. Pharmaceutics, 2021. **14**(1): p. 86.
53. Jiang, Y., et al., *Apolipoprotein E peptide-directed chimeric polymersomes mediate an ultrahigh-efficiency targeted protein therapy for glioblastoma*. ACS Nano, 2018. **12**(11): pp. 11070–11079.
54. Wagner, S., et al., *Uptake mechanism of ApoE-modified nanoparticles on brain capillary endothelial cells as a blood-brain barrier model*. PloS One, 2012. **7**(3): p. e32568.
55. Michaelis, K., et al., *Covalent linkage of apolipoprotein e to albumin nanoparticles strongly enhances drug transport into the brain*. Journal of Pharmacology and Experimental Therapeutics, 2006. **317**(3): pp. 1246–1253.
56. Tamaru, M., et al., *Application of apolipoprotein E-modified liposomal nanoparticles as a carrier for delivering DNA*

and nucleic acid in the brain. International journal of Nanomedicine, 2014: pp. 4267–4276.
57. Neves, A.R., et al., *Solid lipid nanoparticles as a vehicle for brain-targeted drug delivery: Two new strategies of functionalization with apolipoprotein E.* Nanotechnology, 2015. **26**(49): p. 495103.
58. Topal, G.R., et al., *ApoE-targeting increases the transfer of solid lipid nanoparticles with donepezil cargo across a culture model of the blood–brain barrier.* Pharmaceutics, 2020. **13**(1): p. 38.
59. Zhang, Z.-A., et al., *Novel brain-targeted nanomicelles for anti-glioma therapy mediated by the ApoE-enriched protein corona in vivo.* Journal of Nanobiotechnology, 2021. **19**(1): pp. 1–24.
60. Kato, N., et al., *Development of an apolipoprotein E mimetic peptide–lipid conjugate for efficient brain delivery of liposomes.* Drug Delivery, 2023. **30**(1): p. 2173333.
61. Gao, J.-Q., et al., *Glioma targeting and blood–brain barrier penetration by dual-targeting doxorubincin liposomes.* Biomaterials, 2013. **34**(22): pp. 5628–5639.
62. Li, M., et al., *Synergistic tumor microenvironment targeting and blood–brain barrier penetration via a pH-responsive dual-ligand strategy for enhanced breast cancer and brain metastasis therapy.* Nanomedicine: Nanotechnology, Biology and Medicine, 2018. **14**(6): pp. 1833–1843.
63. Kuo, Y.-C. and C.-H. Lee, *Inhibition against growth of glioblastoma multiforme in vitro using etoposide-loaded solid lipid nanoparticles with p-aminophenyl-α-D-mannopyranoside and folic acid.* Journal of Pharmaceutical Sciences, 2015. **104**(5): pp. 1804–1814.
64. Lockman, P.R., et al., *Brain uptake of thiamine-coated nanoparticles.* Journal of Controlled Release, 2003. **93**(3): pp. 271–282.
65. Lewicky, J.D., et al., *Mannosylated glycoliposomes for the delivery of a peptide kappa opioid receptor antagonist to the brain.* European Journal of Pharmaceutics and Biopharmaceutics, 2020. **154**: pp. 290–296.
66. Kuo, Y.-C. and C.-Y. Lin, *Targeting delivery of liposomes with conjugated p-aminophenyl-α-d-manno-pyranoside and apolipoprotein E for inhibiting neuronal degeneration insulted with β-amyloid peptide.* Journal of Drug Targeting, 2015. **23**(2): pp. 147–158.
67. Rip, J., et al., *Glutathione PEGylated liposomes: Pharmacokinetics and delivery of cargo across the blood–brain barrier in rats.* Journal of Drug Targeting, 2014. **22**(5): pp. 460–467.
68. Maussang, D., et al., *Glutathione conjugation dose-dependently increases brain-specific liposomal drug delivery in vitro and in vivo.* Drug Discovery Today: Technologies, 2016. **20**: pp. 59–69.
69. Kuo, Y.-C., H.-C. Tsai, and R. Rajesh, *Glutathione liposomes carrying ceftriaxone, fk506, and nilotinib to control overexpressed dopamine markers and apoptotic factors in neurons.* ACS Biomaterials Science & Engineering, 2021. **7**(7): pp. 3242–3255.
70. Grover, A., et al., *Brain-targeted delivery of docetaxel by glutathione-coated nanoparticles for brain cancer.* AAPS PharmSciTech, 2014. **15**: pp. 1562–1568.
71. Qin, Y., et al., *In vitro and in vivo investigation of glucose-mediated brain-targeting liposomes.* Journal of Drug Targeting, 2010. **18**(7): pp. 536–549.
72. Tamaru, M., et al., *Leptin-derived peptide, a targeting ligand for mouse brain-derived endothelial cells via macropinocytosis.* Biochemical and Biophysical Research Communications, 2010. **394**(3): pp. 587–592.
73. Kuo, Y.-C., I.-H. Wang, and R. Rajesh, *Use of leptin-conjugated phosphatidic acid liposomes with resveratrol and epigallocatechin gallate to protect dopaminergic neurons against apoptosis for Parkinson's disease therapy.* Acta Biomaterialia, 2021. **119**: pp. 360–374.
74. Kuo, Y.-C., I.-Y. Chen, and R. Rajesh, *Astragaloside IV-and nesfatin-1-encapsulated phosphatidylserine liposomes conjugated with wheat germ agglutinin and leptin to activate anti-apoptotic pathway and block phosphorylated tau protein expression for Parkinson's disease treatment.* Materials Science and Engineering: C, 2021. **129**: p. 112361.
75. Kuo, Y.-C. and C.-C. Lin, *Rescuing apoptotic neurons in Alzheimer's disease using wheat germ agglutinin-conjugated and cardiolipin-conjugated liposomes with encapsulated nerve growth factor and curcumin.* International Journal of Nanomedicine, 2015: pp. 2653–2672.
76. Ordóñez-Gutiérrez, L., et al., *Repeated intraperitoneal injections of liposomes containing phosphatidic acid and cardiolipin reduce amyloid-β levels in APP/PS1 transgenic mice.* Nanomedicine: Nanotechnology, Biology and Medicine, 2015. **11**(2): pp. 421–430.
77. Lu, Y.-m., et al., *Targeted therapy of brain ischaemia using Fas ligand antibody conjugated PEG-lipid nanoparticles.* Biomaterials, 2014. **35**(1): pp. 530–537.
78. Kim, J.S., D.H. Shin, and J.-S. Kim, *Dual-targeting immunoliposomes using angiopep-2 and CD133 antibody for glioblastoma stem cells.* Journal of Controlled Release, 2018. **269**: pp. 245–257.
79. Bies, C., C.-M. Lehr, and J.F. Woodley, *Lectin-mediated drug targeting: History and applications.* Advanced Drug Delivery Reviews, 2004. **56**(4): pp. 425–435.
80. Wen, Z., et al., *Odorranalectin-conjugated nanoparticles: Preparation, brain delivery and pharmacodynamic study on Parkinson's disease following intranasal administration.* Journal of Controlled Release, 2011. **151**(2): pp. 131–138.

18 Nanocarriers for Brain Delivery
Safety, Stability Concerns, and Regulatory Aspects

Asmaa Ashraf, Mohamed Dahab, and Rahma Soliman*

18.1 INTRODUCTION

Our brains are protected by the blood–brain barrier physically and biologically. The blood–brain barrier (BBB) controls the movement of chemicals into the brain, hence maintaining hemostasis in the brain. The primary defense mechanism of the BBB is linked to the prevalence of tight junctions (TJs) and efflux pumps (Mehrabian et al., 2022). The BBB has a crucial role in a healthy brain's defense mechanism against toxins, but it also plays a role in brain disorders. Neurodegenerative disorders (NDDs) are a group of heterogeneous disorders which are mostly distinguished by progressive deterioration of the central nervous system's (CNS) structure and function. The likelihood of getting an NDD rises significantly with age. Recent research and findings recommend that NDDs may affect many people in the not-too-distant future (Muheem et al., 2020). More research in this area is urgently needed to determine the cause of this situation pathophysiologically and behind drug delivery strategies for improved drug effectiveness. Passing through the BBB is one of the biggest obstacles in brain cancer treatment. The BBB's endothelial cells feature extremely tight connections and fenestrations, which prevent the passage of both hydrophilic medications and more than 95% of hydrophobic drugs (Garanti et al., 2021). The BBB is the main factor that may present serious obstacles in the effective early diagnosis and treatment of NDDs since the majority of medications and diagnostics have inadequate penetration.

Brain drug delivery has been one of the most difficult problems faced by researchers in developing successful therapies for brain diseases (Ayub & Wettig, 2022). Nano-drug delivery systems become more effective tools for better diagnosis and treatment of NDDs. The effective molecules could be adsorbed, enveloped, or attached to the nanodelivery system according to the nanoparticle surface, size, drug-loading methods, and zeta potential modifications. Cells range between 6–9 and 10–20 μm and blood capillaries, respectively; almost all types of nano-drug delivery systems could reach and deliver drugs into the desired organs and cells. Particularly, targeting of blood–brain barrier receptors by surface-functionalized nano-drug delivery systems make them suitable candidates for brain disorders. The more downsizing the nanoparticles could take to cross the BBB, the more utilization for diagnosis and therapy of NDDs.

Despite significant development in treatment protocols, drug delivery is still an ongoing challenge; the number of available treatments is limited. The aqueous environment of the brain's interstitial fluid needs molecule that can pass through the BBB to be somewhat lipid-soluble to go through the brain. A molecule with a higher level of lipophilicity may enter the BBB more readily; however, this may also result in less of the molecule reaching the target location. Consequently, the chance of a molecule traversing the BBB and reaching the brain is best when it has a more lipophilic content. The chances of molecule permeation are also impacted by its mass. Even though lighter molecules have a higher chance of passing through the barrier and entering the brain, the maximum molecular weight of the compounds relies on the mechanism of transportation. The development of an effective drug circumvents the limitations forced by the BBB, requiring understanding of the biological firewall, the drug's physicochemical characteristics, then how it interacts with the BBB to prevent undesired side effects (Guillama Barroso et al., 2020). Researchers can create fascinating prospective ND-related nanomaterials and nanomedicines for delivering pharmaceuticals or imaging probes to the brain thanks to the amount of understanding about the BBB. However, neurotoxicity is now a significant problem. The CNS becomes exposed to nanomaterials and nanomedicines that may result in a little neurotoxicity because nanoparticles could enter the brain through a variety of pathways. Numerous different types of nanocarriers, like metal nanoparticle, carbon nanotubes, graphene, and cationic polymers, have been created with the ability to enter the CNS and cause neurotoxicity. Oxidative stress changes the BBB integrity by triggering a particular signaling route, inducing cell death, autophagy, immunological response, and inflammation, which are some of the potential mechanisms that lead to common neurotoxicity. Particle size, shape, crystal type, charge, surface morphology, the release of ions, and the method of administering nanoparticles are among the several contributing factors responsible for neurotoxicity. By using multiple delivery methods inside

* Corresponding author: asmaaashraf2382000@gmail.com

DOI: 10.1201/9781003383376-18

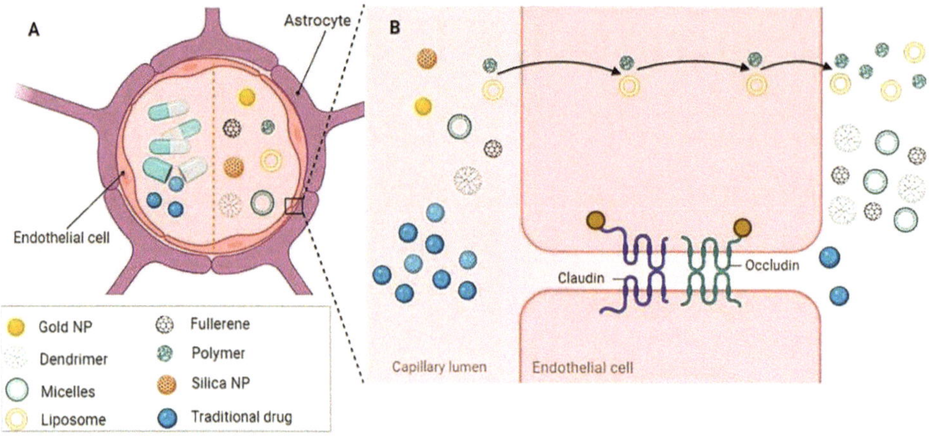

FIGURE 18.1 This figure represents some nano-drug delivery systems and traditional drug crossing the tight junction of the blood–brain barrier (BBB): (A) A blood vesicle containing traditional drug for the brain, on the left, and nano-drug delivery carrier, on the right. (B) Nanocarriers highly efficiently crossing the tight junction due to the small size of the carrier, with some of it crossing the BBB by the lipophilicity across the membrane bilayer, so a higher concentration of the carrier goes through the BBB. Traditional drugs cannot cross the BBB due to their large size.

the subject's body, such as ingestion, inhalation, and skin penetration, nanoparticles may accumulate in numerous tissues or organs, including the brain tissue. Therefore, an in-depth study of the neurotoxicity of nanocarriers can end up with safer treatment, reducing NP-related side effects.

18.2 SAFETY OF NANOPARTICLES

Nanocarriers can control the release of drugs under regulated conditions and prevent deterioration of the payload. Recent advances in structured nanoemulsions and solid lipid nanoparticles can be used to promote the development of safer nanocarriers(NCs). According to studies, numerous harmful effects of NCs have also been linked to solid- or metal-containing NCs. There are some types of nanocarriers for which cytotoxic effects have regularly been observed. Since the field of nanomedicine exhibits great promise, it is crucial to assess the safety of nanocarriers in order to achieve biocompatibility and required activity and establish safety tests that can precisely predict the potential toxicity of nanotherapeutics. We discuss this in the following sections.

18.2.1 Toxicity of Nanoparticles

Nanomaterials are materials at the nanoscale (1–1,000 nm), which makes them highly active, so neurotoxicological studies are important to ensure the safety of nanomaterials (Patel et al., 2022). The safety of nanomaterials depends on their toxicity and biocompatibility. The pharmacokinetic and pharmacodynamic behaviors of nanoparticles can help in understanding biocompatibility, bioavailability, and biodegradability. The size of nanoparticles can affect their safety, so that nanoparticles with a size less than 100 nm are found to be nontoxic and biocompatible for many types of cell lines in the optimum dose. Nanoparticle charge is one of the factors that cause toxicity, in which metal ions' presence increases the toxic effect of nanocarriers. Also, an increased concentration of nanocarriers causes toxicity compared to less concentration (Ahmad & Imran, 2022). Increased surface area ratio of many nanocarriers can cause oxidative stress by interacting with cellular macromolecules. Metal-based nanocarriers and carbon-based nanocarriers are potentially neurotoxic as they cause cytokines release, leading to neuroinflammation, causing neural cell death (Sartaj et al., 2022). In some cases, nanocarriers may induce neuroinflammation, DNA damage, and neurodegeneration, preventing them from being used as medications, but they can be used alone, without the need for carrying a drug to a targeted region to eliminate an exact disease (Patel et al., 2022).

18.2.2 Safety Assessment of Nanocarriers

Safety assessment is important to evaluate the risk for humans or the environment, that toxicity can be evaluated in silico, *in vitro*, or *in vivo*. Safety assessments of nanocarriers have no standards, but most studies perform toxicological assays categorized into (1) physicochemical characterization, (2) cell line assay, and (3) *in vivo* models.

18.2.2.1 Physicochemical Characterization

Physicochemical properties such as shape, size, chemical composition, charge, stability, aggregation tendency, and surface coating are all parameters that define the surface reactivity of nanocarrier and can help predict the ability of nanocarriers to generate oxidative stress (Scimeca & Verron, 2022). Nanoparticles have a large surface area due to their tiny size, which increases their carcinogenic potential. Coating metallic or inorganic nanocarriers with organic or natural polymers makes them less toxic and increases biocompatibility for use as medications. Nanocarriers can be coated with polyethylene glycol (PEG), gelatin, lipid coatings, cetyltrimethylammonium bromide (CTAB), chitosan, etc., to protect the loaded drug and increase the stability of the nanocarrier in addition to minimizing toxicity and maximizing biocompatibility for many cell lines and different tissues (Ahmad & Imran, 2022).

18.2.2.2 Cell Line Assay

Cell line assay or *in vitro* assessment is considered the most economical, reliable, accessible, and free from animal models. Cell line assay is used to evaluate toxicity, oxidative stress, and gene expression changes (Ahmad & Imran, 2022). Some precautions are important for performing *in vitro* assessments, since nanocarriers have a small size and can aggregate sediment, which means that their behavior will be different, so adequate test is required. Methods used to evaluate toxicity must be able to detect low levels of toxicity biomarkers to have accurate results. Also, handling the sample and nanoparticle separation is important for performing a suitable analytical procedure. The formulation of nanocarriers and laboratory procedures can affect results. Nanoparticle separation can be performed by centrifugation (Patel et al., 2022). To optimize the stability of nanocarriers and prevent aggregation, samples need to be prepared freshly, nanocarriers may include serum proteins, and sonication may be performed. Contamination of nanocarriers with endotoxins is a serious issue in that endotoxins cause an inflammatory reaction by activating macrophages and dendritic cells, so the level of contamination should be carefully considered (Scimeca & Verron, 2022). Also, using the appropriate dose of nanocarrier is one of the important points to consider, as increased concentrations will increase toxicity. In some situations, nanocarriers can interact with *in vitro* assay, so it is recommended to perform multiple toxicity assays in parallel to validate the safety assessment results (Patel et al., 2022).

Initial tests are carried out to identify changes in cytoskeletal integrity, organelle structure, and cell membrane integrity. MTT, trypan blue dye assay, comet assay, and 2,7-dichlorofluorescein diacetate assay are all *in vitro* toxicity tests. By measuring the amount of tetrazolium salts that are converted to formazan, the MTT assay evaluates the metabolic activity of the cell. A preliminary experiment, such as adding tetrazolium salts to untreated cells, extracting the resulting formazan, quantification by spectrophotometry, or incubating the extract in the presence of nanocarriers before a second quantification by spectrophotometry, is required to demonstrate that there are no interactions between the NPs and the salts. Changes in optical density readings suggest that nanocarriers are interfering with the MTT assay.

Additionally, genotoxicity is crucial for identifying chromosomal and gene mutations. The Ames test, based on the survival of bacterial strains that have previously been genetically altered to be more sensitive by altering their DNA repair system and making them permeable to connections, is used to assess the mutagenicity of substances. However, because bacteria and nanocarriers are similar in size, *in vitro* micronucleus assay is used because it is less accurate for nanocarriers. It finds micronuclei in the cytoplasm of cells that are in the interphase. The Ames test evaluates the mutagenicity of substances by examining the persistence of bacterial strains that have previously undergone genetic modification to become more sensitive. This modification affects their DNA repair system and renders the bacteria permeable to connections. But it' is not accurate for nanocarriers due to the size similarity of bacteria and nanocarriers, so *in vitro* micronucleus assay is used. It finds micronuclei in the cytoplasm of cells that are in the interphase. To stop cytokinesis and avoid the separation of daughter cells after mitosis, cytochalasin B should be added. Since cytochalasin B prevents NP uptake, cells must first be treated with NPs before cytochalasin B is added. Due to its simplicity, dependability, and sensitivity, the comet assay is a technique used to examine particular DNA lesions or single/double-strand breaks (Scimeca & Verron, 2022). Increased production of reactive oxygen and nitrogen species, inflammation, and excessive production of cytokines that trigger cell death mechanisms like apoptosis, autophagy, and necrosis are the main indicators of nanomaterial-induced toxicity (Ahmad & Imran, 2022).

18.2.2.3 *In Vivo* Models

In vivo assessment is better assigned to humans as it considers many crucial factors, such as hormonal changes and cell matrix interactions. Also, long-term toxicity effects are not measured by *in vitro* studies. Nanocarriers are introduced into the animal body and monitored pharmacodynamically. These studies involve real-time analysis, where the results are consistent with the functioning of the human body. Identifying the risks of an infusion reaction at the earlier stages of drug development phases can help mitigate the vital safety impact when advancing nanocarriers to clinical trials and save money and time and protect patients from significant deleterious consequences (Ahmad & Imran, 2022). *In vivo* models enable documenting the toxicokinetic parameters of nanoparticles as absorption, distribution, metabolism, and elimination of nanocarriers. The kinetic study considers the route of administration, the properties of the nanomaterial, and the animal species. Small rodents and rabbits are preferred over large animal models for the study of nanocarrier toxicity. Following the implantation of the material, blood samples are taken at various intervals to monitor the distribution of the nanocarriers. If the analytical methods are validated, chemical measurements are also taken. Blood samples are also used to calculate the hemoglobin concentration and complete blood count, and these analyses are complemented by tissue impregnation measurements. The animals are put to death after medium or prolonged delays from implantation, and various organs are removed to look for potential bioaccumulation and conduct histological research. Histological studies on bone tissue after stains are required to assess systemic toxicity as well as the local response at the implantation site (Scimeca & Verron, 2022).

18.2.3 Safety of Polymeric-Based Nanoparticles

Polymeric nanocarriers have important properties that include stability, nontoxicity, biodegradability, biocompatibility, and ease of fabrication processes. Two common polymer NPs are poly(butyl cyanoacrylate) (PBCA) and poly(lactic acid) (PLA), with PBCA offering rapid biodegradability. Polymeric nanocarriers have been tested and shown to improve survival in mouse models (Hersh et al., 2022).

18.2.4 Safety of Carbon-Based Nanoparticles

Due to their many benefits, such as high biocompatibility, low toxicity, small size, and high surface area, carbon nanostructures are strongly advised. The capacity of carbon nanostructures to interact with biological molecules like amyloid and protein molecules is well known. These structures can pass through the BBB and possibly become surface-functionalized because of their small size. They can also successfully restore damaged neurons. The development of a therapeutic strategy for neurodegenerative diseases like Parkinson's and Alzheimer's is based on the activity of carbon structures. To ensure the safety of carbon nanostructures in biological systems, *in vitro* and *in vivo* studies are required. These carbon nanostructures' special qualities are what make them useful in treating brain disorders. Through conjugation, the surface of carbon nanotubes can hold more than one unit. By crossing the BBB, this property can be used to deliver a variety of medications to the brain. To create new medications for neurodegenerative diseases, fullerene's combination of antioxidant and anti-aggregation effects is crucial. But their biomedical application is constrained by poor solubility and aggregation in aqueous solvents. The surface functionalization of fullerenes can solve this issue. Targeted drug delivery and decreased toxicity can be achieved through the attachment of ligand molecules to fullerene surfaces. Additional study is required to assess these nanomaterials' biosafety and address the problems with BBB delivery (Henna et al., 2020).

18.2.5 Safety of Lipid-Based Nanoparticles

A composition for a liposome-based preparation is usually developed using a well-known drug substance that has passed safety tests according to the existing rules and guidelines of relevant organizations of manufacturing countries. One should pick the formulation's lipid composition to ensure safety. The majority of a liposome's components are phospholipids (Nanomedicine in Brain Diseases, 2019). They have a high level of safety for pharmaceutical use because they are a natural building material and a component of all biological membranes. Considering the safety of using nanoliposomes as a drug delivery system may be affected by their potential properties. One is phosphatidylserine, which has a strong affinity for calcium and carries a negative charge. Other phospholipids such as phosphatidylinositol or phosphatidic acid are also applied to create a negative charge on the liposomal surface. When given intravenously to guinea pigs, phosphatidic acid causes an increase in platelet aggregation. Cationic liposomes also have other drawbacks. The incorporated drug may aggregate or release as a result of their interaction with serum proteins and intercellular matrix elements, resulting in systemic toxicity. By acting on protein kinase C (PKC), cationic liposomes can accumulate in macrophages, influence the creation of immunoinflammatory modulators and T-lymphocyte functions, and suppress the synthesis of nitric oxide mediators. Cationic liposome toxicity has been noted to vary with dose. When phosphatidylcholine is substituted for phosphatidylethanolamine, *in vitro* tests have shown reduced toxic effects (Tereshkina et al., 2022). The composition of lipid nanocarriers, excipients used, particle size, surface coating, shape, and slow removal of NPs from the body are some of the factors that contribute to their toxicity. The fact that nanoparticles can cause cellular reactions and degrade in the cellular environment is a significant problem. Due to their small size, NPs can spread from the site of entry into the lymphatic, circulatory, and other systems, and eventually into body tissues and organs. Some of them can irreparably harm cells by increasing oxidative stress or by harming organs because of their composition, charge, and size. Cytotoxicity results from the cellular uptake of NPs and their stay in organelles. Directly or indirectly, NPs can physically damage biological membranes by releasing reactive oxygen species (ROS) that can result in cell death. The damage increases as more nanoparticles gather. Organ accumulation of NPs results in ROS production and inflammation. The most common lipids used in the formulation of lipid-based nanocarriers are waxes, phospholipids, fatty acids, cholesterol, and triglycerides. The materials are obtained or derived from natural sources, which makes them biocompatible and biodegradable when administered *in vivo*. Lipids are thought to be safe, but they can be toxic when consumed in high doses and show dose-dependent toxicity. Lipid nanocarriers for drug delivery result in sustained drug release, which eventually reduces drug dosage frequency, variations in plasma drug concentration, and ultimately, drug toxicity. Additionally, the immunogenicity of the drug carrier complex is decreased by the surface modification of nanocarriers. Additionally, by lessening phagocytes' ability to opsonize and destroy nanocarriers, as well as by reducing the possibility of nanocarrier interaction with constantly circulating blood components, stealth properties are imparted. Lipid nanocarriers have great potential and are a promising tool to formulate pharmaceutical products with safety and efficacy in light of growing awareness of drug safety in healing patients (Patel et al., 2022). Amphiphilic phospholipids naturally form bilayers and regulate the overall characteristics of the liposome, such as permeability and steric hindrance. They are biocompatible, adaptable, and low toxic, and they can be conjugated with ligands like PEG for steric stabilization to slow phagocytosis. To track drug delivery or visualize tumors, imaging agents can also be encapsulated in liposomes. In numerous human clinical trials, it has been discovered that liposomes increase mouse survival after being used to treat gliomas in rat and mouse models. (Hersh et al., 2022).

18.2.6 Safety of Metal-Based Nanoparticles

Inorganic nanomaterials with unique physicochemical properties and variable size and structure have been synthesized, such as silver nanocarriers, titanium oxide nanocarriers, and porous silicon nanocarriers. Gram-positive and Gram-negative bacteria can both be eliminated by silver nanoparticles (AgNPs), which have effective antibacterial characteristics. AgNPs are frequently utilized in a wide range of goods, including cosmetics and healing lotions for daily usage, lab coats, and medical goods, such as surgical gowns and bandages. In many investigations, the effects of AgNP's neurotoxicity on zebra fish have been examined. AgNPs' neurotoxic effects depend on their magnitude. Additionally, the effects of AgNPs on neural behavior profiling have been identified. Different AgNP formulations could produce distinct patterns of developmental neurotoxicity, depending on the particle coating and size (Zhao et al., 2014). Due to their distinctive electrical and optical characteristics, titanium oxide nanocarriers are utilized in many commercial fields. For NPs in adult zebra fish, it has been found that even low doses of TiO_2-NPs can harm zebra fish by impairing spatial recognition memory, which attracted attention to the neurotoxicity of TiO_2. Additionally, wimmingg speed and rotation times can be decreased by TiO_2 NPs. The loss of dopaminergic neurons in zebra fish increases the possibility that Parkinson's disease may develop as a result of exposure to TiO_2 NPs, which can also cause neurotoxicity. Zebra fish larvae's brains may become clogged with TiO_2 NPs, which could cause the death of hypothalamic cells and the production of reactive oxygen species (ROS). Additionally, they stop motor neurons' axonal growth, which might help explain the behavioral change. Nano-TiO_2 can increase triphenyl phosphate accumulation, which can cause abnormal neuronal function, serotonin system dysfunction, and worsened abnormal locomotion (Zhao et al., 2014). Porous silicon nanoparticles (pSiNPs) have a high specific surface area, high porosity, high charging efficiency, and quick decomposition into non-toxic products. To deliver nucleic acids, SNPs can be created (Zhang et al., 2022). Numerous metal-based nanoparticles can be neurotoxic, depending on their concentration and physicochemical characteristics. The exposure system, including concentration and exposure conditions, had an impact on the effects of iron ions and iron oxide nanoparticles on neurotoxicity and cardiotoxicity (Zhao et al., 2014).

18.3 STABILITY

Nanocarriers ideally have to be maintained stable, resistant to accumulation, and capable of retaining the medicine until it reaches the specified target site. The stability concerns may affect *in vivo* efficiency. As a result, assessing the stability of nanocarriers is a significant characterization step. According to the source and nature of the nanocarriers, there are numerous methodologies for forecasting their stability in physical, chemical, and physiological environments.

18.3.1 Effect of Dosage Form on Stability

Drug nanoparticles' special features have enabled their broad use in a variety of forms of doses, including oral, parenteral, ophthalmic, pulmonary, cutaneous, and other specified delivery systems all over the body. Although various forms of doses may have some of the same stability issues, such as particle agglomeration, sedimentation, or crystal growth, the impacts on medicinal products are obviously different. Particle agglomeration, for example, could be a serious concern for the drug delivery pulmonary system since it influences deposition quantity site and, hence, therapeutic efficacy. Agglomeration in intravenous formulations, on the other hand, might induce blood capillary obstruction and restrict blood flow. Furthermore, the choice of stabilizers is intimately related to the dispersion medium, the dosage form, and is carefully restricted by FDA regulations. There are currently a wide range of approved stabilizers for oral dosage form; however, the excipients permitted for inhalation are quite limited. Drug nanoparticles can be found in final drug formulations in the form of dry powder or solution. Dry powder inhalers, lyophilized powder for injection, and oral tablets or capsules are all examples of dry powder forms. Because solid dosage forms often have strong storage stability profiles, converting the suspension to solid form is a frequent technique for improving nanosuspension stability. The majority of documented stability difficulties stem from nanosuspensions, in which therapeutic nanoparticles are disseminated in a media with or without stabilizers (Wu et al., 2011). Furthermore, the mechanisms involved in the stability of small and big biomolecule formulations differ due to changes in molecular structure. A small molecule drug is a non-polymeric organic chemical with a low molecular weight, whereas big biomolecules are huge bioactive molecules, such as protein/peptide. One of the key challenges with protein/peptide stability is maintaining the three-dimensional molecular shape, such as secondary and tertiary structure, in order to retain biological functions, whereas tiny chemical molecules have no such concern.

18.3.2 Chemical Stability

Because medication nano-crystals are often dispersed in nanosuspensions with lower limit of solubility, the likelihood of chemical interaction is not as significant as in solution-based formulations. As a result, the chemical stability prospective of nanosuspension is often greater than the solution. Paclitaxel is a nice example of this, which displays a high-performance liquid chromatography (HPLC) diagram of paclitaxel nanosuspensions stabilized with Pluronic F68 (non-ionic surfactant) after four years of storage at a refrigerator at 4–8°C. With a recovery rate of more than 99%, no obvious degradation product was found. Paclitaxel solution containing methanol as a co-solvent, on the other hand, exhibited obvious deterioration after only 48 h at ambient temperature. Paclitaxel nanosuspensions' exceptional chemical stability was attributed to a process similar to an oxidized layer on the aluminum surface. When exposed to water and oxygen, the surface of the nanocrystals degraded into monolayers. This monolayer may protect the interior section of drug crystals from further degradation, hence improving the chemical stability of the nanosuspensions. Chemical stability is drug-specific, as opposed to physical stability, which is a common concern for nanosuspensions. Every molecule of the molecules has unique reaction mechanisms and functional groups that influence stability of this molecule. Chemical functions such as amides and ester, for example, are sensitive to hydrolytic degradation, whereas amino groups are susceptible to oxidative degradation (Mitchell et al., 2021). Although chemical stability of nanosuspensions is usually not a major concern, therapeutic molecules with solubility more than 1 mg/mL or low concentration in suspension should be given special consideration. To improve chemical stability, it is usual practice to convert nanosuspensions into dry solid dosage form, which is significantly more stable than nanosuspensions, or to raise the concentration of the nanosuspensions.

18.3.3 Physical Stability

18.3.3.1 Thermal Stability of Polymeric Nanocarrier

Thermosensitive polymer has a specific temperature at which it undergoes phase transition, known as the low critical solution temperature (LCST). Lower than the LCST, the polymer becomes amphiphilic, allowing the medicine to remain enveloped, but higher than the LCST, the polymer reaches the hydrophobic state, allowing the drug to be released. The thermal stability of nanocarriers is measured by LCST. Filippov et al. found that increasing the maleic anhydride concentration and molecular weight raised the LCST of a poly (N-isopropylacrylamide-co-maleic acid) copolymer from 31°C to 45°C. Na and Bae, on the other hand, confirmed that the pluronics and poly(N-isopropylacrylamide) LCST-reduced when mixed with saccharides.

18.3.3.2 Physical Stability of Self-Assembled Nanocarrier Systems

Polymeric and surfactant micelle systems are examples of self-assembled micelles. There is a concentration that self-assembled micelles associate, which could also be known as critical micelle concentration (CMC), which provides a quantitative assessment of nanocarrier physical stability. Previously, it has been stated that a low CMC makes stability increase. A dramatic polarity contrast between a super hydrophobic lipid field and a highly hydrophilic zwitterionic polymeric field is used to create ultra-low-CMC micelles (Alshawwa et al., 2022). They have the ability to stabilize hydrophobic medicines in diluted settings as well as in serum, where traditional micelles failed. CMC is determined by conductance surface-tension chromatography, light scattering, and fluorescence probes. These procedures are appropriate for the synthesis of self-assembled nanocarriers and micelles, forming of which is affected by constituent concentrations.

18.3.3.3 Stability of Nanocarrier Suspensions and Nanoemulsions

If the nanocarriers are within a given size range, deviations from the mean size could indicate aggregation or dissociation. Turbidity variations in the suspension of the nanocarrier may also be caused by instability. The cloudiness of a sulfonamide-containing nanocarrier dispersion was measured at different pH levels. The nanocarriers had a consistent particle size and turbidity at pH 7 or higher, which increased when the pH was reduced below 7. This response demonstrates pH-induced agglomeration caused by the deionized polymer's hydrophobic interaction. Nanoemulsions are thermodynamically undesirable arrangements; because the emulsified form has a larger free energy than the phase of separated oil and water, they are prone to deterioration by time. As a result, studies of thermodynamic stability for coalescence, gravitational separation, Ostwald ripening, and flocculation or phase inversion are critical for determining the stability of nanoemulsions.

18.3.3.4 Stability of Nanocarriers in Biological Matrices

When the nanoformulations' physicochemical properties are exposed to some complex biological matrices, such as the cerebrospinal fluid (CSF), tears, blood, aqueous humor, synovial fluid, and vitreous humor could be altered substantially. The biological environment's potential impact on nanoformulations includes covering of surface of the nanoparticles at most by some of complex multiple layers of proteins known as the "protein corona." When they come with the biological matrix, the physicochemical properties of it change dramatically, potentially triggering an immunological response and altering targeting, destiny, and toxicity. Analytical methods are necessary to understand how nanosurfaces interact and their interactions' behavior in biological systems. Yet due to their downsizing, the composition variability and the chemistry of particles' surface, and the exceedingly complex nature of biological matrices, nanoparticle characterization and identification are problematic. Proteins that are adsorbed on nanoparticle surfaces could be identified directly or indirectly. To investigate the adsorbed proteins on the surface of the nanoparticle, direct techniques such as spectrometric analysis (e.g., NMR, FT-IR, and MS), microscopy (e.g., AFM, TEM), gel permeation chromatography (GPC), and circular dichroism are used (Alshawwa et al., 2022). On the other hand, protein that is adsorbed is indirectly estimated by assessing changes in nanoparticle properties, such as aggregation, density, Förster resonance energy transfer (FRET), size, absorbance, surface charge and mass; these parameters are measured then connected with the adsorbed protein amount. Both methodologies allow for the identification of adsorbed proteins in situ, whereas other techniques may require removal of unbound proteins before the measurements.

Factors such as phagocytic cells, blood flow, excretion, and coronas can affect nanoparticle stability and delivery while in circulation. The specific effects of each of these environmental elements are reliant on the physiochemical properties of the nanoparticle platform, leading to broad design concepts targeted at manipulating these features to produce positive results. To avoid quick excretion based on surface features, several NP formulations use PEG as a stealth coating. NPs with a diameter smaller than 10 nm, for example, have generally been demonstrated to be promptly removed by the kidneys, but NPs larger than 200 nm risk activating the complement system. PEGylation increases circulation time by altering NP size and solubility while protecting the NP surface from enzymes and antibodies that may induce degradation, secretion, and clearance; however, this physical barrier does not completely prevent recognition by macrophages or other immune system cells.

Another prospect for stability concerns (1) van der Waals force (the force that is formed between nanoparticle interaction in solvents) and (2) growth of nanoparticle mechanism and nucleation mechanism (Sultana et al., 2020).

18.3.4 Some Mechanisms for Stabilization

18.3.4.1 Electrostatic or Charge Stabilization

In this technique, charged particles in polar liquids can induce stability with an electric double layer (EDL). Nanoparticles can be charged or electrostatically stabilized. Adsorption of ions on the surface of particles produces EDL, which counteracts the VDWL attraction forces between colloidal particles. EDL consists of two parallel-charged particle layers, and it is electrically neutral. The first layer is created or formed by adsorption of negative or positive charges on the surface of the particles, known as the stern layer. The second layer, known as the diffuse layer, is generated by the attraction of ions to adsorbed surface charges via the Coulomb interaction, resulting in the formation of a repulsive barrier. Ions in the diffuse layer are loosely linked to the charged surface and, hence, free to migrate. The maximum electric potential value of the Stern layer decreases with increasing distance from the surface and reaches 0 at the EDL's boundary. The mutual repulsion of the particles by the double layer encircling them provides overall stability to the system.

18.3.4.2 Steric Stabilization

This occurrence includes the surface of a particle being coated with polymer, causing a repulsive force, and isolates the particle from another particle. The particles are coated with polymer and surfactant and have water-loving chains that project lengthy loops and chains into the fluid. The surface colloidal particles are sterically stabilized by grafting or attaching a polymer to them. The phenomena are unaffected by high salt concentrations or zeta potential values close to zero. The water-loving chain section remains associated with water rather than interacting with any other surface.

18.3.4.3 Surfactants

Surfactants are naturally capable of controlling formation of the crystals of nanomaterials to obtain specific shapes. Because of the cooperative effect, surface-active biomolecules can be employed in the shape-controlled synthesis of nanomaterials. The quality of nanoparticle dispersion can be increased by covering it with anionic surfactants (SDS), non-ionic surfactants (Triton X 100, PEG-6000), and cationic surfactants (CTAB, PEI). The mechanisms of grafting the surfactant may involve noncovalent adsorption and covalent assembly. In covalent assembly, the surfactants are covalently attached to the end groups of nanoparticles. Shifting zeta potential and optimal stability results from mechanisms of non-covalent adsorption, such as ion pairing, ion exchange, and hydrogen bonding. Anand and Sibi, in 2015, studied the role of different surfactants on the stability of nano-zinc oxide dispersions. Surfactants that create monolayers at the interface are more capable of producing stabilized small-size nanoparticles than surfactants that form bilayers. Among the different surfactants (CTAB, PEG 6000, SDS, and Triton X 100), SDS combination affected stability by increasing it (in terms of settling behavior) with zeta potential value of 28.9 mV, which could be attributed to SDS molecules' higher level of adsorption on ZnO surface (Sultana et al., 2020). Some researchers found that SLS reduced particle aggregation significantly. When distributed in water, rod-shaped iron oxide nanoparticles produced by co-precipitation, such as a chemical method with a surfactant SDBS, showed a 14% increase in thermal conductivity. Via the observation approach, the stability of carbon nanotube–based nanofluids with surfactant is better than that of nanofluids without surfactant. At 0.01% carbon nanotube, the base fluid ethylene glycol/water mixture containing hexadecyltrimethylammonium bromide had the highest zeta potential value. When 0.5 wt% carbon nanotube–based nanofluids containing gum arabic were compared to various surfactants, they displayed the highest thermal conductivity (25.7%).

18.4 BRAIN REGULATORY FRAMEWORK

Nanomaterials utilization for diseases like cancer and central nervous system disorders (CNS) are becoming increasingly popular. Despite the approval of several nanomaterial-based pharmaceuticals, the regulations governing their use in humans are still insufficient. Nanostructures are often not intended to move through the blood–brain barrier. Characterizing nanomaterials biodistribution and the potential toxic effect that could arise from their interaction with the central nervous system is urgently needed because there is not a comprehensive set of standardized procedures to assess their *in vivo* fate. Although there are other ways to administer NMs, including intranasal delivery. They continue to be primarily delivered intravenously. As a result, to contact the brain, NMs must pass through the BBB. Between 98% and 100% of micro- and macromolecules are blocked by the BBB; this covers 99% of the capillary surface in the brain. Several factors that also affect how well NMs cross the BBB have an impact on NM biodistribution. Given the growing and diversifying range of applications of nanocarriers, it is still necessary to further explore the BBB access and *in vivo* fate of several NMs, despite their approval in a few instances. Efforts to offer formal standards of evaluation for this novel pharmacological category have been done throughout the past few years (Szabat-Iriaka & Le Borgne, 2021). Although these programs have built the foundational regulatory and methodological frameworks, they are still insufficient and focused greatly on lipophilic nanocarriers. Further characterization of both the pathways utilized to enter through the brain and the impact of the biological environment on nanomaterial behavior is necessary due to the significance of NC distribution in the organism and the potential of those systems to contact the blood–brain barrier.

18.5 CRITERIA FOR REGULATORY APPROVAL

According to Y. Hu, P. Gaillard, J. Rip, et al.'s analysis of the FDA's Guidelines for Industry documents, a nanocarrier must satisfy numerous requirements before being authorized for use in clinical trials.

18.5.1 PK/PD PK/PD

The pharmacokinetic for the released drug, both systemic PK and disposition in brain, pharmaco-dynamic exposure, and response correlations, and preferably PK of the nanocarrier itself, are criteria. To respond to some relevant concerns, it is critical to ascertain nanocarrier fate, drug linked with the nanocarrier, and the released drug in the body after administration. One of the most important questions is how to identify the transport of drugs formulated in nanocarriers across the BBB. It is possible to assess uptake of nanoformulated drugs in brain and their limitations using some *in vivo* and *in vitro* techniques.

18.5.1.1 The *In Vitro* Models

Since it is hard to control reliable experimental evidence in such studies, *in vivo* methods are typically less appropriate for deep mechanistic research. Accurate *in vitro* models could provide more understanding of the brain uptake system than *in vivo* models do. Nanocarriers like liposomes could nevertheless demonstrate a non-specific *in vitro* cell uptake, complicating in these *in vitro* experiments and possibly hiding the effect of certain drug release pathways that occur *in vivo*. The *in vitro* BBB models, which include co-culture models where endothelial cells are cultured with astrocytes and/or pericytes, and monoculture models of brain capillary endothelial cells, are frequently used to study brain uptake *in vitro*.

18.5.1.2 The *In Vivo* Models

18.5.1.2.1 Brain Distribution Method

Brain distribution is the most used technique for determining the concentrations of drug in plasma and brain when studying nanocarriers for brain drug delivery. Although this approach has some potential applications, it also has significant drawbacks. Unlike blood sampling, many brain samples cannot be taken from a single person, making it impossible to study the delivery process's timing considerations without using a lot of animals. The risk of blood contamination in brain is the second restriction, which may result in false-positive results, particularly if long-circulating nanomaterials are utilized. The drugs might still be present in the endothelial cells or adhere to them in released or nanoencapsulated forms, even if all the blood that was still left in the brain was removed after perfusion. Moreover, this approach does not distinguish between the unbound drug that is attached to plasma proteins or brain tissue and the drug that is linked with nanocarriers; it merely measures overall concentrations. Hence, it is challenging to determine how much of the drug is present in plasma and brain in its released or encapsulated form.

18.5.1.2.2 In Vivo Brain Microdialysis

Microdialysis is the best method for determining influx and efflux transport functions across the BBB under normal and diseased circumstances *in vivo*. A semi-permeable membrane on a microdialysis probe enables passive diffusion of tiny water-soluble solutes over it. When examining nanocarrier-mediated brain delivery, the advantage of microdialysis is that it only permits continuous measurement of the unbound drug concentrations in plasma and brain interstitial fluid (ISF) with time. This enables the separation of released pharmacologically relevant drug from the nanocarrier-associated drug.

The rate and extent of release and transport of the *in vivo* drug through the BBB can be objectively evaluated in conjunction with routine blood samples. The brain microdialysis method is useful and could be the superior tool for the quantitative determination of brain nanodelivery since it offers particular and in-depth knowledge that cannot be attained with other methods. Although microdialysis has been used in preclinical and clinical settings for many years, its application in drug development remains unclear methodologically and from regulatory perspective. The ability to correlate microdialysis data with clinical outcomes will ultimately determine whether microdialysis is accepted as a regulatory method.

18.5.1.2.3 Cerebral Open-Flow Microperfusion Method

Open-flow microperfusion has more recently been developed for cutaneous medication collection as well as continuous glucose and lactate monitoring. It is currently employed in brain ISF for continuous sampling. Cerebral open-flow microperfusion (cOFM), which works similarly to microdialysis, evaluates compound concentrations in the brain using a probe. Instead of a semi-permeable membrane for microdialysis, the cOFM probe features microscopic perforations, making it ideal for bigger and lipophilic nanocarriers. The cOFM sample that is unfiltered and undialyzed would contain both unbound drugs and nanocarrier-delivered drugs when crossing the BBB. To assess the drug concentrations in various entities, they need to be further separated using, for example, ultrafiltration. As it is impossible to directly detect total brain concentrations, cerebrospinal fluid (CSF) sampling is also employed to assess brain uptake, particularly in humans. It is essential to realize that the unbound drug concentration in the brain's ISF does not always correlate with the drug concentration in the CSF (978-3-030-88773-5.Pdf, 2022; Hu et al., 2022). When using long-circulating nanocarriers, contaminated blood with CSF samples becomes a particular issue, as the high concentration of drug in plasma results in high variability and a false-positive result regarding drug concentrations in CSF samples, even if they are carried out at specialized contract laboratories.

To cover all *in vivo* models that is currently available would be excessive. However, we would like to emphasize that in order to compare the efficacy of the examined nanotherapeutics, models should be verified, and the appropriate controls should be used.

18.5.2 Nanotoxicity

The field of nanotoxicology is expanding quickly, especially as more and more nanomaterials are being applied; thus, it is essential to create studies that are consistent across the board. Nanotoxicology is also required because, due to their distinct physicochemical features, nanoparticle interactions with live cells and tissues are frequently unanticipated. Additionally, because the size of synthetic nanoparticles is similar to those of cells, their impact may disrupt important biological functions. Regrettably, the physicochemical properties that give nanoparticles their distinctive qualities and explain their positive impacts in applications to medicine also cause their toxic effect. Via any exposure method, nanoparticles that have successfully crossed the BBB possess a propensity to assemble brain regions in which they are able to pass neural cells. Many processes, such as oxidative stress, that induce cell death, autophagy, immunological responses, and inflammation and impair the function of the BBB, can cause neurotoxicity to appear (Teleanu et al., 2018).

Yet there is a serious absence of knowledge and understanding of neuro-nanotoxicity, which makes risk assessments difficult (Muheem et al., 2020). Hence, a quick, accurate, and effective action plan is required to evaluate the nanocarrier toxic effect to neurons.

18.5.3 Therapeutic Index

The primary concern that must be taken into consideration while developing novel drug formulations is neurotoxicity. Three key factors must be taken into account when dealing with bioactive-loaded nanocarriers: bioactive agents, excipients, and release profile (Babazadeh et al., 2020). The biological activity of their metabolites and their biotransformation may be the main causes for concern in this section. The second stage requires that the excipients used to create the nano-drug delivery to the brain be kept in a minimum and at low levels to decrease neurocytotoxicity. The inclusion components made from food will offer a significant safety benefit to address these problems. The profile of bioactive drugs' release once they reach the intended location of action in the brain is another crucial factor. The method of nanoengineering enables the control of the rate at which encapsulants are released from nanocarriers; however, when it comes to brain delivery, the situation may differ due to distinct behaviors of neuronal networks in the brain. Further research is required to conduct clinical trials and provide appropriate safety rules, and the benefit–risk balance for nanotechnology-based systems promised for brain drug delivery and diagnostics should be carefully considered (Dwivedi et al., 2019).

18.5.4 Effect of Nanoparticle Size and Shape in the Central Nervous System

18.5.4.1 Size

The initial step for particles in circulation to reach the target location is extravasation, which could be influenced by nanocarrier size and shape, since smaller NCs pass capillary walls more readily than bigger ones. Nanocarrier distribution in organs depends on their size. If size is the primary factor, research have shown that nanoparticles (NPs) between 100 and 200 nm in size are preferable because of their longer circulation in blood, low absorption by phagocytosis, and capacity to pass the blood–brain barrier (BBB). Single-walled carbon nanotubes of 30 nm demonstrated more nuclear and cellular internalization than their 50 nm counterparts. No matter their surface charge or composition, most *in vitro* studies show that nanocarriers of a size range from 10 to 60 nm possess high cell uptake in those nonphagocytic ones.

18.5.4.2 Shape

NPs' shape has a big impact on how they are biodistributed, cleared, and internalized by cells. The circulation periods of NPs with worm shapes are significantly longer than those of NPs with other shapes, according to research. The report suggested that elongated NPs have a higher cell adhesion efficiency due to the high surface area, which promotes multivalent interaction with cell surface. This contrasts with spherical NCs, which possess a limited number of binding sites and a smaller contact surface because of their curved shape. *In vitro* and *in vivo* investigations have shown that rod-shaped NPs have a better efficacy and affinity for endothelial cells. Moreover, compared to spherical NPs, ellipsoidal NPs show reduced cellular uptake. It was discovered that the permeation of particles with 200 nm spheres is three times more than that of 100 nm spheres, and that the permeability of 500 nm NPs is 100 and ten times less than that of 200 and 100 nm spheres, respectively (Annu et al., 2022). Rods that were not targeted had less adherence to the artificial surface than spheres. The research also showed that rod-shaped NPs travel more readily across the brain endothelium than sphere-shaped NPs. When compared to their soft counterparts, stiff particles demonstrated more transport.

18.6 CHALLENGES AND LIMITATIONS OF NANOCARRIER CHARACTERIZATION

Many nanoparticles are designed recently as targeting drug delivery systems for therapeutic purposes. Significantly, their physicochemical characteristics, such as charge and hydrodynamic diameter, are influenced by physiological factors, like temperature, pH, and ionic strength; therefore, complications for preclinical characterization arose. There may be certain challenges in developing the basic principles and subsequent clinical application of nanocarriers. Surfactants are frequently used into nanocarrier formulations to encourage dispersion, which conflict with traditional characterization techniques. Cellular membranes can be crossed by cationic nanocarriers more efficiently than by neutral or anionic ones. This also applies to surfactants attached to nanocarriers. After administration, binding to plasma proteins may impact how nanocarriers are distributed and cleared, and the drug release profile in physiological fluids may entirely change. Moreover, changes in polydispersity brought on by contact with bodily fluids may significantly alter the toxicity and biocompatibility profiles. As will the route of administration have a significant impact on characterization techniques. Compared to sophisticated formulations for nanocarrier intended for oral, nasal, or topical administration, intravenous injections require less characterization procedures (Alshawwa et al., 2022). Hence, during *in vitro* characterization, it is crucial to consider any potential influencing of *in vivo* conditions. Control and analytical samples having observable characteristics are typically included in *in vitro* tests to ensure data reliability. There is a need for establishing *in vitro* and *in vivo* models that accurately depict the clinical case which propose standards to enhance the evaluation of nanomaterials' quality. The National Cancer Institute's Nanotechnology Characterization Laboratory (NCL), the Food and Drug Administration (FDA), and the National Institute of Standards and Technology (NIST)—all these regulatory organizations attempt to develop and validate a routinely revised and updated standardized characterization protocol for nanocarriers, including a broad range of nanotherapeutics.

18.7 CONCLUSION

Safety, stability, and regulation of nanoparticles are a major challenge during pharmaceutical nano-drug delivery development in the biomedical field. It is obvious that nanoparticles possess unique physicochemical properties, justifying their broad-spectrum use in applications for the central nervous system. In this chapter, we discussed the three main aspects: safety concerns in characterizations that should be considered while using the nanomaterials, including some studies for safety assessment proposing some safety improvement suggestions; stability issues, discussing physical and chemical stability and proposing some mechanisms for stabilization preferred; regulatory aspect specified in the nano-drug delivery regulation for central nervous system and trying to illustrate FDA criteria for essential clinical trials.

REFERENCES

Ahmad, A., Imran, M., & Sharma, N. (2022). Precision Nanotoxicology in Drug Development: Current Trends and Challenges in Safety and Toxicity Implications of Customized Multifunctional Nanocarriers for Drug-Delivery Applications. Pharmaceutics, 14(11), 2463

Alshawwa, S. Z., Kassem, A. A., Farid, R. M., Mostafa, S. K., & Labib, G. S. (2022). Nanocarrier drug delivery systems: Characterization, limitations, future perspectives and implementation of artificial intelligence. Pharmaceutics, 14(4). https://doi.org/10.3390/pharmaceutics14040883

Annu Sartaj, A., Qamar, Z., Md, S., Alhakamy, N. A., Baboota, S., & Ali, J. (2022). An insight to brain targeting utilizing polymeric nanoparticles: Effective treatment modalities for neurological disorders and brain tumor. Frontiers in Bioengineering and Biotechnology, 10(February), 1–18. https://doi.org/10.3389/fbioe.2022.788128

Ayub, A., & Wettig, S. (2022). An overview of nanotechnologies for drug delivery to the brain. Pharmaceutics, 14(2). https://doi.org/10.3390/pharmaceutics14020224

Babazadeh, A., Mohammadi Vahed, F., & Jafari, S. M. (2020). Nanocarrier-mediated brain delivery of bioactives for treatment/prevention of neurodegenerative diseases. Journal of Controlled Release, 321, 211–221. https://doi.org/10.1016/j.jconrel.2020.02.015

Dwivedi, N., Shah, J., Mishra, V., Tambuwala, M., & Kesharwani, P. (2019). Nanoneuromedicine for management of neurodegenerative disorder. Journal of Drug Delivery Science and Technology, 49(2018), 477–490. https://doi.org/10.1016/j.jddst.2018.12.021

Garanti, T., Alhnan, M. A., & Wan, K. W. (2021). The potential of nanotherapeutics to target brain tumors: Current challenges and future opportunities. Nanomedicine, 16(21), 1833–1837. https://doi.org/10.2217/nnm-2021-0134

Guillama Barroso, G., Narayan, M., Alvarado, M., Armendariz, I., Bernal, J., Carabaza, X., Chavez, S., Cruz, P., Escalante, V., Estorga, S., Fernandez, D., Lozano, C., Marrufo, M., Ahmad, N., Negrete, S., Olvera, K., Parada, X., Portillo, B., Ramirez, A., . . .& Viel, S. (2020). Nanocarriers as potential drug delivery candidates for overcoming the blood-brain barrier: Challenges and possibilities. ACS Omega, 5(22), 12583–12595. https://doi.org/10.1021/acsomega.0c01592

Hersh, A. M., Alomari, S., & Tyler, B. M. (2022). Crossing the blood-brain barrier: advances in nanoparticle technology for drug delivery in neuro-oncology. International journal of molecular sciences, 23(8), 4153.

Henna, T. K., Raphey, V. R., Sankar, R., Shirin, V. A., Gangadharappa, H. V., & Pramod, K. (2020). Carbon nanostructures: The drug and the delivery system for brain disorders. International Journal of Pharmaceutics, 587, 119701

Hu, Y., Gaillard, P. J., Rip, J., & Hammarlund-Udenaes, M. (2022). Blood-to-brain drug delivery using nanocarriers. AAPS Advances in the Pharmaceutical Sciences Series, 33, 501–526. https://doi.org/10.1007/978-3-030-88773-5_16

Mehrabian, A., Mashreghi, M., Dadpour, S., Badiee, A., Arabi, L., Hoda Alavizadeh, S., Alia Moosavian, S., & Reza Jaafari, M. (2022). Nanocarriers call the last shot in the treatment of brain cancers. Technology in Cancer Research and Treatment, 21, 1–21. https://doi.org/10.1177/15330338221080974

Mitchell, M. J., Billingsley, M. M., Haley, R. M., Wechsler, M. E., Peppas, N. A., & Langer, R. (2021). Engineering precision nanoparticles for drug delivery. Nature Reviews Drug Discovery, 20(2), 101–124. https://doi.org/10.1038/s41573-020-0090-8

Moos, T., Gudbergsson, J. M., & Johnsen, K. B. (2022). Transport of Transferrin Receptor-Targeted Antibodies Through the Blood-Brain Barrier for Drug Delivery to the Brain. In AAPS Advances in the Pharmaceutical Sciences Series (Vol. 33, pp. 527–549). Springer. https://doi.org/10.1007/978-3-030-88773-5_17

Muheem, A., Jahangir, M. A., Jaiswal, C. P., Jafar, M., Ahmad, M. Z., Ahmad, J., & Warsi, M. H. (2020). Recent patents, regulatory issues, and toxicity of nanoparticles in neuronal disorders. Current Drug Metabolism, 22(4), 263–279. https://doi.org/10.2174/1389200221999201210213036

Nanomedicine in Brain Diseases. (2019). Nanomedicine in brain diseases. Springer. https://doi.org/10.1007/978-981-13-8731-9

Patel, M., Patel, R. and Rai, M. (2021). Nanomedicines. In Nanotechnology in Medicine (eds M. Rai, M. Patel and R. Patel). https://doi.org/10.1002/9781119769897.ch1

Scimeca, J. C., & Verron, E. (2022). Nano-engineered biomaterials: Safety matters and toxicity evaluation. Materials Today Advances, 15, 100260.

Sultana, S., Alzahrani, N., Alzahrani, R., Alshamrani, W., Aloufi, W., Ali, A., Najib, S., & Siddiqui, N. A. (2020). Stability issues and approaches to stabilised nanoparticles based drug delivery system. Journal of Drug Targeting, 28(5), 468–486. https://doi.org/10.1080/1061186X.2020.1722137

Szabat-Iriaka, B., & Le Borgne, M. (2021). Brain safety concerns of nanomedicines: The need for a specific regulatory framework. Drug Discovery Today, 26(11), 2502–2507. https://doi.org/10.1016/j.drudis.2021.06.011

Teleanu, D. M., Chircov, C., Grumezescu, A. M., Volceanov, A., & Teleanu, R. I. (2018). Impact of nanoparticles on brain health: An up to date overview. Journal of Clinical Medicine, 7(12). https://doi.org/10.3390/jcm7120490

Tereshkina, Y. A., Torkhovskaya, T. I., Tikhonova, E. G., Kostryukova, L. V., Sanzhakov, M. A., Korotkevich, E. I., . . . & Kolesanova, E. F. (2022). Nanoliposomes as drug delivery systems: Safety concerns. Journal of Drug Targeting, 30(3), 313-325.

Wu, L., Zhang, J., & Watanabe, W. (2011). Physical and chemical stability of drug nanoparticles. In Advanced Drug Delivery Reviews (Vol. 63, Issue 6, pp. 456–469). https://doi.org/10.1016/j.addr.2011.02.001

Zhang, L., & Zhao, D. (2014). Applications of nanoparticles for brain cancer imaging and therapy. Journal of biomedical nanotechnology, 10(9), 1713-1731.

Zhang, Z., Conniot, J., Amorim, J., Jin, Y., Prasad, R., Yan, X., . . . & Conde, J. (2022). Nucleic acid-based therapy for brain cancer: Challenges and strategies. Journal of Controlled Release, 350, 80–92.

19 Advancements of Nanocarriers for Brain Targeting and Limitations in Clinical Translation

Nermeen Z. Abuelezz, Fayza Eid Nasr, María-Salud García-Ayllón, and Javier Sáez-Valero*

19.1 INTRODUCTION

The presence of a blood–brain barrier (BBB) as a biological interface that separates the brain from the bloodstream was first noted when Paul Ehrlich (1885, 1906) observed that a peripherally used dye infused from blood did not stain the brain tissue. However, the presence of a structurally different barrier around the brain was not confirmed until 1903 by Goldman, Paul Ehrlich's student (Kadry et al., 2020).

The BBB comprises the blood capillaries that supplement profound regions of the brain and regulate the transport of nutrients and other compounds needed for proper neural functioning, in and outside the CNS. Simultaneously, the BBB protects the brain from foreign neuro-hazardous plasma components, such as plasminogen, thrombin, blood cells, and microbial agents, by preventing their entrance to the brain tissue. Nonetheless, due to the complexity of its structure, the BBB encounters a major challenge against the delivery of therapeutic agents to the brain.

19.1.1 Structure of the Blood–Brain Barrier (BBB)

The blood–brain barrier is primarily composed of endothelial cells (ECs) that line the blood capillaries walls. The ECs are surrounded by the basement membrane, together with pericytes and astrocyte cell projections (astrocyte feet), to provide extra support to the ECs. Two types of basement membranes are present at the BBB: (1) the endothelial (vascular) basement membrane, which is contributed by pericytes and ECs, and (2) the glial basement membrane (parenchymal basement membrane), which is made from molecules secreted by astrocytes (Hall et al., 2014; Peppiatt et al., 2006). Both basement membranes accommodate different fluid secretions and cells. ECs exhibit a unique phenotype when compared to cells from other vascular regions; they are flatter, with fewer luminal caveolae and greater numbers of mitochondria. Furthermore, ECs are deeply enriched with a junction complex that is formed of tight junctions (TJs) and adherent junction (AJs). The TJs are fusion sites of endothelial cells which are closer up to 100 times than those in peripheral capillaries. They comprise three integral membrane proteins, claudin, occludin, and junction adhesion molecules (JAMS), besides various cytoplasmic accessory proteins like Zonula occludens-1, 2, and 3 (ZO-1, ZO-2, ZO-3) and cingulin, whereas the AJs include cadherin–catenin complex in addition to associated proteins. These cytoplasmic proteins link membrane proteins to the cytoskeleton protein actin, which is crucial for maintaining structural and functional integrity of the endothelium (Achar et al., 2021).

19.1.2 Transport of Molecules across the BBB

Although the BBB acts as a barrier of the transport between the bloodstream and the brain tissue, several pathways work to pass peptides and other molecules to keep brain homeostasis. Indeed, passage of molecules across the BBB primarily depends on their structure, chemical composition, and the properties of their surfaces. Generally, an intact BBB allows the entrance of only small-molecular-weight (<400–500 Da) compounds and small lipophilic molecules to the brain. These small lipophilic molecules are transported through transcellular diffusion.

The maintenance of the brain homeostasis depends on efflux transporters to expel foreign molecules outside the brain and nutrient carrier transporters to regulate the entrance of principal nutrients to different brain cells. ATP-binding cassette (ABC) efflux transporters include Mdr-1 (P-glycoprotein), breast cancer resistance protein (BCRP), and multidrug resistance–associated proteins (MRPs) (Daneman & Prat, 2015; Zlokovic, 2008), whereas nutrient transporters include slc2a1GLUT1 for glucose transport, slc7a1 for cationic amino acids, slc7a5 for neutral amino acids transport and slc16a1, as well as L-DOPA for providing lactate and pyruvate supply (Alajangi et al., 2022).

Additionally, receptor-mediated transcytosis (RMT) is another principal method to transfer macromolecules across

* Corresponding author: nermeenzak01@gmail.com/nhassan@umh.es

DOI: 10.1201/9781003383376-19

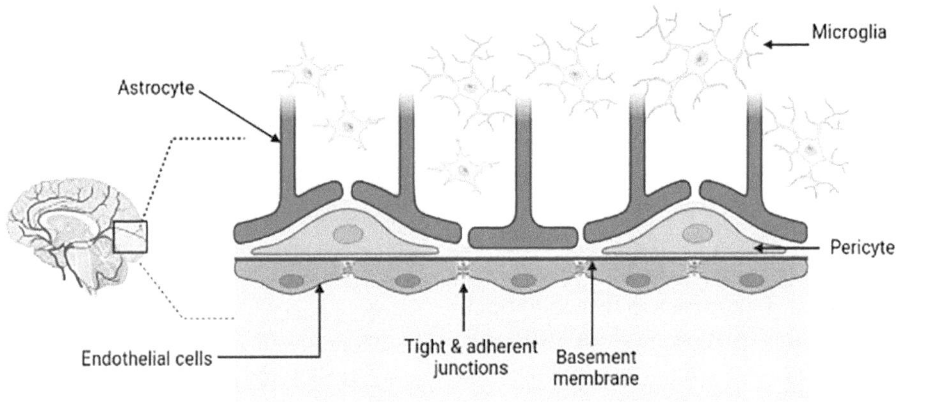

FIGURE 19.1 Structure of the blood–brain barrier. (Created in Biorender.com.)

the BBB. The most well-defined receptors in RMT pathway are insulin receptor (IR), transferrin receptor (TR), and low-density lipoprotein receptor–related protein 1 (LRP1) (Zhang et al., 2020). As a protective strategy, low levels of leukocyte adhesion molecules (LAMs) are also expressed by ECs of BBB to restrict immune cells entry into the CNS (Wu et al., 2021), whereas increasing reports show that B cells upregulate activated leukocyte cell adhesion molecule (ALCAM) expression during inflammation and enhance CNS recruitment of monocytes and CD4+ T cells (Huang et al., 2020).

19.2 ADVANCED STRATEGIES TO CROSS THE BLOOD–BRAIN BARRIER

The BBB provides an obstacle to the delivery of brain therapeutics. In general, crossing the BBB to deliver different therapeutic molecules to the brain has always been a crucial challenge due to several key factors, which are listed as follows:

Selective Permeability of the BBB. Attributed to the complex structure of the specialized endothelial cells of BBB and the included tight junctions, which restrict the movement of large molecules and only allows entrance of small lipophilic molecules.

Physical Barrier. Encountered by the basement membrane and astrocytic end feet, which limit the diffusion of substances into the brain.

Active Efflux Transporters. Trials to maintain effective drug concentrations in the brain are usually hindered by P-glycoprotein transporters that actively pump out foreign molecules back into the bloodstream via the ABC.

Limited Transport Mechanisms. Where transport mechanisms across the BBB are basically designed to maintain homeostasis rather than facilitate diverse molecules delivery.

Variability. Updated studies now confirm that BBB permeability varies, in addition to the physiological individual variations, according to diversities across age and CNS diseases stages (Knox et al., 2022). Such variability adds on more complexity against targeted drug delivery to the brain and urges more research on personalized delivery approaches.

Throughout the last decade, preclinical studies have explored different techniques to tackle the challenging hindrance of BBB passage for enhanced drug delivery to the brain. While some techniques achieved positive results, others did not accomplish the required outcomes. The following section highlights the most updated used techniques, with special focus on strategies that are also currently being investigated in clinical trials.

19.2.1 Hyperosmolar Infusions

Intracarotid arterial hyperosmolar mannitol (ICAHM) infusion is one of the earliest methods used to deliver therapeutic agents to the CNS for treatment of glioblastoma and lymphoma. Osmotically, ICAHM modulates the integrity of BBB endothelial cells and tight junctions, allowing higher accumulated concentrations of drugs or immunomodulatory agents inside the brain, while diminishing systemic toxic side effects. ICAHM has been highlighted as a safe technique; however, emerging proteomic analysis and deeper explorative studies reveal that ICAHM infusion can trigger microglial inflammatory responses and molecular alterations that might develop unwanted BBB damage, especially with prolonged high-dose administration (Achar et al., 2021; Burks et al., 2021).

19.2.2 Radiation Therapy and Focused Ultrasound

Theoretically, magnetic resonance (MR)–guided low-intensity focused ultrasound can temporarily open BBB and its inclusive tight junctions to facilitate targeted delivery of brain therapeutics. Several preclinical studies explored this hypothesis in treating CNS disorders like epilepsy and depression. On the clinical level, this technique was recently explored against Alzheimer's disease (AD). AD patients were subjected to magnetic resonance–guided (220 KHz) focused ultrasound treatment of the hippocampus and microbubble injection. Treatment was then followed by magnetic resonance imaging (MRI) to evaluate BBB opening progression. The study showed that ultrasound usage caused a short-term opening of the BBB through acoustic cavitation, approaching 14% to 71% (average of 29%) of the overall hippocampus volume, with no observed post-treatment side effects (Rezai et al., 2020).

While this non-invasive technique showed an unprecedented improvement in approaching complex and deep brain structures as the hippocampus, the wide window range of BBB opening percentage, the brief opening interval, and the lack of long follow-up information regarding repeated use are limitations that yet need to be investigated.

19.2.3 Chemical/Biological Manipulation of the BBB Nuclear Receptors

Various BBB nuclear receptors are upregulated in CNS diseases, causing disruptions in drug metabolizing enzymes, as well as diminished drug penetration across the BBB (Zhang et al., 2022). Emerging preclinical trials are currently exploring the influence of manipulating these receptors to increase drug bioavailability. For example, silencing the glucocorticoid receptor (GR) in human epileptic brain microvascular endothelial cells led to increased penetration of phenytoin across a dynamic *in vitro* BBB system (Ghosh et al., 2018). Interestingly, GR dysregulation has also been spotted in AD; thus, ongoing clinical trials sponsored by Bristol-Myers Squibb (NCT00860275) and Pfizer (NCT00931073) are now exploring the feasibility of using mifepristone and ketoconazole as potential supplemental therapy for AD.

Yet again, although manipulating nuclear receptors in the BBB can be promising, these receptors are ubiquitously expressed in other body tissues, and therefore, exploring novel means for selective targeting of BBB nuclear receptors is required to avoid adverse effects.

19.2.4 Drugs Inhibiting ABC Transporters and Drug-Metabolizing Enzymes

The third generation of inhibitors for the efflux membrane transporter P-glycoprotein (P-gp) are a group of drugs that selectively inhibit P-gp transporters without impacting other ABC family transporters. Recent preclinical trials have shown that intravenous administration of tariquidar and elacridar (P-gp inhibitors) succeeded in increasing BBB permeability to verapamil and topotecan in mice, respectively (Bauer et al., 2012). Clinically, effects achieved in preclinical trials were not equally successful in human trials (Zhang et al., 2022). Conversely, manipulating drug-metabolizing enzymes showed more promising clinical success, where concomitant administration of DOPA with decarboxylase enzyme inhibitor led to an improvement of clinical response in Parkinson's disease (PD) patients, with a reduction in the used therapeutic dose and consequent adverse effects (Johansson, 1990).

More recently, Kort et al. (2017) found in mice that the oral availability and brain drug accumulation of ponatinib and its active N-desmethyl metabolite (DMP) were increased by inhibiting the drug metabolizing enzyme CYP3A (Kort et al., 2017). However, to date, there remains a lot of unelucidated information about metabolizing enzymes in the BBB, so this field is still widely available for further investigation.

19.2.5 Nanocarriers

While the aforementioned strategies enhance brain-targeted drug delivery, most of the currently used bulk drugs cannot cross the BBB on their own. Nanotechnology-based drug transport systems present a distinguished platform to deliver drugs to the brain. Throughout the last decade, huge progress has been achieved regarding both formulations and selective targeting strategies of nanoparticles (NPs) to address different disorders. For CNS disorders, the distinguished minimal size, biocompatibility, prolonged retention time, and lesser toxicities of emerging nanomaterials offer a great opportunity for crossing the BBB and integrating innovative brain-targeted drug delivery.

The following section presents a comprehensive update of the nanocarriers intended for targeted drug delivery to the brain, with analysis of the adopted strategies for passing to the brain, ongoing clinical trials, and administration routes. We will also discuss the current limitations and challenges that urge future addressing to promote further success for this promising drug delivery intervention.

19.3 IMPLEMENTED STRATEGIES FOR TARGETED DELIVERY OF NANOPARTICLES TO THE BRAIN

The transport of drugs into the brain based on nanotechnology can employ specific and non-specific strategies to target brain cells. To serve as homing tools at the desired target site, inactive therapeutic molecules like antibodies, antibody fragments, and peptides bind to the bioactive substances and transport systems. One example of such a system is the use of polymeric NPs loaded with drugs that are specifically targeted to mitochondria, to control disorders such as cancer, AD, and obesity (Marrache & Dhar, 2012). To deliver therapeutic agents to the brain, nanocarriers mostly adopt one of the following transport systems:

1. *Adsorption-Mediated Transcytosis (AMT).* This approach primarily depends on the electrostatic interaction between the negatively charged BBB endothelial cells and the positively charged surface of the nanocarriers, or cationized proteins surrounding nanomolecules. Although this strategy achieved some success on the preclinical level, it does not ensure cell-specific targeting, since positively charged molecules can rapidly and indiscriminately be adsorbed by different negatively charged cell membranes. Moreover, the excessive introduction of positively charged molecules can manipulate the membrane potential of the ECs and, therefore, can cause adverse alterations in the regulatory signaling pathways (Ashique et al., 2023).
2. *Manipulation of Receptor-Mediated Transcytosis (RMT).* Recent nanocarrier formulations use this emerging strategy. In this mechanism, receptors on the luminal side of the BBB endothelial cells recognize the drug as a ligand. The drug-bound receptor conjugate enters the cell via endocytosis, where it moves to the abluminal side and fuses with the plasma membrane, releasing the drug to the receptor conjugate (Teleanu et al., 2022).

To achieve this, the RMT approach utilizes well-designed molecules to form conjugates with biological or chemical ligands and drugs through linkers or NPs. Used ligands are usually cell-penetrating peptides (CPPs) or monoclonal antibodies (mAbs) that facilitate binding to the BBB receptors present in ECs. Then, binding is followed by merging of the conjugate to plasma membrane and release of the drugs loaded with the transcytosed conjugates to exert their therapeutic effects in the brain. Transferrin receptor (TfR), low-density lipoprotein receptor (LDLR), and insulin receptor (InsR) are the main targets for the ongoing preclinical studies. Interestingly, updated results show promising outcomes that encourage introducing this strategy to further clinical studies (Tashima, 2020).

RMT constitutes a promising delivery strategy, as the nanocarriers utilize an innate transport system without involved damage to the BBB. Consequently, they avoid the chemical complexity of ECs through bypassing AMT, which can be easily compromised by fusion of small entities to the nanocarriers. Additionally, the tiny inter-capillary distances of the brain permit the perfusion of each neuron by its specific surrounding capillaries, facilitating selective drug delivery (Sonoda et al., 2022). Moreover, RMT enables high specificity in trafficking the targeted drug to which specific antibodies are fused.

Recently, a mouse anti-transferrin receptor nanobody that can cross the BBB via RMT was used to deliver a bioactive peptide to the mouse brain (Wouters et al., 2020). The nanobodies were administered peripherally via intraperitoneal, intravenous, and subcutaneous injections.

Clinically, trials using RMT drug delivery via transferrin receptor (Okuyama et al., 2021) and insulin receptor (Giugliani et al., 2018) are currently ongoing against neurodegeneration in mucopolysaccharidosis type I disease.

19.4 TYPES OF NANOCARRIERS FOR TARGETED BRAIN DRUG DELIVERY

19.4.1 Polymers

Polymers are increasingly utilized as vehicles for brain drug delivery due to their favorable biocompatible nature, which makes them relatively much safer than inorganic/metallic NPs, and their efficient encapsulation potential. Moreover, they have a wide spectrum for structuring and make physicochemical modifications that enable for a diversity of manufacturing characteristics according to the targeted disorder (La Barbera et al., 2022). Polymer nanocarriers can originate from natural sources, such as chitosan, cellulose, and starch, or they can be synthetic, like poly(lactic-co-glycolic acid) (PLGA), poly(lactic acid), and poly(glycolic acid). A classic structure of polymer nanocarrier is a hydrophobic core which binds to the drug chemically, surrounded by a hydrophilic nanoshell. The versatility of the external nanoshell facilitates further modifications to enable smart targeted delivery to the brain and minimizes off-target side effects. Hydrophilic drugs can be released directly from the carrier surface, whereas hydrophobic drugs are solubilized in the polymer core and released to the site through diffusion from micropores or via carrier degradation.

Many factors should be considered when formulating polymer NPs for brain drug delivery. For instance, the size of the nanoformula affects both the release and accumulation of the drug in the required region. Moreover, the stiffness of the final formula can affect BBB crossing efficiency and drug accumulation (Zheng et al., 2021). Despite the advantages of polymers that enhance drug bioavailability with reduced toxicity, certain challenges urge further formulating improvements, such as carriers that can be degraded either too rapidly or too slowly and limited modification of binding sites (Wu et al., 2023).

One of the earlier examples that demonstrated the successful usage of polymer NPs for drug delivery was the inclusion of anti-Alzheimer's drug tacrine hydrochloride in poly(n-butyl cyanoacrylate) NPs coated with 1% polysorbate 80 (Tween 80). The used nanopolymer not only improved half-life and pharmacokinetic properties of the drug but also decreased its serious side effects (Wilson et al., 2008). More recently, Chen et al. (2015) encapsulated antiepileptic drug lacosamide in PLGA shell polymer and demonstrated its promising potential as injectable microspheres or local delivery implants for epilepsy control (Chen et al., 2015).

In the last few years, polymeric NPs have been developed as significant drug delivery platform through the intranasal route, which bypasses BBB. This is attributed to the excellent capability of polymeric nanoparticles to modify their composition to exhibit mucoadhesiveness, reduce mucociliary movement, and consequently extend their residence time in the nasal mucosa and delivery to the brain, in comparison to other routes (Awad et al., 2023). In a recent study, chitosan–lecithin NPs loaded with phenytoin (PHT) achieved remarkably high encapsulation efficiency and successfully reached the brain through the intranasal route, achieving sustained release as well (Yousfan et al., 2020). In the same context, encapsulation of clonazepam in mixed polymeric micelles of Pluronic® P123 and Pluronic® L121 exhibited a notable increase in drug concentration in the brain and a more prolonged protection against epileptic seizures compared to the free intranasally administered solution (Nour et al., 2016). Furthermore, another study showed in rats that the non-ergoline agonist of dopamine receptors, rotigotine loaded in polymeric micelles of 88 nm size, enhanced (CNS) its bioavailability as a Parkinson's drug when administered intranasally compared to intravenous injection (Wang et al., 2020).

These promising preclinical results persuade interest in following with clinical trials based on intranasal polymeric NPs against different neurological and neurodegenerative disorders. However, to date, actual clinical trials using nanoformulas for brain drug delivery, and especially through the intranasal route, are still limited. Some promising trials are ongoing, while others failed or were terminated due to adverse effects or unfeasibility tackling the challenging neuro disorders (Huang et al., 2020). Table 19.1 highlights an updated summary of nanoparticles with different administration route which are employed in ongoing CNS disorders clinical trials that are not terminated or declared a failure.

19.4.2 Liposomes

Liposomes are small vesicles composed of a phospholipid bilayer that encloses an aqueous core. They mimic cell plasma membranes in having an outer shell of phospholipids and cholesterol which facilitates their merge with cell membranes and release of either hydrophilic or hydrophobic drugs into cells (Wu et al., 2023). In this context, liposomes exhibit remarkable biocompatibility and safety characteristics as nanocarrier for brain-targeted drug delivery. Recently, PEGylated liposomes were conjugated with HAIPRH (T7), a peptide that targets transferrin receptors and ZL006, an uncoupling agent used to neutralize ischemia-induced effects. After intravenous administration, the study showed that coumarin-6-labeled T7-PEGylated liposomes were widely distributed throughout brain regions. Furthermore, the penetration and concentration of T7-PEGylated liposomes was greater compared to PEGylated liposomes without T7. Such findings spot the benefit of utilizing transferrin receptor-RMT through structural and chemical modifications by versatile liposomes (Wang et al., 2015).

Interestingly, Al-Ahmady et al. (2019) demonstrated the promising potential of selective liposomal transport to enhance anti-ischemic drug delivery and localization in

TABLE 19.1
Updated summary of ongoing clinical trials using drugs at nanoscale/incorporated in nanocarriers for targeted brain drug delivery

Study	Drug/Formula	Disease	Route of administration	Country	Reference
APH-1105 in patients with mild to moderate AD	APH-1105 (alpha-secretase modulator) in nanopolymer carrier	Dementia & AD	Intranasal	UK	(Awad et al., 2023)
Study of the safety, tolerability, and pharmacokinetics of PRV-002 in healthy volunteers	PRV-002 (synthetic neuro steroid) nasal spray	Traumatic brain injury (TBI)	Intranasal	Australia	(Awad et al., 2023)
NCT03790709	ANAVEX2–73 anti-tau anti-amyloid	AD	Oral	Australia, Germany, Canada	(Huang et al., 2020)
NCT01760005	Gantenerumab and solanezumab monoclonal antibodies reduce amyloid production	AD	Subcutaneous/ intravenous infusion	United States, European countries	(Huang et al., 2020)
NCT02956486	Elenbecestat BACE inhibitor	Early AD	Oral	United States	Clinicaltrial.gov
NCT02293915	GV-971 (sodium oligomannurarate) Aβ aggregation inhibitor	AD	Oral	China	Clinicaltrial.gov
NCT03491150 2017-002702-12 BN40031	Crenezumab amyloid monoclonal antibody	AD	Intravenous	Multicentral	Clinicaltrial.gov
00998764	Bapineuzumab targeting N-terminal amyloid	AD	Intravenous	United States	(Ivanoiu et al., 2016)

mice brain, with powerful blocking of delayed inflammatory responses and shifting the polarization of microglia/macrophages toward brain repair status (Al-Ahmady et al., 2019).

Translation of this potential to clinical is achieved by few FDA-approved and ongoing clinical studies that use liposomal drug delivery in the therapy of neurological disorders (Table 19.2). However, the high preparation cost poses a significant drawback toward wider clinical applications. Another important point to note is that the phospholipid structure of liposomes is relatively unstable and can therefore be exposed to dynamic dissociations in biological environments, causing unintended distribution of drugs, which can affect efficacy and result in off-target side effects as well.

The basic structure of liposomes is under progressive development, generating updated derivatives as nanoemulsions, dispersed systems of water, oil, and a co-surfactant that maintains successful dispersion. Nanoemulsions can increase drugs bioavailability and solubility; however, their relatively low stability is a remaining challenge. Nanostructured lipid carriers (NLPs) are the most modernized form of lipid nanoparticles. In NLPs, liquid and solid lipids are mixed, disrupting the lattice structure of solid lipids, and therefore increasing the drug loading capacity. Moreover, it reduces drug precipitation, prevents particle aggregation, and provides more stability with lesser toxicity (Wu et al., 2023). Such advantages can open new avenues for NLPs to targeted brain drug delivery.

19.4.3 Inorganic Nanoparticles

Inorganic NPs comprise a variety of materials that are generally inert, are non-immunogenic, and can be easily modified. Inorganic NPs are powerful tools in size-based electrochemistry and therefore have been utilized successfully for theranostic purposes. Gold and magnetic nanoparticles are the most explored for targeted drug delivery and diagnostic purposes in a variety of CNS disorders (Ayub & Wettig, 2022). More recently, quantum dots and carbon tubes emerged due to their excellent conductivity and efficient ability to identify single or small receptor binding sites for probed ligands. In the last few years, preclinical trials demonstrated the remarkable potential of carbon dots and zero-dimensional nanostructures named carbon onions to cross the BBB and deliver entrapped therapeutic agents in AD models (Ashique et al., 2023).

At a clinical level, CNM-Au8 is an experimental therapy that is undergoing multiple clinical trials (NCT03536559, NCT04098406, NCT04414345, and NCT04081714) as a potential treatment for amyotrophic lateral sclerosis (ALS) and multiple sclerosis. It is made of gold nanocrystals designed to support the energetic needs of nerve cells, remyelination, and enhanced removal of toxic metabolites to reduce neuronal damage (Vucic et al., 2023). Despite the benchmark success of inorganic nanoparticles in theranostic applications, the biosafety of these carriers is still a concern that slows its wider clinical use. A list of inorganic NPs approved by FDA for delivery of drugs in CNS disorders is illustrated in Table 19.3.

19.4.4 Biocarriers, Exosomes, and Biomimetic Nanocarriers

Despite the encouraging potential of synthesized nanomaterials, they are still limited by their abiotic nature, which exposes them to mononuclear phagocytosis, short duration of blood circulation, and the consequent reduction in delivery efficacy (Hernando et al., 2023). The use of bionanocarriers is a smart emerging approach to tackle these challenges. While exosomes are paving the way for a new

TABLE 19.2
FDA-approved and ongoing clinical trials of liposomal-incorporated drugs for brain drug delivery

Drug/Clinical trial details	Disease	Year of approval	Route of administration	Reference
Lomustine	Brain tumor	2010	Oral	(Gilert & Machluf, 2010)
Riluzole	Amyotrophic lateral sclerosis	2019	Intravenous	(Wang et al., 2020)
NCT00734682 CPT-11 (NL CPT-11) camptothecin derivative	High-grade glioma	Ongoing phase I	Intravenous	Clinicaltrials.gov

TABLE 19.3
FDA-approved inorganic nanoparticles for drug brain delivery

Nanoparticle type	Drug name	Disease	Approval year	Reference
Gold nanocrystals	Molecular surgery	AD	2006	(Kogan et al., 2006)
Magnetic iron oxide	Nanotherm®(MagForce)	Glioblastoma	2010	(Nguyen et al., 2021)
Cerium oxide	Photothermal therapy	Stroke	2021	(Tian et al., 2021)
Gold nanocrystals	CNM-Au8	Multiple sclerosis	2021	(Jagaran & Singh, 2021)

era of precision medicine and diagnosis, their use as a therapeutic, biocompatible carrier for targeted drug brain delivery is achieving increasing attention.

Exosomes are a great strategy for targeting different CNS disorders as they are biologically natural, are stable, and their small size enables penetration of the BBB. In PD, both *in vitro* and *in vivo* studies demonstrated the efficacy of dopamine-loaded exosomes to cross the BBB, with remarkable decrease in systemic toxicity and 15-fold more drug accumulation in the neuronal cells (Qu et al., 2018). Meanwhile, exosomes loaded with natural agents such as curcumin and quercetin showed a significant increase in targeted brain drug accumulation. Such effect resulted in efficient suppression of tau phosphorylation and distinguished improvement of cognitive functions in mice models of AD (Qi et al., 2020; Wang et al., 2019). Drug encapsulation through exosome was also investigated in mice models of traumatic brain injury, with treated mice manifesting cognitive improvements upon intravenous infusion of MSC-derived CD63+CD81+ exosomes (Kim et al., 2016). Similar promising preclinical results have been reported for various neurological disorders (Harrell et al., 2021).

Anyhow, clinical trials using exosomes for targeting CNS disorders are still in their early stages. An ongoing phase II clinical trial (NCT03384433) with stem cell–derived exosomes aims to promote neurogenesis in acute ischemic stroke.

Despite the reported success in preclinical studies, wider clinical applications of natural exosomes are hindered by low-scale production and isolation yields, decreased encapsulation efficiency, and basically absence of standardization methods for isolation, enrichment, and purification processing.

Focusing on coating nanomaterials with a variety of biological membranes is an innovative strategy to address the aforementioned drawbacks of both natural exosomes and synthetic nanocompounds. Bio-coating can provide the combined advantages of greater encapsulation and loading capacity, prolonged systemic circulation, better biocompatibility, and immune passage, together with better controlled release of the therapeutic load. Gold nanoparticles functionalized with neuron-directed exosomes conjugated with neuron-focused ligands such as the rabies virus glycoprotein (RVG) have recently demonstrated higher targeting and accumulation in the brain cells (Khongkow et al., 2019). The establishment of successful processing of bio-coated nanocarriers can launch new avenues for broader applications in both diagnosis and treatment of CNS disorders. However, the interesting results in the preclinical studies of these biocarriers/bio-coated nanocarriers are still far from direct clinical applications due to lack of general standardization, isolation, and purification methods for biological membranes and exosomes (Lopes et al., 2023). Moreover, the complex heterogeneity of biological microvesicles and the absence of scientific agreement on optimized manufacturing procedure for NPS coating still need a well-defined establishment.

19.4.5 Nucleic Acid–Loaded Nanocarriers for Gene Targeting

RNA-based therapies are a rapidly developing field that has a strong potential for modulating different diseases. They comprise antisense oligonucleotides (ASOs), small interfering RNAs (siRNAs), and micro-RNAs mimics and inhibitors. To

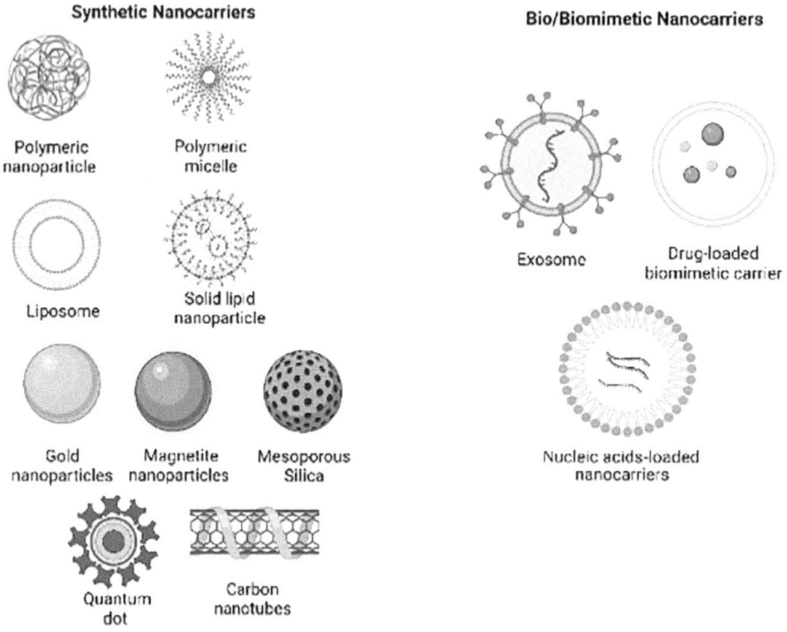

FIGURE 19.2 Illustration of the different types of nanocarriers used in diagnostic and targeted therapeutic delivery to the brain. (Created by Biorender.com.)

date, siRNAs are the most established RNA platforms, and multiple candidates are either FDA-approved or are being tested in ongoing clinical trials (Holm et al., 2022). Regarding CNS disorders, promising preclinical studies are taking place to incorporate the RNA cargo in nanocarrier systems.

Spherical nucleic acids (SNAs) are nanocarriers created by attaching siRNA or ASOs through chemical linkers to a hydrophobic core made of gold or liposomes; such arrangement facilitates RNA exposure and resistance to nucleases cleavage. Most importantly, it enables efficient BBB crossing and uptake by targeted cells through scavenger receptors (Narasipura et al., 2023). SNAs have been successfully tested in glioblastoma treatment; however, they remain to be tested in other CNS disorders (Sita et al., 2017).

Other strategies have been recently investigated to deliver siRNAs to the brain in preclinical AD models. A glycosylated nanodelivery system that manipulate CNS glucose transporters was successfully used to deliver siRNA against the enzyme with β-secretase activity (BACE1), decreasing generation β-amyloid peptides and restoring cognitive abilities in AD mice with minimal side effects (Zhou et al., 2020). Moreover, decorating siRNA-loaded dendrimers nanoparticles with novel cell-penetrating peptides has recently showed in AD mice models efficient targeting of BBB and neurons, resulting in a reduction of the β-amyloid plaques and neurotoxicity with a cognitive improvement (Cai et al., 2020).

Clinically, patisiran is currently an FDA-approved siRNA therapy for the treatment of polyneuropathy associated with hereditary transthyretin-mediated amyloidosis (hATTR amyloidosis). Patisiran comprises DLin-MC3-DMA, as an ionizable lipid; distearoylphosphatidylcholine (DSPC), as a phospholipid; DMG-PEG-2000, as polyethylene glycol coating; and cholesterol. DLin-MC3-DMA incorporates a lipid tail derived from linoleic acid and is connected to a tertiary amine, forming the key component of the LNP formulation (Adams et al., 2018).

The aforementioned trials spot the high potential of nucleic acid–loaded nanocarriers to target CNS disorders, with remarkable ability to cross the BBB. However, again, the high cost of processing and the risk of manipulating nucleic acids that have ubiquitous roles in other different organs are still challenges that will take some time before we can expand and evaluate their efficiency and safety in CNS disorders.

19.5 EMERGING POSSIBILITIES FOR ENHANCED TARGETED NANO-DRUG DELIVERY TO THE BRAIN

19.5.1 Blood–Brain Barrier Leakage

Increasing evidence confirms that BBB is exposed to leakage during the progression of different neurodegenerative disorders, as well as in traumatic brain injuries (Al-Bachari et al., 2020). For instance, in AD, apolipoprotein E4, a stronger genetic risk factor associated with sporadic AD, is known to contribute to BBB breakdown (Raulin et al., 2022). Meanwhile, in PD, vascular dysfunction occurs in the basal ganglia, leading to BBB disruption and dysfunction. In the same context, in Huntington's disease mice models, the detection of the macromolecule albumin in the brain indicates BBB leakage (Drouin-Ouellet et al., 2015). Some recent studies explored the opportunity of using this transient BBB damage to deliver nanoparticles in a more enhanced and precise way (Bharadwaj et al., 2016). However, this approach is challenged with the heterogeneity and grade of BBB leakage in different CNS disorders, together with the physiological variability and the differential response and time required among patients for endogenous BBB repair (Achar et al., 2021).

19.5.2 Intranasal Drug Delivery

Despite the achieved progress, drug delivery to the CNS using oral or parenteral route is enormously hindered by the diminished permeability of BBB to these drugs, together with their reduced bioavailability passing the cerebral parenchyma, rather than using the double-edged sword of BBB leakage. Intranasal drugs administration is another promising solution for nano-drug delivery to the CNS. Histologically, the nasal epithelium has a highly porous endothelial membrane, with large surface area and blood flow that bypasses first-pass metabolism. Moreover, the superior section of the nasal cavity is directly attached to the brain via the olfactory nerves (Bahadur et al., 2020).

Intranasal drugs initially cross the mucociliary clearance in the vestibular part, then pass to the interior section of the nasal cavity to come in direct contact with the neural network and blood vessels of the brain. Consequently, the intranasal route bypasses the BBB and enhances the brain bioavailability for intranasally administered drugs. This provides a direct and minimally invasive, painless pathway with rapid drug absorption and increased accumulation in the brain, while decreasing the chances of systemic adverse effects (Wu et al., 2023).

Intranasal polymeric and liposome nanocarriers have specifically demonstrated in animal models of different CNS disorders success in drug delivery to the brain (Awad et al., 2023; Wu et al., 2023). Moreover, a nanolipid chitosan hydrogel incorporating temozolomide was recently developed for intranasal therapy against brain tumors. The study reported more than twofold increased permeability compared to normal drug, with no reported histological damage to the nasal mucosa (Khan et al., 2018).

With the increasing evidence of the potential of intranasal administration for brain drug delivery, this therapeutic approach still faces some concerns that are witnessed in the limited ongoing clinical trials. A main issue is the different anatomical structure of the commonly used animal models with humans, which jeopardizes the correlation between preclinical and clinical results. Moreover, the dosage volume that should be used through this route could cause damage to the nasal mucosal membrane upon prolonged administration. Another important concern is the lack of validated studies unraveling the cells and brain regions directly targeted by this route, and how the nanocarrier properties and administration via different areas of the nostril affect the drug biodistribution in the CNS.

19.6 CONCLUSION

19.6.1 Current Concerns, Future Prospects, and Recommendations

Progressive advances in industrial nanotechnology and selective drug delivery are opening new avenues to address CNS disorders. Ongoing preclinical studies help explore and evaluate the efficacy of innovative therapeutic–nanocarrier conjugates. However, as commented earlier, translating these breakthroughs into actual clinical success is still hindered by multiple challenges.

A major point is the safety of administration, as there are still limited studies that explore degradation pathways and the fate of different used nanomaterials, not to mention the toxic side effects in the brain, especially over long-term administration. Particularly, regarding intranasal administration, some drugs carriers can cause irritation and/or damage to the mucosal nasal membrane. Simultaneously, the bioavailability of intranasal nanoformulations can be hugely affected by the health condition of the nasal mucosa, where even flu or rhinitis can alter the response to the administered drugs.

Another important concern is attributed to the biological complexity of the human brain structure and the lack of validated information regarding its intermingling signaling pathways, crucial regulatory molecules, innervations, and microenvironment. Additionally, the heterogeneous and multifactorial complexity of the neurological disorders vary with the progressive stages of the disease, age, and even the environmental conditions surrounding the affected patients. The failure of some of the clinical trials may be due to this complex scenario, which is far different from animal models that most often fell short of mimicking all human CNS pathologies and microenvironment. Supplementary studies of discontinued clinical trials are necessary for the redesign of new and more effective trials, based on derivative compounds or new formulations.

Conversely, the wide physicochemical nature of the available nanocarriers, synthetic, and innovative biological therapeutics can affect the membrane potential and electrochemical signaling in the vulnerable brain cells and therefore requires deeper investigation. Furthermore, elevated industrial/processing costs, especially of biological nanocarriers, hinder wider validation and applications in clinical trials.

These challenges can be addressed with deeper studies that elucidate the safety and bioavailability of promising nanoformulations in both healthy and diseased human CNS conditions. New technologies and models, like 3D organoids, induced pluripotent stem cells (IPSCs) and in silico experimentation can be useful to create closer models to human CNS pathologies. Further pharmacokinetics and dynamic studies of the different nanoformulations can optimize the morphologies and physicochemical parameters in accordance with the targeted disorder. They can also incorporate different therapeutic agents to tackle multiple causative receptors/signaling pathways.

In the same context, incorporation of biologically active therapeutic agents or nanocarriers as natural microvesicles, antibodies, and small nucleic acids demands essential standardization steps for isolation, characterization, and manufacturing processes.

Addressing these challenges that face brain-targeted nano-drug delivery can definitely be a game changer against the debilitating burden of the resistant CNS disorders to achieve optimized efficacy with minimal adverse toxicity.

DECLARATION OF INTEREST

Authors declare no conflict of interest.

ACKNOWLEDGMENTS

This collaborative work is partially supported by the "Mujeres Por Africa" Foundation, Spain, 2023.

REFERENCES

Achar, A., Myers, R., & Ghosh, C. (2021). Drug delivery challenges in brain disorders across the blood–brain barrier: Novel methods and future considerations for improved therapy. *Biomedicines*, *9*(12), 1834. https://doi.org/10.3390/biomedicines9121834

Adams, D., Gonzalez-Duarte, A., O'Riordan, W. D., Yang, C.-C., Ueda, M., Kristen, A. V., Tournev, I., Schmidt, H. H., Coelho, T., Berk, J. L., Lin, K.-P., Vita, G., Attarian, S., Planté-Bordeneuve, V., Mezei, M. M., Campistol, J. M., Buades, J., Brannagan, T. H., Kim, B. J., . . . Suhr, O. B. (2018). Patisiran, an RNAi therapeutic, for hereditary transthyretin amyloidosis. *New England Journal of Medicine*, *379*(1), 11–21. https://doi.org/10.1056/NEJMoa1716153

Al-Ahmady, Z. S., Jasim, D., Ahmad, S. S., Wong, R., Haley, M., Coutts, G., Schiessl, I., Allan, S. M., & Kostarelos, K. (2019). Selective liposomal transport through blood brain barrier disruption in ischemic stroke reveals two distinct therapeutic opportunities. *ACS Nano*, *13*(11), 12470–12486. https://doi.org/10.1021/acsnano.9b01808

Alajangi, H. K., Kaur, M., Sharma, A., Rana, S., Thakur, S., Chatterjee, M., Singla, N., Jaiswal, P. K., Singh, G., & Barnwal, R. P. (2022). Blood–brain barrier: Emerging trends on transport models and new-age strategies for therapeutics intervention against neurological disorders. *Molecular Brain*, *15*(1), 49. https://doi.org/10.1186/s13041-022-00937-4

Al-Bachari, S., Naish, J. H., Parker, G. J. M., Emsley, H. C. A., & Parkes, L. M. (2020). Blood–brain barrier leakage is increased in Parkinson's disease. *Frontiers in Physiology*, *11*. www.frontiersin.org/articles/10.3389/fphys.2020.593026

Ashique, S., Afzal, O., Yasmin, S., Hussain, A., Altamimi, M. A., Webster, T. J., & Altamimi, A. S. A. (2023). Strategic nanocarriers to control neurodegenerative disorders: Concept, challenges, and future perspective. *International Journal of Pharmaceutics*, *633*, 122614. https://doi.org/10.1016/j.ijpharm.2023.122614

Awad, R., Avital, A., & Sosnik, A. (2023). Polymeric nanocarriers for nose-to-brain drug delivery in neurodegenerative diseases and neurodevelopmental disorders. *Acta Pharmaceutica Sinica B*, *13*(5), 1866–1886. https://doi.org/10.1016/j.apsb.2022.07.003

Ayub, A., & Wettig, S. (2022). An overview of nanotechnologies for drug delivery to the brain. *Pharmaceutics*, *14*(2), 224. https://doi.org/10.3390/pharmaceutics14020224

Bahadur, S., Pardhi, D. M., Rautio, J., Rosenholm, J. M., & Pathak, K. (2020). Intranasal nanoemulsions for direct nose-to-brain delivery of actives for CNS disorders. *Pharmaceutics*, *12*(12), 1230. https://doi.org/10.3390/pharmaceutics12121230

Bauer, M., Zeitlinger, M., Karch, R., Matzneller, P., Stanek, J., Jäger, W., Böhmdorfer, M., Wadsak, W., Mitterhauser, M., Bankstahl, J. P., Löscher, W., Koepp, M., Kuntner, C., Müller, M., & Langer, O. (2012). Pgp-mediated interaction between I-[11C]verapamil and tariquidar at the human blood–brain barrier: A comparison with rat data. *Clinical Pharmacology & Therapeutics*, *91*(2), 227–233. https://doi.org/10.1038/clpt.2011.217

Bharadwaj, V. N., Lifshitz, J., Adelson, P. D., Kodibagkar, V. D., & Stabenfeldt, S. E. (2016). Temporal assessment of nanoparticle accumulation after experimental brain injury: Effect of particle size. *Scientific Reports*, *6*(1), Article 1. https://doi.org/10.1038/srep29988

Burks, S. R., Kersch, C. N., Witko, J. A., Pagel, M. A., Sundby, M., Muldoon, L. L., Neuwelt, E. A., & Frank, J. A. (2021). Blood–brain barrier opening by intracarotid artery hyperosmolar mannitol induces sterile inflammatory and innate immune responses. *Proceedings of the National Academy of Sciences of the United States of America*, *118*(18), e2021915118. https://doi.org/10.1073/pnas.2021915118

Cai, L., Yang, C., Jia, W., Liu, Y., Xie, R., Lei, T., Yang, Z., He, X., Tong, R., & Gao, H. (2020). Endo/lysosome-escapable delivery depot for improving BBB transcytosis and neuron targeted therapy of Alzheimer's disease. *Advanced Functional Materials*, *30*(27), 1909999. https://doi.org/10.1002/adfm.201909999

Chen, Y., Yue, Z., Moulton, S. E., Hayes, P., Cook, M. J., & Wallace, G. G. (2015). A simple and versatile method for microencapsulation of anti-epileptic drugs for focal therapy of epilepsy. *Journal of Materials Chemistry. B*, *3*(36), 7255–7261. https://doi.org/10.1039/c5tb00675a

Daneman, R., & Prat, A. (2015). The blood-brain barrier. *Cold Spring Harbor Perspectives in Biology*, *7*(1), a020412. https://doi.org/10.1101/cshperspect.a020412

Drouin-Ouellet, J., Sawiak, S. J., Cisbani, G., Lagacé, M., Kuan, W.-L., Saint-Pierre, M., Dury, R. J., Alata, W., St-Amour, I., Mason, S. L., Calon, F., Lacroix, S., Gowland, P. A., Francis, S. T., Barker, R. A., & Cicchetti, F. (2015). Cerebrovascular and blood–brain barrier impairments in Huntington's disease: Potential implications for its pathophysiology. *Annals of Neurology*, *78*(2), 160–177. https://doi.org/10.1002/ana.24406

Ghosh, C., Hossain, M., Mishra, S., Khan, S., Gonzalez-Martinez, J., Marchi, N., Janigro, D., Bingaman, W., & Najm, I. (2018). Modulation of glucocorticoid receptor in human epileptic endothelial cells impacts drug biotransformation in an in vitro blood–brain barrier model. *Epilepsia*, *59*(11), 2049–2060. https://doi.org/10.1111/epi.14567

Gilert, A., & Machluf, M. (2010). Nano to micro delivery systems: Targeting angiogenesis in brain tumors. *Journal of Angiogenesis Research*, *2*, 20. https://doi.org/10.1186/2040-2384-2-20

Giugliani, R., Giugliani, L., de Oliveira Poswar, F., Donis, K. C., Corte, A. D., Schmidt, M., Boado, R. J., Nestrasil, I., Nguyen, C., Chen, S., & Pardridge, W. M. (2018). Neurocognitive and somatic stabilization in pediatric patients with severe Mucopolysaccharidosis Type I after 52 weeks of intravenous brain-penetrating insulin receptor antibody-iduronidase fusion protein (valanafusp alpha): An open label phase 1–2 trial. *Orphanet Journal of Rare Diseases*, *13*, 110. https://doi.org/10.1186/s13023-018-0849-8

Hall, C. N., Reynell, C., Gesslein, B., Hamilton, N. B., Mishra, A., Sutherland, B. A., O'Farrell, F. M., Buchan, A. M., Lauritzen, M., & Attwell, D. (2014). Capillary pericytes regulate cerebral blood flow in health and disease. *Nature*, *508*(7494), 55–60. https://doi.org/10.1038/nature13165

Harrell, C. R., Volarevic, A., Djonov, V., & Volarevic, V. (2021). Mesenchymal stem cell-derived exosomes as new remedy for the treatment of neurocognitive disorders. *International Journal of Molecular Sciences*, *22*(3), 1433. https://doi.org/10.3390/ijms22031433

Hernando, S., Santos-Vizcaíno, E., Igartua, M., & Hernandez, R. M. (2023). Targeting the central nervous system: From synthetic nanoparticles to extracellular vesicles—focus on Alzheimer's and Parkinson's disease. *WIREs Nanomedicine and Nanobiotechnology*, e1898. https://doi.org/10.1002/wnan.1898

Holm, A., Hansen, S. N., Klitgaard, H., & Kauppinen, S. (2022). Clinical advances of RNA therapeutics for treatment of neurological and neuromuscular diseases. *RNA Biology*, *19*(1), 594–608. https://doi.org/10.1080/15476286.2022.2066334

Huang, L.-K., Chao, S.-P., & Hu, C.-J. (2020). Clinical trials of new drugs for Alzheimer disease. *Journal of Biomedical Science*, *27*(1), Article 1. https://doi.org/10.1186/s12929-019-0609-7

Ivanoiu, A., Pariente, J., Booth, K., Lobello, K., Luscan, G., Hua, L., Lucas, P., Styren, S., Yang, L., Li, D., Black, R. S., Brashear, H. R., & McRae, T. (2016). Long-term safety and tolerability of bapineuzumab in patients with Alzheimer's disease in two phase 3 extension studies. *Alzheimer's Research & Therapy*, *8*(1), 24. https://doi.org/10.1186/s13195-016-0193-y

Jagaran, K., & Singh, M. (2021). Nanomedicine for neurodegenerative disorders: Focus on Alzheimer's and Parkinson's diseases. *International Journal of Molecular Sciences*, *22*(16), 9082. https://doi.org/10.3390/ijms22169082

Johansson, B. B. (1990). The physiology of the blood-brain barrier. In J. C. Porter & D. Ježová (Eds.), *Circulating regulatory factors and neuroendocrine function* (pp. 25–39). Springer US. https://doi.org/10.1007/978-1-4684-5799-5_2

Kadry, H., Noorani, B., & Cucullo, L. (2020). A blood–brain barrier overview on structure, function, impairment, and biomarkers of integrity. *Fluids and Barriers of the CNS*, *17*(1), 69. https://doi.org/10.1186/s12987-020-00230-3

Khan, A., Aqil, M., Imam, S. S., Ahad, A., Sultana, Y., Ali, A., & Khan, K. (2018). Temozolomide loaded nano lipid based chitosan hydrogel for nose to brain delivery: Characterization, nasal absorption, histopathology and cell line study. *International Journal of Biological Macromolecules*, *116*, 1260–1267. https://doi.org/10.1016/j.ijbiomac.2018.05.079

Khongkow, M., Yata, T., Boonrungsiman, S., Ruktanonchai, U. R., Graham, D., & Namdee, K. (2019). Surface modification of gold nanoparticles with neuron-targeted exosome for enhanced blood–brain barrier penetration. *Scientific Reports*, *9*(1), Article 1. https://doi.org/10.1038/s41598-019-44569-6

Kim, D., Nishida, H., An, S. Y., Shetty, A. K., Bartosh, T. J., & Prockop, D. J. (2016). Chromatographically isolated CD63+CD81+ extracellular vesicles from mesenchymal stromal cells rescue cognitive impairments after TBI. *Proceedings of the National Academy of Sciences of the United States of America*, *113*(1), 170–175. https://doi.org/10.1073/pnas.1522297113

Knox, E. G., Aburto, M. R., Clarke, G., Cryan, J. F., & O'Driscoll, C. M. (2022). The blood-brain barrier in aging and neurodegeneration. *Molecular Psychiatry*, *27*(6), Article 6. https://doi.org/10.1038/s41380-022-01511-z

Kogan, M. J., Bastus, N. G., Amigo, R., Grillo-Bosch, D., Araya, E., Turiel, A., Labarta, A., Giralt, E., & Puntes, V. F. (2006). Nanoparticle-Mediated Local and Remote Manipulation of Protein Aggregation. *Nano Letters*, *6*(1), 110–115. https://doi.org/10.1021/nl0516862

Kort, A., van Hoppe, S., Sparidans, R. W., Wagenaar, E., Beijnen, J. H., & Schinkel, A. H. (2017). Brain accumulation of ponatinib and its active metabolite, N-desmethyl ponatinib, is limited by P-glycoprotein (P-GP/ABCB1) and breast cancer resistance protein (BCRP/ABCG2). *Molecular Pharmaceutics*, *14*(10), 3258–3268. https://doi.org/10.1021/acs.molpharmaceut.7b00257

La Barbera, L., Mauri, E., D'Amelio, M., & Gori, M. (2022). Functionalization strategies of polymeric nanoparticles for drug delivery in Alzheimer's disease: Current trends and future perspectives. *Frontiers in Neuroscience*, *16*. www.frontiersin.org/articles/10.3389/fnins.2022.939855

Lopes, D., Lopes, J., Pereira-Silva, M., Peixoto, D., Rabiee, N., Veiga, F., Moradi, O., Guo, Z.-H., Wang, X.-D., Conde, J., Makvandi, P., & Paiva-Santos, A. C. (2023). Bioengineered exosomal-membrane-camouflaged abiotic nanocarriers: Neurodegenerative diseases, tissue engineering and regenerative medicine. *Military Medical Research*, *10*(1), 19. https://doi.org/10.1186/s40779-023-00453-z

Marrache, S., & Dhar, S. (2012). Engineering of blended nanoparticle platform for delivery of mitochondria-acting therapeutics. *Proceedings of the National Academy of Sciences of the United States of America*, *109*(40), 16288–16293. https://doi.org/10.1073/pnas.1210096109

Narasipura, E. A., VanKeulen-Miller, R., Ma, Y., & Fenton, O. S. (2023). Ongoing clinical trials of nonviral siRNA therapeutics. *Bioconjugate Chemistry*, *34*(7), 1177–1197. https://doi.org/10.1021/acs.bioconjchem.3c00205

Nguyen, T. T., Dung Nguyen, T. T., Vo, T. K., Tran, N.-M.-A., Nguyen, M. K., Van Vo, T., & Van Vo, G. (2021). Nanotechnology-based drug delivery for central nervous system disorders. *Biomedicine & Pharmacotherapy*, *143*, 112117. https://doi.org/10.1016/j.biopha.2021.112117

Nour, S. A., Abdelmalak, N. S., Naguib, M. J., Rashed, H. M., & Ibrahim, A. B. (2016). Intranasal brain-targeted clonazepam polymeric micelles for immediate control of status epilepticus: *In vitro* optimization, ex vivo determination of cytotoxicity, in vivo biodistribution and pharmacodynamics studies. *Drug Delivery*, *23*(9), 3681–3695. https://doi.org/10.1080/10717544.2016.1223216

Okuyama, T., Eto, Y., Sakai, N., Nakamura, K., Yamamoto, T., Yamaoka, M., Ikeda, T., So, S., Tanizawa, K., Sonoda, H., & Sato, Y. (2021). A phase 2/3 trial of pabinafusp Alfa, IDS fused with anti-human transferrin receptor antibody, targeting neurodegeneration in MPS-II. *Molecular Therapy*, *29*(2), 671–679. https://doi.org/10.1016/j.ymthe.2020.09.039

Peppiatt, C. M., Howarth, C., Mobbs, P., & Attwell, D. (2006). Bidirectional control of CNS capillary diameter by pericytes. *Nature*, *443*(7112), 700–704. https://doi.org/10.1038/nature05193

Qi, Y., Guo, L., Jiang, Y., Shi, Y., Sui, H., & Zhao, L. (2020). Brain delivery of quercetin-loaded exosomes improved cognitive function in AD mice by inhibiting phosphorylated tau-mediated neurofibrillary tangles. *Drug Delivery*, *27*(1), 745–755. https://doi.org/10.1080/10717544.2020.1762262

Qu, M., Lin, Q., Huang, L., Fu, Y., Wang, L., He, S., Fu, Y., Yang, S., Zhang, Z., Zhang, L., & Sun, X. (2018). Dopamine-loaded blood exosomes targeted to brain for better treatment of Parkinson's disease. *Journal of Controlled Release: Official Journal of the Controlled Release Society*, *287*, 156–166. https://doi.org/10.1016/j.jconrel.2018.08.035

Raulin, A.-C., Doss, S. V., Trottier, Z. A., Ikezu, T. C., Bu, G., & Liu, C.-C. (2022). ApoE in Alzheimer's disease: Pathophysiology and therapeutic strategies. *Molecular Neurodegeneration*, *17*(1), 72. https://doi.org/10.1186/s13024-022-00574-4

Rezai, A. R., Ranjan, M., D'Haese, P.-F., Haut, M. W., Carpenter, J., Najib, U., Mehta, R. I., Chazen, J. L., Zibly, Z., Yates, J. R., Hodder, S. L., & Kaplitt, M. (2020). Noninvasive hippocampal blood–brain barrier opening in Alzheimer's disease with focused ultrasound. *Proceedings of the National Academy of Sciences of the United States of America*, *117*(17), 9180–9182. https://doi.org/10.1073/pnas.2002571117

Sita, T. L., Kouri, F. M., Hurley, L. A., Merkel, T. J., Chalastanis, A., May, J. L., Ghelfi, S. T., Cole, L. E., Cayton, T. C., Barnaby, S. N., Sprangers, A. J., Savalia, N., James, C. D., Lee, A., Mirkin, C. A., & Stegh, A. H. (2017). Dual bioluminescence and near-infrared fluorescence monitoring to evaluate spherical nucleic acid nanoconjugate activity in vivo. *Proceedings of the National Academy of Sciences of the United States of America*, *114*(16), 4129–4134. https://doi.org/10.1073/pnas.1702736114

Sonoda, H., Takahashi, K., Minami, K., Hirato, T., Yamamoto, T., So, S., Tanizawa, K., Schmidt, M., & Sato, Y. (2022). Treatment of neuronopathic mucopolysaccharidoses with blood–brain barrier-crossing enzymes: Clinical application of receptor-mediated transcytosis. *Pharmaceutics*, *14*(6), 1240. https://doi.org/10.3390/pharmaceutics14061240

Tashima, T. (2020). Smart strategies for therapeutic agent delivery into brain across the blood–brain barrier using receptor-mediated transcytosis. *Chemical and Pharmaceutical Bulletin*, *68*(4), 316–325. https://doi.org/10.1248/cpb.c19-00854

Teleanu, R. I., Preda, M. D., Niculescu, A.-G., Vladâcenco, O., Radu, C. I., Grumezescu, A. M., & Teleanu, D. M. (2022). Current strategies to enhance delivery of drugs across the blood–brain barrier. *Pharmaceutics*, *14*(5), 987. https://doi.org/10.3390/pharmaceutics14050987

Tian, X., Fan, T., Zhao, W., Abbas, G., Han, B., Zhang, K., Li, N., Liu, N., Liang, W., Huang, H., Chen, W., Wang, B., & Xie, Z. (2021). Recent advances in the development of nanomedicines for the treatment of ischemic stroke. *Bioactive Materials*, *6*(9), 2854–2869. https://doi.org/10.1016/j.bioactmat.2021.01.023

Vucic, S., Menon, P., Huynh, W., Mahoney, C., Ho, K. S., Hartford, A., Rynders, A., Evan, J., Evan, J., Ligozio, S., Glanzman, R., Hotchkin, M. T., & Kiernan, M. C. (2023). Efficacy and safety of CNM-Au8 in amyotrophic lateral sclerosis (RESCUE-ALS study): A phase bilityrised, double-blind, placebo-controlled trial and open label extension. *eClinicalMedicine*, *60*. https://doi.org/10.1016/j.eclinm.2023.102036

Wang, F., Yang, Z., Liu, M., Tao, Y., Li, Z., Wu, Z., & Gui, S. (2020). Facile nose-to-brain delivery of rotigotine-loaded polymer micelles thermosensitive hydrogels: *In vitro* characterization and in vivo behavior study. *International Journal of Pharmaceutics*, *577*, 119046. https://doi.org/10.1016/j.ijpharm.2020.119046

Wang, G. Y., Rayner, S. L., Chung, R., Shi, B. Y., & Liang, X. J. (2020). Advances in nanotechnology-based strategies for the treatments of amyotrophic lateral sclerosis. *Materials Today Bio*, *6*, 100055. https://doi.org/10.1016/j.mtbio.2020.100055

Wang, H., Sui, H., Zheng, Y., Jiang, Y., Shi, Y., Liang, J., & Zhao, L. (2019). Curcumin-primed exosomes potently ameliorate cognitive function in AD mice by inhibiting hyperphosphorylation of the Tau protein through the AKT/GSK-3β pathway. *Nanoscale*, *11*(15), 7481–7496. https://doi.org/10.1039/c9nr01255a

Wang, Z., Zhao, Y., Jiang, Y., Lv, W., Wu, L., Wang, B., Lv, L., Xu, Q., & Xin, H. (2015). Enhanced anti-ischemic stroke of ZL006 by T7-conjugated PEGylated liposomes drug delivery system. *Scientific Reports*, *5*(1), Article 1. https://doi.org/10.1038/srep12651

Wilson, B., Samanta, M. K., Santhi, K., Kumar, K. P. S., Paramakrishnan, N., & Suresh, B. (2008). Targeted delivery of tacrine into the brain with polysorbate 80-coated poly(n-butylcyanoacrylate) nanoparticles. *European Journal of Pharmaceutics and Biopharmaceutics: Official Journal of Arbeitsgemeinschaft Fur Pharmazeutische Verfahrenstechnik e*, *70*(1), 75–84. https://doi.org/10.1016/j.ejpb.2008.03.009

Wouters, Y., Jaspers, T., De Strooper, B., & Dewilde, M. (2020). Identification and in vivo characterization of a brain-penetrating nanobody. *Fluids and Barriers of the CNS*, *17*, 62. https://doi.org/10.1186/s12987-020-00226-z

Wu, D.-D., Salah, Y. A., Ngowi, E. E., Zhang, Y.-X., Khattak, S., Khan, N. H., Wang, Y., Li, T., Guo, Z.-H., Wang, Y.-M., & Ji, X.-Y. (2023). Nanotechnology prospects in brain therapeutics concerning gene-targeting and nose-to-brain administration. *iScience*, *26*(8), 107321. https://doi.org/10.1016/j.isci.2023.107321

Wu, Y.-C., Sonninen, T.-M., Peltonen, S., Koistinaho, J., & Lehtonen, Š. (2021). Blood–brain barrier and neurodegenerative diseases—modeling with iPSC-derived brain cells. *International Journal of Molecular Sciences*, *22*(14), Article 14. https://doi.org/10.3390/ijms22147710

Yousfan, A., Rubio, N., Hakim Natouf, A., Daher, A., Al-Kafry, N., Venner, K., & Kafa, H. (2020). Preparation and characterisation of PHT-loaded chitosan lecithin nanoparticles for intranasal drug delivery to the brain. *RSC Advances*, *10*(48), 28992–29009. https://doi.org/10.1039/D0RA04890A

Zhang, S., Gan, L., Cao, F., Wang, H., Gong, P., Ma, C., Ren, L., Lin, Y., & Lin, X. (2022). The barrier and interface mechanisms of the brain barrier, and brain drug delivery. *Brain Research Bulletin*, *190*, 69–83. https://doi.org/10.1016/j.brainresbull.2022.09.017

Zhang, W., Liu, Q. Y., Haqqani, A. S., Leclerc, S., Liu, Z., Fauteux, F., Baumann, E., Delaney, C. E., Ly, D., Star, A. T., Brunette, E., Sodja, C., Hewitt, M., Sandhu, J. K., & Stanimirovic, D. B. (2020). Differential expression of receptors mediating receptor-mediated transcytosis (RMT) in brain microvessels, brain parenchyma and peripheral tissues of the mouse and the human. *Fluids and Barriers of the CNS*, *17*, 47. https://doi.org/10.1186/s12987-020-00209-0

Zheng, M., Du, Q., Wang, X., Zhou, Y., Li, J., Xia, X., Lu, Y., Yin, J., Zou, Y., Park, J. B., & Shi, B. (2021). Tuning the elasticity of polymersomes for brain tumor targeting. *Advanced Science*, *8*(20), 2102001. https://doi.org/10.1002/advs.202102001

Zhou, Y., Zhu, F., Liu, Y., Zheng, M., Wang, Y., Zhang, D., Anraku, Y., Zou, Y., Li, J., Wu, H., Pang, X., Tao, W., Shimoni, O., Bush, A. I., Xue, X., & Shi, B. (2020). Blood-brain barrier-penetrating siRNA nanomedicine for Alzheimer's disease therapy. *Science Advances*, *6*(41), eabc7031. https://doi.org/10.1126/sciadv.abc7031

Zlokovic, B. V. (2008). The blood-brain barrier in health and chronic neurodegenerative disorders. *Neuron*, *57*(2), 178–201. https://doi.org/10.1016/j.neuron.2008.01.003

20 Toward Applicability of Exosomal-ncRNA Precision vs. Nanoparticles to the Brain

Nadia M. Hamdy, Emad B. Basalious, Mona G. El-Sisi, Rana A. Youness, and Ahmed S. Sultan*

20.1 INTRODUCTION

20.1.1 Problem Definition

Due to a variety of causes, including environmental pollution, inflammation, stress, disease micro-environment activation with immune cells involvement (Hamdy, 2011), and others, the prevalence of neurological disorders such as amyotrophic lateral sclerosis (ALS), Huntington's disease (HD), Parkinson's disease (PD), and Alzheimer's disease (AD) has been skyrocketing (Wang et al., 2020). These disorders are known as neurodegenerative diseases (NDDs) which cause a significant medical, social, and economic impact on the global community.

Despite the extensive research into the NDD pathogenic processes or the therapeutic approaches for treatment, there are still very few suitably effective treatments available (Heemels, 2016). This is attributed to the blood–brain barrier (BBB) preventing the majority of medications from reaching the brain-injured parts, significantly reducing the effectiveness of the available pharmacological therapies (ElShagea et al., 2023).

20.1.2 Background

The optimum drug delivery system (DDS) should keep the drug concentration steady as long as feasible within the therapeutic range (Wang, Cheng, Li, Wang, & Chen, 2019). One of the innovative DDSs are synthetic nanocarriers, which can be engineered to overcome drug problems of poor solubility or inability to cross the BBB with poor bioavailability as well as fast hepatic clearance (Mazayen et al., 2022). Various nanocarriers with various innovative engineered formulation approaches were attempted for brain cell treatment using either the systematic route or the olfactory pathway (Rashed, Shamma, & Basalious, 2017).

Exosomes cargo-carrying iron oxide NPs (IONPs) are an option to pass the brain and overcome the possible IONPs toxicities (Hamdy et al., 2022a). Additionally, targeting and activation are easily employed via magnetic resonance or any other activation means (Wang et al., 2022).

According to the Nanotechnology (NT) Products Database (DB), there are 399 pharmaceuticals products by the FDA https://product.statnano.com/industry/medicine (Accessed May 2023). Nanomaterials database is NanoWerk https://www.nanowerk.com/nanomaterial-database.php For nanomaterials characterization, caNanoLab https://cananolab.cancer.gov/#/.

20.2 DIFFERENT TYPES OF SYNTHETIC NANOCARRIERS SUITABLE FOR BRAIN DELIVERY

The issue of brain therapy, particularly the requirements for (1) treatment that successfully crosses the BBB, (2) multitarget therapy (MTT), (3) enhanced pharmacokinetics, and (4) more precise administration, is met by nanotherapeutics and NPs. Additionally, the ability to customize the biophysical and biological target-specific ligands characteristics of NPs opened the door to highly specialized delivery systems. NPs are able to lessen the side effects of the symptomatic therapies FDA approved, which have shown that they offer greater pharmacokinetics and brain uptake than free pharmaceuticals (Bukhari, 2022).

20.2.1

Organic liposomes containing lipid or inorganic substances with the FDA-approved iron oxide and silver or gold NPs (Huang et al., 2015).

20.2.2

Polylactic-co-glycolic acid (PLGA), polyethylene glycol (PEG), hyaluronic acid, chitosan, and alginate are all polymeric NPs. The short peptide sequences (5–30 amino acids) that help impermeable compounds pass the BBB, employed for NDDs, are cell-penetrating peptides (CPPs) (Tsou et al., 2017).

* Corresponding author: nadia_hamdy@pharma.asu.edu.eg www.orchid.org/0000-0003-2105-107X

DOI: 10.1201/9781003383376-20

20.2.3

Functionally modified polymeric NPs with hydrophobic, amphiphilic, and cationic peptide CPPs increase drug delivery effectiveness and therapeutic success. Bifunctional liposomal NPs containing mannose and CPPs showed improved BBB transport performance. Although their manufacturing method presents difficulties, due to their high targeting and high transmembrane effectiveness, multifunctionally modified NPs can therefore be exploited as prospective drug delivery vehicles for NDDs (Duan et al., 2023). A liposomal preparation of cytarabine made with Depofoam® technology is DepoCyt®, licensed by the FDA for fatal lymphomatous meningitis, administered intrathecally (Farjadian et al., 2019). Dipalmitoyl phosphatidyl glycerol, dioleoyl phosphatidylcholine phosphatidylcholine, triolein, and cholesterol comprise liposomal formulation. DepoCyt is a sustained-release medication with the active ingredient cytarabine, intended to be injected directly into the cerebrospinal fluid. This anticancer drug can influence cells while in the cell cycle's S phase and stop DNA polymerase from working.

20.2.4

The capacity of multitarget nanotherapeutics to address many pathogenic variables at once makes them the most promising approach among the abundance of potential brain nano-drugs, while nano-nutraceuticals are emerging as intriguing tools in the treatment/prevention of NDDs.

20.2.5

The natural nanocarriers, namely, exosomes, are single-membrane, tiny, secreted structures with the same design as cells and range in size from 30 to 200 nm (Pegtel & Gould, 2019). They are abundant in particular proteins, lipids, nucleic acids, and glycoconjugates produced by budding at both plasma and endosome membranes. They exhibit remarkable molecular heterogeneity and contain a variety of membrane-associated, high-order oligomeric protein complexes (Rizk et al., 2022). Exosomes usually move to particular destinations after being released into plasma, such as the CNS, to have various pleiotropic effects (Pei et al., 2020).

20.3 EXOSOMES AND EXOSOMAL CARGOS

20.3.1 Exosomes Physiology

Specific biomolecular characteristic markers confined to exosomes aid in distinguishing them from various extracellular vesicles (ECVs: EVs), beginning with the membrane cluster of differentiation (CD)9, CD63, and CD81 spanins (Toh, Lai, Zhang, & Lim, 2018), as well as heat shock protein (HSP) family within exosomes (Kalluri & LeBleu, 2020).

Most EVs are produced through plasma membrane fission and straight outward budding, which is different from the exosome production involving a three-step procedure. First, endosomal vesicles (endocytic) are created from the plasma membrane. Second, some small vesicles are produced from endosomal membranes by reversed-budding (inward), then one sizeable multivesicular body (MVB) is formed from these small ones, aided with protein and nuclear acid combination; therefore, exosomes are formed, which are released to the extracellular environment via membrane fusion and outward extrusion (Ha, Yang, & Nadithe, 2016; Huang-Doran, Zhang, & Vidal-Puig, 2017). Mesenchymal stem cells (MSCs) produce exosomes that directly target nearby cells and have a paracrine effect (Baglio, Pegtel, & Baldini, 2012) or are discharged into the blood to distant target cells. Exosomes have fortunately been shown to have the ability to penetrate the BBB and enter the target brain cell to achieve its target activity (Alvarez-Erviti et al., 2011; Huang-Doran et al., 2017; Williams et al., 2020). This later advantage of exosomes highlights their therapeutic promise for CNS illness treatment.

It is noteworthy to mention that the formed exosome size, source, and membrane composition, as well as what they carry, may affect how target/recipient cells recognize and/or take them up (Mathieu, Martin-Jaular, Lavieu, & Théry, 2019; Mulcahy, Pink, & Carter, 2014). Selectivity of exosomes recipient cell's recognition and uptake is responsible

TABLE 20.1
Different types of synthetic nanocarriers nanoparticles, shape, and properties suitable for brain delivery

NP type	NP shape	NP properties	For (Ref.)
Polymeric	Nanospheres or nanocapsules	Different routes of administration; penetration of cells and tissues	Various multifunctional drugs and CHO or proteins (Tsou et al., 2017)
Liposomes	Unilamellar and multilamellar	Lipid bilayers of less toxicity; drugs passively enter	Lipid-soluble drugs (bilayer) and biomolecules (Huang et al., 2015)
Micelles co-polymer	Spherical amphiphilic	Drugs protected from degradation	Water-insoluble drugs
Metal NPs	Spherical	—	Iron oxide, silver (Zhang et al., 2020), and gold (Zhang et al., 2021)

Abbreviations: NPs: nanoparticles; CHO: carbohydrates.

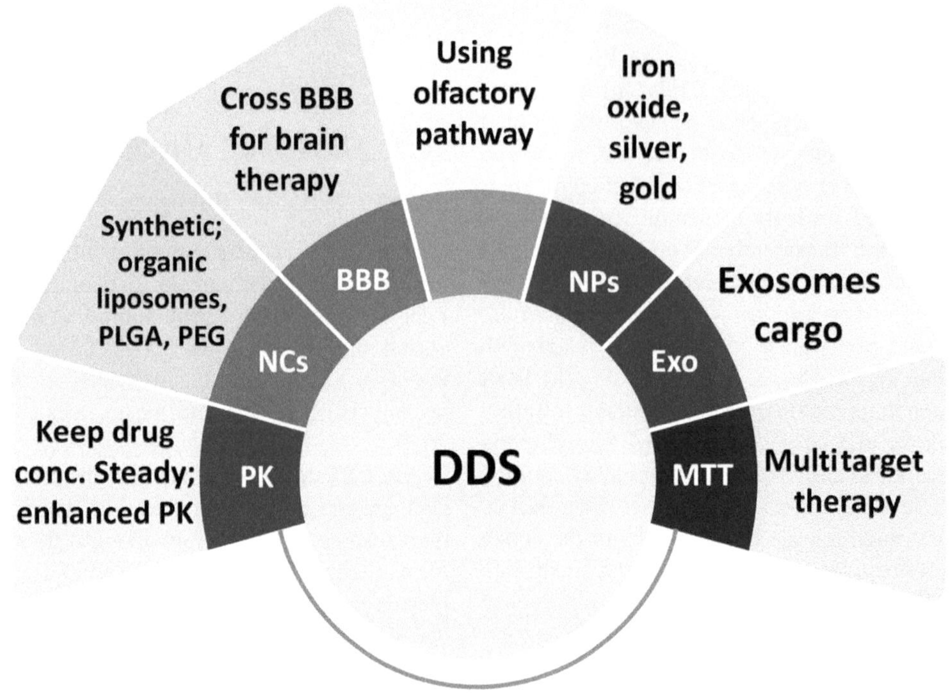

FIGURE 20.1 Drug delivery system (DDS). DDS: drug delivery system; PK: pharmacokinetics; NCs: nanocarriers; BBB: blood–brain barrier; NPs: nanoparticles; PLGA: polylactic-co-glycolic acid; PEG: polyethylene glycol; Exo: exosomes: MTT: multi-target therapy.

for the nanovesicle–content transmission mechanism to the recipient cell's cytosol (Chivet et al., 2014, Fitzner et al., 2011). All these constitute "factors related to the exosomes' internal environment to influence exosomes' external environment of exosome recipient cell's recognition."

20.3.2 Exosomal Cargos

The molecular makeup of the exosomal payload depends on the type of producing cells (Skryabin, Komelkov, Savelyeva, & Tchevkina, 2020). Exosomes transport active versions of numerous biomolecules and carry different cargos, like proteins, lipids, messenger RNA (mRNA), micro-RNAs (miRs), or non-coding RNAs (ncRNAs) in general, such as long ncRNAs (lncRNAs) or circular RNAs (circRNAs) (Wang et al., 2020, Rizk et al., 2022).

Interestingly, endogenous exosomes are released by brain cells and reflect brain injury (Pei et al., 2020), highlighting their potential as biomarkers for NDDs. Exogenous exosomes penetrate the BBB to target brain tissue with the specifically carried treatment (Ha et al., 2016).

Undoubtedly, a substantial amount of proof links exosomes to NDDs, but to what degree exosomes support or inhibit the removal of causative molecular and/or metabolic derangements or dysfunction is a crucial question. Clarifying the role played by secreted exosomes by various brain cells will help us better understand how NDDs occur, spread, and treated.

20.4 EXOSOME'S ROLE IN THE CNS

Exosomes support intercellular communication in the brain (Zappulli, Friis, Fitzpatrick, Maguire, & Breakefield, 2016). They perform their function as cell-to-cell mediators by acting via several different processes, activating target cellular receptors, or transporting their contents. Exosomes have the ability to start signal transduction cascades that have functional effects on recipient cells. All CNS cells release exosomes, including neuroglia–glial cells, microglia, the specialized glial oligodendrocyte cells, neuro-epithelium-derived astrocyte cells, and neurons (Frühbeis, Fröhlich, & Krämer-Albers, 2012). During synaptic activation, cortical neurons release exosomes that only bind to other neurons and not to glial cells (Chivet et al., 2014). However, microglia grab oligodendrocyte exosomes (Fitzner et al., 2011).

Exosome release is often regulated by brain neurotransmitter transmission. Exosome release by mature inter-cortical neurons and projection-cortical neurons and hippocampal pyramidal excitatory neurons grown *in vitro* is stimulated by exposure to the excitatory neurotransmitter glutamate. Contrarily, excitatory activity is reversed with a decreased exosome output by α-amino-3-hydroxyl-5-methyl-4-isoxazole-propionate (AMPA) receptor or N-methyl-D-aspartate (NMDA) receptor antagonists. These findings imply that excitatory synaptic activity controls exosome release, with calcium influx promoting exosome release (Lachenal et al., 2011).

TABLE 20.2
Cell-mediated brain cell exosome subtypes function in CNS

Exosome cell type	Functions in CNS	Ref.
Astrocytic	• Neurotoxic chemicals and inflammatory cytokines, chemokines released by reactive astrocytes, and exosome release aid in the performance of astrocyte functions. • Role in astrocyte's neuroprotective function. • In axonal regeneration, neurogenesis, synaptic plasticity, and enrichment in FGF2 and VEGF.	(Proia et al., 2008; Schänzer et al., 2004; Woodbury & Ikezu, 2014)
Microglial	• Communicate with nearby and distant cells. • Energy source for neurons during synaptic activity via lactate synthesis and anaerobic glycolysis enzymes. • Control synaptic transmission, boosting ceramide and sphingolipids synthesis in neurons. • Pro-inflammatory microglia (M1 phenotype) aid neuroinflammation.	(Antonucci et al., 2012; Jablonski, Gaudet, Amici, Popovich, & Guerau-de-Arellano, 2016; Potolicchio et al., 2005; Tang, Wu, Fang, Qu, & Gao, 2016)
Neuronal	• Contribute to synaptic plasticity, eliminating old proteins, RNA. • Internalized type increases synaptic strength, speeding up the translation of post-synaptic mRNAs, modifying neurotransmission. • Influence neuron–neuron, neuron–glial communication.	(Goldie et al., 2014; Lee et al., 2018; Liu, Huang, Hung, & Hsueh, 2015)
Oligodendrocyte	• Neurons internalize these exosomes for axons trophic support. • During cell stress, improve neuronal survival. • In the conservation of neuronal homeostasis.	(Frühbeis et al., 2013; Frühbeis et al., 2020; Krämer-Albers et al., 2007)

Abbreviations: FGF2: fibroblast growth factor 2; VEGF: vascular endothelial growth factor.

Synaptic adenosine triphosphate (ATP) release activates glial purinergic receptors as a co-transmitter. Exosome release by astrocytes and microglial cells is stimulated by ATP (Kaur et al., 2021; Kim, Otgontenger, Jamsranjav, & Kim, 2020; Pistono, Bister, Stanová, & Malm, 2020; Schnatz, Müller, Brahmer, & Krämer-Albers, 2021). The role of different subtypes of cell-specific exosomes is summarized in Table 20.2.

It is noteworthy to mention that NPs are unable to cross the BBB unless they are functionalized with surface ligand to cross the BBB (Pardridge, 2022), which is a limitation or disadvantage in comparison to exosomes use.

20.5 EXOSOME'S ROLE IN NEURODEGENERATIVE DISEASES

Progressive neuronal and glial dysfunction is brought on by the buildup of abnormal, faulty-folded proteins/un-folded proteins in many NDDs. After release into the extracellular milieu, these abnormally misfolded proteins cause the erroneous folding of their endogenous counterpart, normal proteins, which are then transmitted to normal cells (Frost & Diamond, 2010). This "infective" process causes pathology to intensify and disease to spread to more extensive parts of the brain.

Intranasal NPs have been employed to deliver drugs that are rapidly metabolized if systematically administered (Fouad et al., 2018). Despite intranasal NPs, DDS for brain disease as nose-to-brain (NtB) approach showed promising successful drug delivery rates (Mohsen et al., 2020). The search into the exosomes field would be more promising, as all brain cells secrete exosomes. Exosomes are now considered a crucial component of neuroglia connections and as neurodegenerative mediators between cells or from one cell to another (Kumari & Anji, 2022) that constitutes the promise for treatment from within the diseased cell itself.

Exosomes transport, with protection, medications to the CNS, targeting specific brain cells, while overcoming the BBB drug–efflux mechanisms in favor of progressive brain diseases (Lakhal & Wood, 2011; Malm, Loppi, & Kanninen, 2016; Niu, Chen, & Gao, 2019). Moreover, to target the inflammatory cells, macrophage-derived exosomes (MDEs) bind brain endothelial cell lectin receptors without being fluxed out, namely, carbohydrate-binding C-type lectin receptors (CBCLR) (Yuan et al., 2017). Therefore, exosomes, fortunately, would constitute the ideal delivery method of treatment for brain malignancies as glioblastoma, various cerebrovascular disorders, and NDDs (Haney et al., 2015).

20.6 NEURODEGENERATIVE DISEASES (NDDs)

20.6.1 Dementia

20.6.1.1 Alzheimer's Disease (AD)

The most prevalent form of dementia is AD, which affects parts of the CNS that are crucial for learning and memory processes, whose impairment or changes affecting synapse and brain energy are important interrelated hallmarks of AD (Reza-Zaldivar et al., 2018).

20.6.1.1.1 The Pathogenesis of AD

Amyloid plaques, extracellular aggregates of beta-amyloid (Aβ), and neurofibrillary tangles of tau are key components critical to the pathogenesis of AD.

The successive proteolytic processing of amyloid precursor protein (APP) results in Aβ peptides. Given that APP is an intracellular protein, it has been hypothesized that abnormally folded protein trans-location between neurons causes the pathological lesions of NDDs (Frost & Diamond, 2010). The mystery surrounding the mechanisms through which misfolded proteins are transmitted persists, nevertheless.

20.6.1.1.2 Autophagy

A process where Aβ within endosomes is eliminated by lysosomes to prevent the buildup of Aβ inside neurons (Mizushima & Komatsu, 2011). Phospholipid phosphatidylinositol-3-phosphate (PI3P), mainly produced by PI3-kinase class III Vps34, is a crucial regulator of this system (Jaber et al., 2016).

20.6.1.1.3 Neurofibrillary Tangles

NFTs with improperly phosphorylated tau protein are another pathogenic sign of AD (Wang et al., 2017). Microtubules are known to be stabilized by the cytoplasmic protein tau. The pathogenic tau moves intracellularly and draws more aggregates (Chai, Dage, & Citron, 2012; Saman et al., 2012), hyperphosphorylated tau monomers, and oligomers (Shi et al., 2016, Saman et al., 2012; Saman et al., 2014).

Neurons internalize abnormal protein aggregates and route them to the mitochondria, causing mitochondrial clustering via raised fission protein Drp-1 level, adenosine diphosphate/adenosine triphosphate (ADP/ATP) transporter, to activate caspases for neurite fragmentation activation and more neuronal cell death (Elsherbini et al., 2020).

20.6.1.1.4 Exosomes in the Pathophysiology of AD: Hypes

Alteration of neuronal class III PI3-kinase activities affects lipid metabolism, autophagy, and lysosomal degradation (Miranda et al., 2018). This encourages the brain to release distinct *exosomes rich in lysosomal substrates* that have not been digested, such as plaque proteins and aggregated amyloids, and their processing enzymes (Malm et al., 2016).

The ceramide formation regulatory enzyme sphingomyelinase-2 inhibition causes decreased exosome biogenesis in the brain and serum with decreased Aβ plaque formation (Dinkins et al., 2016). AD spread can be achieved via exosomes produced by injured neuronal cells transmission to neighboring neuronal cells, with their cargos of APP, Aβ peptides, partially digested proteins, enzymes, and tau protein (Chen, Zhao, Zhao, & Li, 2017; Miranda et al., 2018; Yuyama & Igarashi, 2017). Moreover, cellular prion protein (PrP^C) and the amyloid peptides produce more aggregates (Rajendran et al., 2006; Yuyama & Igarashi, 2017; Zappulli et al., 2016; Zheng et al., 2017). Exosomes may serve as foci for the nucleation of amyloid plaques, as evidenced by histological investigation, which revealed Alix and flotillin-1 exosomal markers elevation around plaques (Xiao et al., 2017). PrP^C is substantially concentrated on exosome, constituting a mechanism of exosomal targeted delivery (Falker et al., 2016). This binding induces Aβ fibrillation and could contribute to Aβ extracellular deposition (Kessels, Nguyen, Nabavi, & Malinow, 2010; Laurén, Gimbel, Nygaard, Gilbert, & Strittmatter, 2009).

Exosomes generated within the brain cells operate to engulf, soaking up soluble plaques with final abnormal aggregates destruction (An et al., 2013; Dinkins, Dasgupta, Wang, Zhu, & Bieberich, 2014; Yuyama et al., 2015). The anchoring protein PrP^c and glycosphingolipids, ceramide, all interact with the Aβ aggregates (An et al., 2013; Yuyama & Igarashi, 2017). Due to the abundance of ganglioside and sialylated glycosphingolipids on their surface, neuronal exosomes bind Aβ aggregates more effectively than astrocytic or microglial exosomes (Laurén et al., 2009; Yamamoto, Matsubara, Sato, & Yanagisawa, 2008; Yuyama & Igarashi, 2017). Sphingolipid ceramide is more abundant in astrocyte exosomes (Yuyama & Igarashi, 2017).

20.6.1.1.5 Exosomes in the Treatment of AD: Hopes

During AD, MSCs exosomes provide micro-RNAs to the brain, encouraging neuritic remodeling and plasticity, preventing brain cells and neuronal apoptosis, therefore supporting functional recovery in different brain injuries (Cheng et al., 2018; Xin et al., 2013; Xin et al., 2017).

But which exosomal cargo element is responsible for these effects is the question that remains unanswered as yet.

MiR-133b increasing the secondary release of exosomes from astrocytes, miR-98, miR-155, and miR-125a all suppress programmed brain cell death (Ma, Zang, Xi, Yang, & Zou, 2016).

MiR-21 levels drop significantly during chronic inflammation and apoptosis, yet MSCs in these circumstances produce the hot gene *miR-21* to inhibit apoptosis (Cheng et al., 2018). Yet, miR-21 as being upregulated in various cancer types approaching approval as conventional molecular-biomarker (as for hepatocellular carcinoma) (Hammad et al., 2023; Eldosoky et al., 2023), therefore, needs to be examined extensively in AD and NDDs as well.

In order to avoid neuronal death and alleviate AD symptoms, several researchers used curcumin cargoes into exosomes to block tau phosphorylation and activate energy metabolism from the glycogen pathway, glycogen synthase kinase-3/Protein kinase B (GSK-3/AKT) (Wang et al., 2019).

Exosomes from various cell types injected into mice models have diverse therapeutic effects, first preventing synaptic dysfunction and controlling the inflammatory response via modulating miR-21 to enhance AD memory performance (Cui et al., 2018). Second, adipose-derived stem cell exosomes (ADSC-Exo) treatment reduced Aβ42 and Aβ40 levels, increased levels of pro-apoptotic molecules p53, pro-caspase 3, and lowered B-cell lymphoma 2 (Bcl-2) protein with reduced neuronal apoptosis. Moreover, during treatment, neurite development also increased (Lee, Ban, Yang, Im, & Kim, 2018). Finally, exosomes from human umbilical cord MSCs demonstrated the capacity to treat cognitive dysfunction, to assist in removing Aβ deposits from the brain, reduce neuroinflammation, and alleviate AD symptoms (Ding et al., 2018), therefore controlling the activity of mice brain microglia cells.

20.6.1.1.6 Insight

Future treatments for AD are anticipated to benefit from the growing understanding of miR-regulated molecular pathways and their effects within brain cells. These miRs constitute treatment hits by themselves or via their downstream genes, after being identified, first, computationally and, second, to be approved experimentally, constituting the bases for "exosomal-ncRNA precision therapeutic code for brain diseases."

20.6.1.1.7 NP-Loaded Drugs for AD Treatment

Donepezil®-loaded PLGA-b-PEG NPs has high BBB penetration capacity to cause amyloid fibril disintegration. The pharmacological activity of galantamine (GAL) was enhanced when being carried on PLGA due to their high encapsulation efficiency and the prolonged drug release from these cargos (Baysal, et al., 2017).

Memantine-loaded PLGA-PEGylated NPs were able to cross the BBB to the antagonist of the N-methyl-D-aspartate receptor and decrease Aβ plaques to treat mild to moderate Alzheimer's-type dementia (Matsunaga et al., 2015)

20.6.2 Parkinson's Disease (PD)

The pathological signs and symptoms are nigrostriatal dopaminergic neuron degeneration with Lewy bodies of α-synuclein misfolded protein. PD is a movement condition manifesting as resting tremor, stiffness, bradykinesia, and motor instability (Leverenz et al., 2009). Cognitive deterioration, depression, and psychosis are additional symptoms of this illness (Weintraub & Mamikonyan, 2019).

20.6.2.1 Exosomes in the Pathophysiology of PD: Hypes

α-synuclein protein entered the extracellular area without an extracellular sorting signal; this points a finger toward exosomes having a hand in PD pathophysiology. The first evidence that exosomes might be implicated in PD pathology was provided by an *in vitro* calcium-dependent extracellular production of exosomal α-synuclein (Emmanouilidou et al., 2010).

A closer look found that neuroglioma cells that express synuclein produce oligomers via two different mechanisms, one via exosomes, and the other by the direct release of free synuclein (Beatriz, Vilaça, & Lopes, 2021). PD patients' plasma and cerebrospinal fluid contain free α-synuclein (Borghi et al., 2000; El-Agnaf et al., 2003, Alamri, Vogel, MacAskill, & Anderson, 2016; Tomlinson et al., 2015; Yang et al., 2015).

Exosomes are more effective at exerting their harmful effects than does the free α-synuclein, although exosomes carry lower amounts of α-synuclein (Danzer et al., 2012). How exosomes convey the α-synuclein-damaging effects is via *encouraging α-synuclein aggregation* through their lipid and/or protein content, which makes it easier for cells to take up more α-synuclein. α-synuclein oligomerization is brought on by exosomal–ganglioside lipids mono-sialotetrahexosylganglioside (GM1) or exosomal–ganglioside lipids tri-sialotetrahexosylganglioside (GM3) to hasten α-synuclein oligomerization (Grey et al., 2015). Moreover, ceramides and neurodegeneration-related tau proteins within exosomes have the ability to cause α-synuclein to aggregate as well (Kurzawa-Akanbi et al., 2021). Neuronal exosomes carrying α-synuclein stimulate the aggregation of intracellular proteins in astrocytes, and more synucleinopathies result (Surgucheva, Sharov, & Surguchov, 2012) as well as pathological α-synuclein transfer intra-neuronally (Alvarez-Erviti et al., 2011) or from neurons to microglia, and vice versa (Kordower, Chu, Hauser, Freeman, & Olanow, 2008; Li et al., 2008).

Micro-RNA-containing exosomes from PD animal models, cultured cells, and patients' plasma or cerebrospinal fluid (Gui et al., 2015; Rajkumar et al., 2021) with the following pathways may be affected (Ruf et al., 2021) according to *in silico* pathway analyses, namely, gap junction, dopaminergic, and glutamatergic pathways by miR-23b-3p repression with α-synuclein mRNA upregulation (Cai et al., 2021) with an end result of an increased α-synuclein protein expression in PD.

20.6.2.2 Exosomes in the Treatment of PD: Hopes

20.6.2.2.1

When exosomes from bone marrow MSCs were implanted *in vivo* into a PD mouse model, altered-repaired-damage occurred, preventing the pathogenic symptoms of PD, with an enhanced Wnt family member 5a (Wnt5a) and Wnt pathway in these exosomes (Li et al., 2022).

20.6.2.2.2

Exosomes play a significant role in mediating α-synuclein transmission among brain cells, which is partly controlled by miRs. Again, exosomal–ncRNA could be utilized as a therapeutic hit for PD.

20.6.2.3 NP-Loaded Drugs for PD Treatment

Chitosan NPs loaded with selegiline exerted good therapeutic efficacy as an anti-PD (Sridhar et al., 2018).

20.6.3 Huntington's Disease (HD)

20.6.3.1 The Pathological Signs and Symptoms

The autosomal-dominant progressive NDD HD is characterized by gradual motor impairments from accumulating autophagosomes with overexpressed mutant huntingtin protein (mHtt) (Erie, Sacino, Houle, Lu, & Wei, 2015; Heng et al., 2010). The brain striatum and cortex become cytotoxic as a result of polyglutamine (polyQ) repeats in exon 1, with more protein mHtt aggregates production taken by phagosomes (Maat-Schieman et al., 1999; Takeuchi & Nagai, 2017) as well as autophagic vacuoles accumulation (Nagata, Sawa, Ross, & Snyder, 2004; Rudnicki, Pletnikova, Vonsattel, Ross, & Margolis, 2008).

20.6.3.2 Exosomes in the Pathophysiology of HD: Hypes

A critical axis in HD pathology is miR-124 being repressed in HD, which increases the neuron-restrictive silencing factor (*NRSF*) gene to depress brain-derived neurotrophic genes (Cheng, Pastrana, Tavazoie, & Doetsch, 2009; Das, Jana, & Bhattacharyya, 2013; Hwang & Zukin, 2018). Polyglutamine and amyloid buildup is one theory for HD development (Duan et al., 2023).

20.6.3.3 Exosomes in the Treatment of HD: Hopes

Few researches focused on the relationship between ncRNAs and reactive oxygen species (ROS) burden in HD (Engedal et al., 2018). MiR-124 expression was seen to be downregulated in HD (Johnson et al., 2008), with rise in the expression of its target Cyclin A2, controlling the brain cells' cell cycle (Das et al., 2013, Lee et al., 2017). Exosomal miR-124 constitutes a potential treatment hit for HD.

For autophagy, an increased mHtt aggregation was seen when autophagy was blocked through treatment with an autophagy inhibitor or genetic modification via exosomes (Fu et al., 2017; Korac et al., 2013).

20.6.3.4 NP-Loaded Drugs for HD Treatment

Epigallocatechin-3-gallate (EGCG) or selenium-loaded NPs are used for HD treatment (Debnath et al., 2016).

20.6.4 Amyotrophic Lateral Sclerosis (ALS)

20.6.4.1 Exosomes in the Pathophysiology of ALS: Hypes

Superoxide dismutase (SOD-1) and trans-activating response element (TAR) deoxyribonucleic acid (DNA)–binding protein 43 (TDP-43) misfolding are the main causes of this deadly late-onset NDD (Bosco et al., 2010; Neumann et al., 2006). Knowing that exosomes contain both TDP-43 and SOD-1 (Grad, Pokrishevsky, Silverman, & Cashman, 2014; Nonaka et al., 2013), abnormal SOD-1 folding transported intracellularly by the aid of exosomes and micropinocytosis of SOD protein aggregates via overexpressing both the wild-antimutant-anti-oxidantein and the mutant-antimutant-anti-oxidantein types *in vitro* testing (Grad et al., 2014).

Exosomes from both neurons and astrocytes were enriched in mutant SOD-1, therefore aiding in ALS expansion (Silverman et al., 2019). Exosomes extracted from the cerebral fluid of ALS patients contained TDP-43, a protein involved in the ALS disease's etiology (Feneberg et al., 2014), confirming the hypothesis that exosomes aid in illness spread. Cerebrospinal fluid enriched with exosomes containing the toxic TDP-43 protein was able to build up TDP-43 in glial cell lines (Ding et al., 2015), as well as being transferred between cells (Feiler et al., 2015).

20.6.4.2 Exosomes in the Treatment of ALS: Hopes

For the treatment of oxidative stress–related NDD as ALS (Ha et al., 2016) exosomes could be employed for gene

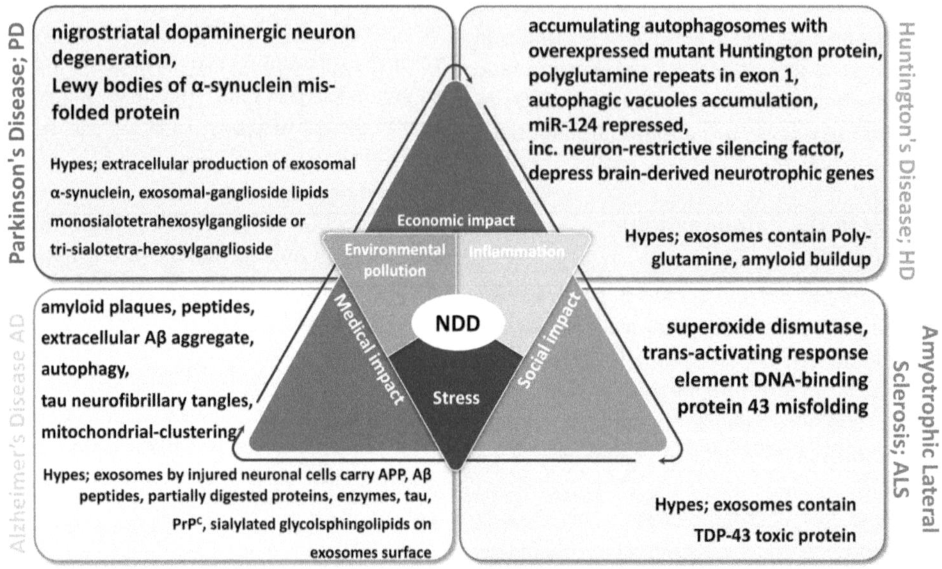

FIGURE 20.2 Exosomes in the pathophysiology of NDDs: "hypes" sketch addressing causes, impact, and the pathogenesis of various NDDs—AD, PD, HD, and ALS.

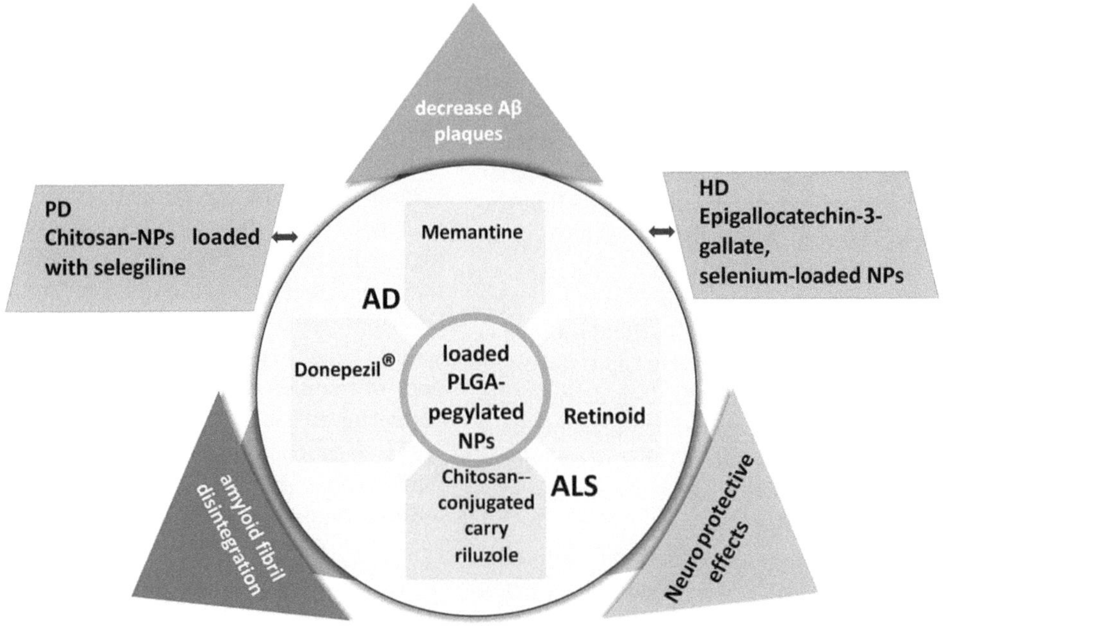

FIGURE 20.3 NP-loaded drugs sketch for NDD treatment: AD treatment to decrease amyloid peptide aggregates and their fibril disintegration when using PLGA-loaded NPs with potential or FDA-approved drugs. ALS treatments have neuroprotective effects via retinoid- or chitosan-conjugated PLGA NPs as for PD.

therapy via transporting miRs and silencing RNA (siRNA) to knock down target genes (Alvarez-Erviti et al., 2011). For example, TDP-43 silencing with downstream abnormalities in the autophagy process (Huang et al., 2020; Ling, Pletnikova, Troncoso, & Wong, 2015; Torres et al., 2018). ALS patients would survive better after autophagy activation via administered exosomes loaded with TDP-43-small interfering RNA (siRNA) (Barmada et al., 2014).

Interestingly, astrocytic–exosomes (ASC-Exo), via preventing mitochondrial malfunction, will control ALS, being a prospective treatment for ALS-enhanced motor function (Lee et al., 2016, Calabria et al., 2019, Bonafede et al., 2020).

20.6.4.3 NP-Loaded Drugs for ALS Treatment

There is no specific treatment for ALS until now, and approaches like using Clopidogrel (Khalil et al., 2016) with other drugs to improve neurological impairment, therefore, going through nanomedicine for treatment or cell-based therapy is crucial.

Retinoic acid signaling pathway dysfunction plays a role in ALS development; therefore, retinoid-activating NPs improved ALS motor function with neuroprotection (Duan et al., 2023).

Chitosan-conjugated NPs could carry large amounts of riluzole across the BBB, producing significant neuroprotective effects for severe motor neuron disorders treatment in ALS (Verma et al., 2016).

A hypothesis to consider: The opioid antagonist's naloxone and naltrexone "nano-drugs" or NPs both exhibited reduced oxidative stress, attenuation of morphine- or oxycodone-induced alterations of tight junction (TJ) protein expression (Peyravian et al., 2022) that may benefit ALS patients. Moreover, neurotrophic factors were employed for ALS treatment in preclinical trials (Pardridge, 2022).

20.7 EXPERT OPINION

20.7.1 Consorting Nano-Biotech for Various NDDs

Due to the complexity of the BBB, innovative strategies to be designed using nanotechnology rather than conventional DDS is a must. Methotrexate, the conventional anti-cancer (El-Mesallamy et al., 2014) has neuroprotective benefit targeting high systemic inflammation, but with no effect toward Alzheimer's-specific pathways.

The brain microenvironment (BME) consists of signaling molecules, endothelial receptors, secreted factors, vasculature, and immune cells with the involvement of natural killer (NK) cells or neutrophils. siRNA therapeutics using a novel lipid nanoparticle (LNP) delivery system.

Nano-therapy targeting the BME would be a promising approach, first, to overcome drug resistance, as well as offering a versatile tool by enabling specific delivery of either single or combinations, along with targeting overexpressed receptors or enzymes of the endothelial linings that would contribute to the BBB.

In an attempt to enhance drug absorption for poorly permeable drugs bioavailability, a self-nanoemulsifying system as well as the integrated nanovesicular/self-nanoemulsifying system were designed.

Numerous NPs approaches have been reported to target cells in the brain tumor microenvironment condition of acidosis, redox potential, hypoxia, hyperthermia, oxidative stress, and high interstitial fluid pressure, based on the concept of "nano-by-design" (NbD); targeted stimuli-responsive nanomedicines is one approach. The second application of the concept of NbD is based on the Internet-of-Nanothings (IoNTs) for personalized medical (PM) treatment.

A hypothesis to consider: Preparing a physically stable nanoemulsion system loaded with a full drug dose inhaler NPs formula to treat pulmonary hypertension and brain cells affection or nano-based intranasal drug delivery formula. Moreover, addressing next-generation nanomedicine (NGNM) to target epigenetics (ncRNA) that contribute to NDDs as one of the BME milieu or to deliver a wide range of genetic therapies effectively.

Sublingual and intranasal transmucosal deliveries would constitute better acceptable delivery routes better than the oral route (Fouad et al., 2016) for exosomes cargo delivery.

It is crucial to deliver vitamins E, C, and D conjugated to the exosomes membrane to target the brain endothelial cells or to be delivered within exosomes to overload the brain cells with these vitamins, to reduce endothelial complicated involvement in NDDs, increase the acidic milieu in the brain cells, antioxidant vitamins, and finally, the anti-inflammatory vitamin D effect (Hamdy et al., 2009; el-Derany et al., 2016).

20.7.2 Recommendations

Engineering exosomes with drug-loaded cargo is an exciting new development in the fight against off-target treatment. By using a design-of-experiment (DOE) approach to identify key variables, as well as the NbD approach, we can create highly targeted and effective drug-loaded exosomes. Moreover, the exosomes-by-design (EbD) approach would be a hero in the fight against cancer treatment-resistance with minimal side effects. Therefore, these innovative approaches -by-design (-bD) are potential game changers in the field of PM and offer hope for a brighter future using ncRNA for PM.

20.7.3 Future Perspectives

Several proteins and enzymes in brain-derived exosomes have been identified and presented according to the Exocarta DB http://exocarta.org/index.html that could be targeted for drug repurposing or drug design, such as tetraspanins, GTPase, fibronectin, heat shock proteins, MVB proteins, cytoskeleton-related proteins, prominin-1 (CD133), basigin (CD147), and B7-H3 (CD276). By targeting these proteins, we can potentially develop new drugs to treat various NDDs. This approach shows great promise and could lead to significant advancements in the field.

20.7.4 Limitations

When considering the use of exosomes for nano-drug delivery, it is important to weigh the potential disadvantages. One of the main drawbacks is the limited quantity of naturally extracted exosomes, which could make sustainable/continuous *in vivo* use challenging. Additionally, the cost associated with using natural exosomes can be prohibitively high. Fortunately, there are new alternatives emerging in the form of artificial exosomes, also known as nanovesicles (NVs), exosome–mimetic (EM) particles, and hybrid exosomes (HEs). These options can help overcome the limitations of traditional exosomes, making them a more viable option for PM drug delivery (Li et al., 2021) for various NDDs mentioned as well as for medulloblastoma brain tumors (Sokolov et al., 2024) or glioblastoma multiforme (GBM) (Hamdy et al., 2024).

20.8 SUMMARY AND CONCLUSION

Neurodegenerative diseases or neurodevelopmental disorders (NDDs) are complex and multifactorial. From the hallmarks that can trigger NDDs are the degradation or aggregation of heat-sensitive brain proteins, the formation of inclusion bodies in the brain, issues with intracellular protein trafficking, and endocytic sorting, mitochondrial dysfunction, iron buildup, imbalances in antioxidants, and oxidative stress, inflammation of nervous tissue, and epigenetic dysregulation such as micro-RNA deregulation (Mahmoud et al., 2021). These factors contribute, either all together or dual, to NDD development, making the condition/disorder a challenge to manage.

It is crucial to find more effective ways to treat NDDs, since new cases are increasing every year, due to epigenetic causes related to the environment, or post-COVID-19 (Hamdy et al., 2022b). The current traditional treatment methods of systemic drug administration have many drawbacks and side effects; therefore, options to use synthetic nanocarriers appeared mandatory, but they also have their downsides. Recent experiments have shown that direct intranasal as NtB drug administration can be successful in targeting the brain, as evidenced by drug targeting effectiveness and direct drug transport percentage indices to the tumor as brain-to-tumor (BtT) approach. One promising natural nanocarrier option for treating NDDs is the use of exosomes. These tiny structures can be found in many different body fluids, and their membranes offer protection for the biological payload they carry. Exosomes are a better option than conventional synthetic nanocarriers and could be a game changer in the treatment of NDDs, a step toward ncRNA precision in brain diseases.

Human cells cultured *in vitro* have been utilized extensively in various exosome *ex vivo* or *in vivo* experiments, along with exosomes derived from plants (as orange juice) or milk. These innovative DDSs have greatly benefited from NT and nanocarriers. However, there is a growing demand for natural, eco-friendly, and safe self-cell-derived nanocarriers that can accurately and specifically target the desired cell with minimal or no adverse effects. Moreover, these nanocarriers do not induce resistance or immunogenicity and, therefore, to be considered a promising solution for drug delivery, considered as one of the next-generation DDSs.

Exosomes fulfil almost all these criteria, expected as the next-generation treatment, and constitute a step for ncRNA precision application.

It is fascinating to learn about the various components carried out within exosomes and their potential roles in NDD pathophysiology. The hype surrounding exosomes is somehow now more understandable, given their potential to provide hope in the form of diagnosis or prognostic follow-up. Furthermore, their biomolecular contents could hold promising treatment options in the near-future PM clinical setting achieving "better health" (SDG3). However, it is important to strike a balance between hype and hope and continue to research and understand the complexities of exosomes.

REFERENCES

Alamri, Y., Vogel, R., MacAskill, M., & Anderson, T. (2016). Plasma exosome concentration may correlate with cognitive impairment in Parkinson's disease. *Alzheimer's & Dementia: Diagnosis, Assessment & Disease Monitoring*, *4*(1), 107–108. https://doi.org/10.1016/j.dadm.2016.08.002

Alvarez-Erviti, L., Seow, Y., Schapira, A. H., Gardiner, C., Sargent, I. L., Wood, M. J. A., & Cooper, J. M. (2011). Lysosomal dysfunction increases exosome-mediated alpha-synuclein release and transmission. *Neurobiology of Disease*, *42*(3), 360–367. https://doi.org/10.1016/j.nbd.2011.01.029

Alvarez-Erviti, L., Seow, Y., Yin, H., Betts, C., Lakhal, S., & Wood, M. J. (2011). Delivery of siRNA to the mouse brain by systemic injection of targeted exosomes. *Nature Biotechnology*, *29*(4), 341–345.

An, K., Klyubin, I., Kim, Y., Jung, J. H., Mably, A. J., O'Dowd, S. T., . . . Kim, J.-H. (2013). Exosomes neutralize synaptic-plasticity-disrupting activity of Aβ assemblies in vivo. *Molecular Brain*, *6*(1), 47. https://doi.org/10.1186/1756-6606-6-47

Antonucci, F., Turola, E., Riganti, L., Caleo, M., Gabrielli, M., Perrotta, C., . . . Verderio, C. (2012). Microvesicles released from microglia stimulate synaptic activity via enhanced sphingolipid metabolism. *EMBO Journal*, *31*(5), 1231–1240. https://doi.org/10.1038/emboj.2011.489

Baglio, S. R., Pegtel, D. M., & Baldini, N. (2012). Mesenchymal stem cell secreted vesicles provide novel opportunities in (stem) cell-free therapy. *Frontiers in Physiology*, *3*, 359.

Barmada, S. J., Serio, A., Arjun, A., Bilican, B., Daub, A., Ando, D. M., . . . Finkbeiner, S. (2014). Autophagy induction enhances TDP43 turnover and survival in neuronal ALS models. *Nature Chemical Biology*, *10*(8), 677–685. https://doi.org/10.1038/nchembio.1563

Baysal, I., Ucar, G., Gultekinoglu, M., Ulubayram, K., & Yabanoglu-Ciftci, S. (2017). Donepezil loaded PLGA-b-PEG nanoparticles: Their ability to induce destabilization of amyloid fibrils and to cross blood brain barrier *in vitro*. *Journal of Neural Transmission*, *124*, 33–45.

Beatriz, M., Vilaça, R., & Lopes, C. (2021). Exosomes: Innocent bystanders or critical culprits in neurodegenerative diseases. *Frontiers in Cell and Developmental Biology*, *9*. https://doi.org/10.3389/fcell.2021.635104

Bonafede, R., Turano, E., Scambi, I., Busato, A., Bontempi, P., Virla, F., . . . Mariotti, R. (2020). ASC-exosomes ameliorate the disease progression in SOD1(G93A) murine model underlining their potential therapeutic use in human ALS. *International Journal of Molecular Sciences*, *21*(10), 3651.

Borghi, R., Marchese, R., Negro, A., Marinelli, L., Forloni, G., Zaccheo, D., . . . Tabaton, M. (2000). Full length α-synuclein is present in cerebrospinal fluid from Parkinson's disease and normal subjects. *Neuroscience Letters*, *287*(1), 65–67. https://doi.org/10.1016/S0304-3940(00)01153-8

Bosco, D. A., Morfini, G., Karabacak, N. M., Song, Y., Gros-Louis, F., Pasinelli, P., . . . Brown, R. H. (2010). Wild-type and mutant SOD1 share an aberrant conformation and a common pathogenic pathway in ALS. *Nature Neuroscience*, *13*(11), 1396–1403. https://doi.org/10.1038/nn.2660

Bukhari SNA. (2022). Nanotherapeutics for Alzheimer's disease with preclinical evaluation and clinical trials: Challenges, promises and limitations. *Current Drug Delivery*, *19*(1), 17–31. https://doi.org/10.2174/1567201818666210910162750

Cai, M., Chai, S., Xiong, T., Wei, J., Mao, W., Zhu, Y., . . . Xiong, N. (2021). Aberrant expression of circulating microRNA leads to the dysregulation of alpha-synuclein and other pathogenic genes in Parkinson's disease. *Frontiers in Cell and Developmental Biology*, *9*. https://doi.org/10.3389/fcell.2021.695007

Calabria, E., Scambi, I., Bonafede, R., Schiaffino, L., Peroni, D., Potrich, V., . . . Mariotti, R. (2019). ASCs-exosomes recover coupling efficiency and mitochondrial membrane potential in an *in vitro* model of ALS. *Frontiers in Neuroscience*, *13*. https://doi.org/10.3389/fnins.2019.01070

Chai, X., Dage, J. L., & Citron, M. (2012). Constitutive secretion of tau protein by an unconventional mechanism. *Neurobiology of Disease*, *48*(3), 356–366. https://doi.org/10.1016/j.nbd.2012.05.021

Chen, J.-J., Zhao, B., Zhao, J., & Li, S. (2017). Potential roles of exosomal microRNAs as diagnostic biomarkers and therapeutic application in Alzheimer's disease. *Neural Plasticity*, *2017*.

Cheng, L.-C., Pastrana, E., Tavazoie, M., & Doetsch, F. (2009). miR-124 regulates adult neurogenesis in the subventricular zone stem cell niche. *Nature Neuroscience*, *12*(4), 399–408.

Cheng, X., Zhang, G., Zhang, L., Hu, Y., Zhang, K., Sun, X., . . . Zhao, J. (2018). Mesenchymal stem cells deliver exogenous miR-21 via exosomes to inhibit nucleus pulposus cell apoptosis and reduce intervertebral disc degeneration. *Journal of Cellular and Molecular Medicine*, *22*(1), 261–276.

Chivet, M., Javalet, C., Laulagnier, K., Blot, B., Hemming, F. J., & Sadoul, R. (2014). Exosomes secreted by cortical neurons upon glutamatergic synapse activation specifically interact with neurons. *Journal of Extracellular Vesicles*, *3*(1), 24722.

Cui, G.-H., Wu, J., Mou, F.-F., Xie, W.-H., Wang, F.-B., Wang, Q.-L., . . . Guo, H.-D. (2018). Exosomes derived from hypoxia-preconditioned mesenchymal stromal cells ameliorate cognitive decline by rescuing synaptic dysfunction and regulating inflammatory responses in APP/PS1 mice. *FASEB Journal*, *32*(2), 654–668. https://doi.org/10.1096/fj.201700600R

Danzer, K. M., Kranich, L. R., Ruf, W. P., Cagsal-Getkin, O., Winslow, A. R., Zhu, L., . . . McLean, P. J. (2012). Exosomal cell-to-cell transmission of alpha synuclein oligomers. *Molecular Neurodegeneration*, *7*(1), 42. https://doi.org/10.1186/1750-1326-7-42

Das, E., Jana, N. R., & Bhattacharyya, N. P. (2013). MicroRNA-124 targets CCNA2 and regulates cell cycle in STHdhQ111/HdhQ111 cells. *Biochemical and Biophysical Research Communications*, *437*(2), 217–224.

Debnath, K., Shekhar, S., Kumar, V., Jana, N. R., & Jana, N. R. (2016). Efficient inhibition of protein aggregation, disintegration of aggregates, and lowering of cytotoxicity by green tea polyphenol-based self-assembled polymer nanoparticles. *ACS Applied Materials & Interfaces*, *8*, 20309–20318.

Ding, M., Shen, Y., Wang, P., Xie, Z., Xu, S., Zhu, Z., . . . Yang, H. (2018). Exosomes isolated from human umbilical cord mesenchymal stem cells alleviate neuroinflammation and reduce amyloid-beta deposition by modulating microglial activation in Alzheimer's disease. *Neurochemical Research*, *43*(11), 2165–2177. https://doi.org/10.1007/s11064-018-2641-5

Ding, X., Ma, M., Teng, J., Teng, R. K., Zhou, S., Yin, J., . . . Wang, X. (2015). Exposure to ALS-FTD-CSF generates TDP-43 aggregates in glioblastoma cells through exosomes and TNTs-like structure. *Oncotarget*, *6*(27), 24178–24191. https://doi.org/10.18632/oncotarget.4680

Dinkins, M. B., Dasgupta, S., Wang, G., Zhu, G., & Bieberich, E. (2014). Exosome reduction in vivo is associated with lower amyloid plaque load in the 5XFAD mouse model of Alzheimer's disease. *Neurobiology of Aging*, *35*(8), 1792–1800. https://doi.org/10.1016/j.neurobiolaging.2014.02.012

Dinkins, M. B., Enasko, J., Hernandez, C., Wang, G., Kong, J., Helwa, I., . . . Bieberich, E. (2016). Neutral sphingomyelinase-2 deficiency ameliorates Alzheimer's disease pathology and improves cognition in the 5XFAD mouse. *Journal of Neuroscience*, *36*(33), 8653–8667.

Duan, L., Li, X., Ji, R., Hao, Z., Kong, M., Wen, X., Guan, F., & Ma, S. (2023). Nanoparticle-based drug delivery systems: An inspiring therapeutic strategy for neurodegenerative diseases. *Polymers*, *15*, 2196. https://doi.org/10.3390/polym15092196

El-Agnaf, O. M. A., Salem, S. A., Paleologou, K. E., Cooper, L. J., Fullwood, N. J., Gibson, M. J., . . . Allsop, D. (2003). α-synuclein implicated in Parkinson's disease is present in extracellular biological fluids, including human plasma. *FASEB Journal*, *17*(13), 1–16. https://doi.org/10.1096/fj.03-0098fje

El-Derany, M. O., Hamdy, N. M., Al-Ansari, N. L., & El-Mesallamy, H. O. (2016). Integrative role of vitamin D related and Interleukin-28B genes polymorphism in predicting treatment outcomes of Chronic Hepatitis C. *BMC Gastroenterology*, *16*(1), 1–12.

Eldosoky, M. A., Hammad, R., Elmadbouly, A. A., Aglan, R. B., AbdelHamid, S. G., Alboraie, M., . . . Lambert, C. (2023). Diagnostic significance of hsa-miR-21-5p, hsa-miR-192-5p, hsa-miR-155-5p, hsa-miR-199a-5p panel and ratios in hepatocellular carcinoma on top of liver cirrhosis in HCV-infected patients. *International Journal of Molecular Sciences*, *24*, 3157.

El-Mesallamy, H. O., Anwar, W., Hamdy, N. M., & Hamdy, M. (2014). High-dose methotrexate in Egyptian pediatric acute lymphoblastic leukemia: The impact of ABCG2 C421A genetic polymorphism on plasma levels, what is next? *Journal of Cancer Research and Clinical Oncology*, *140*(8), 1359–1365.

ElShagea, H. N., Makar, R. R., Salama, A. H., Elkasabgy, N. A., & Basalious, E. B. (2023). Investigating the targeting power to brain tissues of intranasal rasagiline mesylate-loaded transferosomal in situ gel for efficient treatment of Parkinson's disease. *Pharmaceutics*, *15*, 533.

Elsherbini, A., Kirov, A. S., Dinkins, M. B., Wang, G., Qin, H., Zhu, Z., . . . Bieberich, E. (2020). Association of Aβ with ceramide-enriched astrosomes mediates Aβ neurotoxicity. *Acta Neuropathologica Communications*, *8*(1), 60. https://doi.org/10.1186/s40478-020-00931-8

Emmanouilidou, E., Melachroinou, K., Roumeliotis, T., Garbis, S. D., Ntzouni, M., Margaritis, L. H., . . . Vekrellis, K. (2010). Cell-produced α-synuclein is secreted in a calcium-dependent manner by exosomes and impacts neuronal survival. *Journal of Neuroscience*, *30*(20), 6838–6851. https://doi.org/10.1523/jneurosci.5699-09.2010

Engedal, N., Žerovnik, E., Rudov, A., Galli, F., Olivieri, F., Procopio, A. D., . . . Albertini, M. C. (2018). From oxidative stress damage to pathways, networks, and autophagy via microRNAs. *Oxidative Medicine and Cellular Longevity*, *2018*.

Erie, C., Sacino, M., Houle, L., Lu, M. L., & Wei, J. (2015). Altered lysosomal positioning affects lysosomal functions in a cellular model of Huntington's disease. *European Journal of Neuroscience*, *42*(3), 1941–1951. https://doi.org/10.1111/ejn.12957

Falker, C., Hartmann, A., Guett, I., Dohler, F., Altmeppen, H., Betzel, C., . . . Joshi, P. (2016). Exosomal cellular prion protein drives fibrillization of amyloid beta and counteracts amyloid beta-mediated neurotoxicity. *Journal of `Neurochemistry*, *137*(1), 88–100.

Farjadian, F., Ghasemi, A., Gohari, O., Roointan, A., Karimi, M., & Hamblin, M. R. (2019). Nanopharmaceuticals and nanomedicines currently on the market: Challenges and opportunities. *Nanomedicine*, *14*(1), 93–126. https://doi.org/10.2217/nnm-2018-0120

Feiler, M. S., Strobel, B., Freischmidt, A., Helferich, A. M., Kappel, J., Brewer, B. M., . . . Weishaupt, J. H. (2015). TDP-43 is intercellularly transmitted across axon terminals. *Journal of Cell Biology*, *211*(4), 897–911. https://doi.org/10.1083/jcb.201504057

Feneberg, E., Steinacker, P., Lehnert, S., Schneider, A., Walther, P., Thal, D. R., . . . Otto, M. (2014). Limited role of free TDP-43 as a diagnostic tool in neurodegenerative diseases. *Amyotrophic Lateral Sclerosis and Frontotemporal Degeneration*, *15*(5–6), 351–356. https://doi.org/10.3109/21678421.2014.905606

Fitzner, D., Schnaars, M., Van Rossum, D., Krishnamoorthy, G., Dibaj, P., Bakhti, M., . . . Simons, M. (2011). Selective transfer of exosomes from oligodendrocytes to microglia by macropinocytosis. *Journal of Cell Science*, *124*(3), 447–458.

Fouad, S. A., Shamma, R. N., Basalious, E. B., El-Nabarawi, M. M., & Tayel, S. A. (2018) Novel instantly-dispersible nanocarrier powder system (IDNPs) for intranasal delivery of dapoxetine hydrochloride: In-vitro optimization, ex-vivo permeation studies, and in-vivo evaluation. *Drug Development and Industrial Pharmacy*, *44*(9), 1443–1450.

Fouad, S. A., Shamma, R. N., Basalious, E. B., El-Nabarawi, M. M., & Tayel, S. A. (2016) Novel instantly-soluble transmucosal matrix (ISTM) using dual mechanism solubilizer for sublingual and nasal delivery of dapoxetine hydrochloride: In-vitro/in-vivo evaluation. *International Journal of Pharmaceutics*, *505*(1–2), 212–222. https://doi.org/10.1016/j.ijpharm.2016.04.006

Frost, B., & Diamond, M. I. (2010). Prion-like mechanisms in neurodegenerative diseases. *Nature Reviews Neuroscience*, *11*(3), 155–159. https://doi.org/10.1038/nrn2786

Frühbeis, C., Fröhlich, D., & Krämer-Albers, E. M. (2012). Emerging roles of exosomes in neuron-glia communication.

Frontiers in Physiology, *3*, 119. https://doi.org/10.3389/fphys.2012.00119

Frühbeis, C., Fröhlich, D., Kuo, W. P., Amphornrat, J., Thilemann, S., Saab, A. S., . . . Krämer-Albers, E.-M. (2013). Neurotransmitter-triggered transfer of exosomes mediates oligodendrocyte–neuron communication. *PLOS Biology*, *11*(7), e1001604. https://doi.org/10.1371/journal.pbio.1001604

Frühbeis, C., Kuo-Elsner, W. P., Müller, C., Barth, K., Peris, L., Tenzer, S., . . . Krämer-Albers, E.-M. (2020). Oligodendrocytes support axonal transport and maintenance via exosome secretion. *PLOS Biology*, *18*(12), e3000621. https://doi.org/10.1371/journal.pbio.3000621

Fu, Y., Wu, P., Pan, Y., Sun, X., Yang, H., Difiglia, M., & Lu, B. (2017). A toxic mutant huntingtin species is resistant to selective autophagy. *Nature Chemical Biology*, *13*(11), 1152–1154. https://doi.org/10.1038/nchembio.2461

Goldie, B. J., Dun, M. D., Lin, M., Smith, N. D., Verrills, N. M., Dayas, C. V., & Cairns, M. J. (2014). Activity-associated miRNA are packaged in Map1b-enriched exosomes released from depolarized neurons. *Nucleic Acids Research*, *42*(14), 9195–9208. https://doi.org/10.1093/nar/gku594

Grad, L. I., Pokrishevsky, E., Silverman, J. M., & Cashman, N. R. (2014). Exosome-dependent and independent mechanisms are involved in prion-like transmission of propagated Cu/Zn superoxide dismutase misfolding. *Prion*, *8*(5), 331–335. https://doi.org/10.4161/19336896.2014.983398

Grey, M., Dunning, C. J., Gaspar, R., Grey, C., Brundin, P., Sparr, E., & Linse, S. (2015). Acceleration of alpha-synuclein aggregation by exosomes. *Journal of Biological Chemistry*, *290*(5), 2969–2982. https://doi.org/10.1074/jbc.M114.585703

Gui, Y., Liu, H., Zhang, L., Lv, W., & Hu, X. (2015). Altered microRNA profiles in cerebrospinal fluid exosome in Parkinson disease and Alzheimer disease. *Oncotarget*, *6*(35), 37043–37053. https://doi.org/10.18632/oncotarget.6158

Ha, D., Yang, N., & Nadithe, V. (2016). Exosomes as therapeutic drug carriers and delivery vehicles across biological membranes: Current perspectives and future challenges. *Acta Pharmaceutica Sinica B*, *6*(4), 287–296.

Hamdy, N. M. (2011). Relationship between pro-anti-inflammatory cytokines, T-cell activation and CA 125 in obese patients with heart failure. *Medical Science Monitor*, *25*, 17(3), CR174–CR179.

Hamdy, N. M., Suwailem, S.M., El-Mesallamy, H. O. (2009). Influence of vitamin E supplementation on endothelial complications in type 2 diabetes mellitus patients who underwent coronary artery bypass graft. *Journal of Diabetes and its Complications*, *23*(3), 167–173.

Hamdy, N. M., Boseila, A. A., Ramadan, A., Basalious, E. B. (2022a). Iron oxide nanoparticles-plant insignia synthesis with favorable biomedical activities and less toxicity, in the "era of the-green": A systematic review. *Pharmaceutics*, *14*(4), 844.

Hamdy, N. M., Shaker, F. H., Zhan, X., & Basalious, E. B. (2022b). Tangled quest of post-COVID-19 infection-caused neuropathology and what 3P nano-bio-medicine can solve? *EPMA Journal*, *13*, 261–284.

Hamdy, N. M., Gabr, M. M., AbdelHamid, S. G., & Swellam, M. M. (2024). *Scrutinizing the landscape of DNA methylation epigenetic face(s) in glioblastomas, in nutraceutical fruits and foods for neurodegenerative disorders*. Elsevier, Chapter 24; pp. 467–504.

Hammad, R., Eldosoky, M. A., Elmadbouly, A. A., Aglan, R. B., AbdelHamid, S. G., Zaky, S., . . . Hamdy, N. M. (2023). Monocytes subsets altered distribution and dysregulated plasma hsa-miR-21-5p and hsa-miR-155-5p in HCV-linked liver cirrhosis progression to hepatocellular carcinoma. *Journal of Cancer Research and Clinical Oncology*, *149*, 15349–15364.

Haney, M. J., Klyachko, N. L., Zhao, Y., Gupta, R., Plotnikova, E. G., He, Z., . . . Kabanov, A. V. (2015). Exosomes as drug delivery vehicles for Parkinson's disease therapy. *Journal of Controlled Release*, *207*, 18–30.

Heemels, M.-T. (2016). Neurodegenerative diseases. *Nature*, *539*(7628), 179–180.

Heng, M. Y., Duong, D. K., Albin, R. L., Tallaksen-Greene, S. J., Hunter, J. M., Lesort, M. J., . . . Detloff, P. J. (2010). Early autophagic response in a novel knock-in model of Huntington disease. *Human Molecular Genetics*, *19*(19), 3702–3720. https://doi.org/10.1093/hmg/ddq285

Huang-Doran, I., Zhang, C.-Y., & Vidal-Puig, A. (2017). Extracellular vesicles: Novel mediators of cell communication in metabolic disease. *Trends in Endocrinology & Metabolism*, *28*(1), 3–18.

Huang, C. L., Hsiao, I. L., Lin, H. C., Wang, C. F., Huang, Y. J., Chuang, C. Y. (2015). Silver nanoparticles affect on gene expression of inflammatory and neurodegenerative responses in mouse brain neural cells. *Environmental Research*, *136*, 253–263.

Huang, S.-L., Wu, L.-S., Lee, M., Chang, C.-W., Cheng, W.-C., Fang, Y.-S., . . . Shen, C.-K. J. (2020). A robust TDP-43 knock-in mouse model of ALS. *Acta Neuropathologica Communications*, *8*(1), 3. https://doi.org/10.1186/s40478-020-0881-5

Hwang, J.-Y., & Zukin, R. S. (2018). REST, a master transcriptional regulator in neurodegenerative disease. *Current Opinion in Neurobiology*, *48*, 193–200.

Jaber, N., Mohd-Naim, N., Wang, Z., DeLeon, J. L., Kim, S., Zhong, H., . . . Du, G. (2016). Vps34 regulates Rab7 and late endocytic trafficking through recruitment of the GTPase-activating protein Armus. *Journal of Cell Science*, *129*(23), 4424–4435.

Jablonski, K. A., Gaudet, A. D., Amici, S. A., Popovich, P. G., & Guerau-de-Arellano, M. (2016). Control of the inflammatory macrophage transcriptional signature by miR-155. *PLoS One*, *11*(7), e0159724. doi: 10.1371/journal.pone.0159724

Johnson, R., Zuccato, C., Belyaev, N. D., Guest, D. J., Cattaneo, E., & Buckley, N. J. (2008). A microRNA-based gene dysregulation pathway in Huntington's disease. *Neurobiology of Diseas*, *29*(3), 438–445.

Kalluri, R., & LeBleu, V. S. (2020). The biology, function, and biomedical applications of exosomes. *Science*, *367*(6478), eaau6977.

Kaur, S., Verma, H., Dhiman, M., Tell, G., Gigli, G. L., Janes, F., & Mantha, A. K. (2021). Brain exosomes: Friend or foe in Alzheimer's disease? *Molecular Neurobiology*, *58*(12), 6610–6624. https://doi.org/10.1007/s12035-021-02547-y

Kessels, H. W., Nguyen, L. N., Nabavi, S., & Malinow, R. (2010). The prion protein as a receptor for amyloid-β. *Nature*, *466*(7308), E3–E4.

Kim, E., Otgontenger, U., Jamsranjav, A., & Kim, S. S. (2020). Deleterious alteration of glia in the brain of Alzheimer's disease. *International Journal of Molecular Sciences*, *21*(18). https://doi.org/10.3390/ijms21186676

Khalil, B. M., Shahin, M. H., Solayman, M. H. M., Langaee, T., Schaalan, M. F., Gong, Y., . . . Johnson, J. A. (2016). Genetic and nongenetic factors affecting clopidogrel response in the Egyptian population. *Clinical and Translational Science*, *9*(1), 23.

Korac, J., Schaeffer, V., Kovacevic, I., Clement, A. M., Jungblut, B., Behl, C., . . . Dikic, I. (2013). Ubiquitin-independent function of optineurin in autophagic clearance of protein aggregates. *Journal of Cell Science*, *126*(2), 580–592. https://doi.org/10.1242/jcs.114926

Kordower, J. H., Chu, Y., Hauser, R. A., Freeman, T. B., & Olanow, C. W. (2008). Lewy body–like pathology in long-term embryonic nigral transplants in Parkinson's disease. *Nature Medicine*, *14*(5), 504–506. https://doi.org/10.1038/nm1747

Krämer-Albers, E.-M., Bretz, N., Tenzer, S., Winterstein, C., Möbius, W., Berger, H., . . . Trotter, J. (2007). Oligodendrocytes secrete exosomes containing major myelin and stress-protective proteins: Trophic support for axons? *Proteomics – Clinical Applications*, *1*(11), 1446–1461. https://doi.org/10.1002/prca.200700522

Kumari, M., & Anji, A. (2022). Small but mighty—exosomes, novel intercellular messengers in neurodegeneration. *Biology*, *11*(3), 413.

Kurzawa-Akanbi, M., Tammireddy, S., Fabrik, I., Gliaudelytė, L., Doherty, M. K., Heap, R., . . . Morris, C. M. (2021). Altered ceramide metabolism is a feature in the extracellular vesicle-mediated spread of alpha-synuclein in Lewy body disorders. *Acta Neuropathologica*, *142*(6), 961–984. https://doi.org/10.1007/s00401-021-02367-3

Lachenal, G., Pernet-Gallay, K., Chivet, M., Hemming, F. J., Belly, A., Bodon, G., . . . Sadoul, R. (2011). Release of exosomes from differentiated neurons and its regulation by synaptic glutamatergic activity. *Molecular and Cellular Neuroscience*, *46*(2), 409–418. https://doi.org/10.1016/j.mcn.2010.11.004

Lakhal, S., & Wood, M. J. (2011). Exosome nanotechnology: An emerging paradigm shift in drug delivery: Exploitation of exosome nanovesicles for systemic in vivo delivery of RNAi heralds new horizons for drug delivery across biological barriers. *Bioessays*, *33*(10), 737–741.

Laurén, J., Gimbel, D. A., Nygaard, H. B., Gilbert, J. W., & Strittmatter, S. M. (2009). Cellular prion protein mediates impairment of synaptic plasticity by amyloid-β oligomers. *Nature*, *457*(7233), 1128–1132.

Lee, M., Ban, J.-J., Kim, K. Y., Jeon, G. S., Im, W., Sung, J.-J., & Kim, M. (2016). Adipose-derived stem cell exosomes alleviate pathology of amyotrophic lateral sclerosis in vitro. *Biochemical and Biophysical Research Communications*, *479*(3), 434–439. https://doi.org/10.1016/j.bbrc.2016.09.069

Lee, M., Ban, J.-J., Yang, S., Im, W., & Kim, M. (2018). The exosome of adipose-derived stem cells reduces β-amyloid pathology and apoptosis of neuronal cells derived from the transgenic mouse model of Alzheimer's disease. *Brain Research*, *1691*, 87–93. https://doi.org/10.1016/j.brainres.2018.03.034

Lee, S.-T., Im, W., Ban, J.-J., Lee, M., Jung, K.-H., Lee, S. K., . . . Kim, M. (2017). Exosome-based delivery of miR-124 in a Huntington's disease model. *Journal of Movement Disorders*, *10*(1), 45.

Lee, S. H., Shin, S. M., Zhong, P., Kim, H.-T., Kim, D.-I., Kim, J. M., . . . Liu, Q.-S. (2018). Reciprocal control of excitatory synapse numbers by Wnt and Wnt inhibitor PRR7 secreted on exosomes. *Nature Communications*, *9*(1), 3434. https://doi.org/10.1038/s41467-018-05858-2

Leverenz, J. B., Quinn, J. F., Zabetian, C., Zhang, J., Montine, K. S., & Montine, T. J. (2009). Cognitive impairment and dementia in patients with Parkinson disease. *Current Topics in Medicinal Chemistry*, *9*(10), 903–912.

Li, J.-Y., Englund, E., Holton, J. L., Soulet, D., Hagell, P., Lees, A. J., . . . Brundin, P. (2008). Lewy bodies in grafted neurons in subjects with Parkinson's disease suggest host-to-graft disease propagation. *Nature Medicine*, *14*(5), 501–503. https://doi.org/10.1038/nm1746

Li, Y., Li, Z., Gu, J., Xu, X., Chen, H., & Gui, Y. (2022). Exosomes isolated during dopaminergic neuron differentiation suppressed neuronal inflammation in a rodent model of Parkinson's disease. *Neuroscience Letters*, *771*, 136414. https://doi.org/10.1016/j.neulet.2021.136414

Li, Y. J., Wu, J. Y., Liu, J., Xu, W., Qiu, X., Huang, S., Hu, X. B., Xiang, D. X. (2021 Aug 12). Artificial exosomes for translational nanomedicine. *Journal of Nanobiotechnology*, *19*(1), 242. https://doi.org/10.1186/s12951-021-00986-2

Ling, J. P., Pletnikova, O., Troncoso, J. C., & Wong, P. C. (2015). TDP-43 repression of nonconserved cryptic exons is compromised in ALS-FTD. *Science*, *349*(6248), 650–655. https://doi.org/10.1126/science.aab0983

Liu, H.-Y., Huang, C.-M., Hung, Y.-F., & Hsueh, Y.-P. (2015). The microRNAs Let7c and miR21 are recognized by neuronal toll-like receptor 7 to restrict dendritic growth of neurons. *Experimental Neurology*, *269*, 202–212. https://doi.org/10.1016/j.expneurol.2015.04.011

Ma, J. F., Zang, L. N., Xi, Y. M., Yang, W. J., & Zou, D. (2016). MiR-125a Rs12976445 polymorphism is associated with the apoptosis status of nucleus pulposus cells and the risk of intervertebral disc degeneration. *Cellular Physiology and Biochemistry*, *38*(1), 295–305.

Maat-Schieman, M. L., Dorsman, J. C., Smoor, M. A., Siesling, S., Van Duinen, S. G., Verschuuren, J. J., . . . Roos, R. A. (1999). Distribution of inclusions in neuronal nuclei and dystrophic neurites in Huntington disease brain. *Journal of Neuropathology & Experimental Neurology*, *58*(2), 129–137. https://doi.org/10.1097/00005072-199902000-00003

Mahmoud, M. M., Sanad, E. F., & Hamdy, N. M. (2021). MicroRNAs role in the environment-related non-communicable diseases and link to multidrug resistance, regulation or alteration. *Environmental Science and Pollution Research*, *28*, 36984–37000.

Malm, T., Loppi, S., & Kanninen, K. M. (2016). Exosomes in Alzheimer's disease. *Neurochemistry International*, *97*, 193–199.

Mathieu, M., Martin-Jaular, L., Lavieu, G., & Théry, C. (2019). Specificities of secretion and uptake of exosomes and other extracellular vesicles for cell-to-cell communication. *Nature Cell Biology*, *21*(1), 9–17.

Matsunaga, S., Kishi, T., & Iwata, N. (2015). Memantine monotherapy for Alzheimer's disease: A systematic review and meta-analysis. *PLoS One*, *10*, e0123289.

Mazayen, Z. M., Ghoneim, A. M., Elbatanony, R. S., Basalious, B. B., & Bendas, E. R. (2022). Pharmaceutical nanotechnology: From the bench to the market. *Future Journal of Pharmaceutical Sciences*, *8*, 12.

Miranda, A. M., Lasiecka, Z. M., Xu, Y., Neufeld, J., Shahriar, S., Simoes, S., . . . Di Paolo, G. (2018). Neuronal lysosomal dysfunction releases exosomes harboring APP C-terminal fragments and unique lipid signatures. *Nature Communications*, *9*(1), 291.

Mizushima, N., & Komatsu, M. (2011). Autophagy: Renovation of cells and tissues. *Cell*, *147*(4), 728–741.

Mohsen, K., Azzazy, H. M. E., Allam, N. K., & Basalious, E. B. (2020). Intranasal lipid nanocapsules for systemic delivery of nimodipine into the brain: *In vitro* optimization and in vivo pharmacokinetic study. *Materials Science and Engineering: C*, *116*, 111236.

Mulcahy, L. A., Pink, R. C., & Carter, D. R. F. (2014). Routes and mechanisms of extracellular vesicle uptake. *Journal of Extracellular Vesicles*, *3*(1), 24641.

Nagata, E., Sawa, A., Ross, C. A., & Snyder, S. H. (2004). Autophagosome-like vacuole formation in Huntington's disease lymphoblasts. *NeuroReport*, *15*(8), 1325–1328. https://doi.org/10.1097/01.wnr.0000127073.66692.8f

Neumann, M., Sampathu, D. M., Kwong, L. K., Truax, A. C., Micsenyi, M. C., Chou, T. T., . . . Lee, V. M.-Y. (2006). Ubiquitinated TDP-43 in frontotemporal lobar degeneration and amyotrophic lateral sclerosis. *Science*, *314*(5796), 130–133. https://doi.org/10.1126/science.1134108

Niu, X., Chen, J., & Gao, J. (2019). Nanocarriers as a powerful vehicle to overcome blood-brain barrier in treating neurodegenerative diseases: Focus on recent advances. *Asian Journal of Pharmaceutical Sciences*, *14*(5), 480–496.

Nonaka, T., Masuda-Suzukake, M., Arai, T., Hasegawa, Y., Akatsu, H., Obi, T., . . . Hasegawa, M. (2013). Prion-like properties of pathological TDP-43 aggregates from diseased brains. *Cell Reports*, *4*(1), 124–134. https://doi.org/10.1016/j.celrep.2013.06.007

Pardridge, W. M. A. (2022). Historical review of brain drug delivery. *Pharmaceutics*, *14*, 1283. https://doi.org/10.3390/pharmaceutics14061283

Pegtel, D. M., & Gould, S. J. (2019). Exosomes. *Annual Review of Biochemistry*, *88*, 487–514. https://doi.org/10.1146/annurev-biochem-013118-111902

Pei, X., Li, Y., Zhu, L., & Zhou, Z. (2020). Astrocyte-derived exosomes transfer miR-190b to inhibit oxygen and glucose deprivation-induced autophagy and neuronal apoptosis. *Cell Cycle*, *19*(8), 906–917.

Peyravian, N., Sun, E., Dikici, E., Deo, S., Daunert, S., Toborek, M. (2022 Jul 4). Opioid antagonist nanodrugs successfully attenuate the severity of ischemic stroke. *Molecular Pharmaceutics*. *19*(7), 2254–2267. https://doi.org/10.1021/acs.molpharmaceut.2c00079

Pistono, C., Bister, N., Stanová, I., & Malm, T. (2020). Glia-derived extracellular vesicles: Role in central nervous system communication in health and disease. *Frontiers in Cell and Developmental Biology*, *8*, 623771. https://doi.org/10.3389/fcell.2020.623771

Potolicchio, I., Carven, G. J., Xu, X., Stipp, C., Riese, R. J., Stern, L. J., & Santambrogio, L. (2005). Proteomic analysis of microglia-derived exosomes: Metabolic role of the aminopeptidase CD13 in neuropeptide catabolism1. *Journal of Immunology*, *175*(4), 2237–2243. https://doi.org/10.4049/jimmunol.175.4.2237

Proia, P., Schiera, G., Mineo, M., Ingrassia, A. M. R., Santoro, G., Savettieri, G., & Di Liegro, I. (2008). Astrocytes shed extracellular vesicles that contain fibroblast growth factor-2 and vascular endothelial growth factor. *International Journal of Molecular Medicine*, *21*(1), 63–67. https://doi.org/10.3892/ijmm.21.1.63

Rajendran, L., Honsho, M., Zahn, T. R., Keller, P., Geiger, K. D., Verkade, P., & Simons, K. (2006). Alzheimer's disease β-amyloid peptides are released in association with exosomes. *Proceedings of the National Academy of Sciences*, *103*(30), 11172–11177.

Rajkumar, A. P., Hye, A., Lange, J., Manesh, Y. R., Ballard, C., Fladby, T., & Aarsland, D. (2021). Next-generation RNA-sequencing of serum small extracellular vesicles discovers potential diagnostic biomarkers for dementia with lewy bodies. *American Journal of Geriatric Psychiatry*, *29*(6), 573–584. https://doi.org/10.1016/j.jagp.2020.10.012

Rashed, H. M., Shamma, R. N., & Basalious, E. B. (2017). Contribution of both olfactory and systemic pathways for brain targeting of nimodipine-loaded lipo-pluronics micelles: *In vitro* characterization and in vivo biodistribution study after intranasal and intravenous delivery, *Drug Delivery*, *24*(1), 181–187.

Reza-Zaldivar, E. E., Hernández-Sapiéns, M. A., Minjarez, B., Gutiérrez-Mercado, Y. K., Márquez-Aguirre, A. L., & Canales-Aguirre, A. A. (2018). Potential effects of MSC-derived exosomes in neuroplasticity in Alzheimer's disease. *Frontiers in Cellular Neuroscience*, *12*, 317. https://doi.org/10.3389/fncel.2018.00317

Rizk, N. I., Abulsoud, A. I., Kamal, M. M., Kassem, D. H., & Hamdy, N. M. (2022). Exosomal-long non-coding RNAs journey in colorectal cancer: Evil and goodness faces of key players. *Life Sciences*, *292*, 120325.

Rudnicki, D. D., Pletnikova, O., Vonsattel, J.-P. G., Ross, C. A., & Margolis, R. L. (2008). A comparison of Huntington disease and Huntington disease-like 2 neuropathology. *Journal of Neuropathology & Experimental Neurology*, *67*(4), 366–374. https://doi.org/10.1097/NEN.0b013e31816b4aee

Ruf, W. P., Freischmidt, A., Grozdanov, V., Roth, V., Brockmann, S. J., Mollenhauer, B., . . . Danzer, K. M. (2021). Protein binding partners of dysregulated miRNAs in Parkinson's disease serum. *Cells*, *10*(4), 791.

Saman, S., Kim, W., Raya, M., Visnick, Y., Miro, S., Saman, S., . . . Lee, N. C. (2012). Exosome-associated tau is secreted in tauopathy models and is selectively phosphorylated in cerebrospinal fluid in early Alzheimer disease. *Journal of Biological Chemistry*, *287*(6), 3842–3849.

Saman, S., Lee, N. C., Inoyo, I., Jin, J., Li, Z., Doyle, T., . . . Hall, G. F. (2014). Proteins recruited to exosomes by tau overexpression implicate novel cellular mechanisms linking tau secretion with Alzheimer's disease. *Journal of Alzheimer's Disease*, *40*(s1), S47–S70.

Schänzer, A., Wachs, F.-P., Wilhelm, D., Acker, T., Cooper-Kuhn, C., Beck, H., . . . Kuhn, H. G. (2004). Direct stimulation of adult neural stem cells *in vitro* and neurogenesis in vivo by vascular endothelial growth factor. *Brain Pathology*, *14*(3), 237–248. https://doi.org/10.1111/j.1750-3639.2004.tb00060.x

Schnatz, A., Müller, C., Brahmer, A., & Krämer-Albers, E.-M. (2021). Extracellular vesicles in neural cell interaction and CNS homeostasis. *FASEB BioAdvances*, *3*(8), 577–592. https://doi.org/10.1096/fba.2021-00035

Shi, M., Kovac, A., Korff, A., Cook, T. J., Ginghina, C., Bullock, K. M., . . . Aro, P. (2016). CNS tau efflux via exosomes is likely increased in Parkinson's disease but not in Alzheimer's disease. *Alzheimer's & Dementia*, *12*(11), 1125–1131.

Silverman, J. M., Christy, D., Shyu, C. C., Moon, K. M., Fernando, S., Gidden, Z., . . . Cashman, N. R. (2019). CNS-derived extracellular vesicles from superoxide dismutase 1 (SOD1)(G93A) ALS mice originate from astrocytes and neurons and carry misfolded SOD1. *Journal of Biological Chemistry*, *294*(10), 3744–3759. https://doi.org/10.1074/jbc.RA118.004825

Skryabin, G. O., Komelkov, A. V., Savelyeva, E. E., & Tchevkina, E. M. (2020). Lipid rafts in exosome biogenesis. *Biochemistry (Mosc)*, *85*(2), 177–191. https://doi.org/10.1134/s0006297920020054

Sokolov, D., Sharda, N., Banerjee, A., Denisenko, K., Basalious, E. B., Shukla, H., . . . Banerjee, A. (2024). Differential signaling pathways in medulloblastoma: Nano-biomedicine targeting non-coding epigenetics to improve current and future therapeutics. *Current Pharmaceutical Design*, *30*(1), 31–47.

Sridhar, V., Gaud, R., Bajaj, A., Wairkar, S. (2018). Pharmacokinetics and pharmacodynamics of intranasally administered selegiline nanoparticles with improved brain delivery in Parkinson's disease. *Nanomedicine*, *14*, 2609–2618.

Surgucheva, I., Sharov, V. S., & Surguchov, A. (2012). γ-synuclein: Seeding of α-synuclein aggregation and transmission between cells. *Biochemistry*, *51*(23), 4743–4754. https://doi.org/10.1021/bi300478w

Takeuchi, T., & Nagai, Y. (2017). Protein misfolding and aggregation as a therapeutic target for Polyglutamine diseases. *Brain Sciences*, *7*(10). https://doi.org/10.3390/brainsci7100128

Tang, L.-l., Wu, Y.-b., Fang, C.-q., Qu, P., & Gao, Z.-l. (2016). NDRG2 promoted secreted miR-375 in microvesicles shed from M1 microglia, which induced neuron damage. *Biochemical and Biophysical Research Communications*, *469*(3), 392–398. https://doi.org/10.1016/j.bbrc.2015.11.098

Toh, W. S., Lai, R. C., Zhang, B., & Lim, S. K. (2018). MSC exosome works through a protein-based mechanism of action. *Biochemical Society Transactions*, *46*(4), 843–853.

Tomlinson, P. R., Zheng, Y., Fischer, R., Heidasch, R., Gardiner, C., Evetts, S., . . . Tofaris, G. K. (2015). Identification of distinct circulating exosomes in Parkinson's disease. *Annals of Clinical and Translational Neurology*, *2*(4), 353–361. https://doi.org/10.1002/acn3.175

Torres, P., Ramírez-Núñez, O., Romero-Guevara, R., Barés, G., Granado-Serrano, A. B., Ayala, V., . . . Portero-Otín, M. (2018). Cryptic exon splicing function of TARDBP interacts with autophagy in nervous tissue. *Autophagy*, *14*(8), 1398–1403. https://doi.org/10.1080/15548627.2018.1474311

Tsou, Y. H., Zhang, X. Q., Zhu, H., Syed, S., Xu, X. Y. (2017). Drug delivery to the brain across the blood-brain barrier using nanomaterials. *Small*, *13*, 1701921.

Verma, S. K., Arora, I., Javed, K., Akhtar, M., Samim, M. (2016). Enhancement in the neuroprotective power of riluzole against cerebral ischemia using a brain targeted drug delivery vehicle. *ACS Applied Materials & Interfaces*, *8*, 19716–19723.

Wang, H., Sui, H., Zheng, Y., Jiang, Y., Shi, Y., Liang, J., & Zhao, L. (2019). Curcumin-primed exosomes potently ameliorate cognitive function in AD mice by inhibiting hyperphosphorylation of the Tau protein through the AKT/GSK-3β pathway. *Nanoscale*, *11*(15), 7481–7496. https://doi.org/10.1039/C9NR01255A

Wang, J., Zhu, X., Li, C., Cai, L., Pei, W., Ni, M., He, J., Jiang, H., Chen, J. (2022). Efficient exosome extraction through the conjugation of superparamagnetic iron oxide nanoparticles for the targeted delivery in rat brain. *Materials Today Chemistry*, *23*, 100637, https://doi.org/10.1016/j.mtchem.2021.100637

Wang, N., Cheng, X., Li, N., Wang, H., & Chen, H. (2019). Nanocarriers and their loading strategies. *Advanced Healthcare Materials*, *8*(6), e1801002. https://doi.org/10.1002/adhm.201801002

Wang, X., Zhou, Y., Gao, Q., Ping, D., Wang, Y., Wu, W., . . . Shao, A. (2020). The role of exosomal microRNAs and oxidative stress in Neurodegenerative diseases. *Oxidative Medicine and Cellular Longevity*, *2020*, 3232869. https://doi.org/10.1155/2020/3232869

Wang, Y., Balaji, V., Kaniyappan, S., Krüger, L., Irsen, S., Tepper, K., . . . Mandelkow, E. (2017). The release and trans-synaptic transmission of Tau via exosomes. *Molecular Neurodegeneration*, *12*(1), 1–25.

Weintraub, D., & Mamikonyan, E. (2019). The neuropsychiatry of Parkinson disease: A perfect storm. *American Journal of Geriatric Psychiatry*, *27*(9), 998–1018. https://doi.org/10.1016/j.jagp.2019.03.002

Williams, A. M., Bhatti, U. F., Brown, J. F., Biesterveld, B. E., Kathawate, R. G., Graham, N. J., . . . Andjelkovic, A. (2020). Early single-dose treatment with exosomes provides neuroprotection and improves blood-brain barrier integrity in swine model of traumatic brain injury and hemorrhagic shock. *Journal of Trauma and Acute Care Surgery*, *88*(2), 207–218.

Woodbury, M. E., & Ikezu, T. (2014). Fibroblast growth factor-2 signaling in neurogenesis and neurodegeneration. *Journal of Neuroimmune Pharmacology*, *9*(2), 92–101. https://doi.org/10.1007/s11481-013-9501-5

Xiao, T., Zhang, W., Jiao, B., Pan, C.-Z., Liu, X., & Shen, L. (2017). The role of exosomes in the pathogenesis of Alzheimer'disease. *Translational Neurodegeneration*, *6*, 1–6.

Xin, H., Li, Y., Liu, Z., Wang, X., Shang, X., Cui, Y., . . . Chopp, M. (2013). MiR-133b promotes neural plasticity and functional recovery after treatment of stroke with multipotent mesenchymal stromal cells in rats via transfer of exosome-enriched extracellular particles. *Stem Cells*, *31*(12), 2737–2746.

Xin, H., Wang, F., Li, Y., Lu, Q.-E., Cheung, W. L., Zhang, Y., . . . Chopp, M. (2017). Secondary release of exosomes from astrocytes contributes to the increase in neural plasticity and improvement of functional recovery after stroke in rats treated with exosomes harvested from microRNA 133b-overexpressing multipotent mesenchymal stromal cells. *Cell Transplantation*, *26*(2), 243–257.

Yamamoto, N., Matsubara, T., Sato, T., & Yanagisawa, K. (2008). Age-dependent high-density clustering of GM1 ganglioside at presynaptic neuritic terminals promotes amyloid β-protein fibrillogenesis. *Biochimica et Biophysica Acta (BBA)—Biomembranes*, *1778*(12), 2717–2726. https://doi.org/10.1016/j.bbamem.2008.07.028

Yang, Y., Keene, C. D., Peskind, E. R., Galasko, D. R., Hu, S.-C., Cudaback, E., . . . Montine, T. J. (2015). Cerebrospinal fluid particles in Alzheimer disease and Parkinson disease. *Journal of Neuropathology & Experimental Neurology*, *74*(7), 672–687. https://doi.org/10.1097/nen.0000000000000207

Yuan, D., Zhao, Y., Banks, W. A., Bullock, K. M., Haney, M., Batrakova, E., & Kabanov, A. V. (2017). Macrophage exosomes as natural nanocarriers for protein delivery to inflamed brain. *Biomaterials*, *142*, 1–12.

Yuyama, K., & Igarashi, Y. (2017). Exosomes as carriers of Alzheimer's amyloid-ss. *Frontiers in Neuroscience*, *11*, 229.

Yuyama, K., Sun, H., Usuki, S., Sakai, S., Hanamatsu, H., Mioka, T., . . . Igarashi, Y. (2015). A potential function for neuronal exosomes: Sequestering intracerebral amyloid-β peptide. *FEBS Letters*, *589*(1), 84–88. https://doi.org/10.1016/j.febslet.2014.11.027

Zappulli, V., Friis, K. P., Fitzpatrick, Z., Maguire, C. A., & Breakefield, X. O. (2016). Extracellular vesicles and intercellular communication within the nervous system. *Journal of Clinical Investigation*, *126*(4), 1198–1207. https://doi.org/10.1172/jci81134

Zhang, X. J., Li, Y. Y., Hu, Y. N. (2020). Green synthesis of silver nanoparticles and their preventive effect in deficits in recognition and spatial memory in sporadic Alzheimer's rat model. *Colloids and Surfaces A: Physicochemical and Engineering Aspects*, *605*, 125288.

Zhang, J., Liu, R., Zhang, D., Zhang, Z., Zhu, J., Xu, L., & Guo, Y. (2021). Neuroprotective effects of maize tetrapeptide-anchored gold nanoparticles in Alzheimer's disease. *Colloids and Surfaces B: Biointerfaces*, *200*, 111584.

Zheng, T., Pu, J., Chen, Y., Mao, Y., Guo, Z., Pan, H., . . . Zhang, B. (2017). Plasma exosomes spread and cluster around β-amyloid plaques in an animal model of Alzheimer's disease. *Frontiers in Aging Neuroscience*, *9*, 12.

21 Current Advancements of Nanocarriers for Neurodegenerative Disorders Linked to Toxoplasmosis

Mohamed Y. Zaky and Nahed S. Lamloum*

21.1 INTRODUCTION

Toxoplasma gondii (*T. gondii*), a member of the phylum Apicomplexa, is a neurotropic protozoan parasite that has a global presence and infects a range of warm-blooded animals as intermediate hosts. It is worth noting that approximately one-third of the human population serves as intermediate hosts for this parasite. Additionally, cats serve as the definitive hosts for *T. gondii* (Dubey, 1998). Intermediate hosts acquire *T. gondii* infection through multiple routes, such as the consumption of vegetables or water that have been contaminated with oocysts excreted in the feces of infected felids. Other means of infection include the ingestion of tissue cysts found in undercooked meat, solid organ transplantation, blood transfusion, and vertical transmission from an infected mother to her fetus (Hill and Dubey, 2002). The clinical manifestations, such as encephalitis, chorioretinitis, and pneumonitis, associated with *T. gondii* infection are observed in individuals with impaired immune systems. The majority of initial infections often exhibit a benign nature in hosts with intact immune systems. Various systematic review and meta-analysis studies have indicated a potential association between *T. gondii* infection and psychiatric as well as neurological problems (Flegr, 2013). These findings indicated that there were elevated seroprevalence rates of toxoplasmosis in a range of mental disorders, such as schizophrenia (Sutterland et al., 2015), obsessive-compulsive disorder (Nayeri Chegeni et al., 2019), bipolar disorder (de Barros et al., 2017), addiction (Sutterland et al., 2015), epilepsy (Sadeghi et al., 2019), suicide attempts (Sutterland et al., 2019), traffic accidents (Sutterland et al., 2019), and headache. The potential etiology of neurological problems induced by *T. gondii* may be attributed to the enduring presence of parasitic cysts inside the central nervous system (CNS), the host's immunological response to the parasite, and the detrimental effects stemming from the initial infection (McConkey et al., 2013). *T. gondii* exhibits a distinct affinity for the CNS and has the capability to infect many types of brain cells, such as neurons, Purkinje cells, and microglial cells (da Silva and Langoni, 2009). As the cysts proliferate, the host cells undergo degeneration. Moreover, it is worth noting that these cysts have the potential to rupture, leading to the release of bradyzoites that have the ability to transform into tachyzoites. When the immune system is not present, the tachyzoites infiltrate and eliminate the neighboring cells (Frenkel, 1988). The ability of bradyzoites to suppress cellular apoptosis facilitates the establishment of long-term infection by *T. gondii* through the creation of tissue cysts (McConkey et al., 2013). Chronic infection is associated with elevated production of interferon-gamma (IFN-γ), which serves as the primary cytokine involved in immunological defense against *T. gondii*. This heightened production of IFN-γ leads to a reduction in tryptophan and serotonin levels. Furthermore, there is a correlation between reduced serotonin levels and various cognitive deficits observed in both rodents and humans (Bouscaren et al., 2018). IFN-γ is generated as a result of *T. gondii* infection, which subsequently induces the degeneration of neurons responsible for dopamine production (Çelik et al., 2010). Furthermore, it has been observed that infection with *T. gondii* leads to an augmentation in the enzymatic activity of tyrosine hydrolase, a protein encoded by genes within the parasite's genome. This enzymatic activity is responsible for the liberation of dopamine (Bouscaren et al., 2018).

The olfactory route for many viral and/or toxic chemicals may initiate or worsen classical neurodegenerative and autoimmune disorders, particularly in people with a hereditary predisposition (Mattson, 2004). Many researchers have demonstrated that a number of neurologic and neurodegenerative disorders begin to manifest in the olfactory system, resulting in anosmia, or loss of smell, up to ten years before cognitive or motor dysfunction manifests (Doty, 1997). A common characteristic of these illnesses is neuroinflammation, which primarily affects the elderly (Mattson, 2004) and is identified by activated glial cells that release a variety of pro- and anti-inflammatory cytokines as well as other neurobiomediators (Perricone et al., 2013). For instance, a systemic infection that was followed by an aggravation of Alzheimer's symptoms that lasted for several months was also able to raise serum interleukin (IL)

* Corresponding author: mohamedzaki448@science.bsu.edu.eg

DOI: 10.1201/9781003383376-21

1β (Holmes et al., 2003). It has been proposed recently that the primary infectious agent responsible for initiating and developing a number of neurodegenerative illnesses linked to elevated production of pro- and anti-inflammatory cytokines, such as IL-1β, may be chronic *T. gondii* infection (Prandota, 2010).

Nanocarriers have been utilized for the purpose of developing a wide range of drug delivery systems (DDS) (Ghadiri et al., 2017), as well as gene delivery systems (Abedini et al., 2012). The targeting capabilities of nanocarriers can be enhanced through the conjugation of targeting moieties (Ghadiri et al., 2017). Additionally, nanocarriers can be utilized as optical biological probes for imaging by including fluorescent quantum dots (QDs) (Alibolandi et al., 2016). Furthermore, several types of magnetic nanoparticles have been developed to be utilized as an MRI contrast agent and to identify implanted cells in vivo (He et al., 2013). Therefore, it becomes essential to use NPs in drug delivery and toxoplasmosis treatment due to their particular features and size. This book chapter highlights the current advancements of nanocarriers for neurodegenerative disorders linked to toxoplasmosis.

21.2 NEURODEGENERATIVE DISORDERS AND TOXOPLASMOSIS

T. gondii has the capacity to cause persistent infection and can linger in the central nervous system, infecting neurons and functioning glial cells, which can result in a variety of neurological and behavioral diseases (Ortiz-Guerrero et al., 2020). *T. gondii* can cross the blood–brain barrier (BBB) by a number of different routes, including extravasation of the parasite from capillaries into the brain by monocytes and other infected myeloid-derived cells, trans-endothelial migration via attachment of the parasite to CD11b/ICAM1 integrins, paracellular entry of the parasite into the central nervous system via actin–myosin motors, and endothelial lysis (Ortiz-Guerrero et al., 2020). Patients with toxoplasmosis have been found to have a number of neurological and mental conditions, including depression, dysphoria, Alzheimer's disease (AD), schizophrenia, bipolar disorders, generalized anxiety disorder (GAD), obsessive-compulsive disorder, suicidality, Parkinson's disease, epilepsy, and sexual promiscuity (Alvarado-Esquivel et al., 2021). Regarding the chronic neuropathologic mechanisms in toxoplasmosis, there are a number of postulates. These may include changes to glutamate decarboxylase 67 (GAD67), which subsequently results in a decrease in GABAergic synaptic activities, and downregulation of the glutamate receptor (GLT-1) expression, which raises extracellular glutamate levels, excitatory glutamatergic signaling, and neural damage. Furthermore, in *T. gondii*–infected patients, the infection may cause inflammation triggered by the parasite and overexpress a number of cytokines (Wohlfert et al., 2017). Infected individuals with *T. gondii* may also experience altered catecholamine metabolism, which can lead to psychiatric, behavioral, and motor abnormalities (Fabiani et al., 2013). Numerous research studies have documented a noteworthy correlation between *T. gondii* and a range of neurological and mental conditions. According to a recent systematic study, the most common psychiatric condition associated with *T. gondii* infection is schizophrenia (Virus et al., 2021). According to a meta-analysis on *T. gondii*'s impact on epilepsy, it is the primary risk factor for epilepsy (Ngoungou et al., 2015). There is a positive correlation between *T. gondii* infection and neuropsychiatric illnesses, according to a different systematic review and meta-analysis that assessed the relationship between the parasite and Parkinson's and Alzheimer's diseases (Bayani et al., 2019).

21.3 NANOCARRIERS IN DRUG DELIVERY

Nanocarriers, in the form of colloidal nanoparticles, are extensively employed for the purpose of delivering medicinal agents or other chemicals to specific target sites (Din et al., 2017). The diameter of nanocarriers typically ranges from 1 to 100 nm (nanometers), with a maximum size requirement of 200 nm for therapeutic applications due to the diameter of microcapillaries in the human body (Kumar et al., 2023). The biocompatibility of these nanocarriers is notable due to their well-recognized safety and inert nature. The nanocarriers are designed to exhibit an extended circulation time, allowing for the prolonged release of drugs. This property enables them to overcome the endosome–lysosome process (Kingsley et al., 2006). The manipulation of the physiochemical characteristics of nanocarriers, including their surface, composition, and form, has the potential to improve their efficacy while reducing undesirable side effects (Zhao et al., 2019). Therefore, it generates a multitude of effects in the domain of pharmaceutical administration. While a diverse array of nanocarriers has been developed, only a few number exhibit notable efficacy in delivering drugs to specific target sites. Nanocarriers possess several distinctive attributes, including improved biodistribution and pharmacokinetics, heightened stability, enhanced solubility, less toxicity, and the capacity to sustain and target drug delivery (How et al., 2013).

21.4 TYPES OF NANOCARRIERS

Nanocarriers exhibiting a high surface-to-volume ratio are primarily classified into three categories: organic nanocarriers, inorganic nanocarriers, and hybrid nanocarriers.

21.4.1 Organic Nanocarriers

Organic nanocarriers encompass a variety of nanoparticles, including solid lipid nanocarriers, liposomes, dendrimers, polymeric nanocarriers, micelles, and viral nanocarriers. The organic nanocarriers exhibit a high degree of versatility, demonstrating reduced toxicity and the capacity to conjugate a diverse range of medicines and ligands for the purpose of drug delivery. Within the realm of organic nanocarriers, the use of micelles and liposomes for drug administration exhibits the capacity to accumulate at the intended location through the improved permeability and retention effect (López-Dávila et al., 2012). Polymeric nanocarriers and liposome-mediated drug delivery represent the initial generation of nanocarriers due to their uncomplicated nature as excipients (Bhatia, 2016).

21.4.2 Inorganic Nanocarriers

Inorganic nanocarriers encompass a variety of materials, such as gold, magnetic nanocarriers, quantum dots, and mesoporous silica, among others. Inorganic nanocarriers exploit their inherent tractable features. In the field of biotechnology, the utilization of inorganic nanocarriers has proven to be highly efficient in several applications, such as biosensing, cell labeling, targeting, imaging, and diagnostics. In addition, these inorganic nanocarriers exhibit a synergistic therapeutic impact (Santos et al., 2015).

21.4.3 Hybrid Nanocarriers

Nanocarriers that combine two or more inorganic and organic nanocarriers either singly or collectively are known as hybrid nanocarriers. It consists of multiple inorganic–inorganic and organic–inorganic components. Lipid–polymer hybrids and ceramic–polymer hybrids are a few of instances of hybrid nanocarriers. When two nanoparticles are combined, their dual natures will be present, greatly boosting their respective properties (Qian et al., 2012). Organic nanocarriers, such as liposomes, have poor stability and internal solution leakage problems. This facilitates its easy extraction from the bloodstream. Therefore, it can be used for medication administration with extra stability. The hybrid nanocarrier system compensates for these drawbacks (Peer et al., 2007).

These hybrid nanocarriers are the subject of various research projects. With a high rate of retention, the mesoporous silica nanoparticle lipid bilayer hybrid nanocarrier technology has demonstrated zoledronic acid's unique intracellular transport in breast cancer (Desai et al., 2017). By enabling a stimuli-response release of the medicine, this method stops the drug from entering the body too soon. It has been discovered that novel albumin hybrid nanocapsules are effective at encasing hydrophilic peptides or other tiny therapeutic molecules that are intended to target cancer cells. It permits less toxicity and a uniform dissemination of the tumor cells within their surroundings (Zhou et al., 2013). Similar to this, ferritin's use in drug delivery applications offers a significant benefit in encapsulating medicinal substances, and the medication's stimulus-response release allows for a sustained release of the drug into the intended place (Khoshnejad et al., 2018). Numerous studies have also been conducted on the in vivo delivery of small interfering RNA using poly(amidoamine) hybrid nanocarriers (polyethylene glycol–liposome/small interfering RNA nanoparticles and peptide HAIYPRH, known as T7-liposome/small interfering RNA nanoparticles) that are grafted with lipids and cholesterol. This approach is effective in theranostic use for gene silencing and offers the advantages of long stability and strong cellular uptake (Gao et al., 2014).

21.5 TOXOPLASMOSIS TREATMENT

Infection treatment is almost always taken into consideration in the following cases: (1) primary infection in pregnant women to prevent congenital infection; (2) infants with congenital toxoplasmosis; (3) acute infection exacerbation on top of chronic exacerbation in immunocompromised individuals; and (4) acute chorioretinitis (Jones et al., 2003). Because the parasite cannot consume folates from food, it produces folates on its own (Katlama et al., 1996). Examples of classic therapies include spiramycin, pyrimethamine sulfadiazine, and pyrimethamine–clindamycin (Dard et al., 2018). These medications inhibit the production of parasite folate, hence impeding nucleic acid biosynthesis and subsequent parasite multiplication. However, it is important to note that these medications might have several adverse effects, including the development of severe thrombocytopenia, an increase in blood creatinine and liver enzymes, and the occurrence of severe allergic symptoms. Typically, these medications are used concomitantly with folinic acid (Katlama et al., 1996). The administration of pyrimethamine during the initial trimester of pregnancy should be avoided due to its potential to induce teratogenic effects (Goldstein et al., 2008). Spiramycin has been reported to be safe during the first trimester of pregnancy, and it has been shown to reduce the entry of tachyzoites through the placenta (Desmonts and Couvreur, 1974). Because it cannot penetrate the fetus, spiramycin is not necessary for the treatment of an infected fetus. Due to the theoretical possibility that affection for the fetus could develop at any point during gestation from a previously infected placenta, it is delayed until the end of the pregnancy (Couvreur et al., 1988).

21.6 NANOPARTICLES AS A SOLE TREATMENT

Both in vivo and *in vitro* studies were conducted on the application of metal nanoparticles, including silver, gold, platinum, and selenium, for the treatment of *T. gondii*. Techniques such as chemical, physical, and biotechnological have been used to create nanoparticles. Because plant extract is safe and readily available, it has been investigated for the green manufacturing of nanoparticles (Nafari et al., 2020). Because they contain reductase enzymes, bacteria, yeast, and fungus are typically employed in the biological creation of nanoparticles (Rahman et al., 2019). When biogenic selenium nanoparticles were administered to experimentally infected mice harboring the *T. gondii* Tehran strain, the mean number of brain tissue cysts decreased and the levels of inflammatory cytokines increased in comparison to the control group (Keyhani et al., 2020). The present study reports on the anti-Toxoplasma activity of nanosilver particles, which were synthesized biogenically by *Fusarium oxysporum*. The nanosilver particles had a size of 69 nm. The study focused on evaluating the effects of these particles on HeLa cells that were infected with the *T. gondii* RH strain. The silver nanoparticles did not exhibit a significant impact on the inflammatory cytokines. However, they did result in a drop in interleukin (IL) 8 levels (Machado-Santos et al., 2018). When *T. gondii* oocysts were exposed to silver nanoparticles measuring 10 nm, their ultrastructural changes were assessed. The findings revealed an oocyst wall breach, which was directly connected to the oocyst's vitality (Vergara-Duque et al., 2020). Using optical and electron microscopy, the morphological effects of silver nanoparticles on *T. gondii* tachyzoites and tissue cysts were investigated. Both tachyzoites and tissue cysts displayed structural alterations (Shojaee et al., 2019). Using the ME49 strain of *T. gondii* in human foreskin fibroblast (HFF) cells, metal nanoparticles of gold, silver, and platinum with sizes of 5, 10, and 3 nm, respectively, were assessed. The results showed that the nanoparticles had the ability to alter the formation of the parasite cyst wall and reduce the burden of bradyzoites (Adeyemi et al., 2019). It has been observed that metal nanoparticles with diameters ranging from 5 to 50 nm, such as silver, gold, and platinum, moderately inhibit *T. gondii*'s *in vitro* growth. Additionally, an investigation on the anti-parasitic effect of nanoparticles on *Trypanosoma* spp. yielded encouraging results (Adeyemi et al., 2018). According to Adeyemi et al. (2017), the anti-Toxoplasma properties of commercial inorganic nanoparticles, such as gold (5 nm), silver (10 nm), and platinum (3 nm), were assessed, and their anti-parasitic processes were shown to be connected to redox signaling. It was also reported that the nanoparticles affected Toxoplasma's potential for infectivity, invasion by parasites, replication, and mitochondrial membrane potential (Adeyemi et al., 2017). Another method used to kill tumor cells and parasites was the conjugation of gold nanorods with anti–*T. gondii* antibodies and laser heat therapy (Pissuwan et al., 2007).

Natural and nontoxic polysaccharide chitosan is made up of N-acetyl-D-glucosamine and D-glucosamine units. It is produced when chitin is deacetylated in an alkaline environment (Goy et al., 2009). Research has been conducted on the antibacterial properties of chitosan nanoparticles against a range of microorganisms, including parasites. But in addition to its antimicrobial, anticancer, and antifungal qualities, chitosan nanoparticles are said to be a viable drug delivery method (De Marchi et al., 2017). Chitosan nanoparticles of different molecular weights and concentrations were evaluated in *T. gondii* RH strain tachyzoites, and the results showed a strong anti-Toxoplasma efficacy, especially at low molecular weights (Teimouri et al., 2018). In order to cure toxoplasmosis in mice, silver and chitosan nanoparticles were also created. Laboratory mice's liver and spleen contained significantly less parasites when exposed to silver nanoparticles either by themselves or in combination with chitosan, as compared to the control group. In this study, animals receiving nanoparticles had higher serum levels of IFNγ, and tachyzoites that were secreted from the peritoneum of mice were distorted and immobilized (Gaafar et al., 2014).

21.7 NANOTECHNOLOGY AND NANOMEDICINE

Nanotechnology and nanoscience emerged with the development of the scanning tunnelling microscope in the early 1980s. As a result, carbon nanotubes were discovered, and semiconductor nanomaterials were produced and characterized (Ahsan, 2017). With a confirmed resolution of nanometers, the atomic force microscope is an extremely high-resolution type of scanning probe microscopy that is more than 1,000 times better than the optical diffusion–bound. Furthermore, nanostructures can be employed by using the scanning probe's slope (Blanchard, 1996). The development of technology in the nanoscale, or between 0.1 and 100 nm, is known as nanotechnology. One billionth of a meter, or nanometer, is the smallest size that a conventional microscope cannot view. As a branch of nanotechnology, nanomedicine describes very specific therapeutic interference at the nanoscale (Saha, 2009). Nanoparticles are made up of different components, each of which has a single measurement of less than 100 nm and contains particular compounds (Laurent et al., 2008). The generic form states that those substances may have zero, one, two, or three dimensions (Tiwari et al., 2012). The relevance of these substances was revealed when researchers found that a material's physiochemical characteristics can be influenced by its dimension (Dreaden et al., 2012). The advantages of these compounds appear as soon as they were produced using organic biological materials, such microbes, biodegradable polymers, plant extracts, proteins, lipids, carbohydrates, and vitamins. A few inorganic nanoparticles, primarily metal nanoparticles, various metal oxides, and salts, are produced as a result of these advancements (Chimanuka et al., 2002). They have a substantially higher specific surface area due to their reduced size because a molecule's entire surface area is inversely correlated with its length (Kawaguchi, 2000). Additionally, this smaller size allows for a better ingestion and is more suitable for injectable drug use (Danhier et al., 2012). These materials direct fundamental cellular functions and have dimensions in the same range of units. Moreover, some properties—like ocular reactions, for example—are size-dependent and exclusive to the nanoscale, which permits their application in imaging processes (Ratner et al., 2004). These materials differ from bulk constituents of comparable structures in that they have unique physicochemical properties, such as extreme trivial size, a large surface-area-to-mass ratio, and significant reactivity. Some of the limits seen in traditional therapy approaches can be removed with the use of these characteristics (Zhang et al., 2008).

21.8 APPLICATIONS OF NANOMEDICINE IN TOXOPLASMOSIS

The effectiveness of gold nanoparticles conjugated with anti–*Toxoplasma gondii* antibodies for the antigen of the acute strain (RH) of this parasite and treated with laser light was investigated by Pissuwan et al. (2009). They came to the conclusion that dose-dependent levels of gold nanoparticles in a given laser dose increased the mortality rate of tachyzoites *in vitro*, but that the death rate did not significantly alter when the laser dose was increased to a fixed concentration of nanomaterials (Pissuwan et al., 2009). A study conducted by Kunjachan et al. (2011) assessed the use of chitosan and silver nanoparticles in treating infection in experimental animals and compared their use separately and in combination. When they were combined, the parasite load in the spleen and liver was significantly reduced. Additionally, under microscopy, the tachyzoites showed little movement and had an imperfection in their wall (Kunjachan et al., 2011). According to Azami et al. (2018), mice infected with acute strains of the parasite and treated with the nanoemulsion of curcumin showed a statistically significant increase in survival time when compared to the control group. This study evaluated the therapeutic effect of the nanoemulsion in both acute and chronic strains of the parasite in mice. Furthermore, when compared to mice that were not given any treatment, the emulsion significantly decreased the mean counts of parasites in the peritoneum of acutely infected mice (Azami et al., 2018). According to a related study conducted in 2019, Alajmi et al. discovered that treating liver-infected mice with date palm–produced silver nanoparticles was more effective than traditional therapies in reducing hepatic damage by improving their immune-histological characteristics (Alajmi et al., 2019). Another study by El-Shafey et al. (2020) demonstrated that the mean number of parasite cysts in the rats' brains was significantly decreased when curcumin and nano-curcumin@MOFs were added to metalorganic frameworks used to treat rats infected with the chronic ME49 strain of the parasite (El-Shafey et al., 2020).

21.9 CONCLUSION

In conclusion, the exploration of nanocarriers in the context of neurodegenerative disorders associated with toxoplasmosis presents a promising frontier in biomedical research. The intricate interplay between neuroinflammation, infection, and the progression of neurodegenerative diseases has spurred innovative approaches to treatment, with nanotechnology at the forefront. Through an extensive review of current advancements, it is evident that nanocarriers offer a multifaceted platform for targeted drug delivery, enhanced bioavailability, and improved therapeutic outcomes. The integration of nanotechnology into the field of neurodegenerative disorders not only addresses the challenges of conventional drug delivery but also opens avenues for personalized and precise treatment strategies. The ability of nanocarriers to traverse the blood–brain barrier, coupled with their capacity to encapsulate and release therapeutic agents in a controlled manner, holds immense potential for optimizing drug efficacy while minimizing adverse effects. Furthermore, the understanding of the role of toxoplasmosis in exacerbating neurodegenerative processes underscores the importance of a holistic approach to treatment. Nanocarriers not only act as carriers of therapeutic payloads but also serve as diagnostic tools, allowing for real-time monitoring and assessment of treatment responses. This synergy between diagnostics and therapeutics represents a paradigm shift in the management of neurodegenerative disorders linked to toxoplasmosis. As we navigate this evolving landscape, collaboration between researchers, clinicians, and industry stakeholders becomes paramount. Future research should delve into refining nanocarrier design, optimizing drug formulations, and conducting rigorous preclinical and clinical studies to bridge the gap between bench and bedside. Additionally, exploring the long-term safety and potential challenges associated with nanocarrier-based therapies will be crucial for their successful translation into clinical practice. In essence, the current body of evidence underscores the immense potential of nanocarriers as a transformative tool in the fight against neurodegenerative disorders complicated by toxoplasmosis. This chapter serves as a testament to the strides made in this interdisciplinary field and as an impetus for future endeavors that aim to harness the full therapeutic potential of nanotechnology for the benefit of those affected by these challenging conditions.

REFERENCES

Abedini, F., Hosseinkhani, H., Ismail, M., Domb, A.J., Omar, A.R., Chong, P.P., Hong, P.-D., Yu, D.-S., Farber, I.-Y., 2012. Cationized dextran nanoparticle-encapsulated CXCR4-siRNA enhanced correlation between CXCR4 expression and serum alkaline phosphatase in a mouse model of colorectal cancer. Int. J. Nanomedicine 4159–4168.

Adeyemi, O.S., Molefe, N.I., Awakan, O.J., Nwonuma, C.O., Alejolowo, O.O., Olaolu, T., Maimako, R.F., Suganuma, K., Han, Y., Kato, K., 2018. Metal nanoparticles restrict the growth of protozoan parasites. Artif. Cells Nanomed. Biotechnol. 46, 86–94.

Adeyemi, O.S., Murata, Y., Sugi, T., Han, Y., Kato, K., 2019. Nanoparticles show potential to retard bradyzoites *in vitro* formation of Toxoplasma gondii. Folia Parasitol. (Praha). 66, 1–6.

Adeyemi, O.S., Murata, Y., Sugi, T., Kato, K., 2017. Inorganic nanoparticles kill Toxoplasma gondii via changes in redox status and mitochondrial membrane potential. Int. J. Nanomedicine 1647–1661.

Ahsan, M.Z., 2017. Nanotechnology: Prospects and challenges. http://hdl.handle.net/123456789/511.

Alajmi, R.A., Al-Megrin, W.A., Metwally, D., Al-Subaie, H., Altamrah, N., Barakat, A.M., Abdel Moneim, A.E., Al-Otaibi, T.T., El-Khadragy, M., 2019. Anti-toxoplasma activity of silver nanoparticles green synthesized with Phoenix dactylifera and Ziziphus spina-christi extracts which inhibits inflammation through liver regulation of cytokines in Balb/c mice. Biosci. Rep. 39, BSR20190379.

Alibolandi, M., Abnous, K., Sadeghi, F., Hosseinkhani, H., Ramezani, M., Hadizadeh, F., 2016. Folate receptor-targeted multimodal polymersomes for delivery of quantum dots and doxorubicin to breast adenocarcinoma: *In vitro* and in vivo evaluation. Int. J. Pharm. 500, 162–178.

Alvarado-Esquivel, C., Estrada-Martínez, S., Ramos-Nevárez, A., Pérez-Álamos, A.R., Beristain-Garcia, I., Alvarado-Félix, Á.O., Cerrillo-Soto, S.M., Alvarado-Félix, G.A., Guido-Arreola, C.A., Sáenz-Soto, L., 2021. Is toxoplasma gondii infection associated with sexual promiscuity? A cross-sectional study. Pathogens 10, 1393.

Azami, S.J., Teimouri, A., Keshavarz, H., Amani, A., Esmaeili, F., Hasanpour, H., Elikaee, S., Salehiniya, H., Shojaee, S., 2018. Curcumin nanoemulsion as a novel chemical for the treatment of acute and chronic toxoplasmosis in mice. Int. J. Nanomedicine 7363–7374.

Bayani, M., Riahi, S.M., Bazrafshan, N., Gamble, H.R., Rostami, A., 2019. Toxoplasma gondii infection and risk of Parkinson and Alzheimer diseases: A systematic review and meta-analysis on observational studies. Acta Trop. 196, 165–171.

Bhatia, S., 2016. Nanoparticles types, classification, characterization, fabrication methods and drug delivery applications. In Natural polymer drug delivery systems: Nanoparticles, plants, and algae. Springer, 33–93.

Blanchard, C.R., 1996. Atomic force microscopy. Chem. Educ. 1, 1–8.

Bouscaren, N., Pilleron, S., Mbelesso, P., Ndamba-Bandzouzi, B., Dartigues, J., Clément, J., Preux, P., Dardé, M., Guerchet, M., Group, E., 2018. Prevalence of toxoplasmosis and its association with dementia in older adults in Central Africa: A result from the EPIDEMCA programme. Trop. Med. Int. Heal. 23, 1304–1313.

Çelik, T., Kamişli, Ö., Babür, C., Çevik, M.Ö., Öztuna, D., Altinayar, S., 2010. Is there a relationship between Toxoplasma gondii infection and idiopathic Parkinson's disease? Scand. J. Infect. Dis. 42, 604–608.

Chimanuka, B., Gabriëls, M., Detaevernier, M.-R., Plaizier-Vercammen, J.A., 2002. Preparation of β-artemether liposomes, their HPLC–UV evaluation and relevance for clearing recrudescent parasitaemia in Plasmodium chabaudi malaria-infected mice. J. Pharm. Biomed. Anal. 28, 13–22.

Couvreur, J., Desmonts, G., Thulliez, P., 1988. Prophylaxis of congenital toxoplasmosis. Effects of spiramycin on placental infection. J. Antimicrob. Chemother. 22, 193–200.

Danhier, F., Ansorena, E., Silva, J.M., Coco, R., Le Breton, A., Préat, V., 2012. PLGA-based nanoparticles: An overview of biomedical applications. J. Control. Release 161, 505–522.

Dard, L., Bellance, N., Lacombe, D., Rossignol, R., 2018. RAS signalling in energy metabolism and rare human diseases. Biochim. Biophys. Acta (BBA)-Bioenergetics 1859, 845–867.

da Silva, R.C., Langoni, H., 2009. Toxoplasma gondii: Host–parasite interaction and behavior manipulation. Parasitol. Res. 105, 893–898.

de Barros, J.L.V.M., Barbosa, I.G., Salem, H., Rocha, N.P., Kummer, A., Okusaga, O.O., Soares, J.C., Teixeira, A.L., 2017. Is there any association between Toxoplasma gondii infection and bipolar disorder? A systematic review and meta-analysis. J. Affect. Disord. 209, 59–65.

De Marchi, J.G.B., Jornada, D.S., Silva, F.K., Freitas, A.L., Fuentefria, A.M., Pohlmann, A.R., Guterres, S.S., 2017. Triclosan resistance reversion by encapsulation in chitosan-coated-nanocapsule containing α-bisabolol as core: Development of wound dressing. Int. J. Nanomedicine 7855–7868.

Desai, D., Zhang, J., Sandholm, J., Lehtimaki, J., Gronroos, T., Tuomela, J., Rosenholm, J.M., 2017. Lipid bilayer-gated mesoporous silica nanocarriers for tumor-targeted delivery of zoledronic acid in vivo. Mol. Pharm. 14, 3218–3227.

Desmonts, G., Couvreur, J., 1974. Congenital toxoplasmosis: A prospective study of 378 pregnancies. N. Engl. J. Med. 290, 1110–1116.

Din, F.U., Aman, W., Ullah, I., Qureshi, O.S., Mustapha, O., Shafique, S., Zeb, A., 2017. Effective use of nanocarriers as drug delivery systems for the treatment of selected tumors. Int. J. Nanomedicine 7291–7309.

Doty, R.L., 1997. Studies of human olfaction from the University of Pennsylvania Smell and Taste Center. Chem. Senses 22, 565–586.

Dreaden, E.C., Alkilany, A.M., Huang, X., Murphy, C.J., El-Sayed, M.A., 2012. The golden age: Gold nanoparticles for biomedicine. Chem. Soc. Rev. 41, 2740–2779.

Dubey, J.P., 1998. Advances in the life cycle of Toxoplasma gondii. Int. J. Parasitol. 28, 1019–1024.

El-Shafey, A.A.M., Hegab, M.H.A., Seliem, M.M.E., Barakat, A.M.A., Mostafa, N.E., Abdel-Maksoud, H.A., Abdelhameed, R.M., 2020. Curcumin@ metal organic frameworks nano-composite for treatment of chronic toxoplasmosis. J. Mater. Sci. Mater. Med. 31, 1–13.

Fabiani, S., Pinto, B., Bruschi, F., 2013. Toxoplasmosis and neuropsychiatric diseases: Can serological studies establish a clear relationship? Neurol. Sci. 34, 417–425.

Flegr, J., 2013. How and why Toxoplasma makes us crazy. Trends Parasitol. 29, 156–163.

Frenkel, J.K., 1988. Pathophysiology of toxoplasmosis. Parasitol. Today 4, 273–278.

Gaafar, M.R., Mady, R.F., Diab, R.G., Shalaby, T.I., 2014. Chitosan and silver nanoparticles: Promising anti-toxoplasma agents. Exp. Parasitol. 143, 30–38.

Gao, L.-Y., Liu, X.-Y., Chen, C.-J., Wang, J.-C., Feng, Q., Yu, M.-Z., Ma, X.-F., Pei, X.-W., Niu, Y.-J., Qiu, C., 2014. Core-shell type lipid/rPAA-Chol polymer hybrid nanoparticles for in vivo siRNA delivery. Biomaterials 35, 2066–2078.

Ghadiri, M., Vasheghani-Farahani, E., Atyabi, F., Kobarfard, F., Mohamadyar-Toupkanlou, F., Hosseinkhani, H., 2017. Transferrin-conjugated magnetic dextran-spermine nanoparticles for targeted drug transport across blood-brain barrier. J. Biomed. Mater. Res. Part A 105, 2851–2864.

Goldstein, E.J.C., Montoya, J.G., Remington, J.S., 2008. Management of Toxoplasma gondii infection during pregnancy. Clin. Infect. Dis. 47, 554–566.

Goy, R.C., de Britto, D., Assis, O.B.G., 2009. A review of the antimicrobial activity of chitosan. Polímeros 19, 241–247.

He, W., Hosseinkhani, H., Hong, P., Chiang, C., Yu, D., 2013. Magnetic nanoparticles for imaging technology. Int. J. Nanotechnol. 10, 930–944.

Hill, D., Dubey, J.P., 2002. Toxoplasma gondii: Transmission, diagnosis and prevention. Clin. Microbiol. Infect. 8, 634–640.

Holmes, C., El-Okl, M., Williams, A.L., Cunningham, C., Wilcockson, D., Perry, V.H., 2003. Systemic infection, interleukin 1β, and cognitive decline in Alzheimer's disease. J. Neurol. Neurosurg. Psychiatry 74, 788–789.

How, C.W., Rasedee, A., Manickam, S., Rosli, R., 2013. Tamoxifen-loaded nanostructured lipid carrier as a drug delivery system: Characterization, stability assessment and cytotoxicity. Colloids Surf B Biointerfaces 112, 393–399. https://doi.org/10.1016/j.colsurfb.2013.08.009

Jones, J., Lopez, A., Wilson, M., 2003. Congenital toxoplasmosis. Am. Fam. Physician 67, 2131–2138.

Katlama, C., Mouthon, B., Gourdon, D., Lapierre, D., Rousseau, F., 1996. Atovaquone as long-term suppressive therapy for toxoplasmic encephalitis in patients with AIDS and multiple drug intolerance. Atovaquone Expanded Access Group. AIDS 10, 1107–1112.

Kawaguchi, H., 2000. Functional polymer microspheres. Prog. Polym. Sci. 25, 1171–1210.

Keyhani, A., Ziaali, N., Shakibaie, M., Kareshk, A.T., Shojaee, S., Asadi-Shekaari, M., Sepahvand, M., Mahmoudvand, H., 2020. Biogenic selenium nanoparticles target chronic toxoplasmosis with minimal cytotoxicity in a mouse model. J. Med. Microbiol. 69, 104–110.

Khoshnejad, M., Parhiz, H., Shuvaev, V. V., Dmochowski, I.J., Muzykantov, V.R., 2018. Ferritin-based drug delivery systems: Hybrid nanocarriers for vascular immunotargeting. J. Control. Release 282, 13–24.

Kingsley, J.D., Dou, H., Morehead, J., Rabinow, B., Gendelman, H.E., Destache, C.J., 2006. Nanotechnology: A focus on nanoparticles as a drug delivery system. J. Neuroimmune Pharmacol. 1, 340–350. https://doi.org/10.1007/s11481-006-9032-4

Kumar, M., Hilles, A.R., Almurisi, S.H.A., Bhatia, A., Mahmood, S., 2023. Micro and nano-carriers-based pulmonary drug delivery system: Their current updates, challenges, and limitations–A review. JCIS Open 100095.

Kunjachan, S., Gupta, S., Dwivedi, A.K., Dube, A., Chourasia, M.K., 2011. Chitosan-based macrophage-mediated drug targeting for the treatment of experimental visceral leishmaniasis. J. Microencapsul. 28, 301–310.

Laurent, S., Forge, D., Port, M., Roch, A., Robic, C., Vander Elst, L., Muller, R.N., 2008. Magnetic iron oxide nanoparticles: Synthesis, stabilization, vectorization, physicochemical characterizations, and biological applications. Chem. Rev. 108, 2064–2110.

López-Dávila, V., Seifalian, A.M., Loizidou, M., 2012. Organic nanocarriers for cancer drug delivery. Curr. Opin. Pharmacol. 12, 414–419.

Machado-Santos, J., Saji, E., Tröscher, A.R., Paunovic, M., Liblau, R., Gabriely, G., Bien, C.G., Bauer, J., Lassmann, H., 2018. The compartmentalized inflammatory response in the multiple sclerosis brain is composed of tissue-resident CD8+ T lymphocytes and B cells. Brain 141, 2066–2082.

Mattson, M.P., 2004. Infectious agents and age-related neurodegenerative disorders. Ageing Res. Rev. 3, 105–120.

McConkey, G.A., Martin, H.L., Bristow, G.C., Webster, J.P., 2013. Toxoplasma gondii infection and behaviour–location, location, location? J. Exp. Biol. 216, 113–119.

Nafari, A., Cheraghipour, K., Sepahvand, M., Shahrokhi, G., Gabal, E., Mahmoudvand, H., 2020. Nanoparticles: New agents toward treatment of leishmaniasis. Parasite Epidemiol. Control 10, e00156.

Nayeri Chegeni, T., Sarvi, S., Amouei, A., Moosazadeh, M., Hosseininejad, Z.A., Aghayan, S., Daryani, A., 2019. Relationship between toxoplasmosis and obsessive compulsive disorder: A systematic review and meta-analysis. PLoS Negl. Trop. Dis. 13, e0007306.

Ngoungou, E.B., Bhalla, D., Nzoghe, A., Dardé, M.-L., Preux, P.-M., 2015. Toxoplasmosis and epilepsy—systematic review and meta analysis. PLoS Negl. Trop. Dis. 9, e0003525.

Ortiz-Guerrero, G., Gonzalez-Reyes, R.E., de-la-Torre, A., Medina-Rincón, G., Nava-Mesa, M.O., 2020. Pathophysiological mechanisms of cognitive impairment and neurodegeneration by Toxoplasma gondii infection. Brain Sci. 10, 369.

Peer, D., Karp, J.M., Hong, S., Farokhzad, O.C., Margalit, R., Langer, R., 2007. Nanocarriers as an emerging platform for cancer therapy. Nat. Nanotechnol. 2, 751–760.

Perricone, C., Shoenfeld, N., Agmon-Levin, N., de Carolis, C., Perricone, R., Shoenfeld, Y., 2013. Smell and autoimmunity: A comprehensive review. Clin. Rev. Allergy Immunol. 45, 87–96.

Pissuwan, D., Valenzuela, S.M., Miller, C.M., Cortie, M.B., 2007. A golden bullet? Selective targeting of Toxoplasma gondii tachyzoites using antibody-functionalized gold nanorods. Nano Lett. 7, 3808–3812.

Pissuwan, D., Valenzuela, S.M., Miller, C.M., Killingsworth, M.C., Cortie, M.B., 2009. Destruction and control of Toxoplasma gondii tachyzoites using gold nanosphere/antibody conjugates. Small 5, 1030–1034.

Prandota, J., 2010. Neuropathological changes and clinical features of autism spectrum disorder participants are similar to that reported in congenital and chronic cerebral toxoplasmosis in humans and mice. Res. Autism Spectr. Disord. 4, 103–118.

Qian, W.-Y., Sun, D.-M., Zhu, R.-R., Du, X.-L., Liu, H., Wang, S.-L., 2012. pH-sensitive strontium carbonate nanoparticles as new anticancer vehicles for controlled etoposide release. Int. J. Nanomedicine 5781–5792.

Rahman, K., Khan, S.U., Fahad, S., Chang, M.X., Abbas, A., Khan, W.U., Rahman, L., Haq, Z.U., Nabi, G., Khan, D., 2019. Nano-biotechnology: A new approach to treat and prevent malaria. Int. J. Nanomedicine 1401–1410.

Ratner, B.D., Hoffman, A.S., Schoen, F.J., Lemons, J.E., 2004. Biomaterials science: An introduction to materials in medicine. Elsevier.

Sadeghi, M., Riahi, S.M., Mohammadi, M., Saber, V., Aghamolaie, S., Moghaddam, S.A., Aghaei, S., Javanian, M., Gamble, H.R., Rostami, A., 2019. An updated meta-analysis of the association between Toxoplasma gondii infection and risk of epilepsy. Trans. R. Soc. Trop. Med. Hyg. 113, 453–462.

Saha, M., 2009. Nanomedicine: Promising tiny machine for the healthcare in future-a review. Oman Med. J. 24, 242.

Santos, H.A., Bimbo, L.M., Peltonen, L., Hirvonen, J., 2015. Inorganic nanoparticles in targeted drug delivery and imaging. Target. Drug Deliv. Concepts Des. 571–613.

Shojaee, S., Firouzeh, N., Keshavarz, H., Azami, S.J.-P., Salimi, M., Mohebali, M., 2019. Nanosilver colloid inhibits Toxoplasma gondii tachyzoites and bradyzoites *in vitro*. Iran. J. Parasitol. 14, 362.

Sutterland, A.L., Fond, G., Kuin, A., Koeter, M.W.J., Lutter, R., Van Gool, T., Yolken, R., Szoke, A., Leboyer, M., De Haan, L., 2015. Beyond the association. Toxoplasma gondii in schizophrenia, bipolar disorder, and addiction: Systematic review and meta-analysis. Acta Psychiatr. Scand. 132, 161–179.

Sutterland, A.L., Kuin, A., Kuiper, B., van Gool, T., Leboyer, M., Fond, G., de Haan, L., 2019. Driving us mad: The association of Toxoplasma gondii with suicide attempts and traffic accidents–a systematic review and meta-analysis. Psychol. Med. 49, 1608–1623.

Teimouri, A., Azami, S.J., Keshavarz, H., Esmaeili, F., Alimi, R., Mavi, S.A., Shojaee, S., 2018. Anti-Toxoplasma activity of various molecular weights and concentrations of chitosan nanoparticles on tachyzoites of RH strain. Int. J. Nanomedicine 1341–1351.

Tiwari, J.N., Tiwari, R.N., Kim, K.S., 2012. Zero-dimensional, one-dimensional, two-dimensional and three-dimensional nanostructured materials for advanced electrochemical energy devices. Prog. Mater. Sci. 57, 724–803.

Vergara-Duque, D., Cifuentes-Yepes, L., Hincapie-Riaño, T., Clavijo-Acosta, F., Juez-Castillo, G., Valencia-Vidal, B., 2020. Effect of silver nanoparticles on the morphology of Toxoplasma gondii and Salmonella braenderup. J. Nanotechnol. 2020, 1–11.

Virus, M.A., Ehrhorn, E.G., Lui, L.M., Davis, P.H., 2021. Neurological and neurobehavioral disorders associated with Toxoplasma gondii infection in humans. J. Parasitol. Res. 2021, 1–18.

Wohlfert, E.A., Blader, I.J., Wilson, E.H., 2017. Brains and brawn: Toxoplasma infections of the central nervous system and skeletal muscle. Trends Parasitol. 33, 519–531.

Zhang, L., Gu, F.X., Chan, J.M., Wang, A.Z., Langer, R.S., Farokhzad, O.C., 2008. Nanoparticles in medicine: Therapeutic applications and developments. Clin. Pharmacol. Ther. 83, 761–769.

Zhao, Z., Ukidve, A., Krishnan, V., Mitragotri, S., 2019. Effect of physicochemical and surface properties on in vivo fate of drug nanocarriers. Adv. Drug Deliv. Rev. 143, 3–21.

Zhou, J., Zhang, X., Li, M., Wu, W., Sun, X., Zhang, L., Gong, T., 2013. Novel lipid hybrid albumin nanoparticle greatly lowered toxicity of pirarubicin. Mol. Pharm. 10, 3832–3841.

22 Toward Nano-Drug Precision for Neurodegenerative Diseases or Glioma

Nadia M. Hamdy, Emad B. Basalious, Al-Aliaa Sallam, Fatma H. Shaker, Sujay Paul, Shih-Min Hsia, and Mai M. Anwar*

22.1 INTRODUCTION

22.1.1 Background

Mitochondria, which are intracellular organelles derived from bacterial symbionts, are best known for producing adenosine triphosphate (ATP) through oxidative phosphorylation (OXPHOS) (Nissanka & Moraes, 2018), reactive oxygen species (ROS) regulation, Ca^{2+} homeostasis and clearance, and contributing to cellular stress responses, such as autophagy and apoptosis (Jeong & Seol, 2008), considering their crucial functions that extend beyond the boundaries of the cell and affect the whole organism's physiology by controlling communication between cells and tissues (Nunnari & Suomalainen, 2012). Being the "powerhouse" generating cellular energy via OXPHOS, they are also an integrative hub for several cellular metabolic and signaling pathways, which is important for connectivity, plasticity, and survival (Spinelli & Haigis, 2018). These special dynamic organelles may modify their functions and morphology according to the surrounding stressors and physiological situation. Numerous evidences suggest that mitochondria can also regulate the survival and death of cells, along with playing an important role in neurodegeneration (Rahman et al., 2023; Shaw, 2021). Sufficient energy is highly required in the central nervous system (CNS) for the excitability and survival of neurons, which is highly based on mitochondrial energy sources; thus, mitochondrial dysfunctions are the most common molecular feature in neurodegenerative diseases (NDDs). In certain brain disorders, it directly leads to neuronal dysfunction, loss of brain homeostasis, and degenerative diseases (Rahman et al., 2023; Shaw, 2021; Uddin et al., 2020). The characterization of NDDs as a type of mitochondrial disorders has recently gained a lot of interest among researchers. Several types of NDDs, such as Alzheimer's disease (AD), have been found to progress due to triggered free radical, DNA mutation, dysregulation of mitochondrial calcium, and mitochondrial respiratory disorders (Anwar, 2021; Anwar, Özkan, & Gürsoy-Özdemir, 2022; Hamdy, Shaker, & Zhan, 2022). Since mitochondrial dysfunctions are known to be highly associated with the neurological disorders' pathophysiological mechanisms, it is suggested that targeting mitochondria dysfunctions can be considered a vital clinical approach against NDDs.

22.1.2 Problem Definition

Per the importance of mitochondria in normal neurons development and neurogenesis, including axon branching, synaptic function, glial cell function, calcium regulation, neuronal repair, therefore, neurons have high energy requirements, which are provided by the high mitochondrial content within the CNS; thus, mitochondrial dysfunction can have crucial effects on neuron development (Wu, Chen, & Jiang, 2019). Therefore, mitochondrial dysfunction emerged as a crucial component of NDDs such as Parkinson's disease (PD), Alzheimer's disease (AD), Huntington's disease (HD), and amyotrophic lateral sclerosis (ALS) (Nunnari & Suomalainen, 2012), as well as glioma and glioblastoma (GBM), where changes in mitochondrial morphology and/or transport are characteristic.

22.1.3 Problem Statement

Accumulation of mitochondrial DNA (mtDNA) mutations, excessive production of ROS, alterations in mitochondrial dynamics, mitophagy, and mitochondrial biogenesis all enhance mitochondrial dysfunction (Lin, Liu, Qin, & Wang, 2022; Wu et al., 2019). Neuronal morphology is very complex, with optimal functioning and distribution of mitochondria as crucial features. In several NDDs, significant changes in mitochondrial transport, alterations in mitochondrial respiration, as well as functional changes in specific genes expressed in specific neuronal tissues with/out signaling function involved into various mitochondrial functions are observed.

* Corresponding author: nadia_hamdy@pharma.asu.edu.eg

DOI: 10.1201/9781003383376-22

22.2 CHAPTER AIM

Via extensive literature search and *in silico* analysis as well as tracing clinical trials (CTs) done, the chapter will identify hallmarks of mitochondrial dysfunction in NDDs or glioma after discussing mitochondrial biogenesis, mitochondria dynamics, and mitophagy. Second, various epigenetics and mitochondrial dysfunction axis in the selected NDDs as well as glioma will be enumerated. Third, autophagy in NDDs is to be detailed, as well as "neuronal autophagic activity" affection in NDDs in comparison to glioma. Finally, glioma/GBM, in brief relation to epigenetics, will be discussed, a step-toward the development of nano-anti-epigenetic drugs or nano-autophagy-modulating drugs and nano-mitochondria-directed drugs precision in precision medicine (PM).

22.3 HALLMARKS OF MITOCHONDRIAL DYS/FUNCTION IN NDDs

22.3.1 Mitochondrial Biogenesis

22.3.1.1 Mitochondrial Biogenesis Process

In the mitochondrial biogenesis process, new mitochondria are created through the growth and division of pre-existing ones, increasing the cellular mitochondrial mass (Johnson et al., 2021) and allowing cells to adapt their energy requirements in response to conditions such as growth, cell division, and changes in oxidative stimuli and hormones (Wang, Xu, Musich, & Lin, 2019). Mitochondrial transcription factor A (TFAM), nuclear respiratory factors 1 and 2 (NRF1 and NRF2), peroxisome proliferator-activated receptors (PPAR) gamma coactivator-1α (PGC-1α) family, and sirtuins are all involved in gene expression control during mitochondrial biogenesis (Wang et al., 2019). PGC-1α, the master regulator of mitochondrial biogenesis, directly coordinates the activity of NRF1–2, which in turn regulates the expression of TFAM and a large number of nuclear-encoded mitochondrial proteins, including those involved in mitochondria-OXPHOS, mitochondria import, and proteins mediating anti-oxidant response (Wareski et al., 2009). Moreover, PGC-1α contributes to the replication of mtDNA as well as the synthesis of proteins that are encoded by both the mitochondrial and nuclear genomes. Sirtuin 1 (SIRT1), a different energy sensor from PGC-1α that is crucial for mitochondrial biogenesis, is a deacetylase activated in response to a rise in the NAD+/NADH ratio caused by energy stress (Hees & Harbauer, 2022).

22.3.1.2 Defects in Mitochondrial Biogenesis

Defects in mitochondrial biogenesis may contribute to mitochondrial dysfunction in many NDDs, such as AD. Although there is no clear mechanism linking AD to mitochondrial biogenesis, PGC-1α, NRF2, or TFAM have been reported to be downregulated in cellular and mouse models of AD (Branca et al., 2017; Hees & Harbauer, 2022). Impaired PGC-1α-mediated signaling pathways may contribute to the pathogenesis of AD at multiple levels by altering amyloid beta (Aβ) production and clearance (Katsouri, Parr, Bogdanovic, Willem, & Sastre, 2011) and controlling the expression of BDNF, one of the most highly distributed neurotrophins implicated in neuroprotection (Xia et al., 2017). Similarly, crosstalk between mitochondrial biogenesis and mitochondrial dysfunction has been reported in PD, whereas impaired mitochondrial biogenesis through lower PGC-1α has been considered as a potential biomarker for PD patients, and PGC-1α polymorphisms are associated with the age of onset and risk of PD (Asghar et al., 2022). It has been suggested that mitochondrial biogenesis mediated by impaired SIRT1/PGC-1α signaling contributes to the selective susceptibility of dopaminergic neurons and explains the neurodegeneration in PD (Liu et al., 2022). Mitochondrial biogenesis is impaired in several NDDs; thus, targeting mitochondrial biogenesis may result in promising therapeutic strategies.

22.3.2 Mitochondrial Dynamics

Mitochondria are highly mobile, dynamic organelles that often undergo fusion and fission in response to various environmental stimuli through a strictly regulated process called "mitochondrial dynamics" (Johnson et al., 2021).

22.3.2.1 Mitochondrial Fusion

In contrast to mitochondrial fusion, which includes the union of two mitochondria resulting in one mitochondrion, *mitochondrial fission* is characterized by the division of one mitochondrion into two daughter mitochondria (Tilokani, Nagashima, Paupe, & Prudent, 2018), as shown in Figure 22.1. This process serves as a homeostatic mechanism and gives the mitochondria a quality control system, enabling the cell to efficiently adapt to rapidly changing energy demands (Tilokani et al., 2018). As a result, loss of either fission or fusion causes disturbances to mitochondrial dynamics, which leads to mitochondrial dysfunction and may contribute to development and progression of the disease (Chan, 2006; Joshi & Mochly-Rosen, 2018). Mitochondrial fusion is essential, as it allows the exchange of intermembrane and matrix contents, including mtDNA, between mitochondria. On the other hand, mitochondrial fission may facilitate the mitochondria's migration to compartments with high-energy demands; it also allows for the efficient clearance of mitochondria through mitophagy and decreases energy production for efficient cell death during necrosis (Johnson et al., 2021). The primary proteins composing the core machinery are large and multi-domain dynamin-related guanosine triphosphate hydrolases (GTPases), including dynamin-related protein 1 (Drp1), which mediates mitochondrial constriction and scission, while mitofusins 1 and 2 (Mfn1 and Mfn2) are responsible for mitochondrial fusion (Tilokani et al., 2018) as shown in Figure 22.1.

Because neuronal function requires high levels of mitochondrial metabolism, any change in mitochondrial function or morphology can have serious consequences for neurons.

22.3.2.2 Changes in Mitochondrial Dynamics

Changes in mitochondrial dynamics cause neurodegeneration by affecting inter-organelle communication, mitochondrial trafficking, and quality control (Corrado, Scorrano, & Campello, 2012). In multiple NDDs models, Drp1 expression is activated, resulting in excessive mitochondrial fission. Thus, Drp1 inhibition has emerged as a potential therapeutic strategy for reducing dysfunctional mitochondrial fission (Johnson et al., 2021). Furthermore, interactions between Drp1 and amyloid evolve time and are related to the degree of mitochondrial fragmentation (Corrado et al., 2012). In neurons with amyloid expression, Mfns are reported to be downregulated, while Drp1 is upregulated (Wang et al., 2009). Overexpression of synuclein alpha (SNCA) in transgenic mice resulted in lower Mfn1–2 protein levels, which correlated with decreased mitochondrial fusion and smaller mitochondria, as SNCA has influenced mitochondrial size independently and dependent on fusion/fission proteins (Valdinocci et al., 2019), again as summarized in Figure 22.1. Accordingly, mitochondrial dysfunction has been reported to be one of the downstream pathogenic effects of PD pathology (Portz & Lee, 2021).

22.3.3 Some of the Mitochondrial Ligands

22.3.3.1 Mitochondrial-Derived Damage-Associated Molecular Patterns (DAMPs)

Mitochondrial dysfunction causes the mitochondrial membrane to burst, releasing mitochondrial ligands into the cytoplasm or outside the cell (Gong, Liu, Jiang, & Zhou, 2020). Inflammation in NDDs is not pathogen-derived; therefore, "neuroinflammation" may be triggered by autologous particles, such as DAMPs, recognized as endogenous molecules, initiating intracellular signaling cascade and inducing the expression of some inflammatory markers (Donnelly, Chen, & Ji, 2020). Because DAMPs could be derived from bacterial infection, the immune system incorrectly identifies DAMPs derived from mitochondria as if bacterial-derived, provoking an innate immune mechanism (Wang et al., 2019). Several pro-inflammatory molecules are found in mitochondria, including mtDNA, mitochondrial unfolded protein response (mtUPR), mitochondrial ROS, ATP, TFAM, cardiolipin as well as cytochrome *c* or mitochondrial Ca^{2+}, and iron, which serve as danger flags for the neuro-inflammatory response (Rodríguez-Nuevo & Zorzano, 2019).

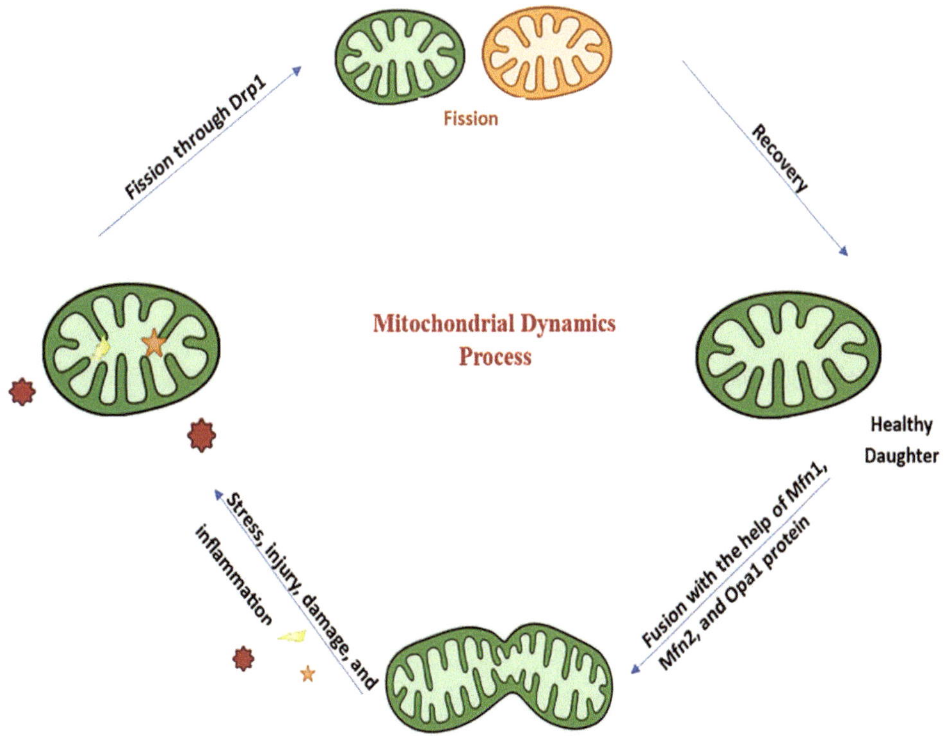

FIGURE 22.1 Simplified overview of mitochondrial dynamics depicting the processes involved in morphological alterations and recycling. Over time, when subjected to stress, a mitochondrion accumulates damage, initiating fission through dynamin-related protein 1 (Drp1) and facilitating the segregation of damaged mitochondria. Subsequent to the fission event, a healthy daughter mitochondrion undergoes recovery and fusion, aided by the proteins GTPases Mitofusin 1 (Mfn1), Mitofusin 2 (Mfn2), and optic atrophy 1 (Opa1). Under normal physiological conditions, a delicate balance between fission and fusion events is maintained, ensuring mitochondria undergo healthy life cycles.

22.3.3.2 mtDNA

mtDNA is a small circular molecule of double-stranded DNA (dsDNA) that encodes respiratory chain proteins as well as the transfer RNA (tRNA) and ribosomal RNA (rRNA) needed for their translation in the mitochondrial matrix. Also, mtDNA is a key signaling molecule that triggers inflammatory responses and regulates energy production and cell metabolism. mtDNA serves as a DAMP to regulate the inflammatory response by activating interferon gene, toll-like receptor (TLR9), and the NLR family pyrin domain–containing 3 (NLRP3) inflammasome (Gan, Moussaud, Jiang, & McLean, 2015). Decreases in neuronal mtDNA copy number and higher mtDNA heteroplasmy have been reported in the postmortem brains of people with AD (Picca, Calvani, & Coelho-Junior, 2020).

22.3.3.3 TFAM

TFAM is an mtDNA-binding protein that plays a crucial role in mtDNA metabolism and is required for mitochondrial genome stability (Kang, Chu, & Kaufman, 2018). By interacting with the receptor for advanced glycation end products (RAGE) or TLR on the membrane surface of glia, TFAM acts as a DAMP for pro-inflammatory microglial activation and initiates the nuclear factor-κB (NF-κB)-mediated nuclear signaling cascade (Lin et al., 2022). TFAM has been implicated in the inflammation of the CNS in NDDs (Anwar et al., 2022; Deus, Tavares, Beatriz, Mota, & Lopes, 2022). In addition to being essential as a fuel source for cellular homeostasis, intracellular ATP is also a fundamental component of extracellular signaling. Under stress conditions, ATP is secreted from various types of cells and activates microglia, which contributes to the pathogenesis of AD (Calabrese & Scuto, 2020). Importantly, extracellular ATP increased intracellular SNCA protein levels in PD neuronal cells, which was associated with an increase in Ca^{2+} influx (Gan et al., 2015). Moreover, SNCA decreases the activity of ecto-ATPase responsible for ATP degradation (Wilkaniec et al., 2017).

22.3.4 Mitophagy

A process controlled by proteins that promote mitochondrial fission and fusion as part of the mitochondrial quality control system to prevent cell damage by removing impaired mitochondria (Wang et al., 2019). Both phosphatase and tensin homolog (PTEN)–induced putative kinase 1 (PINK1) and parkin, critical mediators of mitophagy, identify and tag defective mitochondria for removal, whereas impaired mitochondria engulfed by an autophagosome are transferred to lysosomes for degradation (Dorn & Kitsis, 2015). Dysfunctional lysosomal and proteasomal degradation pathways are widely involved in NDDs. Impaired PINK1–parkin signaling disrupts mitophagy, and the association between mutations in these proteins and mitophagy dysfunction contributes to the pathogenesis of PD (Valdinocci et al., 2019; Wang et al., 2022). It is believed that PINK1 and parkin activity are required for SNCA clearance. Before neurodegeneration, PINK1 or parkin knockout increases the size and number of neuronal inclusions in SNCA transgenic mice that contain mitochondrial remnants and autophagic markers (Chen, Xie, & Turkson, 2015). Moreover, PINK1 guards against SNCA-induced neurodegeneration by inducing the activity of protein phosphatase 2A (Yang et al., 2018). Both Aβ and tau have been associated with mitophagy in AD. Because PINK1 is downregulated in AD patients and transgenic AD models, thus giving rAAV2–PINK1 into the hippocampus of transgenic mice dramatically reduced Aβ, enhancing synaptic function and memory (Du et al., 2017).

TABLE 22.1
The main mitochondrial ligands nature and mechanisms by which they are involved in NDDs

Ligand	DAMPs	mtDNA	TFAM
Nature	Autologous endogenous molecules/particles	Small circular molecule of dsDNA (Picca et al., 2020)	mtDNA-binding protein
Causes	Release of mitochondrial ligands into the cytoplasm or out of the cell (Gong et al., 2020)	Encodes respiratory chain proteins, tRNA, rRNA needed for their translation (Picca et al., 2020)	Role in mtDNA metabolism, mitochondrial genome stability (Kang et al., 2018)
Mechanism	"Neuroinflammation" initiating intracellular signaling cascade, inflammatory mediators' expression (Donnelly et al., 2020)	Triggers inflammatory response, regulates energy production, cell metabolism (Picca et al., 2020)	Interacting with RAGE or TLR on the membrane surface of glia
Remarks	Could be derived from bacterial infection; the immune system incorrectly recognizes DAMPs derived from mitochondria as if bacterial-derived, provoking an innate immune mechanism (Lin et al., 2022)	Decreases in neuronal mtDNA copy number, higher mtDNA heteroplasmy reported in postmortem brains of people with AD (Picca et al., 2020)	-

Abbreviations: TFAM: mitochondrial transcription factor A; mtDNA: mitochondrial DNA; AD: Alzheimer's disease; DAMPs: mitochondria-derived damage-associated molecular patterns; dsDNA, double-stranded DNA; tRNA: transfer RNA; rRNA: ribosomal RNA; TLR: toll-like receptor; RAGE: receptor for advanced glycation end products.

22.3.5 Mitochondrial Dysfunction via the Estrogen Hormone

Mitochondria can use different sex steroid hormones as regulatory factors for its own functions and activity.

22.3.5.1

Estrogen, the sex steroid hormone, may directly act on the mitochondrion via the non-classical pathway, where estrogen receptors have different mitochondrial levels among both males and females (Shaw, 2021; Uddin et al., 2020). The interaction between the mitochondria and estrogen sex hormone is the most studied type of interaction. Female estrogen hormones are released from the ovaries, which may cross the blood–brain barrier (BBB) to act upon brain cells as well as from female gonadal organs.

22.3.5.2

The primary estrogen source in males is obtained from the conversion of testosterone to estrogen using aromatase enzyme.

22.3.5.3

In both genders, estrogen is highly able to act upon the *neural mitochondria*, by binding to both estrogen response elements and estrogen receptors inside this organelle. The influence of sex steroid hormones on AD development as a behavior/cognitive change–related disease highlights the strong crosstalk mediator role of the mitochondrial function (Shaw, 2021; Uddin et al., 2020), evidenced by a notable high connection observed between estrogen and AD prevalence among females.

22.3.5.4

Recent data has highlighted that estrogen may enhance Aβ toxicity in human and animal models. On the other hand, it is suggested that increased estrogen may protect against Aβ aggregations via triggering the actions of B-cell lymphoma (Bcl) (Uddin et al., 2020). These observed data highlight that estrogen signaling cascades can play an important role in the development of AD by regulating mitochondrial activities.

22.3.6 Mitochondrial Energy and OS Abnormalities in NDDs

One of the most important findings in the brain of AD subjects is the accumulative signs of both mitochondrial dysfunctions and oxidative damage. Mitochondria is responsible for ATP generation through the ETC system. Numerous by-products can be produced during this process, such as hydroxyl radicals (OH.) ROS, hydrogen peroxide (H_2O_2), and superoxide (O_2) (Anwar et al., 2022; Hamdy et al., 2022; Hamdy, Gabr, AbdelHamid, & Swellam, 2024). The mitochondrial dysfunctions in AD were found to be related to the direct deposition of Aβ within the brain mitochondria of AD subjects, dysregulation in the mitochondrial OXPHOS, and PGC-1 within the brain mitochondria of AD subjects in addition to increased oxidative stress (Agrawal & Jha, 2020; Bergman & Ben-Shachar, 2016; Haga et al., 2019; Hamdy et al., 2022; McMeekin & Fox, 2021). For PD- and HD-related mitochondrial dysfunctions, suppressed ATP production, enhanced ROS production, intracellular calcium homeostasis impairment, PPAR-c dysregulation, and impaired mitochondrial functions were highly observed. On the other hand, decreased ca^{2+} mitochondrial capacity, altered mitochondrial morphology, increased ROS mitochondrial production, mitochondrial respiratory defects in the brain and muscles, ATP production defects, reduction in mtDNA, PGC-1 deficiency, and PPAR-c dysregulations were all considered as mitochondrial dysfunctions in ALS (Araujo et al., 2020; Bayer et al., 2017; Jiang, Wang, Perry, Zhu, & Wang, 2015; Joardar et al., 2015; Karagas et al., 2022; Keeney & Bennett, 2010; Muyderman & Chen, 2014; Xiao et al., 2018), as illustrated in Table 22.2.

22.3.7 Neuro-Epigenetics vs. Mitochondrial Epigenetics

Despite significant amount of evidence and numerous underlying mechanisms of NDDs proposed, there is still much to learn about how such diseases start and/or progress with deterioration of structure and function of central or peripheral nervous system (Wang et al., 2019). Environment-related factors have a large influence on NDDs; therefore, such influence involves the epigenetics modification pillar.

22.3.7.1 Neuro-Epigenetics

Neuro-epigenetics refers to epigenetic research in neurons that may interfere with both CNS development and function (Johnson et al., 2021). Neuro-epigenetics explores potential mechanistic pathways involved in the genesis of neurodevelopmental and NDDs. Consequently, neuro-epigenetics has the potential to guide the development of novel therapeutic techniques to mitigate the harmful effects of cognitive impairments and neurodegeneration (Gonzalez-Jaramillo & Portilla-Fernandez, 2019).

22.3.7.2 Mitochondrial Epigenetics

Mitochondrial epigenetics was initially introduced through the identification of a mitochondrial DNA methyltransferase (DNMT) variant (mtDNMT1) (Shock, Thakkar, Peterson, Moran, & Taylor, 2011). Per, mtDNA lacks histones, so mtDNA methylation and hydroxy methylation are the most studied mitochondrial epigenetic alterations in patients with NDDs (Coppedè, 2024).

22.3.7.2.1 DNMT3B-Mediated Hypermethylation

DNMT3B-mediated hypermethylation of the *PGC1-α* gene, which regulates mitochondrial biogenesis, in response to the inflammatory signals, reduced its protein expression, thus reducing the *mitochondrial density*. These findings showed that DNA methylation regulates various nuclear-encoded mitochondrial proteins in response to any pathogenic changes (Ambekar et al., 2021). The different mtDNMT variants appear to be tissue-specific, with DNMT3A mostly localized inside excitable tissues as the heart, skeletal tissues, and neurons, whereas DNMT1

TABLE 22.2
Mitochondrial dysfunction or abnormalities in the studied NDDs

NDD	Mitochondrial dysfunction or abnormalities	Ref.
Alzheimer's	Aβ aggregations OXPHOS dysregulation PGC-1 dysregulation Increased OS Decreased enzymes levels	(Agrawal & Jha, 2020; Bergman & Ben-Shachar, 2016; Gibson, Sheu, & Blass, 1998; Hamdy et al., 2024; Hamdy et al., 2022; McMeekin & Fox, 2021; Parker, Filley, & Parks, 1990)
Parkinson's	Suppressed ATP production Enhanced ROS production Intracellular calcium homeostasis impairments PPAR-c dysregulation Impaired function Increased lipid peroxidation	(Andersen, 2004; Anwar & Fathi, 2023; Dickey et al., 2016; Haga et al., 2019; Pérez-Segura & Santiago-Balmaseda, 2023; Schapira, 2008; Scorziello et al., 2020)
Huntington's	Inhibited ETC Inhibited nitropropionic acid (complex II) Increased OS	(Brouillet, Jacquard, Bizat, & Blum, 2005; Giacomello, Oliveros, Naranjo, & Carafoli, 2013; Hariharan, Shetty, Shirole, & Jagtap, 2014; Kumar & Ratan, 2016; Polyzos et al., 2019)
Amyotrophic lateral sclerosis	Decreased Ca^{2+} capacity Altered morphology Increased ROS production Respiratory defects ATP production defects Reduction in mtDNA PGC-1 deficiency PPAR-c dysregulation Mutant SOD1	(Araujo et al., 2020; Bayer et al., 2017; Bergemalm et al., 2006; Jiang et al., 2015; Joardar et al., 2015; Karagas et al., 2022; Keeney & Bennett, 2010; Muyderman & Chen, 2014; Xiao et al., 2018)

Abbreviations: ATP: adenosine triphosphate; Aβ: amyloid beta; mtDNA: mitochondrial DNA; OXPHOS: oxidative phosphorylation; ROS: reactive oxygen species; OS: oxidative stress; PGC-1: PPARγ coactivator-1α; PPAR: peroxisome proliferator-activated receptors; SOD: superoxide dismutase; ETC: electron transport chain.

was only found in the mitochondrial outer membrane in these tissues, but DNMT3B was undetectable (Wong, Gertz, Chestnut, & Martin, 2013).

22.3.7.2.2 Histone Modifications

Histone modifications of acetylation and demethylation are a further epigenetic signature that contributes to mitochondrial epigenomics. Histone lysine deacetylases (HDACs) and histone acetyltransferases (HATs), nuclear-encoded enzymes, remove or add acetyl groups to histones, respectively. In genome-wide investigations, it has been found that acetylation is linked to the availability of acetyl-CoA, which is mainly produced in the mitochondrial matrix (Pietrocola, Galluzzi, Bravo-San Pedro, Madeo, & Kroemer, 2015). Additionally, histone demethylases require various metabolic cofactors dependent on the mitochondrial metabolism, namely, flavin adenine dinucleotide (FAD) and alpha-ketoglutarate, suggesting these Krebs-related metabolic cofactors' involvement in mitochondrial regulation (Ambekar et al., 2021).

22.4 EPIGENETICS PERTURBATION AND MITOCHONDRIAL DYSFUNCTION IN SOME NDDs (AD AND PD)

Epigenetic alterations such as DNA methylation, post-translational histone tail modifications, and non-protein coding RNAs (ncRNA) influence the neuronal genetic function. These epigenetics modifications may either damage mtDNA or impair the mitochondrial biogenesis/bioenergetics, abrupt mitochondrial regulation or dynamics, as well as cause defective mitophagy, with an increased oxidative stress (ROS and free radicals' production) vs. decreased antioxidant/anti-inflammatory capacity (Chen, Rasheed, & Deng, 2022). As a result of these mentioned crucial roles in neurodegeneration, and hence NDDs, it could be hypothesized that DNA methylation, histone modifications, and ncRNA epigenetic alterations may have potential as markers for disease prognosis and treatment outcome measure, as well as constituting a potential promising therapy hit approach for "mitochondria-dependent NDD types." Mitochondrial dysfunction cascades leading to neurodegeneration are presented in Figure 22.2.

22.4.1 Alzheimer's Disease

Alzheimer's disease is a complicated disease characterized by the progressive loss of neural connections and cell death. At the molecular level, AD is distinguished by the formation of extracellular plaques from Aβ filament accumulations, intracellular neurofibrillary tangles of hyperphosphorylated tau, and neuroinflammation (Liu, Jiao, & Shen, 2018).

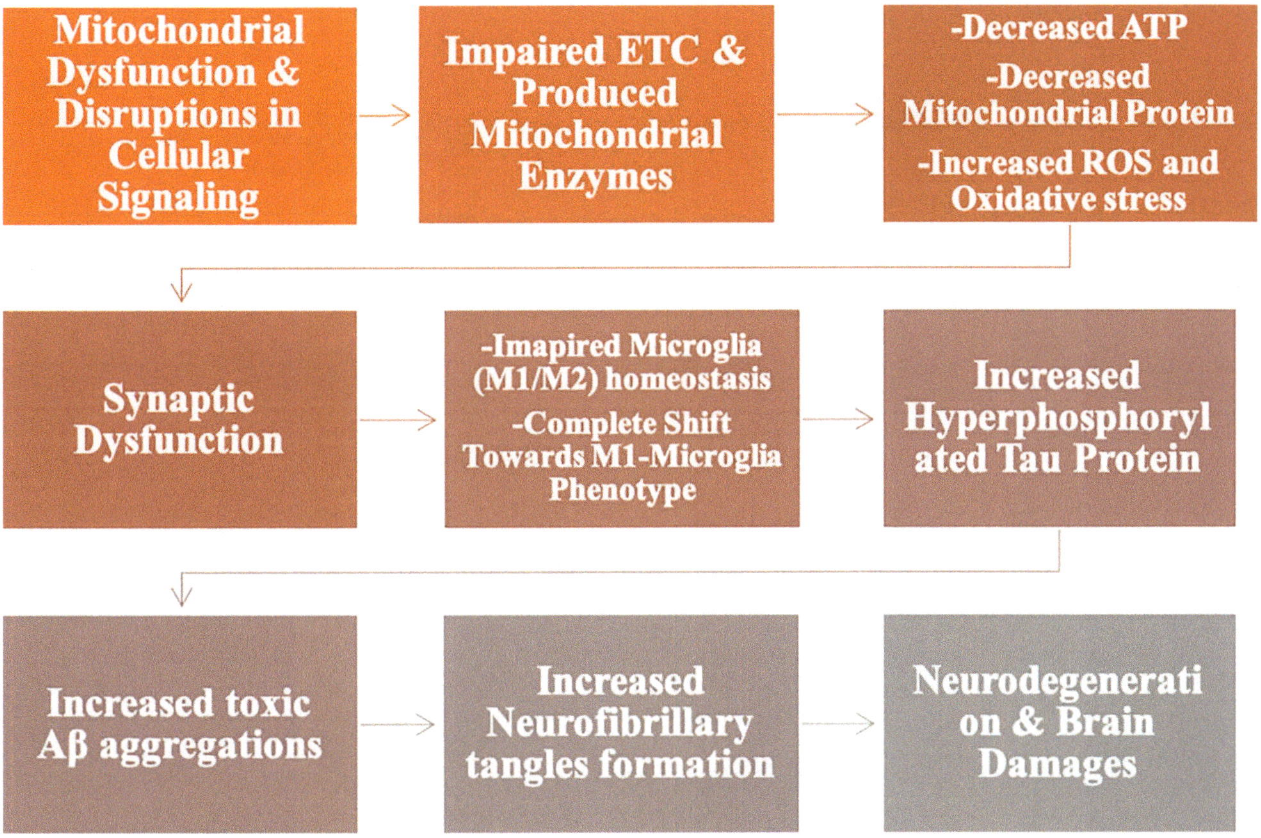

FIGURE 22.2 Hypothetical "mitochondrial dysfunction cascades" leading to neurodegeneration, where impaired electron transport chain (ETC) and mitochondrial enzymes characteristic of mitochondrial dysfunction lead to more Aβ aggregates and tangles via decreased energy with an impaired antioxidant function, synaptic dysfunction, microglia homeostasis impairment, hyperphosphorylation of the tau proteins. All are the typical neurodegeneration hallmarks of the NDDs.

22.4.1.1 ncRNAs

Altered micro-RNAs (miRs) levels have been associated with mitochondrial dysfunction that occurs in AD (Catanesi & d'Angelo, 2020). Several miRs that are specific to the brain, including miR-134, miR-135, let-7g, miR-101, miR-181a-b, miR-191, miR-124, miR-let-7c, let-7a, miR-29a, and miR-107, control synaptic plasticity, neurotransmitter release, and neurite growth (80). In AD brain tissues, many miRs are found to be downregulated, such as miR-107, which is markedly diminished even in patients with early-stage AD. Additionally, miR-107 target β-site amyloid precursor protein-cleaving enzyme-1 (BACE1), which contributes to the accumulation of Aβ peptides (Siedlecki-Wullich & Miñano-Molina, 2021). Also, miR-101 inhibits the expression of the amyloid precursor protein (APP) and prevents the accumulation of Aβ in hippocampal neurons, indicating a potential function for miR-101 in the development of AD (Vilardo, Barbato, Ciotti, Cogoni, & Ruberti, 2010).

22.4.1.1.1 miR Dysregulation and Mitochondrial Dysfunction Crosstalk

Observed in AD through the activation of the Notch signaling system, miR-98 in AD mice decreased the generation of Aβ and alleviated OS and mitochondrial dysfunction (Chen, Zhao, & Chen, 2019). By targeting mfn2, altered miR-195 expression may have contributed to mitochondrial dysfunction. Consequently, miR-195 inhibition may be a potential novel therapeutic approach for the treatment of AD by upregulating the mfn2 protein (Zhang et al., 2016). Several miRNAs, including miR-34a-5p, miR-125b-5p, miR-15b, and miR-149, suppress the expression of BACE1, reducing amyloid buildup and improving neuronal damage (Sayad et al., 2022). BACE1-AS (BACE1 antisense RNA) long non-coding RNAs (lncRNA) regulate BACE1 post-transcriptionally and protect BACE1 transcripts by targeting miR-29b-3p/miR-107/miR-124–3p/miR-485–5p/miR-761 (He et al., 2020). Thus, it is critical to study how some miRNAs can contribute to the onset and progression of other NDDs.

22.4.1.2 LncRNA and NDDs

In silico analysis of human functional lncRNA (list in NDD retrieved from nrStar™ Functional LncRNA PCR Array (human) https://www.arraystar.com/nrstar-functional-lncrna-pcr-array/accessed December 18, 2023) include: 17A, 51A, anti-NOS2A, ATXN8OS, BACE1-AS, BDNF-AS-1, BDNF-AS-2, CDKN2B-AS1, DGCR5, DLX6-AS1, DPY19L2P2–1, DPY19L2P2–2, GDNF-AS1–1, H19–1,

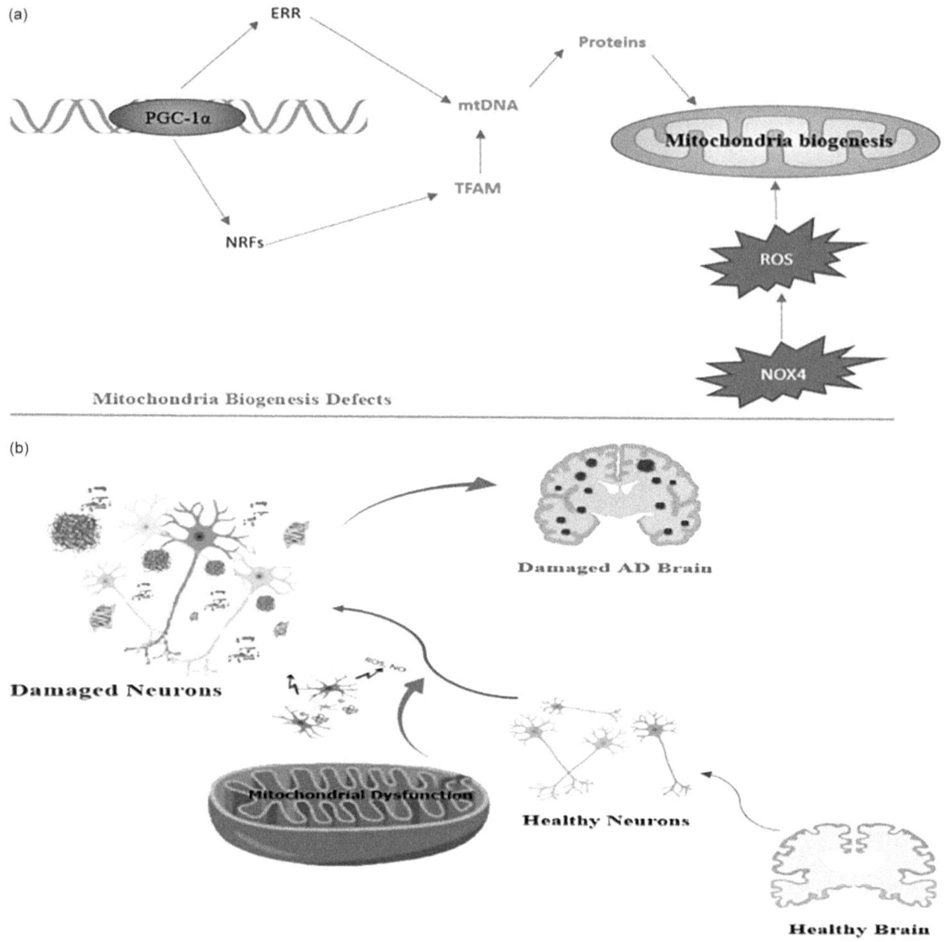

FIGURE 22.3 Illustration of mitochondria biogenesis defects and neuronal mitochondrial dysfunctions in Alzheimer's disease (AD) brain vs. a healthy brain: (a) Mitochondrial biogenesis is a regulated peroxisome proliferator–activated receptor gamma co-activator 1 (PGC-1), along with their key partner, nuclear respiration factors (NRFs), and the estrogen-related receptor (ERR), serving as the central hub. PGC-1α governs the transcription of crucial mitochondrial enzymes and the synthesis of mitochondrial DNA (mtDNA) and mitochondrial transcription factor A (TFAM). Changes in mtDNA have a direct impact on the modification of mitochondrial proteins, thereby influencing mitochondrial biogenesis. Furthermore, the process is regulated by nicotinamide adenine dinucleotide phosphate oxidase 4 (NOX4), with reactive oxygen species (ROS) playing a direct role in this regulatory mechanism. (b) Damaged neurons are torn and are exposed to much sparks of ROS with an impaired antioxidant capacity, leading to protein aggregation and tangles formation in the brain that could be considered as a result of mitochondrial dysfunction.

H19–2, HAR1A, HAR1B (HAR1R), HTT-AS, LINC00299, LINC00599, LINC01262, lincRNA-p21, MALAT1, MEG3–1, MEG3–2, MIAT, NAT-RAD18, NEAT1–1, NEAT1–2, PINK1-AS, PNKY, RMST, SCAANT1, SNHG1, SNHG3, SOX2OT, TUG1–1, TUG1–2, TUNA-1, TUNA-2, U1 spliceosomal lncRNA, UCH1LAS.

22.4.1.3 DNA Methylation

Mediated by a family of DNMT enzymes to regulate gene expression. DNMT1 primarily methylates hemi-methylated DNA and preserves genomic DNA methylation patterns after DNA replication, whereas DNMT3A and DNMT3B are responsible for *de novo* DNA methylation (Xu et al., 2021). The CNS's mitochondria also contain DNMT3A and DNMT3B, which increase the evidence that numerous mtDNA abnormalities, including methylation, are related to NDD pathogenesis (Chen et al., 2022). Generally, DNA demethylation promotes gene expression, while DNA hypermethylation inhibits gene expression. The evidence suggests that mtDNA hypermethylation in AD damages the mitochondrial biogenesis and the normal functioning of the mitochondrial respiratory chain, and demethylation is most likely a compensatory mechanism for this damage (Song, Zhu, Zhang, & Xiong, 2022). In AD patients, specific demethylation of the APP promoter and the presenilin-1 (PS1) promoter regions led to overexpression of PS1 and APP, which might promote the overproduction of Aβ. Thus, folic acid supplementation increases DNA methylation and subsequently lowers levels of APP, PS1, and Aβ through an epigenetic gene-silencing mechanism (Li et al., 2015).

22.4.1.4 Histone Modifications

Histone tails acetylation via HAT vs. deacetylation via HDAC are the most well-studied post-translational modification of core histones over histone methylation or phosphorylation. Class 1 HDACs (HDAC2 and HDAC3) are found in higher concentrations in the brain areas linked with memory. When class 1 HDACs genes are silenced, microglial amyloid phagocytosis is increased in an AD animal model (Broide et al., 2007). HDAC2 reduces histone acetylation in the genes involved in memory and learning. One study reported HDAC2 upregulation in the prefrontal cortex and hippocampus of AD mice (Broide et al., 2007; Lardenoije et al., 2015). HDAC2 dysregulation has recently been linked to cholinergic nucleus basalis of Meynert (nbM) neuronal failure and cognitive impairment during the clinical development of AD (Mahady et al., 2019). These investigations suggest that histone modification has significant role in AD development (Ghosh & Saadat, 2021).

22.4.2 Parkinson's Disease

A multifactorial disease characterized by a complex interplay of genetic and epigenetic elements. PD results from the presence of Lewy bodies that causes the death of dopaminergic neurons containing neuromelanin in the substantia nigra of the brain (Migliore & Coppedè, 2009), impairing PARK, SNCA, DJ-1, Parkin, and leucine-rich receptor kinase 2 genes that cause PD. These genes cause neuroinflammation, oxidative stress, mitochondrial dysfunction, and SNCA protein aggregation (Catanesi & d'Angelo, 2020; Goh, Chao, Dheen, Tan, & Tay, 2019; Titze-de-Almeida, Soto-Sánchez, & Fernandez, 2020) in PD patients.

22.4.2.1 ncRNAs

In the PD model, an elevated miR-124 inhibits mitochondrial apoptotic signaling pathways, maintaining the reduced autophagic activity and suppressing Bcl2-associated X (Bax) translocation to the mitochondria (Wang et al., 2016). miRs downstream target genes and proteins would be related to the mitochondrial function and OS development and, hence, PD development. miR-214 contributed to the neuroprotective effects of resveratrol in PD via inhibiting SNCA aggregation, OS, and mitochondrial impairment. Further, miR-7 enhances mitochondrial function by targeting voltage-dependent anion channel 1 (VDAC1), a component of the mitochondrial permeability transition pore, which may be a potential target for overcoming and reducing mitochondrial dysfunction in PD (Wang et al., 2015).

Small nucleolar RNA host gene 1 (SNHG1) identified as long intergenic long ncRNA 00057 is implicated in molecular and cellular processes directly associated with the PD's phenotype (linc00057), because SNHG1 acts through the SNHG1-miR15b-5p-SIAH1-linc-p21-miR-1277–5p axis to enhance SNCA–ubiquitination and regulate its toxicity (Rasheed & Liang, 2021). Furthermore, SNHG1 increases neuroinflammation via the miR-7/NLRP3 axis, and when overexpressed, 1-methyl-4-phenylpyridinium (MPP+) causes cytotoxicity via miR-153–3p sponging (Rasheed & Liang, 2021).

22.4.2.2 DNA Methylation

Sequestration of DNMT1 results in global DNA hypomethylation, which is associated with the upregulation of the *SNCA* gene to feed-forward Parkinsonism (Desplats et al., 2011). In another study, the *PPARGC1A* (PPARG coactivator 1 alpha) gene was found to be hypermethylated in peripheral blood leukocytes of PD patients with low PGC-α expression. This shows that gene dysregulations may be incorporated in altered cellular energy metabolism and mitochondrial dysfunction involved in PD development (Yang et al., 2020) Remarkably, vulnerability in the substantia nigra of Parkinson's disease has been associated with DNA hypomethylation of the tumor necrosis factor (TNF-α). Additionally, overexpression of TNF-α triggers a caspase cascade that leads to increased mitochondrial malfunction and death (Anwar, 2021, 2023). Therefore, the inhibitors of TNF-α may act as potential neuroprotective agents against SNCA aggregation (Desplats et al., 2011) and mitochondrial dysfunction, suggesting a possible treatment strategy for PD.

22.4.2.3 Histone Modifications

SNCA inhibits histone acetylation, when it binds to histone; hence, HDAC inhibitors would protect neurons from the neurotoxicity caused by SNCA (Ghosh & Saadat, 2021). Several proteins, including SNCA, PINK1, Parkin, protein deglycase DJ1, and leucine-rich repeat kinase2 (LRRK2 or dardarin), are implicated in the development of familial PD and govern mitochondrial dynamics and quality control (Zhu et al., 2012). Inhibiting the histone deacetylase SIRT 2 reduced SNCA-mediated neurotoxicity in a variety of PD models (Outeiro et al., 2007). Another study found that activating SIRT1 protected against SNCA aggregation via stimulating PGC-1α, as well as molecular chaperones (Donmez et al., 2012). Overexpression of SNCA in humans has been shown to downregulate the nine genes that produce histones H1, H2B, and H4 (107). Histone methylation, via H3K27, contributes to PD by increasing the neurodegeneration of DA1 and DA2 neurons by downregulating histone lysine methyltransferases enhancer of zeste homolog1/2 (EZh1/2) (Vartiainen, Pehkonen, Lakso, Nass, & Wong, 2006). Though the evidence suggests epigenetic participation in PD patients, epigenetics relationship to the mitochondrial features remains unknown and has to be clarified.

22.4.3 Huntington's Disease

An autosomal dominant genetic NDD that damages the striatal medium spiny neurons and is manifested clinically by mobility abnormalities, progressive dementia, and psychiatric and behavioral disorders (Landles & Bates, 2004). It is caused by an abnormally enlarged CAG repeats (more than 35 repeats) in the 5′-coding regions of the huntingtin gene (Htt), which promotes the synthesis of a mutant protein (mHtt) with a cytotoxic polyglutamine, leading to gradual motor and cognitive deterioration. Many components of neuronal homeostasis are disrupted in this disease, including improper histone acetylation and chromatin remodeling, as well as inappropriate interactions of the Htt protein. PGC-1α stimulates the clearance of misfolded Htt proteins via the transcription factor EB (TFEB), a regulator of the autophagy–lysosome pathway, hence reducing neurotoxicity and Htt aggregation. Low PGC-1α expression reduces mitochondrial biogenesis, OXPHOS, and mitochondrial malfunction (Ambekar et al., 2021).

22.4.3.1 ncRNA

Taurine upregulated gene-1 (*TUG1*) appears to play a role in NDDs in various physiological processes, such as controlling genes at multiple levels, including epigenetic, post-transcriptional, and post-translational levels. *TUG1* acts as a survival factor in neurons by functioning as a direct downstream transcriptional inhibitor of p53, which is upregulated in HD (Khalil et al., 2009). Other lncRNAs implicated in HD pathogenesis include maternally expressed gene-3 (*MEG3*), nuclear-enriched abundant transcript 1 (*NEAT1*), and X-inactivation-specific transcript (*XIST*). According to some research, various miRNAs, such as miR-9, miR-125b, miR-132, miR-146a, miR-150, miR-221, and miR-222, may interact with these lncRNAs, potentially reducing the efficacy of binding to their target mRNAs and so causing HD development (Riva, Ratti, & Venturin, 2016).

22.4.3.2 DNA Methylation

mHtt expression in many HD model systems and postmortem HD samples has been linked to significant changes in DNA methylation (Xu et al., 2021). Human fibroblasts from HD patients exhibited significant changes in 5-mC DNA methylation, which can be partially alleviated by inhibiting HDAC (Jia, Morris, Williams, Loring, & Thomas, 2015). The DNA methylation pattern was shown to be significantly altered at promoter, proximal, and distal regulatory regions in cells containing either wild-type or mHtt. Furthermore, promoter regions of genes involved in neurogenesis and neuronal development were shown to be hyper-methylated in mutant cells, resulting in lower expression levels (Jia et al., 2015). However, it is unknown how mHtt causes DNA methylation and whether DNMTs and DNA methylation maintenance factors are directly or indirectly implicated in aberrant epigenetic alterations in HD, a question to be answered.

22.4.3.3 Histone Modifications

Acetylation of H3 is reduced in the striatum of HD mice at the promoters of the following genes (*Drd2*, *Penkl*, *Actb*, or *Grin1*), which are normally downregulated for the function of neurons in the striatum; however, histone acetylation induced the expression of these genes (Sadri-Vakili et al., 2007).

Mitochondrial metabolism contributes to chromatin control by supplying intermediary metabolites such as acetyl-CoA for chromatin modifications by HAT. Consequently, changes in mitochondrial metabolism or the mitochondrial metabolic status influence the availability of these cofactors and/or substrates for chromatin-modifying enzymes proper function, resulting in aberrant epigenetic signatures that alter chromatin accessibility and, therefore, gene expression (Hyeon, Kim, & Yano, 2021). Therefore, fingers point to an interplay between mitochondrial function and chromatin dynamics in HD.

22.5 EPIGENETICS PERTURBATION AND MITOCHONDRIAL DYSFUNCTION IN GLIOMA

22.5.1 ncRNA, Micro-RNA in Glioma and Glioblastoma

ncRNA, micro-RNA in glioma and glioblastoma according to *Kyoto Encyclopedia of Genes and Genomes* (KEGG) are *targeted pathways* that are upregulated or downregulated in glial cells leading to tumorigenesis and glioma from increased cell survival, invasion, and metastasis (Figure 22.4).

In GBM, certain miRNAs are upregulated and others are downregulated, leading to glial cell tumorigenesis, survival, invasion, and metastasis. Upregulated miRNAs include miR-21. Inhibition of miR-21 leads to upregulation of the matrix metalloproteinase (MMP) inhibitors, a reversion-inducing cysteine-rich protein with Kazal motifs (RECK), and TIMP metallopeptidase inhibitor 3 (TIMP3) that decreases glioma motility and activates caspases (Rhim, Baek, Seo, & Kim, 2022).

Tumor protein 63 (TP63) and heterogeneous nuclear ribonucleoprotein K (hnRPK) are of the downstream targets of miR-21 involved in cell proliferation, apoptosis, and cell cycle (Li, Liang, Xu, & Zou, 2011). Besides, programmed cell death 4 (PDCD4) is critical in mediating apoptosis in GBM and is downregulated by miR-21, which may mediate the resistance of GBM cells against chemotherapy or radiation via its target genes PDCD4 (Chen et al., 2008). MiR-221 also is an upregulated molecular marker for glioma that targets p27 (Xue, Wang, Yue, & Zhang, 2017). Moreover, downregulation of miR-26a-5p reduced cell growth and tumor metastasis of human glioma via induction of PTEN protein expression (Xu & Xu, 2023). On the other hand, for the downregulated miRNAs, the expression level of both miR-124 and miR-137 is downregulated in GBM. Meanwhile, restoring their expression level leads to increased cell cycle arrest at G0/G1 phase by targeting cyclin-dependent kinase (CDK6) (Sati & Parhar, 2021). Both miR-34a as well as miR-128 act on E2Fa and CDK6 also (Sati & Parhar, 2021). Downregulation of miR-128 may contribute to glioma and GBM, in part, by coordinately upregulating ARP5 (ANGPTL6), Bmi-1, and E2F-3a, resulting in the proliferation of undifferentiated GBM cells (Cui et al., 2010). Further, miR-15b regulates cell cycle progression in glioma cells by targeting cell cycle–related molecules like cyclin E1 (Xia et al., 2009). Similarly, miR-7 downregulation potently and directly regulates epidermal growth factor receptor (EGFR) and IRS-2, forcing tumor invasion (Zhang, Dutta, & Abounader, 2012). miR-34a in GBM was identified as a micro-RNA that regulates GBM stem cells via regulation of Notch signaling

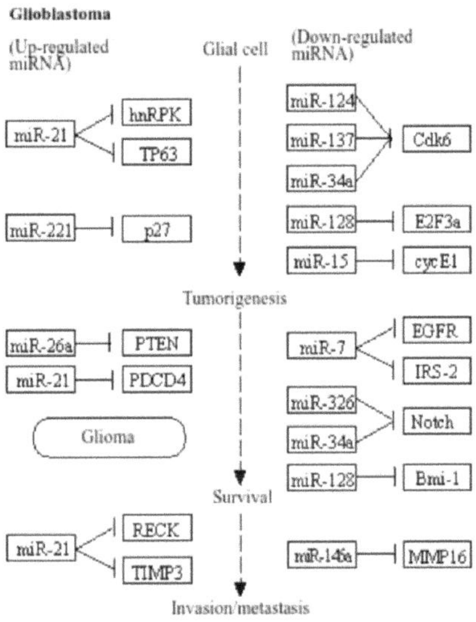

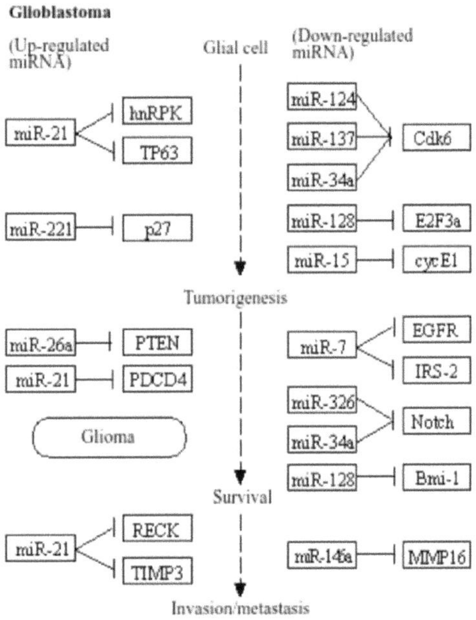

FIGURE 22.4 Micro-RNA in glioblastoma and glioma. (Retrieved from KEGG, accessed December 18, 2023 [www.kegg.jp/pathway/map05206].)

(Abba, Patil, & Allgayer, 2014). miR-146a directly regulated the expression of matrix metalloproteinases-16 (MMP-16), therefore triggering metastasis (Li & Li, 2013).

22.5.2 DNA Methylation

Gliomas are characterized by global changes in DNA methylation that may soon become a diagnostic and prognostic marker (Figure 22.5). They are most frequently triggered by isocitrate dehydrogenase (*IDH)* mutation. The *IDH* mutation not only changes global DNA methylation but also alters histone modifications (Dabrowski & Wojtas, 2019). Notably, the main distinction between glioma subtypes in the classification released by the WHO is based on the *IDH* gene mutation status that has a tremendous effect on global DNA methylation pattern (Louis et al., 2016).

DNA methylation pattern of the promoter of the *MGMT* (O6-methylguanine–DNA methyltransferase) gene was shown to be a prognostic marker for GBM patients treated with temozolomide (TMZ). Patients with silenced *MGMT* gene expression due to high gene promoter methylation were found to have a favorable outcome compared to patients with an unmethylated *MGMT* promoter. In fact, in patients without *MGMT* methylation, TMZ treatment did not give any benefit as compared to the radiotherapy alone (Hegi et al., 2005).

Moreover, in glioma-CpG island methylator phenotype (G-CIMP), hypermethylated phenotype is exhibited,

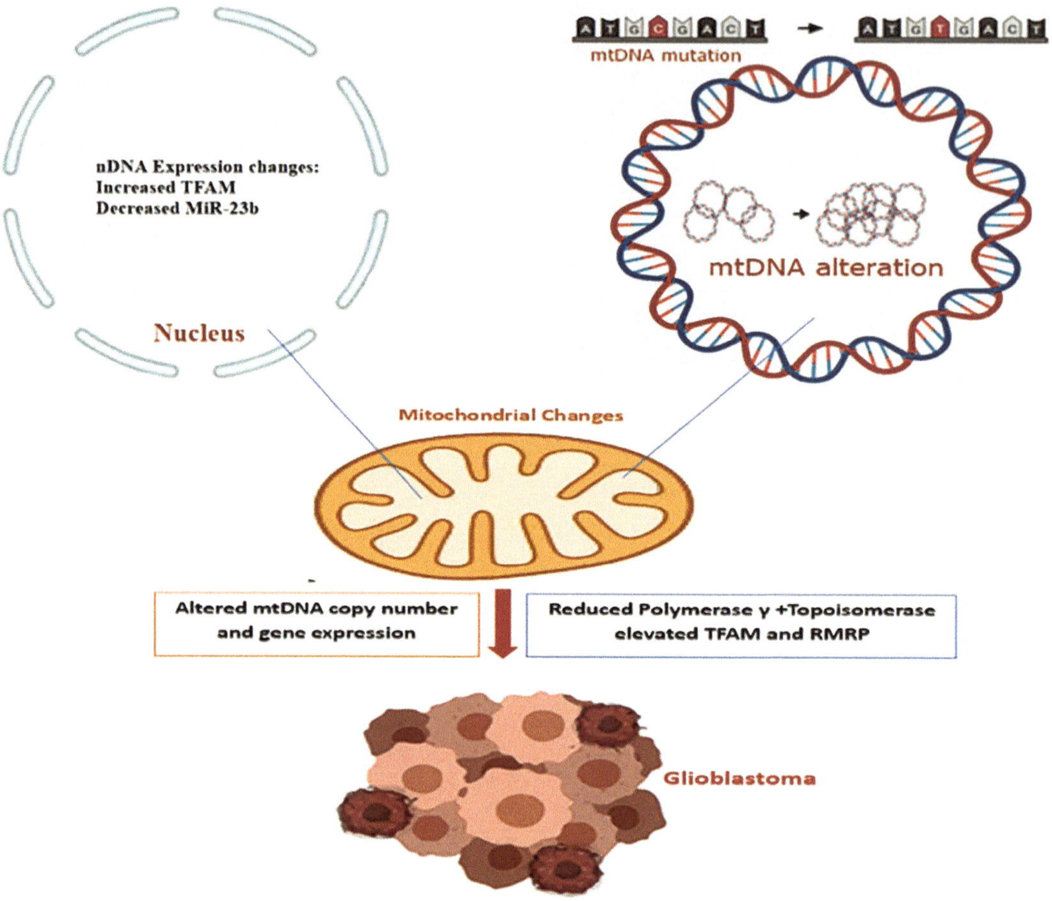

FIGURE 22.5 Epigenetic modifications' impact on the expression of mitochondrial DNA in glioblastoma. Methylation of nuclear DNA genes leads to decreased production of polymerase γ and topoisomerase proteins, resulting in diminished transport and availability in the mitochondria. Gliomas exhibit reduced transcription of MiR-23b, leading to increased TFAM expression within the nucleus. This alteration, coupled with decreased polymerase γ and topoisomerase levels and elevated TFAM and RMRP, affects mitochondrial DNA copy number and gene expression. Reductions in mitochondrial DNA copy number can trigger epigenetic modifications of nuclear DNA within the cell nucleus through retrograde signaling. Ultimately, these epigenetic changes lead to impaired oxidative phosphorylation capabilities in the mitochondria.

causing a repressive epigenetic state and blockage of differentiation. Changes in DNA methylation in gliomas have been shown to alter the binding affinity of several transcription factors, including SRY-related HMG-box (SOX2) (Alonso et al., 2011) and homeobox (HOX) (Kurscheid et al., 2015), as well as the RE1-silencing transcription factor (REST) (Dabrowski et al., 2018) that seems to affect glioma progression. Another remarkable example is CCCTC-binding factor (CTCF), which was described as an insulator, important for the maintenance of chromatin structure. DNA methylation changes within CTCF binding sites lead to global alterations of chromatin structure (Flavahan et al., 2016).

22.5.3 Histone Modifications

Great proof from recent studies confirms that histone modulation through acetylation, phosphorylation, ubiquitination, and methylation can regulate gene expression and may play a role in the pathogenesis of GBM (Azab, 2023). These changes include a widely dysregulated expression of genes involved in histone modifications, as well as changes in the histone modification pattern of individual genes. Overall aberrations at the histone level result from mutations in regulatory genes, as observed in large-scale genomic analyses of GBM samples, including HDAC2 and HDAC9, histone demethylases, and histone methyltransferases (Parsons et al., 2008). Additionally, altered expression levels of HDACs, due to reasons that have yet to be defined, have been linked to tumor recurrence and progression in HDAC1, HDAC2, and HDAC3 (Campos et al., 2011). Histone modifications regulating individual genes have been reported in several studies. For example, a repressed expression of the tumor suppressor RRP22 and the cell cycle regulator p21, combined with an enhanced expression of the pro-proliferative transcription factor HOXA9, has been linked to alterations in histone modification patterns (138). Interestingly, some virulent features characterizing GBM can be targeted using different histone deacetylase inhibitors (Azab, 2023)

22.6 AUTOPHAGY IN NDDs

22.6.1 Autophagy

Autophagy is a catabolic process in which lysosomal lytic enzymes digest cellular contents to protect cells in various physiological and pathological circumstances by enabling the breakdown of damaged cellular elements and subsequent recycling of amino acids, lipids, nutrients, and metabolites (Fujikake, Shin, & Shimizu, 2018). There are at least three types of autophagy: macro-autophagy, micro-autophagy, and chaperone-mediated autophagy. Macro-autophagy is thought to be the primary mechanism for the degradation of cytoplasmic proteins and organelles. Autophagy is a disease common to numerous NDDs, including AD and PD, because it is a cellular mechanism that breaks down aberrant proteins, including misfolded ones (Nakatogawa, Suzuki, Kamada, & Ohsumi, 2009). It is thought that neurons are particularly susceptible to autophagy disorders, since the accumulation of cellular waste would be a significant burden for these cells. The autophagy–lysosome pathway is principally responsible for the degradation of Aβ plaques and mutant SNCA proteins (Park & Kang, 2020) Autophagy induction, phagophore nucleation, elongation, and closure, autophagosome maturation, and fusion with lysosomes to generate autolysosomes are the subsequent stages that make up the entire autophagy process. Autophagy-related gene (ATG) proteins and several kinases, including AMP-activated protein kinase (AMPK), mammalian target of rapamycin (mTOR) complex 1 (mTORC1), and TFEB, play a key role in mediating autophagy complex mechanisms (Ren, Zhai, Lu, & Wang, 2021). There are cross-links between autophagy and PD-related genetic factors, such as SNCA, which is a hallmark of pathological changes in PD and can be destroyed by a variety of clearance mechanisms, including autophagy (Aski et al., 2018). However, wild-type and mutant SNCA, which suppresses autophagosome, activates mTORC1, and sequesters TFEB, is a crucial regulator of autophagosome formation and lysosomal biogenesis and function in the cytoplasm to block autophagic flux (Ren et al., 2021). Many studies have suggested the involvement of autophagy in AD pathogenesis through autophagy-mediated Aβ clearance Autophagy and neuronal cell death in neurological disorders. (Honda et al., 2020; Uddin et al., 2018). Accordingly, regulation of autophagy could be a potential strategy for treating these disorders; moreover, next-generation nano-drug targeting autophagy mechanisms are to be more developed.

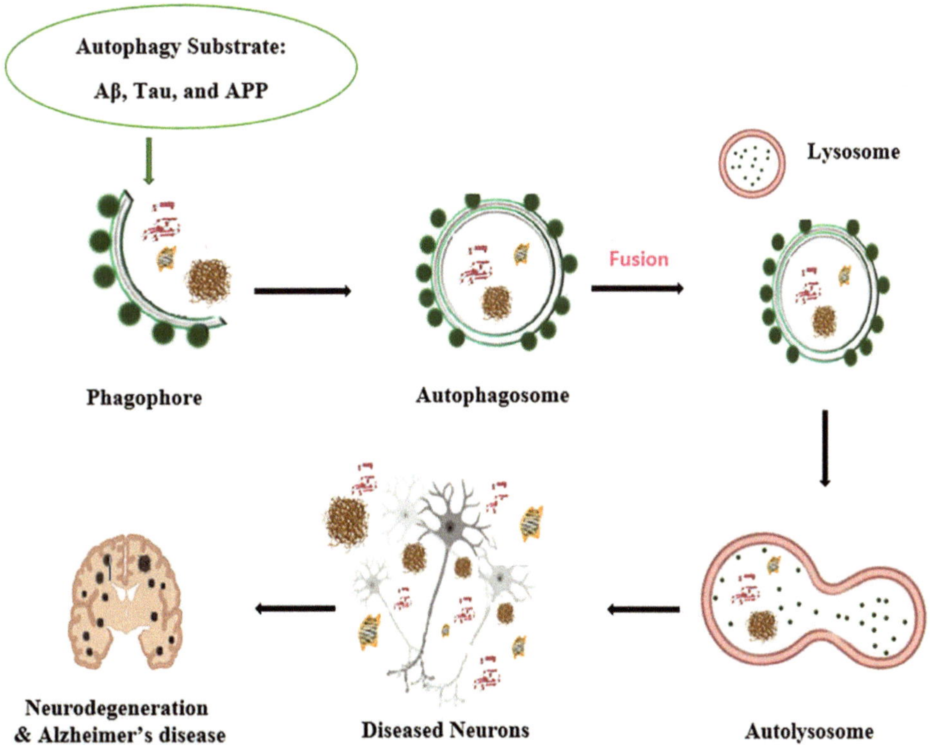

FIGURE 22.6 Dysfunctional autophagy–lysosome pathway in Alzheimer's disease. Autophagy is a well-preserved pathway responsible for breaking down long-lived intracellular proteins, protein aggregates, and organelles, through lysosomes, to uphold homeostasis under normal physiological conditions. Autophagy is generally divided into several steps, which include autophagy initiation, the formation of autophagosomes, the fusion of autophagosomes with lysosomes, and subsequent autophagic cargo degradation. The figure highlights the established direct connection between dysfunctions in the autophagy–lysosome pathway and Alzheimer's disease (AD) with a decreased clearance of Aβ accumulation in the context of AD, indicating impairment in cellular pathway.

22.6.2 "Autophagic-Neuronal Activity" Axis in NDDs

Knockouts of neuron-specific ATG genes show that basal autophagy rates are necessary for normal neuronal survival (Nixon & Yang, 2012). In addition to preserving the viability of the mature neuron, autophagy may play a significant role in early neurogenesis during neural tube formation. Also, it may be crucial in axon outgrowth and guidance, as well as in establishing connectivity in the developing brain (Kulkarni & Maday, 2018) Depletion of ATG9, a critical autophagy protein that regulates membrane growth, results in the aberrant formation of axon tracts and defects in neurite outgrowth (Yamaguchi et al., 2018). Autophagy at the axon's leading edge might promote membrane recycling or signaling information turnover in the migrating growth cone. Autophagy has been implicated in later neurodevelopment by mediating synaptic remodeling and eliminating excess dendritic spines; thus, altered autophagy may be associated with neurodevelopmental disorders. Selective autophagy loss in pyramidal neurons results in increased dendritic spines in the brain and social deficits resembling autism-like behaviors (Tang et al., 2014). Synapses are high-demand sites for cellular degradation pathways because they are dynamic, being remodeled in response to developmental and sensory signals to modify connections and circuits. Levels of autophagy in neurons are controlled by synaptic activity; in turn, autophagy also has an influence on synaptic function at both the pre- and post-synaptic levels (Sumitomo & Tomoda, 2021). Presynaptic autophagy likely modulates neurotransmitter release to control neurotransmission. The neuronal activity triggers autophagy at postsynaptic regions, prompting endocytic turnover of neurotransmitter receptors, such as gamma-aminobutyric acid ($GABA_A$), receptors to be sequestered in the cytoplasm or transported to the autophagosome and lysosome for destruction. Reduced surface $GABA_A$ receptors levels are mediated by elevated p62 levels triggered by reduced autophagy activity (Sumitomo & Tomoda, 2021).

Now, a similar question is raised: What about autophagy in gliomas?

22.6.3 "Autophagic-Neuronal Activity" in Glioma

The influence of autophagy in GBM throughout glioma initiation, promotion, and progression is still divisive, given its regulatory role in both cell death and survival. Autophagy has been anticipated as an inhibitor of tumor initiation by removing damaged proteins and organelles, protecting cells from ROS, necrosis, inflammation, genomic instability, and metabolic alterations (Choi, Ryter, & Levine, 2013). Nevertheless, it has been anticipated that glioma cells promote autophagy under unfavorable conditions, such as nutrient deficiency, acidosis, oxidative, or hypoxic stress, to keep up their survival, escaping the physiological response to cancer and therapy (Kimmelman, 2011; Vial & McKeown-Longo, 2016). Autophagy could affect the prognosis of glioma, either positively or negatively. Astrocytoma and GBM cells show lower levels of autophagy-related proteins with respect to low-grade astrocytic tumors (Magnus et al., 2014; Pirtoli et al., 2009; Vial & McKeown-Longo, 2016); on the other hand, it has been demonstrated that the progression of astrocytic tumors is related to a decrease in autophagic capacity (X. Huang, Bai, Chen, Li, & Lu, 2010). It was demonstrated that Unc-51-like autophagy-activating kinasesl–2 (ULK1–2) mRNA and protein levels are notably decreased in GBM patients in comparison to normal brains, stimulating astrocytic transformation by impairing autophagy (Shukla et al., 2014). On the other hand, it has been reported that the induction of autophagy by stress in tumor cells can result in resistance to the treatment, with the consequent tumor recurrence and progression (Kaza, Kohli, & Roth, 2012). Noor et al. reported that inhibiting autophagy by genetically suppressing ATG7, ATG13, or ULK1 significantly decreased glioma growth and oncogenic progression in a Ki-ras2 Kirsten rat sarcoma viral oncogene homolog (KRAS)–driven GBM mouse model, suggesting that autophagy is essential for the initiation and growth of glioma (Gammoh et al., 2016).

22.6.4 Epigenetic-Autophagic-Mitochondrial Axis Dysfunction in NDDs

Progressive, irreversible loss of neurons hallmarks of the NDD with the progressive accumulation of misfolded protein aggregates, abnormal protein dynamics due to deficiency of the ubiquitin-proteosome-autophagy system, oxidative stress, and free radical formation, endoplasmic reticulum stress, mitochondrial dysfunction and (secondary) disruptions of axonal transport all are addressed in Figure 22.7. *In silico* Pathways of Neurodegeneration—Multiple Diseases via KEGG Accessed Dec. 18th, 2023 www.genome.jp/kegg-bin/show_pathway?hsa05022 According to KEGG, the multiple diseases are H00056 Alzheimer's disease, H00057 Parkinson's disease, H00058 amyotrophic lateral sclerosis, H00059 Huntington's disease, H00063 spinocerebellar ataxia, H00061 prion disease www.genome.jp/network/nt06466+N00953+N00984+N01000+N00967+N00970+N01019+N01021+N01023+N01025+N01029+N01027+N01009+N01011+N01013+N01015.

However, the network pathways of neurodegeneration nt06466 are N00953 mGluR1-TRPC3 signaling pathway, N00984 mGluR5-Ca2+-apoptotic pathway, N01000 mAChR-Ca2+-apoptotic pathway, N00967 VGCC-Ca2+-apoptotic pathway, N00970 transport of calcium, N01019 parkin-mediated ubiquitination, N01021 parkin-mediated ubiquitination, N01023 parkin-mediated ubiquitination, N01025 parkin-mediated ubiquitination, N01029 26S proteasome-mediated protein degradation, N01027 UCHL1-mediated hydrolysis, N01009 PERK-ATF4 signaling pathway, N01011 IRE1a-XBP1 signaling pathway, N01013 IRE1a-JNK signaling pathway, N01015 ATF6-mediated transcription

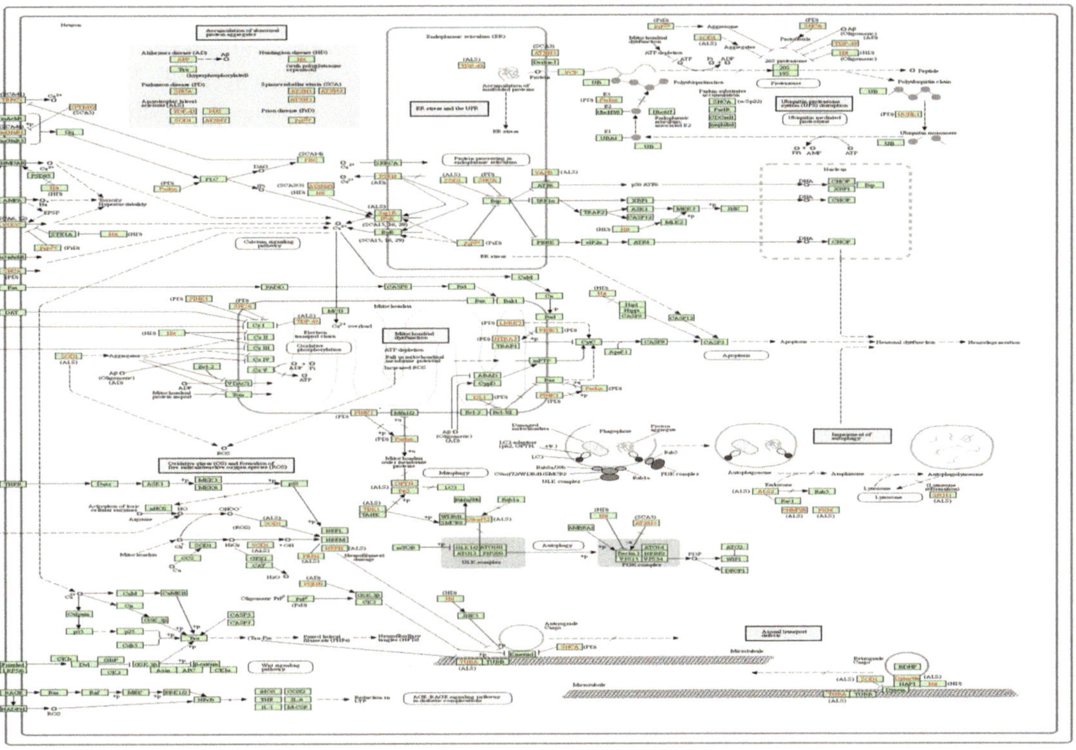

FIGURE 22.7 Pathways of neurodegeneration—multiple diseases. (Retrieved from KEGG. Accessed December 18, 2023 [www.genome.jp/kegg-bin/show_pathway?hsa05022].)

Mitophagy or autophagic-mitochondrial dysfunction is a type of selective autophagy that targets damaged mitochondria to lysosomes for degradation, which suppresses the cellular damage induced by dysfunctional mitochondria while guaranteeing the recycling of substances and energy provision (Um & Yun, 2017). Mitophagy has piqued the interest of researchers due to being a fundamental component of the mitochondrial quality control system, which keeps neurons healthy and functional (Wang, Feng, Wang, Chen, & Zhang, 2021). As a result, impaired mitophagy may be harmful to human health and may have a role in the etiology of several diseases, including neurodegeneration. Mitophagy molecular pathways can be mainly classified into ubiquitin (Ub)–dependent pathways and Ub-independent pathways (Tai, Chen, Tao, & Ren, 2022). Ub-dependent mitophagy is mediated by Parkin and PINK1, which cooperate with Ub to make a feed-forward loop to promote mitophagy. Meanwhile, Ub-independent pathways include receptor-meditated and lipid-mediated mitophagy, which are mostly located at the mitochondrial outer membrane and can interact with light chain3 (LC3) protein, thereby targeting the damaged mitochondria to autophagy vesicle (Shefa et al., 2019).

Alterations of mitophagic function play an important role in the development of many NDDs, such as AD and PD. Multiple studies have shown impaired mitochondrial function and compromised mitophagy in AD patients and models. Reduced mitophagic induction may result in an increase in malonaldehyde, which increases Aβ deposition and limits Beclin1 activity and, consequently, progression in AD pathology (Shefa et al., 2019). Defective mitophagy is commonly reported in AD with high tau protein levels. Excessive accumulation of Aβ and hyperphosphorylated tau accumulation can impair mitophagy by lowering levels of mitophagy-related proteins or forming abnormal interactions with Drp1, PINK1, or parkin in damaged AD neurons (Reddy & Oliver, 2019). Additionally, several PD-associated genes are critical modulators of mitochondrial quality control, and mutations of them can lead to a series of mitochondrial abnormality, including mitophagy, indicating that defective mitophagy plays a key role in the pathogenesis of PD (Fivenson et al., 2017; Wang et al., 2021). Autophagy and mitophagy work together to regulate mitochondrial homeostasis, and dysfunctional mitophagy may hasten the ageing process and the development of age-related NDDs. Autophagy and mitophagy enhancement may thus constitute a treatment strategy for NDDs.

Although the precise molecular mechanisms by which mitophagy processes influence neurodegeneration remain largely unknown, there is mounting evidence that ncRNAs and other epigenetic regulators play important roles in the pathogenesis of NDDs.

PINK1 is genetically linked to PD and performs important roles in mitochondrial homeostasis. According to a recent study, miR-27a/b plays a key role in the autophagic clearance of damaged mitochondria by regulating the *PINK1* gene, whereas miR-27a/b limit *PINK1* stability, preventing mitophagic flux of defective mitochondria (Kim et al., 2016). Further, miR-34a-5p suppresses mitophagy by decreasing

PINK1 through inhibiting phosphorylated Ub accumulation, limiting parkin recruitment to mitochondria, reducing ubiquitination, and delaying the clearance of damaged mitochondria. Suggesting, miR-34a-5p is implicated in promoting the pathogenesis of age-related neurological diseases (Tai et al., 2021). Meanwhile, miR-30b-5p inhibited TFEB transcriptional activity and could downregulate almost all the expression of TFEB-regulated genes involved in lysosomal biogenesis and autophagy (Guo et al., 2021). Modulation of miR-34a/SIRT1 signaling modulates mitophagy and mitochondrial biogenesis, thus protecting against oxidative stress, and delays age-related hearing loss. Additionally, lncRNAs with a length of more than 200 nucleotides are a novel class of crucial regulators of gene activities and various physiological events. The lncRNA NEAT1 was shown to be overexpressed in AD animal models. Furthermore, NEAT1 might promote the ubiquitination and degradation of PINK1, hence impairing PINK1-dependent autophagy and mitochondrial dysfunction (Huang, Zhao, Wang, Zhou, & Zhang, 2020). LncRNA MEG3 improves diabetes-induced cognitive dysfunction via reducing ROS generation, promoting FUN14 domain-containing protein 1 (FUNDC1)–related mitophagy and mitochondrial-derived apoptosis, and eventually provides adequate neuroprotection (Wang et al., 2021). Moreover, mtDNA methylation and mitochondrial DNMT3a levels, regulatory mechanisms for epigenetic modification, are abnormal in skeletal muscle and spinal cord of ALS mice, and these abnormalities occur in parallel with loss of myofiber mitochondria, decreased mitochondrial content, and increased mitophagy (Wong et al., 2013).

22.6.5 Epigenetic-Autophagic-Mitochondrial Axis Dysfunction in Glioma

22.6.5.1

In glioma, mitochondrial function is diminished by noticeable changes in the mitochondrial genome, leading to changed morphology and irregular bioenergetics, with greater-than-before ROS production (Lu & Ho, 2021). Mitochondrial dysfunction has a vital role in the control of numerous cancer key pathways linked to tumor metabolism, survival, proliferation, and cell death in GBM (Lu & Ho, 2021). Triggering mitophagy has been involved in GBM treatment (Cammarata et al., 2023; Wang et al., 2022). It can release stress and defeat tumors by eliminating dysfunctional mitochondria, and mitophagy-mediated clearance of pro-apoptotic mitochondria is responsible for cytoprotective benefits (Wang et al., 2022).

22.6.5.2 Recent Advances in Nano-Biomedicine for GBM/Glioma Treatment

Recently, drugs to act on mitophagy can endorse GBM cell death by encouraging mitochondrial damage, impairing ATP production, and extensively exhausting ATP stores. Therefore, stimulation of lethal autophagy has turned out to be an acceptable approach to remove GBM cells, which is apparently an efficient routine to terminate cancer cell growth (Meng et al., 2022; Rademaker et al., 2022). Through the speedy advances in nano-biomedicine for GBM, a growing number of molecular targets was notorious, providing basis for progress of specific nanotherapeutic systems in the near future. However, developing more effective drugs, precisely, nano-drugs, for GBM treatment targeting various mitochondrial pathways is mandatory. It is noteworthy to mention that the current rise in the number and applications of polymeric nano-agents presents a hope with potential to improve GBM therapeutic efficacy and outcome (Li & Xu, 2023).

22.7 PRECLINICAL AND POSTMORTEM EVIDENCE OF MITOPHAGY AND AUTOPHAGY IN NDDs

Preclinical, clinical, and postmortem brains evidenced several mitochondrial dysfunctions that may lead to the Aβ hallmark depositions and the progression of AD, as illustrated in Table 22.3. These alterations were found to be correlated with disease progression. It was reported that the neurotoxic NH2-tau fragment form harmfully stimulates mitochondrial dysfunctions and degradation in the brain by mitophagy. NH2-tau was also reported to limit mitochondrial fractions, thus leading to the progression of AD (Amadoro et al., 2014). The exogenous application of Aβ on rats' pheochromocytoma cells (PC12) resulted in an impaired mitophagy associated with diminished PINK1, parkin, and Bcl-1 levels. The observed results highlight that autophagy dysfunction may directly leads to harmful protein depositions in AD subjects (Wang, Jiang, Li, Gao, & Zhang, 2018; Wang, Wang, Li, Li, & Zhang, 2019). The treatment of human non-neuronal cells (HEK293T) with oligomeric Aβ1–42 resulted in mitophagy process blockade, associated with increased parkin, p62 mitochondrial fraction, and LC3-II/I levels, where Aβ aggregations induced mitochondrial dysfunctions and fragmentation in the non-neuronal cells (Wang et al., 2020). On the other hand, it was observed in postmortem human brain samples that any alteration or defect in mtDNA results in the progression of neurological diseases and mental illnesses (Arthur, Morton, Dunham, Keeney, & Bennett, 2009; Coskun, Beal, & Wallace, 2004; Janetzky et al., 1996; Melov, Schneider, Coskun, Bennett, & Wallace, 1999; Tatuch et al., 1992; Torrell et al., 2013).

22.8 THERAPIES FOR ENHANCEMENT OF MITOCHONDRIAL FUNCTION IN NDDs

Since mitochondrial dysfunction appears to be highly linked to the development of AD, researchers have thus postulated to target various mitochondrial dysfunctions in the development of potential nano/therapeutic approaches against AD.

TABLE 22.3
Mitochondrial dysfunctions and expected manifestations in preclinical and postmortem findings in NDDs

Brain region affected	Mitochondrial dysfunctions and expected manifestations	Ref.
Hippocampal neurons from rats	Mitochondrial dysfunctions and degradation in brain by mitophagy	(Amadoro et al., 2014)
Aβ-42 peptide on PC12	Impaired mitophagy, reduced levels of PINK1, parkin, Bcl-1, LC3-II/I ratio, p62	(Wang et al., 2019)
Exogenous, prolonged Aβ-24 application on PC12	Induced parkin, Bcl-1, LC3-II/I increase in PC12 cells	(Wang et al., 2018)
Treatment of human non-neuronal cells (HEK293T) with oligomeric Aβ	Increased parkin levels, p62 mitochondrial fractions, and LC3-II/I levels	(Wang et al., 2020)
Postmortem human brain	mtDNA alterations in different brain regions	(Arthur et al., 2009; Coskun et al., 2004; Janetzky et al., 1996; Melov et al., 1999; Tatuch et al., 1992; Torrell et al., 2013)

Abbreviations: PINK1: PTEN-induced putative kinase 1; mtDNA: mitochondrial DNA; Aβ: amyloid beta; Bcl: B cell lymphoma; LC3: light chain 3; PC12: rats' pheochromocytoma cells.

TABLE 22.4
Mode of action of cofactors and supplementary nutrients, as natural therapeutic approaches, for enhancement of mitochondrial dysfunctions and neurodegeneration

Cofactor or supplementary nutrient	Mode of action	Ref.
Acetyl-l-carnitine	OS markers reduction with cognitive abilities restoration	(Hamdy et al., 2024; Kelso et al., 2001)
Coenzyme Q10	Diminishes mitochondrial free radicals and dysfunction, enhances ATP synthesis and cognitive function improvement	(Abdulwahid Arif & Ahmad Khan, 2010)
Creatine	Ameliorates mitochondrial dysfunction and restores motor neuron activity	(Chaturvedi & Beal, 2008)
Curcumin	Reduces oxidative damage and *Aβ* deposits	(Lin & Beal, 2006)
Vitamins E and C	AD amelioration	(Anwar & Mabrouk, 2023)
Gingko biloba		(Anwar, Ali, Rashed, Badawi, & Eltablawy, 2019; Anwar & Mabrouk, 2023)
Probiotics	Antioxidant, anti-inflammatory, and spatial memory function improvement	(Anwar et al., 2018, 2019)
Cineole		(Khan et al., 2014)
Sulforaphane	Upregulates cytoprotective genes and reduces ROS and inflammation	(Carrasco-Pozo et al., 2015)

Abbreviations: ATP: adenosine tri-phosphate; OS: oxidative stress; ROS: reactive oxygen species; Aβ: beta amyloid; AD: Alzheimer's disease.

22.8.1

Based on postmortem AD patient findings, several therapeutic protocols were suggested by researchers to alleviate AD progressions, including antioxidant, anti-inflammatory, immunotherapy, cells therapy, cholinergic therapy, and hormonal therapy (Uddin et al., 2020). Earlier before, the hurdle for developing efficient and effective AD therapy was primary based on the inability to increase mitochondrial redox potentials.

22.8.2

Recently, the use of potential antioxidants that specially target accumulated mitochondrial free radicals offers great free ROS neutralization potentials. Several successful antioxidant therapies showed a promising efficiency against restoring normal mitochondrial and cognitive abilities, including acetyl-L-carnitine, curcumin, creatine, coenzyme Q10, vitamins E and C, ginkgo biloba, and probiotics (Abdulwahid Arif & Ahmad Khan, 2010; Anwar, Ali, Rashed, Badawi, & Eltablawy, 2018, 2019; Chaturvedi & Beal, 2008; Kelso et al., 2001; Khan et al., 2014).

22.8.3

The localized administration of cellular therapy has been recently evaluated preclinically and showed an efficient potential against the progression of AD. Neurons and other associated structures have been successfully restored via using mesenchymal stem cells, healthy microglia, and extracellular vesicles, where they showed antioxidant and anti-inflammatory actions (Anwar, 2019; Anwar & Fathi, 2022; Anwar, Ali, Rashed, Badawi, & Eltablawy, 2021; Anwar et al., 2023). NRF2 pathway activation may lead to antioxidant effects via different downstream products, such as heme-oxygenase, NADPH, and glutathione peroxidase. Thus, sulforaphane and cineole have potential anti-inflammatory and antioxidant actions against AD (Carrasco-Pozo, Tan, & Borges, 2015; Khan et al., 2014), as illustrated in Table 22.4.

22.9 UP-TO-DATE CLINICAL TRIALS (2019 TO DATE)

22.9.1 Bryostatin-1

Bryostatin-1 is an oxygenated macrolide with a polyacetate backbone. Safety, tolerability, and effectiveness of bryostatin in enhancing the cognitive function of 150 individuals with advanced AD were examined in 2019. In this 12-week phase-II, double-blind, randomized, placebo-controlled study, bryostatin was administered to the subjects. Every day, patients (aged 55–85) received intravenous injections of either a placebo or 20 µg or 40 µg bryostatin. The results of this study confirmed that both bryostatin dosages are safe. Compared to the larger dose (40 µg), cognitive performance was improved with 20 µg of bryostatin (Farlow et al., 2019).

22.9.2 Cannabidiol

Cannabidiol, a natural substance, helps PD patients with their tremors, anxiety, and psychosis. A randomized, multicenter, double-blind, placebo-controlled, phase-II study with 13 participants in 2020 examined the effect of cannabidiol on tremors in people with idiopathic PD. For as long as five weeks, cannabidiol was taken orally. Starting at 5 mg/kg/day, the dosage was raised by 2.5–5 mg/kg every three to five days until the desired dose of 20 mg/kg/day was achieved. Dyskinesia and tremor, the two main research variables, were significantly improved (Leehey et al., 2020).

22.9.3 Vitamin B3

Vitamin B3 is another possible therapeutic option for PD because of their antioxidant properties and their ability to regulate gene expression. Safety and tolerability of vitamin B3 were assessed in early- and late-stage PD patients in 2021, as well as the potential benefits of vitamin B3 supplementation for symptom improvement and inflammation reduction. Forty-seven participants were enrolled in a single-center, randomized, double-blind, placebo-controlled study. Participants were given either a placebo or 250 mg of vitamin B3 for a period of six months. Supplementing vitamin B3 greatly enhanced motor functions, decreased fatigue, and regulated serotonin levels, all of which may be beneficial for non-motor symptoms improvement (Wakade et al., 2021).

22.10 EXOSOME STUDIES IN GLIOMA

22.10.1

Almost all cell types, including astrocytes, microglia, oligodendrocytes, and neurons, release exosomes. Bioactive cargos as lipids, metabolites, proteins, and nucleic acids are carried by exosomes, transferred to and taken up by near or distant target cells (Younas, Fernandez Flores, Hopfner, Höglinger, & Zerr, 2022). Exosomes play a key role in a wide variety of biological functions, such as cellular intercommunication, immune responses, synaptic plasticity, and neurotransmission. Moreover, recent data indicates that exosomes can easily pass through the BBB in both directions, connecting the peripheral and central compartments (Kalluri & LeBleu, 2020). The concentration and composition of exosomal cargo, which includes both DNA and RNA, vary depending on the source and target cells. Both genetic materials are important for intracellular communication and gene control (Rastogi et al., 2021). An increasing amount of research has shown the role of exosomes in the cell-to-cell transfer of pathogenic protein aggregation, thereby receiving considerable interest as prospective diagnostic tools and biomarkers for glioma. They are considered to be more reliable than usual specimens, such as pure cerebrospinal fluid (CSF), blood, and urine, because of their capacity to cross the BBB in a bidirectional manner, low immunogenicity, and the fact that their contents (proteins and different RNA species) are changed during NDDs, GBM, and glioma. Finally, the presence of surface markers that uptake exosomes of CNS origin make them eligible to be utilized to identify their cellular origin (Younas et al., 2022) and for targeting cargo nano-drugs.

22.10.2

Potential within nanosized endogenous extracellular vehicles, exosomes, presents enormous promise as immune response stimulants against gliomas. Exosomes derived from malignant glioma cells have been investigated for their potential as robust biomarkers for glioma evaluation (Luo et al., 2023). Many circulating exosomes and exosomal cargos play critical roles in the intricate crosstalk systems in glioma initiation, development, and progression, providing potent opportunities for glioma diagnosis and prognosis (Shao et al., 2015). To increase the precision of diagnosis and prognosis, highly sensitive detection of glioma-derived exosomes and their associated biomarkers are recommended. In this sense, knowledge of glioma growth and metabolic flexibility allows surface proteins of glioma-derived exosomes to be regarded as reliable diagnostic markers as well as flags for next-generation targeted drug delivery. A variety of bioactive molecules, including miRs or lncRNAs, as possible potential suppressor molecules for glioma cell growth, may be found in glioma-derived exosomes, providing a novel theoretical and experimental base foundation for glioma-targeted future therapy.

22.11 EXPERT EXPERIENCE WITH PRECLINICAL NANO-DRUG DEVELOPMENT

Experimental up-to-date preclinical innovative nanoformulas have been developed for various NDDs; these mandates mean rapid leaps forward to their clinical examination phase (recommendation for sustainability). These include either innovative precisely designed intranasal lipid nano-capsules or lipo-pluronic micelles for systematic or olfactory drug delivery to the brain, respectively (Mohsen, Azzazy, Allam, & Basalious, 2020; Rashed, Shamma, & Basalious, 2016); polymeric nanoparticles-bearing drugs for AD treatment (Mohsen et al., 2020); and intranasal transferosomal *in situ* gel (ElShagea, Makar, Salama, Elkasabgy, & Basalious, 2023) or transdermal novasomal gels (ElShagea et al., 2024), both for PD treatment.

Various engineered nano-formulas are developed per their intended use (nano-by-design) as drug delivery systems (DDS) for already FDA-approved drugs to treat CNS, but in a lower dose or in a new route for administration, or for those drugs with problems in their solubility or bioavailability (Sokolov et al., 2024), minimizing off-target side effects (Hamdy, Eskander, & Basalious, 2022) and ensuring safety (Hamdy, Boseila, Ramadan, & Basalious, 2022).

22.12 SUMMARY AND CONCLUSIONS

Because of their diverse phenotypes and complex pathogenesis, there have been no effective drugs for brain tumors, GBM, or NDDs (Sokolov & Sharda, 2024). Mitochondrial dysfunction and autophagy are early pathological phenomenon in NDDs, crucial for drug discovery or repurposing already-known autophagy-modulating drugs. Via extensive literature search and *in silico* analysis, as well as tracing clinical trials done, the current chapter provided integrated mechanisms to understand the fundamental mitochondrial hallmarks, such as energy metabolism, mitochondrial biogenesis, mitophagy, mitochondrial dynamics, mitochondrial ligands, mitochondrial quality control, and metabolism, all of which hold promise to develop NDD-targeted nano-drugs. Moreover, this chapter provided insight into the effects of epigenetic modifications on neural cells and the disorders associated with them. Second, the chapter explored the epigenetic mechanisms contributing to autophagic–mitochondrial axis dysfunction in Alzheimer's disease (AD), Parkinson's disease (PD), Huntington's disease (HD), as well as glioma, providing a brief explanation of the disease pathologies and the role of nanosized extracellular exosome vesicles, a step toward encouraging drug discovery or drug repurposing, better in a nano form, a step toward nano-drugs precision, targeting either pillar of the epigenetic-autophagic-mitochondrial axis dysregulated or not functioning in NDDs or gliomas.

REFERENCES

Abba, M., Patil, N., & Allgayer, H. (2014). MicroRNAs in the regulation of MMPs and metastasis. *Cancers (Basel)*, *6*(2), 625–645. doi: 10.3390/cancers6020625

Abdulwahid Arif, I., & Ahmad Khan, H. (2010). Environmental toxins and Parkinson's disease: Putative roles of impaired electron transport chain and oxidative stress. *Toxicology and Industrial Health*, *26*(2), 121–128. doi: 10.1177/0748233710362382

Agrawal, I., & Jha, S. (2020). Mitochondrial dysfunction and Alzheimer's disease: Role of microglia. *Frontiers in Aging Neuroscience*, *12*. doi: 10.3389/fnagi.2020.00252

Alonso, M. M., Diez-Valle, R., Manterola, L., Rubio, A., Liu, D., Cortes-Santiago, N., . . . Gomez-Manzano, C. (2011). Genetic and epigenetic modifications of Sox2 contribute to the invasive phenotype of malignant gliomas. *PLoS One*, *6*(11), e26740. doi: 10.1371/journal.pone.0026740

Amadoro, G., Corsetti, V., Florenzano, F., Atlante, A., Ciotti, M. T., Mongiardi, M. P., . . . Calissano, P. (2014). AD-linked, toxic NH2 human tau affects the quality control of mitochondria in neurons. *Neurobiology of Disease*, *62*, 489–507. doi: 10.1016/j.nbd.2013.10.018

Ambekar, T., Pawar, J., Rathod, R., Patel, M., Fernandes, V., Kumar, R., . . . Khatri, D. K. (2021). Mitochondrial quality control: Epigenetic signatures and therapeutic strategies. *Neurochemistry International*, *148*, 105095. doi: 10.1016/j.neuint.2021.105095

Andersen, J. K. (2004). Iron dysregulation and Parkinson's disease. *Journal of Alzheimer's Disease*, *6*(6 Suppl), S47–S52. doi: 10.3233/jad-2004-6s602

Anwar, M. (2019). The effect of exaggerated induced inflammatory disorders and related disturbances in microglia–neuron homeostasis on the precipitation and remission of neurological disorders. *IBRO Reports*, *6*, S493. doi: 10.1016/j.ibror.2019.07.1547

Anwar, M., & Fathi, M. (2022). The cross-talk between microglia-derived exosomes and brain printing biometrics as a diagnostic approach for Parkinson's disease prior to motor symptoms implications [abstract]. *Movement Disorders*, *37*(S2). doi: 10.1002/mds.29223.

Anwar, M. M. (2021). Oxidative stress-a direct bridge to central nervous system homeostatic dysfunction and Alzheimer's disease. *Cell Biochemistry & Function*, doi: 10.1002/cbf.3673

Anwar, M. M. (2023). The orchestrating role of deteriorating neurons and TREM-1 in crosstalk with SYK in Alzheimer's disease progression and neuroinflammation. *Inflammopharmacology*, *31*(5), 2303–2310. doi: 10.1007/s10787-023-01270-5

Anwar, M. M., Ali, O. S., Rashed, L. A., Badawi, A. M., & Eltablawy, N. A. (2018). Regulation of miRNA-124, nuclear factor-Kappa B and β-Catenin expression in response to novel therapeutic protocol in LPS induced Alzheimer's disease in rats. *Research in Neuroscience*, *7*(1), 17. doi: 10.5923/j.neuroscience.20180701.03

Anwar, M. M., Ali, O. S., Rashed, L. A., Badawi, A. M., & Eltablawy, N. A. (2019). The neuro engraftment and neuroregenerative effects of hydrogen sulphide donor, intracerebral MSCs, Ginko Biloba and Kefir in attenuating neuropathological hallmarks of lipopolysaccharide induced

Alzheimer's disease rat models. *International Journal of Aging Research, 2*(3), 38. doi: 10.28933/ijoar-2019-06-2005

Anwar, M. M., Ali, O. S. M., Rashed, L. A., Badawi, A. M., & Eltablawy, N. A. (2021). The therapeutic potential and efficiency of intracerebroventricular transplantation and intravenous injection of mesenchymal stem cells in relieving Aß hallmarks and improving cognitive dysfunction in AD induced model. *Gene Reports, 25*, 101323. doi: 10.1016/j.genrep.2021.101323

Anwar, M. M., & Fathi, M. H. (2023). Early approaches of YKL-40 as a biomarker and therapeutic target for Parkinson's disease. *Neurodegenerative Disease Management*. doi: 10.2217/nmt-2022-0010

Anwar, M. M., & Mabrouk, A. A. (2023). Hepatic and cardiac implications of increased toxic amyloid-beta serum level in lipopolysaccharide-induced neuroinflammation in rats: New insights into alleviating therapeutic interventions. *Inflammopharmacology*. doi: 10.1007/s10787-023-01202-3

Anwar, M. M., Özkan, E., & Gürsoy-Özdemir, Y. (2022). The role of extracellular matrix alterations in mediating astrocyte damage and pericyte dysfunction in Alzheimer's disease: A comprehensive review. *European Journal of Neuroscience, 56*(9), 5453–5475. doi: 10.1111/ejn.15372

Anwar, M. M., Özkan, E., Shomalizadeh, N., Sapancı, S., Özler, C., Kesibi, J., & Gürsoy-Özdemir, Y. (2023). Assessing the role of primary healthy microglia and gap junction blocker in hindering Alzheimer's disease neuroinflammatory type: Early approaches for therapeutic intervention. *Frontiers in Neuroscience, 16*. doi: 10.3389/fnins.2022.1041461

Araujo, B. G., Souza, E. S. L. F., de Barros Torresi, J. L., Siena, A., Valerio, B. C. O., Brito, M. D., & Rosenstock, T. R. (2020). Decreased mitochondrial function, biogenesis, and degradation in peripheral blood mononuclear cells from amyotrophic lateral sclerosis patients as a potential tool for biomarker research. *57*(12), 5084–5102. doi: 10.1007/s12035-020-02059-1

Arthur, C. R., Morton, S. L., Dunham, L. D., Keeney, P. M., & Bennett, J. P., Jr. (2009). Parkinson's disease brain mitochondria have impaired respirasome assembly, age-related increases in distribution of oxidative damage to mtDNA and no differences in heteroplasmic mtDNA mutation abundance. *Molecular Neurodegeneration, 4*, 37. doi: 10.1186/1750-1326-4-37

Asghar, M., Odeh, A., Fattahi, A. J., Henriksson, A. E., Miglar, A., Khosousi, S., & Svenningsson, P. (2022). Mitochondrial biogenesis, telomere length and cellular senescence in Parkinson's disease and Lewy body dementia. *Scientific Reports, 12*(1), 17578. doi: 10.1038/s41598-022-22400-z

Aski, M. L., Rezvani, M. E., Khaksari, M., Hafizi, Z., Pirmoradi, Z., Niknazar, S., & Mehrjerdi, F. Z. (2018). Neuroprotective effect of berberine chloride on cognitive impairment and hippocampal damage in experimental model of vascular dementia. *Iranian Journal of Basic Medical Sciences, 21*(1), 53–58. doi: 10.22038/ijbms.2017.23195.5865

Azab, M. A. (2023). The potential role of histone modifications in glioblastoma therapy: Review article. *Journal of Molecular Pathology, 4*(4), 196–212.

Bayer, H., Lang, K., Buck, E., Higelin, J., Barteczko, L., Pasquarelli, N., . . . Witting, A. (2017). ALS-causing mutations differentially affect PGC-1α expression and function in the brain vs. peripheral tissues. *Neurobiology of Disease, 97*(Pt A), 36–45. doi: 10.1016/j.nbd.2016.11.001

Bergemalm, D., Jonsson, P. A., Graffmo, K. S., Andersen, P. M., Brännström, T., Rehnmark, A., & Marklund, S. L. (2006). Overloading of stable and exclusion of unstable human superoxide dismutase-1 variants in mitochondria of murine amyotrophic lateral sclerosis models. *Journal of Neuroscience, 26*(16), 4147–4154. doi: 10.1523/jneurosci.5461-05.2006

Bergman, O., & Ben-Shachar, D. (2016). Mitochondrial oxidative phosphorylation system (OXPHOS) deficits in schizophrenia: Possible interactions with cellular processes. *Canadian Journal of Psychiatry, 61*(8), 457–469. doi: 10.1177/0706743716648290

Branca, C., Ferreira, E., Nguyen, T. V., Doyle, K., Caccamo, A., & Oddo, S. (2017). Genetic reduction of Nrf2 exacerbates cognitive deficits in a mouse model of Alzheimer's disease. *Human Molecular Genetics, 26*(24), 4823–4835. doi: 10.1093/hmg/ddx361

Broide, R. S., Redwine, J. M., Aftahi, N., Young, W., Bloom, F. E., & Winrow, C. J. (2007). Distribution of histone deacetylases 1-11 in the rat brain. *Journal of Molecular Neuroscience, 31*(1), 47–58. doi: 10.1007/bf02686117

Brouillet, E., Jacquard, C., Bizat, N., & Blum, D. (2005). 3-Nitropropionic acid: A mitochondrial toxin to uncover physiopathological mechanisms underlying striatal degeneration in Huntington's disease. *Journal of Neurochemistry, 95*(6), 1521–1540. doi: 10.1111/j.1471-4159.2005.03515.x

Calabrese, V., & Scuto, M. (2020). Hydrogen sulfide and carnosine: Modulation of oxidative stress and inflammation in kidney and brain axis. *9*(12). doi: 10.3390/antiox9121303

Cammarata, F. P., Torrisi, F., Vicario, N., Bravatà, V., Stefano, A., Salvatorelli, L., . . . Cuttone, G. (2023). Proton boron capture therapy (PBCT) induces cell death and mitophagy in a heterotopic glioblastoma model. *Communications Biology, 6*(1), 388. doi: 10.1038/s42003-023-04770-w

Campos, B., Bermejo, J. L., Han, L., Felsberg, J., Ahmadi, R., Grabe, N., . . . Herold-Mende, C. (2011). Expression of nuclear receptor corepressors and class I histone deacetylases in astrocytic gliomas. *Cancer Science, 102*(2), 387–392. doi: 10.1111/j.1349-7006.2010.01792.x

Catanesi, M., & d'Angelo, M. (2020). MicroRNAs dysregulation and mitochondrial dysfunction in neurodegenerative diseases. *21*(17). doi: 10.3390/ijms21175986

Carrasco-Pozo, C., Tan, K. N., & Borges, K. (2015). Sulforaphane is anticonvulsant and improves mitochondrial function. *Journal of Neurochemistry, 135*(5), 932–942. doi: 10.1111/jnc.13361

Chan, D. C. (2006). Mitochondria: Dynamic organelles in disease, aging, and development. *Cell, 125*(7), 1241–1252. doi: 10.1016/j.cell.2006.06.010

Chaturvedi, R. K., & Beal, M. F. (2008). Mitochondrial approaches for neuroprotection. *Annals of the New York Academy of Sciences, 1147*, 395–412. doi: 10.1196/annals.1427.027

Chen, F. Z., Zhao, Y., & Chen, H. Z. (2019). MicroRNA-98 reduces amyloid β-protein production and improves oxidative stress and mitochondrial dysfunction through the Notch signaling pathway via HEY2 in Alzheimer's disease mice. *International Journal of Molecular Medicine, 43*(1), 91–102. doi: 10.3892/ijmm.2018.3957

Chen, L., Xie, Z., & Turkson, S. (2015). A53T human α-synuclein overexpression in transgenic mice induces pervasive mitochondria macroautophagy defects preceding dopamine neuron degeneration. *35*(3), 890–905. doi: 10.1523/jneurosci.0089-14.2015

Chen, Y., Liu, W., Chao, T., Zhang, Y., Yan, X., Gong, Y., . . . Peng, X. (2008). MicroRNA-21 down-regulates the expression of tumor suppressor PDCD4 in human glioblastoma cell T98G. *Cancer Letters, 272*(2), 197–205. doi: 10.1016/j.canlet.2008.06.034

Chen, Z., Rasheed, M., & Deng, Y. (2022). The epigenetic mechanisms involved in mitochondrial dysfunction: Implication for Parkinson's disease. *32*(3), e13012. doi: 10.1111/bpa.13012

Choi, A. M., Ryter, S. W., & Levine, B. (2013). Autophagy in human health and disease. *New England Journal of Medicine*, *368*(7), 651–662. doi: 10.1056/NEJMra1205406

Coppedè, F. (2024). Mitochondrial DNA methylation and mitochondria-related epigenetics in neurodegeneration. *Neural Regeneration Research*, *19*(2), 405–406. doi: 10.4103/1673-5374.379045

Corrado, M., Scorrano, L., & Campello, S. (2012). Mitochondrial dynamics in cancer and neurodegenerative and neuroinflammatory diseases. *International Journal of Cell Biology*, *2012*, 729290. doi: 10.1155/2012/729290

Coskun, P. E., Beal, M. F., & Wallace, D. C. (2004). Alzheimer's brains harbor somatic mtDNA control-region mutations that suppress mitochondrial transcription and replication. *Proceedings of the National Academy of Sciences of the United States of America*, *101*(29), 10726–10731. doi: 10.1073/pnas.0403649101

Cui, J. G., Zhao, Y., Sethi, P., Li, Y. Y., Mahta, A., Culicchia, F., & Lukiw, W. J. (2010). Micro-RNA-128 (miRNA-128) down-regulation in glioblastoma targets ARP5 (ANGPTL6), Bmi-1 and E2F-3a, key regulators of brain cell proliferation. *Journal of Neuro-Oncology*, *98*(3), 297–304. doi: 10.1007/s11060-009-0077-0

Dabrowski, M. J., & Wojtas, B. (2019). Global DNA methylation patterns in human gliomas and their interplay with other epigenetic modifications. *20*(14). doi: 10.3390/ijms20143478

Dabrowski, M. J., Draminski, M., Diamanti, K., Stepniak, K., Mozolewska, M. A., Teisseyre, P., . . . Wojtas, B. (2018). Unveiling new interdependencies between significant DNA methylation sites, gene expression profiles and glioma patients survival. *Scientific Reports*, *8*(1), 4390. doi: 10.1038/s41598-018-22829-1

Desplats, P., Spencer, B., Coffee, E., Patel, P., Michael, S., Patrick, C., . . . Masliah, E. (2011). Alpha-synuclein sequesters Dnmt1 from the nucleus: A novel mechanism for epigenetic alterations in Lewy body diseases. *Journal of Biological Chemistry*, *286*(11), 9031–9037. doi: 10.1074/jbc.C110.212589

Deus, C. M., Tavares, H., Beatriz, M., Mota, S., & Lopes, C. (2022). Mitochondrial damage-associated molecular patterns content in extracellular vesicles promotes early inflammation in neurodegenerative disorders. *11*(15). doi: 10.3390/cells11152364

Dickey, A. S., Pineda, V. V., Tsunemi, T., Liu, P. P., Miranda, H. C., Gilmore-Hall, S. K., . . . Arreola, M. (2016). PPAR-δ is repressed in Huntington's disease, is required for normal neuronal function and can be targeted therapeutically. *22*(1), 37–45. doi: 10.1038/nm.4003

Donmez, G., Arun, A., Chung, C. Y., McLean, P. J., Lindquist, S., & Guarente, L. (2012). SIRT1 protects against α-synuclein aggregation by activating molecular chaperones. *Journal of Neuroscience*, *32*(1), 124–132. doi: 10.1523/jneurosci.3442-11.2012

Donnelly, C. R., Chen, O., & Ji, R. R. (2020). How do sensory neurons sense danger signals? *Trends in Neurosciences*, *43*(10), 822–838. doi: 10.1016/j.tins.2020.07.008

Dorn, G. W., 2nd, & Kitsis, R. N. (2015). The mitochondrial dynamism-mitophagy-cell death interactome: Multiple roles performed by members of a mitochondrial molecular ensemble. *Circulation Research*, *116*(1), 167–182. doi: 10.1161/circresaha.116.303554

Du, F., Yu, Q., Yan, S., Hu, G., Lue, L. F., Walker, D. G., . . . Yan, S. S. (2017). PINK1 signalling rescues amyloid pathology and mitochondrial dysfunction in Alzheimer's disease. *Brain: A Journal of Neurology*, *140*(12), 3233–3251. doi: 10.1093/brain/awx258

ElShagea, H. N., Makar, R. R., Salama, A. H., Elkasabgy, N. A., & Basalious, E. B. (2023). Investigating the targeting power to brain tissues of intranasal rasagiline mesylate-loaded transferosomal in situ gel for efficient treatment of Parkinson's disease. *15*(2). doi: 10.3390/pharmaceutics15020533

ElShagea, H. N., Makar, R. R., Salama, A. H., Elkasabgy, N. A., Salama, A., & Basalious, E. B. (2024). Terpene-augmented novasomal gels for the sustainment of rasagiline mesylate delivery; A new approach for treating Parkinson's disease induced by rotenone in rats. *Journal of Drug Delivery Science and Technology*.

Farlow, M. R., Thompson, R. E., Wei, L. J., Tuchman, A. J., Grenier, E., Crockford, D., . . . Alkon, D. L. (2019). A randomized, double-blind, placebo-controlled, phase II study assessing safety, tolerability, and efficacy of bryostatin in the treatment of moderately severe to severe Alzheimer's disease. *Journal of Alzheimer's Disease, 67*(2), 555–570. doi: 10.3233/jad-180759

Fivenson, E. M., Lautrup, S., Sun, N., Scheibye-Knudsen, M., Stevnsner, T., Nilsen, H., . . . Fang, E. F. (2017). Mitophagy in neurodegeneration and aging. *Neurochemistry International*, *109*, 202–209. doi: 10.1016/j.neuint.2017.02.007

Flavahan, W. A., Drier, Y., Liau, B. B., Gillespie, S. M., Venteicher, A. S., Stemmer-Rachamimov, A. O., . . . Bernstein, B. E. (2016). Insulator dysfunction and oncogene activation in IDH mutant gliomas. *Nature, 529*(7584), 110–114. doi: 10.1038/nature16490

Fujikake, N., Shin, M., & Shimizu, S. (2018). Association between autophagy and neurodegenerative diseases. *Frontiers in Neuroscience*, *12*, 255. doi: 10.3389/fnins.2018.00255

Gammoh, N., Fraser, J., Puente, C., Syred, H. M., Kang, H., Ozawa, T., . . . Jiang, X. (2016). Suppression of autophagy impedes glioblastoma development and induces senescence. *Autophagy*, *12*(9), 1431–1439. doi: 10.1080/15548627.2016.1190053

Gan, M., Moussaud, S., Jiang, P., & McLean, P. J. (2015). Extracellular ATP induces intracellular alpha-synuclein accumulation via P2X1 receptor-mediated lysosomal dysfunction. *Neurobiology of Aging*, *36*(2), 1209–1220. doi: 10.1016/j.neurobiolaging.2014.10.037

Ghosh, P., & Saadat, A. (2021). Neurodegeneration and epigenetics: A review. *Neurologia (Engl Ed)*. doi: 10.1016/j.nrl.2021.01.016

Giacomello, M., Oliveros, J. C., Naranjo, J. R., & Carafoli, E. (2013). Neuronal Ca(2+) dyshomeostasis in Huntington disease. *Prion*, *7*(1), 76–84. doi: 10.4161/pri.23581

Gibson, G. E., Sheu, K. F., & Blass, J. P. (1998). Abnormalities of mitochondrial enzymes in Alzheimer disease. *Journal of Neural Transmission (Vienna)*, *105*(8–9), 855–870. doi: 10.1007/s007020050099

Goh, S. Y., Chao, Y. X., Dheen, S. T., Tan, E. K., & Tay, S. S. (2019). Role of microRNAs in Parkinson's disease. *International Journal of Molecular Sciences*, *20*(22). doi: 10.3390/ijms20225649

Gong, T., Liu, L., Jiang, W., & Zhou, R. (2020). DAMP-sensing receptors in sterile inflammation and inflammatory diseases. *20*(2), 95–112. doi: 10.1038/s41577-019-0215-7

Gonzalez-Jaramillo, V., & Portilla-Fernandez, E. (2019). Epigenetics and inflammatory markers: A systematic review of the current evidence. *2019*, 6273680. doi: 10.1155/2019/6273680

Guo, H., Pu, M., Tai, Y., Chen, Y., Lu, H., Qiao, J., . . . Huang, R. (2021). Nuclear miR-30b-5p suppresses TFEB-mediated lysosomal biogenesis and autophagy. *28*(1), 320–336. doi: 10.1038/s41418-020-0602-4

Haga, H., Matsuo, K., Yabuki, Y., Zhang, C., Han, F., & Fukunaga, K. (2019). Enhancement of ATP production ameliorates motor and cognitive impairments in a mouse model of MPTP-induced Parkinson's disease. *Neurochemistry International*, *129*, 104492. doi: 10.1016/j.neuint.2019.104492

Hamdy, N. M., Boseila, A. A., Ramadan, A., & Basalious, E. B. (2022). Iron oxide nanoparticles-plant insignia synthesis with favorable biomedical activities and less toxicity, in the "era of the-green": A systematic review. *14*(4). doi: 10.3390/pharmaceutics14040844

Hamdy, N. M., Eskander, G., & Basalious, E. B. (2022). Insights on the dynamic innovative tumor targeted-nanoparticles-based drug delivery systems activation techniques. *17*, 6131–6155. doi: 10.2147/ijn.s386037

Hamdy, N. M., Gabr, M. M., AbdelHamid, S. G., & Swellam, M. M. (2024). Chapter 24 – scrutinizing the landscape of DNA methylation epigenetic face(s) in glioblastomas. In R. K. Keservani, R. K. Kesharwani, M. Emerald, & A. K. Sharma (Eds.), *Nutraceutical fruits and foods for neurodegenerative disorders* (pp. 467–504). Academic Press.

Hamdy, N. M., Shaker, F. H., & Zhan, X. (2022). Tangled quest of post-COVID-19 infection-caused neuropathology and what 3P nano-bio-medicine can solve? *13*(2), 261–284. doi: 10.1007/s13167-022-00285-2

Hariharan, A., Shetty, S., Shirole, T., & Jagtap, A. G. (2014). Potential of protease inhibitor in 3-nitropropionic acid induced Huntington's disease like symptoms: Mitochondrial dysfunction and neurodegeneration. *Neurotoxicology, 45*, 139–148. doi: 10.1016/j.neuro.2014.10.004

He, W., Chi, S., Jin, X., Lu, J., Zheng, W., Yan, J., & Zhang, D. (2020). Long non-coding RNA BACE1-AS modulates isoflurane-induced neurotoxicity to Alzheimer's disease through sponging miR-214-3p. *Neurochemical Research*, *45*. doi: 10.1007/s11064-020-03091-2

Hees, J. T., & Harbauer, A. B. (2022). Metabolic regulation of mitochondrial protein biogenesis from a neuronal perspective. *12*(11). doi: 10.3390/biom12111595

Hegi, M. E., Diserens, A. C., Gorlia, T., Hamou, M. F., de Tribolet, N., Weller, M., . . . Stupp, R. (2005). MGMT gene silencing and benefit from temozolomide in glioblastoma. *New England Journal of Medicine*, *352*(10), 997–1003. doi: 10.1056/NEJMoa043331

Honda, S., Arakawa, S., Yamaguchi, H., Torii, S., Tajima Sakurai, H., Tsujioka, M., . . . Shimizu, S. (2020). Association between Atg5-independent alternative autophagy and neurodegenerative diseases. *Journal of Molecular Biology*, *432*(8), 2622–2632. doi: 10.1016/j.jmb.2020.01.016

Huang, X., Bai, H. M., Chen, L., Li, B., & Lu, Y. C. (2010). Reduced expression of LC3B-II and Beclin 1 in glioblastoma multiforme indicates a down-regulated autophagic capacity that relates to the progression of astrocytic tumors. *Journal of Clinical Neuroscience: Official Journal of the Neurosurgical Society of Australasia*, *17*(12), 1515–1519. doi: 10.1016/j.jocn.2010.03.051

Huang, Z., Zhao, J., Wang, W., Zhou, J., & Zhang, J. (2020). Depletion of LncRNA NEAT1 rescues mitochondrial dysfunction through NEDD4L-dependent PINK1 degradation in animal models of Alzheimer's disease. *Frontiers in Cellular Neuroscience*, *14*, 28. doi: 10.3389/fncel.2020.00028

Hyeon, J. W., Kim, A. H., & Yano, H. (2021). Epigenetic regulation in Huntington's disease. *Neurochemistry International*, *148*, 105074. doi: 10.1016/j.neuint.2021.105074

Janetzky, B., Schmid, C., Bischof, F., Frölich, L., Gsell, W., Kalaria, R. N., . . . Reichmann, H. (1996). Investigations on the point mutations at nt 5460 of the mtDNA in different neurodegenerative and neuromuscular diseases. *European Neurology*, *36*(3), 149–153. doi: 10.1159/000117233

Jeong, S.-Y., & Seol, D.-W. (2008). The role of mitochondria in apoptosis. *BMB Reports, 41*, 11–22. doi: 10.5483/BMBRep.2008.41.1.011

Jia, H., Morris, C. D., Williams, R. M., Loring, J. F., & Thomas, E. A. (2015). HDAC inhibition imparts beneficial transgenerational effects in Huntington's disease mice via altered DNA and histone methylation. *Proceedings of the National Academy of Sciences of the United States of America*, *112*(1), E56–E64. doi: 10.1073/pnas.1415195112

Jiang, Z., Wang, W., Perry, G., Zhu, X., & Wang, X. (2015). Mitochondrial dynamic abnormalities in amyotrophic lateral sclerosis. *Translational Neurodegeneration*, *4*, 14. doi: 10.1186/s40035-015-0037-x

Joardar, A., Menzl, J., Podolsky, T. C., Manzo, E., Estes, P. S., Ashford, S., & Zarnescu, D. C. (2015). PPAR gamma activation is neuroprotective in a drosophila model of ALS based on TDP-43. *Human Molecular Genetics*, *24*(6), 1741–1754. doi: 10.1093/hmg/ddu587

Johnson, J., Mercado-Ayon, E., Mercado-Ayon, Y., Dong, Y. N., Halawani, S., Ngaba, L., & Lynch, D. R. (2021). Mitochondrial dysfunction in the development and progression of neurodegenerative diseases. *Archives of Biochemistry and Biophysics*, *702*, 108698. doi: 10.1016/j.abb.2020.108698

Joshi, A. U., & Mochly-Rosen, D. (2018). Mortal engines: Mitochondrial bioenergetics and dysfunction in neurodegenerative diseases. *Pharmacological Research*, *138*, 2–15. doi: 10.1016/j.phrs.2018.08.010

Kalluri, R., & LeBleu, V. S. (2020). The biology, function, and biomedical applications of exosomes. *367*(6478). doi: 10.1126/science.aau6977

Kang, I., Chu, C. T., & Kaufman, B. A. (2018). The mitochondrial transcription factor TFAM in neurodegeneration: Emerging evidence and mechanisms. *FEBS Letters*, *592*(5), 793–811. doi: 10.1002/1873-3468.12989

Karagas, N. E., Gupta, R., Rastegari, E., Tan, K. L., Leung, H. H., Bellen, H. J., . . . Wong, C.-O. (2022). Loss of activity-induced mitochondrial ATP production underlies the synaptic defects in a <em>drosophila</em> model of ALS. *Journal of Neuroscience, 42*(42), 8019–8037. doi: 10.1523/jneurosci.2456-21.2022

Katsouri, L., Parr, C., Bogdanovic, N., Willem, M., & Sastre, M. (2011). PPARγ co-activator-1α (PGC-1α) reduces amyloid-β generation through a PPARγ-dependent mechanism. *Journal of Alzheimer's Disease*, *25*(1), 151–162. doi: 10.3233/jad-2011-101356

Kaza, N., Kohli, L., & Roth, K. A. (2012). Autophagy in brain tumors: A new target for therapeutic intervention. *Brain Pathology*, *22*(1), 89–98. doi: 10.1111/j.1750-3639.2011.00544.x

Keeney, P. M., & Bennett, J. P. (2010). ALS spinal neurons show varied and reduced mtDNA gene copy numbers and increased mtDNA gene deletions. *Molecular Neurodegeneration*, *5*(1), 21. doi: 10.1186/1750-1326-5-21

Kelso, G. F., Porteous, C. M., Coulter, C. V., Hughes, G., Porteous, W. K., Ledgerwood, E. C., . . . Murphy, M. P. (2001). Selective targeting of a redox-active ubiquinone to mitochondria within cells: Antioxidant and antiapoptotic properties. *Journal of Biological Chemistry, 276*(7), 4588–4596. doi: 10.1074/jbc.M009093200

Khalil, A. M., Guttman, M., Huarte, M., Garber, M., Raj, A., Rivea Morales, D., . . . Rinn, J. L. (2009). Many human large intergenic noncoding RNAs associate with chromatin-modifying complexes and affect gene expression. *Proceedings of the National Academy of Sciences of the United States of America, 106*(28), 11667–11672. doi: 10.1073/pnas.0904715106

Khan, A., Vaibhav, K., Javed, H., Tabassum, R., Ahmed, M. E., Khan, M. M., . . . Islam, F. (2014). 1,8-cineole (eucalyptol) mitigates inflammation in amyloid beta toxicated PC12 cells: Relevance to Alzheimer's disease. *Neurochemical Research, 39*(2), 344–352. doi: 10.1007/s11064-013-1231-9

Kim, J., Fiesel, F. C., Belmonte, K. C., Hudec, R., Wang, W.-X., Kim, C., . . . Kim, J. (2016). MiR-27a and miR-27b regulate autophagic clearance of damaged mitochondria by targeting PTEN-induced putative kinase 1 (PINK1). *Molecular Neurodegeneration, 11*(1), 55. doi: 10.1186/s13024-016-0121-4

Kimmelman, A. C. (2011). The dynamic nature of autophagy in cancer. *Genes & Development, 25*(19), 1999–2010. doi: 10.1101/gad.17558811

Kulkarni, V. V., & Maday, S. (2018). Compartment-specific dynamics and functions of autophagy in neurons. *78*(3), 298–310. doi: 10.1002/dneu.22562

Kumar, A., & Ratan, R. R. (2016). Oxidative stress and Huntington's disease: The good, the bad, and the ugly. *Journal of Huntington's Disease, 5*(3), 217–237. doi: 10.3233/jhd-160205

Kurscheid, S., Bady, P., Sciuscio, D., Samarzija, I., Shay, T., Vassallo, I., . . . Hegi, M. E. (2015). Chromosome 7 gain and DNA hypermethylation at the HOXA10 locus are associated with expression of a stem cell related HOX-signature in glioblastoma. *Genome Biology, 16*(1), 16. doi: 10.1186/s13059-015-0583-7

Landles, C., & Bates, G. P. (2004). Huntingtin and the molecular pathogenesis of Huntington's disease. Fourth in molecular medicine review series. *EMBO Reports, 5*(10), 958–963. doi: 10.1038/sj.embor.7400250

Lardenoije, R., Iatrou, A., Kenis, G., Kompotis, K., Steinbusch, H. W., Mastroeni, D., . . . Rutten, B. P. (2015). The epigenetics of aging and neurodegeneration. *Progress in Neurobiology, 131*, 21–64. doi: 10.1016/j.pneurobio.2015.05.002

Leehey, M. A., Liu, Y., Hart, F., Epstein, C., Cook, M., Sillau, S., . . . Bainbridge, J. (2020). Safety and tolerability of cannabidiol in Parkinson disease: An open label, dose-escalation study. *Cannabis and Cannabinoid Research, 5*(4), 326–336. doi: 10.1089/can.2019.0068

Li, L., & Li, H. (2013). Role of microRNA-mediated MMP regulation in the treatment and diagnosis of malignant tumors. *Cancer Biology & Therapy, 14*(9), 796–805. doi: 10.4161/cbt.25936

Li, S., Liang, Z., Xu, L., & Zou, F. (2011). MicroRNA-21: A ubiquitously expressed pro-survival factor in cancer and other diseases. *Molecular and Cellular Biochemistry, 360*, 147–158.

Li, W., & Xu, X. (2023). Advances in mitophagy and mitochondrial apoptosis pathway-related drugs in glioblastoma treatment. *Frontiers in Pharmacology, 14*, 1211719. doi: 10.3389/fphar.2023.1211719

Li, W., Liu, H., Yu, M., Zhang, X., Zhang, M., Wilson, J. X., & Huang, G. (2015). Folic acid administration inhibits amyloid β-peptide accumulation in APP/PS1 transgenic mice. *Journal of Nutritional Biochemistry, 26*(8), 883–891. doi: 10.1016/j.jnutbio.2015.03.009

Lin, M. M., Liu, N., Qin, Z. H., & Wang, Y. (2022). Mitochondrial-derived damage-associated molecular patterns amplify neuroinflammation in neurodegenerative diseases. *Acta Pharmacologica Sinica, 43*(10), 2439–2447. doi: 10.1038/s41401-022-00879-6

Lin, M. T., & Beal, M. F. (2006). Mitochondrial dysfunction and oxidative stress in neurodegenerative diseases. *Nature, 443*(7113), 787–795. doi: 10.1038/nature05292

Liu, J., Jiang, J., Qiu, J., Wang, L., Zhuo, J., Wang, B., . . . Lou, H. (2022). Urolithin a protects dopaminergic neurons in experimental models of Parkinson's disease by promoting mitochondrial biogenesis through the SIRT1/PGC-1α signaling pathway. *Food & Function, 13*(1), 375–385. doi: 10.1039/D1FO02534A

Liu, X., Jiao, B., & Shen, L. (2018). The epigenetics of Alzheimer's disease: Factors and therapeutic implications. *Frontiers in Genetics, 9*. doi: 10.3389/fgene.2018.00579

Louis, D. N., Perry, A., Reifenberger, G., von Deimling, A., Figarella-Branger, D., Cavenee, W. K., . . . Ellison, D. W. (2016). The 2016 world health organization classification of tumors of the central nervous system: A summary. *Acta Neuropathologica, 131*(6), 803–820. doi: 10.1007/s00401-016-1545-1

Lu, R. O., & Ho, W. S. (2021). Mitochondrial dysfunction, macrophage, and microglia in brain cancer. *Frontiers in Cell and Developmental Biology, 8*. doi: 10.3389/fcell.2020.620788

Luo, H., Zhang, H., Mao, J., Cao, H., Tao, Y., Zhao, G., . . . Liu, G. (2023). Exosome-based nanoimmunotherapy targeting TAMs, a promising strategy for glioma. *Cell Death & Disease, 14*(4), 235. doi: 10.1038/s41419-023-05753-9

Magnus, N., Garnier, D., Meehan, B., McGraw, S., Lee, T. H., Caron, M., . . . Rak, J. (2014). Tissue factor expression provokes escape from tumor dormancy and leads to genomic alterations. *Proceedings of the National Academy of Sciences, 111*(9), 3544–3549. doi: 10.1073/pnas.1314118111

Mahady, L., Nadeem, M., Malek-Ahmadi, M., Chen, K., Perez, S. E., & Mufson, E. J. (2019). HDAC2 dysregulation in the nucleus basalis of Meynert during the progression of Alzheimer's disease. *45*(4), 380–397. doi: 10.1111/nan.12518

McMeekin, L. J., & Fox, S. N. (2021). Dysregulation of PGC-1α-dependent transcriptional programs in neurological and developmental disorders: Therapeutic challenges and opportunities. *10*(2). doi: 10.3390/cells10020352

Melov, S., Schneider, J. A., Coskun, P. E., Bennett, D. A., & Wallace, D. C. (1999). Mitochondrial DNA rearrangements in aging human brain and in situ PCR of mtDNA. *Neurobiology of Aging, 20*(5), 565–571. doi: 10.1016/s0197-4580(99)00092-5

Meng, Y., Qiu, L., Zeng, X., Hu, X., Zhang, Y., Wan, X., . . . Han, J. (2022). Targeting CRL4 suppresses chemoresistant ovarian cancer growth by inducing mitophagy. *Signal Transduction and Targeted Therapy, 7*(1), 388. doi: 10.1038/s41392-022-01253-y

Migliore, L., & Coppedè, F. (2009). Genetics, environmental factors and the emerging role of epigenetics in neurodegenerative diseases. *Mutation Research, 667*(1–2), 82–97. doi: 10.1016/j.mrfmmm.2008.10.011

Mohsen, K., Azzazy, H. M. E., Allam, N. K., & Basalious, E. B. (2020). Intranasal lipid nanocapsules for systemic delivery of nimodipine into the brain: *In vitro* optimization and in vivo pharmacokinetic study. *Materials Science and*

Engineering C: Materials for Biological Applications, 116, 111236. doi: 10.1016/j.msec.2020.111236

Muyderman, H., & Chen, T. (2014). Mitochondrial dysfunction in amyotrophic lateral sclerosis - a valid pharmacological target? *British Journal of Pharmacology, 171*(8), 2191–2205. doi: 10.1111/bph.12476

Nakatogawa, H., Suzuki, K., Kamada, Y., & Ohsumi, Y. (2009). Dynamics and diversity in autophagy mechanisms: Lessons from yeast. *Nature Reviews Molecular Cell Biology, 10*(7), 458–467. doi: 10.1038/nrm2708

Nissanka, N., & Moraes, C. T. (2018). Mitochondrial DNA damage and reactive oxygen species in neurodegenerative disease. *FEBS Letters, 592*(5), 728–742. doi: 10.1002/1873-3468.12956

Nixon, R. A., & Yang, D. S. (2012). Autophagy and neuronal cell death in neurological disorders. *Cold Spring Harbor Perspectives in Biology, 4*(10). doi: 10.1101/cshperspect.a008839

Nunnari, J., & Suomalainen, A. (2012). Mitochondria: In sickness and in health. *Cell, 148*(6), 1145–1159. doi: 10.1016/j.cell.2012.02.035

Outeiro, T. F., Kontopoulos, E., Altmann, S. M., Kufareva, I., Strathearn, K. E., Amore, A. M., . . . Kazantsev, A. G. (2007). Sirtuin 2 inhibitors rescue alpha-synuclein-mediated toxicity in models of Parkinson's disease. *Science, 317*(5837), 516–519. doi: 10.1126/science.1143780

Park, H., & Kang, J. H. (2020). Autophagy in neurodegenerative diseases: A hunter for aggregates. *21*(9). doi: 10.3390/ijms21093369

Parker, W. D., Jr., Filley, C. M., & Parks, J. K. (1990). Cytochrome oxidase deficiency in Alzheimer's disease. *Neurology, 40*(8), 1302–1303. doi: 10.1212/wnl.40.8.1302

Parsons, D. W., Jones, S., Zhang, X., Lin, J. C., Leary, R. J., Angenendt, P., . . . Kinzler, K. W. (2008). An integrated genomic analysis of human glioblastoma multiforme. *Science, 321*(5897), 1807–1812. doi: 10.1126/science.1164382

Pérez-Segura, I., & Santiago-Balmaseda, A. (2023). PPARs and their neuroprotective effects in Parkinson's disease: A novel therapeutic approach in α-synucleinopathy? *24*(4). doi: 10.3390/ijms24043264

Picca, A., Calvani, R., & Coelho-Junior, H. J. (2020). Mitochondrial dysfunction, oxidative stress, and neuroinflammation: Intertwined roads to neurodegeneration. *9*(8). doi: 10.3390/antiox9080647

Pietrocola, F., Galluzzi, L., Bravo-San Pedro, J. M., Madeo, F., & Kroemer, G. (2015). Acetyl coenzyme A: A central metabolite and second messenger. *Cell Metabolism, 21*(6), 805–821. doi: 10.1016/j.cmet.2015.05.014

Pirtoli, L., Cevenini, G., Tini, P., Vannini, M., Oliveri, G., Marsili, S., . . . Miracco, C. (2009). The prognostic role of Beclin 1 protein expression in high-grade gliomas. *Autophagy, 5*(7), 930–936. doi: 10.4161/auto.5.7.9227

Polyzos, A. A., Lee, D. Y., Datta, R., Hauser, M., Budworth, H., Holt, A., . . . McMurray, C. T. (2019). Metabolic reprogramming in astrocytes distinguishes region-specific neuronal susceptibility in huntington mice. *Cell Metabolism, 29*(6), 1258–1273.e1211. doi: 10.1016/j.cmet.2019.03.004

Portz, P., & Lee, M. K. (2021). Changes in Drp1 function and mitochondrial morphology are associated with the α-synuclein pathology in a transgenic mouse model of Parkinson's disease. *10*(4). doi: 10.3390/cells10040885

Rademaker, G., Boumahd, Y., Peiffer, R., Anania, S., Wissocq, T., Liégeois, M., . . . Peulen, O. (2022). Myoferlin targeting triggers mitophagy and primes ferroptosis in pancreatic cancer cells. *Redox Biology, 53*, 102324. doi: 10.1016/j.redox.2022.102324

Rahman, M. M., Tumpa, M. A. A., Rahaman, M. S., Islam, F., Sutradhar, P. R., Ahmed, M., . . . Ashraf, G. M. (2023). Emerging promise of therapeutic approaches targeting mitochondria in neurodegenerative disorders. *Current Neuropharmacology, 21*(5), 1081–1099. doi: 10.2174/1570159x21666230316150559

Rasheed, M., & Liang, J. (2021). Epigenetic regulation of neuroinflammation in Parkinson's disease. *22*(9). doi: 10.3390/ijms22094956

Rashed, H. M., Shamma, R. N., & Basalious, E. B. (2016). Contribution of both olfactory and systemic pathways for brain targeting of nimodipine-loaded lipo-pluronics micelles: *In vitro* characterization and in vivo biodistribution study after intranasal and intravenous delivery. *Drug Delivery, 24*(1), 181–187. doi: 10.1080/10717544.2016.1236848

Rastogi, S., Sharma, V., Bharti, P. S., Rani, K., Modi, G. P., Nikolajeff, F., & Kumar, S. (2021). The evolving landscape of exosomes in neurodegenerative diseases: Exosomes characteristics and a promising role in early diagnosis. *International Journal of Molecular Sciences, 22*(1), 440.

Reddy, P. H., & Oliver, D. M. (2019). Amyloid beta and phosphorylated tau-induced defective autophagy and mitophagy in Alzheimer's disease. *Cells, 8*(5). doi: 10.3390/cells8050488

Ren, H., Zhai, W., Lu, X., & Wang, G. (2021). The cross-links of endoplasmic reticulum stress, autophagy, and neurodegeneration in Parkinson's disease. *Frontiers in Aging Neuroscience, 13*. doi: 10.3389/fnagi.2021.691881

Rhim, J., Baek, W., Seo, Y., & Kim, J. H. (2022). From molecular mechanisms to therapeutics: Understanding microRNA-21 in cancer. *11*(18). doi: 10.3390/cells11182791

Riva, P., Ratti, A., & Venturin, M. (2016). The long non-coding RNAs in neurodegenerative diseases: Novel mechanisms of pathogenesis. *Current Alzheimer Research, 13*(11), 1219–1231. doi: 10.2174/1567205013666160622112234

Rodríguez-Nuevo, A., & Zorzano, A. (2019). The sensing of mitochondrial DAMPs by non-immune cells. *Cell Stress, 3*(6), 195–207. doi: 10.15698/cst2019.06.190

Sadri-Vakili, G., Bouzou, B., Benn, C. L., Kim, M. O., Chawla, P., Overland, R. P., . . . Cha, J. H. (2007). Histones associated with downregulated genes are hypo-acetylated in Huntington's disease models. *Human Molecular Genetics, 16*(11), 1293–1306. doi: 10.1093/hmg/ddm078

Sati, I. S. E. E., & Parhar, I. (2021). MicroRNAs regulate cell cycle and cell death pathways in glioblastoma. *International Journal of Molecular Sciences, 22*(24), 13550.

Sayad, A., Najafi, S., Hussen, B. M., Abdullah, S. T., Movahedpour, A., Taheri, M., & Hajiesmaeili, M. (2022). The emerging roles of the β-secretase BACE1 and the long non-coding RNA BACE1-AS in human diseases: A focus on neurodegenerative diseases and cancer. *Frontiers in Aging Neuroscience, 14*. doi: 10.3389/fnagi.2022.853180

Schapira, A. H. (2008). Mitochondria in the aetiology and pathogenesis of Parkinson's disease. *Lancet Neurology, 7*(1), 97–109. doi: 10.1016/s1474-4422(07)70327-7

Scorziello, A., Borzacchiello, D., Sisalli, M. J., Di Martino, R., Morelli, M., & Feliciello, A. (2020). Mitochondrial homeostasis and signaling in Parkinson's disease. *Frontiers in Aging Neuroscience, 12*. doi: 10.3389/fnagi.2020.00100

Shao, H., Chung, J., Lee, K., Balaj, L., Min, C., Carter, B. S., . . . Weissleder, R. (2015). Chip-based analysis of exosomal mRNA mediating drug resistance in glioblastoma. *Nature Communications, 6*(1), 6999. doi: 10.1038/ncomms7999

Shaw, G. A. (2021). Mitochondria as the target for disease related hormonal dysregulation. *Brain, Behavior, & Immunity: Health, 18*, 100350. doi: 10.1016/j.bbih.2021.100350

Shefa, U., Jeong, N. Y., Song, I. O., Chung, H. J., Kim, D., Jung, J., & Huh, Y. (2019). Mitophagy links oxidative stress conditions and neurodegenerative diseases. *Neural Regeneration Research, 14*(5), 749–756. doi: 10.4103/1673-5374.249218

Shock, L. S., Thakkar, P. V., Peterson, E. J., Moran, R. G., & Taylor, S. M. (2011). DNA methyltransferase 1, cytosine methylation, and cytosine hydroxymethylation in mammalian mitochondria. *Proceedings of the National Academy of Sciences of the United States of America, 108*(9), 3630–3635. doi: 10.1073/pnas.1012311108

Shukla, S., Patric, I. R., Patil, V., Shwetha, S. D., Hegde, A. S., Chandramouli, B. A., . . . Somasundaram, K. (2014). Methylation silencing of ULK2, an autophagy gene, is essential for astrocyte transformation and tumor growth. *Journal of Biological Chemistry, 289*(32), 22306–22318. doi: 10.1074/jbc.M114.567032

Siedlecki-Wullich, D., & Miñano-Molina, A. J. (2021). MicroRNAs as early biomarkers of Alzheimer's disease: A synaptic perspective. *10*(1). doi: 10.3390/cells10010113

Sokolov, D., & Sharda, N. (2024). Differential signaling pathways in medulloblastoma: Nano-biomedicine targeting non-coding epigenetics to improve current and future therapeutics. *30*(1), 31–47. doi: 10.2174/0113816128277350231219062154

Sokolov, D., Sharda, N., Banerjee, A., Denisenko, K., Basalious, B. E., Shukla, H., . . . Banerjee, A. (2024). Differential signaling pathways in medulloblastoma: Nano-biomedicine targeting non-coding epigenetics to improve current and future therapeutics. *Current Pharmaceutical Design, 30*(1), 31–47. doi: 10.2174/0113816128277350231219062154

Song, Y., Zhu, X.-Y., Zhang, X.-M., & Xiong, H. (2022). Targeted mitochondrial epigenetics: A New direction in Alzheimer's disease treatment. *International Journal of Molecular Sciences, 23*(17), 9703.

Spinelli, J. B., & Haigis, M. C. (2018). The multifaceted contributions of mitochondria to cellular metabolism. *Nature Cell Biology, 20*(7), 745–754. doi: 10.1038/s41556-018-0124-1

Sumitomo, A., & Tomoda, T. (2021). Autophagy in neuronal physiology and disease. *Current Opinion in Pharmacology, 60*, 133–140. doi: 10.1016/j.coph.2021.07.013

Tai, Y., Chen, J., Tao, Z., & Ren, J. (2022). Non-coding RNAs: New players in mitophagy and neurodegeneration. *Neurochemistry International, 152*, 105253. doi: 10.1016/j.neuint.2021.105253

Tai, Y., Pu, M., Yuan, L., Guo, H., Qiao, J., Lu, H., . . . Ren, J. (2021). MiR-34a-5p regulates PINK1-mediated mitophagy via multiple modes. *Life Sciences, 276*, 119415. doi: 10.1016/j.lfs.2021.119415

Tang, G., Gudsnuk, K., Kuo, S. H., Cotrina, M. L., Rosoklija, G., Sosunov, A., . . . Sulzer, D. (2014). Loss of mTOR-dependent macroautophagy causes autistic-like synaptic pruning deficits. *Neuron, 83*(5), 1131–1143. doi: 10.1016/j.neuron.2014.07.040

Tatuch, Y., Christodoulou, J., Feigenbaum, A., Clarke, J. T., Wherret, J., Smith, C., . . . Robinson, B. H. (1992). Heteroplasmic mtDNA mutation (T----G) at 8993 can cause leigh disease when the percentage of abnormal mtDNA is high. *American Journal of Human Genetics, 50*(4), 852–858.

Tilokani, L., Nagashima, S., Paupe, V., & Prudent, J. (2018). Mitochondrial dynamics: Overview of molecular mechanisms. *Essays in Biochemistry, 62*(3), 341–360. doi: 10.1042/ebc20170104

Titze-de-Almeida, S. S., Soto-Sánchez, C., & Fernandez, E. (2020). The promise and challenges of developing miRNA-based therapeutics for Parkinson's disease. *9*(4). doi: 10.3390/cells9040841

Torrell, H., Montaña, E., Abasolo, N., Roig, B., Gaviria Gómez, A., Vilella, E., & Martorell, L. (2013). Mitochondrial DNA (mtDNA) in brain samples from patients with major psychiatric disorders: Gene expression profiles, MtDNA content and presence of the MtDNA common deletion. *American Journal of Medical Genetics. Part B, Neuropsychiatric Genetics: The Official Publication of the International Society of Psychiatric Genetics, 162*. doi: 10.1002/ajmg.b.32134

Uddin, M. S., Rahman, M. M., Jakaria, M., Rahman, M. S., Hossain, M. S., Islam, A., . . . Ashraf, G. M. (2020). Estrogen signaling in Alzheimer's disease: Molecular insights and therapeutic targets for Alzheimer's dementia. *Molecular Neurobiology, 57*(6), 2654–2670. doi: 10.1007/s12035-020-01911-8

Uddin, M. S., Stachowiak, A., Mamun, A. A., Tzvetkov, N. T., Takeda, S., Atanasov, A. G., . . . Stankiewicz, A. M. (2018). Autophagy and Alzheimer's disease: From molecular mechanisms to therapeutic implications. *Frontiers in Aging Neuroscience, 10*, 04. doi: 10.3389/fnagi.2018.00004

Um, J. H., & Yun, J. (2017). Emerging role of mitophagy in human diseases and physiology. *BMB Reports, 50*(6), 299–307. doi: 10.5483/bmbrep.2017.50.6.056

Valdinocci, D., Simões, R. F., Kovarova, J., Cunha-Oliveira, T., Neuzil, J., & Pountney, D. L. (2019). Intracellular and intercellular mitochondrial dynamics in Parkinson's disease. *Frontiers in Neuroscience, 13*, 930. doi: 10.3389/fnins.2019.00930

Vartiainen, S., Pehkonen, P., Lakso, M., Nass, R., & Wong, G. (2006). Identification of gene expression changes in transgenic C. elegans overexpressing human alpha-synuclein. *Neurobiology of Disease, 22*(3), 477–486. doi: 10.1016/j.nbd.2005.12.021

Vial, D., & McKeown-Longo, P. J. (2016). Role of EGFR expression levels in the regulation of integrin function by EGF. *Molecular Carcinogenesis, 55*(6), 1118–1123. doi: 10.1002/mc.22346

Vilardo, E., Barbato, C., Ciotti, M., Cogoni, C., & Ruberti, F. (2010). MicroRNA-101 regulates amyloid precursor protein expression in hippocampal neurons. *Journal of Biological Chemistry, 285*(24), 18344–18351. doi: 10.1074/jbc.M110.112664

Wakade, C., Chong, R., Seamon, M., Purohit, S., Giri, B., & Morgan, J. C. (2021). Low-dose niacin supplementation improves motor function in US veterans with Parkinson's disease: A single-center, randomized, placebo-controlled trial. *Biomedicines, 9*(12), 1881.

Wang, H., Jiang, T., Li, W., Gao, N., & Zhang, T. (2018). Resveratrol attenuates oxidative damage through activating mitophagy in an in vitro model of Alzheimer's disease. *Toxicology Letters, 282*, 100–108. doi: 10.1016/j.toxlet.2017.10.021

Wang, H., Ye, Y., Zhu, Z., Mo, L., Lin, C., Wang, Q., . . . Zhang, S. (2016). MiR-124 regulates apoptosis and autophagy process in MPTP model of Parkinson's disease by targeting to bim. *Brain Pathology, 26*(2), 167–176. doi: 10.1111/bpa.12267

Wang, H., Zhang, T., Ge, X., Chen, J., Zhao, Y., & Fu, J. (2020). Parkin overexpression attenuates Aβ-induced mitochondrial dysfunction in HEK293 cells by restoring impaired mitophagy. *Life Sciences, 244*, 117322. doi: 10.1016/j.lfs.2020.117322

Wang, J., Qiu, X., Huang, J., Zhuo, Z., Chen, H., Zeng, R., . . . Luo, Y. (2022). Development and validation of a novel mitophagy-related gene prognostic signature for glioblastoma multiforme. *BMC Cancer*, *22*(1), 644. doi: 10.1186/s12885-022-09707-w

Wang, N., Wang, H., Li, L., Li, Y., & Zhang, R. (2019). β-asarone inhibits amyloid-β by promoting autophagy in a cell model of Alzheimer's disease. *Frontiers in Pharmacology*, *10*, 1529. doi: 10.3389/fphar.2019.01529

Wang, X., Su, B., Zheng, L., Perry, G., Smith, M. A., & Zhu, X. (2009). The role of abnormal mitochondrial dynamics in the pathogenesis of Alzheimer's disease. *Journal of Neurochemistry*, *109*(Suppl 1), 153–159. doi: 10.1111/j.1471-4159.2009.05867.x

Wang, X. L., Feng, S. T., Wang, Y. T., Yuan, Y. H., Li, Z. P., Chen, N. H., . . . Zhang, Y. (2022). Mitophagy, a form of selective autophagy, plays an essential role in mitochondrial dynamics of Parkinson's disease. *42*(5), 1321–1339. doi: 10.1007/s10571-021-01039-w

Wang, X. L., Feng, S. T., Wang, Z. Z., Chen, N. H., & Zhang, Y. (2021). Role of mitophagy in mitochondrial quality control: Mechanisms and potential implications for neurodegenerative diseases. *Pharmacological Research*, *165*, 105433. doi: 10.1016/j.phrs.2021.105433

Wang, Y., Xu, E., Musich, P. R., & Lin, F. (2019). Mitochondrial dysfunction in neurodegenerative diseases and the potential countermeasure. *25*(7), 816–824. doi: 10.1111/cns.13116

Wang, Z., Xia, P., Hu, J., Huang, Y., Zhang, F., Li, L., . . . Ye, Z. (2021). LncRNA MEG3 alleviates diabetic cognitive impairments by reducing mitochondrial-derived apoptosis through promotion of FUNDC1-related mitophagy via Rac1-ROS axis. *12*(13), 2280–2307. doi: 10.1021/acschemneuro.0c00682

Wang, Z. H., Zhang, J. L., Duan, Y. L., Zhang, Q. S., Li, G. F., & Zheng, D. L. (2015). MicroRNA-214 participates in the neuroprotective effect of resveratrol via inhibiting α-synuclein expression in MPTP-induced Parkinson's disease mouse. *Biomedicine & Pharmacotherapy*, *74*, 252–256. doi: 10.1016/j.biopha.2015.08.025

Wareski, P., Vaarmann, A., Choubey, V., Safiulina, D., Liiv, J., Kuum, M., & Kaasik, A. (2009). PGC-1{alpha} and PGC-1{beta} regulate mitochondrial density in neurons. *Journal of Biological Chemistry*, *284*(32), 21379–21385. doi: 10.1074/jbc.M109.018911

Wilkaniec, A., Gąssowska, M., Czapski, G. A., Cieślik, M., Sulkowski, G., & Adamczyk, A. (2017). P2X7 receptor-pannexin 1 interaction mediates extracellular alpha-synuclein-induced ATP release in neuroblastoma SH-SY5Y cells. *Purinergic Signal, 13*(3), 347–361. doi: 10.1007/s11302-017-9567-2

Wong, M., Gertz, B., Chestnut, B. A., & Martin, L. J. (2013). Mitochondrial DNMT3A and DNA methylation in skeletal muscle and CNS of transgenic mouse models of ALS. *Frontiers in Cellular Neuroscience*, *7*, 279. doi: 10.3389/fncel.2013.00279

Wu, Y., Chen, M., & Jiang, J. (2019). Mitochondrial dysfunction in neurodegenerative diseases and drug targets via apoptotic signaling. *Mitochondrion*, *49*, 35–45. doi: 10.1016/j.mito.2019.07.003

Xia, D. Y., Huang, X., Bi, C. F., Mao, L. L., Peng, L. J., & Qian, H. R. (2017). PGC-1α or FNDC5 Is involved in modulating the effects of Aβ(1-42) oligomers on suppressing the expression of BDNF, a beneficial factor for inhibiting neuronal apoptosis, Aβ deposition and cognitive decline of APP/PS1 Tg mice. *Frontiers in Aging Neuroscience*, *9*, 65. doi: 10.3389/fnagi.2017.00065

Xia, H., Qi, Y., Ng, S. S., Chen, X., Chen, S., Fang, M., . . . Lin, M. C. (2009). MicroRNA-15b regulates cell cycle progression by targeting cyclins in glioma cells. *Biochemical and Biophysical Research Communications*, *380*(2), 205–210. doi: 10.1016/j.bbrc.2008.12.169

Xiao, Y., Karam, C., Yi, J., Zhang, L., Li, X., Yoon, D., . . . Zhou, J. (2018). ROS-related mitochondrial dysfunction in skeletal muscle of an ALS mouse model during the disease progression. *Pharmacological Research*, *138*, 25–36. doi: 10.1016/j.phrs.2018.09.008

Xu, M., & Xu, L. (2023). Up-regulation of miR-26a-5p promoted cell growth and tumor metastasis of intracranial malignancy through phosphatase and tensin homolog deleted on chromosome ten/phosphatidylinositol3-kinase/protein kinase B signaling pathway. *Journal of Biomaterials and Tissue Engineering, 13*(1), 31–42. doi: 10.1166/jbt.2023.3228

Xu, S., Zhang, X., Liu, C., Liu, Q., Chai, H., Luo, Y., & Li, S. (2021). Role of mitochondria in neurodegenerative diseases: From an epigenetic perspective. *Frontiers in Cell and Developmental Biology*, *9*. doi: 10.3389/fcell.2021.688789

Xue, L., Wang, Y., Yue, S., & Zhang, J. (2017). The expression of miRNA-221 and miRNA-222 in gliomas patients and their prognosis. *Neurological Sciences: Official Journal of the Italian Neurological Society and of the Italian Society of Clinical Neurophysiology*, *38*(1), 67–73. doi: 10.1007/s10072-016-2710-y

Yamaguchi, J., Suzuki, C., Nanao, T., Kakuta, S., Ozawa, K., Tanida, I., . . . Komatsu, M. (2018). Atg9a deficiency causes axon-specific lesions including neuronal circuit dysgenesis. *14*(5), 764–777. doi: 10.1080/15548627.2017.1314897

Yang, W., Wang, X., Liu, J., Duan, C., Gao, G., Lu, L., . . . Yang, H. (2018). PINK1 suppresses alpha-synuclein-induced neuronal injury: A novel mechanism in protein phosphatase 2A activation. *Oncotarget*, *9*(1), 37–53. doi: 10.18632/oncotarget.21554

Yang, X., Xu, S., Qian, Y., He, X., Chen, S., & Xiao, Q. (2020). Hypermethylation of the gene coding for PGC-1α in peripheral blood leukocytes of patients with Parkinson's disease. *Frontiers in Neuroscience*, *14*, 97. doi: 10.3389/fnins.2020.00097

Younas, N., Fernandez Flores, L. C., Hopfner, F., Höglinger, G. U., & Zerr, I. (2022). A new paradigm for diagnosis of neurodegenerative diseases: Peripheral exosomes of brain origin. *Translational Neurodegeneration*, *11*(1), 28. doi: 10.1186/s40035-022-00301-5

Zhang, R., Zhou, H., Jiang, L., Mao, Y., Cui, X., Xie, B., . . . Xu, S. (2016). MiR-195 dependent roles of mitofusin2 in the mitochondrial dysfunction of hippocampal neurons in SAMP8 mice. *Brain Research*, *1652*, 135–143. doi: 10.1016/j.brainres.2016.09.047

Zhang, Y., Dutta, A., & Abounader, R. (2012). The role of microRNAs in glioma initiation and progression. *Frontiers in Bioscience (Landmark Ed)*, *17*(2), 700–712. doi: 10.2741/3952

Zhu, J. H., Gusdon, A. M., Cimen, H., Van Houten, B., Koc, E., & Chu, C. T. (2012). Impaired mitochondrial biogenesis contributes to depletion of functional mitochondria in chronic MPP+ toxicity: Dual roles for ERK1/2. *Cell Death & Disease*, *3*(5), e312. doi: 10.1038/cddis.2012.46

Index

Note: Page numbers in *italics* indicate a figure and page numbers in **bold** indicate a table on the corresponding page.